Direct and Projective Limits of Geometric Banach Structures

This book describes in detail the basic context of the Banach setting and the most important Lie structures found in finite dimension. The authors expose these concepts in the convenient framework which is a common context for projective and direct limits of Banach structures. The book presents sufficient conditions under which these structures exist by passing to such limits. In fact, such limits appear naturally in many mathematical and physical domains. Many examples in various fields illustrate the different concepts introduced.

Many geometric structures, existing in the Banach setting, are "stable" by passing to projective and direct limits with adequate conditions. The convenient framework is used as a common context for such types of limits. The contents of this book can be considered as an introduction to differential geometry in infinite dimension but also a way for new research topics.

This book allows the intended audience to understand the extension to the Banach framework of various topics in finite dimensional differential geometry and, moreover, the properties preserved by passing to projective and direct limits of such structures as a tool in different fields of research.

Monographs and Research Notes in Mathematics

Series Editors:

John A. Burns, Thomas J. Tucker, Miklos Bona, Michael Ruzhansky

About the Series

This series is designed to capture new developments and summarize what is known over the entire field of mathematics, both pure and applied. It will include a broad range of monographs and research notes on current and developing topics that will appeal to academics, graduate students, and practitioners. Interdisciplinary books appealing not only to the mathematical community, but also to engineers, physicists, and computer scientists are encouraged.

This series will maintain the highest editorial standards, publishing well-developed monographs as well as research notes on new topics that are final, but not yet refined into a formal monograph. The notes are meant to be a rapid means of publication for current material where the style of exposition reflects a developing topic.

Fixed Point Results in W-Distance Spaces
Vladimir Rakočević

Analysis of a Model for Epilepsy
Application of a Max-Type Difference Equation to Mesial Temporal Lobe Epilepsy
Candace M. Kent, David M. Chan

Double Sequence Spaces and Four-Dimensional Matrices
Feyzi Başar, Medine Yeşilkayagil Savaşcı

Banach-Space Operators On C*-Probability Spaces Generated by Multi Semicircular Elements
Ilwoo Cho

Applications of Homogenization Theory to the Study of Mineralized Tissue
Robert P. Gilbert, Ana Vasilic, Sandra Klinge, Alex Panchenko, Klaus Hackl

Constructive Analysis of Semicircular Elements
From Orthogonal Projections to Semicircular Elements
Ilwoo Cho

Inverse Scattering Problems and Their Applications to Nonlinear Integrable Equations, Second Edition
Pham Loi Vu

Generalized Notions of Continued Fractions
Ergodicity and Number Theoretic Applications

Juan Fernández Sánchez, Jerónimo López-Salazar Codes, Juan B. Seoane Sepúlveda,
Wolfgang Trutschnig

Aspects of Integration
Novel Approaches to the Riemann and Lebesgue Integrals
Ronald B. Guenther, John W. Lee

Direct and Projective Limits of Geometric Banach Structures
Patrick Cabau and Fernand Pelletier

For more information about this series please visit: https://www.routledge.com/Chapman--HallCRC-Monographs-and-Research-Notes-in-Mathematics/book-series/CRCMONRESNOT

Direct and Projective Limits of Geometric Banach Structures

Patrick Cabau
Fernand Pelletier
With the assistance and partial collaboration
of Daniel Beltiță

CRC Press
Taylor & Francis Group
Boca Raton London New York

CRC Press is an imprint of the
Taylor & Francis Group, an informa business
A CHAPMAN & HALL BOOK

First edition published 2024
by CRC Press
6000 Broken Sound Parkway NW, Suite 300, Boca Raton, FL 33487-2742

and by CRC Press
4 Park Square, Milton Park, Abingdon, Oxon, OX14 4RN

CRC Press is an imprint of Taylor & Francis Group, LLC

ISBN: 978-1-032-56171-4 (hbk)
ISBN: 978-1-032-56450-0 (pbk)
ISBN: 978-1-003-43558-7 (ebk)

DOI: 10.1201/9781003435587

Typeset in CMR10
by KnowledgeWorks Global Ltd.

Publisher's note: This book has been prepared from camera-ready copy provided by the authors.

À nos épouses,
pour leur patience et leurs encouragements.

Contents

Preface

The purpose of this book is to lay down the foundations of both direct and projective limits of Banach geometric structures.

Banach spaces provide a framework for functional analysis, operator theory, probability, optimization and other branches of mathematics. They constitute a natural extension to the study of finite-dimensional linear spaces. Banach manifolds and bundles lead to examples in global analysis.

In many situations, such as global analysis or mathematical physics, the Banach framework is not sufficient. We then have to consider the more general frameworks of direct or projective limits of Banach structures.

Despite their differences, the categories of Banach and Fréchet spaces are connected through countable projective limits. This allows to extend, under suitable conditions, geometrical objects, structures and properties of the Banach framework to the Fréchet one, as described in the book [DGV16], e.g. ILB-Lie groups ([Omo97]) or projective limits of manifolds ([AbbMa99]), Banach-Lie groups ([Gal96]), bundles ([Gal98], [AgSu07]), connections and differential equations ([ADGS09]). This process is extended to other structures such as algebroids ([Cab12]), tensor structures ([CaPe20]), symplectic and Poisson structures ([PeCa19]) as described in Chapter 4.

Note that, more generally, each complete locally convex space is the projective limit of an uncountable sequence of Banach spaces and this approach is useful in C^*-algebra theory.

Around the period 1970–1990, few authors studied some topological properties of countable direct limit of finite-dimensional manifolds (for instance [Han71], [Hei82], [Sak92] and also the recent book [Sak20], Chap. 6). More recently, the theory of countable direct limit of Banach manifolds was developed by H. Glöckner, who applied his results essentially to Banach-Lie groups ([Glo03], [Glo05], [Glo07], [Glo11] and [Glo19]). One can notice that direct limits of Banach manifolds are convenient manifolds which are not Hausdorff in general (cf. [CaPe19]). In fact, the convenient structure described in the book [KrMi97] can be seen as an uncountable direct limit of adequate Banach spaces, and it is possible to overcome some deficiencies of this structure and obtain results analogous to those obtained in the Fréchet framework as described above.
Indeed any LB-space, i.e. direct limit of a sequence of Banach spaces whose bonding maps are closed range, satisfies two fundamental properties:

- it is c^∞-complete;

• its c^∞-topology coincides with the direct limit topology.

Since Fréchet and so Banach spaces are convenient and since the category of convenient spaces with convenient smooth maps is cartesian closed and the core tool of exponential law holds, **we mainly choose the convenient framework as common context for projective and direct limits**.

One of the main objectives of this survey is to show that many geometric structures, existing in the Banach setting, are "stable" by passing to projective and direct limits with adequate conditions. So the contents of this book pave the way for new research topics...

When dealing with projective or direct limits of partial Poisson structures, some interesting results are obtained by combining both techniques: an example of such a combination can be seen in Chapter 7 which is based on the paper [Pel16].

Aiming at the reader's convenience, we give an outline of the structure of the book.

Chapter 1, which is the first basic chapter, introduces the essential notions on Banach structures needed in the remainder of these notes. After having given some fundamental elements about Banach spaces and ordinary differential equations, we focus on the structures of manifolds, bundles, Lie groups, Lie algebras and the fundamental notion of exponential maps in the Banach framework.

Chapter 2 is devoted to some linear operators which provide a lot of examples used in various situations. The notion of G-structures and tensor structures are also studied, and many examples are given to illustrate these notions.

Chapter 3, which is the second basic chapter, first offers a detailed overview of the topology of locally convex topological vector spaces and the links between bornology and convenient space. Then the different convenient structures (manifolds, bundles, Lie groups and algebras) described in the book [KrMi97] are exposed in detail. We end this chapter with classical results in Banach spaces which are not any more true in this setting.

Chapter 4 incorporates results given in the Dodson, Galanis and Vasiliou book ([DGV16]) in the framework of Fréchet structures and develops this setting for G-structures and tensor structures; it is illustrated by examples of projective limits of tensor structures.

Chapter 5 contains a lot of direct limits of Banach structures which give rise to analogous ones in the convenient setting. Some difficulties appear in this framework: e.g. for a direct limit of a sequence of Banach linear bundles $((E_n, \pi_n, M_n))_{n \in \mathbb{N}}$, we have to consider the Fréchet topological group $\mathrm{G}(\mathbb{E})$ which does not correspond to the pathological group $\mathrm{GL}(\mathbb{E})$, where $\mathbb{E}$ is the direct limit of the type fibres $\mathbb{E}_n$.

Chapter 6 is focused on the prolongations of Lie algebroids in the convenient setting. We begin with general results on convenient Lie algebroids which extend classical results on Lie algebroids in finite dimension. In this framework, using similar arguments as in finite dimension, we can generalize the processus of prolongation, along a convenient fibred manifold $\mathbf{p} : \mathcal{M} \to M$, for an anchored convenient bundle $(\mathcal{A}, \pi, M, \rho)$.

Chapter 7 deals with the notion of partial Poisson structure in the convenient setting, which generalizes the core notion of Poisson structure. An almost partial Poisson structure on a convenient manifold M is the datum of a convenient anchored bundle $\left(T^\flat M, p^\flat, M, P\right)$ where $T^\flat M$ is a (weak) subbundle of the kinematic cotangent bundle $T'M$ and P satisfies the skew-symmetric property relative to the dual pairing. In this way, we obtain an almost Lie-Poisson bracket $\{.,.\}_P$ on the algebra of smooth functions defined over any open set of M and whose differentials induce a section of $T^\flat M$. If, moreover, $\{.,.\}_P$ satisfies the Jacobi identity, we get a partial Poisson structure. Such a structure occurs naturally for the canonical symplectic structure on the cotangent bundle of a convenient manifold M. But this type of Poisson structure also appears naturally in some dynamical problems in Physics. The last part of the chapter is devoted to the definition and properties of some direct (resp. projective) limits of Banach partial Poisson manifolds endowed with a convenient (resp. Fréchet) structure.

Chapter 8 is concerned with the integrability of weak distributions (link with involutivity, invariance and Lie invariance) in the Banach framework. This situation is illustrated with Banach-Lie algebroids orbits and symplectic foliations associated with partial Poisson manifolds. This chapter ends with a criterion of integrability of some projective limits (resp. direct limits) of distributions.

Appendix A gives a brief account of some basic results of Fréchet spaces and the link with projective sequences of Banach spaces is highlighted. Finally, general results which are true in the Banach framework but are not anymore true in the Fréchet setting are listed.

In Appendix B, the fundamental notion of category is presented.

In Appendix C, the notion of sheaf is recalled, and some basic results of this notion used in the book are presented.

Appendix D is devoted to the notion of locally convex vector bundles.

In Appendix E, we present different frameworks for the famous KdV equation.

Because (systems of) PDEs are the heart of mathematical physics, in Appendix F, we present the theory of formal integrability which is linked with different geometrical objects on an infinite jet bundle and projective limits of corresponding ones on finite jet bundles.

We end with some open problems that could be addressed by using the different techniques presented and developed in this book.

We have paid particular attention to illustrate the different concepts introduced by examples.

This *opus* is self-contained enough for researchers who have “basic culture in differential geometry in finite dimension” and are interested in various domains of infinite-dimensional differential geometry and mathematical physics.

In order to give adapted frameworks to such domains, more general structures are needed.

The finalization of this book and a lot of notions and proofs were realized in collaboration with the authors and D. Beltiţă.

Reading tree for the cross reader. The tree below gives the most important links between all chapters of this book. It clearly appears that both Chapters 1 and 3 constitute the basic elements of this survey. Chapter 2 is for the reader who is interested in Banach-Lie structures; it can be omitted at first reading. Chapters 4 and 5, founded on these fundamental chapters, give essential materials for the reading and comprehension of the following chapters.

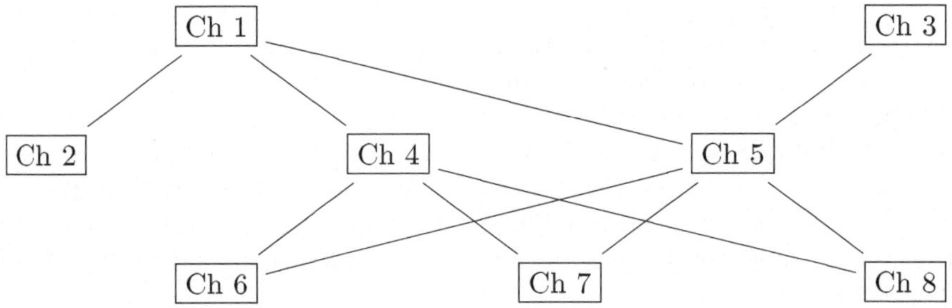

Ch 1 Preliminaries on the Banach setting.

Ch 2 Some Banach-Lie structures.

Ch 3 Convenient structures.

Ch 4 Projective limits.

Ch 5 Direct limits.

Ch 6 Some properties of convenient Lie algebroids and prolongations.

Ch 7 Partial Poisson structures.

Ch 8 Integrability of distributions and foliations.

Authors

Patrick Cabau is an independent researcher. He received his M.Sc. at the University of Toulouse in 1980, l'Agrégation de Mathématiques in 1987 and his Ph.D. from the University of Savoy in 1999. He has taught at the ENSET (University Nord-Madagascar), has been a member of the LIM (Tunisia Polytechnic School) and a teacher in high schools. His primary current interests are in the areas of differential geometry, mechanics and mathematical physics.

Fernand Pelletier began his career as a researcher at the University of Burgundy in 1970. He obtained his third cycle doctorate in 1973 and his habilitation in 1980 at this University. Appointed Professor at the University of Corsica Pasquale Paoli in 1983, he was transferred to the University of Savoy in 1986. He was the Director of the Mathematics Laboratory (LAMA) of the University of Savoy from 1989 to 1996, Director of differential geometry teams in LAMA from 1996 to 2002 and Director of the "South Rodhanian" research group from 2002 to 2006. From his retirement in 2010 until now, he has been Professor Emeritus at the University of Savoy. His main research topics relate to differential geometry, dynamic systems, control theory and mathematical physics in finite and infinite dimensions.

Acknowledgements

The authors are indebted to Daniel Belţiţă for his precious collaboration to this book, for many valuable comments and corrections and for his partial participation for several chapters.

Thanks to Jean-Paul Dufour, Franco Magri, Charles-Michel Marle and Alice Barbora Tumpach for interesting conversations about Poisson geometry.

A special thought for Joseph Grifone, who organized the meeting of both authors at the I.M.T. of Toulouse and initiated their collaboration. Various contents of this book have been greatly influenced by our discussions about different subjects (bihamiltonian structures, PDEs, etc.).

We would not forget the LaMa (Laboratoire de Mathématiques, Université de Savoie Mont-Blanc in Chambéry), its director, Georges Comte, and the geometry team for its material and financial support.

F.P. wishes to begin by personally thanking Daniel Belţiţă with whom he has had the pleasure of collaborating for several years and who is, to a large extent, at the origin of this book. On the other hand, he would like to thank Anatol Odzijewicz, and he deeply regrets his passing. They had fruitful discussions motivating their common interest for differential geometry in infinite dimensions. Professor Anatol Odzijewicz invited him regularly to the W.G.M.P. in Bialowieza workshop. This was the opportunity of meeting many researchers concerned with differential geometry in infinite dimensions and its applications in Mathematics and Physics. In particular, fruitful discussions with Tomasz Goliński, Grzegorz Jakimowicz, Aneta Sliżewska and Alice Barbora Tumpach have been greatly appreciated and have given rise to ongoing collaborations. All these discussions and meetings were the basis of the motivation for this book.

P.C. would also like to thank Daniel Belţiţă for the constructive exchanges, his mathematical rigour as well as his kindness. Second, he would like to acknowledge Eduardo Martínez for interesting and friendly discussions at workshops at I.C.M.A.T. in Madrid. His thanks go to Philippe Monnier for his kind welcome at the I.M.T. and for discussions about Poisson geometry. He also acknowledges Tien Zung Nguyen for interesting exchanges about various subjects at Torus AI. Thanks to Aude for her linguistic support.

Finally, the authors are indebted to Bob Ross for the trust he has placed in them by publishing this *opus*. They are also grateful to Beth Hawkins for her responsiveness and helpful assistance. Thanks to Sumati Agarwal and her team for the finalization of this book.

List of Figures

1

Preliminaries on the Banach Setting

This first chapter gives a brief account of some preliminaries about basic Banach structures, such as linear spaces, manifolds, Lie groups, Lie algebras and vector bundles, with many examples of such structures.
We also fixed the notations which will be used throughout this book.
The relevant structure of finite jet bundles (and their projective limit) appears as an adapted framework to model PDEs and their symmetries and is described at the end of this chapter.

Throughout this book, we will use the following notations:

- $\mathbb{N}$ denotes the set of natural numbers (non-negative integers) $\{0, 1, 2, \dots\}$ and $\mathbb{N}^*$ is the set $\mathbb{N} \setminus \{0\} = \{1, 2, \dots\}$.
- $\mathbb{Z}$ denotes the set of all integers.
- $\mathbb{K}$ is the field $\mathbb{R}$ or $\mathbb{C}$.

1.1 Banach Spaces

In 1906, Maurice Fréchet working on the notion of general metric space endowed the space $C([a, b])$ of continuous functions on the compact set $[a, b]$ with a complete metric which is fact induced by a norm ([Fre1906]). After Frédéric Riesz's works in 1918, one can consider that the general theory of normed spaces was developed independently by Stefan Banach and Hans Hahn in 1922.

Definition 1.1 *A norm on a $\mathbb{K}$-vector space $\mathbb{E}$ is a function $N : \mathbb{E} \to \mathbb{R}+$ satisfying the following conditions:*

(N1) $\forall (u, v) \in \mathbb{E}^2,\ N(u + v) \leq N(u) + N(v);$

(N2) $\forall \lambda \in \mathbb{R}, \forall u \in \mathbb{E},\ N(\lambda.u) = |\lambda| N(u);$

(N3) $N(u) = 0 \Longrightarrow u = \mathbf{0}.$

DOI: 10.1201/9781003435587-1

Two norms $\| \|_1$ and $\| \|_2$ on a $\mathbb{K}$-vector space $\mathbb{E}$ are *equivalent* if there exist some positive numbers $c_1 < c_2$ satisfying

$$\forall v \in \mathbb{E} \setminus \{\mathbf{0}\},\ c_1 \leq \frac{\|v\|_2}{\|v\|_1} \leq c_2.$$

Definition 1.2 *A Banach space is a $\mathbb{K}$-vector space equipped with an equivalence class of complete norms.*

A linear operator $A : \mathbb{E}_1 \to \mathbb{E}_2$ between normed spaces $\left(\mathbb{E}_i, \| \|_{\mathbb{E}_i}\right)_{i \in \{1,2\}}$ is *bounded* if

$$\exists M \geq 0 : \ \forall u \in \mathbb{E}_1,\ \|A.u\|_{\mathbb{E}_2} \leq M. \|u\|_{\mathbb{E}_1}.$$

The smallest such M is called the *operator norm* of A and is denoted by $\|A\|^{\mathrm{op}}$. A linear operator $A : \mathbb{E}_1 \to \mathbb{E}_2$ between normed spaces $\left(\mathbb{E}_i, \| \|_{\mathbb{E}_i}\right)$, $i \in \{1, 2\}$, is *continuous* if

$$\forall \varepsilon > 0,\ \exists \alpha > 0,\ \|u - v\|_{\mathbb{E}_1} < \alpha \Rightarrow \|A(u) - A(v)\|_{\mathbb{E}_2} < \varepsilon.$$

Proposition 1.1 *A linear map $A : \mathbb{E}_1 \to \mathbb{E}_2$ between normed spaces (and so Banach spaces) is continuous if and only if it is bounded.*

The following notations of linear maps will also be used at the convenient setting.

Notation 1.1 *If $\mathbb{E}_1$ and $\mathbb{E}_2$ are Banach spaces, we denote by*

- $\mathrm{L}(\mathbb{E}_1, \mathbb{E}_2)$ *the space of linear maps from $\mathbb{E}_1$ to $\mathbb{E}_2$;*
- $\mathcal{L}(\mathbb{E}_1, \mathbb{E}_2)$ *the space of continuous linear maps from $\mathbb{E}_1$ to $\mathbb{E}_2$.*

If $\mathbb{E}_1 = \mathbb{E}_2 = \mathbb{E}$, we, respectively, use the notations $\mathrm{L}(\mathbb{E})$ and $\mathcal{L}(\mathbb{E})$.
For invertible elements of these spaces, we denote by

- $\mathrm{GL}(E)$ *the space of linear automorphisms of E;*
- $\mathcal{GL}(E)$, *the space of continuous linear automorphisms of E.*

In particular, $\mathcal{L}(\mathbb{E})$ is a Banach algebra. The open subset $\mathcal{GL}(\mathbb{E})$ of invertible elements can be endowed with a structure of group under the composition of automorphisms; it is called the *general Lie group of* $\mathbb{E}$.

Definition 1.3 *A Schauder basis of a Banach space $\mathbb{E}$ is a sequence $(e_n)_{n \in \mathbb{N}}$ of elements of $\mathbb{E}$ such that*

$$\forall v \in \mathbb{E},\ \exists! (\lambda_n) \in \mathbb{K}^{\mathbb{N}} : \ v = \sum_{n=0}^{+\infty} \lambda_n e_n$$

where the convergence of sequences is understood relative to the norm topology:

$$\lim_{n \to +\infty} \left\| v - \sum_{k=0}^{n} \lambda_k e_k \right\| = 0.$$

In a lot of situations, finite-dimensional spaces and Banach spaces behave in a similar way. But one can point out some basic differences between both these types of spaces:

(1) the unit ball (relative to some norm) is not compact;

(2) the topological dual of the topological dual $\mathbb{E}^*$ of a Banach space $\mathbb{E}$ is, in general, not isomorphic to $\mathbb{E}$. However, this is the case for reflexive Banach spaces (cf. § 1.3);

(3) a closed subspace of a Banach space need not split[1], i.e. it need not have a supplemented closed subspace (cf. Remark 1.1);

(4) if $\mathbb{E}_1$ and $\mathbb{E}_2$ are Banach spaces and if U is an open set of $\mathbb{E}_1$, the C^k-differentiability of a map $f : U \to \mathcal{L}(\mathbb{E}_1, \mathbb{E}_2)$ is no more equivalent to the C^k-differentiability of the map $\begin{array}{ccc} U \times \mathbb{E}_1 & \to & \mathbb{E}_2 \\ (x, v) & \mapsto & f(x).v \end{array}$;

(5) any equivalent norm which defines the Banach structure of $\mathbb{E}$ is, in general, not smooth away from zero. This problem is closely related to the existence of a smooth bump function on $\mathbb{E}$. In particular, there does not exist smooth partitions of unity on $\mathbb{E}$, in general, if $\mathbb{E}$ is not a Hilbert space.

1.2 Examples of Banach Spaces

Example 1.1 *Any real or complex finite-dimensional vector space is a Banach space for some norm.*

Example 1.2 *Sequence spaces.— The space of real sequences will be denoted $\mathbb{R}^{\mathbb{N}}$. Let $x = (x_n)_{n\in\mathbb{N}}$ be a real sequence.*

(1) For $p \in]0, +\infty[$, we define

$$\|x\|_p = \left(\sum_{n=0}^{+\infty} |x_n|^p \right)^{1/p} .$$

The vector space $\ell^p_{\mathbb{R}}(\mathbb{N}) = \left\{ x \in \mathbb{R}^{\mathbb{N}} : \|x\|_p \in \mathbb{R} \right\}$ is a Banach space[2].

(2) The space $\ell^\infty_{\mathbb{R}}(\mathbb{N})$ of all bounded real sequences equipped with the norm

$$\|x\|_\infty = \sup_{n\in\mathbb{N}} |x_n|$$

is a Banach space[3].

[1] A Banach space whose every closed subspace is supplemented is necessarily isomorphic to a Hilbert space ([LiTz73]).

[2] Also denoted ℓ^p.

[3] Also denoted ℓ^∞.

(3) The space $\ell^{\infty}_{\mathbb{R}}(\mathbb{N})$ *contains two classical subspaces:*

- *the space* c *of convergent sequences;*
- *the space* c_0 *of sequences whose limit is* 0.

These subspaces are closed in $\ell^{\infty}_{\mathbb{R}}(\mathbb{N})$ *and so are also Banach spaces.*
As sets, for the real numbers p *and* q *such that* $1 < p < q$*, we have the inclusions*

$$\ell^{1}_{\mathbb{R}}(\mathbb{N}) \subset \ell^{p}_{\mathbb{R}}(\mathbb{N}) \subset \ell^{q}_{\mathbb{R}}(\mathbb{N}) \subset c_0 \subset c \subset \ell^{\infty}_{\mathbb{R}}(\mathbb{N})$$

and the inequalities

$$\|x\|_1 \geq \|x\|_p \geq \|x\|_q \geq \|x\|_\infty .$$

Example 1.3 *Let* Ω *be a topological space. The space* $\mathcal{C}_b(\Omega)$ *of continuous and bounded functions from* Ω *to* $\mathbb{R}$ *is endowed with the norm*

$$\|f\|_\infty = \sup_{x\in\Omega} |f(x)| .$$

Then $(\mathcal{C}_b(\Omega), \|.\|_\infty)$ *is a Banach space.*

Example 1.4 *For* $i = (i_1, \dots, i_n) \in \mathbb{N}^n$*, we denote* $|i| = \sum_{j=1}^{n} i_j$*. The space* $C^r_K(\mathbb{R}^n)$ *of* r*-continuously differentiable functions on* $\mathbb{R}^n$ *with support in the compact* K*, equipped with the norm*

$$\|f\| = \sum_{|i|\leq r} \sup_{x\in K} \left| f^{(i)}(x) \right| ,$$

is a Banach space.

Example 1.5 *Spaces of integrable functions.— For* $p \in [1, +\infty[$*, we consider the set* $\mathrm{L}^p([0,1])$ *of equivalence classes (with respect to the relation of equality almost everywhere on the compact* $[0,1]$*) of measurable functions* $f : [0,1] \to \mathbb{R}$ *for which the integral* $\left(\int_0^1 |f(x)|^p \, dx\right)^{1/p}$ *converges.*
Endowed with the norm

$$\|f\|_p = \left(\int_0^1 |f(x)|^p \, dx \right)^{1/p} ,$$

$\mathrm{L}^p([0,1])$ *is a Banach space.*

Example 1.6 *Sobolev spaces.— In this example, we use the notations of [Schr15]; the main part of this example comes from [Schr17] and [AdFo03].*
We denote by $\mathcal{L}^k(\mathbb{R}^n, \mathbb{R}^m)$*, the vector space of continuous* k*-multilinear maps from* $(\mathbb{R}^n)^k$ *to* $\mathbb{R}^m$*. This space is provided with the inner product*

$$A.B = \mathrm{Tr}(B^* A) = \sum_{i=1}^{nk} A(e_i).B(e_i)$$

Let U be a bounded open set in $\mathbb{R}^n$. For $k \in \mathbb{N} \cup \{\infty\}$, we denote by $C^k(U, \mathbb{R}^m)$ the set of maps of class C^k from U to $\mathbb{R}^m$. For $f \in C^\infty(U, \mathbb{R}^m)$ and for any $k \in \mathbb{N}$, we have the continuous (total) derivative $D^k f : U \to \mathcal{L}^k(\mathbb{R}^n, \mathbb{R}^m)$. On the vector space $C^k(U, \mathbb{R}^m)$, for $0 \leq k < \infty$ and $1 \leq p < \infty$, we consider the two norms:

$$|f|_k = \sum_{i=0}^{k} \sup_{x \in U} ||D^i f(x)|| \quad \textit{and} \quad ||f||_{k,p} = \left(\sum_{i=0}^{k} \int_U ||D^i(x) f||^p \right)^{1/p}.$$

We denote by

- $C_b(U, \mathbb{R}^m) = \{f \in C^k(U, \mathbb{R}^m) \; : \; |f|_k < \infty\}$ *(which is a Banach space)*
- $\mathrm{L}^p_k(U, \mathbb{R}^m)$ *the completion of the vector space*

$$\{f \in C^k(U, \mathbb{R}^m)\} \; : \; ||f||_{k,p} < \infty\}.$$

The collection of all spaces $\mathrm{L}^p_k(U, \mathbb{R}^m)$ is collectively called Sobolev spaces. They have an equivalent formulation as spaces of p-integrable functions with p-integrable distributional (or weak) derivatives up to order k (see for instance [Rou05]).
Now, for all $1 < p < \infty$, each space $\mathrm{L}^p_k(U, \mathbb{R}^m)$ is a separable reflexive Banach space (see [Rou05]).
We have the following embeddings (cf. [Belm08])

$$\mathrm{L}^p(U) \subset \mathrm{L}^q(U) \quad \textit{for } p \geq q;$$

$$\mathrm{L}^p_k(U) \subset \mathrm{L}^q_h(U) \quad \textit{for } k \geq h, \; p \geq q.$$

If U is "sufficiently regular", $\mathrm{L}^p_k(U)$ is dense in $\mathrm{L}^p_{k-1}(U)$ for $1 \leq p < \infty$ and so, by induction, $\mathrm{L}^p_k(U)$ is dense in $\mathrm{L}^p(U)$ for $1 \leq p < \infty$.
Now, following [AdFo03], 3.5, for any given integers $n > 1$ and $k > 0$, let $N = N(n, k)$ be the number of multi-indices $i = (i_1, \ldots, i_n)$ such that

$$|i| = \sum_{\alpha=1}^{n} |i_\alpha| \leq k.$$

For each such multi-index i with $|i| \leq k$, let U_i be the disjoint union of U_{i_α} where each U_{i_α} is a copy of $U \subset \mathbb{R}^n$. In this way, we get N domains U_i which are considered as disjoints sets. Let $U^{(k)}$ be the union of these N disjoint domains U_i. Given a function $f \in \mathrm{L}(U, \mathbb{R}^m)$, let $\tilde{f}^{(k)}$ be the map on $U^{(k)}$ which coincides with $D^i f$ in the distributional sense. Then the map $P : f \to \tilde{f}^{(k)}$ is an isometry from $\mathrm{L}^p_k(U, \mathbb{R})$ into $\mathrm{L}^p(U^{(k)}, \mathbb{R})$ whose range is closed. Now since

$$\mathrm{L}^p_k(U, \mathbb{R}^m) = \{(f_1, \ldots, f_m) \; : \; f_1, \ldots, f_m \in \mathrm{L}^p(U^{(k)}, \mathbb{R})\},$$

we obtain an isometry P from $\mathrm{L}^p_k(U, \mathbb{R}^m)$ into $\mathrm{L}^p(U^{(k)}, \mathbb{R})^m$ whose range is also closed.

Let q be the "conjugate" exponent to p, that is $\dfrac{1}{p}+\dfrac{1}{q}=1$. According to [AdFo03], 3.8, we have a "bracket duality" between $\mathrm{L}_k^q(U,\mathbb{R}^m)$ and $\mathrm{L}_k^p(U,\mathbb{R}^m)$ given by

$$\langle f,g\rangle=\sum_{j=1}^{m}\int_U f_j(x)g_j(x)dx.$$

For every $L\in(\mathrm{L}_k^p(U,\mathbb{R}^m))^*$, there exists $g\in\mathrm{L}_k^q(U^{(k)},\mathbb{R}^m)$ such that, if g_i denotes the restriction of g to U_i, then, for all $f\in\mathrm{L}_k^p(U,\mathbb{R}^m)$, we have

$$L(f)=\sum_{0\leq|i|\leq k}\langle D^i f,g_i\rangle.$$

In fact, on $\mathrm{L}_k^q(U,\mathbb{R}^m)$, according to [AdFo03], 3.14, we can define the norm

$$\|g\|_{k,q}=\sup\{|\langle f,g\rangle|\ :f\in\mathrm{L}_k^p(U,\mathbb{R}^m),\|f\|_{k,p}\leq 1\}.$$

Then $(\mathrm{L}_k^p(U,\mathbb{R}^m)^*$ is the completion of $\mathrm{L}_k^q(U,\mathbb{R}^n)$ according to the previous norm. Using the same arguments as in [Schr15], Corollary 4.3.3, we obtain that all the previous results are true by replacing U with a compact manifold without boundary.

Example 1.7 *Space of absolutely continuous paths of $\mathbb{R}^p$.— A path $\gamma\colon[a,b]\to\mathbb{R}^p$ is absolutely continuous if*

$$\exists h\in\mathrm{L}^1\left([a,b],\mathbb{R}^p\right):\forall t\in[a,b],\ \gamma(t)=\gamma(a)+\int_a^t h(s)ds$$

and then we define $\dot{\gamma}:=h$.
The space $\mathcal{W}\left(I,\mathbb{R}^{\mathbb{N}}\right)$ of such paths endowed with the norm

$$\|\gamma\|_{\mathcal{W}}=|\gamma(a)|+\|\dot{\gamma}\|_{\mathrm{L}^1}$$

is a separable Banach space ([Ngu97]).

Example 1.8 *Hilbert spaces.— Hilbert spaces, named after the German mathematician David Hilbert, generalize the notion of finite-dimensional Euclidian spaces. They are the first infinite-dimensional vector spaces whose theory has been studied in the first decade of the XXth century. They play a crucial role in pure and applied Mathematics.*
A Hilbert space over $\mathbb{K}$ is a $\mathbb{K}$-vector space equipped with an inner product $\langle .,.\rangle$ which is complete with respect to the induced norm. In the case $\mathbb{K}=\mathbb{R}$, the scalar product is required to be bilinear, while in the case $\mathbb{K}=\mathbb{R}$, the scalar product is required to be sesquilinear, that is, $\mathbb{C}$-linear with respect to its first variable and antilinear with respect to its second variable.
A Hilbert space is a Banach space.
The spaces ℓ^2 and $\mathrm{L}^2\left([0,1]\right)$ are classical real Hilbert spaces.

1.3 Dual Spaces

Let $\mathbb{E}$ be a $\mathbb{K}$ vector space. The set of all linear functionals, i.e. linear maps from $\mathbb{E}$ to $\mathbb{K}$, endowed with the vector space structure of pointwise addition and multiplication by constants, is called *the algebraic dual space* of $\mathbb{E}$ and is denoted $\mathbb{E}^\sharp$.

If $(\mathbb{E}, \| \ \|)$ is a Banach space and, more generally, a topological vector space (cf. Chapter 3, § 3.1.2), the set of all continuous linear functionals can be endowed with a structure of vector space, called the *topological dual* and denoted by $\mathbb{E}^*$.

For a normed vector space, a linear functional is continuous if it is bounded on the unit ball.

The topological dual $\mathbb{E}^*$ of a Banach space $(\mathbb{E}, \| \ \|)$ can be endowed with a norm $\| \ \|^*$ defined for any $\alpha \in \mathbb{E}^*$ by

$$\|\alpha\|^* = \sup_{x \in \mathbb{E}: \|x\|=1} |\alpha.x|.$$

So $\left(\mathbb{E}^*, \| \ \|^*\right)$ is a Banach space.
By reiterating the process, the topological dual $(\mathbb{E}^*)^*$ of $\mathbb{E}^*$, denoted by $\mathbb{E}^{**}$, can be endowed with a structure of Banach space relatively to the norm $\| \ \|^{**}$. One has the canonical map

$$\begin{array}{rccc} \iota: & \mathbb{E} & \to & \mathbb{E}^{**} \\ & x & \mapsto & x^{**} \end{array} \quad \text{where} \quad \begin{array}{rccc} x^{**}: & \mathbb{E} & \to & \mathbb{K} \\ & \alpha & \mapsto & \alpha.x \end{array}$$

By the Hahn-Banach theorem, ι is injective and is actually an isometry:

$$\forall x \in \mathbb{E}, \ \|\iota(x)\|^{**} = \|x\|.$$

Moreover, the range of ι is closed in $\mathbb{E}^{**}$, but it need not be equal to this space.

Definition 1.4 *A Banach space is called reflexive if the map $\iota : \mathbb{E} \to \mathbb{E}^{**}$ is surjective.*

Example 1.9 *Every finite-dimensional Banach space is reflexive.*

Example 1.10 *Any Hilbert space is reflexive.*

The Banach spaces ℓ^1, ℓ^∞ and c_0 are not reflexive.

1.4 Properties

Proposition 1.2 *A closed subspace of a Banach space is a Banach space.*

Proposition 1.3 *Any quotient of a Banach space by a closed subspace is a Banach space.*

Remark 1.1 *Let $\mathbb{E}$ be a Banach space and let $\mathbb{F}$ be a linear subspace of $\mathbb{E}$. A supplement linear subspace of $\mathbb{F}$ in $\mathbb{E}$ need not exist.*
For example, let $\mathbb{E} = \ell^\infty$ be the Banach space of bounded real sequences; the closed linear subspace $\mathbb{F} = c_0$ (Example 1.2, 3.) has no closed supplement subspace in $\mathbb{E}$ (cf. [Kot69], Ch. VI, 31, p. 427).

Proposition 1.4 *Let $\left(\mathbb{E}_1, \|.\|_{\mathbb{E}_1}\right)$ be a normed vector space and let $\left(\mathbb{E}_2, \|.\|_{\mathbb{E}_2}\right)$ be a Banach space. The space $\mathcal{L}(\mathbb{E}_1, \mathbb{E}_2)$ of continuous linear mappings from $\mathbb{E}_1$ to $\mathbb{E}_2$ equipped with the operator norm $\|.\|_{\mathcal{L}(\mathbb{E}_1,\mathbb{E}_2)}$ defined by*

$$\|A\|_{\mathcal{L}(\mathbb{E}_1,\mathbb{E}_2)} = \sup_{\|x\|_{\mathbb{E}_1} \leqslant 1} \|Ax\|_{\mathbb{E}_2}$$

is a Banach space.

Proposition 1.5 *If $\mathbb{E}$ is a Banach space, the group of invertible elements $\mathcal{GL}(\mathbb{E})$, equipped with the operator norm $\|.\|_{\mathcal{L}(\mathbb{E})}$, is open in $\mathcal{L}(\mathbb{E})$.*

Proposition 1.6 (Open mapping theorem) *Let $\mathbb{E}$ and $\mathbb{F}$ be two Banach spaces and let T be a continuous linear operator from $\mathbb{E}$ to $\mathbb{F}$.*
If T is bijective, then T^{-1} is a continuous linear operator from $\mathbb{F}$ to $\mathbb{E}$.

1.5 Derivatives in Banach Spaces

Let $(\mathbb{E}_1, \|.\|_1)$ and $(\mathbb{E}_2, \|.\|_2)$ be two Banach spaces.

Definition 1.5 *Let U be a not empty open subset of $\mathbb{E}_1$. A map $f : U \to \mathbb{E}_2$ is called differentiable at x, element of U, if there exists a map $Df(x) \in \mathcal{L}(\mathbb{E}_1, \mathbb{E}_2)$, the Fréchet derivative of f at x, such that*

$$\lim_{h \to 0} \frac{\|f(x+h) - f(x) - Df(x)(h)\|_2}{\|h\|_1} = 0.$$

If f is differentiable at any x of U, the *total derivative* or *differential* is the map $\begin{array}{cccc} Df: & U & \to & \mathcal{L}(\mathbb{E}_1, \mathbb{E}_2) \\ & x & \mapsto & Df(x) \end{array}$.
Inductively, we define the derivative of order k ($k \in \mathbb{N}^*$) as

$$D^k f = D\left(D^{k-1} f\right) : U \to \mathcal{L}^k(\mathbb{E}_1, \mathbb{E}_2) \equiv \mathcal{L}\left(\mathbb{E}_1, \mathcal{L}^{k-1}(\mathbb{E}_1, \mathbb{E}_2)\right).$$

The map f is called *smooth* or of class C^∞ if the derivatives D^k exist for every k.

Proposition 1.7 (Inverse function theorem) *Let* $(\mathbb{E}_1, \|.\|_1)$ *and* $(\mathbb{E}_2, \|.\|_2)$ *be two Banach spaces. A* C^1*-mapping* $f : \mathbb{E}_1 \to \mathbb{E}_2$ *with derivative everywhere invertible is a local diffeomorphism at every point.*

1.6 Ordinary Differential Equations

Definition 1.6 *Let* $\mathbb{E}$ *be a Banach space, and let* U *be a non-empty open set of* $\mathbb{E}$.
A C^r*-mapping, where* $r \in \mathbb{N}^*$, $X : U \to \mathbb{E}$ *is called a vector field on* U.
A curve $c :]-\epsilon, \epsilon[\to U$ *which satisfies*

$$\begin{cases} \forall t \in]-\epsilon, \epsilon[, \ \dfrac{d}{dt} c(t) = X(c(t)) \\ c(0) = x_0 \end{cases}.$$

where $x_0 \in U$ *is called an integral curve of* X *with initial condition* x_0.

Theorem 1.1 *Let* $X : U \to \mathbb{E}$ *be a* C^r*-vector field where* $r \in \mathbb{N}^* \cup \{+\infty\}$. *For all* $x_0 \in U$, *there exists a neighbourhood* V *of* x_0, *an open interval* $]-\epsilon, \epsilon[$ *and a* C^r*-mapping* $\mathrm{Fl}^X : V \times]-\epsilon, \epsilon[\to U$ *called flow of the vector field* X, *such that:*

(SolODE1) Fl *is entirely determined by* X;

(SolODE2) *for all* x *in* V, *the curve* $\mathrm{Fl}_x : t \mapsto \mathrm{Fl}^X(x, t)$ *is the integral curve of* X *with initial condition* x;

(SolODE3) *for all* $t \in]-\epsilon, \epsilon[$, *the map* $\mathrm{Fl}_t^X : x \mapsto \mathrm{Fl}^X(x, t)$ *is a* C^r*-diffeomorphism from* V *to* U;

(SolODE4) *any integral curve* X *with initial condition* $x \in V$ *coincides with* Fl_x^X *for all t in the intersection of the definition sets of both curves.*

1.7 Banach Manifolds

The framework of manifolds modelled on Banach spaces leads to interesting examples in global analysis (see, for instance, [Eel66], [Pal68] and [PiTa01]).

Definition 1.7 *Let* M *be a set. For* $p \in \mathbb{N}$, *an atlas of class* C^p *is a family* $\{(U_\alpha, \phi_\alpha)\}_{\alpha \in A}$ *satisfying the following conditions:*

(AtlB1) *each U_α is a nonempty subset of M and $M = \bigcup_{\alpha \in A} U_\alpha$, i.e. $\{U_\alpha\}_{\alpha \in A}$ is a covering of M;*

(AtlB2) *each ϕ_α is a bijection from U_α onto an open subset $\phi_\alpha(U_\alpha)$ of a Banach space $\mathbb{M}_\alpha$;*

(AtlB3) *for any $(\alpha, \beta) \in A^2$ such that $U_\alpha \cap U_\beta \neq \emptyset$, $\phi_\alpha(U_\alpha \cap U_\beta)$ is open in $\mathbb{M}_\alpha$ and the map*

$$\phi_{\alpha\beta} = \phi_\alpha \circ (\phi_\beta)^{-1} : \phi_\beta(U_\alpha \cap U_\beta) \to \phi_\alpha(U_\alpha \cap U_\beta)$$

is a C^p-isomorphism.

Each pair (U_α, ϕ_α) is called a *local chart* on M and the collection $\{(U_\alpha, \phi_\alpha)\}_{\alpha \in A}$ is called a C^p*-atlas* on M.

Given an atlas $\{(U_\alpha, \phi_\alpha)\}_{\alpha \in A}$ on M, there exists a unique topology on M such that each U_α is open and each ϕ_α is a homeomorphism. Note that this topology may not be Hausdorff.

Classically, we have a notion of equivalent C^∞-atlases on M. An equivalence class of C^∞-atlases on M is a maximal C^∞-atlas.

Definition 1.8 *A maximal C^∞-atlas on M is called a not necessary Hausdorff Banach manifold structure on M (n.n.H. Banach manifold M for short); it is called a Hausdorff Banach manifold structure on M (Banach manifold M for short) when the topology defined by this atlas is a Hausdorff topological space.*

From the definition, it follows that all Banach spaces $\mathbb{M}_\alpha$ are topologically isomorphic on connected components of M. If all connected components of M are modelled on a fixed Banach space $\mathbb{M}$ (up to an isomorphism), then we will say that M is a *pure Banach manifold* which is modelled on $\mathbb{M}$.
For the sake of simplicity, unless otherwise stated, a Banach manifold M is assumed to be a pure Banach manifold which is Hausdorff.

Example 1.11 *Space of absolutely continuous paths of a Riemannian manifold.— Let (M, g) be a C^∞-Riemannian manifold of dimension n. According to Nash's theorem (cf. [Nas56]), let $\mathcal{P}$ be an isometric embedding of M in some $\mathbb{R}^N$. The space*

$$\mathcal{W}([a,b], M) = \left\{c\colon\ [a,b] \to M : \mathcal{P} \circ c \in \mathcal{W}\left([a,b], \mathbb{R}^N\right)\right\}$$

can be endowed with a structure of Hausdorff Banach manifold modelled on the Banach space $\mathcal{W}\left([a,b], \mathbb{R}^N\right)$ as defined in Example 1.7.
Let us give an atlas of this manifold. To any (σ, θ) where $\sigma = (t_i)_{0 \leq i \leq m}$ is a subdivision of $[a,b]$: $a = t_0 \leq t_1 \leq \cdots \leq t_m = b$ and $\theta = ((U_i, \phi_i))_{1 \leq i \leq m}$ is a m-uple of charts of M, is associated with the set

$$W_{(\sigma,\theta)} = \{c \in C([a,b], M) : \forall i \in \{1, \dots, m\}, c([t_{i-1}, t_i]) \subset U_i\}$$

A continuous path c of $W_{(c,\theta)}$ is absolutely continuous if and only if

$$\forall i \in \{1,\ldots,m\}, \phi_i \circ c_i \in \mathcal{W}\left([t_{i-1},t_i], \mathbb{R}^N\right).$$

where c_i is the restriction of the path c to the interval $[t_{i-1},t_i]$. If one considers the map

$$\begin{array}{cccc} \phi_{(\sigma,\theta)}: & \Omega_{(\sigma,\theta)} & \to & \mathbb{R}^N \times \mathrm{L}^1\left([t_0,t_1], \mathbb{R}^N\right) \times \mathrm{L}^1\left([t_{m-1},t_m], \mathbb{R}^N\right) \\ & c & \mapsto & (\phi_1 \circ c_1(a), \phi_1 \circ c_1, \ldots, \phi_m \circ c_m) \end{array}$$

where $\Omega_{(\sigma,\theta)} = W_{(\sigma,\theta)} \cap \mathcal{W}([a,b], M)$, the set of all possible charts $\left(\Omega_{(\sigma,\theta)}, \phi_{(\sigma,\theta)}\right)$ is an atlas of the manifold $\mathcal{W}([a,b], M)$ (cf. [Ngu97]).

1.8 Banach Manifold Structures on Sets of Maps

We recall classical structures of Banach manifolds on sets of maps and immersions. We detail a little more the situation of the loop spaces because the elements of construction of such structures will be used later.

1.8.1 Banach Manifold Structure on the Set of Loops of a Manifold

Let M be a finite-dimensional manifold of dimension m. We denote by $C^0(\mathbb{S}^1, M)$ the set of C^0-loops of M, that is continuous maps from $\mathbb{S}^1$ to M. We endow $C^0(\mathbb{S}^1, M)$ with the *compact-open topology*[4].
From [Schr15], it follows that there exists a well-defined subset $\mathsf{L}^p_k(\mathbb{S}^1, M)$ of $C^0(\mathbb{S}^1, M)$, which has a Banach structure modelled on the Banach space $\mathsf{L}^p_k(\mathbb{S}^1, \gamma^*(TM))$ of sections of "class L^p_k" of the pull-back $\gamma^*(TM)$ over $\mathbb{S}^1$ for any $\gamma \in C^\infty(\mathbb{S}^1, M)$. From [Schr15], we get the following result:

Theorem 1.2 *For $k > 0$ and $1 \leq p < \infty$, there exists a subset $\mathsf{L}^p_k(\mathbb{S}^1, M)$ of the set $C^0(\mathbb{S}^1, M)$ of continuous loops in M which has a Banach manifold structure modelled on $\mathsf{L}^p_k(\mathbb{S}^1, \mathbb{R}^m)$, which is a reflexive Banach space. Moreover the topology of this manifold is Hausdorff and paracompact.*

Proof *(sketch) For a complete proof, see [Schr15].*
We only describe how the set $\mathsf{L}^p_k(\mathbb{S}^1, M)$ is built and gives an atlas of this Banach structure since these results will be used latter.
Choose any Riemannian metric on M and denote by exp *the exponential map of its Levi-Civita connection. Then* exp *is defined on a neighbourhood*

[4] If X and Y are topological spaces, given a compact subset K of X and an open subset $\mathcal{O}$, let $\mathcal{V}(K, \mathcal{O})$ be the set of all continuous functions $f \in C(X,Y)$ such that $f(K) \subset \mathcal{O}$. Then the family of all such $\mathcal{V}(K, \mathcal{O})$ is a subbase of the compact-open topology of $C(X,Y)$.

$\mathcal{U}$ in TM of the zero section. In fact, since $\exp$ *is a local diffeomorphism, if $\pi_M : TM \to M$ is the projection, after shrinking $\mathcal{U}$ if necessary, we can assume that the map $\mathcal{F} = (\exp, \pi_M)$ is a diffeomorphism from $\mathcal{U}$ onto an open neighbourhood $\mathcal{V}$ of the diagonal of $M \times M$. For any smooth curve $\gamma : \mathbb{S}^1 \to M$, we consider*

$$\mathcal{O}_\gamma = \{\gamma' \in C^0(\mathbb{S}^1, M) \ : \forall t \in \mathbb{S}^1, \ (\gamma(t), \gamma'(t)) \in \mathcal{V}\}$$

and $\Phi_\gamma(\gamma') = \mathcal{F}^{-1}(\gamma, \gamma')$ for $\gamma' \in \mathcal{O}_\gamma$; of course γ belongs to $\mathcal{O}_\gamma$. If we consider the map $i_\gamma : \gamma' \mapsto (\gamma, \gamma')$ from $\mathcal{O}$ to $C^0(\mathbb{S}^1, M \times M)$, then we have $\Phi_\gamma = \mathcal{F}^{-1} \circ i_\gamma$ and so $\Phi_\gamma(\mathcal{O}_\gamma)$ is an open set in the set $\Gamma^0(\gamma^* TM) \equiv C^0(\mathbb{S}^1, \mathbb{R}^m)$ of continuous sections of $\gamma^* TM$. We set

$$U_\gamma = \{\gamma' \in \mathcal{O}_\gamma, \gamma' \in \mathsf{L}^p_k(\mathbb{S}^1, \mathbb{R}^m) \cap \Phi_\gamma(\mathcal{O}_\gamma)\}.$$

Now, for $k > 0$, $\mathsf{L}^p_k(\mathbb{S}^1, \mathbb{R}^m)$ is continuously embedded in $C^0(\mathbb{S}^1, \mathbb{R}^m)$ (cf. consequence of [Schr15]) and $C^\infty(\mathbb{S}^1, \mathbb{R}^m) \cap \mathsf{L}^p_k(\mathbb{S}^1, \mathbb{R}^m)$ is dense in $\mathsf{L}^p_k(\mathbb{S}^1, \mathbb{R}^m)$ (consequence of the Theorem of Meyers-Serrin, cf. [MeSe64]); it follows that U_γ is a non-empty open set of $\mathsf{L}^p_k(\mathbb{S}^1, \mathbb{R}^m)$ and again, γ belongs to U_γ. Then we set

$$\mathsf{L}^p_k(\mathbb{S}^1, M) = \bigcup_{\gamma \in C^\infty(\mathbb{S}^1, M)} U_\gamma.$$

Then $\{(U_\gamma, \Phi_\gamma)\}_{\gamma \in C^\infty(\mathbb{S}^1, M)}$ defines an atlas for a Banach structure modelled on $\mathsf{L}^p_k(\mathbb{S}^1, \mathbb{R}^m)$.
The proof that the topology of this manifold is Hausdorff and paracompact in our context is, point by point, analogous to the proof of [Sta05], Corollary 3.23.

In the same way, we can define a Banach structure on the set $\mathsf{C}^k(\mathbb{S}^1, M)$ of C^k-maps from $\mathbb{S}^1$ in M.

The subset $\mathsf{I}^p_k(\mathbb{S}^1, M)$ (resp. $\mathsf{I}^k(\mathbb{S}^1, M)$) of maps c in $\mathsf{L}^p_k(\mathbb{S}^1, M)$ (resp. $\mathsf{C}^k(\mathbb{S}^1, M)$) which are immersions from S^1 in M is an open Banach submanifold.

1.8.2 General Case of Banach Structures on Sets of Maps

Let (N, g) be a closed Riemannian, C^∞-manifold of dimension d and let (M, h) be a closed, C^∞-manifold which is isometrically and C^∞ embedded in $\mathbb{R}^n$ and identified with its image. For any integer $d \geq 1$, $k \geq 1$ and any real number $p \in [1, \infty[$, we define:

$$\mathsf{L}^p_k(M, N) = \{f \in \mathsf{L}^p_k(M, \mathbb{R}^n) \ : \ f(x) \in N, \text{for a.e. } x \in M\}.$$

The Theorems 13.5 and 13.6 in [Pal70], proved by Palais, imply that the space $\mathsf{L}^p_k(M, N)$ has a Banach manifold structure for $kp > d$. Moreover, if $p = 2$, this set has a structure of Hilbert manifold. To put it more simply, under the

same assumptions, the set $\mathsf{C}^k(N,M)$ also has a Banach manifold structure (cf. [Bru17]). When $p=2$, $\mathsf{L}_k^p(M,N)$ has a Hilbert manifold structure. This implies that the sets $\mathsf{I}_k^p(M,N) \subset \mathsf{L}_k^p(M,N)$ and $\mathsf{I}^k(M,N) \subset \mathsf{C}^k(M,N)$ of immersions are open Banach submanifolds.
More generally, the reader will find in [PiTa01] a general construction process of the sets $\mathcal{M}(N,M)$ of maps from some set N into a paracompact separable Banach manifold M.

1.9 Banach Submanifolds

Since, in the Banach framework, a subspace need not be supplemented, we introduce different kinds of submanifolds.

Definition 1.9 *Let M be a Banach manifold.*

(1) A weak submanifold is a Banach manifold N endowed with a smooth injective mapping $\iota : N \to M$ such that, for all $x \in N$, the linear mapping $T_x\iota : T_xN \to T_{\iota(x)}M$ is injective.

(2) A weak submanifold N is called a closed submanifold of M if for all $x \in N$, $T\iota(T_xN)$ is closed in $T_{\iota(x)}M$.

(3) A weak submanifold N is called a split submanifold of M if, for all $x \in N$, $T\iota(T_xN)$ is supplemented in $T_{\iota(x)}M$.
In this case, N will be sometimes simply called a submanifold of M.

1.10 Banach-Lie Groups and Lie Algebras

Finite-dimensional Lie theory was created in the late XIXth century by the Norwegian mathematician Marius Sophus Lie ([Lie1880]). Some of the early ideas were developed in collaboration with Felix Klein. This fundamental theory of continuous transformation groups was developed with Friedrich Engel. As remarked in [SiSt65], S. Lie seemed to have two objectives:

(1) to construct a theory for differential equations analogous to algebraic Galois theory;

(2) to investigate groups which leave some geometric structures invariant as described in Erlangen program published by F. Klein in 1892.

In [Bir36] and [Bir38], G. Birkhoff extends the concept of Lie group to the Banach framework. In these articles, he develops the local Lie theory of Banach-Lie groups and Banach-Lie algebras.

Banach-Lie groups play a key role in physical systems both as phase spaces and as symmetry groups of dynamical systems (cf. [Schm04]). But, one can remark that this structure is not so relevant in global analysis because if a Banach-Lie group acts effectively on a finite-dimensional manifold, and it must be finite dimensional itself ([Del72]).

1.10.1 Banach-Lie Groups

Definition 1.10 *A Banach-Lie group* $(G,.)$ *is a Banach manifold endowed with a group structure for which multiplication* $\boldsymbol{m}: G \times G \to G$ *and inversion* $\boldsymbol{i}: G \to G$ *are smooth mappings.*

Remark 1.2 *It follows by an application of the implicit function theorem for Banach manifolds that the smoothness of the inversion mapping* $\boldsymbol{i}$ *results from the smoothness of the multiplication mapping* $\boldsymbol{m}$.

The unit element of the group G will be denoted e, and the inverse of an element g by g^{-1}.
For $g \in G$, $L_g : x \mapsto g.x$ is the *left translation*, and $R_g : x \longmapsto x.g$ is the *right translation*.
Let G_1 and G_2 be two Banach-Lie groups. A continuous map $\phi : G_1 \to G_2$ is called a *morphism of Lie groups* if ϕ is also a group morphism and, in this case, ϕ is a smooth mapping.

1.10.2 The Lie Algebra of a Banach-Lie Group

Let G be a Banach-Lie group.

Definition 1.11 *A vector field* X *on* G *is said to be left invariant if*

$$\forall g \in G,\ (L_g)^* X = X$$

where

$$(L_g)^* X = TL_{g^{-1}} \circ X \circ L_g.$$

Since

$$(L_g)^* [X,Y] = \left[(L_g)^* X, (L_g) * Y \right],$$

the set of all left-invariant vector fields on G forms a Lie subalgebra of $\mathfrak{X}(G)$ denoted by $\mathbf{L}(G)$ or $\mathfrak{g}$.

Definition 1.12 $\mathbf{L}(G)$ *is called the Lie algebra of the Banach-Lie group* G.

$\mathbf{L}(G)$ is in bijective correspondence with the tangent space T_eG *via* the linear isomorphism

$$\begin{array}{ccc} \mathbf{L}(G) & \to & T_eG \\ X & \mapsto & X_e \end{array}.$$

So T_eG becomes a Lie algebra for the bracket still denoted by $[.,.]$

$$[X_e, Y_e] = [X,Y]_e$$

Proposition 1.8 *Let $\varphi : G \to H$ be a morphism of Lie groups. Then $T_e\varphi : \mathbf{L}(G) \to \mathbf{L}(H)$ is a homomorphism of Lie algebras.*

1.10.3 The Exponential Map

An essential tool in the finite-dimensional context is the exponential map.

Definition 1.13 *Let G be a Banach-Lie group with Lie algebra $\mathfrak{g}$. A 1-parameter subgroup of G is a morphism of Lie groups $\gamma : (\mathbb{R}, +) \to (G, .)$.*

So a 1-parameter subgroup of G is a smooth curve γ in G such that

$$\begin{cases} \forall (s,t) \in \mathbb{R}^2, \ \gamma(s+t) = \gamma(s).\gamma(t) \\ \gamma(0) = e \end{cases}.$$

Proposition 1.9 *Let $\gamma : \mathbb{R} \to G$ be a smooth curve where $\gamma(0) = e$ and consider $X \in \mathfrak{g}$. We denote $\hat{X}$ the left-invariant vector field such that $\hat{X}_e = X$. Then the following assertions are equivalent:*

(1) γ is a 1-parameter subgroup of G and $X = \left.\dfrac{\partial}{\partial t}\right|_{t=0} \gamma(t)$;

(2) γ is an integral curve of the left-invariant vector field $\hat{X}$;

(3) $\mathrm{Fl}^{\hat{X}} : (t,x) \mapsto x.\gamma(t)$ is the unique global flow of the vector field $\hat{X}$.

Definition 1.14 *Let G be a Banach-Lie group with Lie algebra $\mathfrak{g}$. The exponential mapping of G is the smooth mapping $\exp : \mathfrak{g} \to G$ such that $t \mapsto \exp(tX)$ is the unique 1-parameter subgroup with tangent vector X at 0.*

We then have the following properties:

(ExpB1) $\exp(0) = e$;

(ExpB2) $T_0 \exp = \mathrm{Id}_{\mathfrak{g}}$

where **(ExpB2)** is a consequence of $\left.\dfrac{d}{dt}\right|_{t=0} \exp(tX) = X$.

A Banach-Lie group is *regular*, i.e. for any smooth curve $\alpha : [0,1] \to \mathfrak{g}$, the Cauchy problem

$$\begin{cases} \forall t \in \mathbb{R}, \ \gamma'(t) = \gamma(t).\alpha(t) \\ \gamma(0) = e \end{cases}$$

has a solution $\gamma_\alpha : [0,1] \to G$ where $\gamma_\alpha(1)$ smoothly depends on α, i.e. the mapping

$$\begin{array}{rccc} \mathrm{evol}_G : & C^\infty([0,1], \mathfrak{g}) & \to & G \\ & \alpha & \mapsto & \gamma_\alpha(1) \end{array}$$

is smooth (cf. [Nee06]).

1.10.4 Lie Subgroups and Lie Subalgebras

As for submanifolds, we have different types of Lie subgroups.
Consider a Banach-Lie group G and recall that we denote by $\mathbf{L}(G)$ its Lie algebra.
We say that a subgroup H of G is an *integral subgroup* of G if H is provided with a structure of connected Banach-Lie group for which H is a split submanifold of G in the sense of Definition 1.9, (3).

Theorem 1.3 *The map $H \mapsto \mathbf{H}$ from the set of integral subgroups of G to the set of supplemented Banach-Lie subalgebras of $\mathbf{L}(G)$ is a bijection.*

Following [Belt06], Corollary 3.7, we have the following result:

Theorem 1.4 *Let H be a closed subgroup of G and denote*

$$\mathfrak{h} = \{X \in \mathbf{L}(G) \mid \forall t \in \mathbb{R},\ \exp(tX) \in H\}.$$

Then $\mathfrak{h}$ is a closed subalgebra of $\mathbf{L}(G)$ and there exists, on H, a uniquely determined topology $\mathfrak{t}$ and manifold structure making H into a Banach-Lie group with its Lie algebra isomorphic to $\mathfrak{h}$, such that H is a closed submanifold of G and the following diagram is commutative:

$$\begin{array}{ccc} \mathfrak{h} & \longrightarrow & \mathbf{L}(G) \\ {\scriptstyle \exp_{|\mathfrak{h}}}\downarrow & & \downarrow{\scriptstyle \exp} \\ H & \longrightarrow & G \end{array}$$

where the horizontal arrows stand for inclusion maps.

Definition 1.15 *A closed subgroup H of G endowed with the Lie group structure from G which satisfies the assumption of Theorem 1.4 is called a weak Lie subgroup of G.*

1.11 Banach Vector Bundles

Let M be a Banach manifold modelled on a Banach space $\mathbb{M}$ and $\pi : E \to M$ be a surjective smooth map from a Banach manifold E to M.

Definition 1.16 *We say that the triple (E, π, M) is a Banach vector bundle with typical fibre a Banach space $\mathbb{E}$, if there exists a covering $\{U_\alpha\}_{\alpha \in A}$ by open subsets U_α of M such that, for each $\alpha \in A$, there exists a diffeomorphism $\tau_\alpha : \pi^{-1}(U_\alpha) \to U_\alpha \times \mathbb{E}$, called a trivialization, satisfying the following conditions:*

(BVB1) $\mathrm{pr}_1 \circ \tau_\alpha = \pi$ *where* $\mathrm{pr}_1 : U_\alpha \times \mathbb{E} \to U_\alpha$ *is the first projection and the restriction $\tau_{\alpha,x}$ of τ_α to the fibre $E_x := \pi^{-1}(x)$ is a Banach space isomorphism onto $\{x\} \times \mathbb{E}$ for all $x \in U_\alpha$;*

(BVB2) *if* (U_α, τ_α) *and* (U_β, τ_β) *are two trivializations such that* $U_{\alpha\beta} := U_\alpha \cap U_\beta \neq \emptyset$, *the map* $\tau_{\alpha,x} \circ (\tau_{\beta,x})^{-1} : \mathbb{E} \to \mathbb{E}$. *is an isomorphism for all* $x \in U_{\alpha\beta}$;

(BVB3) *for any trivializations* (U_α, τ_α) *and* (U_β, τ_β) *as previously, the transition function* $T_{\alpha\beta}$ *is characterized by*

$$\tau_{\alpha,x} \circ (\tau_{\beta,x})^{-1}(x,u) = (x, T_{\alpha\beta}(x,u)).$$

The map $x \mapsto T_{\alpha\beta}(x,.)$ *is a smooth map from* $U_{\alpha\beta}$ *to* $\mathcal{GL}(\mathbb{E})$

Under the previous notations, (U_α, τ_α) is called a *local bundle chart* and $\{(U_\alpha, \tau_\alpha)\}_{\alpha \in A}$ a *trivializing covering.*
Moreover, the transition functions satisfy the *cocycle conditions*:

(BCocy1) $\forall x \in U_\alpha \cap U_\beta \cap U_\gamma, \quad T_{\alpha\beta}(x,.) \circ T_{\beta\gamma}(x,.) = T_{\alpha\gamma}(x,.)$;

(BCocy2) $\forall x \in U_\alpha, \quad T_{\alpha\alpha}(x,.) = \mathrm{Id}_{\mathbb{E}}$.

Conversely, we have (cf. [Lan95]):

Proposition 1.10 *Let* $\pi : E \to M$ *be any mapping, where* M *is a Banach manifold, and let* $\mathbb{E}$ *be a Banach space. Assume that there exists a covering* $\{U_\alpha\}_{\alpha \in A}$ *by open sets* U_α *of* M *and, for each* $\alpha \in A$, *there exists a diffeomorphism* $\tau_\alpha : \pi^{-1}(U_\alpha) \to U_\alpha \times \mathbb{E}$ *such that, for each pair* (α, β) *and* $x \in U_\alpha \cap U_\beta$, *the map* $\tau_{\alpha,x} \circ (\tau_{\beta,x})^{-1}$ *is a Banach isomorphism where the conditions* **(BVB3)**, **(BCocy1)** *and* **(BCocy2)** *are satisfied. Then* E *has a unique structure of Banach manifold for which* (E, π, M) *is Banach vector bundle (with typical fibre* $\mathbb{E}$*) and* $\{(U_\alpha, \tau_\alpha)\}_{\alpha \in A}$ *is a trivializing covering.*

Consider two Banach vector bundles (E_1, π_1, M_1) and (E_2, π_2, M_2) whose typical fibre is $\mathbb{E}_1$ and $\mathbb{E}_2$, respectively.

Definition 1.17 *A Banach vector bundle morphism is a pair of smooth maps* $\phi : M_1 \to M_2$ *and* $\Phi : E_1 \to E_2$ *satisfying the following properties:*

(VBMor1) *the diagram*

$$\begin{array}{ccc} E_1 & \xrightarrow{\Phi} & E_2 \\ {\scriptstyle \pi_1}\downarrow & & \downarrow{\scriptstyle \pi_2} \\ M_1 & \xrightarrow{\phi} & M_2 \end{array}$$

is commutative and, for any $x \in M_1$, *the restriction* Φ_x *of* Φ *to each fibre* $(E_1)_x$ *is a continuous linear map into the fibre* $(E_2)_{\phi(x)}$;

(VBMor2) *for each* $x \in M_1$, *there exists bundle charts* (U_1, τ_1) *around* x *and* (U_2, τ_2) *around* $\phi(x)$ *such that* $\phi(U_1) \subset U_2$ *and the map*

$$x \mapsto (\tau_2)_{\phi(x)} \circ \Phi_x \circ (\tau_1)_x^{-1}$$

is a smooth map from U_1 *to* $\mathcal{L}(\mathbb{E}_1, \mathbb{E}_2)$.

When there is no ambiguity, a Banach vector bundle morphism will be simply called a bundle morphism.

Fix some Banach vector bundle (E, π, M) and consider a smooth map $f : N \to M$ from a Banach manifold N to M. We set

$$f^*E := \{(y, u) \in N \times E \ : \ \pi(u) = \phi(y)\}$$

Let $f^*\pi : f^*E \to N$ be the map defined by $f^*\pi(y, u) := y$ and $\pi^*f : \phi^*(E) \to E$ the map $\pi^*f(y, u) := (f(y), u)$. Then we have (cf. [DGV16], [Lan95]):

Proposition 1.11 *$(f^*E, f^*\pi, N)$ is a Banach vector bundle whose typical fibre is $\mathbb{E}$ and the pair (f, π^*f) is a bundle morphism such that, for each $y \in N$, the restriction π^*f_y of π^*f to $\{f^*E\}_y$ is an isomorphism onto the fibre $E_{f(x)}$.*

Definition 1.18 *The Banach vector bundle $(f^*E, f^*\pi, N)$ is called the pull-back of (E, π, M) over $f : N \to M$.*

1.12 Jets of Sections of a Vector Bundle

The theory of jet bundles occurring in the geometric theory of partial differential equations was introduced by the pioneer of differential topology Charles Ehresman in [Ehr53].

We first define the notion of jet in the Banach field.

Because evolutionary PDEs can be expressed on projective limits of jets on finite-dimensional vector bundles (cf. Appendix F), we shall focus on the case of finite-dimensional bundles (E, π, M). In this framework, we define the Cartan distribution on the bundle of k-jets of sections of π and talk about symmetries.

1.12.1 Jets of Sections of a Banach Vector Bundle

Let (E, π, M) be a Banach vector bundle of typical fibre $\mathbb{E}$ over the Banach manifold M modelled on the Banach vector space $\mathbb{M}$. We denote by $\Gamma(E)$ the $C^\infty(M)$-module of smooth sections of $\pi : E \to M$. Analogously, if U is an open set of M, the module of smooth sections of E over U is denoted by $\Gamma_U(E)$.

For a given section $s \in \Gamma(E)$ and a vector bundle chart (U, ϕ, Φ), we define the local representation of s as the map $\Phi \circ s \circ \phi^{-1} : \phi(U) \to \phi(U) \times \mathbb{E}$ and the corresponding local principal part $s_\phi : \phi(U) \to \mathbb{E}$ as follows

$$\forall x \in \phi(U), \ \Phi \circ s \circ \phi^{-1}(x) = (x, s_\phi(x)).$$

If $\mathcal{L}_{\text{sym}}^k(\mathbb{M},\mathbb{E})$ is the space of continuous symmetric k-linear maps of $\mathbb{M}^k$ in $\mathbb{E}$, then

$$P^k(\mathbb{B},\mathbb{E}) = E \times \mathcal{L}(\mathbb{M},\mathbb{E}) \times \mathcal{L}_{\text{sym}}^2(\mathbb{M},\mathbb{E}) \times \cdots \times \mathcal{L}_{\text{sym}}^k(\mathbb{M},\mathbb{E})$$

is the Banach space of $\mathbb{E}$-valued polynomials of degree k on $\mathbb{M}$.
For an element a of an open set $O \subset \mathbb{M}$, and a map $f : O \to \mathbb{E}$, we denote by $p^k f(a)$ the following polynomial, element of $P^k(\mathbb{M},\mathbb{E})$

$$p^k f(a) = \left(f(a), Df(a), \dots, D^k f(a)\right).$$

Let s_1 and s_2 be two local smooth sections of π defined on a neighbourhood of x in M. We say that s_1 and s_2 have the same k-jet at x if there exists a vector bundle chart (U,ϕ,Φ) where $x \in U$ such that

$$p^k(s_1)_\phi(\phi(x)) = p^k(s_2)_\phi(\phi(x)).$$

We then get an equivalence relation and we denote by $j_x^k s$ the equivalence class of s. If we denote by $J^k(\pi)$ the quotient space and consider the projection

$$\begin{array}{cccc} \pi^k: & J^k(\pi) & \to & M \\ & j_x^k s & \mapsto & x \end{array}$$

then $\left(J^k(\pi), \pi^k, M\right)$ is a Banach vector bundle of typical fibre $P^k(\mathbb{M},\mathbb{E})$ called the k-jet bundle of sections of $\pi : E \to M$.

If (U,τ) is a local trivialization of (E,π,M) with corresponding vector bundle chart (U,ϕ,Φ), we obtain a local trivialization $\left(U,\tau^k\right)$ of $J^k(\pi)$ where

$$\begin{array}{cccc} \tau^k: & \left(\pi^k\right)^{-1}(U) & \to & U \times P^k(\mathbb{M},\mathbb{E}) \\ & j_x^k s & \mapsto & \left(x, p^k s(\phi(x))\right) \end{array}.$$

1.12.2 Jet Bundle of a Finite-Dimensional Vector Bundle

We consider a vector bundle $\pi : E \to M$ above the n-dimensional connected manifold M where the fibre is m-dimensional. The projection π is a smooth surjective submersion.
Let $x \in M$ and $v \in \pi^{-1}(x) \subset E$ and let (U,φ) be a local chart of M around x with $\left(x^i\right)_{1 \leq i \leq n}$. It is possible to choose the open set U in such a way there exists V in $\mathbb{R}^m$ with coordinates $(u^\alpha)_{1 \leq \alpha \leq m}$, an open set U^0 of E such that $v \in U^0$ and a diffeomorphism φ^0 of U^0 onto $U \times V \subset \mathbb{R}^n \times \mathbb{R}^m$ such that the following diagram

$$\begin{array}{ccc} U^0 & \xrightarrow{\varphi^0} & \mathbb{R}^n \times \mathbb{R}^m \\ {\scriptstyle \pi}\downarrow & & \downarrow{\scriptstyle \text{pr}_1} \\ U & \xrightarrow{\varphi} & \mathbb{R}^n \end{array}$$

commutes. (U^0, φ^0) is called an *adapted coordinate chart* on E.

The vertical bundle VE can be seen as the pull-back $\pi^*(\pi)$ or $\pi^*(E)$ where

$$\begin{array}{ccc} \pi^*(E) & \longrightarrow & E \\ \downarrow & & \downarrow \pi \\ E & \xrightarrow{\pi} & M \end{array}$$

$\mathcal{V} = \Gamma(VE)$ denotes the C^∞-module of sections of VE.

Let s_1 and s_2 be local sections of π defined on a neighbourhood of $x \in M$. s_1 and s_2 have the same k-jet at x ($k \in \mathbb{N}$), if with respect to an adapted coordinate chart (U^0, φ^0) on E containing v, all the partial derivatives of $\mathrm{pr}_2 \circ \varphi^0 \circ s_1 \circ \varphi^{-1}$ and $\mathrm{pr}_2 \circ \varphi^0 \circ s_2 \circ \varphi^{-1}$ agree up to order k at $\varphi(x)$.

Denote by $\pi_M^k : J^k(\pi) \to M$ the fibre bundle of k-jets of C^∞ local sections of π endowed with a fibred manifold structure whose fibre above $x \in M$ is the space $\prod_{j=0}^{k} \mathcal{L}^j_{\mathrm{sym}}(\mathbb{R}^n, \mathbb{R}^m)$ where $\mathcal{L}^j_{\mathrm{sym}}(\mathbb{R}^n, \mathbb{R}^m)$ is the space of continuous j-linear symmetric mappings $\mathbb{R}^n \to \mathbb{R}^m$. We then have

$$\dim J^k(\pi) = n + m \begin{pmatrix} n+k \\ k \end{pmatrix}.$$

We use the identification $J^0(\pi) \simeq E$. Each projection $\pi_k^l : J^l(\pi) \to J^k(\pi)$ defined, for $l \geq k$, by $\pi_k^l \left[j^l(s)(x) \right] = j^k(s)(x)$ is a smooth surjection and $\pi_k^{k+1} : J^{k+1}(\pi) \to J^k(\pi)$ is an affine bundle over $J^k(\pi)$ (cf. [KMS93], 12.11).

For $k = 0$, we shall write $\pi_E^l : J^l(\pi) \to E$. We denote by $C^\infty(J^k(\pi))$ the $C^\infty(M)$-module of smooth functions on $J^k(\pi)$.

Let (U^0, φ^0) be an adapted coordinate chart on E whose local coordinates are (x^i, u^α); we then get local charts (U^k, φ^k) on $J^k(\pi)$ as follows: we take $U^k = (\pi_E^k)^{-1}(U^0)$ and $s : U \to U^0$ is written in such coordinates as: $(x^i) \mapsto (x^i, u^\alpha)$; we then get $(x^i, u^\alpha, u_i^\alpha, \ldots, u^\alpha_{i_1 \ldots i_k})$ as local coordinates at $j^k(s)(x) \in J^k(\pi)$ where for $l = 0, 1, \ldots, k$

$$u^\alpha_{i_1 i_2 \ldots i_l} = \frac{\partial^l u^\alpha(x^1, \ldots, x^n)}{\partial x^{i_1} \cdots \partial x^{i_l}}.$$

Notation 1.2 *We then denote* $\left[x, u^{(k)}\right] = (x^i, u^\alpha, u_i^\alpha, \ldots, u^\alpha_{i_1 \ldots i_k})$.

Remark 1.3 *In [Sei10], Seiler uses the notation* $u^\alpha_{[j_1, j_2, \ldots, j_n]}$ *for* $u^\alpha_{i_1 \ldots i_k}$ *where* $j_1 + \cdots + j_n = k$ *and* j_p *corresponds to the number of derivatives of* s^α *with respect to* x^p.

Example 1.12 *The subset* $J^1(\pi)$ *where* $\pi : \mathbb{R} \times N \to \mathbb{R}$ *of elements* $j_0^1 f$ *is a submanifold corresponding to the tangent bundle* TN *([Cha09]).*

Example 1.13 *The subset* $J^1(\pi)$ *where* $\pi : N \to N \times \mathbb{R}$ *of elements* $j_a^1 f$ *where* $f(a) = 0$ *corresponds to the cotangent bundle* T^*N.

1.12.3 Cartan Distributions on $J^k(\pi)$ and Symmetries

Let s be a section of π on a neighbourhood U of $x \in M$. For $\xi = j^k(s)(x) \in J^k(\pi)$, the n-dimensional subspace $R(s, x)$ of $T_\xi J^k(\pi)$ equals to the tangent space at ξ to the submanifold $j^k(s)(U) \subset J^k(\pi)$ is called an *R-plane*.

The Cartan subspace $\mathcal{C}^k(\xi)$ of $T_\xi J^k(\pi)$ is the linear subspace spanned by all R-planes $R(s', x)$ such that $j^k(s')(x) = \xi$. So it is the hull of the union of $\left(j^k(s)\right)_*(x)(T_x M)$ where s is any local section of π around de x.

Cartan subspaces form a smooth distribution on $J^k(\pi)$ called *Cartan distribution* of $J^k(\pi)$ and denoted $\mathcal{C}^k$.

In an adapted coordinate chart,

$$D_{k,i} = \frac{\partial}{\partial x^i} + \sum_{\substack{\alpha=1,\dots,m \\ |\sigma|\leq k-1}} u^\alpha_{\sigma i} \frac{\partial}{\partial u^\alpha_\sigma} \qquad i = 1, \dots, n$$

$$\frac{\partial}{\partial u^{\alpha'}_{\sigma'}} \qquad \begin{array}{c} \alpha' = 1, \dots, m \\ |\sigma'| = k \end{array}$$

is a basis of vector fields for the Cartan distribution where the vector fields $\frac{\partial}{\partial u^{\alpha'}_{\sigma'}}$ ($\alpha' = 1, \dots, m$, $|\sigma'| = k$) are vertical for the projection π^k_{k-1}.

Remark 1.4 *The Cartan distribution also called canonical contact structure can be defined in a dual way via the differential* 1*-forms on* $J^k(\pi)$

$$\omega^\alpha_\sigma = du^\alpha_\sigma - \sum_{i=1}^n u^\alpha_{\sigma i} dx^i \qquad \begin{array}{c} \alpha = 1, \dots, m \\ |\sigma| \leq k-1 \end{array}$$

Because for $\sigma' = \underbrace{(i, i, \dots, i)}_{k \text{ times}}$, we have:

$$\left[\frac{\partial}{\partial u^{\alpha'}_{\sigma'}}, D_{k,i}\right] = \frac{\partial}{\partial u^{\alpha'}_{\sigma''}} \qquad \sigma'' = \underbrace{(i, i, \dots, i)}_{k-1 \text{ times}},$$

the Cartan distribution on $J^k(\pi)$ is not involutive.

Proposition 1.12 *Let* $N \subset J^k(\pi)$ *be a submanifold such that* $\pi_{k|N} : N \to M$ *is a diffeomorphism onto an open subset* U *of* M.
Then N *is an integral submanifold of the Cartan distribution on* $J^k(\pi)$ *if and only if* $N = j^k(s)(U)$ *for some section* $s : U \to E$ *of* π.

Remark 1.5 *The fibres of the projection* π^k_{k-1} *are also integral manifolds of the Cartan distribution.*

Definition 1.19 *A vector field* X *on* $J^k(\pi)$ *is said to be a symmetry if for any vector field* Y *from the Cartan distribution on* $J^k(\pi)$*, the Lie bracket* $[X, Y]$ *also belongs to this distribution.*

Proposition 1.13 *A vector field X on $J^k(\pi)$ is a symmetry if and only if the corresponding* 1*-parameter group of local diffeomorphisms preserves the Cartan distribution on $J^k(\pi)$.*

Theorem 1.5 *(Lie-Bäcklund theorem)*
Let X be a symmetry of $J^k(\pi)$.
For any $p \geq k$, there exists a unique symmetry X^p of $J^p(\pi)$ such that

$$(\pi_k^p)_* (X_p) = X.$$

1.13 Notes

Banach's book ([Ban32]), published in 1932, can be considered as a reference in the field of the general theory of normed spaces.
There exist a lot of books devoted to the Banach structures. For the field of Banach vector spaces the reader is referred to [AbrMa78], [AMR88], [AbRo67], [Bou06], [Caro05], [ChDeDi82], [DGV16], [Meg98], [Tre67] and [RoRo64]. For the more general framework of Banach manifolds (§ 1.7), we use [Bou67] and [Lan95]. Our references for the structures of Banach vector bundles (§ 1.11) are [Pal70], [DGV16] and [Lan95]. For the core notions of Banach-Lie groups and Lie algebras (§ 1.10), we also refer to [Belt06], Chapter 2.
For the notions of jets in the Banach framework, we refer to [Bou67] and [AbRo67] (see also [DGV16]); in the finite-dimensional framework, our basic references are [And92], [IVV04], [Kis12], [KrKe2000], [Sau89], [Vit17] and [Tak79]).

2

Some Banach-Lie Structures

In this chapter, we define and develop some Lie structures on Banach manifolds and bundles which correspond to classical Lie structures in finite dimension. Essentially, these structures are defined by tensors. We begin by a panorama of such linear structures on a Banach space. This permits to define such structures on Banach bundles and on the tangent space of a Banach manifold.
This naturally leads to developing the notion of tensor structure, which is a particular case of the classical notion of G-structure. This context will be illustrated by many examples.

2.1 Linear Tensor Structures

Unless otherwise mentioned, we work in this section only with real Banach spaces.

2.1.1 Operators from a Banach Space to Its Dual and Bilinear Forms

Let $\mathbb{E}$ be a Banach space equipped with a norm $\| \ \|$. We recall that the topological dual is denoted $\mathbb{E}^*$. It will be equipped with the associated norm $\| \ \|^*$ defined by

$$\|\alpha\|^* = \sup_{\|u\|=1} |\alpha.u| . \tag{2.1}$$

If $\mathbb{F}$ is another Banach space provided with a norm $\| \ \|'$, for any continuous injective operator $A : \mathbb{F} \to \mathbb{E}^*$, we consider the norm $\| \ \|_A$ on $\mathbb{F}$ defined by

$$\|u\|_A = \|A.u\|^* = \sup_{\|v\|=1} |\langle A.u, v\rangle| = \sup_{v\in\mathbb{E}\setminus\{0\}} \frac{|\langle A.u, v\rangle|}{\|v\|} \tag{2.2}$$

where $\langle \ , \ \rangle$ is the duality pairing.

The *operator norm* of A is given by

$$\|A\|^{\mathrm{op}} = \sup_{u\in\mathbb{F}\setminus\{0\}} \frac{\|A.u\|^*}{\|u\|'} = \sup_{u\in\mathbb{F}\setminus\{0\}} \sup_{v\in\mathbb{E}\setminus\{0\}} \frac{|\langle A.u, v\rangle|}{\|u\|' \, \|v\|}. \tag{2.3}$$

DOI: 10.1201/9781003435587-2

Because we have

$$\forall u \in \mathbb{E},\ \|u\|_A \leqslant \|A\|^{\mathrm{op}} \|u\|',$$

then the Identity map $\mathrm{Id} : (\mathbb{F}, \| \, \|_A) \to (\mathbb{F}, \| \, \|')$ is continuous.
Denote by $\widehat{\mathbb{F}}_A$ the completion of $\mathbb{F}$ for the norm $\| \, \|_A$.
To such an operator A is associated a canonical continuous bilinear form $\mathcal{B}_A$ on $\mathbb{F} \times \mathbb{E}$ defined by

$$\mathcal{B}_A(u, v) = \langle A.u, v \rangle . \tag{2.4}$$

Conversely, any continuous bilinear form $\mathcal{B}$ on $\mathbb{F}\times\mathbb{E}$ gives rise to the following continuous linear mappings $\mathcal{B}_L^\flat : \mathbb{F} \to \mathbb{E}^*$ and $\mathcal{B}_R^\flat : \mathbb{E} \to \mathbb{F}^*$, respectively, defined by

$$\begin{array}{rccc} \mathcal{B}_L^\flat.u : & \mathbb{F} & \to & \mathbb{R} \\ & v & \mapsto & \mathcal{B}(u, v) \end{array}, \quad \begin{array}{rccc} \mathcal{B}_R^\flat.v : & \mathbb{E} & \to & \mathbb{R} \\ & u & \mapsto & \mathcal{B}(u, v) \end{array}$$

Definition 2.1 *Let $\mathbb{E}$ and $\mathbb{F}$ be two Banach spaces The continuous bilinear form $\mathcal{B} : \mathbb{F} \times \mathbb{E} \to \mathbb{R}$ is said to be left (resp. right) weakly non-degenerate if $\mathcal{B}_L^\flat : \mathbb{F} \to \mathbb{E}^*$ (resp. $\mathcal{B}_R^\flat : \mathbb{E} \to \mathbb{F}^*$) is injective.*
If $\mathcal{B}_L^\flat$ (resp. $\mathcal{B}_R^\flat$) is also surjective, then $\mathcal{B}$ is said to be left (resp. right) strongly non degenerate.

Note that if $\mathbb{F} = \mathbb{E}$ then, when $\mathcal{B}$ is symmetric or skew-symmetric, $\mathcal{B}$ is left weakly (resp. strongly) non-degenerate if and only if $\mathcal{B}$ is right weakly (resp. strongly) non-degenerate. In this case, we will simply say that $\mathcal{B}$ is weakly (resp. strongly) non-degenerate.

If we assume that $\mathcal{B}$ is a continuous bilinear form on $\mathbb{E}$ which is right and left weakly non-degenerate, as previously, we can define two new norms $\|u\|_{\mathcal{B}_L^\flat} = \left\|\mathcal{B}_L^\flat(u)\right\|^*$ and $\|u\|_{\mathcal{B}_R^\flat} = \left\|\mathcal{B}_R^\flat(u)\right\|^*$ and we denote $\mathbb{E}_L$ (resp. $\mathbb{E}_R$) the normed space $\left(E, \| \, \|_{\mathcal{B}_L^\flat}\right)$ (resp. $\left(E, \| \, \|_{\mathcal{B}_R^\flat}\right)$).

Remark 2.1

(1) If $\widetilde{\| \, \|}$ is a norm that is equivalent to $\| \, \|$on $\mathbb{E}$, then the corresponding norms $\widetilde{\| \, \|}^*$ and $\| \, \|^*$ are also equivalent norms on $\mathbb{E}^*$ and so are $\widetilde{\| \, \|}_{\mathcal{B}_L^\flat}$ and $\| \, \|_{\mathcal{B}_L^\flat}$ (resp. $\widetilde{\| \, \|}_{\mathcal{B}_R^\flat}$ and $\| \, \|_{\mathcal{B}_R^\flat}$). Therefore, the completion of $\mathbb{E}_L$ (resp. $\mathbb{E}_R$) only depends on the Banach structure on $\mathbb{E}$.

(2) If $\mathcal{B}$ is symmetric or skew-symmetric, then we have $\| \, \|^L = \| \, \|^R$ and $\mathbb{E}_L = \mathbb{E}_R$. In this case, we set $\| \, \|_{\mathcal{B}^\flat} = \| \, \|_{\mathcal{B}_L^\flat} = \| \, \|_{\mathcal{B}_R^\flat}$ and the associated Banach space will be denoted $\mathbb{E}_{\mathcal{B}^\flat}$.

Proposition 2.1 *Let $A : \mathbb{E} \to \mathbb{E}^*$ be an injective continuous linear operator.*

(1) There exists a unique continuous linear mapping $\widehat{A} : \widehat{\mathbb{E}}_A \to \mathbb{E}^*$ whose restriction to $\mathbb{E}$ is A.
Moreover, the range of $\widehat{A}$ is closed and $\widehat{A}$ is an isometry from $\widehat{\mathbb{E}}_A$ onto its range.

(2) *The associated bilinear form $\mathcal{B}_A$ is a continuous left weakly non-degenerate bilinear form which can be extended to a form of the same type $\widehat{\mathcal{B}}_A^L$ on $\widehat{\mathbb{E}}_A \times \mathbb{E}$.*
Moreover, if $\mathcal{B}_A$ is symmetric (resp. skew-symmetric) then $\mathcal{B}_A$ can be extended to a continuous right weak non-degenerate bilinear form $\widehat{\mathcal{B}}_A^R$ on $\mathbb{E} \times \widehat{\mathbb{E}}_A$ such that $\widehat{\mathcal{B}}_A^L(u,v) = \widehat{\mathcal{B}}_A^R(v,u)$ (resp. $\widehat{\mathcal{B}}_A^L(u,v) = -\widehat{\mathcal{B}}_A^R(v,u)$)

(3) *If $\mathcal{B}_A$ is symmetric (resp. skew-symmetric) and $\mathbb{E}$ is reflexive, then $\widehat{A}$ is an isomorphism from $\widehat{\mathbb{E}}_A$ to $\mathbb{E}^*$, and the adjoint $\widehat{A}^*$ is an isomorphism from $\mathbb{E}^{**} \simeq \mathbb{E}$ to $\widehat{\mathbb{E}}_A^*$.*

Proof *By construction, the linear operator A is an isometry from the normed space $(\mathbb{E}, || \, ||_A)$ into $\mathbb{E}^*$. So A has a unique extension $\widehat{A} : \widehat{\mathbb{E}}_A \to \mathbb{E}^*$ which is also an isometry. This implies that $\widehat{A}$ is also an isometry from the completion $\widehat{E}_A$ onto its range $\mathbb{E}_A^\flat$ which is closed in $\mathbb{E}^*$, which ends the proof of (1).*

Set $\widehat{\mathcal{B}}_A^L(u,v) = \widehat{A}(u)(v)$ for $u \in \mathbb{E}_A$, $v \in \mathbb{E}$. On the one hand, we have:

$$|\widehat{\mathcal{B}}_A^L(u,v)| = |\widehat{A}(u)(v)| \leq ||\widehat{A}(u)||^* ||v|| = ||u||_A ||v|| \tag{2.5}$$

This implies that $\widehat{\mathcal{B}}_A^L$ is a continuous bilinear form on $\widehat{\mathbb{E}}_A \times \mathbb{E}$. Assume that $\widehat{\mathcal{B}}_A^L(u,v) = 0$ for all $v \in \mathbb{E}$, then this implies that $\widehat{A}(u) = 0$ and so $u = 0$ since $\widehat{A}$ is injective. Thus $\widehat{\mathcal{B}}_A^L$ is left non-degenerate.

On the other hand, since the restriction of $\widehat{A}$ to the vector space $\mathbb{E}$ is A and $\mathcal{B}_A$ is symmetric so the restrictions of $\widehat{\mathcal{B}}_A^L$ and $\widehat{\mathcal{B}}_A^R$ to $\mathbb{E} \times \mathbb{E}$ coincide. Therefore, for all $(u,v) \in \mathbb{E}^2$, we have

$$|\widehat{\mathcal{B}}_A^L(u,v)| = |\mathcal{B}_A(u,v)| = |\mathcal{B}_A(v,u)| = |\widehat{A}(v)(u)| \leq ||\widehat{A}(v)||^* ||u|| = ||v||_A ||u||.$$

Note that since $\widehat{A}$ is injective, $\widehat{A}$ is weakly non-degenerate. As previously, $\mathcal{B}_A$ can be also extended to a bilinear continuous form on $\mathbb{E} \times \widehat{\mathbb{E}}_A$ which is right weakly non-degenerate, and this completes the proof of (2) for the symmetric case. It is clear that analog arguments can be applied in the skew-symmetric case using the relation $\mathcal{B}_A(u,v) = -\mathcal{B}_A(v,u)$.
Now if $\mathbb{E}$ is reflexive, from the Hahn-Banach theorem, we must have $E_A^\flat = \mathbb{E}^*$ (cf. [Bam99], Lemma 2.7). This implies that the adjoint $(\widehat{A})^*$ is an isomorphism from $\mathbb{E} = \mathbb{E}^{**}$ to $\mathbb{E}_A^*$. In fact, according to the proof of Lemma 2.8 of [Bam99], if $\mathbb{E}$ is reflexive, A is symmetric, the inverse of the map $u \mapsto \widehat{A}(u, \,)$ is defined by the map $\alpha \mapsto u$ by the following relation:

$$\forall \beta \in \mathbb{E}_A^*, \; u(\beta) := \alpha((\widehat{A})^{-1}(\beta)).$$

If A is skew-symmetric, then this inverse is defined in the same way, replacing the second member with its opposite.

Remark 2.2 *Under the assumptions of Proposition 2.1.(3), if $\mathcal{B}_A$ is symmetric (resp. skew-symmetric), then $\widehat{A}^* = \widehat{A}$ (resp. $\widehat{A}^* = -\widehat{A}$) and so $\widehat{A}$ is an isomorphism from $\mathbb{E}$ to $(\widehat{\mathbb{E}}_A)^*$.*

Remark 2.3 *In [Tro76], Tromba defines a notion of almost inner product on a Banach space $\mathbb{E}$. Specifically, if $|| \; ||$ is a norm on $\mathbb{E}$ which defines the topology of $\mathbb{E}$, assume that $\mathbb{E}$ can be provided with a norm $|| \; ||_w$ such that the identity map from $\mathbb{E}_w := (\mathbb{E}, || \; ||_w)$ to $(\mathbb{E}, || \; ||)$ is continuous. An almost inner product on $\mathbb{E}$ with respect to $|| \; ||_w$ is a continuous bilinear map $B : E \times E_w \to \mathbb{R}$ such that*

(a) $u \mapsto \sup\{|B(u, v)| \; : \; ||u||_w = 1\}$ *is equivalent to* $|| \; ||$;

(b) $v \mapsto \sup\{|B(u, v)| \; : \; ||u|| = 1\}$ *is equivalent to* $|| \; ||_w$;

(c) for any $\alpha \in \mathbb{E}^*$ which belongs to $\mathbb{E}_w^*$, there exists $v_\alpha \in \mathbb{E}$ such that one has $\alpha = B^\flat(v_\alpha)$;

(d) $\forall u \in \mathbb{E} \setminus \{0\}$, $B(u, u) > 0$.

Then the map $u \mapsto \mathcal{B}^\flat(u)$ gives rise to an isomorphism from $\mathbb{E}$ to $(\widehat{\mathbb{E}}_w)^*$ ([Tro76], Proposition 2).

Note that, in fact, given an almost inner product on $\mathbb{E}$, the equivalence class of the norm $|| \; ||_w$ is characterized by (b) for a fixed norm $|| \; ||$ on $\mathbb{E}$. In this way the equivalence class of $|| \; ||_w$ does not depend on the choice of $|| \; ||$ as in Remark 2.1 (1) and, in particular, $|| \; ||_w$ is equivalent to $|| \; ||_A$ for $A = B^\flat$ as defined in Remark 2.1 (2).

2.1.2 Notations for Inner Pre-Hilbert Product and Hilbert Extension

Let g be a continuous inner pre-Hilbert product on a Banach space $\mathbb{E}$. On $\mathbb{E}$ we have a structure of pre-Hilbert space associated with g and we denote by $|| \; ||_g$ the associated norm. Throughout this chapter, we will denote by $\mathbb{E}_g$ the vector space $\mathbb{E}$ with this pre-Hilbert structure while $\mathbb{E}$ will correspond to the vector space $\mathbb{E}$ provided with its Banach structure.

For any norm $|| \; ||$ on $\mathbb{E}$ which defines its Banach structure, recall that the operational norm $||g||^{\mathrm{op}}$ is defined by

$$||g||^{\mathrm{op}} = \sup\{|g(u, v)|, \; ||u|| \leq 1, \; ||v|| \leq 1\}$$

Now, for any u in $\mathbb{E}$, we have

$$||u||_g \leq \sqrt{||g||^{\mathrm{op}}}||u|| \tag{2.6}$$

Thus the identity map $\mathfrak{i} : \mathbb{E}_g \to \mathbb{E}$ is a linear continuous map which is surjective but, of course, which is not an automorphism in general. We will denote by $\widehat{\mathbb{E}_g}$ the Hilbert space, which is the completion of $\mathbb{E}_g$. Then we have an embedding $\iota : \mathbb{E}_g \to \widehat{\mathbb{E}_g}$. If we denote by $\hat{g}$ the extension of g to $\widehat{\mathbb{E}_g}$, then we have $g = \iota^* \hat{g}$.

Let T be a continuous endomorphism of $\mathbb{E}_g$, then $\mathrm{i} \circ T$ is a well-defined continuous endomorphism of $\mathbb{E}$. Note that $\mathrm{i} \circ T$ is injective/surjective/bijective if and only if T is so. In particular, if T is an automorphism of $\mathbb{E}_g$, then $\mathrm{i} \circ T$ is also an automorphism of $\mathbb{E}$. For simplicity, when **no ambiguity is possible, we will identify T with $\mathrm{i} \circ T$.**

Remark 2.4 The reader must be careful, *since from § 2.1.1, to a pre-Hilbert inner product g on a Banach space is associated with a norm $|| \ ||_{g^\flat}$ which should not be confused with $|| \ ||_g$, which is previously defined. Thus, when $\mathbb{E}$ is reflexive, the extension $\widehat{\mathbb{E}}_{g^\flat}$ should not be confused with the Hilbert space $\widehat{\mathbb{E}_g}$.*

2.1.3 Krein Inner Products

In this section, the reader is referred to the book [AzIo89].

Definition 2.2 *Let $\mathbb{E}$ be a real Banach space.*

(1) An indefinite inner product on $\mathbb{E}$ is a continuous, weakly non-degenerate, symmetric bilinear form g on $\mathbb{E}$.

(2) An element u of $\mathbb{E}$ is called

- *positive if $g(u,u) > 0$;*
- *negative if $g(u,u) < 0$;*
- *isotropic if $g(u,u) = 0$.*

We denote by

- $\mathbb{E}^{++}$ the set of positive elements of $\mathbb{E}$;
- $\mathbb{E}^{--}$ the set of negative elements of $\mathbb{E}$;
- $\mathbb{E}^{0}$ the set of neutral elements of $\mathbb{E}$.

We set $\mathbb{E}^{00} = \mathbb{E}^0 \setminus \{0\}$.
For $(u,v) \in \mathbb{E}^2$, we will denote $g(u,v) = (u|v)$.
If $\mathbb{E}^{++}$ and $\mathbb{E}^{--}$ are non-empty sets, g is called an *indefinite metric.*
If $\mathbb{E}^{--} = \emptyset$ (resp. $\mathbb{E}^{++} = \emptyset$), then g is called *positive definite* (resp. *negative definite*). In these cases, $\mathbb{E}^{00} = \emptyset$.

Definition 2.3 *Let $\mathbb{F}$ be a linear subspace of a Banach space $\mathbb{E}$ and let g be an indefinite inner product on $\mathbb{E}$. The orthogonal of $\mathbb{F}$ is the set*

$$\mathbb{F}^\perp = \{u \in \mathbb{E} : \forall v \in \mathbb{F}, g(u,v) = 0\}.$$

$\mathbb{F}^\perp$ is a closed linear subspace of $\mathbb{E}$. If g is indefinite, in general, $\mathbb{F} \cap \mathbb{F}^\perp \neq \{0\}$. If $\mathbb{F}$ is the vector space generated by any vector in $\mathbb{E}^+$ or $\mathbb{E}^-$, then $(\ | \ )_{\mathbb{F}}$ is

positive definite or negative definite.
A subspace $\mathbb{F}$ is *isotropic* if $\mathbb{F} \subset \mathbb{F}^{\perp}$.

Definition 2.4 *Let $\mathbb{E}$ be a Banach space.*

(1) An indefinite inner product g on $\mathbb{E}$ is called an indefinite Krein inner product if there exists a decomposition $\mathbb{E} = \mathbb{E}_1 \oplus \mathbb{E}_2$ of $\mathbb{E}$ such that the restriction of g to $\mathbb{E}_1$ (resp. $\mathbb{E}_2$) is an indefinite inner product where $\mathbb{E}_1$ is the g-orthogonal of $\mathbb{E}_2$.

(2) A Krein inner product is called a neutral inner product if there exists an associated decomposition $\mathbb{E} = \mathbb{E}^+ \oplus \mathbb{E}^-$ of $\mathbb{E}$ such that $\mathbb{E}^+$ and $\mathbb{E}^-$ are isomorphic.

Remark 2.5 *An indefinite inner product g on a Banach space may not have a non-trivial decomposition as in Definition 2.4.*

Consider a Krein inner product g on $\mathbb{E}$ and a decomposition $\mathbb{E} = \mathbb{E}^+ \oplus \mathbb{E}^-$. To g is canonically associated a pre-Hilbert product γ on $\mathbb{E}$ given by

$$\gamma_{\mathbb{E}^+} = g_{\mathbb{E}^+}, \quad \gamma_{\mathbb{E}^-} = -g_{\mathbb{E}^-}$$

where the γ-orthogonal of $\mathbb{E}^+$ is $\mathbb{E}^-$.

Let $\widehat{\mathbb{E}_\gamma}$ the Hilbert space generated by the pre-Hilbert product γ on $\mathbb{E}_\gamma$. We then have an orthogonal decomposition $\widehat{\mathbb{E}_\gamma} = \widehat{\mathbb{E}_\gamma}^+ \oplus \widehat{\mathbb{E}_\gamma}^-$ and $\mathbb{E}^+ = \mathbb{E} \cap \widehat{\mathbb{E}_\gamma}^+$, $\mathbb{E}^- = \mathbb{E} \cap \widehat{\mathbb{E}_\gamma}^-$.
If we consider the map

$$\begin{array}{rccc} J: & \mathbb{E} & \to & \mathbb{E} \\ & u = u^+ + u^- & \mapsto & u^+ - u^-, \end{array}$$

we have $g(u,v) = \gamma(u, Jv)$ and $\gamma(u,v) = g(u, Jv)$ and $\gamma(u, Jv) = \gamma(Ju, v)$.
J is called the *fundamental symmetry* of g, and γ the *fundamental Hilbert product* of g.

Proposition 2.2 *The Krein inner product g can be extended by continuity to a Krein inner product $\hat{g}$ on $\widehat{\mathbb{E}_\gamma}$ and $\hat{g}$ is strongly non-degenerate.*

Proof *In this proof, we simply denote by $\widehat{\mathbb{E}}$ the Hilbert space $\widehat{\mathbb{E}_\gamma}$ generated by the pre-Hilbert space $\mathbb{E}_\gamma$; we can extend the inner product γ to a continuous bilinear form $\hat{\gamma}$ on $\widehat{\mathbb{E}_\gamma}$ which is the inner product on $\widehat{\mathbb{E}}$; in particular, $\hat{\gamma}$ is strongly non-degenerate. According to a decomposition $\mathbb{E} = \mathbb{E}^+ \oplus \mathbb{E}^-$, let P^+ and P^- be the projections of $\mathbb{E}$ onto $\mathbb{E}^+$ and $\mathbb{E}^-$, respectively. Note that P^+ and P^- are continuous according to the norm $|| \; ||_\gamma$. Since $J = P^+ - P^-$, it follows that J can be extended to a map $\widehat{J} : \widehat{\mathbb{E}} \to \widehat{\mathbb{E}}$. Therefore, since $g(u,v) = \gamma(u, Jv)$, it follows that g can be extended to a Krein inner product on $\widehat{\mathbb{E}}$. We denote by $\hat{\gamma}^\flat : \widehat{\mathbb{E}} \to \widehat{\mathbb{E}}^* \equiv \widehat{\mathbb{E}}$ the linear isomorphism $u \mapsto \gamma(u,)$*

and we set $\hat{g}^\flat = \hat{\gamma}^\flat \circ \widehat{J}$. Since $\widehat{J}$ is invertible, it follows that $\hat{g}$ is strongly non-degenerate.

We then have the following results:

Corollary 2.1 *Let $\mathbb{E}$ be a separable Banach space.*

(1) Let g_1 and g_2 be two pre-Hilbert products on $\mathbb{E}$ such that the norms $|| \;||_{g_1}$ and $|| \;||_{g_2}$ are equivalent on $\mathbb{E}$. Then there exists a linear automorphism A of the Banach space $\mathbb{E}$ such that $A^* g_1 = g_2$.

(2) Let g_1 and g_2 be two Krein inner products on $\mathbb{E}$. Assume that the norms $|| \;||_{\gamma_1}$ and $|| \;||_{\gamma_2}$ are equivalent on $\mathbb{E}$ where γ_i is the pre-Hilbert product associated with g_i, for $i \in \{1,2\}$. Then there exists an automorphism A of $\mathbb{E}$ such that $A^* g_1 = g_2$.

Proof

(1) From our assumptions, the pre-Hilbert spaces $\mathbb{E}_{g_1}$ and $\mathbb{E}_{g_2}$ define the same Hilbert space $\widehat{\mathbb{E}}$.
From [Dix53], there exists a countable total orthonormal system $\{e^i_\alpha\}_{\lambda \in \mathbf{\Lambda}}$ in $\mathbb{E}_{g_i}$, for $i \in \{1,2\}$. Since each $\mathbb{E}_{g_i}$ is dense in $\widehat{\mathbb{E}}$, it follows that $\{e^i_\alpha\}_{\lambda \in \mathbf{\Lambda}}$ is also a total orthonormal system in $\widehat{\mathbb{E}}$. So, $x \in \widehat{\mathbb{E}}$ can be written:

$$x = \sum_{\lambda \in \mathbf{\Lambda}_i(u)} g_i(x, e^i_\lambda) e^i_\lambda. \tag{2.7}$$

Moreover, if $|| \;||_{g_i}$ is the norm on $\widehat{\mathbb{E}}$ associated with g_i, we have

$$||x||_{g_i} = \Big(\sum_{\lambda \in \mathbf{\Lambda}_i(u)} |g_i(x, e^i_\lambda)|^2\Big)^{\frac{1}{2}} \tag{2.8}$$

(cf. Theorem 3.6-3 in [Krey78]). In particular, such properties are true in particular for each $x \in \mathbb{E}_{g_i}$.
On the other hand, we can define an automorphism $\hat{T}$ of $\widehat{\mathbb{E}}$ by $\hat{T}(e^1_\lambda) = e^2_\lambda$ for all $\lambda \in \Lambda$ (cf. proof of Theorem 3.6-5 in [Krey78]). Moreover, for any $x \in \widehat{\mathbb{E}}$, we have

$$\hat{T}(x) = \sum_{\lambda \in \mathbf{\Lambda}_1(x)} g_1(x, e^1_\lambda) e^2_\lambda. \tag{2.9}$$

Since each e^i_λ belongs to $\mathbb{E}_{g_i}$, we have $\hat{T}(\mathbb{E}) \subset \mathbb{E}$. But the same argument can be applied to $\hat{T}^{-1}$. If T is the restriction of $\hat{T}$ to E, and since $\hat{T}$ is bijective and $\hat{T}(\mathbb{E}) = \mathbb{E}$, it follows that T is bijective. Moreover, as we have $||\hat{T}(u)||_{g_2} = ||u||_{g_1}$, that is $||\hat{T}|| = 1$, it follows that T is bounded. The same argument applied to $\hat{T}^{-1}$ implies that T is an automorphism of the pre-Hilbert space $\mathbb{E}$ which satisfies $g_1(Tu, Tv) = g_2(u, v)$.

Now, given any norm $|| \;||$ on the Banach space $\mathbb{E}$, the identity from $\mathbb{E}_{g_2}$ into the Banach space $(\mathbb{E}, || \;||)$ is an injective continuous map denoted ι_2.

Then $A = \iota_2 \circ T$ is a bijective bounded linear endomorphism of the Banach space $\mathbb{E}$ and so A belongs to $\mathrm{GL}(\mathbb{E})$ and satisfies $A^* g_2 = g_1$.

(2) From the construction of the weak inner products γ_1 and γ_2, since $\| \ \|_{g_1}$ and $\| \ \|_{g_2}$ are equivalent on $\mathbb{E}$, the result is a consequence of (1).

Remark 2.6

1. *The assumption of separability for $\mathbb{E}$ is in general necessary since any separable pre-Hilbert space need not have a total orthonormal system (cf. [Dix53] for such an example).*
2. *If g is a continuous symmetric coercive positive bilinear form on a Hilbert space $\mathbb{H}$ provided with an inner product $< .,. >$, there exists a symmetric positive automorphism A in $\mathbb{H}$ such that $g(u,v) = \langle Au, v\rangle$. By the way, it is well known that there exists a positive symmetric automorphism B of $\mathbb{H}$ such that $B^2 = A$, and so we have $g(u,v) = \langle Bu, Bv\rangle$. In the context of Corollary 2.1 (1), we give direct proof of such a result. But precisely, in this context, we have proved that this square root keeps the pre-Hilbert space invariant; such a result cannot be deduced directly from the usual construction of the square root of a positive operator in a Hilbert space.*

Definition 2.5 *An isometry of a Krein inner product g on a Banach space is an operator $I : \mathbb{E} \to \mathbb{E}$ such that*

$$\forall\, (u,v) \in \mathbb{E}^2, \ g(Iu, Iv) = g(u,v).$$

From Proposition 2.2, it follows that such an isometry is the restriction to $\mathbb{E}$ of an isometry $\hat{I}$ of $\hat{g}$ on $\widehat{\mathbb{E}_\gamma}$. According to a decomposition $\widehat{\mathbb{E}_\gamma} = \widehat{\mathbb{E}_\gamma}^+ \oplus \widehat{\mathbb{E}_\gamma}^-$, $\hat{I}$ is the sum of an isometry $\hat{I}^+$ and $\hat{I}^-$ of $\widehat{\mathbb{E}_\gamma}^+$ and $\widehat{\mathbb{E}_\gamma}^-$, respectively. It follows that $\hat{I}_\gamma = \hat{I}^+ - \hat{I}^-$ is an isometry of $\hat{\gamma}$. Conversely, each isometry $\hat{I}_\gamma$ of $\hat{\gamma}$ can be decomposed in a sum of an isometry $\hat{I}^+_\gamma$ of $\widehat{\mathbb{E}_\gamma}^+$ and an isometry $\hat{I}^-_\gamma$ of $\widehat{\mathbb{E}_\gamma}^-$ and $\hat{I} = \hat{I}^+_\gamma - \hat{I}^-_\gamma$ is an isometry of $\hat{g}$.

Thus we obtain:

Proposition 2.3 *Let g be a Krein inner product g on $\mathbb{E}$ and γ the associated pre-Hilbert product on $\mathbb{E}$. If $\widehat{\mathbb{E}_\gamma}$ is the associated Hilbert space, we have the dense embedding ι of $\mathbb{E}_\gamma$ into $\widehat{\mathbb{E}_\gamma}$. We consider a decomposition $\widehat{\mathbb{E}_\gamma} = \widehat{\mathbb{E}_\gamma}^+ \oplus \widehat{\mathbb{E}_\gamma}^-$.*

(1) Each element of the isometry group of $\widehat{\mathbb{E}_\gamma}$ which is compatible with the decomposition can be written as a matrix $\begin{pmatrix} \hat{I}^+ & 0 \\ 0 & \hat{I}^- \end{pmatrix}$ where $\hat{I}^+$ (resp. $\hat{I}^-$) is an isometry of $\widehat{\mathbb{E}_\gamma}^+$ (resp. $\widehat{\mathbb{E}_\gamma}^-$).

(2) *The isometry group of the Krein inner product* $\hat{g}$ *is the set of matrices* $\begin{pmatrix} \hat{I}^+ & 0 \\ 0 & -\hat{I}^- \end{pmatrix}$ *where* $\hat{I}^+$ *(resp.* $\hat{I}^-$*) is an isometry of* $\widehat{\mathbb{E}_\gamma}^+$ *(resp.* $\widehat{\mathbb{E}_\gamma}^-$*).*

(3) *If* $\mathbb{E}^+ = \widehat{\mathbb{E}_\gamma}^+ \cap \mathbb{E}$ *and* $\mathbb{E}^- = \widehat{\mathbb{E}_\gamma}^- \cap \mathbb{E}$*, then each isometry which is compatible with the decomposition is of type* $\iota^* \hat{I}$ *where* $\hat{I}$ *is an isometry of* $\hat{g}$*.*

2.1.4 Symplectic structures

In this section, $\mathbb{E}$ is a Banach space equipped with a continuous skew-symmetric bilinear form $\Omega : \mathbb{E} \times \mathbb{E} \to \mathbb{R}$. Consider the associated continuous linear map $\Omega^\flat : \mathbb{E} \to \mathbb{E}^*$ as defined in § 2.1.1.

Definition 2.6 $\Omega^\flat$ *is called weakly non-degenerate if it is injective; the pair* $(\mathbb{E}, \Omega)$ *is called a weak linear symplectic structure.*

Definition 2.7 $\Omega^\flat$ *is called (strongly) non-degenerate if it is an isomorphism; the pair* $(\mathbb{E}, \Omega)$ *is called a (strong) linear symplectic structure.*

Example 2.1 *If* $\mathbb{E}$ *is any Banach space, the product* $\mathbb{E} \times \mathbb{E}^*$ *is naturally equipped with a weak symplectic structure defined by*

$$\Omega\left((u,\eta),(v,\xi)\right) = \langle \eta, v \rangle - \langle \xi, u \rangle .$$

Remark 2.7 *The existence of a linear symplectic structure* $(\mathbb{E}, \Omega)$ *implies not only that* $\mathbb{E}$ *is isomorphic to* $\mathbb{E}^*$ *but also that* $\mathbb{E}$ *is reflexive.*

In finite-dimensional symplectic geometry, Lagrangian spaces play a key role. In [Wei71], Weinstein has studied the problem of Lagrangian subspaces of a symplectic Banach space.

Definition 2.8 *Let* Ω *be a symplectic structure on the Banach space* $\mathbb{E}$*.*
A Banach subspace $\mathbb{F}$ *is isotropic if*

$$\forall (u,v) \in \mathbb{F}^2,\ \Omega(u,v) = 0.$$

If

$$\mathbb{F}^\perp = \{u \in \mathbb{E} : \forall v \in \mathbb{F}, \Omega(u,v) = 0\}$$

is the symplectic orthogonal of the subspace $\mathbb{F}$, then $\mathbb{F}$ is isotropic if and only if $\mathbb{F} \subset \mathbb{F}^\perp$.

Definition 2.9 *Let* Ω *be a symplectic structure on the Banach space* $\mathbb{E}$*.*
A subspace $\mathbb{F}$ *is maximal isotropic if and only if it is a maximal element of the set of all isotropic subspaces of* $\mathbb{E}$ *or, equivalently, if* $\mathbb{F} = \mathbb{F}^\perp$*.*

Unfortunately, in the Banach framework, a maximal isotropic subspace $\mathbb{L}$ need not be supplemented. Following Weinstein's terminology ([Wei71]), we recall the notion of Lagrangian space.

Definition 2.10 *Let Ω be a strong symplectic structure on the Banach space $\mathbb{E}$. An isotropic space $\mathbb{L}$ is called a Lagrangian space if there exists an isotropic space $\mathbb{P}$ such that $\mathbb{E} = \mathbb{P} \oplus \mathbb{L}$.*

Since Ω is (strongly) degenerate, this implies that $\mathbb{L}$ and $\mathbb{P}$ are maximal isotropic and then Lagrangian spaces (see [Wei71]).

Unfortunately, even for a strong symplectic structure, Lagrangian subspaces need not exist (cf. [KaSw82]). Moreover, for a strong symplectic structure on a Banach space which is not Hilbertizable[1], the non-existence of Lagrangian subspaces is an open problem to our knowledge.

In [Swa90], the author introduces the notion of *Darboux symplectic structure on a Banach space.* It means that the symplectic form is of "Darboux type" as precised in the following definition:

Definition 2.11 *A symplectic form Ω on a Banach space $\mathbb{E}$ is a Darboux (linear) form if there exist a Banach subspace $\mathbb{L}$ and an isomorphism $A : \mathbb{E} \to \mathbb{L} \oplus \mathbb{L}^*$ such that $\Omega = A^*\Omega_{\mathbb{L}}$, where $\Omega_{\mathbb{L}}$ is defined by*

$$\Omega_{\mathbb{L}}\left((u,\eta),(v,\xi)\right) = \langle \eta, v\rangle - \langle \xi, u\rangle .$$

Note that, in this case, Ω is a strong symplectic form and $\mathbb{E}$ must be reflexive.

2.1.5 Cotangent Structures

As in finite dimension, a cotangent structure is a particular case of symplectic structure.

Definition 2.12 *A (linear) cotangent structure on a Banach space $\mathbb{E}$ is a weak symplectic form Ω on $\mathbb{E}$ such that there exists a maximal isotropic space $\mathbb{L}$ of $\mathbb{E}$ giving rise to a decomposition $\mathbb{E} = \mathbb{W} \oplus \mathbb{L}$, where $\mathbb{W}$ is isomorphic to $\mathbb{L}$ as a Banach space. Such a space will be called a weak Lagrangian subspace.*

Note that $\mathbb{L}$ is not assumed to have a Lagrangian supplement, even an isotropic supplement, which is equivalent to Lagrangian (cf. [ThSc91]). Nevertheless, if $\mathbb{E}$ is a Hilbert space, then a cotangent structure must be a (linear) Darboux form (see [Wei71]).

We do not know if there exists a cotangent structure which is not a Darboux form.

Proposition 2.4 *If Ω is a linear Darboux form, then Ω is a cotangent structure if we have a Lagrangian decomposition $\mathbb{E} = \mathbb{P} \oplus \mathbb{L}$ where $\mathbb{L}$ and $\mathbb{P}$ are isomorphic.*

[1] i.e. not isomorphic to a Hilbert space.

Then, for any other supplement subspace $\mathbb{V}$ of $\mathbb{L}$, $\mathbb{E} = \mathbb{V} \oplus \mathbb{L}$ is also an admissible decomposition for Ω. Since a Darboux form is always a strong symplectic form, if $\mathbb{E}$ is a Hilbert space that is always true for any Lagrangian decomposition (cf. [Wei71], Proposition 5.1).

2.1.6 Tangent Structures

Definition 2.13 *A (linear) tangent structure on a Banach space $\mathbb{E}$ is an endomorphism J of $\mathbb{E}$ such that*

$$\operatorname{im} J = \ker J$$

and there exists a decomposition $\mathbb{E} = \mathbb{V} \oplus \ker J$.

If J is a tangent structure on $\mathbb{E}$, consider the decomposition

$$\mathbb{E} = \mathbb{V} \oplus \ker J \equiv \mathbb{V} \times \ker J.$$

The restriction $J_{\mathbb{V}}$ of J to $\mathbb{V}$ is an isomorphism onto $\ker J$. According to this decomposition, J can be identified with the matrix

$$\begin{pmatrix} 0 & 0 \\ J_{\mathbb{V}} & 0 \end{pmatrix}.$$

On the product of Banach spaces $\mathbb{E}^2 = \mathbb{E} \times \mathbb{E}$, we have a canonical tangent structure J_{can} defined by $J_{\text{can}}(u_1, u_2) = (0, u_1)$ and the restriction of J_{can} to $\{0\} \times \mathbb{E}$ vanishes. Then J_{can} can be identified with the matrix

$$\begin{pmatrix} 0 & 0 \\ \mathrm{Id}_{\mathbb{E}} & 0 \end{pmatrix}.$$

Proposition 2.5 *If $\mathbb{E} = \mathbb{V} \oplus \ker J$ is a decomposition associated with a tangent structure J on $\mathbb{E}$, then $A : \mathbb{E} \to \ker J \oplus \ker J$ defined by $A(u_1, u_2) = J_{\mathbb{V}}^{-1}(u_1) + u_2$ is an isomorphism such that $A^* J_{\text{can}} = J$ where J_{can} is the canonical tangent structure on $\ker J \times \ker J$.*

Proof *To the decomposition $\mathbb{E} = \mathbb{V} \oplus \ker J$, if $J_{\mathbb{V}}$ is the restriction of J to $\mathbb{V}$, we associate the isomorphism A from $\mathbb{E}$ to $\ker J \oplus \ker J$ defined by $A(u_1, u_2) = (u_1, J_{\mathbb{V}}(u_2))$. Then clearly $A^* J_{\text{can}} = J$ if J_{can} is the canonical tangent structure on $\ker J \oplus \ker J$.*

2.1.7 Complex and Para-Complex Structures

2.1.7.1 Complex Structures

Definition 2.14 *A complex structure on a Banach space $\mathbb{E}$ is an endomorphism $\mathcal{I}$ of $\mathbb{E}$ such that*

$$\mathcal{I}^2 = -\mathrm{Id}_{\mathbb{E}}.$$

If $\mathcal{I}$ is a complex structure on $\mathbb{E}$, then $\mathcal{I}$ is an isomorphism of $\mathbb{E}$ and this space can be provided with a structure of complex Banach space $\mathbb{E}_{\mathbb{C}}$ defined by

$$\forall(\lambda,\mu)\in\mathbb{R}^2,\ \forall u\in\mathbb{E},\ (\lambda+\mathrm{i}\mu)u=\lambda u+\mu\mathcal{I}(u).$$

In this way, $\mathbb{E}_{\mathbb{C}}$, as a real space, is isomorphic to $\mathbb{E}\oplus\mathbb{E}$. Moreover, $\mathcal{I}$ can be extended to a complex isomorphism of $\mathbb{E}_{\mathbb{C}}$, again denoted $\mathcal{I}$, which has two complex eigenvalues i and $-$i. Thus we have the decomposition $\mathbb{E}_{\mathbb{C}}=\mathbb{E}^+\oplus\mathbb{E}^-$ where $\mathbb{E}^+$ (resp. $\mathbb{E}^-$) is the eigenspace associated with the eigenvalue i (resp. $-$i).

According to this decomposition, $\mathcal{I}$ can be identified with the matrix

$$\begin{pmatrix} \mathrm{i}.\,\mathrm{Id}_{\mathbb{E}^+} & 0 \\ 0 & -\mathrm{i}.\,\mathrm{Id}_{\mathbb{E}^-}\end{pmatrix}.$$

As well known, on the product of Banach spaces $\widetilde{\mathbb{E}}=\mathbb{E}\times\mathbb{E}$, we have a canonical complex structure $\mathcal{I}_{\mathrm{can}}(u_1,u_2)=(u_2,-u_1)$. If $\mathbb{E}_1=\mathbb{E}\times\{0\}$, $\mathbb{E}_2=\{0\}\times\mathbb{E}$ and if $\iota_i:\mathbb{E}\to\widetilde{\mathbb{E}}$ is the canonical isomorphism from $\mathbb{E}$ to $\mathbb{E}_i$, for $i\in\{1,2\}$, then we have $\mathcal{I}_{\mathrm{can}}\circ\iota_1=-\iota_2$ and $\mathcal{I}_{\mathrm{can}}\circ\iota_2=\iota_1$. According to the decomposition $\widetilde{\mathbb{E}}=\mathbb{E}_1\oplus\mathbb{E}_2$, $\mathcal{I}_{\mathrm{can}}$ can be identified with the matrix

$$\begin{pmatrix} 0 & \mathrm{Id}_{\mathbb{E}_2} \\ -\mathrm{Id}_{\mathbb{E}_1} & 0\end{pmatrix}.$$

Definition 2.15 *A complex structure $\mathcal{I}$ on a real Banach space is decomposable if there exist supplemented isomorphic subspaces $\mathbb{E}_1$ and $\mathbb{E}_2$ such that $\mathcal{I}$ can be identified with a matrix of type*

$$\begin{pmatrix} 0 & I^{-1} \\ -I & 0\end{pmatrix},$$

where I is an isomorphism from $\mathbb{E}_1$ to $\mathbb{E}_2$.

The canonical complex structure on $\mathbb{E}\oplus\mathbb{E}$ is always decomposable.

Remark 2.8 *There exist non-decomposable complex structures on Banach spaces. Indeed, in [Fer07], Theorem 3, the reader can find an example of a real Banach space with only two complex structures, not isomorphic and not decomposable.*

For an even-dimensional space $\mathbb{E}$, there always exists a complex structure which is decomposable and is isomorphic to the canonical one. However, in the Banach context, the existence or the uniqueness of the isomorphism class of complex structures is much more difficult than in the finite-dimensional framework. Indeed, it is well known that there exist Banach spaces without complex structures, (see, for instance, [Die52]). A real Banach space whose complexification is a primary space has at most one complex structure ([FeGa10], Th. 22).

In particular, the classical spaces of sequences c_0, ℓ_p or functions $\mathrm{L}^p([0,1])$ with $1 \leq p \leq \infty$ and $C([0,1])$ have a unique complex structure. But there exist complex structures which are not necessarily isomorphic to the canonical one. Even more, there can exist infinitely many isomorphic classes of complex structures on Banach spaces (for references about such different situations, see for instance [Carr14]). Now, the existence of a decomposable complex structure on a Banach space $\mathbb{E}$ is equivalent to the existence of a decomposition $\mathbb{E} = \mathbb{E}_1 \oplus \mathbb{E}_2$ where $\mathbb{E}_1$ and $\mathbb{E}_2$ are isomorphic. Such a Banach space will be called *decomposable.*

2.1.7.2 Para-Complex Structures

Definition 2.16 *A product structure on a Banach space $\mathbb{E}$ is an endomorphism $\mathcal{J}$ of $\mathbb{E}$ such that*

$$\mathcal{J}^2 = \mathrm{Id}_{\mathbb{E}}\,.$$

If $\mathcal{J}$ is a product structure we have a decomposition $\mathbb{E} = \mathbb{E}^+ \oplus \mathbb{E}^-$ where $\mathbb{E}^+$ (resp. $\mathbb{E}^-$) is the eigenspace associated with the eigenvalue $+1$ (resp. -1). In general $\mathbb{E}^+$ and $\mathbb{E}^-$ are not isomorphic.

Definition 2.17 *A product structure $\mathcal{J}$ will be called a para-complex structure if in the decomposition $\mathbb{E} = \mathbb{E}^+ \oplus \mathbb{E}^-$, the subspaces $\mathbb{E}^+$ and $\mathbb{E}^-$ are isomorphic.*

In the one hand, a para-complex structure $\mathcal{J}$ on $\mathbb{E}$, can be written as a matrix of type

$$\begin{pmatrix} \mathrm{Id}_{\mathbb{E}^+} & 0 \\ 0 & -\mathrm{Id}_{\mathbb{E}^-} \end{pmatrix}.$$

On the other hand, consider a decomposition $\mathbb{E} = \mathbb{E}_1 \oplus \mathbb{E}_2$ such that $\mathbb{E}_1$ and $\mathbb{E}_2$ are isomorphic subspaces of $\mathbb{E}$, then the operator $\mathcal{J}$ is defined by a matrix of type

$$\begin{pmatrix} 0 & I^{-1} \\ I & 0 \end{pmatrix}$$

where I, isomorphism from $\mathbb{E}_1$ onto $\mathbb{E}_2$. In this case, we have

$$\mathbb{E}^+ = \{u + Iu, \in u \in \mathbb{E}_1\} \quad \text{and} \quad \mathbb{E}^- = \{u - Iu, \in u \in \mathbb{E}_1\}.$$

Now if $\mathcal{S}$ is the symmetry on $\mathbb{E}$ defined by the matrix

$$\begin{pmatrix} \mathrm{Id}_{\mathbb{E}_1} & 0 \\ 0 & -\mathrm{Id}_{\mathbb{E}_2} \end{pmatrix}, \tag{2.10}$$

then $\mathcal{I} = \mathcal{S}\mathcal{J}$ is a decomposable complex structure on $\mathbb{E}$ associated with the decomposition $\mathbb{E} = \mathbb{E}_1 \oplus \mathbb{E}_2$.
Conversely, if $\mathcal{I}$ is a decomposable complex structure and $\mathbb{E} = \mathbb{E}_1 \oplus \mathbb{E}_2$ is an associated decomposition, $\mathcal{J} = \mathcal{S}\mathcal{I}$ is a para-complex structure on $\mathbb{E}$.

Thus from this last argument, on the product of Banach spaces $\widehat{\mathbb{E}} = \mathbb{E} \times \mathbb{E}$, we have a "canonical" para-complex structure $\mathcal{J}_{\text{can}}(u_1, u_2) = (u_2, u_1)$.
If $\mathbb{E}_1 = \mathbb{E} \times \{0\}$, $\mathbb{E}_2 = \{0\} \times \mathbb{E}$ and $\iota_i : \mathbb{E} \to \hat{\mathbb{E}}$ is the canonical isomorphism from $\mathbb{E}$ onto $\mathbb{E}_i$, for $i \in \{1, 2\}$, then we have $\mathcal{J}_{\text{can}} \circ \iota_1 = \iota_2$ and $\mathcal{J}_{\text{can}} \circ \iota_2 = \iota_1$.
According to the decomposition $\widehat{\mathbb{E}} = \mathbb{E}_1 \oplus \mathbb{E}_2$, $\mathcal{J}_{can}$ can be identified with the matrix

$$\begin{pmatrix} 0 & \text{Id}_{\mathbb{E}_1} \\ \text{Id}_{\mathbb{E}_2} & 0 \end{pmatrix}.$$

Now, if $\mathcal{I}$ (resp. $\mathcal{J}$) is a decomposable complex (resp. a para-complex) structure on $\mathbb{E}$ and if $\mathbb{E} = \mathbb{E}_1 \oplus \mathbb{E}_2$ is an associated decomposition, then we can associate to $\mathcal{I}$ (resp. $\mathcal{J}$) a tangent structure $J_{\mathcal{I}}$ (resp. $J_{\mathcal{J}}$) given, for $u = u_1 + u_2$, and $\ker J_{\mathcal{I}} = \mathbb{E}_1$ (resp. $\ker J_{\mathcal{I}} = \mathbb{E}_1$) by $J_{\mathcal{I}}(u) = -\mathcal{I}(u_2)$ (resp. $J_{\mathcal{J}}(u) = -\mathcal{I}(u_2)$). Of course $J_{\mathcal{I}} = J_{\mathcal{J}}$ if $\mathcal{I} = \mathcal{S}\mathcal{J}$.
Conversely, if J is a tangent structure on $\mathbb{E}$ with $\mathbb{E} = \mathbb{V} \oplus \ker J$ and if J is defined by the matrix

$$\begin{pmatrix} 0 & 0 \\ J_{\mathbb{V}} & 0 \end{pmatrix},$$

then, in an evident way, we have a unique decomposable complex (resp. para-complex) structure $\mathcal{I}_J$ (resp. $\mathcal{J}_J$) associated with J; we also have $\mathcal{I}_J = \mathcal{S}\mathcal{J}_J$ with a clear adequate operator $\mathcal{S}$.
In fact, we have the following result about decomposable complex, para-complex and tangent structures:

Proposition 2.6 *Let $\mathcal{I}$ (resp. $\mathcal{J}$) be a decomposable complex (resp. para-complex) structure on a Banach space $\mathbb{E}$. If $\mathbb{E} = \mathbb{E}_1 \oplus \mathbb{E}_2$ is a decomposition for $\mathcal{I}$ and $\mathcal{J}$ and I is the associated isomorphism from $\mathbb{E}_1$ to $\mathbb{E}_2$, then the map $A : \mathbb{E} \to \mathbb{E}_1 \oplus \mathbb{E}_1$ given by $A(u, v) = (u, Iv)$ is an isomorphism such that $A^*\mathcal{I}_{\text{can}} = \mathcal{I}$ (resp. $A^*\mathcal{J}_{\text{can}} = \mathcal{J}$) where $\mathcal{I}_{\text{can}}$ (resp. $\mathcal{J}_{\text{can}}$) is the canonical complex (resp. para-complex) structure on $\mathbb{E}_1 \oplus \mathbb{E}_1$.*

Proof *We only consider the case of a complex structure. The proof for a para-complex structure is analog.*
Let $\mathcal{I}$ be a decomposable complex structure on a Banach space $\mathbb{E}$. If $\mathbb{E} = \mathbb{E}_1 \oplus \mathbb{E}_2$ is a decomposition for $\mathcal{I}$ and J, the associated isomorphism from $\mathbb{E}_2$ to $\mathbb{E}_1$ then $A : \mathbb{E} \to \mathbb{E}_1 \oplus \mathbb{E}_1$ given by $A(u, v) = (u, J(v))$ is an isomorphism such that $A^*\mathcal{I}_{\text{can}} = \mathcal{I}$ where $\mathcal{I}_{\text{can}}$ is the canonical complex structure on $\mathbb{E}_1 \oplus \mathbb{E}_1$.

Remark 2.9 *We have seen that there exists a bijection between decomposable complex structures and tangent structures. In fact, if $\mathcal{I}$ is the complex structure associated with a tangent structure J, then J is the tangent structure associated with $\mathcal{I}$. This implies that if $A^*\mathcal{I}_{\text{can}} = \mathcal{I}$ and if $\mathcal{I}$ is associated with J then one has $A^*J_{\text{can}} = J$ and conversely.*

2.1.8 Compatibility between Different Structures

2.1.8.1 Compatibility between Tangent, Cotangent Structures and Indefinite Inner Products

Definition 2.18 *A cotangent structure* Ω *on the Banach space* $\mathbb{E}$ *is compatible with a tangent structure* J *if* $\ker J$ *is Lagrangian and*

$$\forall\,(u,v)\in\mathbb{E}^2,\ \Omega(Ju,v)+\Omega(u,Jv)=0.$$

Given a decomposition $\mathbb{E}=\mathbb{V}\oplus\mathbb{L}$ associated with a tangent structure J where $\mathbb{L}=\ker J$, the restriction $J_{\mathbb{V}}$ of J to $\mathbb{V}$ is an isomorphism $J_{\mathbb{V}}:\mathbb{V}\to\mathbb{L}$ and, in this decomposition, J can be written as a matrix of the type $\begin{pmatrix}0 & 0\\ J_{\mathbb{V}} & 0\end{pmatrix}$. According to such a decomposition, to a cotangent structure Ω is associated an operator $\Omega^\flat$ which can be written as a matrix $\begin{pmatrix}\Omega_{\mathbb{VV}} & \Omega_{\mathbb{LV}}\\ \Omega_{\mathbb{VL}} & 0\end{pmatrix}$. With these notations, Ω is compatible with J if and only if we have

$$\Omega^\flat\circ J+J^*\circ\Omega^\flat=0.$$

The following proposition gives a link between tangent structures, cotangent structures and indefinite inner products.

Proposition 2.7 *Let* $\mathbb{E}$ *be a Banach space.*

(1) Assume that we have a tangent structure J *on* $\mathbb{E}$ *and an indefinite inner product* g *on* $\ker J$*. Then there exists a cotangent structure* Ω *on* $\mathbb{E}$ *compatible with* J *such that* $\ker J$ *is a weak Lagrangian space for* Ω*.*

(2) Assume that we have a cotangent structure Ω *on* $\mathbb{E}$ *and let* $\mathbb{E}=\mathbb{V}\oplus\mathbb{L}$ *be an associated decomposition. Then there exists a tangent structure* J *on* $\mathbb{E}$ *such that* $\ker J=\mathbb{L}$*.*

(3) Assume that we have a tangent structure J *on* $\mathbb{E}$ *which is compatible with a cotangent structure* Ω*. Then, if* $\mathbb{E}=\mathbb{V}\oplus\ker J$*,* $\ker J$ *is a weak Lagrangian space for* Ω*. Moreover, for any* $u=Ju'$ *and* $v\in\ker J$ *then* $g(u,v)=\Omega(Ju',v)$ *defines an indefinite inner product on* $\ker J$

Proof

(1) Under the assumptions of (1), if J *is a tangent structure with a decomposition* $\mathbb{E}=\ker J\oplus\mathbb{V}$ *and since* $J_{\mathbb{V}}$ *is an isomorphism from* $\mathbb{V}$ *onto* $\ker J$*, we can extend* g *to a Krein inner product on* $\mathbb{E}$*, by*

- $\bar{g}=g$ *on* $\ker J$,
- $\forall(u,v)\in(\ker J)^2,\ \bar{g}(u,v)=g(J_{\mathbb{V}}^{-1}u,J_{\mathbb{V}}^{-1}v)$ *on* $\mathbb{V}$,
- $\ker J$ *and* $\mathbb{V}$ *are* $\bar{g}$ *orthogonal.*

Let $\mathcal{I}$ *be a complex structure associated with* J *as previously and we set*

$$\forall(u,v)\in\mathbb{E}^2,\ \Omega(u,v)=\bar{g}(u,\mathcal{I}(v))-\bar{g}(v,\mathcal{I}(u)).$$

First, we must show that Ω *is non-degenerate. We then prove that* $\bar{g}(\mathcal{I}(u),\mathcal{I}(v)) = \bar{g}(u,v)$. *Note that for* $u \in \ker J$ *and* $v \in \mathbb{V}$, *we have* $\bar{g}(u,v) = 0$ *and since* $\mathcal{I}(u) \in \mathbb{V}$ *and* $\mathcal{I}(v) \in \ker J$, *we also have* $\bar{g}(\mathcal{I}u,\mathcal{I}v) = 0$. *Now assume that* $(u,v) \in (\ker J)^2$, *then* $\mathcal{I}(u) = -J_{\mathbb{V}}^{-1}(u)$ *and* $\mathcal{I}(v) = -J_{\mathbb{V}}^{-1}(v)$. *But, from the definition of* $\bar{g}$ *on* $\mathbb{V}$, *we get* $\bar{g}(\mathcal{I}(u),\mathcal{I}(v)) = \bar{g}(u,v)$.
Finally, for $(u,v) \in \mathbb{V}^2$, *we have* $\mathcal{I}(u) = J_{\mathbb{V}}(u)$ *and* $\mathcal{I}(v) = J_{\mathbb{V}}(v)$. *Thus* $\bar{g}(\mathcal{I}(u),\mathcal{I}(v)) = \bar{g}(J_{\mathbb{V}}(u),J_{\mathbb{V}}(v)) = g(u,v)$ *from the definition of* $\bar{g}$.
Now assume that $\Omega(u,v) = 0$ *for all* $v \in \mathbb{E}$. *This means that*

$$\bar{g}(u,\mathcal{I}(v)) = \bar{g}(v,\mathcal{I}(u)) = -\bar{g}(\mathcal{I}(v),u)$$

since $\mathcal{I}^2 = -\operatorname{Id}_{\mathbb{E}}$. *As* $\bar{g}$ *is symmetric and weak non-degenerate this implies that* $u = 0$ *since* $\mathcal{I}$ *is an isomorphism of* $\mathbb{E}$. *The other properties are obvious because* Ω *is a cotangent structure. Since* $J \circ \mathcal{I}$ *is zero in restriction to* $\mathbb{V}$, *from the definition of* Ω, *we only have to show that*

$$\forall u,v \in \ker J,\ \Omega(u,Jv) + \Omega(Ju,v) = 0.$$

But $\Omega(u,Jv) = g(u,v)$ *and* $\Omega(Ju,v) = -g(v,u)$, *which ends the proof of (1).*

(2) According to the assumptions of (2), from the definition, we have a decomposition $\mathbb{E} = \mathbb{L} \oplus \mathbb{V}$ *and there exists an isomorphism* $J_{\mathbb{V}} : \mathbb{V} \to \mathbb{L}$. *We define* $J : \mathbb{E} \to \mathbb{E}$ *by*

$$\forall u \in \mathbb{V},\ J(u) = J_{\mathbb{V}}(u)$$

$$\forall u \in \mathbb{L},\ J(u) = 0.$$

Then clearly J *is a tangent structure on* $\mathbb{E}$. *If* $\mathcal{I}$ *is the complex structure associated with* J *as previously then for any* $(u,v) \in \mathbb{E}^2$, *we have* $g(u,v) = \Omega(\mathcal{I}u,v) - \Omega(u,\mathcal{I}v)$ *and* g *is symmetric. Assume that* $\forall (u,v) \in \mathbb{E}^2$, $g(u,v) = 0$. *Then we have* $\forall v \in \mathbb{E}$, $\Omega(\mathcal{I}u,v) = \Omega(u,\mathcal{I}v)$. *For* $v = u$, *we get* $\Omega(\mathcal{I}u,u) = 0$; *so* $\mathcal{I}u = \pm u$. *Assume* $u \neq 0$. *If* $\mathcal{I}u = u$, *we must have* $\forall v \in \mathbb{E}$, $\Omega(u,v) = \Omega(u,\mathcal{I}v)$ *and so* $\forall v \in \mathbb{E}$, $\mathcal{I}v = v$, *which contradicts the definition of* $\mathcal{I}$. *In the same way, the case* $\mathcal{I} = -u$ *implies* $\forall v \in \mathbb{E}$, $\mathcal{I}v = -v$, *which contradicts again the definition of* $\mathcal{I}$. *Finally, we obtain that* g *is non-degenerate, which ends the proof.*

(3) According to assumption (3), we must show that $\ker J$ *is a weak Lagrangian space for* Ω, *i.e.* $\ker J = (\ker J)^{\perp_\Omega}$. *The compatibility between* Ω *and* J *means that* $\Omega^\flat \circ J = -J^* \circ \Omega^\flat$. *The annihilator* $\mathbb{L}^0$ *in* $\mathbb{E}^*$ *of* $\mathbb{L}$ *is* $\ker J^*$ *and so* $(\ker J)^{\perp_\Omega} = (\Omega^\flat)^{-1}(\mathbb{L}^0)$. *But also* $\mathbb{L} = \ker J = J(E)$. *Thus from compatibility condition, we have*

$$\mathbb{L} = \ker(\Omega^\flat \circ J) = \ker(J^* \circ \Omega^\flat) = (\Omega^\flat)^{-1}(\ker J^*) = (\ker J)^{\perp_\Omega}$$

We set $g(u,v) = \Omega(Ju,v)$ *for* $(u,v) \in \mathbb{V}^2$. *Since* Ω *is compatible with* J, *this implies that* g *is symmetric. Assume that* $\forall v \in \mathbb{K}$, $g(u,v) = 0$. *This implies that* $\forall v \in \mathbb{V}$, $\Omega(Jv,u) = 0$. *Since we have* $\forall v \in \mathbb{L}$, $\Omega(v,u) = 0$, *this implies that* $\forall v \in \mathbb{E}$, $\Omega(u,v) = 0$*; since* Ω *is non-degenerate, we must have* $u = 0$*. Thus* g *is a weak indefinite metric on* $\mathcal{V}$*. Since* J *is an isomorphism from* $\mathbb{V}$ *to* $\ker J$*, this ends the proof.*

2.1.8.2 Compatibility between Symplectic Structures, Inner Products and Decomposable Complex Structures

Definition 2.19 *Let* $\mathbb{E}$ *be a Banach space.*

(1) *We say that a weak symplectic structure* Ω *and a complex structure* $\mathcal{I}$ *on* $\mathbb{E}$ *are compatible if* $(u,v) \mapsto \Omega(u,\mathcal{I}v)$ *is a pre-Hilbert product on* $\mathbb{E}$ *and*

$$\forall (u,v) \in \mathbb{E}^2,\ \Omega(\mathcal{I}u,\mathcal{I}v) = \Omega(u,v).$$

(2) *We say that a pre-Hilbert product* g *and a complex structure* $\mathcal{I}$ *on* $\mathbb{E}$ *are compatible if*

$$\forall (u,v) \in \mathbb{E}^2,\ g(\mathcal{I}u,\mathcal{I}v) = g(u,v).$$

(3) *We say that a pre-Hilbert product* g *and a weak symplectic structure* Ω *on* $\mathbb{E}$ *are compatible if* $\mathcal{I} = (g^\flat)^{-1} \circ \Omega^\flat$ *is a well-defined complex structure on* $\mathbb{E}$*.*

If $\mathbb{E}$ is isomorphic to a Hilbert space and Ω is a symplectic form, then there exist a compatible complex structure $\mathcal{I}$ on $\mathbb{E}$ and a Hilbert product compatible with $\mathcal{I}$ and Ω (see, for instance, [ChMa74]). In our context, we have:

Proposition 2.8 *Let* $\mathbb{E}$ *be a Banach space.*

(1) *Assume that there exist a decomposable complex structure* $\mathcal{I}$ *on* $\mathbb{E}$ *and a pre-Hilbert product* g *on* $\mathbb{E}$ *which are compatible. Then there exists a weak symplectic structure* Ω *on* $\mathbb{E}$ *given by*

$$\forall (u,v) \in \mathbb{E}^2,\ \Omega(u,v) = g(\mathcal{I}u,v)$$

which is compatible with $\mathcal{I}$*. Moreover, if* $\mathbb{E} = \mathbb{E}_1 \oplus \mathbb{E}_2$ *is a decomposition such that* $\mathbb{E}_1$ *and* $\mathbb{E}_2$ *are* g*-orthogonal, then* Ω *is a Darboux form and* $\mathbb{E}_1$ *and* $\mathbb{E}_2$ *are supplemented Lagrangian subspaces.*

(2) *Assume that there exists a pre-Hilbert inner product* g *on* $\mathbb{E}$*. Let* $\widehat{\mathbb{E}_g}$ *be the Hilbert space which is the completion of the pre-Hilbert space* $\mathbb{E}_g$*. For any weak symplectic structure* Ω *(not necessarily compatible with* g*) such that* $\Omega^\flat(\mathbb{E}) \subset g^\flat(\mathbb{E})$*, assume that* $\Omega^\flat(\mathbb{E})$ *is dense in* $\widehat{\mathbb{E}_g}$*. Then there exist a Hilbert product* $\bar{g}$ *and a decomposable complex structure* $\mathcal{I}$ *on* $\mathbb{E}$ *which are compatible with* Ω*. Moreover, there exists a decomposition* $\mathbb{E} = \mathbb{E}_1 \oplus \mathbb{E}_2$ *associated with* $\mathcal{I}$ *such that* $\mathbb{E}_1$ *and* $\mathbb{E}_2$ *are* $\bar{g}$*-orthogonal, then* Ω *is a Darboux form,* $\mathbb{E}_1$ *and* $\mathbb{E}_2$ *are supplemented Lagrangian subspaces.*

(3) *Assume that there exists a decomposable complex structure* $\mathcal{I}$ *on* $\mathbb{E}$ *compatible with a weak symplectic structure* Ω*. Then* $g(u,v) = \Omega(u,\mathcal{I}v)$ *is a*

neutral inner product on $\mathbb{E}$ which is compatible with $\mathcal{I}$. Moreover, there exists a decomposition $\mathbb{E} = \mathbb{E}_1 \oplus \mathbb{E}_2$ which is associated with $\mathcal{I}$ and to Ω then $\mathbb{E}_1$ and $\mathbb{E}_2$ are g-orthogonal and Lagrangian.

Proof *(1) Let $\mathcal{I}$ be a complex structure compatible with a pre-Hilbert product g. Then if we set $\forall (u,v) \in \mathbb{E}^2$, $\Omega(u,v) = g(\mathcal{I}u, v)$, Ω is a continuous skew-symmetric bilinear form. Since g is weakly non-degenerate, so is Ω. In this proof, simply denote by $\widehat{\mathbb{E}}$ the Hilbert space which the completion of $\mathbb{E}_g$. We can extend g to a Hilbert product $\hat{g}$ on $\widehat{\mathbb{E}}$. From the compatibility of $\mathcal{I}$ with g, it follows that $\mathcal{I}$ is an isometry for the norm $|| \;||_g$; so we can extend $\mathcal{I}$ to a complex structure $\hat{\mathcal{I}}$ on $\widehat{\mathbb{E}}$ and so Ω can also be extended by the same formula to a symplectic form $\hat{\Omega}$ on $\widehat{\mathbb{E}}$. By classical results (cf. [ChMa74]) in a Hilbert space, the announced result is true in $\widehat{\mathbb{E}}$. Now, since $\mathcal{I}$ leaves $\mathbb{E}$ invariant, the announced results are also true in restriction to $\mathbb{E}$.*

(2) Fix some pre-Hilbert product g in $\mathbb{E}$ and some symplectic structure Ω on $\mathbb{E}$ such that $\Omega^\flat(\mathbb{E}) \subset g^\flat(\mathbb{E})$ and $\Omega^\flat(\mathbb{E})$ is dense in the completion, simply denoted $\widehat{\mathbb{E}}$ in this proof, of the pre-Hilbert space $\mathbb{E}_g$, simply denoted $\widehat{\mathbb{E}}$ in this proof. Let $\hat{g}$ be the extension of g to $\widehat{\mathbb{E}}$. Since Ω is bounded on $\mathbb{E}$ (relative to its original norm $|| \;||$) and the inclusion of $\mathbb{E}_g$ in $\mathbb{E}$ is continuous, we can extend Ω to a 2-form $\widehat{\Omega}$ on $\widehat{\mathbb{E}}$. We set $A = (g^\flat)^{-1} \circ \Omega^\flat$. Thus A is an injective endomorphism of $\mathbb{E}$ and we have

$$g(Au, v) = \langle \Omega^\flat(u), v \rangle = \Omega(u, v). \tag{2.11}$$

where $< .,. >$ is the duality pairing between $\mathbb{E}^*$ and $\mathbb{E}$.
Thus, if we had $A^2 = -\operatorname{Id}_{\mathbb{E}}$, then the proof would be completed.
However, in general, that is not true. Therefore, we shall build from A a new endomorphism which will be the announced complex structure. At first, for the sake of simplicity, we denote by $\ll \,,\, \gg$ the Hilbert product on $\widehat{\mathbb{E}}$ defined by g. From our assumptions on Ω, it follows that the range of A is dense. Thus we can define the adjoint $A^* : \widehat{\mathbb{E}}^* \equiv \widehat{\mathbb{E}} \to \widehat{\mathbb{E}}$ and, by definition, for all $u, v \in \mathbb{E}$ we have:

$$\ll A^*u, v \gg = \ll u, Av \gg = \Omega(v, u) = -\Omega(u, v) = \ll -Au, v \gg .$$

Since A is injective with dense range, A^* is an isomorphism. Now the extension $\widehat{A}$ of A to $\widehat{\mathbb{E}}$ is well defined and we have $\widehat{A} = (\hat{g}^\flat)^{-1} \circ \widehat{\Omega}^\flat$. Since the restriction of $\widehat{A}$ to $\mathbb{E}$ is $-A$, we must have $A^* = \widehat{A}^*$ and so $\widehat{A}$ is an isomorphism and, in particular, $\widehat{\Omega}$ is a strong symplectic form.

Now we have

$$\ll \widehat{A}\widehat{A}^*v, u \gg = \ll \widehat{A}^*v, \widehat{A}^*u \gg = \ll v, \widehat{A}\widehat{A}^*u \gg . \tag{2.12}$$

This implies that $\widehat{A}A^*$ is symmetric. From (2.12) we also have

$$\forall u \in \widehat{\mathbb{E}} \setminus \{0\}, \ll \widehat{A}\widehat{A}^*u, u \gg = \ll \widehat{A}^*u, \widehat{A}^*u \gg > 0.$$

Because $\widehat{A}^* = -\widehat{A}$, then $\widehat{A}\widehat{A}^*$ is a symmetric auto-adjoint definite positive operator. Therefore there exits $\widehat{R}$ such that $\widehat{R}^2 = \widehat{A}\widehat{A}^*$ is auto-adjoint. Moreover, since $\widehat{A}$ and $\widehat{A}^*$ leave $\mathbb{E}$ invariant, the same is true for $\widehat{R}$. We set $\widehat{\mathcal{I}} = \widehat{R}^{-1}\widehat{A}$.

In finite dimension (cf. [Can08], 12) as in this Hilbert context (cf. [Wei71], proof of Proposition 5.1), we can show that $\widehat{\mathcal{I}}$ commutes with $\widehat{A}$ and $\widehat{R}$, $\widehat{\mathcal{I}}$ is compatible with $\hat{g}$, and we have the following properties:

$$\widehat{\mathcal{I}}\widehat{\mathcal{I}}^* = \mathrm{Id},\ \widehat{\mathcal{I}}^* = -\widehat{\mathcal{I}},\ \widehat{\mathcal{I}}^2 = -\mathrm{Id} \quad \text{and } \widehat{\Omega}(\widehat{\mathcal{I}}u, \widehat{\mathcal{I}}v) = \widehat{\Omega}(u,v).$$

We can remark that $\widehat{\Omega}(u, \mathcal{I}v) = \hat{g}(\widehat{R}u, v)$ and so $\overline{g}$ defined by $\overline{g}(u,v) = \hat{g}(\widehat{R}u, v)$ is a Hilbert product compatible with $\widehat{\Omega}$. Since $\widehat{R}$ and $\widehat{A}^*$ leave $\mathbb{E}$ invariant, the same is true for $\widehat{\mathcal{I}}$ and so we have the same property in restriction to $\mathbb{E}$.

On the other hand, since $\widehat{\Omega}$ is a symplectic form on the Hilbert space $\widehat{\mathbb{E}}$, we have a decomposition $\widehat{\mathbb{E}} = \widehat{\mathbb{L}_1} \oplus \widehat{\mathbb{L}_2}$ in two orthogonal Lagrangian spaces relative to the Hilbert product defined by $\ll\ ,\ \gg$ (cf. [Wei71], proof of Proposition 5.1). As $\widehat{\mathcal{I}} = (\overline{g}^\flat)^{-1} \circ \widehat{\Omega}^\flat$, the restriction of $\mathcal{I}$ to $\mathbb{L}_2 = \mathbb{E} \cap \widehat{\mathbb{L}_2}$ is an isomorphism onto $\mathbb{L}_1 = \mathbb{E} \cap \widehat{\mathbb{L}_1}$, and so $\mathcal{I} = \widehat{\mathcal{I}}_{|\mathbb{E}}$ is decomposable.

(3) Consider a complex structure $\mathcal{I}$ on $\mathbb{E}$ compatible with a weak symplectic structure Ω on $\mathbb{E}$. We set $g(u,v) = \Omega(u, \mathcal{I}v)$. From the definition of compatibility between $\mathcal{I}$ and Ω, it follows that g is symmetric and positive definite and so is a pre-Hilbert product. Now, we have

$$g(\mathcal{I}u, \mathcal{I}v) = \Omega(\mathcal{I}u, -v) = \Omega(v, \mathcal{I}u) = g(v,u) = g(u,v),$$

and so g is compatible with $\mathcal{I}$. The result is then a consequence of (1).

Let $\mathbb{L}$ be a Hilbert space provided with the Hilbert product $< .,. >_{\mathbb{L}}$. Consider the direct sum $\mathbb{H}_0 = \mathbb{L} \oplus \mathbb{L}$, provided with the Hilbert product

$$\langle u_1 + u_2, v_1 + v_2 \rangle_{\mathbb{H}_0} = \langle u_1, v_1 \rangle_{\mathbb{L}} + \langle u_2, v_2 \rangle_{\mathbb{L}}.$$

We denote by g_{can} the previous Hilbert product, $\mathcal{I}_{\mathrm{can}}$ the unique complex structure on $\mathbb{H}_0$ compatible with g_{can} and Ω_{can} the canonical Darboux form on $\mathbb{H}$ compatible with the previous structures g_{can} and $\mathcal{I}_{\mathrm{can}}$.

Corollary 2.2 *Consider a weak Darboux form Ω, a pre-Hilbert product g and a decomposable complex structure $\mathcal{I}$ on a separable Banach space $\mathbb{E}$. Assume that any pair of such a triple is compatible. Then the third one is compatible with any of the given pairs.*
Moreover, there exists a Banach space $\mathbb{H} = \mathbb{L} \oplus \mathbb{L}$ and an isomorphism $A : \mathbb{E} \to \mathbb{H}$ such that

$$A^*\Omega_{\mathrm{can}} = \Omega,$$
$$A^*g_{\mathrm{can}} = g,$$
$$A^*\mathcal{I}_{\mathrm{can}} = \mathcal{I}.$$

Proof *Proposition 2.8 means exactly that for any pair of such a triple, the third element is compatible with anyone of the given pair.*
From Point (3) of this Proposition, we have a decomposition $\mathbb{E} = \mathbb{E}_1 \oplus \mathbb{E}_2$ *which is associated with* $\mathcal{I}$ *and to* Ω, *so* $\mathbb{E}_1$ *and* $\mathbb{E}_2$ *are g-orthogonal and Lagrangian. Now, let* $\widehat{\mathbb{E}}$ *be the Hilbert space generated by* $\mathbb{E}_g$ *and* $\hat{g}$ *the prolongation of* g. *Then we have a decomposition*

$$\widehat{\mathbb{E}} = \widehat{E}_1 \oplus \widehat{E}_2$$

where $\widehat{E}_1$ *and* $\widehat{E}_2$ *are the Hilbert spaces generated by* $\mathbb{E}_1$ *and* $\mathbb{E}_2$ *and which are* $\hat{g}$ *orthogonal.*

Let $\widehat{\Omega}$ *and* $\widehat{\mathcal{I}}$ *be the extensions to* $\widehat{\mathbb{E}}$ Ω *and* $\mathcal{I}$, *respectively. By density of* $\mathbb{E}_g$, *it follows that* $\widehat{E}_1$ *and* $\widehat{E}_2$ *are Lagrangian relative to* $\hat{\Omega}$ *and so, from Point (2) of Proposition 2.8, the previous decomposition is a decomposition associated with* $\widehat{\mathcal{I}}$. *Note that* $\widehat{E}_1$ *and* $\widehat{E}_2$ *are isomorphic Lagrangian spaces. We set* $\widehat{\mathbb{H}} = \widehat{\mathbb{L}} \oplus \widehat{\mathbb{L}}$ *where* $\mathbb{L} = \mathbb{E}_1$ *and* $\widehat{\mathbb{H}}$ *are provided with the orthogonal inner product* $\hat{g}_{\text{can}}$ *from the inner product on* $\widehat{\mathbb{E}}_1$. *Let* $\widehat{\Omega}_{\text{can}}$ *be the canonical Darboux form on* $\widehat{\mathbb{H}}$ *associated with this decomposition.*
By classical results (cf. [Wei71]), there exists an isomorphism $\widehat{A} : \widehat{\mathbb{E}} \to \widehat{\mathbb{H}}$ *such that* $\hat{A}^*\widehat{\Omega}_{\text{can}} = \hat{\Omega}$ *and note that* $\hat{A}$ *is an isometry; so it follows that* $\hat{A}^*\widehat{\mathcal{I}}_{\text{can}} = \hat{\mathcal{I}}$ *and so, from Definition 2.19 (2), we also have* $\hat{A}^*\hat{g}_{\text{can}} = \hat{g}$. *We set* $\mathbb{H} = \mathbb{L} \oplus \mathbb{L}$. *Then, from Remark 2.6, we must have* $\hat{A}(\mathbb{E}_1) = \mathbb{L}$ *and* $\hat{A}(\mathbb{E}_2) = \mathbb{L}$ *which implies clearly the result.*

Remark 2.10 *Under the assumption of Corollary 2.2, we have a decomposition* $\mathbb{E} = \mathbb{E}_1 \oplus \mathbb{E}_2$ *such that* $\mathbb{E}_1$ *and* $\mathbb{E}_2$ *are isomorphic, Lagrangian and orthogonal subspaces of* $\mathbb{E}$.

Definition 2.20 *A Banach space* $\mathbb{E}$ *is called a weak (resp. strong) Kähler-Banach space if there exists on* $\mathbb{E}$ *a weak (resp. strong) Darboux form, a weak (resp. strong) inner product and a decomposable complex structure such that any pair of such a triple data is compatible.*

2.1.8.3 Compatibility between Symplectic Structures, Neutral Inner Products and Para-Complex Structures

Definition 2.21 *Let* $\mathbb{E}$ *be a Banach space.*

(1) We say that a weak symplectic structure Ω *and a para-complex structure* $\mathcal{J}$ *on* $\mathbb{E}$ *are compatible if* $(u, v) \mapsto \Omega(u, \mathcal{J}v)$ *is a neutral inner product on* $\mathbb{E}$ *and*

$$\forall (u, v) \in \mathbb{E}^2, \ \Omega(\mathcal{J}u, \mathcal{J}v) = -\Omega(u, v).$$

(2) We say that a neutral inner product g *and a para-complex structure* $\mathcal{J}$ *on* $\mathbb{E}$ *are compatible, if*

$$\forall (u, v) \in \mathbb{E}^2, \ g(\mathcal{J}u, \mathcal{J}v) = -g(u, v).$$

(3) *We say that a neutral inner product* g *and a weak symplectic structure* Ω *on* $\mathbb{E}$ *are compatible if* $\mathcal{J} = (g^\flat)^{-1} \circ \Omega^\flat$ *is a well-defined para-complex structure on* $\mathbb{E}$.

Remark 2.11 *Let* $\mathcal{J} = \mathcal{S}\mathcal{I}$ *be the para-complex structure naturally associated with a decomposable complex structure* $\mathcal{I}$. *If* Ω *is a weak symplectic form on* $\mathbb{E}$, *then* Ω *and* $\mathcal{J}$ *are compatible if and only if* Ω *and* $\mathcal{I}$ *are compatible. If* g *is a pre-Hilbert product on* $\mathbb{E}$, *we denote by* $g_{\mathcal{S}}$ *the neutral inner product defined by* $g_{\mathcal{S}}(u,v) = -g(\mathcal{S}u, v)$. *Then* $g_{\mathcal{S}}$ *and* $\mathcal{J}$ *are compatible if and only if* g *and* $\mathcal{I}$ *are compatible.*

According to this Remark, from Proposition 2.8, we obtain:

Proposition 2.9 *Let* $\mathbb{E}$ *be a Banach space.*

(1) *Assume that there exists a para-complex structure* $\mathcal{J}$ *on* $\mathbb{E}$ *and a neutral product* g *on* $\mathbb{E}$ *which are compatible. Then there exists a weak symplectic structure* Ω *on* $\mathbb{E}$ *given by* $\Omega(u,v) = g(\mathcal{J}u, v)$ *for all* $(u,v) \in \mathbb{E}^2$ *which is compatible with* $\mathcal{J}$. *Moreover if* $\mathbb{E} = \mathbb{E}_1 \oplus \mathbb{E}_2$ *is a decomposition associated with* $\mathcal{J}$ *such that* $\mathbb{E}_1$ *and* $\mathbb{E}_2$ *are* g*-orthogonal, then* Ω *is a Darboux form and* $\mathbb{E}_1$ *and* $\mathbb{E}_2$ *are supplemented Lagrangian subspaces.*

(2) *Assume that there exists a neutral inner product* g *on* $\mathbb{E}$. *Consider an inner product* $<.,.>$ *on* $\mathbb{E}$ *canonically associated with a decomposition* $\mathbb{E} = \mathbb{E}^+ \oplus \mathbb{E}^-$ *associated with* g *and let* $\widehat{\mathbb{E}}$ *be the Hilbert space associated do* g. *For any weak symplectic structure* Ω *(not necessarily compatible with* g*) such that* $\Omega^\flat(\mathbb{E}) \subset g^\flat(\mathbb{E})$, *if* $\Omega^\flat(\mathbb{E})$ *is dense in* $\widehat{\mathbb{E}}$, *then there exist a neutral product* $\bar{g}$ *and a para-complex structure* $\mathcal{J}$ *on* $\mathbb{E}$ *which are compatible with* Ω. *Moreover, there exists a decomposition* $\mathbb{E} = \mathbb{E}_1 \oplus \mathbb{E}_2$ *associated with* $\mathcal{J}$ *such that the restriction of* $\bar{g}$ *to* $\mathbb{E}_1$ *(resp.* $\mathbb{E}_2$*) is positive definite (resp. negative definite),* Ω *is a Darboux form and* $\mathbb{E}_1$ *and* $\mathbb{E}_2$ *are supplemented Lagrangian subspaces.*

(3) *Assume that there exists a para-complex structure* $\mathcal{J}$ *on* $\mathbb{E}$ *compatible with a weak symplectic structure* Ω. *Then* $g(u,v) = \Omega(u, \mathcal{I}v)$ *is a neutral product on* $\mathbb{E}$ *which is compatible with* $\mathcal{J}$. *Moreover, there exists a decomposition* $\mathbb{E} = \mathbb{E}_1 \oplus \mathbb{E}_2$ *which is associated with* $\mathcal{J}$ *and to* Ω *such that* $\mathbb{E}_1$ *and* $\mathbb{E}_2$ *are Lagrangian and the restriction of* g *to* $\mathbb{E}_1$ *(resp.* $\mathbb{E}_2$*) is positive definite (resp. negative definite).*

Let $\mathbb{L}$ be a Hilbert space provided with the Hilbert product $<.,.>_{\mathbb{L}}$. Consider the direct sum $\mathbb{H}_0 = \mathbb{L} \oplus \mathbb{L}$, provided with the Hilbert product

$$\langle u_1 + u_2, v_1 + v_2 \rangle_{\mathbb{H}_0} = \langle u_1, v_1 \rangle_{\mathbb{L}} + \langle u_2, v_2 \rangle_{\mathbb{L}}.$$

We denote by g_{can} is the canonical neutral inner product on $\mathbb{H}_0$ defined by $g_{\text{can}} = <.,.>$ on the first factor $\mathbb{L}$ and $g_{\text{can}} = - <.,.>$ on the second factor $\mathbb{L}$ and $\mathcal{J}_{\text{can}}$ is the unique para-complex structure on $\mathbb{H}_0$ compatible with g_{can} and Ω_{can} the Darboux form compatible with the previous structures g_{can} and $\mathcal{J}_{\text{can}}$

Corollary 2.3 *Consider a weak Darboux form Ω, a neutral inner product g and a para-complex structure $\mathcal{J}$ on a separable Banach space $\mathbb{E}$. Assume that any pair of such a triple is compatible. Then the third one is compatible with any of the given pair.*
If $\overline{\mathbb{E}}$ denotes the pre-Hilbert space $\mathbb{E}$ provided with the inner Hilbert product, there exists a Hilbert space $\mathbb{H}$ such that $\mathbb{E}$ is continuously and densely embedded in $\mathbb{H}$ and an isomorphism $A_{\mathbb{H}} : \mathbb{H} \to \mathbb{H}_0 = \mathbb{L} \oplus \mathbb{L}$ such that the restriction of A of $A_{\mathbb{H}}$ to $\mathbb{E}$ has the following properties:

$$\begin{aligned} A^*\Omega_{\text{can}} &= \Omega, \\ A^*g_{\text{can}} &= g, \\ A^*\mathcal{I}_{\text{can}} &= \mathcal{J}. \end{aligned}$$

Remark 2.12 *As in the previous section, under the assumption of Corollary 2.3, we have a decomposition $\mathbb{E} = \mathbb{E}_1 \oplus \mathbb{E}_2$ such that $\mathbb{E}_1$ and $\mathbb{E}_2$ are isomorphic, Lagrangian and the restriction of g to $\mathbb{E}_1$ (resp. $\mathbb{E}_2$) is positive definite (resp. negative definite).*

Definition 2.22 *A Banach space $\mathbb{E}$ is called a weak (resp. strong) para-Kähler Banach space if there exist on $\mathbb{E}$ a weak (resp. strong) Darboux form Ω, a weak (resp. strong) neutral inner product g and a para-complex structure such that any pair of such a triple data is compatible.*

2.2 Banach G-Structures and Tensor Structures

The concept of G-structure provides a unifying framework for a wide variety of interesting geometric structures. This notion is defined in finite dimension as a reduction of the frame bundle (cf [Mol72]).
In order to extend this notion to the Banach framework, following Bourbaki ([Bou67]), the frame bundle $\ell(TM)$ is defined as an open submanifold of the linear maps bundle $L\left(M \times \mathbb{M}, TM\right)$ where the manifold M is modelled on the Banach space $\mathbb{M}$ (cf. [DGV16], 1.6.5).
The notion of tensor structure which corresponds to an intersection of G-structures, where the various groups G are isotropy groups for tensors, is relevant in many domains in differential geometry: Krein metrics (cf. [Bog74]), almost tangent and almost cotangent structures (cf. [ClGo72] and [ClGo74]), symplectic structures (cf. [Vai03] and [Wei71]), inner products and decomposable complex structures (cf. [ChMa74]), etc.

2.2.1 G-Structures and Tensor Structures on a Banach Space

We fix a Banach space $\mathbb{E}_0$. For any space $\mathbb{E}$, the Banach space of continuous linear maps from $\mathbb{E}_0$ to $\mathbb{E}$ is denoted by $\mathcal{L}(\mathbb{E}_0, \mathbb{E})$ and, if it is non-empty, the open subset of $\mathcal{L}(\mathbb{E}_0,\mathbb{E})$ of linear isomorphisms from $\mathbb{E}_0$ onto $\mathbb{E}$ will denoted by $\mathcal{L}is(\mathbb{E}_0,\mathbb{E})$. In this section, we always assume that $\mathcal{L}is(\mathbb{E}_0,\mathbb{E}) \neq \emptyset$. We then have a natural right transitive action of $\mathcal{GL}(\mathbb{E}_0)$ on $\mathcal{L}is(\mathbb{E}_0,\mathbb{E})$ given by $(\phi, g) \mapsto \phi \circ g$. According to [PiTa06], we define the notion of G-structure on $\mathbb{E}$.

Definition 2.23 *Let G be a weak Banach-Lie subgroup of $\mathcal{GL}(\mathbb{E}_0)$.*
A G-structure on $\mathbb{E}$ is a subset $\mathcal{S}$ of $\mathcal{L}is(\mathbb{E}_0,\mathbb{E})$ such that:

(GStrB1) $\forall\, (\phi,\psi) \in \mathcal{S}^2,\ \phi \circ \psi^{-1} \in G;$

(GStrB2) $\forall\, (\phi,g) \in \mathcal{S} \times G,\ \phi \circ g \in \mathcal{S}.$

The set $L_r^s(\mathbb{E}_0)$ (resp. $\mathcal{L}_r^s(\mathbb{E}_0)$) of tensors (resp. continuous tensors) of type (r,s) on $\mathbb{E}_0$ is the space (resp. Banach space) of $(r+s)$-multilinear maps (resp. continuous $(r+s)$-multilinear maps) from $(\mathbb{E}_0)^r \times (\mathbb{E}_0^*)^s$ into $\mathbb{R}$.

Each $g \in \mathcal{GL}(\mathbb{E}_0)$ induces an automorphism g_r^s of $\mathcal{L}_r^s(\mathbb{E}_0)$ which gives rise to a right action of $\mathcal{GL}(E_0)$ on $\mathcal{L}_r^s(\mathbb{E}_0)$ which is $(\mathbb{T},g) \mapsto g_r^s(\mathbb{T})$ where, for any $(u_1,\dots,u_r,\alpha_1,\dots,\alpha_s) \in (\mathbb{E}_0)^r \times (\mathbb{E}_0^*)^s$,

$$g_r^s(\mathbb{T})\,(u_1,\dots,u_r,\alpha_1,\dots,\alpha_s) = \mathbb{T}\left(g^{-1}.u_1,\dots,g^{-1}.u_r,\alpha_1\circ g,\dots,\alpha_s\circ g\right).$$

Definition 2.24 *The isotropy group of a tensor $\mathbb{T}_0 \in \mathcal{L}_r^s(\mathbb{E}_0)$ is the set*

$$G(\mathbb{T}_0) = \{g \in \mathcal{GL}(\mathbb{E}_0)\ :\ g_r^s(\mathbb{T}_0) = \mathbb{T}_0\}.$$

Definition 2.25 *Let $\mathbb{T}$ be a k-uple $(\mathbb{T}_1,\dots,\mathbb{T}_k)$ of tensors on $\mathbb{E}_0$.*
A tensor structure on $\mathbb{E}_0$ of type $\mathbb{T}$ or a $\mathbb{T}$-structure on $\mathbb{E}_0$ is a $G = \bigcap\limits_{i=1}^{k} G(\mathbb{T}_i)$-structure on $\mathbb{E}_0$.

It is clear that such a group $G = \bigcap\limits_{i=1}^{k} G(\mathbb{T}_i)$ is a closed topological subgroup of the Banach-Lie subgroup of $\mathcal{GL}(\mathbb{E}_0)$. Therefore, from Theorem 1.4, it is a weak Lie subgroup of $\mathcal{GL}(\mathbb{E}_0)$.

2.2.2 The Frame Bundle of a Banach Vector Bundle

Let (E,π,M) be a vector bundle of typical fibre the Banach space $\mathbb{E}$, with total space E, projection π and base M (modelled on the Banach space $\mathbb{M}$). Because $\mathbb{E}$ has not necessarily a Schauder basis, it is not possible to define the frame bundle of this vector bundle as it is done in finite dimension.

An extension of the notion of frame bundle in finite dimension to the Banach framework can be found in [Bou67], 7.10.1 and in [DGV16], 1.6.5.

As seen above, the set of linear continuous isomorphisms from $\mathbb{E}$ to E_x (where E_x is the fibre over $x \in M$) is denoted by $\mathcal{L}is\,(\mathbb{E},E_x)$.

The set

$$P\,(E) = \{(x,f) : x \in M,\ f \in \mathcal{L}is\,(\mathbb{E},E_x)\}$$

is an open submanifold of the linear map bundle $\mathcal{L}(M \times \mathbb{E}, E) = \bigcup_{x \in M} \mathcal{L}(\mathbb{E}, E_x)$. The Lie group $\mathcal{GL}(\mathbb{E})$ of the continuous linear automorphisms acts on the right of $P(E)$ as follows:

$$\begin{array}{rccc} \widehat{R}: & P(E) \times \mathcal{GL}(\mathbb{E}) & \to & P(E) \\ & ((x,f),g) & \mapsto & (x,f).g = (x, f \circ g) \end{array}.$$

The quadruple $\ell(E) = (P(E), \pi, M, \mathcal{GL}(\mathbb{E}))$ where $\begin{array}{rccc} \pi: & P(E) & \to & M \\ & (x,f) & \mapsto & x \end{array}$ is a principal bundle, called *the frame bundle of* E.
We can write

$$P(E) = \bigcup_{x \in M} \mathcal{L}is(\mathbb{E}, E_x).$$

Let us describe the local structure of $P(E)$.
Let (U_α, τ_α) be a local trivialization of E with $\tau_\alpha : \pi^{-1}(U_\alpha) \to U_\alpha \times \mathbb{E}$. This trivialization gives rise to a local section $s_\alpha : U_\alpha \to P(E)$ of $P(E)$ as follows:

$$\forall x \in U_\alpha, \; s_\alpha(x) = \left(x, (\overline{\tau}_{\alpha,x})^{-1} \right)$$

where $\overline{\tau}_{\alpha,x} \in \mathcal{L}is(E_x, \mathbb{E})$ is defined by $\overline{\tau}_{\alpha,x} = \mathrm{pr}_2 \circ \tau_\alpha|_{E_x}$.
We then get a local trivialization of $P(E)$:

$$\begin{array}{rccc} \boldsymbol{\tau}_\alpha: & \pi^{-1}(U_\alpha) & \to & U_\alpha \times \mathcal{GL}(\mathbb{E}) \\ & (x,f) & \mapsto & (x, \overline{\tau}_{\alpha,x} \circ f) \end{array}.$$

In particular, we have

$$\boldsymbol{\tau}_\alpha(s_\alpha(x)) = \left(x, \overline{\tau}_{\alpha,x} \circ (\overline{\tau}_{\alpha,x})^{-1} \right) = (x, \mathrm{Id}_{\mathbb{E}}).$$

$\boldsymbol{\tau}_\alpha$ gives rise to $\overline{\boldsymbol{\tau}}_{\alpha,x}$ defined by

$$\overline{\boldsymbol{\tau}}_{\alpha,x}(x,f) = \overline{\tau}_{\alpha,x} \circ f.$$

Moreover, for $g \in \mathcal{GL}(\mathbb{E})$, we have

$$\widehat{R}_g(s_\alpha(x)) = \left(x, (\overline{\tau}_{\alpha,x})^{-1} \circ g \right).$$

Because the local structure of $P(E)$ is derived from the local structure of the vector bundle E, we get

$$\overline{\boldsymbol{\tau}}_{\alpha,x} \circ (\overline{\boldsymbol{\tau}}_{\beta,x})^{-1} (\overline{\tau}_{\beta,x} \circ f) = \overline{\boldsymbol{\tau}}_{\alpha,x}(x,f) = \overline{\tau}_{\alpha,x} \circ f.$$

We then have the following result:

Proposition 2.10 *The transition functions*

$$\begin{array}{rccc} \mathrm{T}_{\alpha\beta}: & U_\alpha \cap U_\beta & \to & \mathcal{GL}(\mathbb{E}) \\ & x & \longmapsto & \overline{\tau}_{\alpha,x} \circ (\overline{\tau}_{\beta,x})^{-1} \end{array}$$

of E *coincide with the transition functions of* $P(E)$.

The transition functions form a cocycle (cf. Chapter 1, § 1.11):

$$\forall x \in U_\alpha \cap U_\beta \cap U_\gamma,\ \mathrm{T}_{\alpha\gamma}(x) = \mathrm{T}_{\alpha\beta}(x) \circ \mathrm{T}_{\beta\gamma}(x).$$

Definition 2.26 *The tangent frame bundle $\ell(TM)$ of a Banach manifold M is the frame bundle of TM.*

Proposition 2.11 *Let $\pi_1 : E_1 \to M$ and $\pi_2 : E_2 \to M$ be two Banach vector bundles with the same typical fibre $\mathbb{E}$ and let $\Phi : E_1 \to E_2$ be a bundle morphism above Id_M. Then Φ induces a unique bundle morphism $\ell(\Phi) : \ell(E_1) \to \ell(E_2)$ which is injective (resp. surjective) if and only if Φ is injective (resp. surjective).*

2.2.3 G-Structures and Tensor Structures on a Banach Vector Bundle

A reduction of the frame bundle $\ell(E)$ of a Banach fibre bundle $\pi : E \to M$ of typical fibre $\mathbb{E}$ corresponds to the data of a weak Banach-Lie subgroup G of $\mathcal{GL}(\mathbb{E})$ and a topological principal subbundle $\left(F, \pi_{|F}, M, G\right)$ of $\ell(E)$ such that $\left(F, \pi_{|F}, M, G\right)$ has its own smooth principal bundle structure, and the inclusion is smooth. In fact, such a reduction can be obtained in the following way.

Assume that there exists a bundle atlas $\{(U_\alpha, \tau_a)\}_{\alpha\in\mathbf{A}}$ whose transition functions

$$\mathrm{T}_{\alpha\beta} : x \longmapsto \overline{\tau}_{\alpha,x} \circ \left(\overline{\tau}_{\beta,x}\right)^{-1}$$

belong to a weak Banach-Lie subgroup G of $\mathcal{GL}(\mathbb{E})$.

To any local trivialization τ_α of E, is associated the following map:

$$\begin{array}{cccc} \phi_\alpha : & U_\alpha \times G & \to & P(E) \\ & (x,g) & \mapsto & s_\alpha(x).g \end{array}$$

where

$$s_\alpha(x).g = \left(x, \left(\overline{\tau}_{\alpha,x}\right)^{-1}\right).g = \left(x, \left(\overline{\tau}_{\alpha,x}\right)^{-1} \circ g\right).$$

If $U_\alpha \cap U_\beta \neq \phi$, we have

$$\forall x \in U_\alpha \cap U_\beta,\ s_\beta(x) = s_\alpha(x).\mathrm{T}_{\alpha\beta}(x).$$

Let $F = \bigcup\limits_{\alpha\in\mathbf{A}} V_\alpha$ be the subset of $P(E)$ where $V_\alpha = \phi_\alpha(U_\alpha \times G)$. Because a principal bundle is determined by its cocycles (cf., for example, [DGV16], 1.6.3), the quadruple $\left(F, \pi_{|F}, M, G\right)$ can be endowed with a structure of Banach principal bundle and is a topological principal subbundle of $\ell(E)$. In fact, moreover, we have

Lemma 2.1 *F is closed in $\ell(E)$ and F is a weak submanifold of $\ell(E)$.*

Proof *Consider a sequence* $((x_n, g_n))_{n\in\mathbb{N}} \in F^{\mathbb{N}}$ *which converges to some* $(x,g) \in \ell(E)$. *Take a trivialization* $\tau_\alpha : P(E) \to U_\alpha \times \mathcal{GL}(\mathbb{E})$ *which contains* (x,g). *Now, for n large enough,* $\tau_\alpha(x_n, g_n)$ *belongs to* $U_\alpha \times G \subset U_\alpha \times \mathcal{GL}(\mathbb{E})$. *But* $U_\alpha \times G$ *is closed in* $U_\alpha \times \mathcal{GL}(\mathbb{E})$; *this implies that* (x,g) *belongs to* F. *Since the inclusion of* G *in* $\mathcal{GL}(\mathbb{E})$ *is smooth, then the inclusion of* F *in* $\ell(E)$ *and* F *is also smooth.*

Definition 2.27 *The weak subbundle* $\left(F, \pi_{|F}, M, G\right)$ *of the frame bundle* $\ell(E) = (P(E), \pi, M, \mathrm{GL}(\mathbb{E}))$ *is called a G-structure on E.*
When $E = TM$, *a G-structure on* TM *is called a G-structure on* M.

Intuitively, a G-structure F on a Banach bundle $\pi : E \to M$ may be considered as a family $\{F_x\}_{x\in M}$ on the fibres E_x which varies smoothly with x in M, in the sense that $F = \bigcup\limits_{x\in M} F_x$ has a principal bundle structure over M with structural group G.

Definition 2.28 *Let* $\pi_1 : E_1 \to M$ *and* $\pi_2 : E_2 \to M$ *be two Banach vector bundles with the same fibre* $\mathbb{E}$ *provided with the G-structures* $\left(F_1, \pi_{1|F_1}, M, G\right)$ *and* $\left(F_2, \pi_{2|F_2}, M, G\right)$, *respectively. A morphism* $\Phi : E_1 \to E_2$ *is said to be G-structure preserving if, for any* $x \in M$, $\Phi\left((F_1)_x\right) = (F_2)_{\Phi(x)}$.

Definition 2.29 *Let* $\mathbb{T}$ *be a k-uple* $(\mathbb{T}_1, \dots, \mathbb{T}_k)$ *of tensors on the typical fibre* $\mathbb{E}$ *of a Banach vector bundle* $\pi : E \to M$. *A tensor structure on* E *of type* $\mathbb{T}$ *is a* $G(\mathbb{T}) = \bigcap\limits_{i=1}^{k} G(\mathbb{T}_i)$ *structure on* E *where* $G(\mathbb{T}_i)$ *is the isotropy group of* $\mathbb{T}_i$ *in* $\mathcal{GL}(\mathbb{E})$.

Definition 2.30 *The Banach fibre bundle* $\pi_{\widehat{\mathcal{L}_r^s}} : \widehat{\mathcal{L}_r^s}(E) \to M$ *whose typical fibre is* $\mathcal{L}_r^s(\mathbb{E})$ *is the tensor Banach bundle of type* (r,s).

Definition 2.31 *Let* $\pi_{\widehat{\mathcal{L}_r^s}} : \widehat{\mathcal{L}_r^s}(E) \to M$ *be the tensor Banach bundle of type* (r,s).

(1) *A smooth (local) section* $\mathscr{T}$ *of this bundle is called a (local) tensor of type (r,s) on* E.

(2) *Let* $\Phi : E \to E'$ *be an isomorphism from a Banach vector bundle* $\pi : E \to M$ *to another Banach vector bundle* $\pi' : E' \to M'$ *over a map* $\phi : M \to M'$. *Given a (local) tensor* $\mathscr{T}$ *of type (r,s) on* E', *defined on an open set* U' *of* M', *the pull-back of* $\mathscr{T}$ *is the (local) tensor* $\Phi^*\mathscr{T}$ *of the same type defined (on* $\phi^{-1}(U')$*) by*

$$\begin{aligned} & (\Phi^*\mathscr{T})_{\phi^{-1}(x')}\left(\Phi^{-1}(u'_1), \dots, \Phi^{-1}(u'_r), \Phi^*(\alpha'_1), \dots, \Phi^*(\alpha'_s)\right) \\ = \; & \mathscr{T}_{x'}(u'_1, \dots, u'_r, \alpha'_1, \dots, \alpha'_s). \end{aligned}$$

(3) *A local tensor* $\mathscr{T}$ *is called locally modelled on* $\mathbb{T} \in L_r^s(\mathbb{E})$ *if there exists a trivialization* $\tau : E_{|U} \to U \times \mathbb{E}$ *such that*

$$\mathscr{T}_{|U}(x) = \overline{\tau}_x^*(\mathbb{T}).$$

Proposition 2.12 *A tensor $\mathscr{T}$ of type (r, s) on a Banach bundle $\pi : E \to M$, where M is a connected manifold, defines a tensor structure on E if and only if there exists a tensor $\mathbb{T} \in \mathcal{L}_r^s(\mathbb{E})$ and a bundle atlas $\{(U_\alpha, \tau_a)\}_{\alpha \in \mathbf{A}}$ such that $\mathscr{T}_\alpha = \mathscr{T}_{|U_a}$ is locally modelled on $\mathbb{T}$ and the transition functions $T_{\alpha\beta}(x)$ belong to the isotropy group $G(\mathbb{T})$ for all $x \in U_\alpha \cap U_\beta$ (where $U_\alpha \cap U_\beta \neq \emptyset$) and all $(\alpha, \beta) \in A^2$.*

Proof *Fix some tensor $\mathbb{T}$ on $\mathbb{E}$ and denote the isotropy group of $\mathbb{T}$ by G. Assume that we have a G-structure on E and let $\{(U_\alpha, \tau_\alpha)\}_{\alpha \in \mathbf{A}}$ be a bundle atlas such that each transition function $T_{\alpha\beta} : U_\alpha \cap U_\beta \to \mathcal{GL}(\mathbb{E})$ takes values in G. Fix some point $x_0 \in M$ and denote an open set of the previous atlas which contains x_0 by U_α. Since we have a trivialization $\tau_\alpha : E_{U_\alpha} \to U_\alpha \times \mathbb{E}$, we consider on U_α the section of $\widehat{\mathcal{L}_r^s}(E)$ defined by $\mathscr{T}(x) = \bar{\tau}_{\alpha,x}^* \mathbb{T}$. If there exists U_β which contains x_0, then, on $U_\alpha \cap U_\beta$, the transition function $T_{\alpha\beta}$ takes values in G; it follows that the restriction of $\mathscr{T}$ on $U_\alpha \cap U_\beta$ is well defined. Therefore, there exists an open set U in M on which is defined a smooth section $\mathscr{T}$ of $\widehat{\mathcal{L}_r^s}(E)$ such that $\mathscr{T}_{|U_\alpha \cap U} = \bar{\tau}_{\alpha,x}^* \mathscr{T}$, for all α such that $U_\alpha \cap U \neq \emptyset$. If x belongs to the closure of U, let U_β be an open set of the previous atlas which contains x. Then, using the previous argument, we can extend $\mathscr{T}$ to $U \cup U_\beta$. Since M is connected, we can define such a tensor $\mathscr{T}$ on M.*

Conversely, if $\mathscr{T}$ is a tensor such that there exists a bundle atlas $\{(U_\alpha, \tau_\alpha)\}_{\alpha \in \mathbf{A}}$ where $\mathscr{T}_{|U_\alpha}$ is locally modelled on $\mathbb{T}$, this clearly implies that each transition function $T_{\alpha\beta}$ must take values in G.

2.3 Examples of Tensor Structures on a Banach Bundle

Let (E, π, M) be a vector bundle whose fibre is the Banach space $\mathbb{E}$ and whose base is modelled on the Banach space $\mathbb{M}$.

2.3.1 Krein Metrics

Definition 2.32 *A weak pseudo-Riemannian metric on (E, π, M) is a smooth field of weak non-degenerate symmetric forms g on E. Moreover, g is called*

(1) a weak Riemannian metric if each g_x is a pre-Hilbert inner product on the fibre E_x;

(2) a weak Krein metric if there exists a decomposition $E = E^+ \oplus E^-$ into a Whitney sum of Banach bundles such that for each fibre E_x, g_x is a Krein inner product associated with the decomposition $E_x = E_x^+ \oplus E_x^-$;

(3) a weak neutral metric if it is a neutral Krein metric such that there exists a

decomposition $E = E^+ \oplus E^-$ *where* E^+ *and* E^- *are isomorphic subbundles of* E*;*

(4) *a weak Krein indefinite metric if there exists a decomposition* $E = E_1 \oplus E_2$ *in a Whitney sum of Banach bundles such that for each fibre* E_x*,* g_x *is a Krein indefinite inner product associated with the decomposition* $E_x = (E_1)_x \oplus (E_2)_x$*.*

Remark 2.13 *In Definition 2.32, if we impose that* g *is a strong non-degenerate symmetric form we will use the terminology strong in place of weak in each item of this definition. Note that this condition implies in each item that the typical fibre* $\mathbb{E}$ *is isomorphic to a Hilbert space and so is reflexive.*

From now on, we adopt the following convention:
unless otherwise mentioned "pseudo-Riemannian metric", "Krein metric", "neutral metric" and "Krein indefinite metric" will always mean "weak pseudo-Riemannian metric", "weak Krein metric", "weak neutral metric" "weak Krein indefinite metric". Otherwise, we will add the adjective weak or strong if such a precision is needed.

Let g be Riemannian metric on a Banach bundle (E, π, M). Let $\widehat{E}_x$ be the Hilbert space which is the completion of the pre-Hilbert space $\check{E}_x := (E_x)_{g_x}$. Following the construction of the Hilbert completion bundle of a weak Riemannian convenient manifold in [MMM12], p. 8, we introduce the following definition:

Definition 2.33 *A Riemannian Banach bundle* (E, π, M, g) *has the property of g-Hilbert extension* E *if the following conditions are satisfied:*

(1) $\widehat{E} = \bigcup\limits_{x \in M} \widehat{E}_x$ *(resp.* $\check{E} = \bigcup\limits_{x \in M} \check{E}_x$*) is a Hilbert bundle (resp. pre-Hilbert bundle) over* M *whose typical fibre is a Hilbert space* $\widehat{\mathbb{E}}$ *(resp.* $\check{\mathbb{E}}$*) which the completion of* $\check{\mathbb{E}}$*.*

(2) *There exists an atlas bundle* $\{(U_\alpha, \tau_\alpha)\}_{\alpha \in \mathbf{A}}$ *(resp.* $\{(U_\alpha, \check{\tau}_\alpha)\}_{\alpha \in \mathbf{A}}$*) with trivializations* $\tau_\alpha : E_U \to U \times \mathbb{E}$ *such that*

   the restriction of τ_α *to* $\check{E}_\alpha$ *is* $\check{\tau}_\alpha$*;*

   $\check{\tau}_\alpha$ *can be extended to a trivialization* $\widehat{\tau}_\alpha : \widehat{E}_{|U_\alpha} \to U_\alpha \times \widehat{\mathbb{E}}$*.*

(3) *The Riemannian metric* g *can be extended to a (strong) Riemannian metric* $\hat{g}$ *on* $\widehat{E}$*.*

Remark 2.14

(1) *As we have seen in Proposition 2.2, for each* $x \in M$ *the inner product defined by* g_x *extends to an inner product on* $\widehat{E}_x$*. Under the notation of Definition 2.33, over each* U_α*, the map* $x \mapsto (\check{\tau}_\alpha^{-1})^* g_x$ *is a smooth map which takes values in* $\mathrm{L}_s(\check{\mathbb{E}})$ *(Banach space of symmetric operators on* $\check{\mathbb{E}}$*). Although* $\check{g}_x^\alpha := (\check{\tau}_\alpha^{-1})^* g_x$ *has a natural extension* $\widehat{g}_x^\alpha$ *to* $\mathcal{L}_s(\widehat{\mathbb{E}})$ *for each fixed*

$x \in U_\alpha$, a priori there is no reason that the map $x \mapsto \widehat{g}_x^\alpha$ from U_α to $\mathcal{L}_s(\widehat{\mathbb{E}})$ would be again a smooth map. Unfortunately, we have no counter-example. This motives the assumption (3) in Definition 2.33.

(2) *From Definition 2.33, the identity map $\mathfrak{i} : \check{E} \to E$ and the inclusion ι : $\check{E} \to \widehat{E}$ are bundle morphisms. By the way, we have $\mathfrak{i}^* g = \iota^* \hat{g}$.*

Definition 2.33 can be illustrated by the following simple Example.

Example 2.2 *Hilbert snake.— (see [PeSa13]). Let $\mathbb{H}$ be a separable infinite-dimensional Hilbert space and $< .,. >$ (resp. $|| \ ||$) the inner product (resp. the norm) on $\mathbb{H}$. We consider a fixed Hilbertian basis $(e_i)_{i \in \mathbb{N}}$ in $\mathbb{H}$. Any $x \in \mathbb{H}$ will be written as a series $x = \sum_{i \in \mathbb{N}} x_i e_i$ where $x_i = \langle x, e_i \rangle$ is the ith coordinate of x. We denote by $\mathbb{S}_{\mathbb{H}} = \{x \in \mathbb{H} : \|x\| = 1\}$ the unit sphere in $\mathbb{H}$. Note that $\mathbb{S}_{\mathbb{H}}$ is a hypersurface of codimension 1 in $\mathbb{H}$.*
Given any metric space (X, d), we denote by $\mathcal{C}([a, b], X)$ the set of continuous curves $u : [a, b] \to X$. Recall that on $\mathcal{C}([a, b], X)$, we have the usual distance d_∞ defined by

$$d_\infty(u_1, u_2) = \sup_{t \in [a,b]} d(u_1(t), u_2(t))$$

and $(\mathcal{C}([a, b], X), d_\infty)$ is a complete metric space.
We fix a real number $L > 0$ and let $\mathcal{P}=\{0 = s_0 < s_1 < ... < s_N = L\}$ be a fixed partition of $[0, L]$. We denote by $\mathcal{C}^k_{\mathcal{P}}([0, L], \mathbb{S}_{\mathbb{H}})$ (resp. $\mathcal{C}^k_{\mathcal{P}}([0, L], \mathbb{H})$) the set of curves $u \in \mathcal{C}([0, L], \mathbb{S}_{\mathbb{H}})$ (resp. $u \in \mathcal{C}([0, L], \mathbb{H})$) which are C^k-piecewise relatively to $\mathcal{P}$ for $k \in \mathbb{N}$.

A Hilbert snake *is a continuous piecewise C^1-curve $S : [0, L] \to \mathbb{H}$, such that*

$$\begin{cases} \forall t \in [0, L], \ ||\dot{S}(t)|| = 1 \\ S(0) = 0 \end{cases}.$$

In fact, a snake is characterized by $u(t) = \dot{S}(t)$ and, of course, we have $S(t) = \int_0^t u(s)ds$ where $u : [0, L] \to \mathbb{S}_{\mathbb{H}}$ is a piecewise C^0-curve associated with the partition $\mathcal{P}$.
The set $\mathcal{C}^L_{\mathcal{P}} = \mathcal{C}^0_{\mathcal{P}}([0, L], \mathbb{S}_{\mathbb{H}})$ is called the configuration space of the snakes *in $\mathbb{H}$ of length L relatively to the partition $\mathcal{P}$. We can also put on $\mathcal{C}^L_{\mathcal{P}}$ the distance d_∞ defined by*

$$d_\infty(u_1, u_2) = \sup_{t \in [a,b]} ||u_1(t) - u_2(t)||.$$

The natural map

$$\begin{array}{rcl} h: & \mathcal{C}^L_{\mathcal{P}} & \to \prod_{i=1}^N \mathcal{C}^0([s_{i-1}, s_i], \mathbb{S}_{\mathbb{H}}) \\ & u & \mapsto (u\,|_{[s_0,s_1]}, ..., u\,|_{[s_i,s_{i+1}]}, u\,|_{[s_{N-1},s_N]}) \end{array} \tag{2.13}$$

is a homeomorphism which provides structure of Banach manifold on $\mathcal{C}^L_{\mathcal{P}}$ *and we will identify these sets.*
The tangent space $T_u\mathcal{C}^L_{\mathcal{P}}$ *can be identified with the set*

$$\{v \in \mathcal{C}^0_{\mathcal{P}}([0,L],\mathbb{H}) : \forall s \in [0,L],\ \langle u(s), v(s)\rangle = 0\}. \tag{2.14}$$

This space is naturally provided with the induced norm $||\ ||_\infty$. *On the other hand, note that any* $v \in \mathcal{C}^0_{\mathcal{P}}([0,L],\mathbb{H})$ *is integrable on* $[0,L]$ *and so we get an inner product on this space given by:*

$$\langle v, w\rangle_{L^2} = \int_0^L \langle v(s), w(s)\rangle ds \tag{2.15}$$

This inner product induces a natural norm $||\ ||_{L^2}$ *on* $T_u\mathcal{C}^L_{\mathcal{P}}$ *given by:*

$$||v||_{L^2} = \left[\int_0^L \langle v(s), v(s)\rangle ds\right]^{\frac{1}{2}}$$

So the inner product defined by (2.15) gives rise only to a Riemannian g metric on $T\mathcal{C}^L_{\mathcal{P}}$.
According to (2.14), the completion $\widehat{T_u\mathcal{C}^L_{\mathcal{P}}}$ *of the pre-Hilbert space* $T_u\mathcal{C}^L_{\mathcal{P}}$ *is isomorphic to the subspace*

$$\left\{v = (v_1, \ldots, v_N) \in \prod_{i=1}^N L^2([s_{i-1}, s_i], \mathbb{H}) : \right.$$
$$\left. \forall\, i \in \{1, \ldots, N\},\ \int_0^L \langle v_i(s), u_i(s)\rangle ds = 0 \right\}. \tag{2.16}$$

If we also identify $\mathcal{C}^0_{\mathcal{P}}([0,L],\mathbb{H})$ *with* $\prod_{i=1}^N \mathcal{C}^0([s_{i-1}, s_i], \mathbb{H})$, *then,* $\widehat{T\mathcal{C}^L_{\mathcal{P}}}$ *can be identified with the closed sub-bundle of* $\prod_{i=1}^N \mathcal{C}^0([s_{i-1}, s_i], \mathbb{H}) \times L^2([s_{i-1}, s_i], \mathbb{H})$ *defined by (2.16). It is easy to see that the assumptions of Definition 2.33 are satisfied for* $T\mathcal{C}^L_{\mathcal{P}}$.

Remark 2.15 *Let* M *be a compact Riemannian manifold. We have also the situation of Definition 2.33 for a* L^2*-metric on the* C^k *loops space* $L^k(\mathbb{S}^1, M)$ *(cf. [Wur95]), on the Banach manifolds* $C^k(M,N)$ *of* C^k *maps from* M *to a manifold* N*, (cf [Bru18]), on the Banach manifold* $\mathcal{M}^k(M)$ *of* C^k*-Riemannian metrics on* M *(cf. [Ebi67]), on the Banach manifold* $\mathcal{S}^k(E,M)$ *of* C^k *sections of a vector bundle* $\pi : E \to M$ *(cf. [Pal68], [Pal70]).*
Note that, in many of these references, such results are proved in some adequate Sobolev context.

Under the assumption of Definition 2.33, given a Krein metric g on E, according to Proposition 2.2, we get the following result.

Theorem 2.1 *Let g be a Krein metric on a Banach bundle (E,π,M) whose typical fibre is separable. Consider a decomposition $E = E^+ \oplus E^-$ in a Whitney sum such that the restriction g^+ (resp. $-g^-$) of g (resp. $-g$) to E^+ (resp. E^-) is a Riemannian metric. We denote by $\bar{g}$ the Riemannian metric defined by*

- $\bar{g}_{|E^+} = g_{|E^+}$
- $\bar{g}_{|E^-} = -g_{|E^-}$
- *E^+ and E^- are $\bar{g}$ -orthogonal.*

Assume that E has the $\bar{g}$-Hilbert extension.
Then g can be extended to a strong Krein metric $\hat{g}$ on $\widehat{E}$ and we have a decomposition $\widehat{E} = \widehat{E}^+ \oplus \widehat{E}^-$ such that the restriction $\hat{g}^+$ (resp. $\hat{g}^-$) of $\hat{g}$ (resp. $-\hat{g}$) to $\widehat{E}^+$ (resp. $\widehat{E}^-$) such that $\hat{g}^+$ (resp. $-\hat{g}^-$) is a strong Riemannian metric. In fact $\widehat{E}^+$ (resp. $\widehat{E}^-$) is the closure of E^+ (resp. E^-) in $\widehat{E}$.
Given any point x_0 in M, we identify the typical fibre of $\widehat{E}$ (resp. E) with $\widehat{E}_{x_0}$ (resp. E_{x_0}). Then $\widehat{\bar{g}} = \widehat{g}^+ - \hat{g}^-$ defines a $\widehat{\bar{g}}_{x_0}$-structure and a $\hat{g}_{x_0}$-structure on $\widehat{E}$. Moreover, g defines a g_{x_0}-structure on E.

Proof *Consider a decomposition $E = E^+ \oplus E^-$ in a Whitney sum such that the restriction g^+ (resp. $-g^-$) of g (resp. $-g$) to E^+ (resp. E^-) is a Riemannian metric.*
The assumption of $\bar{g}$-Hilbert extension implies on the one hand that E is a subset of $\widehat{E}$ and, on the other hand, that we have a bundle atlas $\{(U_\alpha,\tau_\alpha)\}_{\alpha\in\mathbf{A}}$ (resp. $\{(U_\alpha,\widehat{\tau_\alpha})\}_{\alpha\in\mathbf{A}}$) for E (resp. for $\widehat{E}$) such that, for each $\alpha\in\mathbf{A}$, the restriction of $\widehat{\tau_\alpha}$ to $\check{E}_{U_\alpha}$ is $\check{\tau}_\alpha$ and also $\tau_\alpha = \check{\tau}_\alpha$ as a map. In other words, on E, we have a structure of normed manifold modelled on $\mathbb{M}\times\check{\mathbb{E}}$ where $\check{\mathbb{E}}$ denotes the pre-Hilbert space $(\mathbb{E},<.,.>)$ which is also a normed vector bundle over M. By the way, the identity map $\mathfrak{i}$ from $\check{E}$ to E and the natural inclusion $\iota:\check{E}\to\widehat{E}$ are bundle morphisms of normed bundles over M. In particular, we have $\widehat{\tau}_\alpha\circ\iota = \check{\tau}_\alpha$, and $\tau_\alpha = \mathfrak{i}\circ\check{\tau}_\alpha$. Moroever, if $U_\alpha\cap U_\beta\neq\emptyset$, the associated transition map $\check{T}_{\alpha\beta}$ for $\check{E}$ (resp. $\widehat{T}_{\alpha\beta}$ for $\widehat{E}$) also satisfies $\widehat{T}_{\alpha\beta}\circ\iota = \check{T}_{\alpha\beta}$.
Since $\bar{g}$ is a Riemannian metric, from our assumption, we can extend $\bar{g}$ to a smooth symmetric tensor $\widehat{\bar{g}}$ which, in fact, is a strong Riemannian metric on $\widehat{E}$.
Via the trivializations over U_α, we obtain

- *a weak Riemannian metric g^α on $U^\alpha\times\check{E}$;*
- *a strong Riemannian metric $\hat{g}^\alpha$ on $U^\alpha\times\widehat{E}$;*
- *a injective morphisms $\mathfrak{i}^\alpha : U_\alpha\times\check{\mathbb{E}}\to U_\alpha\times\mathbb{E}$;*
- *a injective morphism $\iota^\alpha : U_\alpha\times\check{\mathbb{E}}\to U_\alpha\times\widehat{\mathbb{E}}$.*

Therefore, from to Remark 2.14 we have

$$(\mathfrak{i}^\alpha)^* g^\alpha = (\iota^\alpha)^* \hat{g}^\alpha. \tag{2.17}$$

Now, according to proof of Theorem 3.1, chap VII in [Lan95], there exists a unique smooth field of positive definite symmetric operators $x \mapsto \hat{G}^\alpha_x$ *on* $\widehat{\mathbb{E}}$ *such that* $\hat{g}^\alpha(u,v) = \langle \hat{G}^\alpha(u), v\rangle$. *If* $\hat{T}^\alpha_x$ *is a square root of* $\hat{G}^\alpha(u)$, *then* $x \mapsto \hat{T}^\alpha_x$ *is a smooth field from* U_α *to* $\mathcal{GL}(\widehat{\mathbb{E}})$ *such that*

$$\hat{g}^\alpha(u,v) = \langle \hat{T}^\alpha(u), \hat{T}^\alpha(v)\rangle$$

Note that from Remark 2.6 the restriction $\check{T}^\alpha_x$ *of* $\hat{T}^\alpha_x$ *to* $\check{\mathbb{E}}$ *is an automorphism of* $\check{\mathbb{E}}$. *Thus* $T^\alpha_x = \mathfrak{i} \circ \check{T}^\alpha_x$ *belongs to* $\mathcal{GL}(\mathbb{E})$ *and we have*

$$(T^\alpha_x)^* g^\alpha_x = (g_0)_x \alpha \tag{2.18}$$

where g^α_0 *is the Riemannian metric on* $U_\alpha \times \mathbb{E}$ *defined by* $(g^\alpha_0)_x = <.,.>$.

We will show that the maps $\begin{array}{ccc} U_\alpha & \to & \mathcal{GL}(\mathbb{E}) \\ x & \mapsto & T^\alpha_x \end{array}$ are smooth.

Since $x \mapsto \hat{T}^\alpha_x$ *is smooth, for any* $u \in \widehat{\mathbb{E}}$, *the map* $x \mapsto \hat{T}^\alpha(u)$ *is a smooth map from* U_α *to* $\widehat{\mathbb{E}}$. *Let* $c : \mathbb{R} \to U_\alpha$ *be a smooth curve. For any* $\xi \in \mathbb{E}^*$, *we consider the curve* $c_\xi(t) = \xi(T^\alpha_{c(t)}(u))$ *from* $\mathbb{R}$ *to* $\mathbb{R}$ *and we have* $c_\xi(t) = (\mathfrak{i}^\alpha)^*\xi(\hat{T}^\alpha_{c(t)}(u))$. *Thus for any* $u \in \mathbb{E}$ *and* $\xi \in \mathbb{E}^*$ *the map* $t \mapsto T^\alpha_{c(t)}$ *is weakly smooth on* $\mathbb{R}$. *Thus, by Lemma 5.A.2 (a) in [ABG96], it follows that the map* $t \mapsto T^\alpha_{c(t)}$ *is a smooth map from* $\mathbb{R}$ *to* $\mathcal{GL}(\mathbb{E})$. *Now this result is true for any smooth curve* $c : \mathbb{R} \to U_\alpha$; *from the characterization of a smooth map in the convenient setting, it follows that* $x \mapsto T^\alpha_x$ *is a smooth map from* U_α *to* $\mathcal{GL}(\mathbb{E})$.
Now, as in the proof of [Lan95], Chap. VII, Theorem 3.1, the atlas $\{(U_\alpha, \tau_\alpha \circ T^\alpha)\}_{\alpha \in \mathbf{A}}$, *satisfies the assumptions of Proposition 2.12.*
Finally, we can apply the previous result to E^+ *and* E^- *and for the Riemannian metric* g^+ *and* $-g^-$, *respectively. The result is then a consequence of Proposition 2.3.*

2.3.2 Almost Tangent, Decomposable Complex and Para-complex Structures

Let (E, π, M) be a Banach bundle such that M is connected. As in finite dimension, we introduce the notions of almost tangent and almost complex structures.

Definition 2.34 *An endomorphism* J *of* E *is called an almost tangent structure on* E *if*

$$\left\{ \begin{array}{c} \operatorname{im} J = \ker J \\ \ker J \textit{ is a supplemented subbundle of } E \end{array} \right. .$$

Definition 2.35 *An endomorphism $\mathcal{I}$ (resp. $\mathcal{J}$) is called an almost complex (resp. para-complex) structure on E if*

$$\mathcal{I}^2 = -\operatorname{Id}_E \ \textit{(resp. } \mathcal{J}^2 = \operatorname{Id}_E \textit{).}$$

According to § 2.1.6 and § 2.1.7.1, if J is an almost tangent structure, there exists a decomposition $E = \ker J \oplus K$ in a Whitney sum, the restriction J_K of J to K is a bundle isomorphism from K to $\ker J$. Moreover, we can associate to J an almost complex structure $\mathcal{I}$ (resp. $\mathcal{J}$) on E given by $\mathcal{I}(u) = -J_K u$ (resp. $\mathcal{J}(u) = J_K u$) for $u \in K$ and $\mathcal{I}(u) = J_K^{-1}(u)$ (resp. $\mathcal{I}(u) = J_K^{-1}(u)$) for $u \in \ker J$.

Definition 2.36 *An almost complex structure is called decomposable if there exists a decomposition in a Whitney sum $E = E_1 \oplus E_2$ and an isomorphism $I : E_1 \to E_2$ such that $\mathcal{I}$ can be written*

$$\begin{pmatrix} 0 & I^{-1} \\ -I & 0 \end{pmatrix}.$$

Given a Withney decomposition $E = E_1 \oplus E_2$ associated with a decomposable almost complex structure $\mathcal{I}$, let $\mathcal{S}$ be a the isomorphism of E defined by the matrix

$$\begin{pmatrix} -\operatorname{Id}_{E_1} & 0 \\ 0 & \operatorname{Id}_{E_2} \end{pmatrix}.$$

According to § 2.1.7.2, $\mathcal{J} = \mathcal{S}\mathcal{I}$ is an almost para-complex structure. Conversely, if $\mathcal{J}$ is an almost para-complex structure there exists a decomposition in a Whitney sum $E = E_1 \oplus E_2$ and an isomorphism $I : E_1 \to E_2$ such that $\mathcal{I}$ can be written

$$\begin{pmatrix} 0 & I^{-1} \\ I & 0 \end{pmatrix}$$

and, in this way, $\mathcal{I} = \mathcal{S}\mathcal{J}$ is a decomposable almost complex structure.

From Propositions 2.5 and 2.6, we obtain the following result:

Theorem 2.2 *Let (E, π, M) be a Banach vector bundle. For a fixed x_0 in M, we identify the fibre E_{x_0} with the typical fibre of E.*

(1) Let J be an almost tangent structure on (E, π, M).
If J_{x_0} is the induced tangent structure on E_{x_0}, then J defines a J_{x_0}-structure on E.

(2) Let $\mathcal{I}$ be an almost decomposable complex structure on (E, π, M).
If $\mathcal{I}_{x_0}$ is the induced complex structure on E_{x_0}, then $\mathcal{I}$ defines a $\mathcal{I}_{x_0}$-structure on E.

(3) Let $\mathcal{J}$ be an almost para-complex structure on (E, π, M).
If $\mathcal{J}_{x_0}$ is the induced para-complex structure on E_{x_0}, then $\mathcal{J}$ defines a $\mathcal{J}_{x_0}$-structure on E.

Proof *(1) Consider a decomposition* $E = K \oplus \ker J$ *in Whitney sum. Let* $\{(U_\alpha, \tau_\alpha)\}_{\alpha\in\mathbf{A}}$ *be a bundle atlas of* E *which induces an atlas on each subbundle* $\ker J$ *and* K*. For any* $x \in U_\alpha$, $\tau_\alpha(x) : E_x \to E_{x_0}$ *is a an isomorphism and so* $\tau_\alpha(x)^* J_{x_0}$ *is a tangent structure on* E_x*. Moreover, if* τ^1_α *(resp.* τ^2_α*) is the restriction of* τ_α *to* $\ker J_{|U_\alpha}$ *(resp.* $K_{|U_\alpha}$*), then* $\tau^1_\alpha(x)$ *and* $\tau^2_\alpha(x)$ *is an isomorphism of* $\ker J_x$ *and* K_x *onto* $\ker J_{x_0}$ *and* K_{x_0}*, respectively. Therefore* $(\tau_\alpha(x)^{-1})^*(J_x)$ *is a (linear) tangent structure on* E_{x_0} *whose kernel is also* $\ker J_{x_0}$ *with* $K_{x_0} = (\tau^2_\alpha(x))^{-1}(K_x)$ *and* $(\tau_\alpha(x)^{-1})^* J_{|K_x}$ *is an isomorphism from* K_{x_0} *to* $\ker J_{x_0}$*. From Proposition 2.5,* $T_x : E_{x_0} \to E_{x_0}$ *defined by* $T_x(u,v) = (u, (\tau_\alpha(x)^{-1})^* J_{|K_x} \circ J^{-1}_{x_0}$ *is an automorphism of* E_{x_0} *such that* $T^*_x(J_{x_0}) = (\tau_\alpha(x)^{-1}) J_x$ *and so* $(T_x \circ \tau_\alpha(x))^*(J_{x_0}) = J_x$*. Since* τ_α *is smooth on* U*, this implies that* $\tau'_\alpha(x) = T_x \circ \tau_\alpha(x)$ *defines a trivialization* $\tau'_\alpha : E_{U_\alpha} \to U_\alpha \times E_{x_0}$*, so* $J_{|U_\alpha}$ *is locally modelled on* J_{x_0}*. Now, it is clear that each transition map* $T_{\alpha\beta}$ *belongs to the isotropy group of* J_{x_0}*.*

The proofs of (2) and (3) use, step by step, the same type of arguments as in the previous one and is left to the reader.

2.3.3 Compatible Almost Tangent and Almost Cotangent Structures

Let (E, π, M) be a Banach bundle such that M is connected.

Definition 2.37 *A weak non-degenerate 2-form* Ω *on* E *is called an almost cotangent structure on* E *if there exists a decomposition of* E *in a Whitney sum* $L \oplus K$ *of Banach subbundles of* E *such that each fibre* L_x *is a weak Lagrangian subspace of* Ω_x *in* E_x *for all* $x \in M$*. In this case,* L *is called a weak Lagrangian bundle.*

Definition 2.38 *An almost tangent structure* J *on* E *is compatible with an almost cotangent structure* Ω *if* Ω_x *is compatible with* J_x *for all* $x \in M$*.*

Assume that E is a Whitney sum $E_1 \oplus E_2$ of two subbundles of E and there exists a bundle isomorphism $J : E_1 \to E_2$, then as we have already seen, we can associate to J an almost complex structure $\mathcal{I}$ on E. On the other hand, if g is a pseudo-Riemannian metric on E_1, we can extend g to a canonical pseudo-Riemannian metric $\bar{g}$ on E such that E_1 and E_2 are $\bar{g}$ orthogonal and the restriction of $\bar{g}$ on E_2 is given by $\bar{g}(u,v) = g(Ju, Jv)$. Now, according to § 2.1.8.1, by application of Proposition 2.7 on each fibre, we get the following result.

Proposition 2.13

(1) Assume that there exists an almost tangent structure J *on* E *and let* $E = K \oplus \ker J$ *be an associated decomposition in a Whitney sum. Assume that the typical fibre of* E *is separable. If there exists a pseudo-Riemannian*

metric g on $\ker J$*, then there exists a cotangent structure Ω on E compatible with J such that* $\ker J$ *is a weak Lagrangian bundle of E where*

$$\forall (u,v) \in E^2, \Omega(u,v) = \bar{g}(\mathcal{I}u,v) - \bar{g}(u,\mathcal{I}v) \tag{2.19}$$

if $\bar{g}$ is the canonical extension of g and $\mathcal{I}$ is the almost complex structure on E defined by J.

(2) Assume that there exists an almost cotangent structure Ω on E and let $E = K \oplus L$ be an associated Whitney decomposition. Then there exists a tangent structure J on E such that $\ker J = L$.

(3) Assume that there exists a tangent structure J on E which is compatible with a cotangent structure Ω. Then there exists a decomposition $E = K \oplus$ $\ker J$ *where* $\ker J$ *is a weak Lagrangian bundle. Moreover for any $u = Ju'$ and v in* $\ker J$*, then $g(u,v) = \Omega(Ju,v)$ defines a pseudo-Riemannian metric on* $\ker J$.

In the context of the previous proposition, if g is a weak Riemannian metric, so is its extension $\bar{g}$. In this case, as we have already seen in § 2.3.1, each fibre E_x can be continuously and densely embedded in a Hilbert space which will be denoted $\widehat{E}_x$. According to Theorem 2.1, with these notations, we get the following theorem.

Theorem 2.3 *Consider a tangent structure J on E, a weak Riemannian metric g on* $\ker J$ *and Ω the cotangent structure compatible with J as defined in (2.19). Assume that the typical fiber of E is separable. If $\bar{g}$ is the natural extension of g to E, assume that E satisfies the $\bar{g}$-Hilbert extension property. For any x_0 in M, the triple (J, g, Ω) defines a $(J_{x_0}, \bar{g}_{x_0}, \Omega_{x_0})$-structure on E where $\bar{g}$ is the natural extension of g to E.*

Remark 2.16 *According to Proposition 2.13 (3), if there exists a tangent structure J on E which is compatible with a cotangent structure Ω such that $g(u,v) = \Omega(Ju,v)$ is a Riemannian metric on* $\ker J$*, then Ω satisfies the relation (2.19). Therefore we have a corresponding version of Theorem 2.19 with the previous assumption.*

Proof *According to our assumption, as in the proof of Theorem 2.1, we have compatible bundle atlases for $\widehat{E}$ and E. Using such atlases, it is easy to see that we can extend J to an almost tangent structure $\widehat{J}$ and by assumption, we have a strong Riemannian metric $\widehat{g}$ on $\widehat{E}$ which extends $\bar{g}$. By the way, if $\mathcal{I}$ is the almost complex structure on E associated with J, we can also extend $\mathcal{I}$ to an almost complex structure $\widehat{\mathcal{I}}$ on $\widehat{E}$ which is exactly the almost complex structure associated with $\widehat{J}$. Thus this implies by relation (2.19) that we can extend Ω into a 2-form $\widehat{\Omega}$ on $\widehat{E}$ which will satisfy also relation (2.19) on $\widehat{E}$, using $\hat{g}$ and $\widehat{\mathcal{I}}$. Then $\widehat{\Omega}$ is a strong non-degenerate 2-form on $\widehat{E}$. Note that* $\ker \widehat{J}$ *is the closure in $\widehat{E}$ of* $\ker J$ *and since $J : K \to$* $\ker J$ *is an isomorphism, this implies that $\widehat{K} = \widehat{J}(\ker \widehat{J})$ is a Banach subbundle of $\widehat{E}$ such*

that $\widehat{E} = \ker \widehat{J} \oplus \widehat{K}$. *Moreover, each fibre* $\ker \widehat{J}_x$ *of* $\ker J$ *is exactly the Hilbert space which is the closure of* $\ker J_x$ *provided with the pre-Hilbert product* $\hat{g}_x$. *Now, from the construction of the extension* $\bar{g}$ *of* g *to* E, *the restriction of* J *to* K *is an isometry and so the restriction of* $\widehat{J}$ *to* $\widehat{K}$ *is also an isometry. This implies that* $\mathcal{I}$ *(resp.* $\widehat{\mathcal{I}}$*) is an isometry of* $\bar{g}$ *(resp.* $\hat{g}$*). Fix a point* x_0 *in* M *and identify the fibres of* E *and* $\widehat{E}$ *with* E_{x_0} *and* $\widehat{E}_{x_0}$.
According to the assumption of $\bar{g}$*-Hilbert extension, using the same notation as in the proof of Theorem 2.1, there exists an atlas* $\{(U_\alpha, \tau_\alpha)\}_{\alpha \in \mathbf{A}}$ *of* E *and an atlas* $\{(U_\alpha, \widehat{\tau}_\alpha)\}_{\alpha \in \mathbf{A}}$ *of* $\widehat{E}$ *such that if* $\check{\tau}_\alpha = \widehat{\tau_\alpha} \circ \iota$, *then* $\{(U_\alpha, \check{\tau}_\alpha)\}_{\alpha \in \mathbf{A}}$ *is an atlas of the pre-Hilbert bundle* $\check{E}$ *and we have* $\tau_\alpha = \check{\tau}_\alpha \circ \mathfrak{i}$. *For each* α, *there, exists on* U_α *smooth fields* $\hat{T}^\alpha$, $\check{T}^\alpha$ *of automorphisms of the fibre* $\widehat{E}_{x_0}$ *and* $\check{E}_{x_0}$, *respectively, such that the transitions functions associated to the atlases* $\left\{\left(U_\alpha, \hat{T}^\alpha \circ \widehat{\tau}_\alpha\right)\right\}_{\alpha \in \mathbf{A}}$ *and* $\left\{(U_\alpha, \check{T}^\alpha \circ \check{\tau}_\alpha)\right\}_{\alpha \in \mathbf{A}}$ *are isometries for* $(\widehat{E}_{x_0}, \widehat{\bar{g}}_{x_0})$ *and* $(\check{E}_{x_0}, g_{x_0})$.

Since $\mathcal{I}$ *and* $\widehat{\mathcal{I}}$ *are isometries for* $\bar{g}$ *and* $\hat{g}$, *respectively, this implies that* $\mathcal{I}$ *is also a* $\mathcal{I}_{x_0}$*-structure on* E *and* $\widehat{\mathcal{I}}$ *is a,* $\widehat{\mathcal{I}}_{x_0}$*-structure on* $\widehat{E}$. *But, since* J *and* $\widehat{J}$ *are canonically defined from* $\mathcal{I}$ *and* $\widehat{\mathcal{I}}$ *respectively, we obtain a similar result for* J *and* $\widehat{J}$. *Finally, according to the relation (2.19), we also obtain a similar result for* Ω *and* $\widehat{\Omega}$.
Now if $T^\alpha = \mathfrak{i}^\alpha \circ \check{T}^\alpha$, *according to § 2.1.2, then* $T^\alpha \circ \tau_\alpha$ *is also a trivialization of* E *such that* $(T^\alpha \circ \tau_\alpha)^* \mathcal{T}_x = \mathcal{T}_{x_0}$ *for* $\mathcal{T} \in \{J, \bar{g}, \Omega\}$, *which ends the proof.*

2.3.4 Compatible Weak Symplectic Form, Weak Riemannian Metric and Almost Complex Structures

If g (resp. Ω) is a weak Riemannian metric (resp. a weak symplectic form) on a Banach bundle (E, π, M), as in the linear context, we denote by $g^\flat$ (resp. $\Omega^\flat$) the associated morphism from E to E^* where (E^*, π^*, M) is the dual bundle of (E, π, M). Moreover, if $\mathcal{I}$ is an almost complex structure on E, following § 2.1.8, we introduce the following notions.

Definition 2.39

(1) *We say that a weak symplectic form* Ω *and an almost complex structure* $\mathcal{I}$ *on* E *are compatible if* $(u, v) \mapsto \Omega(u, \mathcal{I}v)$ *is a weak Riemannian metric on* E *and*

$$\forall (u, v) \in E^2,\ \Omega(\mathcal{I}u, \mathcal{I}v) = \Omega(u, v).$$

(2) *We say that a weak Riemannian metric* g *and an almost complex structure* $\mathcal{I}$ *on* E *are compatible if*

$$\forall (u, v) \in E^2,\ g(\mathcal{I}u, \mathcal{I}v) = g(u, v).$$

(3) *We say that a weak Riemannian metric g and a weak symplectic structure Ω on E are compatible, if $\mathcal{I} = (g^\flat)^{-1} \circ \Omega^\flat$ is well defined and is a complex structure on E.*

(4) *A weak symplectic form Ω on E will be called a Darboux form if there exists a decomposition $E = E_1 \oplus E_2$ such that for each $x \in M$ each fibre $(E_1)_x$ and $(E_2)_x$ is Lagrangian.*

Now, by application of Proposition 2.8 and Corollary 2.2, we obtain:

Theorem 2.4 *Consider a Darboux form Ω, a weak Riemannian g and a decomposable complex structure $\mathcal{I}$ on a Banach bundle E whose typical fibre is separable. Assume that any pair among such a triple exists on E and is compatible. Then the third one also exists and is compatible with any element of the given pair.*
If E has the g-Hilbert extension property, then we can extend Ω, g and $\mathcal{I}$ to a strong symplectic form $\widehat{\Omega}$, a strong Riemannian metric $\hat{g}$, an almost complex structure $\widehat{\mathcal{I}}$ on $\widehat{E}$, respectively. Moreover, for any $x_0 \in M$, if we identify the typical fibre of E and of $\widehat{E}$ with E_{x_0} and $\widehat{E}_{x_0}$, respectively, then the triple $(\Omega, g, \mathcal{I})$ (resp. $(\widehat{\Omega}, \widehat{g}, \widehat{\mathcal{I}})$) defines a $(\Omega_{x_0}, g_{x_0}, \mathcal{I}_{x_0})$-structure (resp. $(\widehat{\Omega}_{x_0} \widehat{g}_{x_0}, \widehat{\mathcal{I}}_{x_0})$-structure) on E (resp. $\widehat{E}$).

Proof *The first property is a direct application of Proposition 2.8 and Corollary 2.2, the other ones are obtained as in the proof of the corresponding parts of Theorem 2.3.*

Definition 2.40 *A Banach bundle $\pi : E \to M$ has a weak (resp. strong) almost Kähler structure if there exists on E a weak (resp. strong) Darboux form Ω, a weak (resp strong) Riemanian metric g and a decomposable almost complex structure $\mathcal{I}$ such that there exists a compatible pair among these three data.*

From Remark 2.10, to a weak or strong almost Kähler structure on E is associated a decomposition $E = E_1 \oplus E_2$ of isomorphic subbundles which are Lagrangian and orthogonal. Note that Theorem 2.4 can be seen as sufficient conditions under which a weak Kähler structure on a Banach bundle E is a tensor structure on E.

2.3.5 Compatible Weak Symplectic Forms, Weak Neutral Metrics and Almost Para-Complex Structures

If g (resp. Ω) is a weak neutral metric (resp. weak symplectic form) on a Banach bundle (E, π, M), as in the linear context, we denote by $g^\flat$ (resp. $\Omega^\flat$) the associated morphism from E to E^* where (E^*, π^*, M) is the dual bundle of (E, π, M). Moreover, if $\mathcal{J}$ is an almost para-complex structure on E, following § 2.1.8, we introduce the following notions.

Definition 2.41

(1) We say that a weak symplectic form Ω and an almost para-complex structure $\mathcal{J}$ on E are compatible if $(u,v) \mapsto \Omega(u, \mathcal{J}v)$ is a weak neutral metric on E and

$$\forall (u,v) \in E^2,\ \Omega(\mathcal{J}u, \mathcal{J}v) = -\Omega(u,v).$$

(2) We say that a weak neutral metric g and an almost para-complex structure $\mathcal{J}$ on E are compatible if

$$\forall (u,v) \in E^2,\ g(\mathcal{J}u, \mathcal{J}v) = -g(u,v)$$

(3) We say that a weak neutral metric g and a weak symplectic structure Ω on E are compatible, if

$$\forall (u,v) \in E^2,\ \mathcal{J} = (g^\flat)^{-1} \circ \Omega^\flat$$

is well defined and is an almost para-complex structure on E.

According to Remark 2.11 and Theorem 2.4, we have

Theorem 2.5 *Consider a Darboux form Ω, a weak neutral metric g and a para-complex structure $\mathcal{J}$ on a Banach bundle (E, π, M) whose typical fibre is separable. Assume that any pair among such a triple exists on E and is compatible. Then the third one also exists and is compatible with any element of the given pair. Let $\bar{g}$ be a weak Riemaniann metric canonically associated with some decomposition $E = E^+ \oplus E^-$ relative to g and assume that E has the $\bar{g}$-Hilbert extension property. Then we can extend Ω, g and $\mathcal{J}$ to a strong symplectic form $\widehat{\Omega}$, a strong neutral metric $\hat{g}$ and an almost para-complex structure $\widehat{\mathcal{J}}$ on $\widehat{E}$, respectively. Moreover, for any $x_0 \in M$, if we identify the typical fibre of E and of $\widehat{E}$ with E_{x_0} and $\widehat{E}_{x_0}$, respectively, then the triple $(\Omega, g, \mathcal{J})$ (resp. $(\widehat{\Omega}, \widehat{g}, \widehat{\mathcal{J}})$) defines a $(\Omega_{x_0}, g_{x_0}, \mathcal{J}_{x_0})$-structure (resp. $(\widehat{\Omega}_{x_0}\hat{g}_{x_0}, \widehat{\mathcal{J}}_{x_0})$-structure) on E (resp. $\widehat{E}$).*

As in finite dimension (cf. [Lib55] for instance), we introduce the para-Kähler structures.

Definition 2.42 *A Banach bundle $\pi : E \to M$ has a weak (resp. strong) almost para-Kähler structure if there exists on E a weak (resp. strong) Darboux form Ω, a weak (resp strong) neutral metric g and a almost para-complex structure $\mathcal{J}$ such that there is a compatible pair among these three data.*

As for weak or strong almost para-Kähler bundles, from Remark 2.12, to a weak or strong para-Kähler structure on E is associated a decomposition $E = E_1 \oplus E_2$ of isomorphic subbundles which are Lagrangian and the restriction of g to E_1 (resp. E_2) is positive definite (resp. negative definite).
Note also that Theorem 2.5 can be seen as sufficient conditions under which a weak almost para-Kähler structure on a Banach bundle E is a tensor structure on E.

2.4 Examples of Integrable Tensor Structures on a Banach Manifold

The integrability problem for tensor structures on finite-dimensional manifolds consists in finding sufficient conditions under which there exists local coordinates in which the tensor is constant. This situation of "local homogeneity" is the subject of classical theorems as Darboux Theorem for a symplectic form or Newlander-Nirenberg Theorem for an almost-complex structure, but also, for example, the existence of a flat (pseudo-) Riemannian metric. Note that this property has also some interest in the physical context (cf. [Cam10], [EES90] or [HRG85] for instance). In this section, we will give some examples of such integrable tensor structures in the Banach setting.

2.4.1 Integrable Tensor Structures on a Banach Manifold

Definition 2.43 *A tensor $\mathscr{T}$ on a Banach manifold M, modelled on a Banach space $\mathbb{M}$ is called an integrable tensor structure if there exists a tensor $\mathbb{T} \in L_r^s(\mathbb{M})$ and a bundle atlas $\{(U_\alpha, \tau_a)\}_{\alpha\in\mathbf{A}}$ such that $\mathscr{T}_\alpha = \mathscr{T}_{|U_a}$ is locally modelled on $\mathbb{T}$ and the transition functions $T_{\alpha\beta}(x)$ belong to the isotropy group $G(\mathbb{T})$ for all $x \in U_\alpha \cap U_\beta$ (where $U_\alpha \cap U_\beta \neq \emptyset$) and all $(\alpha, \beta) \in A^2$.*

2.4.2 The Darboux Theorem on a Banach Manifold

In finite dimension, from the Darboux theorem, a symplectic form on a manifold defines an integrable tensor structure on M. The extension of such a result to the Banach framework is given in [Bam99] for weak symplectic Banach manifolds.

Definition 2.44 *A weak symplectic form on a Banach manifold M modelled on a Banach space $\mathbb{M}$ is a closed 2-form Ω on M which is non-degenerate.*

If $\Omega^\flat : TM \to T^*M$ is the associated morphism, then non-degeneracy of Ω means that $\Omega^\flat$ is an injective bundle morphism. The symplectic form Ω is weak if $\Omega^\flat$ is not surjective. Assume that $\mathbb{M}$ is reflexive. We denote by $\widehat{T_xM}$ the Banach space which is the completion of T_xM provided with the norm $\|\ \|_{\Omega_x^\flat}$. Recall that $\widehat{T_xM}$ does not depend on the choice of the norm on T_xM (cf. Remark 2.1). Then Ω can be extended to a continuous bilinear map $\widehat{\Omega}_x$ on $T_xM \times \widehat{T_xM}$ and $\Omega_x^\flat$ becomes an isomorphism from T_xM to $(\widehat{T_xM})^*$. We set

$$\widehat{TM} = \bigcup_{x\in M} \widehat{T_xM} \text{ and } (\widehat{TM})^* = \bigcup_{x\in M} (\widehat{T_xM})^*.$$

We have the following Darboux Theorem (see [Bam99], [Pel16] or § 2.5 for a complete version of these results):

Theorem 2.6 (Local Darboux theorem) *Let Ω be a weak symplectic form on a Banach manifold M modelled on a reflexive Banach space $\mathbb{M}$. Assume that the following conditions are satisfied:*

(i) There exists a neighbourhood U of $x_0 \in M$ such that $\widehat{TM}_{|U}$ is a trivial Banach bundle whose typical fibre is the Banach space $(\widehat{T_{x_0}M}, || \ ||_{\Omega^\flat_{x_0}})$;

(ii) via a trivialization, Ω can be extended to a smooth field of continuous bilinear forms on $TM_{|U} \times \widehat{TM}_{|U}$.

Then there exists a chart (V, F) around x_0 such that $F^*\Omega_0 = \Omega$ where Ω_0 is the constant form on $F(V)$ defined by $(F^{-1})^*\Omega_{x_0}$.

Definition 2.45 *The chart (V, F) in Theorem 2.6 will be called a Darboux chart around x_0.*

Remark 2.17 *The assumptions of Darboux theorem in [Bam99] (Theorem 2.1) are formulated in a different way. In fact, the assumption on all those norms $|| \ ||_{\Omega^\flat_x}$ in this Theorem 2.1 on the typical fibre $\widehat{\mathbb{M}}$ is a consequence of assumptions (i) and (ii) of Theorem 2.6 after shrinking U if necessary (cf. [Pel16]).*

Remark 2.18 *If Ω is a strong symplectic form on M, then $\mathbb{M}$ is reflexive and $\Omega^\flat$ is a bundle isomorphism from TM to T^*M. In particular, the norm $|| \ ||_{\Omega^\flat_x}$ is equivalent to any norm $|| \ ||$ on T_xM which defines its Banach structure and so all the assumptions (i) and (ii) of Theorem 2.6 are always locally satisfied. Thus Theorem 2.6 recovers the Darboux Theorem which is proved in [Mars72] or [Wei71].*

Weinstein gives an example of a weak symplectic form Ω on a neighbourhood of 0 of a Hilbert space $\mathbb{H}$ for which the Darboux Theorem is not true. The essential reason is that the operator $\Omega^\flat$ is an isomorphism from T_xU onto T_x^*U on $U \setminus \{0\}$, but $\Omega^\flat_0$ is not surjective.

Finally, from Theorem 2.6, we also obtain the following global version of a Darboux Theorem whose proof is left to the reader:

Theorem 2.7 (Global Darboux Theorem) *Let Ω be a weak symplectic form on a connected Banach manifold modelled on a reflexive Banach space $\mathbb{M}$. Assume that we have the following assumptions:*

(i) $\widehat{TM} \to M$ is a Banach bundle whose typical fibre is the Banach space $(\widehat{T_{x_0}M}, || \ ||_{\Omega^\flat_{x_0}})$ for some $x_0 \in M$.

(ii) $\Omega^\flat$ can be extended to a bundle morphism from $\widehat{TM}$ to T^*M

Then, for any $x_0 \in M$, there exists a Darboux chart (V, F) around x_0. In particular, Ω defines an integrable tensor structure on M if and only if the assumptions (i) and (ii) are satisfied.

Note that, from Remark 2.18, a strong symplectic form on a Banach manifold is always an integrable tensor structure.

2.4.3 Flat Pseudo-Riemannian Metrics on a Banach Manifold

In finite dimension, a pseudo-Riemannian metric on a manifold M defines an integrable tensor structure on M. We give a generalization of this result to the Banach framework.

Definition 2.46 *A pseudo-Riemannian metric (resp. a Krein metric) g on a Banach manifold M is a pseudo-Riemannian metric (resp. a Krein metric) on the tangent bundle TM.*

When g is a Krein metric, we have a decomposition $TM = TM^+ \oplus TM^-$ in a Whitney sum such that the restriction g^+ (resp. $-g^-$) of g (resp. $-g$). to TM^+ (resp. TM^-) is a weak Riemannian metric. To g is associated a canonical Riemannian metric $\gamma = g^+ - g^-$. According to Chapter 2, § 2.1, on each fibre T_xM, we have a norm $|| \; ||_{g_x}$ which is associated with the inner product γ_x; we denote by $\widehat{T_xM}$ the Banach Hilbert space associated with the normed space $(T_xM, || \; ||_{g_x})$.

By application of Theorem 2.1 to TM we have the following theorem.

Theorem 2.8 *Let g be a Krein metric on a Banach M modelled on a separable Banach space $\mathbb{M}$. Consider a decomposition $TM = TM^+ \oplus TM^-$ in a Whitney sum such that the restriction g^+ (resp. $-g^-$) of g (resp. $-g$) to TM^+ (resp. TM^-) is a weak Riemannian metric. If $\bar{g}$ is the weak Riemannian metric on TM associated with g assume that TM has the $\bar{g}$-Hilbert extension property. Then g can be extended to a strong Krein metric $\hat{g}$ on the bundle $\widehat{TM}$ and we have a decomposition $\widehat{TM} = \widehat{TM}^+ \oplus \widehat{TM}^-$ such that the restriction $\hat{g}^+$ (resp. $\hat{g}^-$) of $\hat{g}$ (resp. $-\hat{g}$) to $\widehat{TM}^+$ (resp. $\widehat{TM}^-$) is a strong Riemannian metric. In fact $\widehat{TM}^+$ (resp. $\widehat{TM}^-$) is the closure of TM^+ (resp. TM^+) in $\widehat{TM}$.*

Now, as for a strong Riemannian metric on a Banach manifold, to a strong pseudo-Riemannian metric g is associated a Levi-Civita connection which is a Koszul connection (cf. Chapter 3, § 3.15.4 for the convenient setting) characterized, for all (local) vector fields X, Y, Z on M, by

$$\begin{aligned} 2g(\nabla_X Y, Z) =& X(g(Y,Z)) + Y(g(Z,X)) - Z(g(X,Y)) \\ &+ g([X,Y],Z) - g([Y,Z],X) - g([X,Z],Y) \end{aligned} \tag{2.20}$$

When g is a weak pseudo-Riemannian metric, a Levi-Civita connection may not exist (cf. [BBM12]). But if it exists, this connection is unique.
There are many examples of weak Riemannian metrics for which its Levi-Civita connection is defined. For a complete panorama of such Examples, the reader can see [Bil04] and references inside.

Recall that the curvature R of the connection ∇ is given, for all (local) vector fields X, Y, Z on M, by

$$R(X,Y)Z = \nabla_X \nabla_Y Z - \nabla_Y \nabla_X Z - \nabla_{[X,Y]} Z. \tag{2.21}$$

When the Levi-Civita connection exists, we have then the following criterion, which gives rise to an integrable tensor structure for a weak pseudo-Riemannian metric:

Theorem 2.9 *Let g be a weak pseudo-Riemannian metric on a Banach manifold. Assume that the Levi-Civita connection ∇ of g is defined. Then g defines an integrable tensor structure if and only if the curvature of ∇ vanishes.*

This result is well known in finite dimension (see, for instance, [Val19] for a recent complete proof). However, for this result in the Banach setting, without precise reference to our knowledge and in order to be complete, we give a sketch of the proof.

Proof *On the principal frame bundle $\ell(TM)$, the Levi-Civita connection gives rise to a connection form ω with values in the Lie algebra $\mathfrak{gl}(\mathbb{M})$ of the Banach-Lie group* $\mathrm{GL}(\mathbb{M})$ *(cf. [KrMi97], 8). Let Ω be the curvature of ω. If X and Y are local vector fields on M, let X^h and Y^h their horizontal lifts in $\ell(TM)$ and we have $\Omega(X^h, Y^h) = -\omega([X^h, Y^h])$. Therefore, the horizontal bundle is integrable if the curvature vanishes. Moreover, in this case, over any simply connected open set U of M, the bundle $\ell(TM)_{|U}$ is trivial (consequence of [KrMi97], Theorem 39.2 for instance). Thus, around any point $x_0 \in M$, there exists a chart domain (U, ϕ) such that $\ell(TM)_{|U} \equiv \phi(U) \times \mathrm{GL}(\mathbb{M})$. Without loss of generality, we may assume that $U = \mathbb{M}$, $x_0 = 0 \in \mathbb{M}$ and so $\ell(TM)_{|U} = \mathbb{M} \times \mathrm{GL}(\mathbb{M})$. The horizontal leaves are obtained from $\mathbb{M} \times \mathrm{Id}_{\mathbb{M}}$ by right translation in* $\mathrm{GL}(\mathbb{M})$. *Consider an horizontal section $\psi : \mathbb{M} \to \mathrm{GL}(\mathbb{M})$ such that $\psi(0) = \mathrm{Id}_{\mathbb{M}}$ then $\psi(x) = \mathrm{Id}_{\mathbb{M}}$ for all $x \in \mathbb{M}$. Since the parallel transport from some point x to some point y does not depend on the curve which joins x to y, this implies that $\psi(x)$ is the parallel transport from $T_0\mathbb{M}$ to $T_x\mathbb{M}$. For the same reason, if ∇ is the Levi-Civita connection, then for any constant vector field X and Y on M, we have $\nabla_Y X = 0$. As ∇ is compatible with g, this implies that, for any $X \in T_x\mathbb{M}$, we have $X\{g_x(Y,Z)\} = 0$, for all $x \in \mathbb{M}$, all $X \in T_x\mathbb{M}$ and any local constant vector fields Y, Z around x. Consider the diffeomorphism $\Psi(x) = x + \mathrm{Id}_{\mathbb{M}}$ of $\mathbb{M}$. The previous property implies that $\Psi^* g_0 = g$.*
Let (U_1, ϕ_1) and (U_2, ϕ_2) be two charts such that $U_1 \cap U_2 \neq \emptyset$ and if $x_0 \in U_1 \cap U_2$ then, for $i \in \{1, 2\}$, $\phi_i^* g = g_{x_0}$ on $\phi_i(U_i) \subset \mathbb{M} \equiv T_{x_0}M$ where g_{x_0} is the constant pseudo-Riemannian metric on $\mathbb{M}$ defined by g_{x_0}. Then it is easy to see that the the transition function T_{12} on $U_1 \cap U_2$ associated with $T\phi_1$ and $T\phi_2$ belongs to the isotropy group of g_{x_0}.

Conversely, if g is an integrable tensor structure, locally there exists a chart (U, ϕ) such that $\phi^* g = g_{x_0}$ on $\phi(U) \subset \mathbb{M} \equiv T_{x_0}M$ where g_{x_0} is the *constant pseudo-Riemannian defined on* $\mathbb{M} \equiv T_{x_0}M$ *by* g_{x_0}*. Then it is clear that the curvature of the Levi-Civita connection of the constant metric* g_{x_0} *vanishes. Now since* ϕ *is an isometry from* $(U, g_{|U})$ *onto* $(\phi(U), (g_{x_0})_{|\phi(U)})$*, the curvature of the Levi-Cevita connection of* g *must also vanish on* U*, which ends the proof.*

Corollary 2.4 *Let* g *be a Krein metric on a Banach manifold* M *for which its Levi-Civita connection is defined. Then* g *is an integrable tensor structure if and only if the curvature of the Levi-Civita connection of* g *vanishes.*

2.4.4 Integrability of Almost Tangent, Para-Complex and Decomposable Complex Structures

Definition 2.47 *An almost tangent structure* J *(resp. para-complex structure* $\mathcal{J}$*, resp. decomposable complex) structure on a Banach manifold* M *is an almost tangent (resp. para-complex structure* $\mathcal{J}$*, resp. decomposable complex) structure structure on the tangent bundle* TM *of* M*.*

If J is an almost tangent structure, we have a decomposition $TM = L \oplus \ker J$, such that L and $\ker J$ are isomorphic subbundles of TM. Moreover, there exists an isomorphism $J_L : L \to \ker J$ such that J can be written as a matrix of type $\begin{pmatrix} 0 & 0 \\ J_L & 0 \end{pmatrix}$.

If $\mathcal{J}$ is an almost para-complex structure, there exists a Whitney decomposition $TM = E^+ \oplus E^-$ where E^+ and E^- are the eigen-bundles associated with the eigenvalues $+1$ and -1 of $\mathcal{J}$, respectively.

Now, in order to obtain a classical criterion of the integrability of such structures, we need the classical notion of Nijenhuis tensor.

Definition 2.48 *If* A *is an endomorphism of* TM*, the Nijenhuis tensor* N_A *of* A *is defined, for all (local) vector fields* X *and* Y *on* M*, by:*

$$N_A(X,Y) = [AX, AY] - A[AX, Y] - A[X, AY] + A^2[X, Y].$$

Theorem 2.10 *An almost tangent (resp. para-complex) structure* J *(resp.* $\mathcal{J}$*) structure on* M *is integrable if and only its Nijenhuis tensor is null.*

Proof *For any almost tangent structure* J*, since* $J^2 = 0$*, we have:*

$$N_J(X,Y) = [JX, JY] - J[JX, Y] - J[X, JY].$$

Moreover, if X *or* Y *is a section of* $\ker J$*, we always have* $N_J(X,Y) = 0$*. Therefore we only have to consider* N_J *in restriction to* L*. In particular, we can note that if* $N_J \equiv 0$*, then we have*

$$[JX, JY] = J[JX, Y] + J[X, JY].$$

Therefore, when we restrict this relation to a section of L, since J_L is an isomorphism, this implies that $\ker J$ *is an involutive supplemented subbundle of* TM.

Fix some $x_0 \in M$ and consider a chart (U, ϕ) around x_0 such that $(\ker J)_{|U}$ *and $L_{|U}$ are trivial. Therefore,* $T\phi(TM) = \phi(U) \times \mathbb{M}$, $T\phi(\ker J) = \phi(U) \times \mathbb{V}$ *and* $T\phi(L) = \phi(U) \times \mathbb{L}$.

If $\mathbb{V} \subset \mathbb{M}$ and $\mathbb{L} \subset \mathbb{M}$ are the typical fibers of $\ker J$ *and L, respectively, we have $\mathbb{M} = \mathbb{L} \oplus \mathbb{V}$. Therefore, without loss of generality, we may assume that U is an open subset of $\mathbb{M} \equiv \mathbb{L} \times \mathbb{V}$ and $TM = U \times \mathbb{M}$,* $\ker J = U \times \mathbb{V}$ *and $L = U \times \mathbb{L}$. Thus $x \mapsto (J_L)_x$ is a smooth field which takes values in the set* $\mathrm{Iso}(\mathbb{L}, \mathbb{V})$ *of isomorphisms from $\mathbb{L}$ to $\mathbb{V}$ and $x \mapsto J_x$ is a smooth field from M to $L(\mathbb{M})$ such that* $\ker J_x = \mathbb{V}$.

As in [Belt105], Proposition 2.1, we obtain in our context:

Lemma 2.2 *With the previous notations, we have*

$$N_J(X,Y) = J'(Y,JX) - J'(JX,Y) + J'(X,JY) - J'(X,JY)$$

where J' stands for the differential of J as a map from M to $\mathrm{L}(\mathbb{M})$.

At first, assume that J is an integrable almost tangent structure. This means that, for any $x \in M$, the value J_x in $\mathrm{L}(\mathbb{M})$ *(resp. $(J_L)_x$ in* $\mathrm{Iso}(\mathbb{L}, \mathbb{V})$*) is a constant. So from Lemma 2.2. it follows that $N_J = 0$.*

Conversely, assume that $N_J = 0$ and, as we have already seen, $\ker J$ *is involutive. If X and Y are sections of L, in the expression of N_J given in Lemma 2.2, we can replace J by J_L and this implies that J_L satisfies the assumption of [Lan95], Chapter IV, Theorem 1.2. By the same arguments as in the proof of the Frobenius Theorem in [Lan95], we produce a local diffeomorphism Ψ from a neighbourhood $U_0 \times V_0 \in \mathbb{L} \times \mathbb{V}$ of x_0 such that $\Psi^* J_L$ is a smooth field from $U_0 \times V_0$ to* $\mathrm{Iso}(\mathbb{L}, \mathbb{V})$ *which is constant. Thus the same is true for J, which ends the proof in this case.*

Now consider a para-complex structure $\mathcal{J}$. Recall that we have a canonical Whitney decomposition $TM = E^+ \oplus E^-$ where E^+ (resp. E^-) is the eigen-bundle associated with the eigenvalue $+1$ *(resp.* -1*) of $\mathcal{J}$. Note that if $P^\pm = \frac{1}{2}(\mathrm{Id} \pm \mathcal{J})$, then $E^\pm = im\ P^\pm$,* $\ker P^\pm = E^\mp$ *and $TM = E^+ \oplus E^-$. Now, we have $N_{\mathcal{J}} = 0$ if and only if $N_{P^+} = N_{P^-} = 0$ and E^+ and E^- are integrable subbundles of TM. Note that since $\mathcal{J}^2 = Id$, Lemma 2.2 is also valid for $\mathcal{J}$; so if $\mathcal{J}$ is integrable, this implies that $N_{\mathcal{J}} = 0$.*

Conversely, assume that $N_{\mathcal{J}} = 0$. Fix some $x_0 \in M$ and denote by $\mathbb{E}^\pm$ the typical fibre of $E^\pm$. If the subbundles $E^\pm$ are integrable, from Frobenius Theorem, there exists a chart $(U^\pm, \phi_\pm)$ around x_0 where $\phi_\pm$ is a diffeomorphism from $U^\pm$ onto an open neighbourhood $V_1^\pm \times V_1^\pm$ in $\mathbb{E}^+ \times \mathbb{E}^-$ such that $\phi_+^* P^+$

and $\phi_-^* P^-$ *are the fields of constant matrices* $\begin{pmatrix} 0 & \mathrm{Id}_{\mathbb{E}^+} \\ 0 & 0 \end{pmatrix}$ *and* $\begin{pmatrix} 0 & 0 \\ 0 & \mathrm{Id}_{\mathbb{E}^-} \end{pmatrix}$. *Now the transition map* $\phi_- \circ \phi_+^{-1}$ *is necessarily of type* $(\bar{x}, \bar{y}) \mapsto (\alpha(\bar{x}), \beta(\bar{y}))$. *So the restriction of* ϕ_+ *to* $U^+ \cap U^-$ *is such that* $\phi_+^* P^+$ *and* $\phi_+^* P^-$ *are also the previous matrices, which ends the proof since* $\mathcal{J} = P^+ + P^-$.

Unfortunately, the problem of integrability of an almost complex structure $\mathcal{I}$ on a Banach manifold M is not equivalent to the relation $N_{\mathcal{I}} \equiv 0$. The reader can find in [Pat2000] an example of an almost complex structure $\mathcal{I}$ on a smooth Banach manifold M such that $N_{\mathcal{I}} \equiv 0$ for which there exists *no holomorphic chart* in which $\mathcal{I}$ is isomorphic to a linear complex structure. However, if M is analytic and if $N_{\mathcal{I}} \equiv 0$, there exists an holomorphic chart in which $\mathcal{I}$ is isomorphic to a linear complex structure (Banach version of Newlander-Nirenberg theorem, see [Belt105] and [Pen70]). In particular, there exists a structure of holomorphic manifold on M.

Definition 2.49 *An almost complex structure* $\mathcal{I}$ *on a Banach manifold* M *is called formally integrable if* $N_{\mathcal{I}} \equiv 0$.
$\mathcal{I}$ *is called integrable if there exists a structure of complex manifold on* M

2.4.5 Flat Kähler and Para-Kähler Banach Manifolds

Following [Tum07], we introduce the following notions.

Definition 2.50 *A Kähler (resp. formal-Kähler) Banach structure on a Banach manifold* M *is an almost Kähler structure* $(\Omega, g, \mathcal{I})$ *on* TM *such that the almost complex structure* $\mathcal{I}$ *is integrable (resp. formally integrable).*

Note that if M is analytic and $(\Omega, g, \mathcal{I})$ is an almost Kähler structure on TM if and only if $N_{\mathcal{I}} \equiv 0$, then we have a Kähler structure on M. The reader will find many examples of weak and strong formal-Kähler Banach manifolds in [Tum05].

In the same way, a *weak (resp. strong) para-Kähler Banach structure* on a Banach manifold M is an almost para-Kähler structure $(\Omega, g, \mathcal{J})$ on TM (cf. Definition 2.40) *such that the almost para-complex structure* $\mathcal{J}$ *is integrable.*

In what follows, we use the terminology Kähler or para-Kähler manifold instead of weak Kähler or weak para-Kähler manifold.

According to § 2.3.4, if we have a Kähler or para-Kähler manifold structure on M, we have a decomposition $TM = E_1 \oplus E_2$ where E_1 and E_2 are isomorphic subbundles, Lagrangian and mutually orthogonal subbundles of TM.

Definition 2.51 *A formal-Kähler (resp. para-Kähler) Banach structure* $(\Omega, g, \mathcal{I})$ *(resp.* $(\Omega, g, \mathcal{J})$*) is said to be flat if the Levi-Civita connection* ∇ *of* g *exists and is flat.*

Lemma 2.3 *Let* $(\omega, g, \mathcal{I})$ *(resp.* $(\Omega, g, \mathcal{J})$*) be a Kähler (resp. para-Kähler) Banach structure such that the Levi-Civita connection* ∇ *of* g *exists. Then we have* $\nabla\mathcal{I} \equiv 0$ *(resp.* $\nabla\mathcal{J} \equiv 0$*).*

Proof *cf. proof of Proposition 91 in [Tum05].*

Theorem 2.11 *If a formal-Kähler (resp. para-Kähler) Banach structure* $(\Omega, g, \mathcal{I})$ *(resp.* $(\Omega, g, \mathcal{J})$*) on* M *is flat, for each* $x_0 \in M$*, there exists a chart* (U, ϕ)*, a linear Darboux form* Ω_0*, a linear decomposable complex structure* $\mathcal{I}_0$ *and an inner product* g_0 *(resp. a linear Darboux form* Ω_0*, a linear para-complex structure* $\mathcal{J}_0$ *and a neutral inner product* g_0*) on* $\mathbb{M}$ *such that* $\phi^*\Omega_0 = \Omega$*,* $\phi^*\mathcal{I}_0 = \mathcal{I}$ *and* $\phi^*g_0 = g$ *(resp.* $\phi^*\Omega_0 = \Omega$*,* $\phi^*\mathcal{J}_0 = \mathcal{J}$ *and* $\phi^*g_0 = g$*).*

Proof *We only consider the case of an almost para-Kähler manifold, the case of an almost Kähler manifold is similar.*
Consider the decomposition $TM = E_1 \oplus E_2$ *as recalled previously. Since the Levi-Civita connection is a metric connection, then* $\nabla g = 0$*. Since* E_1 *and* E_2 *are orthogonal, the connection* ∇ *induces a connection* ∇^i *on* E_i *which preserves the restriction* g_i *of* g *to* E_i *for* $i \in \{1, 2\}$ *and* $\nabla = \nabla^1 + \nabla^2$*. Thus, for* $i \in \{1, 2\}$*, if* X *and* Y *are sections of* E_i *we have*

$$[X, Y] = \nabla_X Y - \nabla_Y X = \nabla^i_X Y - \nabla^i_Y X$$

and so E_i *is integrable. Moreover, the parallel transport from* $T_{\gamma(0)}M$ *to* $T_{\gamma(1)}M$ *along a curve* $\gamma : [0, 1] \to M$ *induces an isometry of the fibre* $(E_i)_{\gamma(0)}$ *into* $(E_i)_{\gamma(1)}$ *for* $i \in \{1, 2\}$*. Fix some* $x_0 \in M$*. Since* ∇ *is flat, according to the proof Theorem 2.9, we may assume that* $M = \mathbb{M}$ *and* $x_0 = 0 \in \mathbb{M}$*. Moreover, if* $\mathbb{E}_1$ *(resp.* $\mathbb{E}_2$*) is the typical fibre of the integrable subbundle* E_1 *(resp.* E_2*), we may assume that* $\mathbb{M} = \mathbb{E}_1 \oplus \mathbb{E}_2$*. Recall that under the flatness assumption, the diffeomorphism* $\Psi(x) = x + Id_{\mathbb{M}}$ *is such that* $T_0\Psi$ *is the parallel transport from* $T_0\mathbb{M}$ *to* $T_x\mathbb{M}$ *and* $\Psi^*g_0 = g$*. But since* E_1*,* E_2 *and* $\mathcal{J}$ *are invariant by parallel transport, we must have* $\mathcal{J} \circ T\Psi = T\Psi \circ \mathcal{J}$ *and so* $\Psi^*\mathcal{J}_0 = \mathcal{J}$*. Now as* $\Omega(u, v) = g(u, \mathcal{J}v)$*, we also get* $\Psi^*\Omega_0 = \Omega$*.*

Remark 2.19 *Under the assumption of flatness of* ∇ *in Theorem 2.11, we obtain the integrability of* $\mathcal{I}$*. But as we have already seen, in general, the condition of nullity of* $N_{\mathcal{I}}$ *is not sufficient to ensure the integrability of* $\mathcal{I}$*.*

2.5 Darboux Theorem for Symplectic Forms on a Banach Manifold

In this section, we essentially follow the paper [Pel18] and use the notions and notations of § 2.1.

For any finite-dimensional symplectic manifold, the Darboux Theorem asserts that, around each point, there exists a (Darboux) chart in which the 2-form can be written as a constant one, named classical linear Darboux form. Such a result can be proved by induction on the dimension of the manifold. However, using a Moser's idea for volume form on compact manifold ([Mos65]), the Darboux theorem can also be proved using an isotopy obtained by the local flow of a time dependent vector field. Such a method is classically called the *Moser's method* and works in many other frameworks.

2.5.1 Moser's Method and Darboux Theorem

In the Banach context, it is well known that a symplectic form can be strong (cf. Definition 2.7) or weak (cf. Definition 2.6). The Darboux Theorem was firstly proved for strong symplectic Banach manifolds by Weinstein ([Wei71]). Then Marsden ([Mars72]) showed that the Darboux theorem fails for a weak symplectic Banach manifold. However, in [Bam99], Bambusi found necessary and sufficient conditions for the validity of Darboux theorem for a weak symplectic Banach manifold (**Darboux-Bambusi Theorem**). The proof of all these versions of Darboux Theorem relies on Moser's method.

In this section, we prove a generalization of Moser's Lemma (see, for instance, [Lesf14]) to the Banach framework, and, as a corollary, we obtain the result of [Bam99] for a weak symplectic Banach manifold.

Let M be a manifold modelled on a reflexive Banach space $\mathbb{M}$. Consider a weak symplectic form ω on M. Then $\omega^\flat : TM \to T^*M$ is an injective bundle morphism. As in § 2.4.2, we denote by $\widehat{T_xM}$ the Banach space which is the completion of T_xM provided with the norm $|| \; ||_{\omega_x^\flat}$ associated with some norm $|| \; ||$ on T_xM. The Banach space $\widehat{T_xM}$ does not depend on this choice. Then ω_x can be extended to a continuous bilinear form $\hat{\omega}_x$ on $T_xM \times \widehat{T_xM}$ and $\omega_x^\flat$ becomes an isomorphism from T_xM to $(\widehat{T_xM})^*$. We set

$$\widehat{TM} = \bigcup_{x \in M} \widehat{T_xM} \text{ and } (\widehat{TM})^* = \bigcup_{x \in M} (\widehat{T_xM})^*.$$

Theorem 2.12 (Moser's Lemma) *Let ω be a weak symplectic form on a Banach manifold M modelled on a reflexive Banach space $\mathbb{M}$. Assume the following properties:*

(i) There exists a neighbourhood U of $x_0 \in M$ such that $\widehat{TM}_{|U}$ is a trivial Banach bundle whose typical fibre is the Banach space $(\widehat{T_{x_0}M}, || \; ||_{\omega_{x_0}^\flat})$;

(ii) via some trivialization, ω can be extended to a smooth field of continuous bilinear forms on $TM_{|U} \times \widehat{TM}_{|U}$.

Consider a family $\{\omega^t\}_{0 \leq t \leq 1}$ of closed 2-forms on M, depending smoothly on t with the following properties:

- $\omega^0 = \omega$ *and* $\forall t \in [0,1], \omega^t_{x_0} = \omega_{x_0}$;
- ω^t *can be extended to a smooth field of continuous bilinear forms on* $TM_{|U} \times \widehat{TM}_{|U}$.

Then there exists a neighbourhood V *of* x_0 *such that each* ω^t *is a symplectic form on* V *and there exists a family* $\{F_t\}_{0\leq t\leq 1}$ *of diffeomorphisms* F_t *from a neighbourhood* $V_0 \subset V$ *of* x_0 *to a neighbourhood* $F_t(V_0) \subset V$ *of* x_0 *such that* $F_0 = \mathrm{Id}$ *and* $F_t^*\omega^t = \omega$, *for all* $0 \leq t \leq 1$.

Proof *Let* (U, ϕ) *be a chart around* x_0 *such that* $\phi(x_0) = 0 \in \mathbb{M}$ *and consider that all the assumptions of the theorem are true on* U. *Let* $\Psi : \widehat{TM}_{|U} \to \phi(U) \times \widehat{\mathbb{M}}$ *be a trivialization. Without loss of generality, we may assume that* U *is an open neighbourhood of* 0 *in* $\mathbb{M}$ *and* $\widehat{TM}_{|U} = U \times \widehat{\mathbb{M}}$. *Therefore,* $U \times \widehat{\mathbb{M}}$ *is a trivial Banach bundle whose fibre is the Banach space* $(\widehat{\mathbb{M}}, || \, ||_{\omega_0^\flat})$. *Since* ω *can be extended to a non-degenerate skew-symmetric bilinear form (again denoted* ω*) on* $U \times (\mathbb{M} \times \widehat{\mathbb{M}})$ *then* $\omega^\flat$ *is a Banach bundle isomorphism from* $U \times \mathbb{M}$ *to* $U \times \widehat{\mathbb{M}}^*$.
We set $\dot{\omega}^t = \frac{d}{dt}\omega^t$. *Since each* ω^t *is closed for* $0 \leq t \leq 1$, *we have :*

$$d\dot{\omega}^t = \frac{d}{dt}(d\omega^t) = 0$$

and so $\dot{\omega}^t$ *is closed. After shrinking* U *if necessary, from the Poincaré Lemma, there exists a* 1*-form* α^t *on* U *such that* $\dot{\omega}^t = d\alpha^t$ *for all* $0 \leq t \leq 1$. *In fact* α_t *can be given by*

$$\alpha^t_x = \int_0^1 s.(\dot{\omega}^t_{sx})^\flat(x)ds.$$

Now, at $x = 0$, $(\omega^t_{x_0})^\flat$ *is an isomorphism from* $\mathbb{M}$ *to* $\widehat{\mathbb{M}}^*$. *Since, for all* $0 \leq t \leq 1$, $x \mapsto (\omega^t_x)^\flat$ *is a smooth field from* U *to* $\mathcal{L}(\mathbb{M}, \widehat{\mathbb{M}}^*)$, *there exists a neighbourhood* V *of* 0 *such that* $(\omega^t_x)^\flat$ *is an isomorphism from* $\mathbb{M}$ *to* $\widehat{\mathbb{M}}^*$ *for all* $x \in V$ *and* $0 \leq t \leq 1$. *In particular,* ω^t *is a symplectic form on* V. *Moreover* $x \mapsto (\dot{\omega}^t_x)^\flat$ *is smooth and takes values in* $\mathcal{L}(\mathbb{M}, \widehat{\mathbb{M}}^*)$. *Therefore* $x \mapsto \alpha^t_x$ *can be extended to a smooth field on* V *into* $\widehat{\mathbb{M}}^*$. *We set* $X^t_x := -((\omega^t_x)^\flat)^{-1}(\alpha^t_x)$. *It is a well-defined time-dependent vector field and let* Fl_t *be the flow generated by* X^t *defined on some neighbourhood* $V_0 \subset V$ *of* 0. *Note that, for all* $t \in [0,1]$, *since* $\dot{\omega}^t_{x_0} = 0$, *then* $X^t_{x_0} = 0$. *Thus, for all* $t \in [0,1]$, $F_t(x_0) = x_0$. *As classically, we have*

$$\frac{d}{dt}\mathrm{Fl}_t^*\,\omega^t = \mathrm{Fl}_t^*(L_{X^t}\omega^t) + \mathrm{Fl}_t^*\,\frac{d}{dt}\omega^t = \mathrm{Fl}_t^*(-d\alpha^t + \dot{\omega}^t) = 0.$$

Thus $\mathrm{Fl}_t^*\,\omega^t = \omega$.

Remark 2.20 *If* ω *is a strong symplectic form on* M, *then* $\mathbb{M}$ *is reflexive and* $\omega^\flat$ *is a bundle isomorphism from* TM *to* T^*M. *In particular, the norm* $|| \, ||_{\omega_x^\flat}$

is equivalent to any norm $|| \ ||$ on T_xM which defines its Banach structure and so all the assumptions (i) and (ii) of Theorem 2.12 are locally always satisfied. Moreover, if we consider a family $\{\omega^t\}_{0\leq t\leq 1}$ as in Theorem 2.12, the assumption

ω^t can be extended to a smooth field of continuous bilinear forms on $TM_{|U} \times \widehat{TM}_{|U}$, for all $0 \leq t \leq 1$

is always satisfied.

Therefore we get the same conclusion only with the assumptions $\omega^0 = \omega$ and $\omega^t_{x_0} = \omega_{x_0}$ for all $0 \leq t \leq 1$.

Example 2.3 *We consider a symplectic form ω on M such that the assumptions (i) and (ii) of Theorem 2.12 are satisfied on some neighbourhood U of x_0. We consider the family $\omega^{s,t}_x = e^{st}\omega_x$ where $(s,t) \in [0,1]^2$. Then for $s = 0$, we have $\omega^{s,t}_{x_0} = \omega_{x_0}$ for all $t \in [0,1]$ and $\omega^0 = \omega$ for $t = 0$. But $(\omega^{s,t}_{x_0})^\flat \equiv \omega^\flat_{x_0}$ belongs to $\mathcal{L}(\mathbb{M}, \widehat{\mathbb{M}}^*)$. Since the map $\begin{array}{ccc} [0,1]^2 \times U & \to & \mathcal{L}(\mathbb{M}, \widehat{\mathbb{M}}^*) \\ (s,t,x) & \mapsto & (\omega^{s,t}_x)^\flat \end{array}$ is smooth, there exists an open neighbourhood $\Sigma \times V \subset [0,1]^2 \times U$ of $\{0\} \times [0,1] \times \{x_0\}$ such that $(\omega^{s,t}_x)^\flat$ belongs to $\mathrm{GL}(\mathbb{M}, \widehat{\mathbb{M}}^*)$. Therefore the assumptions of Theorem 2.12 are fulfilled for each s fixed such that $(s,t) \in \Sigma$ on an open neighbourhood V_s of x_0. This implies that, for any fixed s such that $\{s\} \times [0,1] \subset \Sigma$, there exists a family $\{F_{s,t}\}_{0\leq t\leq 1}$ of diffeomorphisms $F_{s,t}$ from a neighbourhood W_s of x_0 to $F_{s,t}(W_s) \subset V_s$ is a neighbourhood x_0 and such that $F_0 = \mathrm{Id}$ and $F^*_{s,t}\omega^{s,t} = \omega$ for all $0 \leq t \leq 1$.*

Now as a Corollary of Theorem 2.12, we obtain Bambusi's version of Darboux Theorem ([Bam99], Theorem 2.1).

2.5.2 A Symplectic Form on $\mathsf{L}^p_k(\mathbb{S}^1, M)$

Let (M, ω) be a symplectic manifold of dimension $m = 2m'$.
As in [Kum15], for any $\gamma \in \mathsf{L}^p_k(\mathbb{S}^1, M)$ and $X, Y \in T_\gamma \mathsf{L}^p_k(\mathbb{S}^1, \mathbb{R}^m) \equiv \mathsf{L}^p_k(\gamma^*TM)$, we set

$$\Omega_\gamma(X,Y) = \int_{\mathbb{S}^1} \omega_{\gamma(t)}(X(t), Y(t))dt. \tag{2.22}$$

Theorem 2.13 *For $k > 0$ and $1 < p < \infty$, Ω is a symplectic weakly non-degenerate form. Moreover Ω is a strong symplectic form if and only if $p = 2$.*

Proof *At first, it is clear that Ω_γ is a bilinear skew-symmetric 2-form on $T_\gamma \mathsf{L}^p_k(\mathbb{S}^1, \mathbb{R}^m)$. Since the smoothness of $\gamma \mapsto \Omega_\gamma$ is a local property, we consider a chart (U, Φ) around γ on $\mathsf{L}^p_k(\mathbb{S}^1, M)$. Note that over U, the Banach bundle of bilinear skew-symmetric forms is trivial. Therefore, without loss of generality, we may assume that U is an open set of $\mathsf{L}^p_k(\mathbb{S}^1, \mathbb{R}^m)$ and the previous bundle is*

$$U \times \Lambda^2 \left(\mathsf{L}^p_k\left(\mathbb{S}^1, \mathbb{R}^m\right)\right)^* \equiv U \times \mathsf{L}^p_k\left(\mathbb{S}^1, \Lambda^2\mathbb{R}^m\right).$$

Then the set of smooth sections of this bundle can be identified with the set of smooth maps from U *to* $\mathsf{L}^p_k\left(\mathbb{S}^1, \Lambda^2\mathbb{R}^m\right)$. *First, we prove that the map* $(\gamma, X, Y) \mapsto \Omega_\gamma(X,Y)$ *is a smooth map from* $U \times \left(\mathsf{L}^p_k\left(\mathbb{S}^1, \mathbb{R}^m\right)\right)^2$ *into* $\mathbb{R}$. *From [KrMi97], Corollary 4.13, it is sufficient to prove that for any smooth curve* $\delta : \mathbb{R} \to U \times \left(\mathsf{L}^p_k(\mathbb{S}^1, R^m)\right)^2$ *the map*

$$\tau \mapsto \int_{\mathbb{S}^1} \omega_{\delta_0(\tau)(t)}(X_{\delta_1(\tau)(t)}, Y_{\delta_2(\tau)(t)})dt$$

is a smooth map from $\mathbb{R}$ *to* $\mathbb{R}$, *where we have the decomposition*

$$\delta(\tau) = (\delta_0(\tau), \delta_1(\tau), \delta_2(\tau)) \in U \times \left(\mathsf{L}^p_k(\mathbb{S}^1, \mathbb{R}^m)\right)^2.$$

Indeed, for any $t \in \mathbb{S}^1$, *the map* $\tau \mapsto \omega_{\delta(\tau)(t)}(X_{\delta(\tau)(t)}, Y_{\delta(\tau)(t)})$ *is smooth and so from the properties of the integral of a function depending on a parameter, we obtain the smoothness of* $\tau \mapsto \int_{\mathbb{S}^1} \omega_{\delta(\tau)(t)}(X_{\delta(\tau)(t)}, Y_{\delta(\tau)(t)})dt$. *This implies the smoothness of* $\gamma \mapsto \Omega_\gamma$ *from [KrMi97], Theorem 3.12.*
Now we show that Ω *is closed. From Cartan formulae, we have*

$$\begin{aligned} d\Omega\left(U_0, U_1, U_2\right) &= \sum_{j=0}^{2}(-1)^j U_j\left(\Omega(U_0, \widehat{U}_j, U_2)\right) \\ &\quad + \sum_{0 \le l < j \le 2} (-1)^{l+j}\Omega\left([U_l, U_j], U_0, \widehat{U}_l, \widehat{U}_j, U_2\right). \end{aligned}$$

Fix some $\gamma \in \mathsf{L}^p_k(\mathbb{S}^1, \mathbb{R}^m)$. *Consider a map* $\sigma : (]-\epsilon, \epsilon[)^3 \times \mathbb{S}^1 \to M$ *with the following properties:*

(a) $\sigma(0,0,0,t) = \gamma(t)$;

(b) $\forall j \in \{0,1,2\}, \dfrac{\partial \sigma}{\partial u_j}(0,0,0,t) = U_j(t)$.

We set $\gamma' = \dfrac{\partial \sigma}{\partial t}$ *and for all* $j \in \{0,1,2\}$, $U_j = \dfrac{\partial \sigma}{\partial u_j}$.
Note that by construction, we have

$$\forall j \in \{0,1,2\},\ [U_l, U_j] = 0\ .$$

and so

$$\begin{aligned} U_0\left(\Omega(U_1, U_2)\right) &= \frac{\partial \sigma}{\partial u_0}\int_{\mathbb{S}^1} \omega_{\sigma(0,0,0,t)}\left(\frac{\partial \sigma}{\partial u_1}(0, u_1, 0, t), \frac{\partial \sigma}{\partial u_2}(0,0,u_2,t)\right)dt \\ &= \int_{\mathbb{S}^1} \frac{\partial \sigma}{\partial u_0}\left\{\omega_{\sigma(0,0,0,t)}\left(\frac{\partial \sigma}{\partial u_1}(0,u_1,0,t), \frac{\partial \sigma}{\partial u_2}(0,0,u_2,t)\right)\right\}dt. \end{aligned} \tag{2.23}$$

Using the same calculus for the other terms in the first sum, we obtain

$$d\Omega(U_0, U_1, U_2) = \int_{\mathbb{S}^1} d\omega_{\gamma(t)}(U_0(t), U_1(t), U_2)(t)dt.$$

Therefore, since ω is closed so is Ω.
It remains to show that Ω is weakly non-degenerate. Assume that there exists $X \in T_\gamma \mathsf{L}^p_k(\mathbb{S}^1, \mathbb{R}^m)$ not identically zero such that, for any $Y \in T_\gamma \mathsf{L}^p_k(\mathbb{S}^1, \mathbb{R}^m)$, we have $\Omega_\gamma(X, Y) = 0$.
Since $k > 0$, then we can assume that γ and X are continuous. (cf. [Schr15], consequence of Theorem 4.1.1). Now since X is not identically zero, there exists some interval $]\alpha, \beta[\subset \mathbb{S}^1$ on which $\omega_{\gamma(t)}(X(t), \ )$ is a field of non-zero 1-forms. Let $J \subsetneqq]\alpha, \beta[$ be a closed sub-interval. Since ω is symplectic, it must exist a field of smooth vector fields Y_J on J such that $\omega_{\gamma(t)}(X(t), Y_j(t)) > 0$. We can extend Y_J to a smooth vector field Y on $\mathbb{S}^1$ so that the support of Y is contained in $]\alpha, \beta[$. Since Y is smooth then Y belongs to $T_\gamma \mathsf{L}^p_k(\mathbb{S}^1, \mathbb{R}^m)$ and, by construction, we have

$$\Omega(X, Y) = \int_{\mathbb{S}^1} \omega_{\gamma(t)}(X(t), Y(t))dt > 0.$$

and we obtain a contradiction.
Now, if Ω is a strong symplectic form, this implies that the Banach space $\mathsf{L}^p_k(\mathbb{S}^1, \mathbb{R}^m)$ is isomorphic to its dual which is only true for $p = 2$. Conversely, for $p = 2$, the space $\mathsf{L}^2_k(\mathbb{S}^1, \mathbb{R}^m)$ is a Hilbert space and so $T^*_\gamma \mathsf{L}^2_k(\mathbb{S}^1, M)$ is isomorphic to $\mathsf{L}^2_k(\mathbb{S}^1, \mathbb{R}^m)$. Since ω is symplectic, $\omega^\flat$ is a smooth isomorphism from TM to T^*M. Therefore, if $\eta \in T^*_\gamma \mathsf{L}(\mathbb{S}^1, M)$, $X = (\omega^\flat)^{-1}(\eta)$ belongs to $T_\gamma \mathsf{L}^2_k(\mathbb{S}^1, M)$ and, for any $Y \in T_\gamma \mathsf{L}^2_k(\mathbb{S}^1, M)$, we have

$$\omega_{\gamma(t)}(X(t), Y(t)) = \langle \eta(t), Y(t) \rangle_M$$

where $< ., . >_M$ is the duality bracket between TM and T^*M. Now the duality bracket between $T^*_\gamma \mathsf{L}^2_k(\mathbb{S}^1, M)$ and $T_\gamma \mathsf{L}^2_k(\mathbb{S}^1, M)$ is then given by

$$\int_{\mathbb{S}^1} \langle \xi(t), Z(t) \rangle dt$$

for any $\xi \in T^*_\gamma \mathsf{L}^2_k(\mathbb{S}^1, M)$ and any $Z \in T_\gamma \mathsf{L}^2_k(\mathbb{S}^1, M)$. Therefore we have

$$\Omega^\flat(X)(Y) = \eta(Y),$$

which ends the proof.

2.5.3 Darboux Charts on $\mathsf{L}^p_k(\mathbb{S}^1, M)$

We are now in situation to apply Theorem 2.6 for the symplectic form on $\mathsf{L}^p_k(\mathbb{S}^1, M)$ defined in the previous section.

Theorem 2.14 *There exists a Darboux chart (V, F) around any $\gamma \in \mathsf{L}^p_k(\mathbb{S}^1, M)$ for the weak symplectic manifold $(\mathsf{L}^p_k(\mathbb{S}^1, M), \Omega)$.*

Proof *At first, note that if $p = 2$, then Ω is a strong symplectic form and then the result is an application of the result of [Mars81] and [Wei71] or also Theorem 2.6 (local version of Darboux theorem). Thus, from now on, we assume $1 < p < \infty$ and $p \neq 2$.*
Fix some $\gamma \in \mathsf{L}^p_k(\mathbb{S}^1, M)$ and consider a chart (U, Φ) around γ. It is well known that the maps $T\phi : T\mathsf{L}^p_k(\mathbb{S}^1, M)_{|U} \to \Phi(U) \times \mathsf{L}^p_k(\mathbb{S}^1, \mathbb{R}^m)$ and $T^*\Phi^{-1} : T^*\mathsf{L}^p_k(\mathbb{S}^1, M)_{|U} \to \Phi(U) \times (\mathsf{L}^p_k(\mathbb{S}^1, \mathbb{R}^m))^*$ are the natural trivializations associated with (U, Φ). Without loss of generality, we may assume that U is an open subset of $\mathsf{L}^p_k(\mathbb{S}^1, \mathbb{R}^m)$. Then $TU = U \times \mathsf{L}^p_k(\mathbb{S}^1, \mathbb{R}^m)$ and $T^*U = U \times (\mathsf{L}^p_k(\mathbb{S}^1, \mathbb{R}^m))^*$.
According to [Pel18], end of § 4.1 and Remark 29, it follows that $\mathsf{L}^p_k(\mathbb{S}^1, \mathbb{R}^m)$ is dense in $\mathsf{L}^p(\mathbb{S}^1, \mathbb{R}^m)$. Either $1 < p < 2$ and then for $p \leq q$, $\mathsf{L}^q(\mathbb{S}^1, \mathbb{R}^m)$ is a dense subspace of $\mathsf{L}^p(\mathbb{S}^1, \mathbb{R}^m)$ and so $\mathsf{L}^p_k(\mathbb{S}^1, \mathbb{R}^m) \cap \mathsf{L}^q(\mathbb{S}^1, \mathbb{R}^m)$ is dense in $\mathsf{L}^q(\mathbb{S}^1, \mathbb{R}^m)$. Otherwise, if $p > 2$, then $\mathsf{L}^p(\mathbb{S}^1, \mathbb{R}^m)$ is a dense subspace of $\mathsf{L}^q(\mathbb{S}^1, \mathbb{R}^m)$ and so $\mathsf{L}^p_k(\mathbb{S}^1, \mathbb{R}^m)$ is dense in $\mathsf{L}^q(\mathbb{S}^1, \mathbb{R}^m)$. Thus we will only consider the case $p > 2$ since the other one is similar by replacing $\mathsf{L}^p_k(\mathbb{S}^1, \mathbb{R}^m)$ by $\mathsf{L}^p_k(\mathbb{S}^1, \mathbb{R}^m) \cap \mathsf{L}^q(\mathbb{S}^1, \mathbb{R}^m)$. Since $TU = U \times \mathsf{L}^p_k(\mathbb{S}^1, \mathbb{R}^m)$, and $\mathsf{L}^p_k(\mathbb{S}^1, \mathbb{R}^m)$ is dense in $\mathsf{L}^q(\mathbb{S}^1, \mathbb{R}^m)$ we can consider the trivial bundle

$$L^q(U) = U \times \mathsf{L}^q(\mathbb{S}^1, \mathbb{R}^m),$$

and then the inclusion of TU in $L^q(U)$ is a bundle morphism with dense range. On the other hand, as noted at the end of section 4.1 and Remark 29 in [Pel18], $\mathsf{L}^q(\mathbb{S}^1, \mathbb{R}^m)$ provided with the norm $|| \;||_{-k,p}$ is dense in $(\mathsf{L}^p_k(\mathbb{S}^1, \mathbb{R}^m))^*$. So we have an inclusion of the trivial bundle $L^q(U)$ in T^*U whose range is dense.
Since ω is symplectic, $\omega^\flat_\gamma : \gamma^*(TM) \to \gamma^*(T^*M)$ is an isomorphism and the map $X \in \mathsf{L}^p_k(\mathbb{S}^1, \mathbb{R}^m) \mapsto \alpha = \omega^\flat_{\gamma(t)}(X(t))$ is nothing but that $\Omega^\flat_\gamma$. For each fixed γ, we can extend the operator $\Omega^\flat_\gamma : \mathsf{L}^q_k(\mathbb{S}^1, \mathbb{R}^m) \to (\mathsf{L}^p_k(\mathbb{S}^1, \mathbb{R}^m))^*$ to a continuous operator $(\bar{\Omega}^\flat_\gamma)^q : \mathsf{L}^q(\mathbb{S}^1, \mathbb{R}^m) \to (\mathsf{L}^p_k(\mathbb{S}^1, \mathbb{R}^m))^*$. In fact, this operator is given by

$$\bar{\Omega}^\flat_\gamma(X(t)) = \omega^\flat_{\gamma(t)}(X(t))$$

and $\bar{\Omega}^\flat(X)$ belongs to $\mathsf{L}^q(\mathbb{S}^1, \mathbb{R}^m)$. Therefore the range of $\bar{\Omega}^\flat_\gamma$ is $\mathsf{L}^q(\mathbb{S}^1, \mathbb{R}^m) \subset (\mathsf{L}^p_k(\mathbb{S}^1, \mathbb{R}^m))^*$ and is an isomorphism whose inverse is given by

$$(\bar{\Omega}^\flat)^{-1}(\alpha(t)) = (\omega^\flat_{\gamma(t)})^{-1}(\alpha(t)).$$

Note that $\bar{\Omega}^\flat_\gamma$ is the natural morphism associated with the skew-symmetric bilinear form on $\mathsf{L}^q(\mathbb{S}^1; \mathbb{R}^m)$ defined by

$$\bar{\Omega}_\gamma(X, Y) = \int_{\mathbb{S}^1} \omega_{\gamma(t)}(X(t), Y(t))dt$$

which is an extension of the initial 2-form Ω_γ on $\mathsf{L}^p(\mathbb{S}^1; \mathbb{R}^m)$. Therefore, on $\mathsf{L}^q(\mathbb{S}^1; \mathbb{R}^m)$, considered as a subset of $(\mathsf{L}^p_k(\mathbb{S}^1, \mathbb{R}^m))^*$, we can consider the induced norm $|| \;||_{-k,q}$. Since $\bar{\Omega}_\gamma$ is bijective, then

$$||X||_{\bar{\Omega}_\gamma} = ||\bar{\Omega}^\flat(X)||_{-k,p}$$

defines a norm on $\mathsf{L}^q(\mathbb{S}^1;\mathbb{R}^m)$ *for all* $\gamma \in U$. *This implies that* $\bar{\Omega}^\flat_\gamma$ *is a continuous isomorphism and an isometry from the normed space* $(\mathsf{L}^q(\mathbb{S}^1;\mathbb{R}^m), || \; ||_{\bar{\Omega}_\gamma})$ *to the normed space* $(\mathsf{L}^q(\mathbb{S}^1;\mathbb{R}^m), || \; ||_{-k,q})$. *Therefore the norms* $|| \; ||_{\bar{\Omega}_\gamma}$ *and* $|| \; ||_{-k,q}$ *are equivalent on* $\mathsf{L}^q(\mathbb{S}^1;\mathbb{R}^m$ *for any* $\gamma \in U$. *Now, note that* $|| \; ||_{\bar{\Omega}_\gamma}$ *induces the norm* $|| \; ||_{\Omega_\gamma}$ *on* $T_\gamma U = \{\gamma\} \times \mathsf{L}^p_k(\mathbb{S}^1;\mathbb{R}^m)$ *as defined in § 2.1.1. Since each norm* $|| \; ||_{\bar{\Omega}_\gamma}$ *is equivalent to* $|| \; ||_{-k,p}$, *then each norm* $|| \; ||_{\bar{\Omega}^\flat_\gamma}$ *is equivalent to* $|| \; ||_{\bar{\Omega}_{\gamma_0^\flat}}$; *it follows that the same result is true for the restriction of this norm to* $\mathsf{L}^p_k(\mathbb{S}^1,\mathbb{R}^m)$. *Therefore, the completion of the normed space* $(\mathsf{L}^p_k(\mathbb{S}^1,\mathbb{R}^m), || \; ||_{\Omega^\flat_\gamma})$ *does not depend on* γ. *We denote by* $\widehat{\mathsf{L}^p_k}(\mathbb{S}^1,\mathbb{R}^m)_\gamma$ *this completion. According to the notations of Theorem 2.12, we have* $\widehat{TU} = U \times \widehat{\mathsf{L}^p_k}(\mathbb{S}^1,\mathbb{R}^m)_{\gamma_0}$. *This implies that the condition (i) of the assumptions of Theorem 2.6 is satisfied.*
As we have seen in Proposition 2.1, for each γ, *we can extend the operator*

$$\Omega^\flat : \mathsf{L}^p_k(\mathbb{S}^1,\mathbb{R}^m) \to (\mathsf{L}^p_k(\mathbb{S}^1,\mathbb{R}^m))^*$$

to an isometry $\widehat{\Omega}^\flat_\gamma$ *from* $\widehat{\mathsf{L}^p_k}(\mathbb{S}^1,\mathbb{R}^m)_\gamma$ *to* $(\mathsf{L}^p_k(\mathbb{S}^1,\mathbb{R}^m))^*$. *It remains to show that* $\gamma \mapsto \Omega^\flat_\gamma$ *is a smooth field from* U *to* $\mathcal{L}(\mathsf{L}^p_k(\mathbb{S}^1,\mathbb{R}^m), (\widehat{\mathsf{L}^p_k}(\mathbb{S}^1,\mathbb{R}^m)_{\gamma_0})^*)$.
This property can be shown by the same arguments used in Theorem 2.13, for the smoothness of $\gamma \mapsto \Omega_\gamma$. *Therefore, the assumptions of Theorem 2.6 are satisfied on* U *and so the proof is completed.*

2.6 Notes

A classical reference for Krein spaces is [AzIo89].
Some Banach G-structures defined by a tensor occur, for example, on the loop spaces (cf. [Wur95], [Kum13a] and [Kum13b] among many others) and, more generally, the existence of a (strong) Riemannian metric on a Banach bundle (cf [Lan95], Theorem 3.1). A general result on some G-structures on a Banach principal bundle can be found in [Klo11].

3

Convenient Structures

Differential calculus in infinite dimension has already a long history which goes back to the beginnings of variational calculus in the XVIIIth century. This calculus developed by Jacques Bernoulli, Leonhard Euler and Joseph-Louis Lagrange deals with finding *extrema* of functionals.

During the last decades, a lot of theories of differentiation have been proposed in order to differentiate in spaces more general than Banach ones.

The setting of convenient differential calculus introduced by A. Frölicher and A. Kriegl (see [FrKr88]) is well adapted to a lot of situations and suits perfectly the framework of direct limits. The category of convenient spaces with convenient smooth maps is cartesian closed and the exponential law holds.

In this chapter, we introduce convenient geometric structures (vector spaces, manifolds, Lie groups, Lie algebras, (linear and principal) bundles and algebroids which will be used throughout this book.

3.1 Locally Convex Topological Vector Spaces

The purpose of this section is to present basic definitions on (locally convex) topological vector spaces, duals and bornology which will be used in the convenient setting.
The power set of a set $\mathscr{X}$ will be denoted $\mathscr{P}(\mathscr{X})$.

3.1.1 Topological Spaces

The notion of topology has been introduced to provide a framework to the notions of continuity (of maps) and convergence (of filters).

Definition 3.1 *A topology on a set* $\mathscr{X}$ *is a family* $\mathscr{O} = \{\mathcal{O}_i\}_{i\in I} \subset \mathscr{P}(\mathscr{X})$ *satisfying the following axioms:*

(Top1) $\emptyset \in \mathscr{O}$ *and* $\mathscr{X} \in \mathscr{O}$*;*

DOI: 10.1201/9781003435587-3

(Top2) *The union of elements of an arbitrary subfamily of $\mathscr{O}$ still belongs to $\mathscr{O}$:*

$$(\forall i \in I,\ \mathcal{O}_i \in \mathscr{O}) \Longrightarrow \bigcup_{i \in I} \mathcal{O}_i \in \mathscr{O};$$

(Top3) *The intersection of elements of a finite set of elements of $\mathscr{O}$ still belongs to $\mathscr{O}$:*

$$(\forall j \in J:\ \mathrm{card}(J) < +\infty,\ \mathcal{O}_j \in \mathscr{O}) \Longrightarrow \bigcap_{j \in J} \mathcal{O}_j \in \mathscr{O}.$$

The pair $(\mathscr{X}, \mathscr{O})$ is called a topological space. An element of $\mathscr{O}$ is called an open set.

It can be convenient to consider an equivalent approach by defining a topology *via* the system of the neighborhoods of the points.

Definition 3.2

(1) A family $\mathscr{V} = \{\mathscr{V}_x\}_{x \in \mathbb{E}}$ of subsets of $\mathscr{X}$ is called a basis of neighbourhoods if it satisfies the followings assumptions:

(BN1) $\forall V \in \mathscr{V}_x,\ x \in V;$

(BN2) $U \in \mathscr{V}_x$ *and* $V \in \mathscr{V}_x \Longrightarrow U \cap V \in \mathscr{V}_x;$

(BN3) $\forall U \in \mathscr{V}_x,\ U \subset V \Longrightarrow V \in \mathscr{V}_x;$

(BN4) $\forall U \in \mathscr{V}_x,\ \exists V \in \mathscr{V}_x : \forall y \in V,\ U \in \mathscr{V}_y.$

(2) A subset $\{\mathscr{U}_x\}_{x \in \mathbb{E}}$ is a called a basis of neighborhoods of x if

$$\forall V \in \mathscr{V}_x,\ \exists U \in \mathscr{U}_x :\ U \subset V.$$

A point x in a subset X of $\mathscr{X}$ is called an *interior point* if there exists $V \in \mathscr{V}_x$ such that $V \subset X$.

Notation 3.1 *We denote by $\mathring{X}$ the set of interior points of X.*

As usually, a set O is open if and only if $\mathring{O} = O$.
It is well known that the set $\mathcal{O}$ of such open sets define a topology on $\mathscr{X}$.

The primitive notion of filter (cf. [Tre67], 1.) appears useful in different situations (e.g. Cauchy sequences).

Definition 3.3 *A filter $\mathfrak{F}$ on a set $\mathscr{X}$ is a family $\mathfrak{F} \subset \mathscr{P}(\mathscr{X})$ satisfying the following axioms:*

(Filt1) $\emptyset \notin \mathfrak{F};$

(Filt2) $\forall (A_1, A_2) \in \mathfrak{F}^2,\ A_1 \cap A_2 \in \mathfrak{F};$

(Filt3) $\forall A \in \mathfrak{F}, \forall B \in \mathscr{P}(\mathscr{X}), B \supset A \Longrightarrow B \in \mathfrak{F}$.

Example 3.1 *Let $(\mathscr{X}, \mathscr{O})$ be a topological space. The set of neighbourhoods $\mathscr{N}(x)$ of $x \in \mathscr{X}$ is a filter.*

A family $\mathcal{B} \subset \mathscr{P}(\mathscr{X})$ is a *basis* of a filter $\mathfrak{F}$ on $\mathscr{X}$ if it fulfills the following conditions:

(BF1) $\mathcal{B} \subset \mathfrak{F}$;

(BF2) Every subset of $\mathscr{X}$ belonging to $\mathfrak{F}$ contains some subset of $\mathscr{X}$ which belongs to $\mathcal{B}$:

$$\forall F \in \mathfrak{F},\ \exists B \in \mathcal{B} :\ F \supset B.$$

Example 3.2 *The family of open intervals $]-a, a[$ with $a > 0$ is a basis of the filter of the neighborhoods of 0 on $\mathbb{R}$.*

Continuity of maps is one of the core concepts of topology.

Definition 3.4 *Let $(\mathscr{X}_1, \mathscr{O}_1)$ and $(\mathscr{X}_2, \mathscr{O}_2)$ be topological spaces. A map $f : \mathscr{X}_1 \to \mathscr{X}_2$ is continuous (relatively to the topologies $\mathscr{O}_1$ and $\mathscr{O}_2$) if*

$$\forall \mathcal{O}_2 \in \mathscr{O}_2,\ \exists \mathcal{O}_1 \in \mathscr{O}_1 :\ f(\mathcal{O}_1) \subset \mathcal{O}_2.$$

A homeomorphism[1] $f : \mathscr{X}_1 \to \mathscr{X}_2$ is a continuous bijection whose inverse map is continuous.

3.1.2 Topological Vector Spaces

This concept goes back to A.N. Kolmogoroff ([Kolm34]) and J. von Neumann ([VonN35]).

Definition 3.5 *Let E be a set endowed with a $\mathbb{K}$-vector space structure and a topology $\mathscr{O}$. These structures are said to be compatible if we have the following properties:*

(TVS1) *the vectorial sum* $\begin{array}{ccc} E \times E & \to & E \\ (x, y) & \mapsto & x + y \end{array}$ *is continuous;*

(TVS2) *the external multiplication* $\begin{array}{ccc} \mathbb{K} \times E & \to & E \\ (\lambda, x) & \mapsto & \lambda.x \end{array}$ *is continuous.*

E endowed with both structures is called a $\mathbb{K}$ topological vector space .

Definition 3.6 *Let E be a $\mathbb{K}$ topological vector space. A subset A of E is called balanced if*

$$\forall x \in A, \forall \lambda \in K :\ |\lambda| \leq 1,\ \lambda.x \in M.$$

[1]This word built on the greek words *homeo* (like, resembling) and *morphê* (shape) was introduced by Henri Poincaré.

Every topological vector space has a base of balanced neighborhoods ([RoRo64], Chapter I, Proposition 3.).

We illustrate the notion of topological vector space with the following examples (see also, for instance, [RoRo64], Chapter I, Supplement, for more examples).

Example 3.3 *The space* $C([0,1],\mathbb{K})$ *endowed with the topology of uniform convergence is a topological vector space.*

Example 3.4 *Let* $p \in]0,1[$. *Let* $\mathrm{L}^p([0,1])$ *be the quotient space of measurable functions from* $[0,1]$ *to* $\mathbb{R}$ *such that the integral* $\int_0^1 |f|^p(t)\,dt$ *is finite with respect to the equivalence relation almost everywhere. This space endowed with the metric*

$$d(f,g) = \int_0^1 |f-g|^p(t)\,dt$$

is a topological vector space.

The notion of convex subset is fundamental in our framework.

Definition 3.7 *Let* E *be a* $\mathbb{K}$ *topological vector space. A subset* C *of* E *is called convex if*

$$\forall (x,y) \in C^2, \forall t \in [0,1], t.x + (1-t).y \in C.$$

Definition 3.8 *Let* E *be a* $\mathbb{K}$ *topological vector space. A subset of* E *is called absolutely convex if it is convex and balanced.*

A subset is absolutely convex if and only if

$$\forall (x,y) \in A^2, \forall (\lambda,\mu) \in \mathbb{K}^2 : |\lambda| + |\mu| \leq 1,\ \lambda.x + \mu.y \in A.$$

The cardinal notion of bounded set is defined *via* absorbing subsets.

Definition 3.9 *Let* E *be a* $\mathbb{K}$ *topological vector space and let* A *and* B *two subsets of* E. *We say that* A *absorbs* B *if*

$$\exists \alpha > 0, \forall \lambda : |\lambda| \geq \alpha,\ B \subset \lambda.A.$$

A subset is called absorbing if it absorbs any singleton.

Definition 3.10 *A subset* B *of a topological vector space is said to be bounded if it is absorbed by any neighbourhood of* 0.

A compact set[2] of a topological vector space is bounded ([Tre67], Proposition 14.1).

[2] A topological space is said to be *compact* if it is Hausdorff and if every open covering $\{O_i\}_{i\in I}$ contains a finite covering.

3.1.3 Seminorms

The notion of seminorm is essential to define a locally convex topological vector space.
It is also used in the definition of the weak topology (cf. [RoRo64], Chapter II, 3.).

Definition 3.11 *A map $p : E \to \mathbb{R}$ is a seminorm if it fulfils the properties:*

(SN1) $\forall x \in E, p(x) \geq 0;$

(SN2) $\forall x \in E, \forall \lambda \in \mathbb{K}, p(\lambda.x) = |\lambda| p(x);$

(SN3) $\forall (x, y) \in E^2, p(x + y) \leq p(x) + p(y).$

Note that from definition, we must have $p(0) = 0$ and also, for any $x \in E$, $p(-x) = p(x)$.

Definition 3.12 *Let $\mathcal{P} = \{p_i\}_{i \in I}$ be a family of seminorms on the $\mathbb{K}$-vector space E and let $x \in E$, $r > 0$ and J be a finite subset of I. The $\mathcal{P}$ open ball $B_J(x, r)$ with center x is the convex set*

$$B_J(x, r) = \bigcap_{j \in J} B_{p_j}(x, r) = \{y \in E : \forall j \in J, p_j(y - x) < r\}.$$

Let $\mathcal{P} = \{p_i\}_{i \in I}$ be a family of seminorms on a $\mathbb{K}$ topological vector space E. An open set of the $\mathcal{P}$-topology $\mathcal{O}_{\mathcal{P}}$ associated with the family $\mathcal{P}$ is a subset O of E such that for any $x \in O$, there exists $r > 0$ and a finite subset J of I such that $B_J(x, r) \subset O$.
So the $\mathcal{P}$-topology $\mathcal{O}_{\mathcal{P}}$ is the one whose $\mathcal{P}$-open balls centered at x is a fundamental system of neighborhoods of x.

Let E be a topological vector space and C an absolutely convex and absorbing subset. The *gauge* or *Minkowski functional* p_C defined, for all x of E, by

$$p_C(x) = \inf\{\lambda > 0 : x \in \lambda.C\}$$

is a fundamental example of seminorm where

$$\{x \in E : p_C(x) < 1\} \subset C \subset \{x \in E : p_C(x) \leq 1\}.$$

In particular, p_C is continuous.
A family $\mathfrak{N}$ of seminorms on $\mathbb{E}$ is said to be *separating* if

$$\forall x \in E : x \neq 0, \exists \nu \in \mathfrak{N} : \nu(x) > 0.$$

3.1.4 Locally Convex Topological Vector Spaces

Since there exist topological vector spaces on which there are no continuous linear functionals except the trivial one (cf. Example 3.4), in order to get a meaningful and useful framework for many branches of functional analysis, we mainly consider topological vector spaces whose topologies are locally convex (cf. [Kot69]).

Definition 3.13 *A topological vector space E is called locally convex (l.c.t.v.s. for short) if every neighbourhood of any $x \in E$ contains a convex neighbourhood of x.*

We then have the following characterization of l.c.t.v.s. ([RoRo64], Chapter I, Theorem 2):

Proposition 3.1 *Let E be a $\mathbb{K}$ locally convex topological vector space. There exists a base of 0-neighbourhoods $\mathcal{B}$ such that:*

(BN0LCTVS1) $\forall (U,V) \in \mathcal{B}^2, \exists W \in \mathcal{B} : W \subset U \cap V$;

(BN0LCTVS2) $\forall \lambda \in \mathbb{K} \backslash \{0\}, \forall V \in \mathcal{B}, \lambda.V \in \mathcal{B}$;

(BN0LCTVS3) *Any set V of $\mathcal{B}$ is absolutely convex and absorbent.*

Conversely, given a non-empty set $\mathcal{B}$ of subsets of a vector space E fulfilling the properties **(BN0LCTVS1)** - **(BN0LCTVS3)**, *there exists a topology making E a locally convex topological vector space with $\mathcal{B}$ as a base of neighborhoods.*

Corollary 3.1 *Let $\mathscr{V}$ be any set of absolutely convex absorbent sets in a vector space $\mathbb{E}$. Then the set*

$$\mathscr{U} = \left\{ \bigcap_{1 \leq i \leq n} tV_i, \ \ n \in \mathbb{N}, \ \ V_i \in \mathscr{V}, \ \ t > 0 \right\}$$

defines a base of neighborhoods of 0 which provides $\mathbb{E}$ with a structure of locally convex space. The associated topology is the coarsest topology which provides $\mathbb{E}$ with a structure of topological vector space.

E endowed with the $\mathcal{P}$-topology is a l.c.t.v.s.
Conversely, a topology on E is locally convex if it is defined by a family of seminorms.

Example 3.5 *Topology of pointwise convergence.— Let S be a set. The vector space of all $\mathbb{K}$ valued functions on S can be endowed with a structure of locally convex topological vector space with the topology of pointwise convergence determined, for each $t \in S$, by the seminorms $p_t : x \mapsto |x(t)|$.*

The $\mathcal{P}$-topology $\mathscr{O}_{\mathcal{P}}$ of E is *Hausdorff* if and only if the family $\mathcal{P}$ of seminorms $\{p_i\}_{i \in I}$ separates the points of E:

$$x = 0 \ \Leftrightarrow \forall i \in I, p_i(x) = 0.$$

Let E be a Hausdorff topological vector space. E is metrizable (i.e. its topology is defined by a metric) if and only if the family of seminorms $\mathcal{P}$ is countable.

Example 3.6 *The space* $C^{\infty}([0,1],\mathbb{K})$ *of infinitely differentiable functions on the compact set* $[0,1]$ *taking values in* $\mathbb{K}$ *is a l.c.t.v.s. for the sequence of seminorms* $\mathcal{P} = (p_n)_{n\in\mathbb{N}}$, *where, for all integers* n, *we have*

$$p_n(f) = \sup_{t\in[0,1]} \left| f^{(n)}(t) \right|.$$

On the contrary, for $p \in]0,1[$, the topological vector space $L^p([0,1])$ of Example 3.4 is not locally convex. For this space, the only non-empty convex set is the space itself.

3.1.5 Bornology

Definition 3.14 *A bornology on a set* S *is a family* $\mathfrak{B}$ *of subsets of* S *satisfying the following axioms:*

(Born1) $\mathfrak{B}$ *covers* S.

(Born2) *Every subset of a set of* $\mathfrak{B}$ *belongs to* $\mathfrak{B}$;

(Born3) *The union of elements of a finite set of elements of* $\mathfrak{B}$ *belongs to* $\mathfrak{B}$.

Example 3.7 *On any topological space* $(S,\mathcal{O})$, *the set* $\mathfrak{B}$ *of the subsets of* $\mathcal{O}$ *on which any continuous map is bounded is a bornology on* S.

Example 3.8 *On any topological vector space locally convex* $(E,\mathcal{O})$, *the family of subsets absorbed by any neighborhood of* 0 *is a bornology called Von Neumann's bornology of* E *or canonical bornology.*

3.1.6 Linear Maps

The continuity of a linear map between topological vector spaces is obviously defined.
A linear map is continuous if and only if it continuous at 0.
We can characterize the continuity of a linear map between l.c.t.v.s. *via* families of seminorms.

Proposition 3.2 *Let* E *and* F *be two topological vector spaces locally convex respectively defined by the families of seminorms* $\mathcal{P}$ *and* $\mathcal{Q}$. *A linear map* $f : E \to F$ *is continuous if and only if*

$$\begin{aligned} &\forall q \in \mathcal{Q}, \exists C \geq 0, \exists k \in \mathbb{N}^*, \exists (p_1,\dots,p_k) \in \mathcal{P}^k : \\ &\forall x \in E, q(f(x)) \leq C \sup_{i\in\{1,\dots,k\}} p_i(x). \end{aligned} \tag{3.1}$$

Definition 3.15 *Let* E *and* F *be two topological vector spaces. A linear map* $f : E \to F$ *is bounded or bornological if the range of a bounded subset of* E *is a bounded subset of* F.

Notation 3.2 *In this chapter, we will use the following notations (cf. analogous notations 1.1 for Banach spaces):*

- $\mathrm{L}(E, F)$ *denotes the linear space of linear maps from* E *to* F;
- $L(E, F)$ *denotes the space of bounded linear maps from* E *to* F;
- $\mathcal{L}(E, F)$ *denotes the space of continuous linear maps from* E *to* F.

Theorem 3.1 *Let* E *and* F *be two topological vector spaces. Every continuous linear map from* E *to* F *is bounded.*

The converse is true when E is a bornological l.c.t.v.s. It is true in particular when E is metrizable.
Let us give an example of a bounded linear map which is not continuous.

Example 3.9 *Let* E *be an infinite-dimensional Banach space. Consider the same space endowed with its weak topology (cf. § 3.1.7.3) and denote it by* F. *The identity* $\mathrm{Id} : F \to E$ *is bounded (every weakly bounded set is strongly bounded by Mackey's theorem) but not continuous (the strong topology is strictly finer than the weak topology).*

3.1.7 Duality and Topologies

3.1.7.1 Topological Dual

Definition 3.16 *A linear map from a vector space* E *to* $\mathbb{K}$ *is called a linear form or a linear functional on* E.

Definition 3.17

(1) The set of the linear forms on the vector space E *is the algebraic dual space of* E *and is denoted by* $E^\sharp$.

(2) The set of the continuous linear forms on the topological vector space E *is the topological dual of* E *and is denoted by* E^*.

For a topological vector space E, E^* is a linear subspace of $E^\sharp$.

Notation 3.3 *The space of bounded linear forms on the l.c.t.v.s. space* E *will be denoted* E'.

3.1.7.2 Hahn-Banach Theorems

The Hahn-Banach theorem is a central tool in functional analysis.

Theorem 3.2 (Analytic form of the Hahn-Banach theorem) *Let* p *be a seminorm on the vector space* E *and let* f *be a linear form on a linear subspace* G *of* E *such that:*

$$\forall x \in G, \ |f(x)| \leq p(x).$$

Then there exists a linear form $\widehat{f}$ on E extending f such that:

$$\forall x \in E, \ \left|\widehat{f}(x)\right| \leq p(x).$$

Theorem 3.3 (Geometric form of the Hahn-Banach theorem) *Let E be a topological vector space, G a linear subspace of E and Ω an open convex set of E such that $G \cap \Omega = \emptyset$.*
There exists a hyperplane H of E such that G is a subset of H and $H \cap \Omega = \emptyset$.

Proposition 3.3 *Let E be a Haussdorff topological vector space locally convex. We then have*

$$(\forall f \in E^*, \ f(a) = 0) \Longrightarrow (a = 0).$$

This proposition is frequently used in analysis for problems of approximations, existence of solutions of a functional equation or separation on convex subsets (cf. [Tre67], 18-6).

3.1.7.3 Strong and Weak Topologies

Definition 3.18 *Let E be a l.c.t.v.s. and E^* its topological dual. The strong topology on E^* is the topology defined by the family of seminorms $\{p_B\}_{B \text{ bounded}}$ where*

$$\forall f \in E^*, \forall B \subset E, B \text{ bounded}, \ p_B(f) = \sup_{x \in B} |f(x)|.$$

Definition 3.19 *Let E be a l.c.t.v.s. The weak topology on E denoted $\sigma(E, E^*)$ is the initial topology with respect to the family E^*, i.e. the coarsest topology on E such that each element of E^* remains a continuous function.*

The weak topology $\sigma(E, E^*)$ is defined by the family of seminorms p_φ where, for all $\varphi \in E^*$ and all $x \in E$

$$p_\varphi(x) = |\varphi(x)|.$$

This topology endows E with a structure of l.c.t.v.s.
It is Haussdorf if and only if E^* separates the elements of E:

$$\forall x \in E \backslash \{0\}, \exists \varphi \in E^* : \varphi(x) \neq 0.$$

In particular, a sequence $(u_n)_{n \in \mathbb{N}}$ of E weakly converges towards an element u of E if

$$\forall \varphi \in E^*, \ \lim_{n \longrightarrow +\infty} \varphi(u_n) = \varphi(u).$$

3.1.7.4 Bornology and Dual

In any l.c.t.v.s. $(E, \mathscr{O})$ exists a natural notion of bounded set (cf. § 3.10). The associated bornology (Von Neumann's bornology, cf. Example 3.8) is little related to the topology $\mathscr{O}$ of E: according to a Mackey's theorem (cf. [Tre67], 36), if one varies the topology $\mathscr{O}$ while keeping the same dual space, the bornology remains the same. The bounded sets can be seen relatively to the weak topology rather than to the topology $\mathscr{O}$. For the weak topology, a subset of E, whose range by any linear functional is bounded, is itself bounded.

3.1.7.5 Algebraic Tensor Product

In this section, we define tensor products from a pure algebraic viewpoint.
For vector spaces E and F, we denote by $\mathrm{B}(E \times F)$ the space of bilinear forms of $E \times F$.
We then define the *algebraic tensor product* $E \otimes F$ of the vector spaces E and F as the space of linear functionals on $\mathrm{B}(E \times F)$ constructed as follows.
For $(x, y) \in E^2$, the element $x \otimes y$ is the functional given by the evaluation at the point (x, y) :

$$x \otimes y : \begin{array}{ccc} \mathrm{B}(E \times F) & \rightarrow & \mathbb{K} \\ A & \mapsto & (x \otimes y)(A) = A(x, y) \end{array}.$$

The *tensor product* of the spaces E and F is the subspace of the algebraic dual $\mathrm{B}(E \times F)^\sharp$ spanned by the elements of the form $x \otimes y$. So a tensor of $E \otimes F$ can be written in the (non-necessarily unique) form

$$\sum_{i=1}^{n} \lambda_i x_i \otimes y_i.$$

The algebraic tensor product $E \otimes F$ and the associated bilinear map $\varphi : E \times F \rightarrow E \otimes F$ are characterized, up to an isomorphism, by a universal property making, for any vector space G, the following diagram commutative

$$\begin{array}{ccc} E \times F & \xrightarrow{\ \varphi\ } & E \otimes F \\ {\scriptstyle A}\downarrow & \swarrow {\scriptstyle \tilde{A}} & \\ G & & \end{array}$$

i.e. for any bilinear map $A : E \times F \rightarrow G$, there exists a unique linear map $\tilde{A} : E \otimes F \rightarrow G$ such that $A = \tilde{A} \circ \varphi$.

3.1.7.6 Bornological Tensor Product and Bornological Approximation

In this section, we will refer to [KrMi97]. The solution to the universal problem of linearizing bounded bilinear maps is given by the *bornological tensor product* $E \otimes_\beta F$ for the convenient vector spaces E and F (cf. § 3.3), i.e. the algebraic tensor product provided with the finest locally convex topology such that the map $E \times F \to E \otimes F$ is bounded. This topology is bornological. The *projective tensor product* $E \otimes_\pi F$ is the algebraic tensor product equipped with the finest locally convex topology such that the map $E \times F \to E \otimes F$ is continuous. If E and F are two metrizable spaces, then $E \otimes_\pi F = E \otimes_\beta F$.

If we denote $B(E \times F, G)$ the space of bounded bilinear maps from $E \times F$ to G, we have the following important results (cf. [KrMi97], Theorem 5.7):

Theorem 3.4 *We have the following bornological isomorphisms:*

$$\begin{aligned} &L(E \otimes_\beta F, G) \simeq B(E \times F, G) \\ &E \otimes_\beta F \simeq F \otimes_\beta E \\ &(E \otimes_\beta F) \otimes_\beta G \simeq E \otimes_\beta (F \otimes_\beta G)\,. \end{aligned} \tag{3.2}$$

The *wedge product* is generated from the *alternator*: alt : $\otimes^k E \to \otimes^k E$ defined by

$$\operatorname{alt}(x_1 \otimes \cdots \otimes x_k) = \frac{1}{k!} \sum_{\sigma \in \boldsymbol{\sigma}_k} \operatorname{sign}(\sigma) x_{\sigma(1)} \otimes \cdots \otimes x_{\sigma(k)}$$

where $\boldsymbol{\sigma}_k$ is the set of permutations of $\{1, \dots, k\}$. Then alt is a bounded map whose range is denoted $\bigwedge^k E$. Let $L^k_{\text{alt}}(E; F)$ be the subset of alternating k-multilinear maps from E^k to F, then $L^k_{\text{alt}}(E; F)$ is a closed supplemented subspace of $L^k(E; F)$ and so is a convenient space and we have:

$$L(\bigwedge^k E; F) \simeq L^k_{\text{alt}}(E; F) \tag{3.3}$$

Definition 3.20 *A convenient space E has the bornological approximation property if $E' \otimes_\beta E$ is dense in $L(E, E)$ relatively to the bornological topology.*

We summarize the essential properties for a convenient space E which has the bornological approximation property (cf. [KrMi97], 6.9).

- For any convenient space F, $E' \otimes_\beta F$ is dense in $L(E, F)$ relatively to the bornological topology.
- If E is a reflexive convenient space, then E has the bornological approximation property if and only if E' is so.

3.2 Differential Calculus on Locally Convex Topological Vector Spaces

The differentiation on Banach spaces (cf. Chapter 1, § 1.5) has been extended to more general locally convex topological vector spaces in a variety of

ways depending on the type of derivatives used (cf. [Mica38], [Bas64], [Kel74], [Lesl82]). As a consequence, one obtains non-equivalent differential calculi on l.c.t.v.s. and so on manifolds, bundles and Lie groups.
In the framework of Chapter 4, the differential calculus proposed by J.A. Leslie will be used.
One corner stone in these different approaches is A. Kriegl and P.W. Michor's book which develops differential calculus in so-called convenient vector spaces.

In this section, the reader is referred to [Nee06] and [KrMi97], 13.3 (see also [SzLo07]).

Let E and F be two Hausdorff l.c.t.v.s.
For any $x \in E$, there exists a base of neighborhoods of convex open sets for x. Thus if x belongs to an open set set U, for any $u \in E$ and for any $t \in \mathbb{R}$, there exists an interval $[0, \varepsilon]$ such that the curve $t \mapsto x + tu$ is contained in U for $t \in [0, \varepsilon]$.

The following notion of C^k-map corresponds to the context of C^k_{MB}-map in the Michal-Bastiani sense or C^k_c-map in Keller's context. The main advantage of this notion is that it does not take into account the space of linear mappings; recall that, in general, the space of linear mappings from a Fréchet space to another one is not anymore a Fréchet space.

Definition 3.21 *Let $f : U \to F$ be a mapping defined on an open subset U of E.*
The directional derivative of f at $x \in U$ in the direction v is defined as

$$df(x)(v) = \lim_{t \to 0} \frac{f(x+tv) - f(x)}{t}.$$

(1) The map f is called differentiable at x if $df(x)(v)$ exists for all $v \in E$.

(2) f is called of class C^1 or continuously differentiable if it is differentiable at any point of U and

$$\begin{array}{rccc} df: & U \times E & \to & F \\ & (x,v) & \mapsto & df(x)(v) \end{array}$$

is a continuous map.

(3) By induction, for $k \geq 2$, the directional derivative or order k is defined by

$$\begin{aligned} & d^k f(x)(v_1, \dots, v_k) \\ = & \lim_{t \to 0} \frac{d^{k-1} f(x + tv_k)(v_1, \dots, v_{k-1}) - d^{k-1} f(x)(v_1, \dots, v_{k-1})}{t}. \end{aligned}$$

provided all the limits involved exist.
f is said to be of class C^k if $d^k f : U \times E^k \to F$ is continuous.

(4) *f is said to be of class C^∞ or smooth if it is of class C^k for all k.*

We then have the following basic properties.

Proposition 3.4 *Let f be a map of class C^k with $k \geq 1$ from an open set U in E to F.*

(1) *For $h \in \{1, \dots, k\}$, $df^h(x)(.) : E^h \to F$ is a symmetric h-linear continuous map.*

(2) *Chain rule.– If g is a map of class C^k from an open set V in F into a locally convex topological vector space G with $f(U) \subset V$, the $g \circ f$ is of class C^k and we have*

$$d(g \circ f)(x)(u) = dg\left(f(x)\right)\left(df(x)(u)\right)$$

(3) *Partial derivatives.– Let E_1 and E_2 be locally convex topological vector spaces and $E = E_1 \times E_2$.*
A continuous map f from an open set U of E into a l.c.t.v.s. F is of class C^1 if and only the partial derivatives

$$(\partial_1)_{(x_1,x_2)} f(u_1) = \lim_{t \to 0} \frac{f(x_1 + tu_1, x_2) - f(x_1, x_2)}{t}$$
$$(\partial_2)_{(x_1,x_2)} f(u_2) = \lim_{t \to 0} \frac{f(x_1, x_2 + tu_2) - f(x_1, x_2)}{t}$$

exist in F for all $(x_1, x_2) \in U$ and $(u_1, u_2) \in E$ and if $\partial_1 f : U \times E_1 \to F$ and $\partial_2 f : U \times E_2 \to F$ are continuous maps. Moreover, we have

$$d_{(x_1,x_2)} f(u_1, u_2) = (\partial_1)_{(x_1,x_2)} f(u_1) + (\partial_2)_{(x_1,x_2)} f(u_2)$$

(4) *If there exists an open cover $\{U_\alpha\}_{\alpha \in \mathbf{A}}$ of U such that $f_{|U_\alpha}$ is of class C^k for each $\alpha \in \mathbf{A}$, then f is of class C^k on U.*

Let us remark that, in the beginning of Definition 3.21, even if $df(x)(v)$ exists for all $v \in E$, the mapping $v \mapsto df(x)(v)$ may not be linear in general, and if it is linear it will not be bounded in general.
f is called *Gâteaux-differentiable* at x, if the directional derivatives $df(x)(v)$ exist for all $v \in E$ and the mapping $v \mapsto df(x)(v)$ is a bounded linear mapping from E to F.
The notion of Fréchet differentiability used in the definition of differentiability in Banach spaces (cf. Definition 1.5) is a particular case of Gâteaux differentiability.

Definition 3.22 *$f : U \to F$ is said to be Fréchet-differentiable at $x \in U$ if f is Gâteaux-differentiable at x and $\dfrac{f(x + tv) - f(x)}{t} - df(x)(v) \to 0$ uniformly for v in any bounded set.*

The following remark establishes a relationship between the Michal-Bastiani C^k_{MB}-differentiability and the Fréchet differentiability (cf. [Nee06], Remark 1.2.2).

Remark 3.1 *If E and F are Banach spaces, then the C^1_{MB}-notion is weaker than continuous Fréchet differentiability which requires the continuity of the map $x \mapsto df(x)$ where $\mathcal{L}(E,F)$ is endowed with the operator norm (cf. [Mil82]).*
Nevertheless, we have:

$$C^{k+1}_{MB}\text{-differentiability} \Longrightarrow C^k\text{-Fréchet differentiability} \Longrightarrow C^k_{MB}\text{-differentiability}.$$

So these different concepts of differentiability lead to the same class of smooth maps.

3.3 Convenient Calculus

In order to define the smoothness on locally convex topological vector spaces (l.c.t.v.s.) E, the basic idea is to test it along smooth curves (cf. Definition 3.29), since this notion in this realm is a concept without problems.

Definition 3.23 *Let E be a locally convex topological vector space.*

(1) A curve $c : \mathbb{R} \to E$ is differentiable if for all t, the derivative $c'(t)$ exists, where

$$c'(t) = \lim_{h \to 0} \frac{1}{h} (c(t+h) - c(t)).$$

(2) A curve is smooth (or C^∞) if all iterative derivatives exist.

Definition 3.24 *Let E be a locally convex topological vector space.*

(1) If J is an open subset of $\mathbb{R}$, we say that c is a Lipschitz curve on J if the set $\left\{ \dfrac{c(t_2) - c(t_1)}{t_2 - t_1},\ (t_1, t_2) \in J^2,\ t_1 \neq t_2 \right\}$ is a bounded set in E.

(2) The curve c is locally Lipschitz if every point in $\mathbb{R}$ has a neighbourhood on which c is Lipschitz.

(3) For $k \in \mathbb{N}$, the curve c is of class Lip^k if c is derivable up to order k, and if the kth-derivative $c : \mathbb{R} \to E$ is locally Lipschitz.

We then have the following link between both these notions ([KrMi97], 1.2):

Proposition 3.5 *Let E be a l.c.t.v.s, and let $c : \mathbb{R} \to E$ be a curve. Then c is C^∞ if and only if c is Lip^k for all $k \in \mathbb{N}$.*

The space $C^\infty(\mathbb{R}, E)$ of smooth curves from $\mathbb{R}$ to E does not depend on the locally convex topology on E but only on its associated *bornology* (i.e. the system of bounded sets). Note that the topology can vary considerably without changing the bornology; the *bornologification* E_{born} of E is the finest locally convex structure having the same bounded sets.

The set $C^\infty(\mathbb{R}, E)$ can be endowed with the locally convex topology of uniform C^∞-convergence on compact subsets. In this topology, pointwise addition and scalar multiplication of curves are continuous and hence we get a structure of a locally convex space on $C^\infty(\mathbb{R}, E)$ (cf. [Spa14], Remark 3.14).

Remark 3.2 *One can note that the link between continuity and smoothness in infinite dimension is not as tight as in finite dimension: there are smooth maps which are not continuous for the given topology on E! This situation cannot be avoided if we want that the chain rule holds: the evaluation map $ev : E \times E^* \to \mathbb{R}$ (where E^* is the topological dual of E) is smooth but is continuous only if E is normable.*

Definition 3.25 *Let E be a l.c.t.v.s. A curve $c : \mathbb{R} \to E$ is weakly smooth if $\ell \circ c$ is C^∞ for all $\ell \in E^*$.*

Definition 3.26 *The c^∞-topology on a l.c.t.v.s. E is the final topology with respect to all smooth curves $\mathbb{R} \to E$; it is denoted by $c^\infty E$.*

An open set of this topology will be called *c^∞-open*.

For every absolutely convex closed bounded set B, the linear span E_B of B in E is equipped with the Minkowski functional

$$p_B(v) = \inf \{\lambda > 0 : v \in \lambda.B\}$$

which is a norm on E_B.

We then have the following characterization of c^∞-open sets ([KrMi97], Theorem 2.13):

Proposition 3.6 *$U \subset E$ is c^∞-open if and only if $U \cap E_B$ is open in E_B for all absolutely convex bounded subsets $B \subset E$.*

Remark 3.3 *The c^∞-topology is in general finer than the original topology and E is not a topological vector space when equipped with the c^∞-topology. For Fréchet spaces and so Banach spaces, this topology coincides with the given locally convex topology.*

Definition 3.27 *A linear mapping between locally convex vector spaces is bounded if it maps every bounded set into a bounded one.*

Definition 3.28 *A locally convex vector space E is called bornological if any bounded linear mapping $f : E \to F$ (where F is any Banach space) is continuous.*

Proposition 3.7 *Let E be a bornological vector space. The c^∞-topology and the locally convex topology coincide (i.e. $c^\infty E = E$) if the closure of subsets in E is formed by all limits of sequences in the subset.*

Definition 3.29 *Let E and F be l.c.t.v.s. A mapping $f : E \to F$ is called conveniently smooth if it maps smooth curves into smooth curves, i.e. if $f \circ c \in C^\infty(\mathbb{R}, F)$ for all $c \in C^\infty(\mathbb{R}, E)$.*

Note that in finite-dimensional spaces E and F, this corresponds to the usual notion of smooth mappings as proved by Boman (see [Bom67]).

In finite-dimensional analysis, we use the Cauchy condition, as a necessary condition for the convergence of a sequence, to define the completeness of the space. In the infinite-dimensional framework, completeness can be obtained in various forms (cf. [Jon66]). In the convenient setting, we use the notion of Mackey-Cauchy sequence (cf. [KrMi97], 2).

Definition 3.30 *A sequence $(x_n)_{n\in\mathbb{N}}$ in E is called Mackey-Cauchy if there exists a bounded absolutely convex subset B of E such that (x_n) is a Cauchy sequence in the normed space E_B.*

The following definition is fundamental.

Definition 3.31 *A locally convex vector space is said to be c^∞-complete or convenient if any Mackey-Cauchy sequence converges (c^∞-completeness).*

An equivalent definition requires the concept of Mackey-Cauchy net defined *via* the notion of directed set.

Definition 3.32 *A directed set*[3] *is a pair $(I, \succcurlyeq)$ where I is a non-empty set and $\succcurlyeq$ is a reflexive and transitive binary relation such that*

$$\forall (i,j) \in I^2,\ \exists k \in I : \begin{cases} k \succcurlyeq i \\ k \succcurlyeq j \end{cases}$$

Definition 3.33 *Let $(I, \succcurlyeq)$ be a directed set and X a topological space. A net is a function $\xi : I \to X$. The net ξ is often denoted $\{\xi_i\}_{i\in I}$.*

The notion of Mackey-Cauchy net which generalizes the concept of Mackey-Cauchy sequence can be defined as follows (cf. [KeTa18]).

Definition 3.34 *Let E be a l.c.t.v.s. A Mackey-Cauchy net in E is a net $\{\xi_i\}_{i\in I}$ such that there exists a net of scalars $\{\lambda_{i,j}\}_{(i,j)\in I^2}$ decreasing towards 0 and a bounded set B such that*

$$\forall (i,j) \in I^2,\ \xi_j - \xi_i \in \lambda_{i,j}.B.$$

[3] $(I, \succcurlyeq)$ can be considered as a category I where the objects are the elements of I and the morphisms are the arrows $i \to j$ if and only if $j \succcurlyeq i$.

So a space where every Mackey-Cauchy net converges is Mackey-complete. Because a bounded function is not continuous in general, it does not preserve converging sequences but it does preserve Mackey-Cauchy nets.

We have the following characterizations of a convenient space (cf. [KrMi97], Theorem 2.14):

Proposition 3.8 *A locally convex vector space E is convenient if one of the following equivalent conditions is satisfied:*

(CVS1) *Any Lipschitz curve in E is locally Riemann integrable.*

(CVS2) *For any $c \in C^\infty(\mathbb{R}, E)$, there exists $\gamma \in C^\infty(\mathbb{R}, E)$ such that $\gamma' = c$.*

(CVS3) *Any weakly smooth curve*[4] *is smooth.*

(CVS4) *If B is bounded closed absolutely convex, the associated normed space E_B is a Banach space.*

(CVS5) *Any continuous linear mapping from a normed space into E has a continuous extension to the completion of the normed space.*

Example 3.10 *The direct sum $\mathbb{R}^\infty = \bigoplus\limits_{n\in\mathbb{N}^*} \mathbb{R}$, also denoted $\mathbb{R}^{(\mathbb{N})}$ or c_{00} or Φ, is the set of all sequences of real numbers with only a finite number of non-zero terms. It can be seen as the union $\bigcup\limits_{n\in\mathbb{N}^*} \mathbb{R}^n$ where each $\mathbb{R}^n$ is identified with the hyperplane $\mathbb{R}^n \times \{0\}$ of $\mathbb{R}^{n+1}$ via the inclusion $\iota_n^{n+1} : \mathbb{R}^n \to \mathbb{R}^{n+1}$. The space $\mathbb{R}^\infty$ is endowed with the direct limit topology, i.e. the finest topology for which the inclusions $\iota_n : \mathbb{R}^n \to \mathbb{R}^\infty$ are continuous. A subset U of $\mathbb{R}^\infty$ is open for this topology if and only if $U \cap \mathbb{R}^n$ is open in $\mathbb{R}^n$ for all $n \in \mathbb{N}^*$.*
Endowed with point addition and scalar multiplication, $\mathbb{R}^\infty$ is a non-metrizable locally convex vector space (cf. [Spa14], Example 3.10).
The subsets

$$B_0^\infty(\epsilon_n) = \left\{ (x_n)_{n\in\mathbb{N}^*} \in \mathbb{R}^\infty : |x_n| < \epsilon_n \right\},$$

where $(\epsilon_n)_{n\in\mathbb{N}^*}$ is a sequence of positive real numbers, form a basis of open neighbourhoods of 0 *([Jar81], 4.1.4).*
$\mathbb{R}^\infty$ is a regular space*: every bounded set B of $\mathbb{R}^\infty$ is contained in some $\mathbb{R}^n$.*
According to **(CVS 4)**, *$\mathbb{R}^\infty$ is a convenient vector space.*
$\mathbb{R}^\infty$ is also a strict direct limit of closed linear embeddings of the Banach spaces $\mathbb{R}^n$.

Theorem 3.5 *Let U be a c^∞-open set of a convenient vector space E and let F and G be convenient vector spaces.*

(1) The space $C^\infty(U, F)$ may be endowed with a structure of convenient vector space. The subspace $L(E, F)$ of all bounded linear mappings from E to F is closed in $C^\infty(E, F)$.

---

[4] cf. Definition 3.25.

(2) *The category is cartesian closed, i.e. we have the natural identification:*

$$C^{\infty}(E \times F, G) \simeq C^{\infty}(E, C^{\infty}(F, G)).$$

(3) *The differential operator*

$$d : C^{\infty}(E, F) \to C^{\infty}(E, L(E, F))$$

$$df(x)\,v = \lim_{t \to 0} \frac{f(x + tv) - f(x)}{t}$$

exists and is linear and smooth.

(4) *The chain rule holds:*

$$d(f \circ g)(x)\,v = df(g(x))\,dg(x)\,v.$$

Remark 3.4 *Consider a convenient space E. It is well known that the c^{∞}-topology on E is finer then its locally convex topology. Thus any open set U of the locally convex topology is a c^{∞}-open set. Given a smooth map f from a locally convex open set U in E to any locally convex space F, if $c : \mathbb{R} \to U$ is a smooth curve, then $f \circ c$ is a smooth map from $\mathbb{R} \to F$.*

According to this Remark and Definition 3.21, we introduce:

Definition 3.35

(1) *Let $\mathbb{E}$ be a convenient space and U a c^{∞}-open set in U. We say that a map f from U to a locally convex space $\mathbb{F}$ is convenient smooth, if for any smooth curve $c : \mathbb{R} \to U$, the curve $f \circ c : \mathbb{R} \to \mathbb{F}$ is smooth.*

(2) *Let $\mathbb{E}$ be a locally convex space and a continuous map f from an open set U of $\mathbb{E}$ into a c^{∞}-open set V in a convenient space $\mathbb{F}$. We say that f is convenient smooth if for any smooth curve $c : \mathbb{R} \to U$, the curve $f \circ c : \mathbb{R} \to V$ is smooth.*

Proposition 3.9 *The following constructions:*

- *limits,*
- *direct sums,*
- *strict direct limits of sequences of closed embeddings*

preserve c^{∞}-completeness.

In general, an inductive limit of c^{∞}-complete spaces need not be c^{∞}-complete (cf. [KrMi97], 2.15, Example).

3.4 Convenient Manifolds

Definition 3.36 *A chart (U, ϕ) on a set M is a bijection $\phi : U \to \phi(U)$ where U is a non-empty set of M and $\phi(U)$ is an open set of a convenient vector space $\mathbb{M}$.*

Definition 3.37 *Let (U_α, ϕ_α) and (U_β, ϕ_β) be two charts of M such that $U_{\alpha\beta} = U_\alpha \cap U_\beta$ is non-empty. The mapping*

$$\phi_{\alpha\beta} = \phi_\alpha \circ (\phi_\beta)^{-1} : \phi_\beta(U_{\alpha\beta}) \to \phi_\alpha(U_{\alpha\beta})$$

is called the chart changing.

Definition 3.38 *A family $\{(U_\alpha, \phi_\alpha)\}_{\alpha\in A}$ of charts on a set M, where $\phi_\alpha(U_\alpha)$ is an open set of a convenient vector space $\mathbb{M}_\alpha$, is called c^∞-atlas of M if*

(AtlC1) $M = \bigcup_{\alpha\in A} U_\alpha$ *;*

(AtlC2) *For all α and β such that $U_{\alpha\beta} \neq \emptyset$, the chart changing*

$$\phi_{\alpha\beta} = \phi_\alpha \circ (\phi_\beta)^{-1} : \phi_\beta(U_{\alpha\beta}) \to \phi_\alpha(U_{\alpha\beta})$$

is smooth.

Definition 3.39 *Two c^∞-atlases are called c^∞-equivalent if their union is still a c^∞-atlas.*
An equivalence class of c^∞-atlases on M is called a maximal atlas.

The natural topology on a manifold M is the identification topology with respect to some smooth atlas $\{(U_\alpha, \phi_\alpha)\}_{\alpha\in A}$, i.e. the final topology with respect to all $(\phi_\alpha)^{-1}$.
This topology is not necessarly Hausdorff (cf. [KrMi97], 27.4).

Definition 3.40 *A maximal atlas $\mathcal{A}$ on M is called a structure of not-necessarily-Hausdorff convenient manifold (or c^∞-manifold) (n.n.H. convenient manifold for short) on M.*
$\mathcal{A}$ is called a structure of Hausdorff convenient manifold if the topology defined by this atlas is Hausdorff.

When $\mathbb{M}$ is a Banach (resp. Fréchet) space, a structure of Hausdorff convenient manifold on M is nothing but a Banach (resp. Fréchet) manifold in the usual sense.

Remark 3.5 *If M is a n.n.H. Banach manifold, then its corresponding topology is T_1 hence each finite subset of M is closed. Moreover, since M is locally Hausdorff, by Zorn's Lemma, there exists a maximal open dense subset M_0 of M which is an open Banach manifold in M.*

Example 3.11 *The sphere* S^∞ *([KrMi97], 47.2).— We endow the convenient vector space* $\mathbb{R}^\infty$ *of example 3.10 with the bounded (equivantely smooth) bilinear form*

$$\langle .,. \rangle : (x,y) \longmapsto \sum_{n=1}^{+\infty} x_n y_n.$$

The associated linear mapping $\mathbb{R}^\infty \to (\mathbb{R}^\infty)' = \mathbb{R}^{\mathbb{N}}$ *is injective but not surjective. The infinite-dimensional sphere* S^∞ *is the subset of* $\mathbb{R}^\infty$

$$S^\infty = \{x \in \mathbb{R}^\infty : \langle x, x \rangle = 1\}.$$

S^∞ *corresponding to the direct limit* $\varinjlim S^n$ *(cf. Chapter 5, § 5.6) is endowed with the* c^∞*-topology which coincides with the direct limit topology (*DL*-topology) (cf. [CaPe19]). Endowed with the stereographic atlas,* S^∞ *is a convenient manifold modelled of the convenient vector space* $\mathbb{R}^\infty$*.*

Definition 3.41 *A mapping* $f : M \to N$ *between convenient manifolds is called* c^∞ *or smooth if for each* $x \in M$ *and each chart* (V, ψ) *on* N *with* $f(x) \in V$ *there exists a chart* (U, ϕ) *on* M *where* $x \in U$ *and* $f(U) \subset V$*, such that* $\psi \circ f \circ \phi^{-1}$ *is smooth.*

Let f be a smooth map from a convenient Hausdorff manifold M into $\mathbb{R}$. The *support* of f is the closure of the set $\{x \in M : f(x) \neq 0\}$.
For $k \in \mathbb{N} \cup \{\infty\}$, a C^k*-bump function* on M at x_0 is a map $f : M \to \mathbb{R}$ of class C^k, such that $0 \leq f(x) \leq 1$ whose support is contained in a neighbourhood of x and $f(x_0) = 1$.

Definition 3.42 *A smooth Hausdorff manifold* M *is said to be*

(1) C^k*-regular* $(k \in \mathbb{N} \cup \{\infty\})$ *if for any* $x \in M$ *and any neighbourhood* V *of* x *there exists a* C^k*-bump function at* x *whose support is contained in* V*;*[5]

(2) paracompact if for any open covering $\{U_\alpha\}_{\alpha \in A}$ *there exists a covering* $\{V_\beta\}_{\beta \in B}$ *which is a locally finite refinement*[6] *of* $\{U_\alpha\}_{\alpha \in A}$*;*

(3) having a C^k*-partition of unity* $(k \in \mathbb{N} \cup \{\infty\})$ *or is* C^k*-paracompact, if* M *is paracompact and for each open covering, there exists a locally finite refinement* $\{V_\beta\}_{\beta \in B}$ *and a collection of* C^k*-bump functions* $v_\beta : M \to [0,1]$ *whose support is contained in* V_β *and*

$$\forall x \in M, \sum_{\beta \in B} v_\beta(x) = 1.$$

[5]For a (C^0-)regular manifold we have: for any neighbourhood V of any $a \in M$ there exists a neighbourhood $V' \subset M$ of a such that its closure $\bar{V}'$ is a subset of V.

[6]i.e. for each $\beta \in B$, there exists $\alpha \in A$ such that $V_\beta \subset U_\alpha$ and every point of M has a neighbourhood that intersects only a finite number of open sets V_β.

The regularity condition is satisfied if, for example, M is locally compact or is a topological group (cf. [Mil82]). If the topology of M is metrizable, then M is paracompact and regular. In particular, a Banach space and a Fréchet space are C^0-regular and C^0-paracompact but not necessarily C^k-regular or C^k-paracompact if[7] $k \geq 1$. Thus, any metrizable Banach or Fréchet manifold is paracompact. A Hilbert space is always smoothly paracompact (cf. [Bourl19]). Thus, any paracompact Hilbert manifold is in fact smoothly paracompact. Note that a smoothly regular and paracompact manifold is smoothly paracompact.

3.5 Tangent Vectors

In the convenient setting, different kinds of tangent vectors can be defined, either as derivations or kinematically. Both these concepts differ as it is proved in [KrMi97], 28, where one can find derivations which are not tangent to any smooth curve.

Let M be a convenient manifold. A *smooth curve at* $x \in M$ is a smooth map $\alpha : I \to M$ such that $\alpha(0) = x$, where I is an open interval of $\mathbb{R}$ containing 0. Two curves α and β are said *tangent* at x if there exists a chart (U, ϕ) at x such that

$$(\phi \circ \alpha)'(0) = (\phi \circ \beta)'(0).$$

The corresponding equivalence class $[\alpha, x]$ is a *tangent vector at* x. The set of all tangent vectors at x is the (kinematic) *tangent space* at x, denoted by $T_x M$.

3.6 Tangent Mappings

Let M_1 and M_2 be two convenient manifolds. If $f : M_1 \to M_2$ is a smooth map, then the *differential* or *tangent map* at x is the map

$$\begin{array}{rccc} T_x f: & T_x M_1 & \to & T_{f(x)} M_2 \\ & [\alpha, x] & \mapsto & [f \circ \alpha, f(x)]. \end{array}$$

This definition is independent of the choice of the representatives of the tangent vector at x.

[7] Every reflexive (not necessarily separable) Banach space is C^1-paracompact (cf. [Tor73]).

3.7 Immersions

Definition 3.43 *Let M_1 and M_2 be convenient manifolds, respectively, modelled on the convenient spaces $\mathbb{M}_1$ and $\mathbb{M}_2$.*
A smooth map $f : M_1 \to M_2$ is called

(1) *a weak immersion if, for any $x \in M_1$, there exist a chart (U_1, ϕ_1) around $x \in M_1$ and a chart (U_2, ϕ_2) around $f(x) \in M_2$ such that*

$$\phi_2 \circ f \circ \phi_1^{-1} = \iota_{\phi_1(U)} \tag{3.4}$$

where $\iota : \mathbb{M}_1 \to \mathbb{M}_2$ is a bounded linear injective map.
$\iota_{\phi_1(U)} : \phi_1(U) \times \mathbb{M}_1 \to \phi_1(U) \times \mathbb{M}_2$ is the map $\iota_{\phi_1(U)}(z, u) = (z, \iota(u))$. The pair (M_1, f) is called a weak immersed submanifold of M_2;

(2) *a closed immersion if f is a weak immersion such that $\mathbb{M}_1$ is a closed convenient subspace of $\mathbb{M}_2$.*
The pair (M_1, f) is called a closed immersed submanifold of M_2;

(3) *a split immersion if f is a closed immersion such that $\mathbb{M}_1$ is supplemented in $\mathbb{M}_2$[8]. The pair (M_1, f) is called a split immersed submanifold of M_2.*

Remark 3.6 *Note that if f is a closed immersion, this implies that the tangent map $T_{x_1} f$ is a bounded injective linear map from $T_{x_1} M_1$ in $T_{f(x_1)} M_2$ for any $x_1 \in M_1$. When $\mathbb{M}_1$ and $\mathbb{M}_2$ are Banach spaces then the converse is true. However, this converse is not true in general if one of the spaces is not a Banach space (cf. § 3.19 (2)).*

3.8 Convenient Submanifolds

Let M be a n.n.H. convenient manifold modelled on a convenient space $\mathbb{M}$ (cf. [KrMi97], 28.1).

Definition 3.44 *Let M be a convenient manifold.*

(1) *A weak submanifold is a subset M_1 of M_2 provided with its* **own structure of convenient manifold** *such that such that the inclusion $\iota_{M_1} : M_1 \to \mathbb{M}_2$ is a weak immersion.*

(2) *A closed submanifold of M_2 is a weak submanifold such that the inclusion $\iota_{M_1} : M_1 \to M_2$ is a closed immersion.*

[8] i.e. $\mathbb{M}_2 = \iota(\mathbb{M}_1) \oplus \mathbb{F}$

(3) A split submanifold M_1 of M_2 if the inclusion $\iota_{M_1} : M_1 \to M_2$ is a split immersion.
In this case, M_1 will be simply called a submanifold of M_2.

Notice that for a weak submanifold $\iota_N : M_1 \to M_2$, on N of M, there are two topologies:

1. the induced topology $\mathscr{O}_{|M_2}$ by the one of M_2;
2. the topology $\mathscr{O}_{M_1}$ associated with its convenient structure manifold.

Let M_2 be a convenient manifold modelled on a convenient space $\mathbb{M}_2$ and a weak submanifold M_1 of M_2 modelled on a convenient space $\mathbb{M}_1$, according to [Nee05], we introduce:

Definition 3.45 *M_1 is an initial submanifold of M_2 if the following property is true:*

(ISM) *For any continuous map f from a convenient manifold Z to M_1, then $f : Z \to M_1$ is smooth if and only if $f : Z \to M_2$ is smooth.*

Proposition 3.10 *A closed submanifold M_1 of M_2 has the following properties:*

(1) Around each $x \in M_1$, there exists a chart (U_1, ϕ_1) in M_1 (resp. (U_2, ϕ_2)) in M_2 such that $M_1 \cap U_2 = U_1$. In particular the inclusion ι_{M_1} is continuous and if M_2 is Hausdorff so is M_1

(2) Each closed submanifold M_1 of M_2 is an initial submanifold of M_2.

Proof
(1) Since Point (1) is a local problem, according Definition 3.43 (1), we may assume that M_i is a c^∞-open set in $\mathbb{M}_i$ for $i \in \{1, 2\}$. From (3.4), since ι_{M_1} is a closed immersion, we may assume that $\mathbb{M}_1$ is a c^∞-closed subspace of $\mathbb{M}_2$. Let U be a c^∞-open set in M_1. Let B_2 be any absolutely convex bounded subset of $\mathbb{M}_2$ then $U \cap \{\mathbb{M}_2\}_{B_2}$[9] is open in $\{\mathbb{M}_2\}_{B_2}$. But according to the Hahn-Banach Theorem, any absolutely convex bounded subset B_1 in $\mathbb{M}_1$ is also an absolutely convex bounded subset of $\mathbb{M}_2$. As $\{\mathbb{M}_1\}_{B_1} = \{\mathbb{M}_2\}_{B_1} \cap \mathbb{M}_1$, it follows that $U \cap \{\mathbb{M}_1\}_{B_1}$ is also a c^∞-open set in $\{\mathbb{M}_1\}_{B_1}$ and so $U \cap \mathbb{M}_1$ is a c^∞-open set of $\mathbb{M}_1$ and so ι is C^∞ continuous. From (3.4) it follows that ι_{M_1} is continuous according the topology manifolds of M_1 and M_2. Thus if M_2 is Hausdorff so is M_1.

(2) Under the context of Point (2), fix some continuous map f from Z in some closed submanifold M_1 of M_2. It is clear the property **(ISM)** *is a local problem. Since $\mathbb{M}$ and $\mathbb{N}$ are convenient spaces, according the definition of smoothness in the convenient setting (cf. [KrMi97]), it is sufficient to prove the following property:*

[9] $\{\mathbb{M}_2\}_{B_2}$ is the Banach space generated by B_2.

(ISMloc) *For $z \in Z$ and any smooth curve $\gamma :]-\varepsilon, \varepsilon[\to Z$ where ε is small enough, $f \circ \gamma$ is a smooth curve in M_1 if and only if it is a smooth curve in M_2.*

Thus $x \in M_1$ being fixed, without loss, according to (1), we may assume that, M_1 is a c^∞-open set in $\mathbb{M}_1$ around x and M_2 is a c^∞-open neighbourhood of $\iota(x)$ in $\mathbb{M}_2$ such that $M_1 \cap M_2 = M_2$. According to this context, for ε small enough, $f \circ \gamma$ is contained in $M_1 \subset M_2$. Since ι is smooth, it follows that if $f \circ \gamma$ is smooth in M_1, then it is smooth in M_2. Now, for $i \in \{1, 2\}$, $f \circ \gamma$ is smooth in $\mathbb{M}_i$ if and only $\lambda \circ f \circ \gamma$ is smooth for any continuous linear functional λ on $\mathbb{M}_i$. Then the converse is a consequence of the Hahn-Banach theorem.

3.9 Convenient Bundles

3.9.1 Bundles

Definition 3.46 *A convenient fibration is a triple (F, π, M) where F and M are convenient manifolds and where $\pi : F \to M$ is a smooth surjection for which, for all $x \in M$, all fibres $\pi^{-1}(x)$ are diffeomorphic manifolds.*

Definition 3.47 *A convenient bundle is a quadruple (F, π, M, S) where F, M and S are convenient manifolds, $\pi : F \to M$ is a smooth mapping fulfilling the condition of local trivialization:*

(BLT) *Every $x \in M$ has an open neighbourhood U such that $F_{|U} = \pi^{-1}(U)$ is diffeomorphic with $U \times S$ via a diffeomorphism τ such that $\mathrm{pr}_1 \circ \tau = \pi$ where $\mathrm{pr}_1 : U \times \mathbb{F} \to U$ is the first projection.*

The map π is the projection of the total space F on the base M and S is the standard fibre or fibre type.
The pair (U, τ) is called a trivialization of the bundle F.
A family of trivializations $\{(U_\alpha, \tau_\alpha)\}_{\alpha \in A}$ where $\{U_\alpha\}_{\alpha \in A}$ is an open covering of M is called a trivializing cover of F.

Let (U_α, τ_α) and (U_β, τ_β) be two trivializations such that $U_\alpha \cap U_\beta \neq \emptyset$. We define the transition function $\psi_{\alpha\beta} : (U_\alpha \cap U_\beta) \times S \to S$ by

$$\tau_\alpha \circ (\tau_\beta)^{-1}(x, s) = (x, \psi_{\alpha\beta}(x, s)).$$

For any $x \in U_\alpha \cap U_\beta$, the map $\psi_{\alpha\beta}(x, .)$ is a diffeomorphism of S.
Moreover, the transition functions satisfy the *cocycle conditions*:

(CCocy1) $$\forall x \in U_\alpha \cap U_\beta \cap U_\gamma,\ \psi_{\alpha\beta}(x, .) \circ \psi_{\beta\gamma}(x, .) = \psi_{\alpha\gamma}(x, .)\ ;$$

(CCocy2) $$\forall x \in U_\alpha, \psi_{\alpha\alpha}(x, .) = \mathrm{Id}_S\ .$$

3.9.2 Pull-Back Bundles

In this subsection, the reader is referred to [KrMi97], 37.4.
Let (F, π, M, S) be a convenient bundle. Consider a convenient manifold N and a smooth map $f : N \to M$.
The pull-back

$$N \times_{(f,M,\pi)} F = \{(n, u) : \ f(n) = \pi(u)\}$$

is denoted by f^*F.
We introduce some notations in the following commutative diagram:

$$\begin{array}{ccc} f^\star F & \xrightarrow{\pi^\star f} & F \\ {\scriptstyle f^\star \pi}\downarrow & & \downarrow{\scriptstyle \pi} \\ N & \xrightarrow{f} & M \end{array} \tag{3.5}$$

So $(f^\star F, f^\star \pi, N, S)$ is a convenient bundle and $\pi^* f$ is a fibrewise diffeomorphism.
If $\{(U_\alpha, \tau_\alpha)\}_{\alpha \in A}$ is a bundle atlas of F, then $\left\{\left(f^{-1}\left(U_\alpha\right), \xi_\alpha\right)\right\}_{\alpha \in A}$ where $\xi_\alpha = (f^*\pi, \mathrm{pr}_2 \circ \tau_\alpha \circ \pi^* f)$ is a bundle atlas for $(f^\star F, f^\star \pi, N, S)$.

3.9.3 G-Bundle Structures

When the standard fibre S is endowed with an action of a Lie group G (cf. 3.61), we get the notion of G-bundle.

Definition 3.48 *Let (F, π, M, S) be a convenient bundle and let G be a convenient Lie group. A G-bundle structure is given by:*

(GBS1) *a left action* $\begin{array}{llll} \ell : & G \times S & \to & S \\ & (g, s) & \mapsto & \ell(g, s) \end{array}$ *of the Lie group on the standard fibre;*

(GBS2) *a trivialization cover $\{(U_\alpha, \tau_\alpha)\}_{\alpha \in A}$ of F, where the transition function $\theta_{\alpha\beta}$ acts on S via the action ℓ: there exists a family of smooth functions $\{\psi_{\alpha\beta} : U_\alpha \cap U_\beta \to G\}_{(\alpha,\beta) \in A^2}$ fulfilling the cocycle conditions* **(CCocy 1)** *and* **(CCocy 2)** *with*

$$\theta_{\alpha\beta}(x, s) = \ell\left(\psi_{\alpha\beta}(x), s\right).$$

A bundle endowed with a G-bundle structure is called a G-bundle.

3.9.4 Vertical Bundle

Let (F, π, M, S) be a bundle.
The kernel of the linear tangent map $T\pi : TF \to TM$ is called *the vertical bundle* of E and denoted by VE.

3.10 Convenient Vector Bundles

Let $(E, \pi, M, \mathbb{E})$ be a bundle where the fibre type $\mathbb{E}$ is a convenient vector bundle such that, for all $x \in U_\alpha \cap U_\beta$, the smooth map $\psi_{\alpha\beta}(x, .)$ belongs to the group $GL(\mathbb{E})$ of bounded automorphisms of $\mathbb{E}$.

Two trivializations (U_α, τ_β) and (U_β, τ_β) are said to be *compatible* if the mapping $\tau_\alpha \circ (\tau_\beta)^{-1}$is a fibre isomorphism, i.e.

$$\tau_\alpha \circ (\tau_\beta)^{-1}(x, v) = (x, \theta_{\alpha\beta}(x) v)$$

where $\theta_{\alpha\beta} : U_{\alpha\beta} = U_\alpha \cap U_\beta \to GL(\mathbb{E})$. The mapping $\theta_{\alpha\beta}$ is then unique and smooth in the set $L(\mathbb{E}, \mathbb{E})$ of bounded endomorphisms of $\mathbb{E}$. It is called a *transition function* between both trivializations.

Definition 3.49 *A trivializing cover $\mathcal{A} = \{(U_\alpha, \tau_\alpha)\}_{\alpha \in A}$ for $\pi : E \to M$ is a set of compatible ranges of trivializations where $\{U_\alpha\}_{\alpha \in A}$ is an open covering of M.*

Two trivializing covers of E are *equivalent* if their union is still a trivializing cover of E.

Definition 3.50 *A convenient vector bundle $(E, \pi, M, \mathbb{E})$ is the datum of convenient manifolds E (total space), M (base) with a smooth mapping (projection) $\pi : E \to M$ and an equivalence class of trivializing covers.*

If we assume that the open sets U_α of the trivializing cover are the domains of charts (U_α, ϕ_α) of the base M, we may consider the maps:

$$\Phi_\alpha = (\phi_\alpha \times \mathrm{Id}_\mathbb{E}) \circ \tau_\alpha : \pi^{-1}(U_\alpha) \to \phi_\alpha(u_\alpha) \times \mathbb{E}. \tag{3.6}$$

Definition 3.51 *The triple $(U_\alpha, \phi_\alpha, \Phi_\alpha)$ is called a vector bundle chart.*

Because a convenient vector space is Hausdorff, a vector bundle modelled on a convenient space has a natural structure of n.n.H. convenient manifold; this one is Hausdorff if the base M is Hausdorff.

3.11 Convenient Tangent Bundles

Let M be a convenient manifold endowed with an atlas $\{(U_\alpha, \phi_\alpha)\}_{\alpha \in A}$ where $\phi_\alpha : U_\alpha \to \mathbb{M}$.

Two notions of tangent bundles can be defined on M (cf. [KrMi97], 28.12):

1. the operational tangent bundle DM;

2. the kinematic tangent bundle TM.

We are only interested in the last structure. The bundle DM can be defined from bounded derivations at a point.

In order to build the bundle TM, we first introduce, on the disjoint union $\bigcup_{\alpha\in A} \{\alpha\} \times U_\alpha \times \mathbb{M}$, the following equivalence relation:

$$(\alpha, x, v) \sim (\beta, y, w) \Longleftrightarrow \begin{cases} x = y \\ (d\phi_{\alpha\beta})_{\phi_\beta(x)} w = v \end{cases}.$$

The quotient space TM is called the *kinematic tangent bundle* of M.

Consider the projection $p_M : TM \to M$ defined by $p_M\left(\widetilde{(\alpha, x, v)}\right) = x$ where $\widetilde{(\alpha, x, v)}$ is the equivalence class of (α, x, v). For $TU_\alpha = p_M^{-1}(U_\alpha) \subset TM$, consider $\tau_\alpha : TU_\alpha \to \phi_\alpha(U_\alpha) \times \mathbb{M}$, defined by $\tau_\alpha\left(\widetilde{(\alpha, x, v)}\right) = (\phi_\alpha(x), v)$; we then have:

$$\tau_\alpha\left(\widetilde{(\beta, x, w)}\right) = \left(\phi_a(x), (d\phi_{\alpha\beta})_{\phi_\beta(x)} w\right).$$

The family of charts $\{(TU_\alpha, \tau_\alpha)\}_{\alpha\in A}$ is an atlas of TM because the chart changing is given by:

$$\begin{array}{rcccc} \tau_\alpha \circ (\tau_\beta)^{-1}: & \phi_\beta(U_{\alpha\beta}) \times \mathbb{M} & \to & \phi_\alpha(U_{\alpha\beta}) \times \mathbb{M}. \\ & (x_\beta, v) & \mapsto & \left(\phi_{\alpha\beta}(x_\beta), (d\phi_{\alpha\beta})_{x_\beta} v\right) \end{array}.$$

So the smooth structure of TM only depends on the equivalence class of smooth atlases of M.

Then the projection $p_M : TM \to M$ is smooth. The natural topology is Hausdorff.

For all $x \in M$, the set $T_xM = p_M^{-1}(x)$ is called *kinematic tangent space in x at M*. It is endowed with a structure of convenient vector space.

Example 3.12 *For the convenient manifold S^∞ of Example 3.11, we have: $TS^\infty = DS^\infty$ and this fibre is diffeomorphic to $\{(x, v) \in S^\infty \times \mathbb{R}^\infty : \langle x, v\rangle = 0\}$ ([KrMi97], 47.2).*

3.12 Convenient Vector Fields

Let M be a convenient manifold.

Definition 3.52 *A kinematic tangent vector field X on M is a smooth section of the kinematic tangent bundle $TM \to M$.*

Notation 3.4 *The $c^\infty(M)$-module of kinematic vector fields is denoted $\mathfrak{X}(M)$.*

Definition 3.53 *An operational vector field X on M is a bounded derivation of the sheaf $c^\infty(., \mathbb{R})$, i.e. that, for any open set $U \subset M$, $X_U : c^\infty(U, \mathbb{R}) \to c^\infty(U, \mathbb{R})$ commute with restrictions.*

Example 3.13 *We will indicate the way to produce an example of an operational vector on the Banach space ℓ^p for $1 \leq p \leq 2$ which is not a kinematic vector field. This construction comes from [BGT18] where the reader will find details about this problem.*
Given a Banach space $\mathbb{E}$, let $L_2(\mathbb{E}, \mathbb{R})$ be the set of bounded bilinear operators on $\mathbb{E}$ with values in $\mathbb{R}$. For instance, the second derivative f_0'' at $0 \in \mathbb{E}$ of a smooth map $f : \mathbb{E} \to \mathbb{R}$ belongs to $L_2(\mathbb{E}, \mathbb{R})$. We can consider the natural inclusion of $\mathbb{E}' \times \mathbb{E}'$ into $L_2(\mathbb{E}, \mathbb{R})$ given by

$$(f, g) \mapsto (f \otimes g : (x, y) \mapsto f(x)g(y)) \tag{3.7}$$

A functional $\phi \in (L_2(\mathbb{E}, \mathbb{R}))'$ defines an operational tangent vector of order 2 at $0 \in \mathbb{E}$ by

$$\delta_\phi f = \phi(f_0'') \tag{3.8}$$

if and only if it vanishes on $\mathbb{E}' \times \mathbb{E}'$ (as a subspace of $L_2(\mathbb{E}, \mathbb{R})$).
Consider the Banach space ℓ^p of p-summable real sequences. According to the inclusion (3.7), we consider $(\ell^p)' \times (\ell^p)'$ as a subspace of $L_2(\ell^p, \mathbb{R})$. Since $1 \leq p \leq 2$, the inclusion of ℓ^p in $(\ell^p)^*$ is not compact; according to the Hahn-Banach Theorem, we can find a non-zero functional ϕ on $L_2(\ell^p, \mathbb{R})$ which vanishes on $(\ell^p)' \times (\ell^p)'$. This implies the existence of non-zero operational tangent vectors δ_ϕ of the second order 2 on ℓ^p for $1 \leq p \leq 2$.

Notation 3.5 *The space of operational vector fields is denoted $\mathfrak{D}(M)$. It corresponds to the set of smooth sections of the bundle DM.*

Proposition 3.11 *There exists a natural embedding $\mathfrak{X}(M) \to \mathfrak{D}(M)$.*

Let X and Y be two operational vector fields. The map $f \mapsto X(Y(f)) - Y(X(f))$ is a bounded derivation of the sheaf $c^\infty(\,, \mathbb{R})$; it is denoted $[X, Y]$ and is called *Lie bracket*. This bracket fulfils the usual properties (bilinearity, antisymmetry and Jacobi identity).

Definition 3.54 *A smooth curve $\gamma : I \to M$ defined on a not empty interval I of $\mathbb{R}$ is called an integral curve of the kinematic vector field X if*

$$\forall t \in I, \gamma'(t) = X_{\gamma(t)}.$$

Remark 3.7 *For a given kinematic vector field, integral curves need not exist locally ([KrMi97], 32.12, Example 1.) and even if they exist, they need not be unique for a given initial value ([KrMi97], 32.12, Example 2.).*

Definition 3.55 *Let X be a kinematic vector field. A local flow Fl^X for X is a smooth mapping $U \to M$ where U is a c^∞-open set of $M \times \mathbb{R}$ fulfilling the following properties:*

(Fl1) $U \cap (\{x\} \times \mathbb{R})$ *is an open interval;*

(Fl2) *if* $\mathrm{Fl}_s^X(x)$ *exists, then* $\mathrm{Fl}_{t+s}^X(x)$ *exits if and only if* $\mathrm{Fl}_t^X\left(\mathrm{Fl}_s^X(x)\right)$ *exists; in this case, we have:* $\mathrm{Fl}_t^X\left(\mathrm{Fl}_s^X(x)\right) = \mathrm{Fl}_{t+s}^X(x)$*;*

(Fl3) $\mathrm{Fl}_0^X = \mathrm{Id}_M$*;*

(Fl4) $\dfrac{d}{dt}\left(\mathrm{Fl}_t^X(x)\right) = X_{\mathrm{Fl}_t^X(x)}$.

3.13 Convenient Cotangent Bundles

Let M be a convenient manifold endowed with an atlas $\{(U_\alpha, \phi_\alpha)\}_{\alpha \in A}$ where $\phi_\alpha : U_\alpha \to \mathbb{M}$.
As for the framework of tangent bundle, it is possible to define two types of cotangent bundles ([KrMi97], 33.1):

1. the operational cotangent bundle;
2. the kinematic cotangent bundle $T'M$.

We only are interested in this last structure. $T'M = \bigcup\limits_{x \in M} (T_xM)'$, where $(T_xM)'$ is the set of bounded linear forms on the kinematic tangent bundle T_xM, can be endowed with a structure of fibre bundle where the cocycle of the transition functions is given, for all x in M, by $d\left(\phi_\beta \circ (\phi_\alpha)^{-1}\right)(\phi_\alpha(x))^* \in GL(\mathbb{M})$.

3.14 Convenient Differential Forms

Let $\mathbb{E}_1, \dots, \mathbb{E}_n$ and $\mathbb{F}$ be convenient vector spaces. $L(\mathbb{E}_1, \dots, \mathbb{E}_n; \mathbb{F})$ denotes the space of bounded n-linear mappings $\mathbb{E}_1 \times \cdots \times \mathbb{E}_n \to \mathbb{F}$ endowed with the topology of uniform convergence on bounded sets of $\mathbb{E}_1 \times \cdots \times \mathbb{E}_n$. For $k \in \mathbb{N}^*$ and a convenient vector space $\mathbb{E}$, $L_{\mathrm{alt}}^k(\mathbb{E}; \mathbb{F})$ denotes the sub vector space of antisymmetric maps and $L_{\mathrm{alt}}^k(\mathbb{E})$ denotes the space of bounded antisymmetric k-forms.
For a convenient manifold M, we consider the space

$$L_{\mathrm{alt}}^k(TM, M \times \mathbb{R}) = \bigcup_{x \in M} L_{\mathrm{alt}}^k(T_xM)$$

endowed with a structure of convenient bundle. More generally, we can define

$$L_{\text{alt}}^k (TM, M \times \mathbb{F}) = \bigcup_{x \in M} L_{\text{alt}}^k (T_x M, \mathbb{F}) .$$

Definition 3.56 *A differential k-form is a smooth section of* $L_{\text{alt}}^k(TM, M \times \mathbb{R})$.

Notation 3.6 $\Omega^k (M)$ *will denote the space of differential k-forms on* M.

$\Omega^k (M)$ can be endowed with a structure of convenient vector space ([KrMi97], 33.22).

A *differential* 1*-form* is then a smooth section of the kinematic cotangent bundle T^*M.

More generally, one can define the notion of vector valued differential *k-form.*

Definition 3.57 *Let* $\mathbb{V}$ *be a convenient vector space. A differential k-form with values in* $\mathbb{V}$ *is a smooth section of the bundle* $L_{\text{alt}}^k (TM, M \times \mathbb{V})$.

Notation 3.7 $\Omega^k (M, \mathbb{V})$ *denotes the space of differential k-forms with values in* $\mathbb{V}$.

3.15 Connections on a Convenient Bundle

3.15.1 Double Structure on a Bundle

Let $\pi : F \to M$ be a bundle where the manifold M is modelled on the convenient vector space $\mathbb{M}$ and where the type fibre is the manifold S modelled on the convenient vector space $\mathbb{S}$. The tangent space TF can endowed with two fibre bundle structures:

1. The structure of classical tangent bundle $p_F : TF \to F$;
2. The tangent structures $T\pi : TF \to TM$.

We have the following diagram:

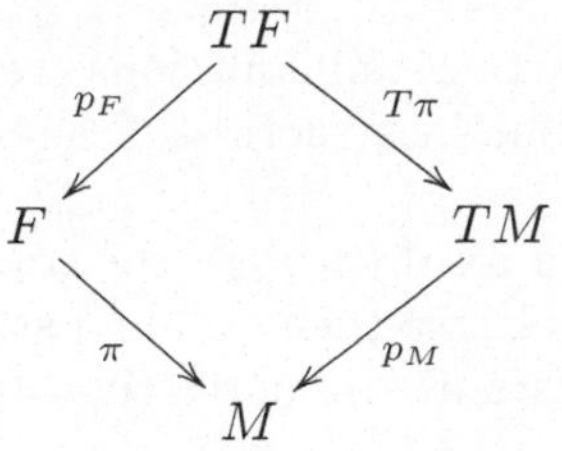

Consider the mapping

$$\begin{array}{rccc} (p_F, T\pi): & TF & \to & F \times_M TM \\ & w & \mapsto & (p_F(w), T\pi(w)) \end{array}$$

where $\pi(p_F(w)) = p_M(T\pi(w)) = x \in M$.

The vertical space at $u \in F$ is then $V_u F = (p_F, T\pi)^{-1} \left(u, 0_{\pi(u)}\right)$. There exists an atlas $\{(U_\alpha, \phi_\alpha)\}_{\alpha \in A}$ of M for which $F_{|U_\alpha}$ is trivial; we then get a local chart $(U_\alpha^F, \tau_\alpha^F)$ of F, where $U_\alpha^F = \pi^{-1}(U_\alpha)$ and such that τ_α^F is a diffeomorphism from U_α^F onto $\phi_\alpha(U_\alpha) \times \mathbb{S}$.

So the family $\left\{\left(T\left(F_{|U^\alpha}\right), T\tau_\alpha^F\right)\right\}_{\alpha \in A}$ where

$$T\tau_\alpha^F : T\left(F_{|U^\alpha}\right) \to T\left(\phi_\alpha(U_\alpha) \times \mathbb{S}\right) = \phi_\alpha(U_\alpha) \times \mathbb{S} \times \mathbb{M} \times \mathbb{S}$$

is an atlas of (TF, p_F, F).

We also have a local chart $(U_\alpha^{TM}, \tau_\alpha^{TM})$ of TM where $U_\alpha^{TM} = p_M^{-1}(U_\alpha)$ and $\tau_\alpha^{TM} = (\phi_\alpha, Tp_M)$.

3.15.2 Tangent Bundle of a Vector Bundle

Let (E, π, M) be a vector bundle.
Given an element (x, u) of the vector fibre $E_x = \pi^{-1}(x)$ where $x \in U_{\alpha\beta} = U_\alpha \cap U_\beta \neq \emptyset$ and let (x, u, y, v) be an element of $T_{(x,u)}E$. For $(x_\alpha, u_\alpha, y_\alpha, v_\alpha) = T\tau_\alpha^E(x, u, y, v)$, we have the following transition functions:

$$\begin{aligned} & \left(T\left((\phi_\alpha \times \mathrm{Id}_{\mathbb{E}}) \circ \tau_\alpha^E\right) \circ \left(T\left((\phi_\beta \times \mathrm{Id}_{\mathbb{E}}) \circ \tau_\beta^E\right)\right)^{-1}\right)((x_\beta, u_\beta, y_\beta, v_\beta)) \\ = & \left(\phi_{\alpha\beta}(x_\beta), \tau_{\alpha\beta}^E\left((\phi_\beta)^{-1}(y_\beta)\right) u_\beta, d\phi_{\alpha\beta}(x_\beta) y_\beta,\right. \\ & \left(d\left(\tau_{\alpha\beta}^E \circ (\phi_\beta)^{-1}\right)(x_\beta) y_\beta\right) u_\beta\Big) \\ & \left.+\tau_{\alpha\beta}^E\left((\phi_\beta)^{-1}(x_\beta) v_\beta\right)\right) \end{aligned}$$

where $\phi_{\alpha\beta} = \phi_\alpha \circ (\phi_\beta)^{-1}$ and $\left(x_\beta, \tau_{\alpha\beta}^E(x_\beta) u_\beta\right) = \left(\tau_\alpha^E \circ \left(\tau_\beta^E\right)^{-1}\right)(x_\beta, u_\beta)$ for $x_\beta \in \phi_\beta(U_{\alpha\beta})$.
So, for a fixed $\left(x^\beta, u^\beta\right)$, the transition functions are linear in $(y_\beta, v_\beta) \in \mathbb{M} \times \mathbb{E}$. This describes the tangent bundle structure of the tangent bundle (TE, p_E, E).

On the other hand, for a fixed (x_β, y_β), the transition functions of TE are also linear in $(u_\beta, v_\beta) \in \mathbb{E} \times \mathbb{E}$; we then obtain a structure of vector bundle on $(TE, T\pi, TM)$ which appears as the derivative structure of the original one on E.

3.15.3 Local Identifications and Expressions in a Convenient Bundle

Consider a convenient bundle $\pi : E \to M$. We denote, respectively by $\Gamma(E)$ and $\Gamma_U(E)$ the module of global sections and sections defined on an open set U.
Consider a chart (U, ψ) on M such that we have a trivialization $\Phi : E_U \to \psi(U) \times \mathbb{E}$ Then $T\psi$ is a trivialization of TM_U on $\psi(U) \times \mathbb{M}$ and $T\Phi$ is a trivialization of TE_U on $\psi(U) \times \mathbb{E} \times \mathbb{M} \times \mathbb{E}$.
For simplicity, we will denote these trivializations

- $E_U = E_{|U} \equiv \psi(U) \times \mathbb{E}$;
- $TM_U = TM_{|U} \equiv \psi(U) \times \mathbb{M}$;
- $TE_U = TE_{|E_{(U)}} \equiv (\psi(U) \times \mathbb{E}) \times (\mathbb{M} \times \mathbb{E})$;
- $T^*E_{|E_{(U)}} \equiv (\psi(U) \times \mathbb{E}) \times (\mathbb{M}^* \times \mathbb{E}^*)$;
- $T^*E^*_{|E^*_U} \equiv \psi(U) \times \mathbb{E}^* \times \mathbb{M}^* \times \mathbb{E}^{**}$;
- $TE^*_{|E^*_U} \equiv \psi(U) \times \mathbb{E}^* \times \mathbb{M} \times \mathbb{E}^*$.

where $U \subset M$ is identified with $\psi(U)$.

We will also use the following associated local coordinates where $\equiv$ stands for the representation in the corresponding trivialization.

$\mathfrak{a} = (x, a) \equiv (\mathsf{x}, \mathsf{a}) \in U \times \mathbb{E}$

$(x, v) \equiv (\mathsf{x}, \mathsf{v}) \in U \times \mathbb{M}$

$(\mathfrak{a}, \mathfrak{b}) \equiv (\mathsf{x}, \mathsf{a}, \mathsf{v}, \mathsf{b}) \in U \times \mathbb{E} \times \mathbb{M} \times \mathbb{E}$.

$(\mathfrak{a}, \eta) \equiv (\mathsf{x}, \mathsf{a}, \eta_1, \eta_2) \in (U \times \mathbb{A}) \times (\mathbb{M}^* \times \mathbb{E}^*)$

$\sigma = (x, \xi) \equiv (\mathsf{x}, \xi) \in U \times \mathbb{E}^*$

$(\sigma, w\eta) \equiv (\mathsf{x}, \xi, \mathsf{w}_1, \eta_2) \in U \times \mathbb{E}^* \times \mathbb{M} \times \mathbb{E}^*$

$(\sigma, \eta w) \equiv (\mathsf{x}, \xi, \eta_1, \mathsf{w}_2) \in U \times \mathbb{E}^* \times \mathbb{M}^* \times \mathbb{E}^{**}$.

3.15.4 Koszul Connections on a Convenient Bundle

Let M be a convenient manifold modelled on a convenient space $\mathbb{M}$ and let $\pi : E \to M$ be a convenient vector bundle on M whose typical fibre is a convenient space $\mathbb{E}$. Let $p_E : TE \to E$ and $p_M : TM \to M$ be the canonical projections of each tangent bundle.

The vertical bundle VE over E is the kernel of $T\pi : TE \to TM$. It is well known that VE is isomorphic to the pull-back $\pi^*E \overset{\widehat{\tau}}{\to} E$ of the bundle $\pi : E \to M$ over π.

We have a canonical isomorphism $E \times_M E \to VE$ called the *vertical lift* vl_E defined by

$$vl_E(x, e, z) = \dot{\gamma}(0)$$

where $\gamma(t) = e + tz$. This map is fibre linear over M.

Let $\iota : VE \to TE$ be the canonical inclusion.

According to [Vil67] we have:

Definition 3.58 *A (non-linear) connection on E is the existence of a decomposition $TE = HE \oplus VE$ of the Banach bundle E. This condition is equivalent to the datum of a bundle morphism $V : TE \to VE$ such that $V \circ \iota = \mathrm{Id}_{VE}$.*

We then have the following diagram:

$$\begin{array}{ccc} VE & \xrightarrow{\ V\ } & TE \\ {\scriptstyle vl_E}\uparrow & & \downarrow{\scriptstyle D} \\ \pi^* E & \xrightarrow{\ \widehat{\tau}\ } & E \end{array}$$

Remark 3.8 *Note that the previous Definition is also equivalent to the datum of an endomorphism N of TE which has two eigenvalues ± 1 with VE (resp. HE) is the bundle of all eigenspaces associated with -1 (resp. 1).*

The bundle morphism $D = \widehat{\tau} \circ vl_E^{-1} \circ V : TE \to E$ is called the *connection map* or *connector* which is a smooth morphism of fibrations. Note that, in each fibre $T_{(x,e)}E$, the kernel of D is exactly the subspace $H_{(x,e)}E$ of HE in $T_{(x,e)}E$. Therefore, the datum of D is equivalent to the datum of V. We then have, modulo the identification $VE \simeq \pi^* E$ *via* vl_E:

$$\begin{array}{rccc} D: & TE & \to & E \\ & (x,e,v,z) & \mapsto & (x, z + \omega(x,e)(v)) \end{array}$$

where $\omega(x,e) \in L(T_x M, E_x)$, where $L(\mathbb{M},\mathbb{E})$ is the set of bounded linear operators from $\mathbb{M}$ to $\mathbb{E}$.
If moreover, ω is linear on each fibre E_x, then the connection D is called a *linear connection*.

Given a chart (U,ψ) on M, we have the following complementary identification:

- $VE_U := V_{|U} \equiv (\psi(U) \times \mathbb{E}) \times \mathbb{E}$ *via* the trivialization $T\Psi$.

In order to give local expressions in the different charts, the local coordinates corresponding to the element x (resp. e, v, z) will be x (resp. e, v, z), the local expression of the mapping ω being ϖ; moreover, $(\mathfrak{e},\mathfrak{z})$ will correspond to (x,e,v,z).

Thus, we obtain the following characterizations of V and D:

$$V(\mathfrak{e},\mathfrak{z}) \equiv (\mathsf{x},\mathsf{e},0,\mathsf{z} + \varpi(\mathsf{x},\mathsf{e})\mathsf{v}) \tag{3.9}$$

$$D(\mathfrak{e},\mathfrak{z}) \equiv (\mathsf{x},\mathsf{z} + \varpi(\mathsf{x},\mathsf{e})\mathsf{v}) \tag{3.10}$$

where ϖ is a smooth map from $\psi(U) \times \mathbb{E}$ to the space $L(\mathbb{M},\mathbb{E})$.
This connection is *linear* if and only if ϖ is linear in the second variable.

In this case, the relation $\digamma(\mathsf{x})(\mathsf{e},\mathsf{v}) := \varpi(\mathsf{x},\mathsf{e})\mathsf{v}$ gives rise to a smooth map $\digamma$ from $\psi(U)$ to the space of bilinear maps $L^2(\mathbb{E},\mathbb{M};\mathbb{E})$ called *local Christoffel components* of the connection.

Definition 3.59 *A Koszul connection on the convenient linear bundle $\pi : E \to M$ is a $\mathbb{R}$-bilinear map $\nabla : \mathfrak{X}(M) \times \Gamma(E) \to \Gamma(E)$ which fulfills, for any smooth function f on M, $X \in \mathfrak{X}(M)$ and $s \in \Gamma(E)$, the following properties:*

(KosC1) $$\nabla_X(f.s) = df(X).s + f\nabla_X s$$

(KosC2) $$\nabla_{fX} s = f\nabla_X s$$

Given a linear connection D on a convenient bundle $\pi : E \to M$, we obtain a covariant derivative $\nabla : \mathfrak{X}(M) \times \Gamma(E) \to \Gamma(E)$ which is a Koszul connection. Since any (linear) connection induces naturally a (linear) connection on the restriction E_U of E to any open set U of M, we also obtain a covariant derivative $\nabla^U : \mathfrak{X}(U) \times \Gamma(E_U) \to \Gamma(E_U)$ with the correspondent previous properties for any function f on U, $X \in \mathfrak{X}(U)$ and $\mathfrak{u} \in \Gamma(E_U)$.

Unfortunately, if M is not smooth regular, then a Koszul connection may be not localizable in the following sense.
Since ∇ is defined on **global** sections of E and since any local section of E (resp. any local vector field on M) can not always be extended to a global section of E (resp. a global vector field on M), the operator ∇ need not always induce a (local) operator ∇^U as previously. Unfortunately, we have no example for which a local section can not be extended to a global one.

Therefore, in this work, a *Koszul connection will always be considered in the sense of the covariant derivative associated with a linear connection D on E*. This implies in particular that

Proposition 3.12 *For any $x \in M$, the value $\nabla_X s(x)$ only depends on the value of X at x and the 1-jet of the section s at x.*

In a local trivialization $E_U \equiv \psi(U) \times \mathbb{E}$, according to the local expressions used above, a local section $\mathfrak{e}$ of E, defined on U, can be identified with a map $\mathsf{e} : U \to \mathbb{E}$. Then ∇ has the local expression:

$$\nabla_X \mathfrak{e} \equiv T\mathsf{e}(\mathsf{X}) + \digamma(\mathsf{e},\mathsf{X}) \tag{3.11}$$

where $\digamma$ is a smooth map from U to $L^2(\mathbb{E},\mathbb{M};\mathbb{E})$ which corresponds to the *local Christoffel components* of the connection D.

Therefore, under this complementary assumption, *there is a one-to-one correspondence between such covariant derivatives and linear connections on E* as in the finite-dimensional framework.

Remark 3.9 *Let $\pi : E \to M$ be a convenient linear bundle.*

(1) Let ∇ be Koszul connection on E. The fibre $(HE)_{(x,e)}$ of the associated horizontal bundle HE is characterized by

$$\begin{aligned} H_{(x,e)}E &= \{T\mathfrak{e}(X) \mid \mathfrak{e}(x) = e,\ X \in T_xM, \nabla_X \mathfrak{e} = 0\} \\ &\equiv \{(\mathsf{v},\mathsf{z}) \in \{(\mathsf{x},\mathsf{e})\} \times \mathbb{M} \times \mathbb{E} \ : \ \mathsf{z} + F(\mathsf{x})(\mathsf{e},\mathsf{v}) = 0\}. \end{aligned} \tag{3.12}$$

(2) Given any connection D on E, from its local expression, the horizontal space $HE_{(x,e)}$ has the following characterization:

$$H_{(x,e)}E \equiv \{(\mathsf{v},\mathsf{z}) \ : \ \mathsf{z} + \varpi(\mathsf{x},\mathsf{e})\mathsf{v} = 0\}.$$

If C is the Liouville section of $(\psi(U) \times \mathbb{E}) \times (\mathbb{M} \times \mathbb{E})$, then D is linear if and only if $L_{\mathsf{C}}\varpi = \varpi$.

Now, as in [AgSu10] in the Banach context, we have

Theorem 3.6 *Let $\pi : E \to M$ be a convenient vector bundle. If there exists a linear connection D on E, there exists a convenient isomorphism bundle*

$$\kappa = (\kappa_1, \kappa_2) : TE \to \pi^*TM \oplus \pi^*E.$$

Proof *The proof of this Theorem is an adaptation, point by point, of the same theorem 3.1 in [AgSu10] in the Banach framework. At first, consider the trivialization*

$$\begin{array}{cccc} \tilde{\Psi}: & \pi^*E_U & \to & \Psi(\mathcal{M}_U) \times \mathbb{E} \\ & (x,e,z) & \mapsto & (\psi(x), \Phi_x(e), \Phi_x(z)) = (\mathsf{x},\mathsf{e},\mathsf{z}) \end{array}$$

In the same way, consider the trivialization

$$\begin{array}{cccc} \widetilde{T\psi}: & \pi^*TM_U & \to & \Psi(\mathcal{M}_U) \times \mathbb{M} \\ & (x,e,v) & \mapsto & (\psi(x), \Phi_x(e), T_x\psi(v)) = (\mathsf{x},\mathsf{e},\mathsf{v}) \end{array}.$$

According to the previous notations, we define κ_1 and κ_2 in local coordinates in the following way:

$$\begin{array}{cccc} \kappa_1: & TE_U & \to & \pi^*TM_U \\ & (\mathsf{x},\mathsf{e},\mathsf{v},\mathsf{z}) & \mapsto & (\mathsf{x},\mathsf{e},\mathsf{v}) \end{array}.$$

$$\begin{array}{cccc} \kappa_2: & TE_U & \to & \pi^*E_U \\ & (\mathsf{x},\mathsf{e},\mathsf{v},\mathsf{z}) & \mapsto & (\mathsf{x},\mathsf{e},\mathsf{z}+F\,(\mathsf{x})\,(\mathsf{e},\mathsf{v})) \end{array}$$

It is clear that, locally, κ_1 and κ_2 are local bundle morphisms which are smooth and so the pair (κ_1, κ_2) is a local convenient bundle morphism.

Let (U', ψ') be another chart on M and consider the corresponding trivializations $T\psi'$, Ψ' and $T\Psi'$. Following our convention, in local coordinates, we set

$$(\mathsf{x}',\mathsf{e}',\mathsf{v}',\mathsf{z}') := T\mathfrak{e}\Psi'(\mathfrak{z}) = (\psi'(x), \Phi'_x(e), T_x\psi'(v), \Phi'_x(z)).$$

Assume that $U \cap U' \neq \emptyset$. For the change of coordinates, we set $\theta_{\mathsf{x}} = \psi' \circ \psi^{-1}(\mathsf{x})$, and each other associated transition map gives rise to a smooth field of isomorphisms of convenient spaces as follows:

$$T_x\theta(\mathsf{v}) = T_x\psi' \circ \psi^{-1}(\mathsf{v}),$$
$$\Theta_x(\mathsf{e}) = (\Psi' \circ \Psi^{-1})_x(\mathsf{e}),$$
$$T_{(x,e)}\Theta(\mathsf{v},\mathsf{z}) = T_{(x,e)}(\Psi' \circ \Psi^{-1})(\mathsf{v},\mathsf{z}).$$

The corresponding change of coordinates $\widetilde{\Psi'} \circ \widetilde{\Psi}^{-1}$ *in* π^*E *is:*

$$(\mathsf{x}',\mathsf{e}',\mathsf{z}') = (\theta(\mathsf{x}), \Theta_x(e), \Theta_x(\mathsf{z})).$$

and the corresponding change of coordinates $\widetilde{T\psi'} \circ \widetilde{T\psi}^{-1}$ *in* π^*TM *is:*

$$(\mathsf{x}',\mathsf{e}',\mathsf{v}') = (\theta(\mathsf{x}), \Theta_x(e), T_x\theta(\mathsf{v}))$$

Using formally the same arguments as in the proof of Theorem 3.1 in [AgSu10], we have the following relations:

$$\begin{array}{rl} & (\kappa_1',\kappa_2')(\mathsf{x}',\mathsf{e}',\mathsf{v}',\mathsf{z}') \\ = & (\kappa_1,\kappa_2)(\theta(\mathsf{x}),\Theta_x(\mathsf{e}),T_x\theta(\mathsf{v}),T_{(x,e)}\Theta(\mathsf{z})) \\ = & (\theta(\mathsf{x}),\Theta_x(\mathsf{e}),T_x\theta(\mathsf{v})) \oplus (\theta(\mathsf{x}),\Theta_x(\mathsf{e}),\Theta_x(\mathsf{z}) + T_{(x,e)}\Theta(\mathsf{v},\mathsf{z}) \\ & \quad + F(\theta(\mathbf{x})\{\Theta_x(\mathsf{e}),T_x\theta(\mathsf{v})\}) \\ = & (\theta(\mathsf{x}),\Theta_x(\mathsf{e}),T_x\theta(\mathsf{v})) \oplus (\theta(\mathsf{x}),\Theta_x(\mathsf{e}),\Theta_x(\mathsf{z}) + \Theta_x(F(\mathsf{x})\{\mathsf{e},\mathsf{v}\})) \\ = & \widetilde{T\psi'} \circ \widetilde{T\psi}^{-1}(\kappa_1(\mathsf{x},\mathsf{e},\mathsf{v},\mathsf{z})) \oplus \widetilde{\Psi}' \circ \widetilde{\Psi}^{-1}(\kappa_2(\mathsf{x},\mathsf{e},\mathsf{v},\mathsf{z})). \end{array}$$

This shows that the local definition of (κ_1,κ_2) *gives rise to a global convenient morphism* κ *over the identity of* E. *Since* κ *is locally a convenient isomorphism, this implies that* κ *is a convenient bundle isomorphism.*

3.15.5 SODE and Exponential Map of a Connection

We consider the tangent bundle $p_M : TM \to M$ of a convenient manifold modelled on a convenient space $\mathbb{M}$ and the fibre bundle $p_{TM} : TTM \to TM$.

Definition 3.60 *A second-order differential equation on* M *(SODE for short) is a (global) vector field* S *on* TM.

Let D be a connection on TM and $VTM \oplus HTM$ be the associated Whitney decomposition. Thus, as in [AgSu13], we have:

Proposition 3.13 *To any connection* D *on* TM *is associated a SODE* S_D *defined by*

$$S_D(x,y) = \{Tp_M\}^{-1}_{|H_{(x,y)}TM}(x,y).$$

Proof *Consider a chart* (U,ψ) *in* M. *Then* $\left(TU = TM_{|U}, \Psi = T\psi\right)$ *is a chart in* TM *and* $\left(p_M^{-1}(TU), T\Psi\right)$ *is a bundle chart for* $TTM \to TM$. *Then* Ψ *is an isomorphism from* TU *onto* $\psi(U) \times \mathbb{M}$ *and* $T\Psi$ *is an isomorphism of* $p_M^{-1}(TU)$ *onto* $(\psi(U) \times \mathbb{M}) \times \mathbb{M} \times \mathbb{M}$. *Let* $\varpi : \Psi(TU) \times \mathbb{M} \times \mathbb{M} \to \mathcal{L}(\mathbb{M} \times \mathbb{M}, \mathbb{M} \times \mathbb{M})$ *the local field associated with* D *such that*

$$H_{(x,y)}TM \equiv \{(\dot{\mathsf{x}},\dot{\mathsf{y}}) : \dot{\mathsf{y}} + \varpi(\mathsf{x},\mathsf{y})\dot{\mathsf{x}} = 0\}$$

for all $(x,y) \in TU$ *(cf. Remark 3.9). On the other hand, in local coordinates,* Tp_M *is the map* $(\mathsf{x},\mathsf{y},\dot{\mathsf{x}},\dot{\mathsf{y}}) \mapsto (\mathsf{x},\dot{\mathsf{x}})$. *Therefore, for any* $(x,u) \in T_xM$, *we have*

$$Tp_M^{-1}(x,u) \cap T_{(x,u)}TM = \{(x,u,u,v) \in T_{(x,u)}TM\}.$$

Therefore, in local coordinates, $Tp_M^{-1}(x,u) \cap H_{(x,u)}TM$ *corresponds exactly to* $(\mathsf{x},\mathsf{u},\mathsf{u},-\varpi(\mathsf{x},\mathsf{u})\mathsf{u})$. *This implies that* S_D *is a well-defined smooth vector field on* TM.

Assume now that D is a linear connection and let ∇ be its covariant derivation. If $c : [a,b] \to M$ is a smooth curve, we can define a derivation along γ as the Koszul connection ∇^c associated with the linear connection c^*D on c^*TM. Therefore, a vector field V along c which is a smooth section of c^*TM, is called *parallel* if $\nabla^c_{c'}V = 0$. In local coordinates, according to (3.11), V satisfies the differential equation:

$$\frac{d\mathsf{V}}{dt}(t) + \mathsf{F}(c(t))(V(t),c'(t)) = 0.$$

In particular, if c' is parallel along c, then c is called *a geodesic of* D *(or* ∇*)*. In local coordinates, according to (3.11), c is a geodesic if and only if we have

$$\mathsf{c}''(t) + \mathsf{F}(\mathsf{c}(t)(c'(t),c'(t)) = 0.$$

According to Proposition 3.13, if S_D is the SODE associated with D we have:

Proposition 3.14 *c is a geodesic of D (or ∇) if and only if c is an integral curve of S_D.*

Assume now that S_D has a flow on TM. Note that this situation occurs at least in the Banach context. Denote by Fl the flow of S_D. This means that there exists an open set O_D in $\mathbb{R} \times TM$ such that Fl is defined. Thus, for any $(x,y) \in TM$, there exists an open interval $]\alpha,\beta[$ such that $t \mapsto \mathrm{Fl}(t,x,y)$ is an integral curve of S_D such that $\mathrm{Fl}(0,x,y) = (x,y)$. As usually, if we set $\mathrm{Fl}_t(x,y) = \mathrm{Fl}(t,x,y)$, we have

- $\mathrm{Fl}_{t+t'} = \mathrm{Fl}_t \circ \mathrm{Fl}_{t'}$ when $\mathrm{Fl}_{t+t'}$ Fl_t and $\mathrm{Fl}_{t'}$ are defined;
- Fl_t is a local diffeomorphism in TM whose inverse is Fl_{-t}.

Moreover, since, in local coordinates, we have

$$S_D(x,y) \equiv (x,y,y,-\mathsf{F}(x)(y,y))$$

and because $\mathsf{F}(x)$ is bilinear, it follows that S_D is zero on the zero section of TM and so Fl_t is the identity in restriction to the zero section of TM.
In this way, as classically, we can define the exponential map:

$$\exp^D : \mathcal{U} \subset TM \to M$$

where $\mathcal{U}$ is an open set of the zero section of TM. Moreover, $\mathcal{U}_x = T_xM \cap \mathcal{U}$ is a star-shaped open neighbourhood of $0 \in T_xM$. If $\exp_x^D$ denotes the restriction of $\exp^D$ to $\mathcal{U}_x$, then $T\exp_x^D(0) = \mathrm{Id}$.

In the Banach context, this implies that $\exp_x^D$ is a local diffeomorphism of an open neighbourhood of $0 \in T_xM$ onto an open neighbourhood of x in M.

3.16 Convenient Lie Groups and Lie Algebras

The framework of Banach-Lie groups (cf. Chapter 1, § 1.10.1) was extended to complete locally convex spaces in a paper of Marsden and Abraham in the context of hydrodynamics. The first attempt to develop a general theory of Lie groups modelled on complete locally convex spaces can be found in Milnor's lectures notes [Mil82].

Some infinite-dimensional Lie groups are adapted frameworks for symmetries depending on an infinite number of parameters. These symmetries play a fundamental role in Mathematical Physics (see for example [Schm04]).

The core of Lie theory consists in solving some integration problems listed in [Nee98] (FP1)..(FP10). In this Neeb's survey, where one can also find historical development on infinite-dimensional Lie theory, four classes of infinite-dimensional Lie groups are described:

1. linear Lie groups;
2. mapping groups and gauge groups;
3. direct limit groups;
4. groups of diffeomorphisms.

3.16.1 Lie Groups

Definition 3.61 *A convenient Lie group G is a convenient manifold endowed with a group structure for which multiplication $\mathbf{m} : G \times G \to G$ and inversion $\mathbf{i} : G \to G$ are smooth mappings.*

The unit element of the group G will be denoted e, and the inverse of an element g by g^{-1}.
For $g \in G$, $L_g : x \mapsto g.x$ is the *left translation* and $R_g : x \mapsto x.g$ is the *right translation*.

3.16.2 The Lie Algebra of a Lie Group

Let G be a convenient Lie group. The tangent bundle to G that we consider is the kinematic tangent bundle TG and the vector fields, smooth sections of this bundle, are the kinematic vector fields whose set is denoted $\mathfrak{X}(G)$.

Definition 3.62 *A vector field X on G is said to be left invariant if*

$$\forall g \in G, (L_g)^* X = X$$

where $(L_g)^* X = TL_{g^{-1}} \circ X \circ L_g$.

Since $(L_g)^* [X, Y] = \left[(L_g)^* X, (L_g)^* Y\right]$, the set of all left-invariant vector fields on G forms a Lie subalgebra of $\mathfrak{X}(G)$ denoted by $\mathbf{L}(G)$ or $\mathfrak{g}$.

Definition 3.63 $\mathbf{L}(G)$ *is called the Lie algebra of the Lie group G.*

$\mathbf{L}(G)$ is in bijective correspondence with the tangent space T_eG *via* the linear isomorphism

$$\begin{array}{ccc} \mathbf{L}(G) & \to & T_eG \\ X & \mapsto & X_e \end{array}$$

So T_eG becomes a Lie algebra for the bracket still denoted by $[.,.]$

$$[X_e, Y_e] = [X, Y]_e$$

Proposition 3.15 *Let $\varphi : G \to H$ be a smooth homomorphism of Lie groups. Then $T_e\varphi : \mathbf{L}(G) \to \mathbf{L}(H)$ is a Lie algebra homomorphism.*

3.16.3 The Exponential Map

An essential tool in the finite-dimensional and Banach contexts is the exponential map.

Definition 3.64 *Let $(G, .)$ be a Lie group with Lie algebra $\mathfrak{g}$. A 1-parameter subgroup of G is a smooth Lie group homorphism $\alpha : (\mathbb{R}, +) \to (G, .)$.*

So a 1-parameter subgroup of G is a smooth curve α in G such that

$$\left\{ \begin{array}{c} \forall (s,t) \in \mathbb{R}^2, \ \alpha(s+t) = \alpha(s) . \alpha(t) \\ \alpha(0) = e \end{array} \right. .$$

Proposition 3.16 *Let $\alpha : \mathbb{R} \to G$ be a smooth curve where $\alpha(0) = e$ and let $X \in \mathfrak{g}$. We denote $\hat{X}$ the left-invariant vector field such that $\hat{X}_e = X$. Then the following assertions are equivalent:*

(1) α is a 1-parameter subgroup of G and $X = \left.\dfrac{\partial}{\partial t}\right|_{t=0} \alpha(t)$;

(2) α is an integral curve of the left-invariant vector field $\hat{X}$;

(3) $\mathrm{Fl}^{\hat{X}}(t, x) = x.\alpha(t)$ *is the unique global flow of the vector field $\hat{X}$.*

Definition 3.65 *Let G be a Lie group with Lie algebra $\mathfrak{g}$. We say that G admits an exponential map if there exists a smooth mapping* $\exp : \mathfrak{g} \to G$ *such that* $t \mapsto \exp(tX)$ *is the unique 1-parameter subgroup with tangent vector X at 0.*

As in the Banach framework, we then have the following properties:

(ExpC1) $\exp(0) = e$;

(ExpC2) $T_0 \exp = \mathrm{Id}_{\mathfrak{g}}$

where **(ExpC2)** is a consequence of $\left.\frac{d}{dt}\right|_{t=0} \exp(tX) = X$.

Unfortunately, as kinematic vector fields on convenient manifolds need not possess integral curves, there is no general theorem that guarantees the existence of such a map.

3.16.4 Regular Lie Group and Local Exponential Lie Group

Consider a convenient Lie group G and its Lie algebra $\mathbf{L}(G)$. The notion of regular Lie group is originally due to Omori et al. (cf. [OMY80] .. [OMY83b]). They defined a notion of regular Fréchet Lie group and studied the properties of such Lie groups; in particular, they showed that any strong **ILB**-Lie group is a Fréchet Lie group (in their sense) and proved that many "classical" Fréchet Lie groups are regular. Their definition of regular Lie group was weakened and made more transparent by Milnor ([Mil82]), and this definition can be adapted to the convenient Lie groups setting (see [KrMi97], 38).

Definition 3.66 *A convenient Lie group G is called regular if, for each curve ξ in $C^\infty(\mathbb{R}, \mathbf{L}(G))$, the following Cauchy problem*

$$\begin{cases} \forall t \in \mathbb{R}, \ \gamma'(t) = T_e R_{\gamma(t)}(\xi(t)) \\ \gamma(0) = e \end{cases}$$

has a solution $\gamma_\xi : \mathbb{R} \to G$ where $\gamma_\xi(1)$ smoothly depends on ξ, i.e. the mapping

$$\begin{array}{rccc} \mathrm{evol}_G: & C^\infty(\mathbb{R}, \mathbf{L}(G)) & \to & G \\ & \xi & \mapsto & \gamma_\xi(1) \end{array}$$

is smooth.

In this way, the exponential map is

$$\begin{array}{rccc} \exp_G: & \mathbf{L}(G) & \to & G \\ & \xi & \mapsto & \mathrm{evol}_G(\xi) \end{array}$$

where $\xi \in \mathbf{L}(G)$ is considered as a constant function from $\mathbb{R}$ to $\mathbf{L}(G)$. Thus, the exponential function is smooth and satisfies the property of Definition 3.65.
A Banach-Lie group is regular (cf. [KrMi97]). Note that, in the context of locally convex Lie-group, in [Glo02a], the reader will find an explicit example of a Lie group which is not regular in the sense of Milnor.
In the convenient setting, all known convenient Lie groups are regular. (cf. [KrMi97], 38.4)

Remark 3.10 *Usually, the definition of a regular Lie group is for curves $\xi \in C^\infty([0,1], \mathbf{L}(G)$. But from Lemma 38.3 of [KrMi97], for any interval $I \subset \mathbb{R}$ if the Cauchy problem in Definition 3.66 has a local solution, this solution is unique. It follows that if G is regular, the map evol_G is unique. Moreover, by the same Lemma, such a solution can be defined on the whole interval I. Thus, for instance in the Banach setting, for any $\xi \in C^\infty(\mathbb{R}, \mathbf{L}(G))$ such a solution can be always defined locally and so any Banach-Lie group is always regular in the sense of Definition 3.66.*

A very complete exposition of problems and all properties of infinite-dimensional Lie groups can be found in [Nee06]. In particular, the author introduces the following notion:

Definition 3.67 *A convenient Lie group G is called locally exponential if there exists an exponential map $\exp : \mathbf{L}(G) \to G$ which is a local smooth diffeomorphism from a neighbourhood of $0 \in \mathbf{L}(G)$ onto a neighbourhood of $e \in G$.*

Of course, any Banach-Lie group is locally exponential. Although a regular convenient Lie group has an exponential map, in general, such a Lie group is not locally exponential. For instance, $\mathbf{Diff}(\mathbb{S}^1)$ is regular but not locally exponential (cf. [Nee06]). But among the set of pro-Lie groups, there exist Fréchet Lie groups which are regular and locally exponential (cf. Example 4.12). In § 4.7.2, we give some necessary conditions under which a projective limit of Banach-Lie groups is a regular locally exponential Fréchet Lie group.

3.16.5 The Adjoint Representation

Let G be a Lie group with Lie algebra $\mathfrak{g}$.

Definition 3.68 *For any $g \in G$, we define the conjugation or inner automorphism*

$$\begin{array}{rccc} \mathrm{conj}_g : & G & \to & G \\ & x & \longmapsto & g.x.g^{-1}. \end{array}$$

Definition 3.69 *The adjoint representation is the mapping*

$$\begin{array}{rccc} \mathrm{Ad} : & G & \to & \mathrm{GL}(\mathfrak{g}) \\ & g & \longmapsto & T_e\left(\mathrm{conj}_g\right) \end{array}$$

The adjoint representation of the Lie algebra $\mathfrak{g}$ is the map $\mathrm{ad} = T_e \mathrm{Ad}$. We then have:

$$\forall (X,Y) \in \mathfrak{g}^2, \ \mathrm{ad}(X)Y = [X,Y]. \tag{3.13}$$

3.16.6 Group Actions

Definition 3.70 *Let E be a set and let $(G,.)$ be a group with identity e. A left action of G on E is a mapping*

$$\begin{array}{cccc} \phi: & G\times E & \to & E \\ & (g,x) & \longmapsto & \phi(g,x) \overset{\text{denoted}}{=} g\cdot x \end{array}$$

such that:

(GA1) $\forall x\in E,\ e\cdot x=x;$

(GA2) $\forall x\in E, \forall (g,g')\in G^2,\ g\cdot(g'\cdot x)=(g.g')\cdot x.$

For $g\in G$, we denote $\begin{array}{cccc} \phi_g: & E & \to & E \\ & x & \mapsto & g\cdot x \end{array}.$

Definition 3.71 *For any $x\in E$, we define the stabilizer of x or isotropy group of x as*

$$G_x=\{g\in G:\phi_g(x)=x\}.$$

Definition 3.72 *An action of the group G on the set E will be called*

(1) effective if $g\mapsto\phi_g$ is injective;

(2) free if $\forall x\in E,\ G_x=\{e\}$.

When G is a Lie group and E is a smooth convenient manifold, we will assume that ϕ is smooth.

3.17 Convenient Principal Bundles

3.17.1 Definition. Actions

Definition 3.73 *A convenient principal bundle is a bundle (P,π,M,G) where the type fibre G is a Lie group acting on P on the right* $\begin{array}{cccc} \widehat{R}: & P\times G & \to & P \\ & (p,g) & \longmapsto & p\cdot g \end{array}$ *in the following way:*

Any $x\in M$ has an open neighbourhood U such that $P_{|U}=\pi^{-1}(U)$ is diffeomorphic to $U\times G$ via a diffeomorphism τ such that $\mathrm{pr}_1\circ\tau=\pi$ where $\mathrm{pr}_1: U\times G\to U$ is the first projection and satisfies the property:

(PB) $\forall y\in U,\forall (g,g')\in G^2, \left(\tau^{-1}(y,g)\right)\cdot g'=\tau^{-1}(y,g.g').$

The property **(PB)** corresponds to the *equivariance* of τ^{-1} with respect to the action $\widehat{R_g}$ of G. If $p\in\pi^{-1}(x)$ and $g\in G$, then $\widehat{R_g}(p)=p\cdot g\in\pi^{-1}(x)$;

so G acts on the right on each fibre and this action is free and transitive. In particular, if (U, τ) is a local trivialization of P, then

$$\forall x \in U, \tau\left(\pi^{-1}(x)\right) = \{x\} \times G.$$

In relation with the right action of the Lie group G on the bundle P, is associated a left action of G on itself, seen as the type fibre.
Let U_α and U_β be two trivialization open sets such that $U_\alpha \cap U_\beta \neq \emptyset$ and let $x \in U_\alpha \cap U_\beta$; fix some p in the fibre$P_x = \pi^{-1}(x)$. For the local trivialization $\tau_\alpha : \pi^{-1}(U_\alpha) \to U_\alpha \times G$, we denote $\left(\widetilde{\tau_\alpha}\right)_x : P_x \to G$ the bijection defined by $\left(\widetilde{\tau_\alpha}\right)_x = \mathrm{pr}_2 \circ \tau_\alpha|_{P_x}$. We then set $g_\alpha = \left(\widetilde{\tau_\alpha}\right)_x(p)$ and in an analogous way $g_\beta = \left(\widetilde{\tau_\beta}\right)_x(p)$. Then there exists an element $\gamma_{\alpha\beta}(x,p)$ of G such that $g_\alpha = \gamma_{\alpha\beta}(x,p).g_\beta$. Hence, thanks to the equivariance of the right action of G on each fibre, $\gamma_{\alpha\beta}(x,p)$ only depends on p: for $p \cdot g \in P_x$, $\left(\widetilde{\tau_\alpha}\right)_x(p \cdot g) = g_\alpha.g$ and $\left(\widetilde{\tau_\beta}\right)_x(p \cdot g) = g_\beta.g$ and so $\gamma_{\alpha\beta}(x,p.g) = \gamma_{\alpha\beta}(x,p)$. We will denote $\gamma_{\alpha\beta}(x,p) = \psi_{\alpha\beta}(x)$. The mappings $\psi_{\alpha\beta} : U_\alpha \cap U_\beta \to G$ are called *transition functions* of the bundle P relatively with the trivializations family $\{(U_\alpha, \tau_a)\}_{\alpha \in A}$. We then have the following relation:

(TF) $\quad \forall x \in U_\alpha \cap U_\beta, \forall g \in G, \ \tau_a \circ (\tau_\beta)^{-1}(x,g) = (x, \psi_{\alpha\beta}(x).g).$

Because $g_\alpha = \left(\widetilde{\tau_\alpha}\right)_x \circ \left(\left(\widetilde{\tau_\beta}\right)_x\right)^{-1}(g_\beta)$ and $\psi_{\alpha\beta}(x)$ does not depend on p, we can write $\psi_{\alpha\beta}(x) = \left(\widetilde{\tau_\alpha}\right)_x \circ \left(\left(\widetilde{\tau_\beta}\right)_x\right)^{-1}(e)$ where e is the identity of the group G.

So a principal bundle is a G-bundle where the type fibre is the Lie group G and where the action ℓ is the left translation of the Lie group.

Proposition 3.17 *A convenient principal bundle is determined by its cocycles.*

For a given cocycle $\{\psi_{\alpha\beta} : U_\alpha \cap U_\beta \to G\}_{(\alpha,\beta)\in A^2}$ where $(U_\alpha)_{\alpha\in A}$ is an open covering of the manifold M, on the set $S = \bigcup\limits_{\alpha \in A} (\{\alpha\} \times U_\alpha \times G)$, we define the equivalence relation

$$(\alpha, x, g) \sim (\beta, y, h) \Longleftrightarrow \left\{ \begin{array}{c} x = y \\ g = \psi_{\alpha\beta}(x).h \end{array} \right. .$$

The quotient set $P = S/\sim$ can be endowed with a structure of principal bundle (P, π, M, G) where $\pi\left(\widetilde{(\alpha, x, g)}\right) = x$ and where the local transitions functions $\tau_\alpha : \pi^{-1}(U_\alpha) \to U_\alpha \times G$ are given by

$$\tau_\alpha\left(\widetilde{(\beta, x, g)}\right) = (x, \psi_{\alpha\beta}(x).g).$$

The bundle thus constructed is unique up to isomorphism.
For this principal bundle a right action

$$\begin{array}{cccc} R: & P \times G & \to & P \\ & (p, g) & \mapsto & p \cdot g \end{array}$$

can be defined as follows. For a chart (U_α, τ_α) of the bundle where $p \in \pi^{-1}(x)$, $x \in U_\alpha$, we have $(\tau_\alpha)^{-1}(p) = (x, g_\alpha)$; so we have

$$\tau_\alpha (p \cdot g) = (x, g_\alpha . g) .$$

Since we have

$$(\tau_\beta)^{-1}(x, g_\beta . g) = (\tau_\beta)^{-1}\left(x, \left((\psi_{\alpha\beta}(x))^{-1} . g_\alpha\right) . g\right) = (\tau_a)^{-1}(x, g_\alpha . g),$$

this right action is well-defined, i.e. independent on the chosen local trivialization.
This action is free and transitive and the elements v of the fibre P_x containing p can be written $v = p \cdot g$ where $g \in G$; the orbits for the right action are the fibres of $\pi : P \to M$.

3.17.2 Vertical Bundle of a Principal Bundle

We have the following proposition (cf. [KMS93], 10.18):

Proposition 3.18 *Let (P, π, M, G) be a convenient principal bundle. The vertical bundle $(VP, p_P, P, \mathfrak{g})$ is the range by a vector bundle isomorphism of the trivial bundle $(P \times \mathfrak{g}, \mathrm{pr}_1, P, \mathfrak{g})$.*

3.17.3 Morphisms of Principal Bundles

Definition 3.74 *A morphism between the convenient principal bundles $\mathcal{P}_1 = (P_1, \pi_1, B_1, G_1)$ and $\mathcal{P}_2 = (P_2, \pi_2, B_2, G_2)$ is a triple $(\lambda, \varepsilon, \varphi)$ where $\lambda : P_1 \to P_2$ and $\varepsilon : M_1 \to M_2$ are smooth mappings and $\varphi : G_1 \to G_2$ is a convenient Lie groups morphism satisfying the following conditions:*

(MPB1) $$\pi_2 \circ \lambda = \varepsilon \circ \pi_1 ;$$

(MPB2) $$\forall g \in G_1, \forall p \in P_1,\ \lambda (p \cdot g) = (\lambda (p)) \cdot \varphi (g)$$
(commutativity with the right actions).

3.17.4 Sections of a Principal Bundle

The sections of a convenient principal bundle are defined in an usual way.

Notation 3.8 $\Gamma(U, P)$ *denotes the set of sections of the principal bundle (P, π, M, G) over the open set U of M.*

If $\mathcal{C} = \{(U_\alpha, \tau_\alpha)\}_{\alpha \in A}$ is a trivializing covering of the principal bundle, the *natural sections* of P, with respect to $\mathcal{C}$, are the maps

$$\begin{array}{rccc} s_\alpha : & U & \to & P \\ & x & \longmapsto & (\tau_\alpha)^{-1}(x, e) \end{array}.$$

3.18 Convenient Lie Algebroids

The notion of Lie algebroid [*algébroïde de Lie* in the French terminology] was first introduced by J. Pradines in [Pra66] in relation with Lie groupoids, then extended to the Banach framework in [Ana11].

Lie algebroids are generalizations of both Lie algebras and tangent vector bundles.

This notion is an adapted framework for different problems one can meet in mechanics ([Wei71], [LMM10], [CLMM09]), symplectic geometry ([Kar87]), and optimal control theory ([Mar01a], [Mar01b]), ...

3.18.1 Convenient Anchored Bundle

Let $\pi : \mathcal{A} \to M$ be a convenient vector bundle whose fibre is a convenient space $\mathbb{A}$.

Definition 3.75 *A morphism of vector bundles $\rho : \mathcal{A} \to TM$ is called an anchor and the quadruple $(\mathcal{A}, \pi, M, \rho)$ is called a convenient anchored bundle.*

Notation 3.9 *For the sake of simplicity, in this section, the anchored bundle $(\mathcal{A}, \pi, M, \rho)$ will be fixed.*
For any open subset $U \subset M$, we denote $\Gamma(\mathcal{A}_U)$ the $C^\infty(U)$-module of smooth sections of the restriction $\mathcal{A}_U$ of $\mathcal{A}$ over U and $\Gamma(\mathcal{A})$ when $U = M$. The morphism ρ gives rise to a $C^\infty(U)$-morphism of modules $\rho_U : \Gamma(\mathcal{A}_U) \to \mathfrak{X}(U)$ defined for any $x \in M$ and any smooth section $\mathfrak{a}$ of $\mathcal{A}_U$ by:

$$(\rho_U(\mathfrak{a}))(x) = \rho(\mathfrak{a}(x))$$

is still denoted by ρ.

3.18.2 Almost Lie Bracket

Definition 3.76 *An almost Lie bracket on an anchored bundle $\mathcal{A}$ is a sheaf of antisymmetric bilinear maps*

$$[.,.]_{\mathcal{A}_U} : \Gamma(\mathcal{A}_U) \times \Gamma(\mathcal{A}_U) \to \Gamma(\mathcal{A}_U)$$

for any open set $U \subseteq M$ and which satisfies the following properties

(AL1) *the Leibniz identity:*

$$\forall (\mathfrak{a}_1, \mathfrak{a}_2) \in \Gamma(\mathcal{A}_U)^2, \forall f \in C^\infty(M),\ [\mathfrak{a}_1, f\mathfrak{a}_2]_{\mathcal{A}} = f.[\mathfrak{a}_1, \mathfrak{a}_2]_{\mathcal{A}} + df(\rho(\mathfrak{a}_1)).\mathfrak{a}_2.$$

(AL2) *For any open set $U \subseteq M$ and any $(\mathfrak{a}_1, \mathfrak{a}_2) \in \Gamma(\mathcal{A}_U)^2$, the map*

$$(\mathfrak{a}_1, \mathfrak{a}_2) \mapsto [\mathfrak{a}_1, \mathfrak{a}_2]_{\mathcal{A}_U}$$

only depends on the 1*-jets of* $\mathfrak{a}_1$ *and* $\mathfrak{a}_2$*.*

By abuse of notation, such a sheaf of almost Lie brackets will be denoted $[.,.]_{\mathcal{A}}$.

Remark 3.11

(1) Note that, in finite dimension, it is classical that a Lie bracket $[.,.]_{\mathcal{A}}$ *on an anchored bundle* $(\mathcal{A}, \pi, M, \rho)$ *respects the sheaf of sections of* $\pi : \mathcal{A} \to M$ *and depends on the* 1*-jets of sections (cf. [Marl08]) and so satisfies the conditions* **(AL2)***. From condition* **(AL2)** *and (1) before, it follows that the bracket* $[\mathfrak{a}_1, \mathfrak{a}_2]_{\mathcal{A}_U}$ *also only depends on the* 1*-jets of* $\mathfrak{a}_1$ *and* $\mathfrak{a}_2$ *at any* $x \in U$*.*

(2) In the Banach context, the definition of an almost bracket given in [Pel12] and [CaPe12] imposes the compatibility of the sheaf of sections of π *:* $\mathcal{A} \to M$ *called the "localization property". But, under these conditions, we cannot show that the bracket of two sections only depends on the* 1*-jets of these sections. This is why we impose here the condition* **(AL2)***. Note that the Lie bracket of vector fields satisfies* **(AL1)** *and* **(AL2)***. More generally, a bracket on the Lie algebroid of a Banach-Lie groupoid satisfies the conditions required in Definition 3.76 (cf. [BGJP19]). Therefore, these conditions seem natural.*

(3) The existence of Poisson brackets which depends on jets of order greater than 1 *(cf. [BGT18]) leads naturally to build some examples of localizable Lie brackets for which the assumption* **(AL2)** *in Definition 3.76 is not satisfied.*

By the same arguments as the ones used in finite dimension, when M is smoothly regular, any almost Lie bracket $[.,.]_{\mathcal{A}}$ on an anchored bundle (E, π, M, ρ) satisfies condition **(AL1)** (cf. [CaPe12]) but not necessarily **(AL2)**. If M is not smoothly regular, we cannot prove that any almost bracket on the whole space $\Gamma(\mathcal{A}) \times \Gamma(\mathcal{A})$ is compatible with the sheaf of sections of $\mathcal{A}$ and satisfies the condition **(AL2)**.

Example 3.14 *Assume that there exists a linear connection on the anchored bundle* $\mathcal{A}$ *and let* ∇ *be the associated Koszul covariant derivative. Then*

$$[\mathfrak{a}_1, \mathfrak{a}_2]_{\nabla} = \nabla_{\mathfrak{a}_1} \mathfrak{a}_2 - \nabla_{\mathfrak{a}_2} \mathfrak{a}_1$$

is an almost Lie bracket.

Example 3.15 *If* M *satisfies the Von Neumann bornology approximation property* **(VNAP)***, the condition* **(AL2)** *is always satisfied and so any sheaf of almost Lie brackets on* $\mathcal{A}$ *is an almost Lie bracket. In particular, if* M *is finite-dimensional or a nuclear Fréchet space or a nuclear (LF) space, this situation occurs.*

Example 3.16 *Assume that $\mathcal{A}$ is a closed supplemented sub-bundle of TM and ρ is the natural inclusion of $\mathcal{A}$ in TM. Choose any supplemented bundle $\mathcal{B}$ of $\mathcal{A}$ in TM and let $q_{\mathcal{A}}$ be the natural map from TM to $\mathcal{A}$ whose kernel is $\mathcal{B}$. If $[.,.]$ denotes the Lie bracket of vector fields on M then*

$$[\mathfrak{a}_1, \mathfrak{a}_2]_{\mathcal{A}} = q_{\mathcal{A}}([\mathfrak{a}_1, \mathfrak{a}_2])$$

is an almost Lie bracket on $\mathcal{A}$.

Notation 3.10 *Local expression of an almost Lie bracket.— Consider an open set $U \subset M$ which is a chart domain and over which $\mathcal{A}$ is trivializable. We can identify $J^1(\mathcal{A})$ with $U \times \mathbb{A} \times L(L(\mathbb{M}, \mathbb{A}), \mathbb{A})$. If $L_{\text{alt}}(\mathbb{E}, \mathbb{F})$ denotes the convenient space of bounded skew-symmetric maps on $\mathbb{E}$ with values in $\mathbb{F}$, consider*

$$\text{Hom}_{\text{alt}}\left(j^1(\mathcal{A}), \mathcal{A}\right) := \bigcup_{x \in M} L_{\text{alt}}\left(J^1_x(\mathcal{A}), \mathcal{A}_x\right).$$

Then we have a structure of convenient bundle of $\text{Hom}_{\text{alt}}\left(j^1(\mathcal{A}), \mathcal{A}\right)$ over M whose typical fibre is $L_{\text{alt}}\left(\mathbb{A} \times L\left(L(\mathbb{M}, \mathbb{A}), \mathbb{A}\right), \mathbb{A}\right)$ and which is trivializable over U. Thus, over U, a section of this bundle can be identified with a pair of smooth fields $x \mapsto A_x$ over U where A_x belongs to $L_{alt}\left(\mathbb{A} \times L\left(L(\mathbb{M}, \mathbb{A}), \mathbb{A}\right), \mathbb{A}\right)$. Moreover, such a field A_x satisfies **(AL1)** *if and only if we have*

$$\begin{aligned} & A_x\left((\mathfrak{a}_1, \omega_1 \otimes s_1), (\mathfrak{a}_2, \omega_2 \otimes \mathfrak{a}_2)\right) \\ = \;& \omega_2\left(\rho_x(\mathfrak{a}_1)\right)\mathfrak{a}_2 - \omega_1\left(\rho_x(\mathfrak{a}_2)\right)s_1 + A_x\left((\mathfrak{a}_1, 0), (\mathfrak{a}_2, 0)\right) \end{aligned} \tag{3.14}$$

for all $\omega_i \in \mathbb{M}^*_x$, $\mathfrak{a}_i \in \mathbb{A}_x$ and $i \in \{1, 2\}$. In this way, without loss of generality, we may assume that U is an open set of $\mathbb{M}$ from the local expression of a linear differential endomorphism, we deduce that the local expression of an almost Lie bracket can be written:

$$[\mathfrak{a}_1, \mathfrak{a}_2]_{\mathcal{A}}(x) = d_x\mathfrak{a}_2(\rho(\mathfrak{a}_1(x)) - d_x\mathfrak{a}_1(\rho(\mathfrak{a}_2(x))) + C_x(\mathfrak{a}_1(x), \mathfrak{a}_2(x)) \tag{3.15}$$

for all $\mathfrak{a}_i \in \Gamma(\mathcal{A}_U)$ and $i \in \{1, 2\}$ where and $\mathsf{x} \mapsto \mathsf{C}_\mathsf{x}$ is a smooth map where C_x belongs to $L_{\text{alt}}(\mathbb{A}, \mathbb{A})$. The Leibnitz property is then always satisfied.

3.18.3 Almost Lie Algebroid and Lie Algebroid

Definition 3.77 *The quintuple $(\mathcal{A}, \pi, M, \rho, [.,.]_{\mathcal{A}})$ where $(\mathcal{A}, \pi, M, \rho)$ is an anchored bundle and $[.,.]_{\mathcal{A}}$ an almost Lie bracket is called a convenient almost Lie algebroid.*

In this way, the *Jacobiator* is the $\mathbb{R}$-trilinear map $J_{\mathcal{A}_U} : \Gamma(\mathcal{A}_U)^3 \to \Gamma(\mathcal{A}_U)$ defined, for any open set U in M and any $(\mathfrak{a}_1, \mathfrak{a}_2, \mathfrak{a}_3) \in \Gamma(\mathcal{A}_U)^3$ by

$$J_{\mathcal{A}_U}(\mathfrak{a}_1, \mathfrak{a}_2, \mathfrak{a}_3) = [\mathfrak{a}_1, [\mathfrak{a}_2, \mathfrak{a}_3]_{\mathcal{A}}]_{\mathcal{A}} + [\mathfrak{a}_2, [\mathfrak{a}_3, \mathfrak{a}_1]_{\mathcal{A}}]_{\mathcal{A}} + [\mathfrak{a}_3, [\mathfrak{a}_1, \mathfrak{a}_2]_{\mathcal{A}}]_{\mathcal{A}}$$

Definition 3.78 *A convenient Lie algebroid is a convenient almost Lie algebroid $(\mathcal{A}, \pi, M, \rho, [.,.]_{\mathcal{A}})$ such that the associated Jacobiator $J_{\mathcal{A}_U}$ vanishes identically on each module $\Gamma(\mathcal{A}_U)$.*

Proposition 3.19 *Let $(\mathcal{A}, \pi, M, \rho, [.,.]_{\mathcal{A}})$ be a convenient almost Lie algebroid.*

(1) For any open set $U \subseteq M$ and any $(\mathfrak{a}_1, \mathfrak{a}_2) \in \Gamma(\mathcal{A}_U)^2$, the map

$$(\mathfrak{a}_1, \mathfrak{a}_2) \mapsto \rho([\mathfrak{a}_1, \mathfrak{a}_2]_{\mathcal{A}}) - [\rho(\mathfrak{a}_1), \rho(\mathfrak{a}_2)]$$

only depends on the 1-jet of ρ at any $x \in U$ and the values of $\mathfrak{a}_1$ and $\mathfrak{a}_2$ at x.

(2) If the Jacobiator $J_{\mathcal{A}_U}$ vanishes identically, then we have:

$$\forall (\mathfrak{a}_1, \mathfrak{a}_2) \in \Gamma(\mathcal{A}_U)^2, \ \rho([\mathfrak{a}_1, \mathfrak{a}_2]_{\mathcal{A}}) = [\rho(\mathfrak{a}_1), \rho(\mathfrak{a}_2)]. \tag{3.16}$$

(3) If property (3.16) is true, then $J_{\mathcal{A}_U}$ is a bounded trilinear $C^\infty(U)$ morphism from $\Gamma(\mathcal{A}_U)^3$ to $\Gamma(\mathcal{A}_U)$ which takes values in $\ker \rho$ *over U.*

Proof

(1) Since Point (1) is local, we may assume that M is a c^∞-open set in $\mathbb{M}$ and $\mathcal{A} = M \times \mathbb{A}$. In this context, according to Notation 3.10

$$[\mathfrak{a}_1, \mathfrak{a}_2]_{\mathcal{A}} = d\mathfrak{a}_2(\rho(\mathfrak{a}_1)) - d\mathfrak{a}_1(\rho(\mathfrak{a}_2)) + C(\mathfrak{a}_1, \mathfrak{a}_2). \tag{3.17}$$

On the other hand, the Lie bracket of the vector fields $\rho(\mathfrak{a}_1)$ and $\rho(\mathfrak{a}_2)$ is:

$$[\rho(\mathfrak{a}_1), \rho(\mathfrak{a}_2)] = d\rho(\rho(\mathfrak{a}_1), \mathfrak{a}_2) - d\rho(\rho(\mathfrak{a}_2), \mathfrak{a}_1) + \rho \circ d\mathfrak{a}_2(\rho(\mathfrak{a}_1)) - \rho \circ d\mathfrak{a}_1(\rho(\mathfrak{a}_2)).$$

We then obtain:

$$\rho([\mathfrak{a}_1, \mathfrak{a}_2]_{\mathcal{A}}) - [\rho(\mathfrak{a}_1), \rho(\mathfrak{a}_2)] = \rho(C(\mathfrak{a}_1, \mathfrak{a}_2)) - d\rho(\rho(\mathfrak{a}_1), \mathfrak{a}_2) + d\rho(\rho(\mathfrak{a}_2), \mathfrak{a}_1).$$

From the last member, it follows that $\rho([\mathfrak{a}_1, \mathfrak{a}_2]_{\mathcal{A}}) - [\rho(\mathfrak{a}_1), \rho(\mathfrak{a}_2)]$ depends only on the 1-jet of ρ at any $x \in U$ and the values of $\mathfrak{a}_1$ and $\mathfrak{a}_2$ at x.

(2) For any triple $(\mathfrak{a}_1, \mathfrak{a}_2, \mathfrak{a}_3) \in (\Gamma(\mathcal{A}_U))^3$ and any $f \in C^\infty(U)$, we have (cf. for instance [DuZu05], Lemma 8.1.4):

$$\begin{aligned} J_{\mathcal{A}_U}(\mathfrak{a}_1, \mathfrak{a}_2, f\mathfrak{a}_3) &= [\mathfrak{a}_1, [\mathfrak{a}_2, f\mathfrak{a}_3]_{\mathcal{A}}]_{\mathcal{A}} + [\mathfrak{a}_2, [f\mathfrak{a}_3, \mathfrak{a}_1]_{\mathcal{A}}]_{\mathcal{A}} + [f\mathfrak{a}_3, [\mathfrak{a}_1, \mathfrak{a}_2]_{\mathcal{A}}]_{\mathcal{A}} \\ &= \rho(\mathfrak{a}_1)(\rho(\mathfrak{a}_2)(f))\mathfrak{a}_3 + \rho(\mathfrak{a}_2)(f)[\mathfrak{a}_1, \mathfrak{a}_3]_{\mathcal{A}} \\ &\quad + \rho(\mathfrak{a}_1)(f)[\mathfrak{a}_2, \mathfrak{a}_3]_{\mathcal{A}} + f[\mathfrak{a}_1, [\mathfrak{a}_2, \mathfrak{a}_3]_{\mathcal{A}}]_{\mathcal{A}} \\ &\quad + \rho(\mathfrak{a}_2)(f)[\mathfrak{a}_3, \mathfrak{a}_1]_{\mathcal{A}} + f[\mathfrak{a}_2, [\mathfrak{a}_3, \mathfrak{a}_1]_{\mathcal{A}}]_{\mathcal{A}} \\ &\quad + f[\mathfrak{a}_3, [\mathfrak{a}_1, \mathfrak{a}_2]_{\mathcal{A}}]_{\mathcal{A}} - \rho([\mathfrak{a}_1, \mathfrak{a}_2]_{\mathcal{A}})(f)\mathfrak{a}_3 \\ &= ([\rho(\mathfrak{a}_1), \rho(\mathfrak{a}_2)] - \rho([\mathfrak{a}_1, \mathfrak{a}_2]_{\mathcal{A}}))(f)\mathfrak{a}_3 + fJ_{\mathcal{A}_U}(\mathfrak{a}_1, \mathfrak{a}_2, \mathfrak{a}_3) \end{aligned} \tag{3.18}$$

Assume that $J_{\mathcal{A}_U}$ vanishes on $\Gamma(\mathcal{A}_U)^3$.
It follows that

$$\forall (\mathfrak{a}_1, \mathfrak{a}_2, \mathfrak{a}_3) \in \left(\Gamma(\mathcal{A}_{|U})\right)^3, \forall f \in C^\infty(U), \ J_{\mathcal{A}_U}(\mathfrak{a}_1, \mathfrak{a}_2, f\mathfrak{a}_3) = fJ_{\mathcal{A}_U}(\mathfrak{a}_1, \mathfrak{a}_2, \mathfrak{a}_3)$$

if and only if

$$\forall (\mathfrak{a}_1, \mathfrak{a}_2, \mathfrak{a}_3) \in \Gamma(\mathcal{A}_U)^3, \forall f \in C^\infty(U),\ ([\rho(\mathfrak{a}_1), \rho(\mathfrak{a}_2)] - \rho([\mathfrak{a}_1, \mathfrak{a}_2]_\mathcal{A})(f)\mathfrak{a}_3 = 0$$

which is equivalent to

$$\forall (\mathfrak{a}_1, \mathfrak{a}_2) \in \Gamma(\mathcal{A}_U)^2, \forall f \in C^\infty(U),\ ([\rho(\mathfrak{a}_1), \rho(\mathfrak{a}_2)] - \rho([\mathfrak{a}_1, \mathfrak{a}_2]_\mathcal{A})(f) = 0$$

which is equivalent to

$$\forall (\mathfrak{a}_1, \mathfrak{a}_2) \in \Gamma(\mathcal{A}_U)^2,\ ([\rho(\mathfrak{a}_1), \rho(\mathfrak{a}_2)] - \rho([\mathfrak{a}_1, \mathfrak{a}_2]_\mathcal{A}) = 0.$$

(3) If the property (3.16) is true, then, from (3.18), we have, for any $(\mathfrak{a}_1, \mathfrak{a}_2, \mathfrak{a}_3) \in \left(\Gamma(\mathcal{A}_{|U})\right)^3$ *and any* $f \in C^\infty(U)$,

$$J_{\mathcal{A}_U}(\mathfrak{a}_1, \mathfrak{a}_2, f\mathfrak{a}_3) = f J_{\mathcal{A}_U}(\mathfrak{a}_1, \mathfrak{a}_2, \mathfrak{a}_3).$$

But, we have $\rho \circ J_{\mathcal{A}_U} \equiv 0$ *according to the Jacobi property of the Lie bracket of vector fields and property (3.16). Finally, on the one hand, from the second member of (3.17) since the differential is a bounded morphism of convenient space, (cf. [KrMi97], Theorem 3.18) and* $x \mapsto C_x(.,.)$ *is a smooth field of a bounded linear operators and so is a bounded convenient operator from* $\Gamma(\mathcal{A}_U)^2$ *to* $\Gamma(\mathcal{A}_U)$ *and, on the other hand, from its definition, it follows that* $J_{\mathcal{A}_U}$ *is a convenient bounded operator from* $\Gamma(\mathcal{A}_U)^3$ *to* $\Gamma(\mathcal{A}_U)$*. This completes the proof according to the uniform boundedness principle in [KrMi97], Proposition 30.1.*

Corollary 3.2 *If* $(\mathcal{A}, \pi, M, \rho, [.,.]_\mathcal{A})$ *is a convenient Lie algebroid, then* ρ *induces a morphism of Lie algebras from* $\Gamma(\mathcal{A}_{|U})$ *into* $\mathfrak{X}(U)$*.*

Remark 3.12 *An almost Banach-Lie algebroid such that*

$$\forall (\mathfrak{a}_1, \mathfrak{a}_2) \in \Gamma(U)^2,\ [\rho(\mathfrak{a}_1), \rho(\mathfrak{a}_2)] = \rho([\mathfrak{a}_1, \mathfrak{a}_2]_\mathcal{A},$$

need not be a Banach-Lie algebroid because the Jacobi identity is not necessarily satisfied. In finite dimension, such an almost Lie algebroid is called a pre-Lie algebroid in [Yud16] and [Pel16], but there does not exist a common denomination and some authors use other names (e.g. [PoPo01], [PopP92], [GrJo11]). Again in finite dimension, such types of Lie algebroids fit into the structure of a two-term L_∞*-algebra (cf [Gui75]) and also in the setting of twisted Poisson manifolds as introduced in [SeWe01].*

In the context of local trivializations (§ 3.15.3), there exists a field

$$\begin{array}{cccc} \mathsf{C}: & U & \to & \mathcal{L}(\mathbb{A} \times \mathbb{A}, \mathbb{A}) \\ & \mathsf{x} & \mapsto & \mathsf{C}_\mathsf{x} \end{array}$$

such that for $\mathfrak{a}(x) \equiv (\mathsf{x}, \mathsf{a}(\mathsf{x}))$ and $\mathfrak{a}'(x) \equiv (\mathsf{x}, \mathsf{a}'(\mathsf{x}))$, we have:

$$[\mathfrak{a}, \mathfrak{a}']_U(x) \equiv (\mathsf{x}, \mathsf{C}_\mathsf{x}(\mathsf{a}(\mathsf{x}), \mathsf{a}'(\mathsf{x})) + d\mathsf{a}'(\rho_\mathsf{x}(\mathsf{a}(\mathsf{x}))) - d\mathsf{a}(\rho_\mathsf{x}(\mathsf{a}'(\mathsf{x})))) \quad (3.19)$$

Definition 3.79 *A convenient Lie algebroid* $(\mathcal{A}, \pi, M, \rho, [.,.]_{\mathcal{A}})$ *will be called split if for each* $x \in M$*, the kernel of* $\rho_x = \rho_{|\pi^{-1}(x)}$ *is supplemented in* $\pi^{-1}(x)$*.*

For example, if $\ker \rho_x$ is finite dimensional or finite codimensional for all $x \in M$ or if $\mathbb{A}$ is a Hilbert space, then $(\mathcal{A}, \pi, M, \rho, [.,.]_{\mathcal{A}})$ is split. Another particular situation is the case where the anchor $\rho \equiv 0$ and then $(\mathcal{A}, \pi, M, \rho, [.,.]_{\mathcal{A}})$ is a *Lie algebra Banach bundle.*

3.18.4 Structure of Convenient Partial Lie Algebroid

Given a convenient bundle $\mathcal{A}$ over M, for any c^∞-open set U of M, the set $\Gamma(\mathcal{A}_U)$ has a structure of module over the ring $C^\infty(U)$ of smooth functions on U and also a structure of convenient space (cf. [KrMi97]). If $\mathcal{A}$ is provided with an anchor ρ, we have seen that a Lie bracket on $\mathcal{A}$ is a sheaf of antisymmetric bilinear maps

$$[.,.]_{\mathcal{A}_U} : \Gamma(\mathcal{A}_U) \times \Gamma(\mathcal{A}_U) \to \Gamma(\mathcal{A}_U)$$

for any open set $U \subseteq M$, which satisfies the Leibniz and Jacobi properties and depends on the 1-jets of (local) sections of $\mathcal{A}$. By the way, provided with this Lie bracket, $\Gamma(\mathcal{A}_U)$ has a Lie algebra structure. Moreover, the restriction of ρ to $\Gamma(\mathcal{A}_U)$ gives rise to a Lie algebra morphism into the Lie algebra $\mathfrak{X}(U)$ of vector fields defined on U provided with the Lie bracket of vector fields.

In a more precise termilogy, (i.e. sheaf terminology as recalled in Appendix C), given an anchored bundle $(\mathcal{A}, \pi, M, \rho)$ on M, a Banach-Lie algebroid structure $(\mathcal{A}, \pi, M, \rho, [.,.]_{\mathcal{A}})$ is equivalent to the datum of a sheaf of Lie algebras on the sheaf

$$\Gamma_M(\mathcal{A}) := \{\Gamma(\mathcal{A}_U),\ U \text{ open in } M\}$$

such that, if $\mathfrak{X}_M := \{\mathfrak{X}(U),\ U \text{ open in } M\}$ is the $C^\infty(M)$-sheaf of vector fields on M, the Lie bracket $[.,.]_U$ and ρ satisfy the following conditions:
for any $(\mathfrak{a}, \mathfrak{a}') \in \Gamma(\mathcal{A}_U)^2$, any $f \in C^\infty(U)$ and any open set U in M:

(i) the Lie bracket $[\mathfrak{a}, \mathfrak{a}']_U$ only depends on the 1-jets of $\mathfrak{a}$ and $\mathfrak{a}'$

(ii) $[\mathfrak{a}, f\mathfrak{a}']_U = df(\rho(\mathfrak{a}))\mathfrak{a}' + f[\mathfrak{a}, \mathfrak{a}']_U$

(iii) ρ induces a Lie algebra morphism from $\Gamma(\mathcal{A}_U)$ to $\mathfrak{X}(U)$.

This motives the following Definition:

Definition 3.80 *Let* $(\mathcal{A}, \pi, M, \rho)$ *be a convenient anchored bundle. Given a sheaf* $\mathfrak{C}_M$ *of subalgebras of the sheaf* C^∞_M *of smooth functions on* M*, let* $\mathfrak{P}_M$ *be a sheaf of* $\mathfrak{C}_M$ *modules of sections of* $\mathcal{A}$*. Assume that* $\mathfrak{P}_M$ *can be provided with a structure of Lie algebras sheaf which satisfies, for any open set U in* M*:*

(CPLA1) *the Lie bracket* $[.,.]_{\mathfrak{P}(U)}$ *on* $\mathfrak{P}(U)$ *only depends on the* 1*-jets of sections of* $\mathfrak{P}(U)$*;*

(CPLA2) *for any* $(\mathfrak{a}, \mathfrak{a}') \in (\mathfrak{P}(U))^2$ *and any* $f \in \mathfrak{E}(U)$, *we have the compatibility conditions:*

$$[\mathfrak{a}, f\mathfrak{a}']_{\mathfrak{P}(U)} = df(\rho(\mathfrak{a}))\mathfrak{a}' + f[\mathfrak{a}, \mathfrak{a}']_{\mathfrak{P}(U)}; \tag{3.20}$$

(CPLA 3) ρ *induces a Lie algebras morphism from* $\mathfrak{P}(U)$ *to* $\mathfrak{X}(U)$, *for any open set* U *in* M.

Then $(\mathcal{A}, \pi, M, \rho, \mathfrak{P}_M)$ *is called a convenient partial Lie algebroid. The family* $\{[.,.]_{\mathfrak{P}(U)}, U$ *open set in* $M\}$ *is called a sheaf bracket and is denoted* $[.,.]_{\mathcal{A}}$.
A convenient partial Lie algebroid $(\mathcal{A}, \pi, M, \rho, \mathfrak{P}_M)$ *is called strong if for any* x *in* M, *the stalk*

$$\mathfrak{P}_x = \varinjlim\{\mathfrak{P}(U),\ \varrho_V^U,\ U \text{ open neighbourhood of } x\}$$

is equal to $\pi^{-1}(x)$ *for any* $x \in M$.

Note that this definition is more general than the definition of convenient partial Lie algebroid structure given in [PeCa19] but both definitions are compatible.

As we have seen in the introduction of this paragraph, any convenient Lie algebroid is a convenient partial Lie algebroid. More generally, if $(\mathcal{A}, \pi, M, \rho)$ is a convenient anchored bundle, any convenient subbundle $\mathcal{B}$ of $\mathcal{A}$ such that $(\mathcal{B}, \pi_{\mathcal{B}} = \pi_{|\mathcal{B}}, M, \rho_{\mathcal{B}} = \rho_{|\mathcal{B}}, [.,.]_{\mathcal{B}})$ is a convenient Lie algebroid provided with a structure of convenient partial Lie algebroid on $\mathcal{A}$ which is not strong in general.

Another type of examples of convenient partial Lie algebroids will be described in the context of the projective limit of Banach-Lie algebroids in Chapter 4, § 4.18, the prolongation of a convenient Lie algebroid in Chapter 6, § 6.4.5 and Chapter 7, § 7.2. In fact, both types of examples correspond to convenient strong partial Lie algebroids, but that is not a general situation.

Remark 3.13 *From the definition of a partial Lie algebroid, it follows that in local coordinates, the sheaf of Lie brackets* $[.,.]_{\mathcal{A}}$ *have the same expression as in (3.19).*

We end this section by a result on a general procedure of construction of a Lie bracket on a sheaf $\mathfrak{P}_M$ of $\mathfrak{E}_M$ modules of sections of $\mathcal{A}$, so that $\mathfrak{P}_M$ can be provided with a structure of Lie algebras sheaf which satisfies the assumptions of Definition 3.80. This result will be used in Chapter 6, § 4.5.

Theorem 3.7 *Under the context of Definition 3.80, let* $\mathfrak{P}_M$ *be a sheaf of* $\mathfrak{E}_M$*-modules of sections of* $\mathcal{A}$ *such that, for any open set* U *in* M *and for each* $x \in U$, *there exists an open neighbourhood* $V \subset U$ *with the following properties:*

(CPLAG1) *there exists a subset* $\mathfrak{S}_V$ *of* $\mathfrak{P}_V$ *which is a generating set of* $\mathfrak{P}_V$ *over* $\mathfrak{E}_V$;

(CPLAG2) *each set $\mathfrak{S}_V$ is provided with a Lie bracket $[.,.]_V$ which satisfies the assumptions* **(CPLA1)** *and* **(CPLA2)***;*

(CPLAG3) *for each $\mathfrak{a}$ and $\mathfrak{b}$ in $\mathfrak{S}_V$, we have*

$$\rho([\mathfrak{a},\mathfrak{b}]_V) = [\rho(\mathfrak{a}), \rho(\mathfrak{b})].$$

Then we can extend the Lie bracket $[.,.]_V$ to $\mathfrak{P}_U \times \mathfrak{P}_U$ so that the assumptions of Definition 3.80 are satisfied.

Proof *Fix some $x \in U$ and choose V as in assumptions in Theorem 3.7. We may assume that V is a simply connected chart domain around x and so $\mathcal{A}_V$ is trivilizable. By the way, without loss of generality, we can suppose that V is a c^∞ open set of $\mathbb{M}$ and $\mathcal{A}_V = V \times \mathbb{A}$. Therefore, each section of $\mathcal{A}_V$ is a smooth map from V to $\mathbb{A}$. From the assumption* **(CPLAG2)***, for any $\mathfrak{a}$ and $\mathfrak{b}$ in $\mathfrak{S}_V$, we have*

$$[\mathfrak{a},\mathfrak{b}]_V(z) = d_z\mathfrak{b}(\rho(\mathfrak{a})) - d_z\mathfrak{a}(\rho(\mathfrak{b})) - C_z(\mathfrak{a},\mathfrak{b}) \tag{3.21}$$

where $z \mapsto C(.,.)$ is a smooth map from V to the set of bounded skew-symmetric maps from $\mathbb{A}$ to $\mathbb{A}$.

Choose $\mathfrak{a} \in \mathfrak{P}_V$. On V, we have a decomposition

$$\mathfrak{a} = \sum_{i=1}^{m} f_i \mathfrak{a}_i \tag{3.22}$$

where f_i belongs to $\mathfrak{E}_V$ and $\mathfrak{a}_i$ to $\mathfrak{S}_V$.

For any $\mathfrak{b} \in \mathfrak{S}_U$ we define

$$\begin{aligned}
[\mathfrak{b},\mathfrak{a}]_V(z) &= \sum_{i=1}^{m} \{d_z f_i(\rho(\mathfrak{b}))\mathfrak{a}_i + f_i\left(d_z\mathfrak{a}_i(\rho(\mathfrak{b})) - d_z\mathfrak{b}(f_i\rho(\mathfrak{a}_i)) - C_z(\mathfrak{b},\mathfrak{a}_i)\right)\} \\
&= \sum_{i=1}^{m} \{d_z f_i(\rho(\mathfrak{b}))\mathfrak{a}_i + f_i[\mathfrak{b},\mathfrak{a}_i]_V\} \\
&= d_z\mathfrak{a}(\rho(\mathfrak{b})) - d_z\mathfrak{b}(\rho(\mathfrak{a})) - C_z(\mathfrak{b},\mathfrak{a})
\end{aligned} \tag{3.23}$$

where the last member is a consequence of Remark 3.13. We can note that in the previous relation, the third member is well defined and belongs to $\mathfrak{P}_V$ and also the fourth member only depends on the 1-jets of $\mathfrak{a}$ and $\mathfrak{b}$ at $z \in V$. Moreover, we also have:

from the fourth member for any $g \in \mathfrak{E}_V$, we get the Leibniz property*:* $[g\mathfrak{b},\mathfrak{a}] = -dg(\rho(a))\mathfrak{b} + g[\mathfrak{b},\mathfrak{a}]_V$,

from the third member and assumption **(CPLAG2)***, we obtain* the compatibility of ρ with this bracket*:* $\rho([\mathfrak{b},\mathfrak{a}]_V) = [\rho(\mathfrak{b}), \rho(\mathfrak{a})]$.

In conclusion, the bracket $[\mathfrak{b}, \mathfrak{a}]_V$ *is well defined and takes values in* $\mathfrak{P}_V$*, only depends on the* 1*-jets of* $\mathfrak{a}$ *and* $\mathfrak{b}$ *on* V*, satisfies the Leibniz property and is compatible with* ρ*.*

Now, any $\mathfrak{b} \in \mathfrak{P}_V$ *can be also decomposed as in (3.22). Using analog arguments as previously and the previous property of* $[\mathfrak{b}, \mathfrak{a}]_V$ *for* $b \in \mathfrak{S}_V$*, one can show easily that, for any* $\mathfrak{a}$ *and* $\mathfrak{b}$ *in* $\mathfrak{P}_V$*, we can define the bracket* $[\mathfrak{b}, \mathfrak{a}]_V$ *by*

$$[\mathfrak{b}, \mathfrak{a}]_V = d_z\mathfrak{a}(\rho(\mathfrak{b})) - d_z\mathfrak{b}(\rho(\mathfrak{a})) - C_z(\mathfrak{b}, \mathfrak{a}), \tag{3.24}$$

which only depends on the 1*-jets of* $\mathfrak{a}$ *and* $\mathfrak{b}$ *on* V*, satisfies the Leibniz property and is compatible with* ρ*. It follows that, around any* $x \in U$*, the bracket in the previous relation defines an almost Lie bracket and so we can extend it to a an almost Lie bracket* $[.,.]_U$ *on* $\mathfrak{P}_U$*.*

It remains to show that this bracket satisfies the Jacobi identity. But since ρ *is compatible with the bracket on* $\mathfrak{P}_U$*, according to Proposition 3.19, the Jacobiator* J_U *on* $\mathfrak{P}_U$ *is trilinear relatively to the ring* $\mathfrak{E}_U$*. Thus, for any* $x \in U$ *and any neighbourhood* V *of* x *in* U *(cf. assumptions as in the Theorem 3.7), the Jacobiator is identically zero in restriction to* $\mathfrak{S}_V$*. From the trilinearity of* J_U*, this implies that* $J_U = 0$ *on* $\mathfrak{P}_V$ *around each* $x \in U$*, which implies the result.*

3.18.5 Examples of Lie Algebroids

Example 3.17 *Let* E *be an integrable subbundle of the tangent bundle* $p_M : TM \to M$ *of a Banach manifold* M*. Denote by* $\iota_E : E \to TM$ *the inclusion and by* $[.,.]_E$ *the restriction to* E *of the Lie Bracket of vector fields to* E*. Then if* $\pi = p_{M|E}$ *then* $(E, \pi, M, \iota_E, [.,.]_E)$ *is a Banach-Lie algebroid. In particular* $(TM, p_M, M, Id_{TM}, [.,.])$ *is a Banach-Lie algebroid.*

Example 3.18 *Let* $\psi : M \times G \to M$ *be a right action of a convenient connected Lie group* G *on a convenient manifold* M*. We denote* $\mathfrak{g}$ *the Lie algebra of* G*. We then have a natural Lie algebras morphism* ξ *from* $\mathfrak{g}$ *to* $\mathfrak{X}(M)$ *defined by:*

$$\xi_X(x) = T_{(x,e)}\psi(X, 0).$$

We then have

$$\forall (X, Y) \in \mathfrak{g}^2, \ \xi_{\{X,Y\}} = [\xi_X, \xi_Y]$$

where $\{.,.\}$ *is the Lie bracket of the Lie algebra* $\mathfrak{g}$ *(cf. [KrMi97], 36.12).*
On the trivial bundle $M \times \mathfrak{g}$*, every section can be identified with a mapping* $\sigma : M \to \mathfrak{g}$*. Then a Lie bracket* $\{\{.,.\}\}$ *can be defined on the set of such sections by:*

$$\{\{\sigma_1, \sigma_2\}\}(x) = \{\sigma_1(x), \sigma_2(x)\} + d\sigma_1\left(\xi_{\sigma_2(x)}\right) - d\sigma_2\left(\xi_{\sigma_1(x)}\right).$$

An anchor $\Psi : M \times \mathfrak{g} \to TM$ *is then defined by* $\Psi(x, X) = \xi_X(x)$*.*
So $(M \times \mathfrak{g}, pr_1, M, \Psi, \{\{.,.\}\})$ *is a convenient Lie algebroid.*

Moreover, if G_x is the closed isotropy subgroup of $x \in M$ and if $\mathcal{G}_x \subset \mathfrak{g}$ is its Lie subalgebra, then $\ker \Psi_x = \mathcal{G}_x$.

Example 3.19 *In finite dimension, a Poisson manifold M is defined by the datum of a 2-contravariant antisymmetric tensor Λ whose Schouten bracket with itself vanishes, i.e. $[\Lambda, \Lambda] = 0$. The cotangent bundle T^*M can be endowed with a Lie algebroid structure where*

- *the bracket $\{.,.\}$ is given by:*

$$\forall(\alpha, \beta) \in \Gamma(T^*M)^2, \{\alpha, \beta\} = L_{P(\alpha)}\beta - L_{P(\beta)}\alpha - d(P(\alpha, \beta))$$

  where $P(\alpha)$ is defined by $< \beta, P(\alpha) >= \Lambda(\alpha, \beta)$;

- *The anchor ρ is exactly P previously defined.*

The reader will find in [CaPe12] more examples in the Banach setting.

3.18.6 Derivative Operators

3.18.6.1 Preliminaries

If U is a c^∞-open subset of a convenient space $\mathbb{E}$, the space $C^\infty(U, \mathbb{F})$ of smooth maps from U to a convenient space $\mathbb{F}$ is a convenient space (cf. [KrMi97], 3.7 and 3.11).
The space $L(\mathbb{E}, \mathbb{F})$ of bounded linear maps from $\mathbb{E}$ to $\mathbb{F}$ endowed with the topology of uniform convergence on bounded subsets in $\mathbb{E}$ is a closed subspace of $C^\infty(\mathbb{E}, \mathbb{F})$ and so is a convenient space.
More generally, the set $L^k_{\text{alt}}(\mathbb{E}, \mathbb{F})$ of all bounded k-linear alternating mappings from $\mathbb{E}^k$ to $\mathbb{F}$ endowed with the topology of uniform convergence of bounded sets is a closed subspace of $C^\infty(\mathbb{E}^k, \mathbb{F})$ (cf. [KrMi97], Corollary 5.13) and so $L^k_{\text{alt}}(\mathbb{E}, \mathbb{F})$ is a convenient space.
On the other hand, if $\bigwedge^k(\mathbb{E})$ is the set of alternating k-tensors on $\mathbb{E}$ then $L^k_{\text{alt}}(\mathbb{E}) := L^k_{\text{alt}}(\mathbb{E}, \mathbb{R})$ is isomorphic as a locally convex space to $L^k_{\text{alt}}(\mathbb{E})$ (cf. [KrMi97], Corollary 5.9) and so has a natural structure of convenient space.
Recall that bounded linear maps are smooth (cf. [KrMi97], Corollary 5.5).

Let us consider a convenient vector bundle $\pi : \mathcal{A} \to M$ with typical fibre $\mathbb{A}$. We study the bundle

$$\begin{array}{rccc} \pi^k : & L^k_{\text{alt}}(\mathcal{A}) = \displaystyle\bigcup_{x \in M} L^k_{\text{alt}}(\mathcal{A}_x) & \to & M \\ & (x, \omega) & \mapsto & x \end{array}$$

Using any atlas for the bundle structure of $\pi : \mathcal{A} \to M$, it is easy to prove that $\pi^k : L^k_{\text{alt}}(\mathcal{A}) \to M$ is a convenient vector bundle. The vector space of local sections of $L^k_{\text{alt}}(\mathcal{A}_U)$ is denoted by $\bigwedge^k \Gamma^*(\mathcal{A}_U)$ and is called the set of

k-exterior differential forms on $\mathcal{A}_U$. We denote by $\bigwedge^k \Gamma^*(\mathcal{A})$ the sheaf of sections of $\pi^k : L^k_{\text{alt}}(\mathcal{A}) \to M$ and $\bigwedge \Gamma^*(\mathcal{A}) = \bigcup_{k=0}^{\infty} \bigwedge^k \Gamma^*(\mathcal{A})$ the sheaf of associated graded exterior algebras.

In this section, we assume that $(\mathcal{A}, \pi, M, \rho)$ is an anchored bundle and that $(\mathcal{A}, \pi, M, \rho, \mathfrak{P}_M)$ is a fixed strong partial Lie algebroid. Note that this situation is always satisfied if $(\mathcal{A}, \pi, M, \rho, [.,.]_{\mathcal{A}})$ is a Lie algebroid. However this situation will also occur in the situation of a partial Poisson manifold and the prolongation of a Lie algebroid

3.18.6.2 Insertion Operator

Let $\mathfrak{a}$ be a local section of $\mathcal{A}$ defined on an open set U. As in [KrMi97], 33.10 we have

Proposition 3.20 *The insertion operator $i_{\mathfrak{a}}$ is the graded endomorphism of degree -1 defined by:*

(1) (i) For any function $f \in C^\infty(U)$

$$i_{\mathfrak{a}}(f) = 0 \tag{3.25}$$

(ii) For any $k+1$-form ω (where $k \geq 0$)

$$(i_{\mathfrak{a}}\omega)(\mathfrak{a}_1, \ldots, \mathfrak{a}_k)(x) = \omega(\mathfrak{a}(x), \mathfrak{a}_1(x), \ldots, \mathfrak{a}_k(x)). \tag{3.26}$$

(2) We have

$$i_{\mathfrak{a}}(\omega \wedge \omega') = i_{\mathfrak{a}}(\omega) \wedge \omega' + (-1)^{\deg \omega} \omega \wedge i_{\mathfrak{a}}(\omega').$$

3.18.6.3 Lie Derivative

Proposition 3.21 *For $k \geq 0$, let ω be a local k-form that is an element of $\bigwedge^k \Gamma^*(\mathcal{A}_U)$ for some open set U of M. Given any section $\overline{\mathfrak{a}} \in \mathfrak{P}(U)$, the* Lie derivative *with respect to $\overline{\mathfrak{a}}$ on sections of $\mathfrak{P}(U)$, denoted by $L^\rho_{\overline{\mathfrak{a}}}$, is the graded endomorphism, with degree* 0 *defined in the following way:*
For any function $f \in C^\infty(U)$

$$L^\rho_{\overline{\mathfrak{a}}}(f) = i_{\rho \circ \overline{\mathfrak{a}}}(df) \tag{3.27}$$

For any k-form ω (where $k > 0$)

$$\begin{aligned}\left(L^\rho_{\overline{\mathfrak{a}}}\omega\right)(\mathfrak{a}_1, \ldots, \mathfrak{a}_k)(x) =& L^\rho_{\overline{\mathfrak{a}}}(\omega(\mathfrak{a}_1, \ldots, \mathfrak{a}_k))(x) \\ &- \sum_{i=1}^{k} \omega\left(\mathfrak{a}_1, \ldots, \mathfrak{a}_{i-1}, [\overline{\mathfrak{a}}, \overline{\mathfrak{a}}_i]_{\mathcal{A}}, \mathfrak{a}_{i+1}, \ldots, \mathfrak{a}_k\right)(x)\end{aligned} \tag{3.28}$$

where $\overline{\mathfrak{a}}_i$ is any section of $\mathfrak{P}(U)$ such that $\overline{\mathfrak{a}}_i(x) = \mathfrak{a}_i(x)$ for $i \in \{1, \ldots, k\}$.

Proof *Since the problem is local, we may assume that U is a c^∞-open set in $\mathbb{M}$ over which $\mathcal{A}$ is trivial. Fix some section $\mathfrak{a}$ of $\mathcal{A}$ on U and fix some $x \in U$. After shrinking U if necessary, if f is a smooth function on U, it is clear that (3.27) is well defined.*
For $k > 0$, consider $\omega \in \bigwedge^k \Gamma^*(\mathcal{A}_U)$. Since we have a strong partial Lie algebroid, this implies that for any k-uple $(\mathfrak{a}_1, \dots, \mathfrak{a}_k)$ of local sections on U of $\mathcal{A}$, the value $\omega(\mathfrak{a}_1(x), \dots, \mathfrak{a}_k(x))$ is well defined. Let $\overline{\mathfrak{a}}_i \in \mathfrak{P}(U)$ be such that $\mathfrak{a}_i(x) = \overline{\mathfrak{a}}_i(x)$ for $i \in \{1, \dots, k\}$, we apply the formula (3.28) to $(\overline{\mathfrak{a}}, \overline{\mathfrak{a}}_1, \dots, \overline{\mathfrak{a}}_k)$. In our context, ω is a smooth field over U with values in $L^k_{\text{alt}}(\mathbb{A})$ and each $\overline{\mathfrak{a}}_i$ is a smooth map from U to $\mathbb{A}$. In this way, we have

$$
\begin{aligned}
L^\rho_{\overline{\mathfrak{a}}}\omega\left(\overline{\mathfrak{a}}_1, \dots, \overline{\mathfrak{a}}_k\right)) =& d_x(\omega)(\rho(\overline{\mathfrak{a}}))\left(\overline{\mathfrak{a}}_1, \dots, \overline{\mathfrak{a}}_k\right) \\
&+ \sum_{i=1}^k \omega\left(\overline{\mathfrak{a}}_1(x), \dots, d_x\overline{\mathfrak{a}}_i\left(\rho(\mathfrak{a})\right), \dots, \overline{\mathfrak{a}}_k(x)\right)
\end{aligned}
$$

Since $[.,.]_{\mathfrak{P}(U)}$ only depends on the 1-jets of sections, as for an almost Lie bracket (cf. (3.19)), we have:

$$
\left[\overline{\mathfrak{a}}, \overline{\mathfrak{a}}_i\right]_{\mathcal{A}}(x) = d_x\overline{\mathfrak{a}}_i\left(\rho(\overline{\mathfrak{a}})\right) - d_x\overline{\mathfrak{a}}\left(\rho(\overline{\mathfrak{a}}_i)\right) + C_x\left(\overline{\mathfrak{a}}(x), \overline{\mathfrak{a}}_i(x)\right).
$$

It follows that we have

$$
\begin{aligned}
L^\rho_{\overline{\mathfrak{a}}}\omega\left(\mathfrak{a}_1, \dots, \mathfrak{a}_k\right)(x) = & d_x\omega(\rho(\mathfrak{a}))\left(\overline{\mathfrak{a}}_1, \dots, \overline{\mathfrak{a}}_k)\right)) \\
&+ \sum_{i=1}^k \omega(\overline{\mathfrak{a}}_1(x), \dots, \overline{\mathfrak{a}}_{i-1}(x), d_x\overline{\mathfrak{a}}(\rho(\overline{\mathfrak{a}}_i)), \\
&C_x\left(\overline{\mathfrak{a}}(x), \overline{\mathfrak{a}}_i(x), \overline{\mathfrak{a}}_{i+1}(x)\right), \dots, \overline{\mathfrak{a}}_k(x))
\end{aligned}
$$

which implies that $L^\rho_{\overline{\mathfrak{a}}}\omega$ is a well defined k-skew-symmetric form on $\mathcal{A}_x$ on U since its value only depends on the $1-$jet of $\overline{\mathfrak{a}}$ and of ω and the values of $(\mathfrak{a}_1, \dots, \mathfrak{a}_k)$ at x. Now, as $x \mapsto C_x$ is a smooth map from U to $L^2_{\text{alt}}(\mathbb{A})$ and so is bounded, the differential of functions is a bounded morphism of convenient spaces (cf. [KrMi97], § 3), and ρ is a bounded morphism of convenient spaces, this completes the proof according to the uniform boundedness principle in [KrMi97], Proposition 30.1.

Remark 3.14 *From the relation (3.27), it is clear that the Lie derivative of a function is defined for any section of $\mathcal{A}_U$. Of course this is also true for any k-form on a Lie algebroid. But, for a strong partial Lie algebroid, this is not true for any k-form with $k > 0$, since the last formulae in the previous proof shows clearly that $\left(L^\rho_{\overline{\mathfrak{a}}}\omega\right)$ also depends on the 1-jet of $\overline{\mathfrak{a}}$.*

More generally, as in Proposition 3.21, consider a tensor Q of type $(r, 0)$, that is a section of convenient fibre bundle $\pi_{\widehat{\mathrm{L}^0_r}:} \widehat{\mathrm{L}^0_r}(\mathcal{A}) \to M$ whose typical fibre is $\mathrm{L}_r^0(\mathbb{A})$. For any section local $\overline{\mathfrak{a}}$ of $\mathcal{P}$, *the Lie derivative* $L^{\rho}_{\overline{\mathfrak{a}}}$ *of* Q is the tensor of type $(r, 0)$ defined by

$$\begin{aligned} & L^{\rho}_{\overline{\mathfrak{a}}} Q\left(\mathfrak{a}_1, \ldots, \mathfrak{a}_r, \alpha_1, \ldots, \alpha_s\right)(x) \\ = {} & L^{\rho}_{\overline{\mathfrak{a}}}\left(Q(\mathfrak{a}_1, \ldots, \mathfrak{a}_r,\right)(x) \\ & - \sum_{i=1}^{r} Q\left(\mathfrak{a}_1(x), \ldots, \mathfrak{a}_{i-1}(x), [\overline{\mathfrak{a}}, \overline{\mathfrak{a}}_i]_{\mathcal{A}}(x), \mathfrak{a}_{i+1}, \ldots, \mathfrak{a}_r(x)\right) \end{aligned} \tag{3.29}$$

where $\overline{\mathfrak{a}}_i$ is any section of $\mathfrak{P}(U)$ such that $\overline{\mathfrak{a}}_i(x) = \mathfrak{a}_i(x)$ for $i \in \{1, \ldots k\}$.

Remark 3.15 *Assume that* $(\mathcal{A}, \pi, M, \rho)$ *is provided with an almost Lie bracket* $[.,.]_{\mathcal{A}}$. *Then the Lie derivative* $L^{\rho}_{\mathfrak{a}}\omega$ *is again well defined by an evident adaptation of formulae (3.28) for any local section* $\mathfrak{a}$ *and* k*-form* ω *defined on some open set* U. *The same is true for the Lie derivative of a tensor. Moreover, if* $[.,.]_{\mathcal{A}}$ *induces the Lie bracket on* $(\mathcal{A}, \pi, M, \rho, \mathfrak{P}_M)$ *then the Lie derivative defined in Proposition 3.21 and the previous global one are compatible. Such a situation will occurs in the context of convenient partial Poisson manifold (cf. Chapter 7, § 7.1).*

3.18.6.4 Exterior Derivative

At first, for any function f, we can also define the 1-form $d_\rho f$, by

$$d_\rho f = \rho^t \circ df \tag{3.30}$$

where $\rho^t : T'M \to \mathcal{A}'$ is the transposed mapping of ρ.

The Lie derivative with respect to any local section $\mathfrak{a}$ of $\mathcal{A}$ commutes with d_ρ.

The *exterior differential* on $\bigwedge \Gamma^*(\mathcal{A})$ is defined as:

Proposition 3.22

(1) The exterior differential d_ρ *is the graded endomorphism of degree* 1 *on* $\bigwedge \Gamma^*(\mathcal{A})$ *defined in the following way:*

- *(1) For any function* f, $d_\rho f$ *is defined previously*
- *(2) For* $k > 0$ *and any* k*-form* ω, *the exterior differential* $d_\rho\omega$ *is the unique* $(k+1)$*-form such that, for all* $(\mathfrak{a}_0, \ldots, \mathfrak{a}_k) \in (\Gamma(\mathcal{A}))^{k+1}$,

$$\begin{aligned} & (d_\rho\omega)\left(\mathfrak{a}_0, \ldots, \mathfrak{a}_k\right)(x) \\ = {} & \sum_{i=0}^{q} (-1)^i L^{\rho}_{\overline{\mathfrak{a}}_i}\Big(\omega\left(\mathfrak{a}_0, \ldots, \widehat{\mathfrak{a}_i}, \ldots, \mathfrak{a}_k\right)\Big)(x) \\ & + \sum_{0 \le i < j \le k} (-1)^{i+j}\Big(\omega\left([\overline{\mathfrak{a}}_i, \overline{\mathfrak{a}}_j]_{\mathcal{A}}, \mathfrak{a}_0, \ldots, \widehat{\mathfrak{a}_i}, \ldots, \widehat{\mathfrak{a}_j}, \ldots, \mathfrak{a}_k\right)\Big)(x) \end{aligned} \tag{3.31}$$

where $\overline{\mathfrak{a}}_i$ *is any section of* $\mathfrak{P}(U)$ *such that* $\overline{\mathfrak{a}}_i(x) = \mathfrak{a}_i(x)$ *for any* $i \in \{0, \ldots k\}$.

(2) *For any k-form η and l-form ζ where $(k,l) \in \mathbb{N}^2$, we then have the following property,*

$$d_\rho(\eta \wedge \zeta) = d_\rho(\eta) \wedge \zeta + (-1)^k \eta \wedge d_\rho(\zeta). \tag{3.32}$$

$$d_\rho \circ d_\rho = {d_\rho}^2 = 0. \tag{3.33}$$

Proof

(1) Using the same context as in the proof of Proposition 3.21, on the one hand, in local coordinates, we have

$$\begin{aligned} L^\rho_{\overline{\mathfrak{a}}_i}\Big(\omega\left(\mathfrak{a}_0,\dots,\widehat{\mathfrak{a}_i},\dots,\mathfrak{a}_k\right)\Big)(x) \quad &= d_x\omega\left(\rho(\overline{\mathfrak{a}}_i)\right)\Big(\overline{\mathfrak{a}}_1(x),\dots,\widehat{\overline{\mathfrak{a}}_i},\dots,\overline{\mathfrak{a}}_k)\Big) \\ &+\sum_{i=1}^{k}\omega\left(\overline{\mathfrak{a}}_1(x),\dots,d_x\overline{\mathfrak{a}}_j(\rho(\mathfrak{a}_i)),\dots,\overline{\mathfrak{a}}_k(x)\right). \end{aligned}$$

On the other hand, we have

$$\begin{aligned} &\omega\left(\left[\overline{\mathfrak{a}}_i,\overline{\mathfrak{a}}_j\right]_{\mathcal{A}},\mathfrak{a}_0,\dots,\widehat{\mathfrak{a}_i},\dots,\widehat{\mathfrak{a}_j},\dots,\mathfrak{a}_q\right)(x) \\ = \quad &\omega(d_x\overline{\mathfrak{a}}_i(\rho(\overline{\mathfrak{a}}_j)) - d_x\overline{\mathfrak{a}}_j(\rho(\overline{\mathfrak{a}}_i)) \\ &+C_x(\overline{\mathfrak{a}}_i(x),\overline{\mathfrak{a}}_j(x)),\overline{\mathfrak{a}}_i(x)),\overline{\mathfrak{a}}_j,\mathfrak{a}_0,\dots,\widehat{\mathfrak{a}_i},\dots,\widehat{\mathfrak{a}_j},\dots,\mathfrak{a}_k)(x) \end{aligned}$$

Finally we obtain:

$$\begin{aligned} &\left(d_\rho\omega\right)\left(\mathfrak{a}_0,\dots,\mathfrak{a}_q\right)(x) \\ = \quad &\sum_{i=0}^{q}(-1)^i\, d_x\omega(\rho(\overline{\mathfrak{a}}_i))\Big(\overline{\mathfrak{a}}_1(x),\dots,\widehat{\overline{\mathfrak{a}}_i},\dots,\overline{\mathfrak{a}}_k)\Big) \\ &+(-1)^{i+j}\,\omega\Big(C_x\left(\overline{\mathfrak{a}}_i(x),\overline{\mathfrak{a}}_j(x)\right),\overline{\mathfrak{a}}_0(x),\dots,\widehat{\overline{\mathfrak{a}}_i}(x),\dots,\widehat{\overline{\mathfrak{a}}_j}(x),\dots,\overline{\mathfrak{a}}_k(x)\Big) \end{aligned}$$

Since this value only depends on the 1-jet of ω at x and the values of each $\overline{\mathfrak{a}}_i(x)$ for $i \in \{0,\dots k\}$, it follows that $(d_\rho\omega)$ is a well defined $(k+1)$-form by same arguments as at the end of the proof of Proposition 3.21.

(2) According to the definition of the wedge product, the last formulae in local coordinates for $(d_\rho\omega)(\mathfrak{a}_0,\dots,\mathfrak{a}_k)$ implies clearly relation (3.32).
Since the Lie bracket on $\mathfrak{P}(U)$ satisfies the Jacobi identity for any open set U, and since the differential $d_\rho\omega$ only depends on the 1-jet of ω, as in finite dimension, it follows that $d_\rho(d_\rho\omega) = 0$.

As in finite dimension (cf. [Marl08]), we have the following classical relations between Lie derivative, inner product and exterior differential:

Proposition 3.23 *For $k \geq 1$, let ω be a local k-form that is an element of $\bigwedge^k \Gamma^*(\mathcal{A}_U)$ for some open set U of M. Given any sections $\overline{\mathfrak{a}}, \overline{\mathfrak{b}} \in \mathfrak{P}(U)$, we have*

1.

$$i_{[\overline{\mathfrak{a}},\overline{\mathfrak{b}}]_{\mathcal{A}}}\omega = \left(L^\rho_{\overline{\mathfrak{a}}} \circ i_{\overline{\mathfrak{b}}} - i_{\overline{\mathfrak{b}}} \circ L^\rho_{\overline{\mathfrak{a}}}\right)\omega \tag{3.34}$$

2.

$$L^{\rho}_{\overline{\mathfrak{a}}}\omega = i_{\overline{\mathfrak{a}}}(d_\rho\omega) + d_\rho(i_{\overline{\mathfrak{a}}}\omega) \tag{3.35}$$

Proof *Fix some $x \in U$. For $i = 1, \dots, k-1$ and any section $\mathfrak{a}_i$ of $\mathcal{A}_U$ we take some section $\overline{\mathfrak{a}}_i \in \mathfrak{P}(U)$ such that $\mathfrak{a}_i(x) = \overline{\mathfrak{a}}_i(x)$. Note that the value of each member of (3.34) on $(\mathfrak{a}_1, \cdots, \mathfrak{a}_{k-1})$ at x depends only on the values of $(\mathfrak{a}_1, \cdots, \mathfrak{a}_{k-1})$ at x. So it is sufficient to prove this relation for sections $(\overline{\mathfrak{a}}_1, \cdots, \overline{\mathfrak{a}}_{k-1}$. Now we have:*

$$\begin{aligned}
&\left(L^{\rho}_{\overline{\mathfrak{a}}} \circ i_{\overline{\mathfrak{b}}}\right)\omega(\overline{\mathfrak{a}}_1, \cdots, \overline{\mathfrak{a}}_{k-1}) \\
=\ & L^{\rho}_{\overline{\mathfrak{a}}}\left(\omega(\overline{\mathfrak{b}}, \overline{\mathfrak{a}}_1, \cdots, \overline{\mathfrak{a}}_{k-1})\right) - \sum_{i=1}^{k-1} \omega\left(\overline{\mathfrak{b}}, \overline{\mathfrak{a}}_1, \cdots, [\overline{\mathfrak{a}}, \overline{\mathfrak{a}}_i]_{\mathcal{A}}, \cdots, \overline{\mathfrak{a}}_{k-1}\right) \\
=\ & \left(L^{\rho}_{\overline{\mathfrak{a}}}\omega\right)\left(\overline{\mathfrak{b}}, \overline{\mathfrak{a}}_1, \cdots, \overline{\mathfrak{a}}_{k-1}\right) + \omega([\overline{\mathfrak{a}}, \overline{\mathfrak{b}}]_{\mathcal{A}}, \cdots, \overline{\mathfrak{a}}_1, \cdots, \overline{\mathfrak{a}}_{k-1}) \\
=\ & \left((i_{\overline{\mathfrak{b}}} \circ L^{\rho}_{\overline{\mathfrak{a}}} + i_{[\overline{\mathfrak{a}}, \overline{\mathfrak{b}}]_{\mathcal{A}}})\omega\right)(\overline{\mathfrak{a}}_1, \cdots, \overline{\mathfrak{a}}_{k-1})
\end{aligned}$$

which ends the proof of Point 1.

The proof of Point 2 used analog arguments and is similar to the proof of Proposition 4.2.3 in [Marl08]. This proof is left to the reader as an exercise.

3.18.6.5 Nijenhuis Endomorphisms

In this subsection we only consider the case of a convenient Lie algebroid $(\mathcal{A}, \pi, M, \rho, [.,.]_{\mathcal{A}})$.

Let A be an endomorphism of $\mathcal{A}$. The *Lie derivative of A* with respect to a local section $\mathfrak{a}$ is defined by

$$L^{\rho}_{\mathfrak{a}}A(\mathfrak{b}) = [\mathfrak{a}, A(\mathfrak{b})]_{\mathcal{A}} - A([\mathfrak{a}, \mathfrak{b}]_{\mathcal{A}}) \tag{3.36}$$

for all local or global sections $\mathfrak{b}$ with the same domain as $\mathfrak{a}$.
The *Nijenhuis tensor* of A is the tensor of type $(2,0)$ defined by:

$$N_A(\mathfrak{a}, \mathfrak{b}) = [A\mathfrak{a}, A\mathfrak{b}]_{\mathcal{A}} - A[A\mathfrak{a}, \mathfrak{b}]_{\mathcal{A}} - A[\mathfrak{a}, A\mathfrak{b}]_{\mathcal{A}} - [\mathfrak{a}, \mathfrak{b}]_{\mathcal{A}} \tag{3.37}$$

for all local or global sections $\mathfrak{a}$ and $\mathfrak{b}$ with same domain.

Remark 3.16 *Consider a partial $(\mathcal{A}, \pi, M, \rho, \mathfrak{P}_M)$. If A is a endomorphism of sheaves of $\mathfrak{P}_M$ the same formulae are well defined for any local section $\overline{\mathfrak{a}}$ of $\mathfrak{P}_M$. In this way, we can also define the Nijenhuis N_A endomorphism of sheaves of $\mathfrak{P}_M$.*

3.18.7 Lie Morphism and Lie Algebroid Morphisms

Let $(\mathcal{A}_1, \pi_1, M_1, \rho_1, [.,.]_{\mathcal{A}_1})$ and $(\mathcal{A}_2, \pi_2, M_2, \rho_2, [.,.]_{\mathcal{A}_2})$ be two convenient Lie algebroids.
We consider a bundle morphism $\Psi : \mathcal{A}_1 \to \mathcal{A}_2$ over $\psi : M_1 \to M_2$.

In the one hand, according to [HiMa90], in finite dimension, we can introduce:

Definition 3.81 *Consider a section $\mathfrak{a}_1$ of $\mathcal{A}_1$ over an open set U_1 and a section $\mathfrak{a}_2$ of $\mathcal{A}_2$ over an open set U_2 which contains $\psi(U_1)$. We say that the pair $(\mathfrak{a}_1, \mathfrak{a}_2)$ are ψ-related if we have*
(RS) $\Psi \circ \mathfrak{a}_1 = \mathfrak{a}_2 \circ \psi$

Definition 3.82 Ψ *is called a Lie morphism over ψ if it fulfils the following conditions:*
(LM1) $\rho_2 \circ \Psi = T\psi \circ \rho_1$
(LM2) $\Psi \circ [\mathfrak{a}_1, \mathfrak{a}'_1]_{\mathcal{A}_1} = [\mathfrak{a}_2, \mathfrak{a}'_2]_{\mathcal{A}_2} \circ \psi$ *for all ψ-related pairs of sections $(\mathfrak{a}_1, \mathfrak{a}_2)$ and $(\mathfrak{a}'_1, \mathfrak{a}'_2)$.*

Remark 3.17 *Let $(\mathcal{A}_1, \pi_1, M_1, \rho_1, \mathfrak{P}_1)$ and $(\mathcal{A}_2, \pi_2, M_2, \rho_2, \mathfrak{P}_2)$ be two convenient partial Lie algebroids. Let Ψ be a morphism of anchored bundles from $(\mathcal{A}_1, \pi_1, M_1, \rho_1)$ to $(\mathcal{A}_2, \pi_2, M_2, \rho_2)$ over a smooth map $\psi : M_1 \to M_2$ and such that $\Psi(\mathfrak{P}_1(U_1)) \subset \mathfrak{P}_2(U_2)$ for any open sets U_1 in M_1 and U_2 in M_2 such that $\Psi(U_1) \subset U_2$. Then the relation* **(LM 2)** *has a sense for all ψ-related pairs of sections $(\mathfrak{a}_1, \mathfrak{a}_2)$ and $(\mathfrak{a}'_1, \mathfrak{a}'_2)$ of $\mathfrak{P}_1(U_1) \times \mathfrak{P}_2(U_2)$ such that the previous property of inclusion is true. By the way, in this case we say that Ψ is a Lie morphism of partial Lie algebroid.*

For any local k-form ω on $\mathcal{A}_2$ defined on U_2, we denote by $\Psi^*\omega$ the local k-form on $\mathcal{A}_1$ defined on $U_1 = \psi^{-1}(U_2)$ by:

$$(\Psi^*\omega)_{x_1}(\mathfrak{a}_1 \dots \mathfrak{a}_k) = \omega_{\psi(x_1)}(\Psi(\mathfrak{a}_1), \cdots, \Psi(\mathfrak{a}_k)) \tag{3.38}$$

for all $x_1 \in U_1$.
On the other hand, as classically in finite dimension, we can introduce:

Definition 3.83 Ψ *is a Lie algebroid morphism over ψ if and only if we have*
(LAM1) $\Psi^*(d_{\rho_2} f) = d_{\rho_1}(f \circ \psi)$ *for all $f \in C^\infty(U_2)$;*
(LAM2) $\Psi^*(d_{\rho_2}\omega) = d_{\rho_1}\Psi^*(\omega)$ *for any 1-form ω on $\{\mathcal{A}_2\}_{U_2}$.*

It is easy to see that conditions **(LM1)** and**(LAM1)** are equivalent (cf. Proof of Proposition 3.24). Property **(LM2)** implies property **(LAM2)** for ψ-related sections but, in general, a pair of local sections $(\mathfrak{a}_1, \mathfrak{a}_2)$ of $\mathcal{A}_1 \times \mathcal{A}_2$ are not ψ-related while each member (3.38) for any two such pairs is well defined. On the other hand, under the assumption of **(LM 2)**, we have

$$[\Psi(\mathfrak{a}_1), \Psi(\mathfrak{a}_2)]_{\mathcal{A}_2}(\psi(x_1)) = [\mathfrak{a}'_1, \mathfrak{a}'_2]_{\mathcal{A}_2}(\psi(x_1)) \tag{3.39}$$

For any $x_1 \in U_1$. Therefore the relation **(LAM1)** is satisfied for any such pair $(\mathfrak{a}_1, \mathfrak{a}_2)$ of sections of $\mathcal{A}_1$ which is ψ-related to a pair of sections $(\mathfrak{a}'_1, \mathfrak{a}'_2)$ of sections of $\mathcal{A}_2$. Of course, this property is no longer true for any pair $(\mathfrak{a}_1, \mathfrak{a}_2)$ of local sections of $\mathcal{A}_2$ and so the bracket "$[\Psi(\mathfrak{a}_1), \Psi(\mathfrak{a}'_1)]_{\mathcal{A}_2}(\psi(x_1))$" is not defined. Thus, in general, both definitions are not comparable. However, if ψ is a local diffeomorphism, we have:

Proposition 3.24 *Let $(\mathcal{A}_1, \pi_1, M_1, \rho_1, [.,.]_{\mathcal{A}_1})$ and $(\mathcal{A}_2, \pi_2, M_2, \rho_2, [.,.]_{\mathcal{A}_2})$ be two convenient Lie algebroids. We consider a bundle morphism $\Psi : \mathcal{A}_1 \to \mathcal{A}_2$ over a local diffeomorphisms $\psi : M_1 \to M_2$. Then Ψ is a Lie algebroid morphism if and only if it is a Lie morphism.*

Proof *Since the set of differentials df of smooth maps around $x_2 \in M_2$ is a separating family on $T_{x_2}M_2$, we easily obtain the equivalence* **(LM1)** $\Leftrightarrow$ **(LAM1)**.
At first, note that **(LM2)** *and* **(LAM2)** *are properties of germs. Thus the equivalence is in fact a local problem. Fix some $x_1^0 \in M_1$ and $x_2^0 = \psi(x_1^0)$. Since ψ is a local diffeomorphism, the models of M_1 and M_2 are the same convenient space $\mathbb{M}$; we have a chart (U_1, ϕ_1) and (U_2, ϕ_2) around x_1^0 in M_1 and x_2^0 in M_2 such that for $i \in \{1, 2\}$:*

- $\phi_i(x_i) = 0 \in \mathbb{M}$
- $\phi_2 \circ \psi \circ \phi_1^{-1}$ *is a diffeomorphism between the c^∞-open sets $\mathsf{U}_1 := \phi_1(U_1)$ and $\mathsf{U}_2 := \phi_2(U_2)$*
- *a trivialization $\tau_i\{\mathcal{A}_i\}_{U_1} = U_i \times \mathbb{A}_i$.*

Thus, without loss of generality, we may assume that M_i is a c^∞-open neighbourhood of $0 \in \mathbb{M}$, ψ is a diffeomorphism from M_1 to M_2 and $\mathcal{A}_i = M_i \times \mathbb{A}_i$. By the way, the anchor ρ_i is a smooth map from M_i to $\mathrm{L}(\mathbb{M}, \mathbb{A}_i)$ and each section $\mathfrak{a}_i$ of $\mathcal{A}_i$ is a smooth map from M_i to $\mathbb{A}_i$. Under this context, on the one hand, for all $x_1 \in M_1$, we have

$$\Psi^* d_{\rho_2}\omega(\mathfrak{a}, \mathfrak{a}')(x_1) = d_{\psi(x_1)}\left(\omega(\Psi(\mathfrak{a}'))\left(\rho_2 \circ \Psi(\mathfrak{a})\right)\right) - d_{\psi(x_1)}\left(\omega(\Psi(\mathfrak{a}))\left(\rho_2 \circ \Psi(\mathfrak{a}')\right)\right) - \omega\left([\Psi(\mathfrak{a}), \Psi(\mathfrak{a}')]_{\mathcal{A}_2}\right)(\psi(x_1)).$$

On the other hand, we have

$$d_{\rho_1}\Psi^*\omega(\mathfrak{a}, \mathfrak{a}')(x_1) = d_{\psi(x_1)}\omega\left((\Psi(\mathfrak{a}'))\left(T\psi \circ \rho(\mathfrak{a})\right)\right) - d_{\psi(x_1)}\omega\left((\Psi(\mathfrak{a}))\left(T\psi \circ \rho(\mathfrak{a}')\right)\right) - \omega\left(\Psi([\mathfrak{a}, \mathfrak{a}']_{\mathcal{A}_1}\right)(\psi(x_1)).$$

Note that for any two pairs of ψ-related sections $(\mathfrak{a}_1, \mathfrak{a}'_1)$ and $(\mathfrak{a}_2, \mathfrak{a}'_2)$, as ψ is a diffeomorphism, **(LM 2)** *is equivalent to*

$$[\Psi(\mathfrak{a}_1), \Psi(\mathfrak{a}_2)]_{\mathcal{A}_2}(\psi(x_1)) = \Psi([\mathfrak{a}_1, \mathfrak{a}_2]_{\mathcal{A}_1})(\psi(x_1)) \tag{3.40}$$

for all $x_1 \in M_1$.

Thus, if **(LM1)** *and* **(LM2)** *are true, then, in the previous local context,* **(LAM2)** *is equivalent to*

$$\omega\left(\Psi([\mathfrak{a}_1,\mathfrak{a}_2]_{\mathcal{A}_1})\left(\psi(x_1)\right)\right)=\omega\left([\Psi(\mathfrak{a}_1),\Psi(\mathfrak{a}_2)]_{\mathcal{A}'}\right)\left(\psi(x_1)\right)\qquad(3.41)$$

for any 1*-form* ω *on* $\mathcal{A}_2$ *and any* $x_1\in M_1$*. As* ψ *is a diffeomorphism for any pair of sections* $(\mathfrak{a}_1,\mathfrak{a}_2)$ *of* $\mathcal{A}_1$ *if we set* $\mathfrak{a}'_1=\Psi(\mathfrak{a}_1)\circ\psi^{-1}$ *and* $\mathfrak{a}'_2=\Psi(\mathfrak{a}_2)\circ\psi^{-1}$*, then* $(\mathfrak{a}_1,\mathfrak{a}'_1)$ *and* $(\mathfrak{a}_2,\mathfrak{a}'_2)$ *are* ψ*-related and it follows that* **(LM 1)** *and* **(LM2)** *implies* **(LAM1)** *and* **(LAM2)***.*

Conversely, assume that **(LAM1)** *and* **(LAM2)** *are true. Consider any two pairs of* ψ*-related sections* $(\mathfrak{a}_1,\mathfrak{a}'_1)$ *and* $(\mathfrak{a}_2,\mathfrak{a}'_2)$*. In this case the relation (3.41) evaluated on* $(\mathfrak{a}_1,\mathfrak{a}_2)$ *is equivalent to* **(LAM2)** *for any* 1*-form* ω *on* U_2*. Since around each point in* M_2*, the set of germs of* 1*-forms on* $\mathcal{A}_2$ *is separating for germs of sections of* $\mathcal{A}_2$ *and as* ψ *is a diffeomorphism, this implies that (3.41).*

It follows the relation **(LM2)** *evaluated on the two pairs* $(\mathfrak{a}_1,\mathfrak{a}'_1)$ *and* $(\mathfrak{a}_2,\mathfrak{a}'_2)$ *is satisfied, which ends the proof.*

3.18.8 Admissible Curves

Definition 3.84 *Let* $(\mathcal{A},\pi,M,\rho,[.,.]_{\mathcal{A}})$ *be a convenient Lie algebroid.*
A curve $c:[a,b]\subset\mathbb{R}\to M$ *is called an* $\mathcal{A}$*-admissible curve if there exists a piecewise smooth curve* $\alpha:[a,b]\to\mathcal{M}$ *such that*

$$\begin{cases}\text{For all } t \text{ at which } c \text{ is differentiable, } \rho(\alpha(t))=\dot{c}(t)\\ \pi(\alpha)=c\end{cases}.$$

In these conditions, α *will be called an* $\mathcal{A}$*-lift of* c *and, more generally,* α *will be called an* E*-path.*

We can define an equivalence relation on M associated with $(\mathcal{A},\pi,M,\rho,[.,.]_{\mathcal{A}})$ in the following way:

$x\sim y$ if and only if there exists an $\mathcal{A}$-admissible curve $\gamma:[a,b]\subset\mathbb{R}\to M$ such that $\gamma(a)=x$ and $\gamma(b)=y$.

The equivalent class of x is called the $\mathcal{A}$-orbit of x of the convenient Lie algebroid $(\mathcal{A},\pi,M,\rho,[.,.]_{\mathcal{A}})$.

3.19 Classical Results on Banach Structures Which Are Not True in the Convenient Setting

In this section, we point out some results on Banach structures which are not any more valid on convenient ones. Our references are [Schm04] and [KrMi97].

(1) If $\mathbb{E}$ is a Banach space, then the group $\mathcal{GL}(\mathbb{E})$ of invertible linear continuous maps is open in $\mathcal{L}(\mathbb{E})$. Similarly, the set $\text{Inj}(\mathbb{E})$ (resp. $\text{Surj}(\mathbb{E})$) of continuous linear injective (resp. surjective) maps is open in $\mathcal{L}(\mathbb{E})$ when this set is not empty. These results are no more valid in a convenient space in general.

(2) The previous results in the Banach setting implies that if a smooth map f from an open set U of a Banach space $\mathbb{E}$ into a Banach space $\mathbb{F}$ is such that its differential $D_x f$ is injective (resp. surjective, resp. bijective), then there exists a neighbourhood W of $x \in V$ such that this property is true on W. This property is no more true in the convenient setting in general.

(3) Because the fixed point theorem is not anymore valid beyond Banach spaces, the classical results on existence and uniqueness of solutions of equations (inverse function theorem, implicit function theorem and the Picard-Lindelöf theorem) are not any more true in the convenient framework.
As a classic application, for a given kinematic vector field, integral curves need not exist, and even if they exist, they need not be unique for a given initial value (cf. [KrMi97], 32.12, Examples 1., 2. and 3.)

(4) If G is a Banach-Lie group, then exp defines a local diffeomorphism from a neighbourhood of $0 \in \mathbf{L}(G)$ onto a neighbourhood of $e \in G$. So the exponential map defines a local chart about e, hence the differential structure of G (cf. [Schm04]). This result is no more true in the convenient context ([KrMi97], 36.9.).

(5) A symplectic form Ω on a Banach manifolds M, the canonical operator $\Omega^\flat : TM \to T^*M$ is an isomorphism if and only if the Banach model $\mathbb{M}$ is reflexive. In this case, there exists a Darboux chart for Ω (cf. [Mars72] and [Wei71]). This result is no more true in the convenient context.

One main aim of the two next chapters is to get results which are true in the Banach framework on particular convenient structures: projective and direct limits of Banach ones.

3.20 Notes

There are some interesting books treating the fundamental notion of (locally convex) vector space such as [Bou07], [Choq73], [Jar81], [Kot69], [Kot79], [Scha80], [Tre67] and [RoRo64].
For the notion of tensor product, the references are [Rya02] and [AbTo11].
The first notions of differentiability on locally convex topological vector spaces can be found in [Mica38], [Bas64], [Kel74], [Lesl82]. Frölicher ([Fro81]) and Kriegl ([Kri82]) presented a "more general" differential named as convenient.

For the main structures encountered in this convenient setting (vector spaces, manifolds, bundles, Lie groups and Lie algebras) the main reference is the tome [KrMi97] which includes some further results.
For the notion of (partial) Lie algebroid, the reader is referred to [Pel16] and [CaPe19].

4

Projective Limits

The notion of projective system first appeared in the work of P.S. Alexandroff in 1926 and was generalized by S. Lefschetz.
The infinite differential geometry and particularly the Fréchet framework is a mathematical tool used in mathematical physics.
In many situations, the Fréchet manifold structures are obtained as projective limits of Banach manifolds, e.g. spaces of smooth loops, of smooth maps between finite-dimensional manifolds and of infinite jets.
This chapter is devoted to the approach of differential geometry in the Fréchet context in terms of projective limits of Banach spaces, manifolds, fibre bundles and some Lie structures on them. This approach was initiated by G. Galanis, C. T. J. Dodson, E. Vassiliou and their collaborators and fix, in particular, the problem of the structural group of a linear Fréchet bundle of type fibre $\mathbb{F}$ since the pathological group $GL(\mathbb{F})$ cannot be used.
In this chapter, we will recall some of their results without proofs. These results are completed by the study of some projective limits of Banach geometrical and Lie structures under appropriate assumptions.

4.1 Projective Limits in Categories

Definition 4.1 *Let $(I, \succcurlyeq)$ be a directed set*[1] *and C a category. A projective system (or inverse system) is a family $\mathcal{S} = \left\{\left(X_i, \delta_i^j\right)\right\}_{(i,j)\in I^2,\ j\succcurlyeq i}$ where X_i is an object of the category C and $\delta_i^j : X_j \to X_i$ is a morphism (bonding map) where:*

(PS1) $\forall i \in I, \delta_i^i = \mathrm{Id}_{X_i}$;

(PS2) $\forall\, (i,j,k) \in I^3 : \ k \succcurlyeq j \succcurlyeq i,\ \delta_i^j \circ \delta_j^k = \delta_i^k$.

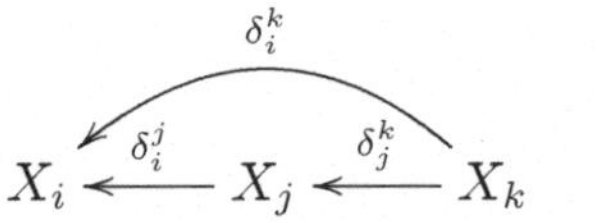

[1] The order relation $\succcurlyeq$ is used in the projective setting; in the direct framework, the order relation will be $\preccurlyeq$.

DOI: 10.1201/9781003435587-4

The projective system $\left\{\left(X_i, \delta_i^j\right)\right\}_{(i,j)\in I^2,\ j\succcurlyeq i}$ is a contravariant functor $I \to \mathsf{C}$ (cf. Appendix B).

When $I = \mathbb{N}$ with the usual order relation, countable projective systems are called *projective sequences*.
The notion of projective limit is defined *via* a universal property.

Definition 4.2 *A cone over $\mathcal{S}$ is a pair $\left(X, \{\delta_i\}_{i\in I}\right)$ where $X \in \mathsf{ob}(\mathsf{C})$ and $\delta_i : X \to X_i$ is such that $\delta_i = \delta_i^j \circ \delta_j$ whenever $j \succcurlyeq i$.*

Definition 4.3 *A cone $\left(X, \{\delta_i\}_{i\in I}\right)$ is a projective limit of $\mathcal{S}$ if, for every cone $\left(Y, \{\psi_i\}_{i\in I}\right)$ over $\mathcal{S}$, there exists a unique morphism $\vartheta : Y \to X$ such that $\psi_i = \delta_i \circ \vartheta$.*
We then write $X = \varprojlim \mathcal{S}$ or $X = \varprojlim X_i$.

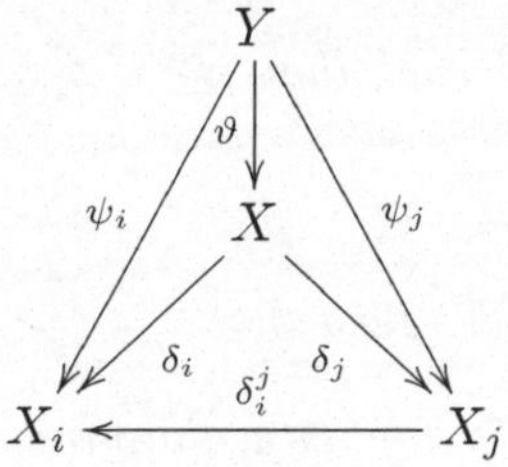

Projective limits in the categories of sets, groups and rings admit the following description:

$$\varprojlim X_i = \left\{ (x_i) \in \prod_{i\in I} E_i : \forall\, (i,j) \in I^2 : j \succcurlyeq i,\ x_i = \delta_i^j\left(x_j\right) \right\}$$

is called the projective limit of the system $\left\{\left(X_i, \delta_i^j\right)\right\}_{(i,j)\in I^2,\ j\succcurlyeq i}$.
In the set category SET, projective limits always exist[2] that is SET is complete.

Example 4.1 *Let p be a prime number. Consider the projective sequence composed of the group factors $\mathbb{Z}/p^n\mathbb{Z}$ and the homomorphisms $\delta_m^n : \mathbb{Z}/p^n\mathbb{Z} \to \mathbb{Z}/p^m\mathbb{Z}$ is the natural projection (cf. [RiZa10]). The inverse limit of this sequence can be identified with*

$$\left\{(x_n)_{n\in\mathbb{N}} : x_n \in \mathbb{Z}, x_n \equiv x_m\ [p^m]\ \textit{if}\ m \leq n\right\}$$

Definition 4.4 *Let $\left\{\left(X_i, \delta_i^j\right)\right\}_{(i,j)\in I^2,\ j\succcurlyeq i}$ and $\left\{\left(Y_i, \eta_i^j\right)\right\}_{(i,j)\in I^2,\ j\succcurlyeq i}$ be two projective families with limit X and Y, respectively. A family $\{f_i\}_{i\in I}$ of mappings $f_i : X_i \to Y_i$ satisfying the coherence condition:*

[2] The projective limit of sets can be empty (cf. [Bou06]).

(CCPM) $\forall (i,j) \in I^2 : j \succcurlyeq i,\ f_i \circ \delta_i^j = \eta_i^j \circ f_j.$

is called a projective family of mappings.
The limit of the projective family of mappings $\{f_i\}_{i\in I}$ *is the map*

$$\begin{array}{rccc} f: & X = \varprojlim X_i & \to & Y = \varprojlim Y_i \\ & \{x_i\}_{i\in I} & \mapsto & \{f_i(x_i)\}_{i\in I} \end{array}$$

4.2 Projective Limits of Topological Spaces

Definition 4.5 *A projective sequence of topological spaces is a sequence* $\left(\left(X_i, \delta_i^j\right)\right)_{(i,j)\in\mathbb{N}^2,\ j\geq i}$ *where*

- **(PSTS1)** *For all* $i \in \mathbb{N}$, X_i *is a topological space;*
- **(PSTS2)** *For all* $(i,j) \in \mathbb{N}^2$ *such that* $j \geq i$, $\delta_i^j : X_j \to X_i$ *is a continuous map;*
- **(PSTS3)** *For all* $i \in \mathbb{N}$, $\delta_i^i = Id_{X_i}$*;*
- **(PSTS4)** *For all* $(i,j,k) \in \mathbb{N}^3$ *such that* $k \geq j \geq i$, $\delta_i^j \circ \delta_j^k = \delta_i^k$.

Notation 4.1 *For the sake of simplicity, the projective sequence* $\left(\left(X_i, \delta_i^j\right)\right)_{(i,j)\in\mathbb{N}^2,\ j\geq i}$ *will be denoted* $\left(X_i, \delta_i^j\right)_{j\geq i}$.

An element $(x_i)_{i\in\mathbb{N}}$ of the product $\prod\limits_{i\in\mathbb{N}} X_i$ is called a *thread* if, for all $j \geq i$, $\delta_i^j(x_j) = x_i$.

Definition 4.6 *The set* $X = \varprojlim X_i$ *of all threads, endowed with the finest topology for which all the projections* $\delta_i : X \to X_i$ *are continuous, is called the projective limit of the sequence* $\left(X_i, \delta_i^j\right)_{j\geq i}$.

A basis of the topology of X is constituted by the subsets $(\delta_i)^{-1}(U_i)$ where U_i is an open subset of X_i (and so δ_i is open whenever δ_i is surjective).

Let $\left\{\left(X_i, \delta_i^j\right)\right\}_{(i,j)\in\mathbb{N}^2,\ j\succcurlyeq i}$ and $\left\{\left(Y_i, \eta_i^j\right)\right\}_{(i,j)\in\mathbb{N}^2,\ j\succcurlyeq i}$ be two projective sequence of topological spaces with limit X and Y, respectively. Consider a family $\{f_i\}_{i\in I}$ of mappings $f_i : X_i \to Y_i$ satisfying the coherence condition **(CCPM)**.
The mapping $f = \varprojlim f_i$ is continuous if all the f_i are continuous (cf. [AbbMa99]).

4.3 Projective Limits of Banach and Normed Spaces

Consider a projective sequence $\left(\mathbb{E}_i, \delta_i^j\right)_{j \geq i}$ of Banach (resp. normed) spaces.

Remark 4.1 *Since we have a countable sequence of Banach (resp. normed) spaces, according to the properties of bonding maps, the sequence* $\left(\delta_i^j\right)_{(i,j)\in\mathbb{N}^2,\, j\geq i}$ *is well defined by the sequence of bonding maps* $\left(\delta_i^{i+1}\right)_{i\in\mathbb{N}}$.

Fix some norm $\| \ \|_i$ on the normed space $\mathbb{E}_i$, for all $i \in \mathbb{N}$. If $x = \varprojlim x_i$, then $p_n(x) = \max\limits_{0\leq i\leq n} \|x_i\|_i$ is a semi-norm on the projective limit $\mathbb{E} = \varprojlim \mathbb{E}_n$ which provides a structure of locally convex topological vector space (cf. for instance [Bell14]). If, moreover, each $\mathbb{E}_i$, is a Banach space for all $i \in \mathbb{N}$, then $\mathbb{E} = \varprojlim \mathbb{E}_n$ have a Fréchet space structure (see [DGV16]). In particular, If each space $\mathbb{E}_i$ is a pre-Hilbert space (resp. Hilbert space) then the projective limit $\mathbb{E} = \varprojlim \mathbb{E}_n$ is called a H-locally convex topological vector space (cf. [Pre69]) (resp. Fréchet Hilbert space (cf. [Zar93]). In this last situation, we have a family of semi-norms on $\mathbb{E}$ which is associated with a semi-inner product on $\mathbb{E}$.

Following [Dub72] we introduce:

Definition 4.7 *A projective sequence* $\left(\mathbb{E}_i, \delta_i^j\right)_{j \geq i}$ *of Banach spaces is called reduced if the range of* δ_i^{i+1} *is dense for all* $i \in \mathbb{N}$.

Example 4.2 *The projective sequence of Banach spaces* $\left(C^i([0,1],\mathbb{R}), \delta_i^{i+1}\right)_{i\in\mathbb{N}}$ *is reduced and* $C^\infty([0,1],\mathbb{R}) = \varprojlim C^i([0,1],\mathbb{R})$.

Definition 4.8 *Two projective sequences* $\left(\mathbb{E}_i^1, \left(\delta_i^j\right)^1\right)_{j\geq i}$ *and* $\left(\mathbb{E}_i^2, \left(\delta_i^j\right)^2\right)_{j\geq i}$ *of Banach spaces are called equivalent if there exist isometries* $A_i : \mathbb{E}_i^1 \to \mathbb{E}_i^2$ *for all* $i \in \mathbb{N}$ *such that*

$$\left(\delta_i^{i+1}\right)^1 = A_i^{-1} \circ \left(\delta_i^{i+1}\right)^2 \circ A_{i+1}.$$

Of course, a projective sequence $\left(\mathbb{E}_i, \delta_i^j\right)_{j \geq i}$ of Banach spaces need not be reduced and, in general, such a sequence is not equivalent to a reduced one. However, by replacing each $\mathbb{E}_i$ by the closure $\overline{\mathbb{E}_i}$ in $\mathbb{E}_i$ of $\delta_i^{i+1}(\mathbb{E}_{i+1})$ and δ_i^{i+1} by the restriction $\overline{\delta_i^{i+1}}$ of δ_i^{i+1} to $\overline{\mathbb{E}_{i+1}}$, we produce a reduced sequence of Banach spaces $\left(\overline{\mathbb{E}_i}, \overline{\delta_i^j}\right)_{j\geq i}$ such that $\varprojlim \mathbb{E}_i = \varprojlim \overline{\mathbb{E}_i}$.
Conversely, any Fréchet space provided with a countable family of semi-norms is topologically isomorphic to the projective limit of a reduced projective sequence.

A particularly important case of projective limit of a reduced projective sequence of Banach spaces corresponds to the case of a decreasing sequence:

$$\mathbb{E}_0 \supset \mathbb{E}_1 \supset \cdots \supset \mathbb{E}_i \supset \mathbb{E}_{i+1} \supset \cdots$$

fulfilling, for any $i \in \mathbb{N}$, the properties:

(DecS1) the inclusion $\iota_i^{i+1} : \mathbb{E}_{i+1} \to \mathbb{E}_i$ is continuous;

(DecS2) $\mathbb{E}_{i+1}$ is dense in $\mathbb{E}_i$.

Then the projective limit $\underleftarrow{\lim}\mathbb{E}_i$ is the intersection $\bigcap\limits_{i\in\mathbb{N}} \mathbb{E}_i$; it is called an inverse limit of Banach spaces or ILB for short (cf. [Omo97]). In fact, any Fréchet space is an ILB space (cf. Appendix A).

4.4 Projective Limits and Linear Functionals

Let $\left(\mathbb{E}_i, \delta_i^j\right)_{j\geq i}$ be a projective sequence of Banach spaces whose projective limit is $\mathbb{E}$.

The topological dual $\mathbb{E}_i^* = \mathcal{L}\left(\mathbb{E}_i, \mathbb{R}\right)$ of $\mathbb{E}_i$ is endowed with the norm $\| \ \|_i^*$ defined (cf. Chapter 1, § 1.3), for any linear functional $\alpha_i \in \mathbb{E}_i^*$ by

$$\|\alpha_i\|_i^* = \sup_{u_i \in \mathbb{E}_i : \|u_i\|_i = 1} |\alpha_i . u_i|$$

Because $\mathbb{E}_i$ and $\mathbb{E}_j$ are normed spaces, the continuity of δ_i^j implies the continuity of $\left(\delta_i^j\right)^*$. Since, for any $(i,j,k) \in \mathbb{N}^3$ such that $k \geq j \geq i$, we have

$$\left(\delta_i^j\right)^* \circ \left(\delta_j^k\right)^* = \left(\delta_i^k\right)^*$$

and so we then get a direct sequence of Banach spaces $\left(\mathbb{E}_i^*, \left(\delta_i^j\right)^*\right)_{i\leq j}$ (cf Chapter 5, § 5.3.1).

Let $(\alpha_i)_{i\in\mathbb{N}}$ be a sequence of continuous linear functionals ($\forall i \in \mathbb{N}$, $\alpha_i \in \mathbb{E}_i^*$) such that the following diagram

$$\begin{array}{ccc}
\mathbb{E}_i & \xleftarrow{\delta_i^j} & \mathbb{E}_j \\
{\scriptstyle \alpha_i}\downarrow & & \downarrow{\scriptstyle \alpha_j} \\
\mathbb{R} & \xleftarrow{\mathrm{Id}} & \mathbb{R}
\end{array}$$

is commutative for any $(i,j) \in \mathbb{N}^2$ such that $j \geq i$.

Since $(\alpha_i : \mathbb{E}_i \to \mathbb{R})_{i\in\mathbb{R}}$ is a projective sequence of maps, its projective limit $\alpha = \varprojlim \alpha_i : \mathbb{E} \to \mathbb{R}$ is well defined (cf. § 4.1). Moreover, since α is linear and continuous, it belongs to $\mathbb{E}^\star$. In order to get another type of dual defined by projective sequences of i-uples of linear functionals, we consider the vector space

$$\mathbb{E}_i^0 = \{\hat{\alpha}_i = (\alpha_0, \dots, \alpha_l, \dots \alpha_i) \in \prod_{l=0}^{i} \mathbb{E}_l : \forall l \in \{0, \dots, i\},\ \alpha_i = \left(\delta_l^i\right)^* (\alpha_l)\}$$

endowed with the norm $\|\ \|_i^0$ defined, for any $\hat{\alpha}_i$, by

$$\|\hat{\alpha}_i\|_i^0 = \sum_{l=0}^{i} \sup_{u_l \in E_l : \|u_l\|_l = 1} |\alpha_l . u_l|$$

For any $(i,j) \in \mathbb{N}^2$ such that $j \geq i$, the projections $\begin{array}{cccc} \pi_i^j : & \mathbb{E}_j^0 & \to & \mathbb{E}_i^0 \\ & \hat{\alpha}_j & \mapsto & \hat{\alpha}_i \end{array}$ are continuous. Since, for any $(i,j,k) \in \mathbb{N}^3$ such that $k \geq j \geq i$, we have $\pi_i^j \circ \pi_j^k = \pi_i^k$, then $\left(\mathbb{E}_i^0, \pi_i^j\right)_{j \geq i}$ is a projective sequence of Banach spaces.

4.5 Projective Limits of Differential Maps

The notion of differentiability used in the framework of Banach spaces has been extended to more general spaces (Fréchet and even locally convex topological vector spaces) in a variety of ways (cf. Chapter 3, § 3.2). In this chapter, the differentiability *à la Leslie* between Fréchet spaces is adopted since it fits well to the setting of projective limits of Banach structures. We then restrict the definitions and results of [Gal96] and [DGV16], 2.2. relative to l.c.t.v.s. to Fréchet spaces.

Let $\mathbb{F}_1$ and $\mathbb{F}_2$ two Fréchet spaces and let U be a non-empty set of $\mathbb{F}_1$.

Definition 4.9 *A continuous map $f : U \to \mathbb{F}_2$ is called differentiable at x if there exists a continuous linear map $df_x : \mathbb{F}_1 \to \mathbb{F}_2$ such that*

$$R(t,v) = \left| \begin{array}{ll} \dfrac{f(x+tv) - f(x) - df_x(tv)}{t} & \text{if } t \neq 0 \\ 0 & \text{if } t = 0 \end{array} \right. .$$

is continuous[3] *at every $(0,v) \in \mathbb{R} \times \mathbb{F}_1$.*

[3]The map $t \mapsto t.R(t,v)$ is tangent to 0 (cf. [Lan95], I, 3 where a definition of differentiability between l.c.t.v.s. is given).

The map f is said to be differentiable if it is differentiable at every $x \in U$. The continuous linear map df_x is called differential of f at x and the map

$$\begin{array}{rccc} df: & U \times \mathbb{F}_1 & \to & \mathbb{F}_2 \\ & (x,v) & \mapsto & df_x(v) \end{array}$$

is called the total differential of f.

As in the differentiation between Banach spaces, the map df_x is uniquely determined.

In contrast to the situation of Banach spaces which involves the Banach space $\mathcal{L}(\mathbb{B}_1, \mathbb{B}_2)$ for the total differential of a map between the Banach spaces $\mathbb{B}_1$ and $\mathbb{B}_2$, the space $\mathcal{L}(\mathbb{F}_1, \mathbb{F}_2)$ cannot be used since it has not any more a structure of Fréchet space.

Definition 4.10 *A map $f : U \to \mathbb{F}_2$ is called C^1-differentiable if it is differentiable at every point $x \in U$ and the total derivative df is continuous.*

The notion of C^n-differentiability ($n \geq 2$) can be defined by induction (cf. [DGV16], Definition 2.2.3).

Definition 4.11 *A map $f : U \to \mathbb{F}_2$ is C^n-differentiable on U if it fulfils the following conditions:*

(1) f is C^{n-1}-differentiable.

(2) For every $x \in U$, there exists a continuous symmetric n-linear map $d^n f_x : \mathbb{F}_1^n \to \mathbb{F}_2$ such that

$$R_n(t,v) = \left| \begin{array}{ll} \dfrac{f(x+tv) - f(x) - df_x(tv) - \cdots - \frac{1}{n!} d^n f_x(t.v, \ldots, t.v)}{t^n} & \text{if } t \neq 0 \\ 0 & \text{if } t = 0 \end{array} \right. .$$

is continuous at every $(0,v) \in \mathbb{R} \times \mathbb{F}_1$.

(3) The differential of order n

$$\begin{array}{rccc} d^n f: & u \times \mathbb{F}_1^n & \to & \mathbb{F}_2 \\ & (x, v_1, \ldots, v_n) & \mapsto & d^n f_x(v_1, \ldots, v_n) \end{array}$$

is continuous.

A map f is called C^∞ or smooth if it is C^n-differentiable for every n.

Remark 4.2 *For Banach spaces, the notion of C^n-differentiability à la Leslie is weaker than the usual Fréchet differentiability but both these notions coincide for smooth functions. Moreover, the differentiation à la Leslie coincides with the usual notion for finite-dimensional topological vector spaces.*

The following proposition (cf. [Gal96], Lemma 1.2 or [DGV16], Proposition 2.3.12) will be used throughout this chapter[4].

[4]In Chapter 5, the differentiability in the convenient setting will be used.

Proposition 4.1 *Let* $\left(\mathbb{E}_i, \delta_i^j\right)_{j\geq i}$ *be a projective sequence of Banach (resp. normed) spaces whose projective limit is the Fréchet space (resp. locally convex space)* $\mathbb{E} = \underleftarrow{\lim}\mathbb{E}_i$. *Given a projective sequence of Banach spaces* $\left(\mathbb{F}_i, \lambda_i^j\right)_{j\geq i}$ *and* $(f_i : \mathbb{E}_i \to \mathbb{F}_i)_{i\in\mathbb{N}}$ *a projective sequence of differential maps whose projective limit* $f = \underleftarrow{\lim} f_i$ *is a map from* $\mathbb{E}$ *to* $\mathbb{F} = \underleftarrow{\lim}\mathbb{F}_i$. *Then the following conditions hold:*

(1) f *is* C^∞ *in the sense of J. Leslie.*

(2) For all $x = (x_i)_{i\in\mathbb{N}}$, $df_x = \underleftarrow{\lim}(df_i)_{x_i}$.

(3) $df = \underleftarrow{\lim} df_i$.

Remark 4.3 *In [Gal96] or [DGV16], this result is proved for projective sequence of Banach space but formally, the proof is the same for projective limit of normed spaces.*

Remark 4.4 *Consider two projective sequences* $\left(\mathbb{E}_i, \delta_i^j\right)_{j\geq i}$ *and* $\left(\mathbb{F}_i, \lambda_i^j\right)_{j\geq i}$ *of Banach spaces and let* $\mathbb{E}$ *and* $\mathbb{F}$ *their respective projective limit. A smooth map* $f : \mathbb{E} \to \mathbb{F}$ *is not in general a projective limit of maps* $f_i : \mathbb{E}_i \to \mathbb{F}_i$ *as shows the following Example.*

Let $\mathbb{R}^\mathbb{N}$ *be the vector space of real valued sequences* $(x_0, x_1, \dots)$. *Provided with the sequence of semi-norms* $(\nu_n)_{n\in\mathbb{N}}$ *where* $\nu_n((x_i)_{i\in\mathbb{N}}) = |x_n|$, $\mathbb{R}^\mathbb{N}$ *has a structure of Fréchet space. If we consider the canonical projective sequence* $(\mathbb{R}^i, \delta_i^j)_{j\geq i}$, *we have* $\mathbb{R}^\mathbb{N} = \underleftarrow{\lim}\mathbb{R}^i$. *Fix some integer* $s > 0$ *and consider the map*

$$\begin{array}{llll} f: & \mathbb{R}^\mathbb{N} & \to & \mathbb{R}^\mathbb{N} \\ & (x_i)_{i\in\mathbb{N}} & \mapsto & (y_i)_{i\in\mathbb{N}} \text{ where } \left| \begin{array}{ll} y_i = 0 & \text{if } 0 \leq i < s \\ y_i = x_{i-s} & \text{if } i \geq s \end{array} \right. \end{array}$$

Since f *is linear and bounded, it is a smooth map (cf. Chapter 3, Theorem 3.5). However, if* f *were the projective limit of a sequence* $(g_i)_{i\in\mathbb{N}}$, *we should have* $g_i \circ \delta_i = \delta_i \circ f$. *Since we have* $g_i \circ \delta_i = \delta_{i+s} \circ f$, *the sequence* $(g_i)_{i\in\mathbb{N}}$ *does not satisfy the compatibility conditions.*

4.6 Projective Limits of Banach Manifolds

Definition 4.12 *The projective sequence* $\left(M_i, \delta_i^j\right)_{j\geq i}$ *is called* projective sequence of Banach manifolds *if*

(PSBM1) *for all* $i \in \mathbb{N}$, M_i *is a manifold modelled on the Banach space* $\mathbb{M}_i$*;*

(PSBM2) $\left(\mathbb{M}_i, \overline{\delta_i^j}\right)_{j \geq i}$ *is a projective sequence of Banach spaces;*

(PSBM3) *for all* $x = (x_i) \in M = \varprojlim M_i$*, there exists a projective sequence of local charts* $(U_i, \varphi_i)_{i \in \mathbb{N}}$ *such that* $x_i \in U_i$ *where one has the relation*

$$\varphi_i \circ \delta_i^j = \overline{\delta_i^j} \circ \varphi_j;$$

(PSBM4) *Under the previous assumptions, if* $U = \varprojlim U_i$ *and* $\phi = \varprojlim \phi_i$*, then* $\phi(U)$ *is an open set of* $\mathbb{M} = \varprojlim \mathbb{M}_i$*.*

Under the assumptions **(PSBM1)** and **(PSBM2)** in Definition 4.12, the pair of assumptions **(PSBM3)**] and **(PSBM4)** around $x \in M$ is called *the projective limit chart property* around $x \in M$ and $(U = \varprojlim U_i, \phi = \varprojlim \phi_i)$ is called a *projective limit chart*.

The projective limit $M = \varprojlim M_i$ has a structure of Fréchet manifold modelled on the Fréchet space $\mathbb{M} = \varprojlim \mathbb{M}_i$ and is called a PLB-*manifold*. The differentiable structure is defined *via* the charts (U, ϕ) where $\phi = \varprojlim \phi_i : U \to (\phi_i(U_i))_{i \in \mathbb{N}}$.
φ is a homeomorphism (projective limit of homeomorphisms) and the charts changings $\left(\psi \circ \varphi^{-1}\right)_{|\varphi(U)} = \varprojlim \left(\left(\psi_i \circ (\phi_i)^{-1}\right)_{|\phi_i(U_i)} \right)$ between open sets of Fréchet spaces are smooth in the sense of convenient spaces[5].

Remark 4.5 *If* M *is the projective limit of the sequence* $\left(M_i, \delta_i^j\right)_{j \geq i}$*, then, as a set,* M *can identified with*

$$\left\{ (x_i)_{i \in \mathbb{N}} \in \prod_{i \in \mathbb{N}} M_i : \forall j \geq i, \; x_i = \delta_i^j(x_j) \right\}.$$

Since each M_i *is a topological space, we can provide* $\prod_{i \in \mathbb{N}} M_i$ *with the product topology and so, since each* δ_i^j *is continuous, it follows that* M *is a closed subset in* $\prod_{i \in \mathbb{N}} M_i$ *which can be provided with the induced topology generated by the open sets of type* $\prod_{i \in \mathbb{N}} V_i \bigcap M$ *where* V_i *is an open set of* M_i *for a finite number of indices* i *and otherwise* $V_i = M_i$*.*
However, under the previous identification, as a Fréchet manifold, the topology on M *is not this projective limit topology according to the assumption* **(PSBM4)***. By the way, this topology is generated by all the sets of projective limits of charts* $(U = \varprojlim U_i)$ *where* U_i *is a chart domain in* M_i *for all* $i \in \mathbb{N}$*.*

[5]They are also differentiable in the sense of Leslie differential calculus (cf. [DGV16], Proposition 3.1.3).

Of course, in general, such a set U is not an open set in the projective limit topology. In fact, if $\delta_i = \varprojlim_{j\geq i}\delta_i^j$, then we have $U = \bigcap_{i\in\mathbb{N}} \delta_i^{-1}(U_i)$. Without the assumption **(PSBM4)**, $\bigcap_{i\in\mathbb{N}} \delta_i^{-1}(U_i)$ *could be reduced to a point.*

Since the composition of two submersions[6] is still a submersion, we have the following result

Proposition 4.2 *Let $\left(M_i, \delta_i^j\right)_{j\geq i}$ and $\left(N_i, \eta_i^j\right)_{j\geq i}$ be two projective sequences of Banach manifolds. Consider a sequence of compatible smooth submersions $(f_i : M_i \to N_i)_{i\in\mathbb{N}}$. Whenever the projective limit of such submersions is defined, it is a smooth submersion in the sense of convenient spaces.*

The sequence $\left(M_i, \delta_i^j\right)_{j\geq i}$ is called *reduced projective sequence of Banach manifolds* if the sequence $\left(\mathbb{M}_i, \bar{\delta}_i^j\right)_{j\geq i}$ is a reduced projective sequence of Banach spaces. Then $\delta_i^j(M_j)$ is dense in M_i for all $j \geq i$. We will say that $\left(M_i, \delta_i^j\right)_{j\geq i}$ is a *reduced projective sequence* and $M = \varprojlim M_i$ is a *reduced* PLB-*manifold*. This situation occurs when the bonding map δ_i^j is a surjective submersion from M_j onto M_i for all $j \geq i$. In this case, we say that $\left(M_i, \delta_i^j\right)_{j\geq i}$ is a *surjective projective sequence* and $M = \varprojlim \mathbb{M}_i$ is a *surjective* PLB-*manifold*. More particular is the situation:

Definition 4.13 *The sequence $\left(M_i, \delta_i^j\right)_{j\geq i}$ is called submersive projective sequence of Banach manifolds if*

(SPSBM1) $\forall (i,j) \in \mathbb{N}^2 : j \geq i$, $\delta_i^j : M_j \to M_i$ *is a surjective submersion;*

(SPSBM2) *Around each $x \in M = \varprojlim \mathbb{M}_i$, there exists a projective limit chart $\left(U = \varprojlim U_i, \varphi = \varprojlim \varphi_i\right)$;*

(SPSBM3) *For all $i \in \mathbb{N}$, there exists a decomposition $\mathbb{M}_i = \ker \bar{\delta}_i^{i+1} \oplus \breve{\mathbb{M}}_i$ such that the following diagram is commutative:*

$$\begin{array}{ccc} U_{i+1} & \xrightarrow{\varphi_{i+1}} & (\ker \bar{\delta}_i^{i+1} \times \breve{\mathbb{M}}_i) \\ {\scriptstyle \delta_i^{i+1}} \downarrow & & \downarrow {\scriptstyle \bar{\delta}_i^{i+1}} \\ U_i & \xrightarrow{\varphi_i} & \mathbb{M}_i \end{array} \tag{4.1}$$

Such a chart is called a submersive projective limit chart around x.

[6]The notion of submersion is defined for convenient maps between convenient manifolds as in [Glo16], replacing open set in a l.c.t.v.s. by c^∞ open set.

The projective limit $M = \underleftarrow{\lim}\mathbb{M}_i$ of a submersive projective sequence $\left(M_i, \delta_i^j\right)_{j \geq i}$ is called a *submersive projective limit of Banach manifolds* or for short a *submersive* PLB-*manifold*. In this case, we have the following results.

Proposition 4.3 *Let $\left(M_i, \delta_i^j\right)_{j \geq i}$ be a surjective (resp. submersive) projective sequence. Then, for each $i \in \mathbb{N}$, the map $\delta_i : M \to M_i$ is surjective (resp. is a submersion).*

Proof *We only prove the case of a submersive projective sequence. Fix some $x \in M = \underleftarrow{\lim}\mathbb{M}_i$. From the definition, for all $i \in \mathbb{N}$, we have a decomposition $\mathbb{M}_i^{i+1} = \ker \bar{\delta}_i \oplus \mathbb{X}_i^{i+1}$ and around each x there exists a projective limit chart with the commutative diagram (4.1).*
This implies that the restriction of $\bar{\delta}_i^{i+1}$ to $\breve{\mathbb{M}}_i$ is an isomorphism. We set $\mathbb{P}_i = \ker \bar{\delta}_i^{i+1}$ and $\mathbb{X}_i = \breve{\mathbb{M}}_i$. Since the kernel of $\bar{\delta}_{i+1}^{i+2}$ is split, we have the decomposition $\mathbb{M}_{i+2} = \ker \bar{\delta}_{i+1}^{i+2} \oplus \breve{\mathbb{M}}_{i+1}$. As the restriction of $\bar{\delta}_{i+1}^{i+2}$ to $\breve{\mathbb{M}}_{i+1}$ is an isomorphism onto $\mathbb{M}_{i+1}$, we have a decomposition

$$\mathbb{M}_{i+2} = \ker \bar{\delta}_{i+1}^{i+2} \oplus_i \breve{\mathbb{X}}_i$$

where the restriction of δ_{i+1}^{i+2} to $\breve{\mathbb{P}}_i$ (resp. $\breve{\mathbb{X}}_i$) is an isomorphism onto $\mathbb{P}_i$ (resp. $\mathbb{X}_i$). Thus $\ker \bar{\delta}_i^{i+2} = \ker \bar{\delta}_{i+1}^{i+2} \oplus \breve{\mathbb{P}}_i$. We set $\mathbb{P}_i^{i+2} = \ker \bar{\delta}_i^{i+2} = \ker \bar{\delta}_{i+1}^{i+2} \oplus \breve{\mathbb{P}}_i$ and $\mathbb{X}_i^{i+2} = \breve{\mathbb{X}}_i$. In particular, the restriction of $\bar{\delta}_i^{i+2}$ to $\mathbb{X}_i^{i+2}$ is an isomorphism onto $\mathbb{M}_i$ with a split kernel and by superposition of diagrams (4.1) for $i \in \{1, 2\}$ we obtain the following diagram

$$\begin{array}{ccc}
U_{i+2} & \xrightarrow{\varphi_{i+2}} & \ker \bar{\delta}_i^{i+2} \times \mathbb{X}_i^{i+2} \\
\delta_i^{i+2} \downarrow & & \downarrow \bar{\delta}_i^{i+2} \\
U_i & \xrightarrow{\varphi_i} & \mathbb{M}_i
\end{array}$$

By induction, we can prove, in the same way, that $\ker \bar{\delta}_i^j$ is split and if $\mathbb{X}_i^j$ is a supplement subspace, we have the following commutative diagram

$$\begin{array}{ccc}
U_j & \xrightarrow{\varphi_j} & \ker \bar{\delta}_i^j \times \mathbb{X}_i^j \\
\delta_i^j \downarrow & & \downarrow \bar{\delta}_i^j \\
U_i & \xrightarrow{\varphi_i} & \mathbb{M}_i
\end{array}$$

It clearly follows that, for a fixed i, by taking the projective limit for such a diagram for $j \geq i$, if we set $\mathbb{X}_i = \underleftarrow{\lim}\mathbb{X}_i^j$, we get the following commutative *diagram around each x*

$$\begin{array}{ccc} U & \xrightarrow{\varphi} & \ker \bar{\delta}_i \times \mathbb{X}_i \\ \delta_i \downarrow & & \downarrow \bar{\delta}_i \\ U_i & \xrightarrow{\varphi_i} & \mathbb{M}_i \end{array}$$

So the proof is completed.

Remark 4.6 *The previous proof shows that in fact each $\delta_i^j : M_j \to M_i$ is a surjective submersion for all $j \geq i$ where $(i,j) \in \mathbb{N}^2$.*

Another important situation of reduced PLB-manifold, is the case of ILB-*manifold* defined as follows:

Definition 4.14 *A* PLB-*manifold $M = \underleftarrow{\lim} \mathbb{M}_i$ is called* ILB-*manifold if*

(ILBM1) $\forall i \in \mathbb{N}$, $M_{i+1} \subset M_i$;

(ILBM2) $\forall i \in \mathbb{N}$, $\delta_i^{i+1} : M_{i+1} \to M_i$ *is the canonical inclusion which is a weak immersion with dense range.*

Note that this definition is stronger than the definition of ILB-manifold in the Omori's sense (see [Omo97] or [Smo06]) since we impose the condition **(PSBM4)**. In this case, $M = \bigcap_{i \in \mathbb{N}} M_i$.

Example 4.3 *Let $\mathbb{F}$ be a Fréchet space (cf. Appendix A) provided with a sequence of semi-norms $(\nu_n)_n \in \mathbb{N}$. The completion $\mathbb{F}_n$ of the normed space $\mathbb{F}/\ker \nu_n$ is a Banach space. We then get a projective sequence $\left(\mathbb{F}_i, \pi_i^j\right)_{j \geq i}$ of Banach spaces whose bonding maps are*

$$\begin{array}{cccc} \pi_i^j : & \mathbb{F}_j & \to & \mathbb{F}_i \\ & [x + \ker \nu_j]_j & \longmapsto & [x + \ker \nu_i]_i \end{array}$$

where $[\ ]_n$ stands for the associated equivalence class. $\mathbb{F}$ will be identified with the projective limit $\underleftarrow{\lim} \mathbb{F}_i$ (cf. [DGV16], Theorem 2.3.8). In fact $\left(\mathbb{F}_i, \pi_i^j\right)_{j \geq i}$ is a reduced projective sequence.

Example 4.4 *Let M be a Banach manifold. We set $M^i = \prod_{k=1}^{i} M_k$ where $M_k = M$ for $k \in \{1, \ldots, i\}$. Then the natural projection*

$$\begin{array}{cccc} \delta_i^j : & M^j & \to & M^i \\ & (x_1, \cdots, x_j) & \mapsto & (x_1, \cdots, x_i) \end{array}$$

where $j \geq i$, is a trivial fibration and so the projective sequence $\left(M^i, \delta_i^j\right)_{j \geq i}$ is submersive.

Example 4.5 *Let* $\mathsf{C}^k(\mathbb{S}^1, M)$ *be the Banach manifold of loops of class* C^k *in a finite-dimensional manifold* M *(cf. Chapter 1, § 1.8.1). The space* $\mathsf{C}^\infty(\mathbb{S}^1, M)$ *of smooth loops in* M *is the projective limit of the sequence* $\left(\mathsf{C}^k(\mathbb{S}^1, M)\right)_{k\in\mathbb{N}}$ *of Banach manifolds. In fact,* $\mathsf{C}^\infty(\mathbb{S}^1, M)$ *is an* ILB*-manifold.*

Example 4.6 *More generally, let* $\mathsf{C}^k(N, M)$ *the Banach manifold of maps of class* C^k *from a compact manifold* N *into a finite-dimensional manifold* M *(cf. Chapter 1, § 1.8.2). The space* $\mathsf{C}^\infty(N, M)$ *of smooth maps from* N *to* M *is the projective limit of the strong projective sequence* $\left(\mathsf{C}^k(N, M)\right)_{k\in\mathbb{N}}$ *of Banach manifolds and is, in fact, an* ILB*-manifold.*

Example 4.7 *Let* D *be a distribution of constant rank* $m + 1$ *on a manifold* M *of dimension* n*. Choose any Riemannian metric* g *on* M*, and we denote by* $S(D, M, g)$ *the unit sphere bundle of* D *associated with the induced Riemannian metric on* D*. So we get a bundle* $\pi : S(D, M, g) \to M$*. On* $S(D, M, g)$*, we consider the antipodal action of* $\mathbb{Z}_2$*. Clearly, the quotient of* $S(D, M, g)$ *by this action can be identified with* $G(D, 1, M)$[7] *and the associated projection* $\tau : S(D, M, g) \to G(D, 1, M)$ *is a bundle morphism over* M*, and also a 2-fold covering. In particular,* τ *is a local diffeomorphism. On* $S(D, M, g)$*, we consider the distribution* $D^{[1]}$ *defined in the following way*

$$D^{[1]}_{(x,\nu)} := \{v \in T_{(x,\nu)}S(D, M, g) : \exists t \in \mathbb{R} : \ d\pi(v) = t\nu \ \}$$

where ν *is a vector in* $D(x)$ *of norm* 1*. For any* $m \geq 2$ *and* $k \geq 1$*, we can define, inductively, a tower of sphere bundles (for a fixed choice of the metric* g *on the manifold* M*):*

$$\hat{P}^0(M) = M \leftarrow \hat{P}^1(M) \leftarrow \cdots \leftarrow \hat{P}^{k-1}(M)\hat{P}^{k-1}(M) \leftarrow \hat{P}^k(M) \leftarrow \cdots \tag{4.2}$$

where, for any $j \in \{0, \cdots, k\}$*,* $\hat{P}^j(M)$ *is a manifold of dimension* $(j+1)m+1$*. Then the tower (4.2) is a sequence of submersive projective Banach manifolds (cf. [PeSl14])*

Example 4.8 *We will use the notations introduced in Chapter 1, § 1.12.2. The sequence* $\left(J^k(\pi), \pi_k^l\right)_{l>k}$ *is a submersive projective sequence and its projective limit is denoted* $J^\infty(\pi)$ *and is called the set of infinite jets of* E*. Let* π_k *be the canonical projection from* $J^\infty(\pi)$ *onto* $J^k(\pi)$ *defined by the sequence* $\left(\pi_k^l\right)_{l>k}$*. A basis of its topology is given by the sets* $\widetilde{W} = \pi_k^{-1}\left(O^k\right)$ *where* O^k *is any open set of* $J^k(\pi)$ *.*

---

[7] The Grassmannian bundle $G(D, 1, M)$ over M is the set

$$G(D, 1, M) := \bigcup_{x\in M} P(D(x), 1), \tag{4.3}$$

where $P(D(x), 1)$ is the projective space of the vector space $D(x)$. Thus, we have a bundle $\pi : G(D, 1, M) \to M$ whose fibre $\pi^{-1}(x)$ is diffeomorphic to the projective space $\mathbb{RP}^m$.

Thus, we provide $J^\infty(\pi)$ of a Fréchet manifold structure where a local chart is a projective limit of local charts $\left(\widetilde{U^k}, \widetilde{\varphi^k}\right)$ where $\widetilde{U^k} = \left(\pi^k\right)^{-1}(U_0)$. We will denote by $\left(x^i, u^\alpha, u_i^\alpha, \dots, u_{i_1\dots i_k}^\alpha, \dots\right)$ the associated local coordinates $J^\infty(\pi)$ which will also be written $[x, u]$.

The reader will find more examples (e.g. the groups $C^0(\mathbb{R}, \mathbb{R}+)$ and $C^\infty(M, G)$ where M is a compact manifold and G a finite-dimensional Lie group) in [DGV16].

4.7 Projective Limits of Banach-Lie Groups

4.7.1 PLB-Lie Group

Definition 4.15 *A group G is called projective limit of Banach-Lie groups modelled on the projective limit $\mathbb{G} = \varprojlim \mathbb{G}_i$ if $G = \varprojlim G_i$ where $\left(G_i, \delta_i^j\right)_{j \geq i}$ is a projective sequence of Banach-Lie groups where G_i is modelled on $\mathbb{G}_i$ and where, for all $i \in \mathbb{N}$, there exists a chart (U_i, φ_i) centered at the unity $e_i \in G_i$ such that:*

(PLBLG1) $\forall (i,j) \in \mathbb{N}^2 : j \geq i,\ \delta_i^j(U_j) \subset U_i$;

(PLBLG2) $\forall (i,j) \in \mathbb{N}^2 : j \geq i,\ \overline{\delta_i^j} \circ \varphi_j = \varphi_j \circ \delta_i^j$;

(PLBLG3) $\varprojlim \varphi_i(U_i)$ *is a non-empty open set of $\mathbb{G}$ and $\varprojlim U_i$ is open in G according to the projective limit topology.*

Under these assumptions, following Galanis' terminology (cf. [Gal96]), we say that G is a PLB-Lie group. When G is an ILB manifold, then G will be called an ILB-Lie group in order to be coherent with our definition of ILB-manifold (cf. Definition 4.14).

Remark 4.7 *Note that our definition of* ILB*-Lie group does not confuse the terminology used in [Omo97] since, in this book, Omori only defines the notion of "strong* ILB*-Lie group". Of course, any strong* ILB*-Lie group is an* ILB *group in the previous sense.*

Following [Gal96], recall that:

Theorem 4.1 *Let $G = \varprojlim G_i$ be a* PLB*-Lie group. Then we have the following properties:*

(1) G is a Fréchet-Lie group.

(2) If $\mathbf{L}(G_i)$ is the Lie algebra of G_i then $\mathbf{L}(G) = \varprojlim \mathbf{L}G_i$.

(3) If $\exp_{G_i}$ is the exponential map for G_i, then $\exp_G = \varprojlim \exp_{G_i}$ is the exponential map of the Fréchet-Lie group G.

Remark 4.8 *For each $i \in \mathbb{N}$, $\exp_{G_i} : \mathbf{L}(G_i) \to G_i$ is a local diffeomorphism from a neighbourhood U_i of $0 \in \mathbf{L}(G_i)$ onto an open neighbourhood V_i of the neutral e_i in G_i. By a good choice of each U_i, the sequence $(U_i)_{i\in\mathbb{N}}$ gives rise to a sequence of projective open sets but, in general, this projective limit is not an open set in $\mathbf{L}(G)$ and it can be reduced to $\{0\}$. Even if this projective limit U is open, then $\exp_{G|U}$ will not be in general a local diffeomorphism around $0 \in \mathbf{L}(G)$.*

Example 4.9 *The space of real sequences $\mathbb{R}^{\mathbb{N}}$, equipped with the product topology is an abelian Lie group, projective limit of the sequence $\left(\mathbb{R}^j, +\right)_{j\in\mathbb{N}}$ of abelian Lie groups.*

Example 4.10 *For any Fréchet space $\mathbb{E}$ there exists a projective sequence $\left(\mathbb{E}_i, \overline{\delta}_i^j\right)_{j\geq i}$ of Banach spaces such that $\mathbb{E} = \varprojlim E_i$; if we consider the sequence of abelian groups $(\mathbb{E}_i, +)_{i\in\mathbb{N}}$, $\mathbb{E}$ is an* ILB *group.*

Example 4.11 *Let $C^\infty(M, G)$ be the set of smooth maps from a compact manifold M into a finite-dimensional Lie group. Then if we provide $C^\infty(M, G)$ with the pointwise operation from the group structure on G, we have*

$$C^\infty(M, G) = \varprojlim C^k(M, G) = \bigcap_{k\in\mathbb{N}} C^k(M, G).$$

Thus $C^\infty(M, G)$ has a ILB*-group structure (cf. [Gal96]).*

Example 4.12 *Of particular interest is the notion of* pro-Lie group *which is a projective limit of a sequence of finite-dimensional Lie groups. From a topological point of view, locally compact groups can be seen as a first generalization of finite-dimensional Lie groups. According to a classical theorem of Yamabe, each locally compact group contains an open subgroup which is a projective limit of finite-dimensional Lie groups. However, not every pro-Lie group is locally compact. The theory of pro-Lie groups has been described in details in [HoMo07]. But any pro-Lie group is not necessarily a* PLB*-Lie group. In [HoNe08], the authors show that a pro-Lie group G is a Fréchet-Lie group (i.e. a* PLB*-Lie group) if and only if G is locally contractible and they also give many characterizations of these* PLB*-Lie groups. Note that such pro-Lie groups are locally exponential (i.e. $\exp_G$ is a local diffeomorphism).*

4.7.2 Regularity of PLB-Lie Group

We have already seen that (cf. Chapter 3, § 3.16.4) that any strong ILB-Lie group is a regular Fréchet Lie group. We will show that so are any PLB-Lie group.

We will say that $\left(\mathfrak{g}_i, \overline{\delta}_i^j\right)_{j\geq i}$ is *a projective sequence of Banach-Lie algebras* if $\left(\mathfrak{g}_i, \overline{\delta}_i^j\right)_{j\geq i}$ is a projective sequence of Banach spaces such that

$\bar{\delta}_i^j$ is a Lie algebra morphism. Given any projective sequence $\left(\mathfrak{g}_i, \bar{\delta}_i^j\right)_{j \geq i}$ of Banach-Lie algebras, it follows that $\mathfrak{g} = \varprojlim \mathfrak{g}_i$ is naturally provided with a Fréchet Lie algebra structure in an evident way (cf. [Las57]) and is called *a* PLB-*Lie algebra*. Note that the Lie algebra of a PLB-Lie group is a PLB-Lie algebra (cf. Theorem 4.1). Since any pro-Lie algebra is a PLB-Lie algebra, according to [HoNe08], the converse is not true in general for enlargeable projective sequence of Banach-Lie algebras $\left(\mathfrak{g}_i, \bar{\delta}_i^j\right)_{j \geq i}$.

Consider a projective sequence $\left(\mathfrak{g}_i, \bar{\delta}_i^j\right)_{j \geq i}$ of Banach-Lie algebras. Denote by $C^\infty(\mathbb{R}, \mathfrak{g}_i)$ (resp. $C^\infty(\mathbb{R}, \mathfrak{g})$) the space of smooth curves from $\mathbb{R}$ to $\mathfrak{g}_i$ (resp. to $\mathfrak{g}$). From [DGV16] (3.1.5), it follows that $C^\infty(\mathbb{R}, \mathfrak{g}) = \varprojlim C^\infty(\mathbb{R}, \mathfrak{g}_i)$. So if we provide it with the pointwise Lie bracket, then the pointwise Lie bracket on $C^\infty(\mathbb{R}, \mathfrak{g})$ is the projective limit of the Lie brackets on $C^\infty(\mathbb{R}, \mathfrak{g}_i)$. Then we obtain:

Theorem 4.2 *Each* PLB-*Lie group is a regular Fréchet-Lie group.*

Proof *Consider a* PLB-*Lie group* $G = \varprojlim G_i$ *and* $\mathbf{L}(G) = \varprojlim \mathbf{L}(G_i)$. *Choose any* $\xi \in C^\infty(\mathbf{R}, \mathbf{L}(G)) = \varprojlim C^\infty(\mathbb{R}, \mathbf{L}(G_i))$. *Then there exists* $\xi_i \in C^\infty(\mathbb{R}, \mathbf{L}(G_i))$ *such that* $\xi = \varprojlim \xi_i$. *Since* G_i *is regular, in the sense of Definition 3.66 (cf. Remark 3.10), it follows that* G *is regular.*

4.8 Projective Limits of Banach and Normed Vector Bundles

Let $\left(M_i, \delta_i^j\right)_{j \geq i}$ be a projective sequence of Banach manifolds where each manifold M_i is modelled on the Banach space $\mathbb{M}_i$.
For any integer i, let (E_i, π_i, M_i) be the Banach (resp. normed) vector bundle whose type fibre is the Banach (resp. normed) vector space $\mathbb{E}_i$ where $\left(\mathbb{E}_i, \lambda_i^j\right)_{j \geq i}$ is a projective sequence of Banach (resp. normed) spaces.

Definition 4.16 $\left((E_i, \pi_i, M_i), \left(\xi_i^j, \delta_i^j\right)\right)_{j \geq i}$, *where* $\xi_i^j : E_j \to E_i$ *is a morphism of vector bundles, is called a projective sequence of Banach vector bundles on the projective sequence of manifolds* $\left(M_i, \delta_i^j\right)_{j \geq i}$ *if there exists a projective sequence of Banach spaces* $\left(\mathbb{E}_i, \lambda_i^j\right)_{j \geq i}$ *such that:*

for all (x_i), *we have a projective sequence of trivializations* (U_i, τ_i) *of* (E_i, π_i, M_i), *where* $\tau_i : (\pi_i)^{-1}(U_i) \to U_i \times \mathbb{E}_i$ *are local diffeomorphisms, where*

U_i is an open neighbourhood of x_i in M_i) and where $U = \varprojlim U_i$ is a non-empty open set in M and, for all $(i,j) \in \mathbb{N}^2$ such that $j \geq i$,, the following compatibility condition is satisfied

(PLBVB) $\left(\delta_i^j \times \lambda_i^j\right) \circ \tau_j = \tau_i \circ \xi_i^j.$

With the previous notations, $(U = \varprojlim U_i, \tau = \varprojlim \tau_i)$ is called a *projective bundle chart limit*. The triple of projective limit $(E = \varprojlim E_i, \pi = \varprojlim \pi_i, M = \varprojlim M_i))$ is called a *projective limit of Banach (resp. normed) bundles* or PLB-bundle (resp. NLB-bundle) for short.

The following proposition generalizes the result of [Gal04] about the projective limit of tangent bundles to Banach manifolds. Its proof can be found in [AgSu10]. Note that for projective sequences of normed bundles the proof is formally the same as for the Banach case.

Proposition 4.4 *Let $\left((E_i, \pi_i, M_i), \left(f_i^j, \delta_i^j\right)\right)_{j \geq i}$ be a projective sequence of Banach (resp. normed) vector bundles.*
Then $\left(\varprojlim E_i, \varprojlim \pi_i, \varprojlim M_i\right)$ is a Fréchet vector bundle (resp. locally convex vector bundle).

Notation 4.2 *From now on and for the sake of simplicity, the projective sequence of vector bundles $\left((E_i, \pi_i, M_i), \left(f_i^j, \delta_i^j\right)\right)_{j \geq i}$ will be denoted $(E_i, \pi_i, M_i)_{\underleftarrow{i}}$.*

If for all $i \in \mathbb{N}$, the typical fibre $\mathbb{E}_i$ of the vector bundle (E_i, π_i, M_i) is a Hilbert space (resp. pre-Hilbert space), then $\left(\varprojlim E_i, \varprojlim \pi_i, \varprojlim M_i\right)$ is a Fréchet Hilbert vector bundle (resp. H-locally convex vector bundle).

Remark that $\mathcal{GL}(\mathbb{E})$ cannot be endowed with a structure of Lie group. So it cannot play the rôle of structural group. We then consider, as in [Gal98], the generalized Lie group $H^0(\mathbb{E}) = \varprojlim H_i^0(\mathbb{E})$ which is the projective limit of the Banach-Lie groups

$$H_i^0(\mathbb{E}) = \left\{ (h_1, \ldots, h_i) \in \prod_{j=1}^{i} \mathcal{GL}(\mathbb{E}_j) : \lambda_k^j \circ h_j = h_k \circ \lambda_k^j, \text{ for } k \leq j \leq i \right\}.$$

Example 4.13 *As a particular case of Proposition 4.4, we can consider the projective sequence of tangent bundles $\left((E_i, \pi_i, M_i), \left(T\delta_i^j, \delta_i^j\right)\right)_{j \geq i}$ of a projective sequence of Banach manifolds $(M_i, \delta_i^j)_{j \geq i}$. Thus, if each M_i is modelled on the Banach space $\mathbb{M}_i$, $\left(\varprojlim TM_i, \varprojlim \pi_i, \varprojlim M_i\right)$ is a Fréchet vector bundle* *whose typical fibre is* $\mathbb{M} = \varprojlim \mathbb{M}_i$ *with structural group* $H^0(\mathbb{M})$*. As we have already seen, this result was firstly proved in [Gal04].*

Remark 4.9 *We have seen in Remark 4.5 that* $M = \varprojlim M_i$ *can be considered as a subset of* $\prod\limits_{i\in\mathbb{N}} M_i$*. In the same way, we can consider* $TM = \varprojlim TM_i$ *as the subset*

$$\left\{(x_i, u_i) \in \prod_{i\in\mathbb{N}} TM_i : \forall j \geq i,\ (x_i, u_i) = (\delta_i^j(x_j), T\delta_i^j(u_j)), \right\}.$$

As in § 4.6, we introduce

Definition 4.17 *A sequence* $(E_i, \pi_i, M_i)_{\underleftarrow{i}}$ *is called a submersive projective sequence of Banach vector bundles if* $(E_i, \pi, M_i)_{\underleftarrow{i}}$ *is a submersive projective sequence of Banach manifolds and if around each* $x \in M$*, there exists a projective limit chart bundle* $(U = \varprojlim U_i, \tau = \varprojlim \tau_i)$ *such that for all* $i \in \mathbb{N}$*, we have a decomposition* $\mathbb{E}_{i+1} = \ker \overline{\lambda}_i^{i+1} \oplus \mathbb{E}'_i$ *such that the condition* **(PLBVB)** *is true.*

The projective limit (E, π, M) of a projective sequence of Banach vector bundles $(E_i, \pi, M_i)_{\underleftarrow{i}}$ is called a *submersive projective limit of Banach bundles* or *submersive* PLB-*bundle* for short.

Now, we have the following result whose proof is similar to the proof of Proposition 4.3:

Proposition 4.5 *Let* $(E_i, \pi_i, M_i)_{\underleftarrow{i}}$ *be a submersive projective sequence of Banach bundles. Then, for each* $i \in \mathbb{N}$*, the map* $\lambda_i : E \to E_i$ *is a submersion.*

Example 4.14 *Let* $\left(E_i^1, \pi_i^1, M_i\right)_{\underleftarrow{i}}$ *and* $\left(E_i^2, \pi_i^2, M_i\right)_{\underleftarrow{i}}$ *be projective sequences of Banach bundles. If* $\mathbb{E}_i^1$ *and* $\mathbb{E}_i^2$ *is the typical fibre of* E_i^1 *and* E_i^2*, respectively, the set* $\mathcal{L}(E_i^1, E_i^2)$ *of morphisms from* E_i^1 *to* E_i^2 *is a Banach bundle over* M_i *with typical fibre* $\mathcal{L}(\mathbb{E}_i^1, \mathbb{E}_i^2)$*. Then it is clear that* $\left(\mathcal{L}(E_i^1, E_i^2), \pi_i, M_i\right)$ *is a projective sequence and* $\left(\mathcal{L}(E^1, E^2) = \varprojlim \mathcal{L}(E_i^1, E_i^2), \pi, M\right)$ *is a Fréchet bundle. Moreover,* $\mathcal{L}(E)$ *is isomorphic to*

$$\left\{(A_n) \in \prod_{n\in\mathbb{N}} \mathcal{L}(E_i^1, E_i^2) :\ \forall m \geq n,\ A_n \circ (\lambda_1)_n^m = (\lambda_2)_n^m \circ A_m, \right\}$$

and so the typical fibre of $\mathcal{L}(E^1, E^2)$ *is* $\mathcal{H}(\mathbb{E}^1, \mathbb{E}^2)$ *where* $\mathbb{E}^1 = \varprojlim \mathbb{E}_i^1$ *and* $\mathbb{E}^2 = \varprojlim \mathbb{E}_i^2$*.*

Example 4.15 *Let* $(M_i, \delta_i^j)_{j \geq i}$ *be a submersive projective sequence of Banach manifolds. Then* $(TM_i, p_{M_i}, M_i)_{\underleftarrow{i}}$ *is a submersive projective sequence of Banach bundles.*

Example 4.16 *Let* $p : E \to M$ *a vector bundle of finite rank over the finite-dimensional manifold* M. *The space* $J^k(E, M)$ *of sections of* E *of class* C^k *has a structure of Banach manifold (cf [Pal68]). The space* $J^\infty(E, M)$ *of infinite jets of sections of* E *is the projective limit of the sequence* $\left(J^k(E, M)\right)_{k \in \mathbb{N}}$ *of Banach manifolds (cf. [Sau89], [AbbMa99]). In fact, it is a submersive projective sequence of Banach bundles (cf. § 4.9).*

4.9 The Infinite Jet Bundle

4.9.1 Infinite Jets of Sections of a Banach Vector Bundle

We use the notations of Chapter 2, § 1.12.1.
For a section s of $\pi : E \to M$ and $l \geq k \geq 0$, we consider the maps

$$\begin{array}{cccc} \pi_k^l : & J^l(\pi) & \to & J^k(\pi) \\ & j_x^l s & \longmapsto & j_x^k s \end{array}.$$

We then have the following result whose proof can be found in [DGV16], 6.3.

Proposition 4.6 $\left(J^k(\pi), \pi_k^l\right)_{\underleftarrow{k}}$ *is a submersive projective sequence of Banach vector bundles and the projective limit*

$$J^\infty(\pi) = \varprojlim J^k(\pi)$$

can be endowed with a structure of Fréchet vector bundle whose fibre is isomorphic to the Fréchet space

$$\mathbb{E} \times \mathcal{L}(\mathbb{M}, \mathbb{E}) \times \mathcal{L}^2_{\mathrm{sym}}(\mathbb{M}, \mathbb{E}) \times \cdots$$

Definition 4.18 *The projective limit* $J^\infty(\pi) = \varprojlim J^k(\pi)$ *is called the infinite jet bundle of* π.

4.9.2 Infinite Jets of Sections of a Finite-Dimensional Vector Bundle

We use the notions and notations of Chapter 2, § 1.12.2.
So we consider a smooth vector bundle $\pi : E \to M$ with total space E and base space M of dimension n with the type fibre is m dimensional.
The local coordinates will be denoted $\left(x^i, u^\alpha, u_i^\alpha, \dots, u_{i_1 \dots i_k}^\alpha, \dots\right)$ in an adapted local chart of $J^\infty(\pi)$.

Using the natural embeddings

$$C^\infty(M) \to \mathcal{F}_0(\pi) = C^\infty(E) \to \mathcal{F}_1(\pi) \to \cdots \mathcal{F}_k(\pi) \to \mathcal{F}_{k+1}(\pi) \cdots$$

where for $k \in \mathbb{N}$, $\mathcal{F}_k(\pi) = C^\infty\left(J^k(\pi)\right)$, we consider the filtered commutative algebra

$$\mathcal{F}(\pi) = \bigcup_{k \in \mathbb{N}} \mathcal{F}_k(\pi)$$

and we identify the algebras $C^\infty(M)$, $C^\infty(E)$ and $\mathcal{F}_k(\pi)$ to the corresponding subalgebras of $\mathcal{F}(\pi)$. So a smooth function on $J^\infty(\pi)$ is a function on $J^k(\pi)$ for some finite but arbitrary k.

Definition 4.19 *Let $\xi = (\xi_k)_{k \in \mathbb{N}}$ be an element of $J^\infty(\pi)$ where $\xi_k \in J^k(\pi)$ and $\forall k_2 \geq k_1$, $\pi_{k_1}^{k_2}(\xi_{k_2}) = \xi_{k_1}$.*
A tangent vector $v \in T_\xi J^\infty(\pi)$ is a sequence $(v_k)_{k \in \mathbb{N}}$ where $v_k \in T_{\xi_k} J^k(\pi)$ and $\forall k_2 \geq k_1$, $\left(\pi_{k_1}^{k_2}\right)_* (\xi_{k_2}) = \xi_{k_1}$

We then have the following result:

Proposition 4.7 *For each $\xi \in J^\infty(\pi)$, $T_\xi J^\infty(\pi)$ is canonically isomorphic with the projective limit of $\left(T_{\pi_k^\infty(\xi)} J^k(\pi), \left(\pi_k^l\right)_*\right)_{(k,l) \in \mathbb{N}^2, l \geq k}$.*

Definition 4.20 *An operational vector field on $J^\infty(\pi)$ is a derivation of the algebra $\mathcal{F}(\pi)$.*

So X is an operational vector field on $J^\infty(\pi)$ if and only if X is a $\mathbb{R}$-linear map $\mathcal{F}(\pi) \to \mathcal{F}(\pi)$ such that

$$\forall (f,g) \in \mathcal{F}(\pi)^2,\ X(f.g) = f.X(g) + X(f).g.$$

4.9.3 The Cartan Distribution on an Infinite Jet Bundle

We use the notions of Chapter 2, § 1.12.3.
The main geometrical object of the infinite jet bundle $J^\infty(\pi)$ is the Cartan distribution $\mathcal{C}(\pi)$ of n-dimensional planes which project without degeneration onto TM via $(\pi^\infty)_*$ and defined as follows.

A tangent vector $(v_k)_{k \in \mathbb{N}}$ belongs to the Cartan subspace $\mathcal{C}(\xi)$ of $T_\xi J^\infty(\pi)$ if and only if

$$\forall k \in \mathbb{N},\ v_k \in \mathcal{C}^k.$$

The spaces $\mathcal{C}(a)$ form the *Cartan distribution* $\mathcal{C}(\pi)$ on $J^\infty(\pi)$.

This distribution is spanned by the total derivatives D_{x^i}, $1 \leq i \leq n$ whose action on the algebra $\mathcal{F}(\pi)$ is defined by restriction onto the infinite jets $j^\infty(s)$ of sections $s \in \Gamma(\pi)$ by

$$\forall s \in \Gamma(\pi), \forall f \in \mathcal{F}(\pi),\ j^\infty(s)\left(D_{x^i}(f)\right) = \frac{\partial}{\partial x^i}\left(j^\infty(s)(f)\right). \tag{4.4}$$

In local coordinates, $\mathcal{C}(\pi)$ is a $\mathcal{F}(\pi)$-module generated by the local vector fields

$$D_i = \frac{\partial}{\partial x^i} + \sum_{\substack{\alpha=1,\ldots,m \\ |\sigma|\leq 0}} u^\alpha_{\sigma i} \frac{\partial}{\partial u^\alpha_\sigma}.$$

Proposition 4.8 *The Cartan distribution is involutive:*

$$[\mathcal{C}(\pi), \mathcal{C}(\pi)] \subset \mathcal{C}(\pi).$$

For any $\xi \in J^\infty(\pi)$, $\dim \mathcal{C}(\xi) = n$ and the n-dimensional submanifold $j^\infty(s)(M) \subset J^\infty(\pi)$ is an integral submanifold of the Cartan distribution.

Recall that an infinite-dimensional smooth Fréchet differentiable manifold equipped with a finite-dimensional involutive distribution corresponds to the notion of *diffiety* (differential variety) as introduced by Vinogradov ([Vin81]). One can find applications of such a framework in non-holonomic mechanics and non-linear control systems (see, for instance, [FLMR97]).

4.9.4 Symmetries of Infinite Jet Spaces

Similarly to the finite-dimensional case, we introduce the notion of symmetry.

Definition 4.21 *A vector field X on $J^\infty(\pi)$ is said to be a symmetry if*

$$[X, \mathcal{C}(\pi)] \subset \mathcal{C}(\pi).$$

According to Lie-Bäcklund theorem, for any symmetry X of $J^k(\pi)$, there is a unique symmetry $\hat{X}$ of $\mathcal{F}(\pi)$ such that

$$\begin{cases} \hat{X}|_{\mathcal{F}_k(\pi)} = X \\ \forall p \geq k, \ \hat{X}|_{\mathcal{F}_p(\pi)} = X_p. \end{cases}$$

Such a symmetry has a flow which preserves the Cartan distribution.

Remark 4.10 *An arbitrary symmetry of $J^\infty(\pi)$ need not have a flow.*

Denote by $\mathrm{Sym}(\pi)$ the Lie algebra of symmetries.
Let us consider the following set

$$S_{(v)}(\pi) = \left\{X \in \mathrm{Sym}(\pi) : X|_{C^\infty(M)} = 0\right\}$$

the structure of which we are going to describe thanks to the vector space $\varkappa(\pi)$ of $\mathcal{F}(\pi)$-values derivations $\varphi : C^\infty(E) \to \mathcal{F}(\pi)$ such that $\varphi|_{C^\infty(M)} = 0$. This $\mathcal{F}(\pi)$-module can be seen as

$$\Gamma(\pi^*_\infty(\pi)) = \Gamma(\pi) \otimes_{C^\infty(M)} \mathcal{F}(\pi).$$

These vertical vector fields on the infinite jet bundle correspond to the infinite prolongations of evolutionary vector fields as defined in [Olv93]. They commute with the generators of the Cartan distribution.
The terminology evolutionary vector field is linked with the notion of *evolution equation*: in the research of the flow $\psi_t = (\psi_t^x, \psi_t^u, \psi_t^{u_x}, \dots)$ one has lead to solve an infinite-dimensional differential system.

Example 4.17 *Heat equation.– To the PDE* $u_t = u_{xx}$ *([Sza07], 1.2.2) is associated the flow of the vector field* $u_{xx}\dfrac{\partial}{\partial u}$.
A solution is given by $u : (x,t) \mapsto \mathrm{e}^{-k^2 t} \times \mathrm{e}^{\mathrm{i}\ kx}$ *and corresponds to a dissipative system, this means that the energy is not conserved and dissipates from the physical system.*

If $\varphi = (\varphi^1, \dots, \varphi^m)$ is the characteristic of an evolutionary vector field Y_φ whose expression in local coordinates is $\varphi^\alpha \left[x, u^{(k)}\right] \dfrac{\partial}{\partial u^\alpha}$, the local expression of the infinite prolongation denoted $\mathrm{pr}^\infty (Y_\varphi)$ in [Olv93] and E_φ in [KiVa11] is

$$E_\varphi = \mathrm{pr}^\infty Y_\varphi = \sum_{\alpha,\sigma} D_\sigma (\varphi^\alpha) \frac{\partial}{\partial u_\sigma^\alpha}$$

$\varkappa(\pi)$ can be endowed with a Lie algebra structure $[.,.]$ defined by

$$[E_{\varphi_1}, E_{\varphi_2}] = E_{\{\varphi_1,\varphi_2\}_0}$$

where

$$\{\varphi_1, \varphi_2\}_0 = E_{\varphi_1}(\varphi_2) - E_{\varphi_2}(\varphi_1).$$

Finally we have

$$\mathrm{Sym}(\pi) = \mathcal{C}(\pi) \oplus \varkappa(\pi).$$

4.10 Projective Limits of Banach Principal Bundles

Definition 4.22 $\mathcal{P} = \left((P_i, \pi_i, M_i, G_i), \left(\lambda_i^j, \mu_i^j, \gamma_i^j\right)\right)_{j \geq i}$ *is called a projective sequence of Banach principal bundles if the following conditions are satisfied:*

(PSBPB1) $\left(M_i, \mu_i^j\right)_{j \geq i}$ *is a projective sequence of Banach* C^∞*-manifolds;*

(PSBPB2) $\left(P_i, \lambda_i^j\right)_{j \geq i}$ *is a projective sequence of Banach* C^∞*-manifolds;*

(PSBPB3) $\left(G_i, \gamma_i^j\right)_{j \geq i}$ *is a projective sequence of Banach-Lie groups;*

(PSBVB4) *For all* $i \leq j$, $\left(\mu_i^j, \lambda_i^j, \gamma_i^j\right)$ *is a principal bundle morphism;*

(PSBVB5) *$(M_i)_{i\in\mathbb{N}}$ has the projective limit chart property at any point relatively to $(U = \varprojlim U_i, \phi = \varprojlim \phi_i)$;*

(PSBVB6) *For any $x \in M = \varprojlim M_i$, there exists a local trivialization $\tau_i : \pi_i^{-1}(U_i) \to U_i \times G_i$ such that the following diagram is commutative:*

$$\begin{array}{ccc} (\pi_i)^{-1}(U_i) & \stackrel{(\pi_i)^{-1}(U_i)}{\longleftarrow} & (\pi_j)^{-1}(U_j) \\ \tau_i \downarrow & & \downarrow \tau_j \\ U_i \times G_i & \xrightarrow{\mu_i^j \times \gamma_i^j} & U_j \times G_j \end{array}$$

We then have the following result:

Theorem 4.3 *Let $\left((P_i, \pi_i, M_i, G_i), \left(\lambda_i^j, \mu_i^j, \gamma_i^j \right) \right)_{j \geq i}$ be a projective sequence of Banach principal bundles. The quadruple $\left(\varprojlim P_n, \varprojlim \pi_n, \varprojlim M_n, \varprojlim G_n \right)$ is a Fréchet principal bundle.*

Notation 4.3 *For the sake of simplicity, the projective sequence of Banach bundles*

$$\left((P_i, \pi_i, M_i, G_i), \left(\lambda_i^j, \mu_i^j, \gamma_i^j \right) \right)_{j \geq i}$$

will be denoted $(P_i, \pi_i, M_i, G_i)_{\underleftarrow{i}}$.

4.11 The Fréchet Space $\mathcal{H}(\mathbb{F}_1, \mathbb{F}_2)$ and the Banach Space $\mathcal{H}_b(\mathbb{F}_1, \mathbb{F}_2)$

Let $(\mathbb{F}_1, \nu_n^1)$ (resp. $(\mathbb{F}_2, \nu_n^2)$) be a graded Fréchet space.
Recall that a linear map $L : \mathbb{F}_1 \to \mathbb{F}_2$ is *continuous* if

$$\forall n \in \mathbb{N}, \exists k_n \in \mathbb{N}, \exists C_n > 0 : \forall x \in \mathbb{F}_1, \nu_2^n (L.x) \leq C_n \nu_1^{k_n} (x).$$

The space $\mathcal{L}(\mathbb{F}_1, \mathbb{F}_2)$ of continuous linear maps between both these Fréchet spaces generally drops out of the Fréchet category. Indeed, $\mathcal{L}(\mathbb{F}_1, \mathbb{F}_2)$ is a Hausdorff locally convex topological vector space whose topology is defined by the family of semi-norms $\{p_{n,B}\}$:

$$p_{n,B}(L) = \sup_{x \in B} \left\{ \nu_n^2 (L.x) \right\}$$

where $n \in \mathbb{N}$ and B is any bounded subset of $\mathbb{F}_1$. This topology is not metrizable since the family $\{p_{n,B}\}$ is not countable.
So $\mathcal{L}(\mathbb{F}_1, \mathbb{F}_2)$ will be replaced, under certain assumptions, by a projective limit

of appropriate functional spaces as introduced in [Gal98].
According to the notations introduced in the previous section, we denote by $\mathcal{L}(\mathbb{B}_1^n, \mathbb{B}_2^n)$ the space of linear continuous maps (or equivalently bounded linear maps because $\mathbb{B}_1^n$ and $\mathbb{B}_2^n$ are normed spaces). We then have the following result ([DGV16], Theorem 2.3.10).

Theorem 4.4 *The space of all continuous linear maps between $\mathbb{F}_1$ and $\mathbb{F}_2$ which can be represented as projective limits*

$$\mathcal{H}(\mathbb{F}_1, \mathbb{F}_2) = \left\{ (L_n) \in \prod_{n \in \mathbb{N}} \mathcal{L}(\mathbb{B}_1^n, \mathbb{B}_2^n) : \varprojlim L_n \text{ exists} \right\}$$

is a Fréchet space.

For this sequence $(L_n)_{n \in \mathbb{N}}$ of linear maps, for any integer $0 \leq n \leq m$, the following diagram is commutative

$$\begin{array}{ccc} \mathbb{B}_1^n & \xleftarrow{(\delta_1)_n^m} & \mathbb{B}_1^m \\ {\scriptstyle L_n}\downarrow & & \downarrow{\scriptstyle L_m} \\ \mathbb{B}_2^n & \xleftarrow{(\delta_2)_n^m} & \mathbb{B}_2^m \end{array}$$

On $\mathcal{H}(\mathbb{F}_1, \mathbb{F}_2)$, the topology can be defined by the sequence of seminorms p_n given by

$$p_n(L) = \max_{0 \leq k \leq n} \sup \left\{ \nu_k^2(L.x), x \in \mathbb{F}_1, \ \nu_k^1(x) = 1 \right\}$$

so that $(\mathcal{H}(\mathbb{F}_1, \mathbb{F}_2), p_n)$ is a graded Fréchet space.

Remark 4.11 *For $l \in \{1,2\}$, given a graduation (ν_n^l) on a Fréchet space $\mathbb{F}_l$, let $\mathbb{B}_l^n$ be the associated local Banach space and $\delta_l^n : \mathbb{F}_l \to \mathbb{B}_l^n$ the canonical projection.*
The quotient norm $\tilde{\nu}_n^l$ associated with ν_n^l is defined by

$$\tilde{\nu}_n^l(\delta_n(z)) = \sup\{\nu_n^l(y) : \ \delta_n(y) = \delta_n(z)\}. \tag{4.5}$$

We denote by $(\tilde{\nu}_n^2)^{\text{op}}$ the corresponding operator norm on $\mathcal{L}(\mathbb{B}_1^n, \mathbb{B}_2^n)$.
If $L = \varprojlim L_n$ where $L_n : \mathbb{B}_1^n \to \mathbb{B}_2^n$, then we have

$$\begin{aligned} (\tilde{\nu}_n^2)^{\text{op}}(L_n) &= \sup\{\tilde{\nu}_n^2(L_n.x), \ \ x \in \mathbb{B}_1^n \ \ \tilde{\nu}_n^1(x) \leq 1\} \\ &= \sup\{\nu_n^2(L.x), x \in \mathbb{F}_1, \nu^1(x) \leq 1\}. \end{aligned}$$

This implies that

$$p_n(L) = \max_{0 \leq i \leq n} (\tilde{\nu}_i^2)^{\text{op}}(L_n).$$

Definition 4.23 *Let $(\mathbb{F}_1, \nu_n^1)$ and $(\mathbb{F}_2, \nu_n^2)$ be graded Fréchet spaces. A linear map $L : \mathbb{F}_1 \to \mathbb{F}_2$ is called a uniformly bounded operator, if*

$$\exists C > O : \forall n \in \mathbb{N}, \ \nu_n(L(x)) \leq C\mu_n(x).$$

We denote by $\mathcal{H}_b(\mathbb{F}_1, \mathbb{F}_2)$ the set of uniformly bounded operators. Of course $\mathcal{H}_b(\mathbb{F}_1, \mathbb{F}_2)$ is contained in $\mathcal{H}(\mathbb{F}_1, \mathbb{F}_2)$ and $L \in \mathcal{H}(\mathbb{F}_1, \mathbb{F}_2)$ belongs to $\mathcal{H}_b(\mathbb{F}_1, \mathbb{F}_2)$ if and only if $\sup_{n\in\mathbb{N}} p_n(L) < \infty$ and so

$$\mathcal{H}_b(\mathbb{F}_1, \mathbb{F}_2) = [\mathcal{H}(\mathbb{F}_1, \mathbb{F}_2)]_b .$$

When $\mathbb{F} = \mathbb{F}_1 = \mathbb{F}_2$ and $\nu_n^1 = \nu_n^2$ for all $n \in \mathbb{N}$, the set $\mathcal{H}(\mathbb{F}, \mathbb{F})$ (resp. $\mathcal{H}_b(\mathbb{F}, \mathbb{F})$) is simply denoted $\mathcal{H}(\mathbb{F})$ (resp. $\mathcal{H}_b(\mathbb{F})$).

We denote by $\mathcal{IH}_b(\mathbb{F}_1, \mathbb{F}_2)$ (resp. $\mathcal{SH}_b(\mathbb{F}_1, \mathbb{F}_2)$) the set of injective (resp. surjective) operators of $\mathcal{H}_b(\mathbb{F}_1, \mathbb{F}_2)$ with closed range. As in the Banach context, according to the context of Theorem 4.4 we have:

Proposition 4.9

(1) Each operator $L \in \mathcal{H}(\mathbb{F}_1, \mathbb{F}_2)$ has a closed range if and only if, for each $n \in \mathbb{N}$, the induced operator $L_n : \mathbb{B}_1^n \to \mathbb{B}_2^n$ has a closed range.

(2) $\mathcal{IH}_b(\mathbb{F}_1, \mathbb{F}_2)$ is an open subset of $\mathcal{H}_b(\mathbb{F}_1, \mathbb{F}_2)$.

(3) $\mathcal{SH}_b(\mathbb{F}_1, \mathbb{F}_2)$ is an open subset of $\mathcal{H}_b(\mathbb{F}_1, \mathbb{F}_2)$.

Proof *(1) For $i \in \{1, 2\}$, consider the representation $\mathbb{F}_i = \varprojlim \mathbb{B}_i^n$ as the projective limit of a projective sequence of Banach spaces $(\mathbb{B}_i^n, (\delta_i)_n^m)_{m\geq n}$ and a sequence of linear operators $L_n : \mathbb{B}_1^n \to \mathbb{B}_2^n$. Assume that $\mathbb{F}_1$ is a closed Fréchet subspace of $\mathbb{F}_2$ and ν_n^1 is the restriction of ν_n^2 to $\mathbb{F}_1$. Since $\mathbb{F}_1$ is closed in $\mathbb{F}_2$, it follows that $\ker \nu_n^1 = \ker \nu_n^2 \cap \mathbb{F}_1$ and so the quotient space $\mathbb{F}_1/\ \ker \nu_n^1$ is a normed subspace of $\mathbb{F}_2/\ \ker \nu_n^2$ whose quotient norm $\tilde{\nu}_n^1$ is the restriction of $\tilde{\nu}_n^2$ to $\mathbb{F}_1/\ \ker \nu_n^1$. It follows that their respective completions $\mathbb{B}_1^n$ and $\mathbb{B}_2^n$ are such that $\mathbb{B}_1^n$ is a closed subspace of $\mathbb{B}_2^n$.*
Conversely, assume that $\mathbb{F}_1$ is a Fréchet space contained in $\mathbb{F}_2$ such that the inclusion $\iota : \mathbb{F}_1 \to \mathbb{F}_2$ is continuous and each operator $\iota_n : \mathbb{B}_1^n \to \mathbb{B}_2^n$ has a closed range. Thus $\mathbb{F}_1$ can be identify with the set

$$\left\{ (x_n) \in \prod_{n\in\mathbb{N}} (\iota_n(\mathbb{B}_1^n)) : \ \forall m \geq n \ x_n = (\delta_2)_n^m (x_m) \right\}$$

which is a closed subspace of

$$\left\{ (x_n) \in \prod_{n\in\mathbb{N}} \mathbb{B}_2^n : \ \forall m \geq n \ x_n = (\delta_2)_n^m (x_m) \right\}$$

which can be identified with $\mathbb{F}_2$. This clearly implies Point 1.

(2) Consider an injective operator $L \in \mathcal{H}(\mathbb{F}_1, \mathbb{F}_2)$. According to the representation $\mathbb{F}_i = \varprojlim \mathbb{B}_i^n$ as a projective limit of a projective Banach sequence

$(\mathbb{B}_i^n, (\delta_i)_n^m)_{m\geq n}$, we have a sequence of linear operators $L_n : \mathbb{B}_1^n \to \mathbb{B}_2^n$ such that $L = \varprojlim L_n$ (cf. Theorem 4.4). Considering each

$$\mathbb{F}_i = \{(x_n) \in \prod_{n\in\mathbb{N}} \mathbb{B}_i^n : \forall m \geq n,\ x_n = (\delta_2)_n^m(x_m)\}$$

as previously, then if $x = (x_n) \in \mathbb{F}_1$ then $L(x) = (L_n(x_n)) \in \mathbb{F}_2$. Thus it is clear that L is injective if and only if L_n is injective for all $n \in \mathbb{N}$.

Now if $L \in \mathcal{IH}_b(\mathbb{F}_1, \mathbb{F}_2)$, then L_n is an isomorphism from $\mathbb{B}_1^n$ onto its range and so we have

$$\frac{1}{\ell_n}.\tilde{\nu}_n^1(x) \leq \tilde{\nu}_n^2(L_n(x) \leq \ell_n.\tilde{\nu}_n^1(x) \tag{4.6}$$

for all $x \in \mathbb{B}_1^n$, all $n \in \mathbb{N}$, where $\tilde{\nu}_n^i$ is the quotient norm of ν_n^i on $\mathbb{B}_i^n$ for $i \in 1, 2$, and

$$\ell_n = (\tilde{\nu}_n^2)^{\mathrm{op}}(L) = \sup\{\frac{\tilde{\nu}_n^2(L(x))}{\tilde{\nu}_n^1(x)},\ \ x \neq 0\}.$$

Since δ_n^2 is the canonical projection of $\mathbb{F}_2$ on $\mathbb{B}_2^n$ and $\nu_n^2 \circ \delta_n = \tilde{\nu}_n^2$, we obtain

$$\frac{1}{\ell_n}\nu_n^1(x) \leq \nu_n^2(L(x)) \leq \ell_n.\nu_n^1(x) \tag{4.7}$$

for all $x \in \mathbb{F}_1$ and $n \in \mathbb{N}$. But we have $\ell_n \leq ||L||_\infty$ and we finally obtain

$$\frac{1}{\ell}\nu_n^1(x) \leq \nu_n^2(L(x)) \leq \ell.\nu_n^1(x) \tag{4.8}$$

for all $x \in \mathbb{F}_1$, all $n \in \mathbb{N}$ and where $\ell = ||L||_\infty$.

Fix some $L \in \mathcal{H}_b(\mathbb{F}_1, \mathbb{F}_2)$ and set $\ell = ||L||_\infty$, we consider the open set

$$W = \{T \in \mathcal{H}_b(\mathbb{F}_1, \mathbb{F}_2),\ : \ ||T - L||_\infty < \frac{\ell}{2}\}$$

Fix some $n \in \mathbb{N}$. For any $x \in \mathbb{F}_1$ and $T \in W$, we have

$$\nu_n^1(x) - \nu_n^1(T(x)) \leq \nu_n^1(T-L)(x) \leq p_n(T-L).\nu_n^1(x) \leq ||T-L||_\infty.\nu_n^1(x) \leq \frac{\ell}{2}\nu_n^1(x).$$

This implies that

$$\nu_n^2(T(x)) \geq \frac{\ell}{2}\nu_n^1(x). \tag{4.9}$$

Since (ν_n^i) is a separating sequence of semi-norms, it follows that L is injective. Now taking in account inequality (4.9) and relation $\tilde{\nu}_n^i = \nu_n^i \circ (\delta_i)_n$, for $T \in W$ and each $n \in \mathbb{N}$, we have

$$\tilde{\nu}_n^2(T_n(x)) \leq \frac{3\ell}{2}\nu_n^1(x) \leq 3\tilde{\nu}_n^2(T_n(x))$$

for all $x \in \mathbb{B}_1^n$. It follows that T_n is closed and so T is closed (cf. 1.). Finally, W is an open neighbourhood of L contained in $\mathcal{IH}_b(\mathbb{F}_1, \mathbb{F}_2)$, which ends the proof of (2).

(3) By same arguments as for the injectivity of an operator $L \in \mathcal{H}(\mathbb{F}_1, \mathbb{F}_2)$ we can show that such an operator L is surjective if and only if each induced operator $L_n : \mathbb{B}_1^n \to \mathbb{B}_2^n$ is surjective. The set of surjective operators in $\mathcal{L}(\mathbb{B}_1^n, \mathbb{B}_2^n)$ is an open set. More precisely the adjoint *operation defines an isometry* $\mathrm{Adj} : T_n \mapsto T_n^*$ *from $\mathcal{L}(\mathbb{B}_1^n, \mathbb{B}_2^n)$ to $\mathbb{L}((\mathbb{B}_2^n)^*, (\mathbb{B}_2^n)^*)$. Since T_n is surjective if and only if T_n^* is a monomorphism*[8] Adj *is an isometry from the set of monomorphisms of $\mathcal{L}(\mathbb{B}_1^n, \mathbb{B}_2^n)$ to $\mathbb{L}((\mathbb{B}_2^n)^*, (\mathbb{B}_2^n)^*)$ into the set surjective operators of $\mathcal{L}(\mathbb{B}_1^n, \mathbb{B}_2^n)$. Fix some surjective operator $L \in \mathcal{H}(\mathbb{F}_1, \mathbb{F}_2)$ and let $(\tilde{\nu}_n^i)^*$ on $(\mathbb{B}_i^n)^*$ be the norm associated with $\tilde{\nu}_n^i$. Fix some $L \in \mathcal{H}_b(\mathbb{F}_1, \mathbb{F}_2)$ and set $\ell = ||L||_\infty$. If $||T - L||_\infty < \dfrac{\ell}{2}$, by using an adequate adaptation of arguments developed between (4.6) and (4.9), we obtain: It follows that each T_n is surjective and so T is also surjective, which ends the proof of (3).*

Remark 4.12 *In general, the set $\mathcal{IH}(\mathbb{F}_1, \mathbb{F}_2)$ of injective continuous operators from $\mathbb{F}_1$ to $\mathbb{F}_2$ which belongs to $\mathcal{H}(\mathbb{F}_1, \mathbb{F}_2)$ is not open in $\mathcal{H}(\mathbb{F}_1, \mathbb{F}_2)$. As in the proof of Lemma 8.2 (1), we need the fact that L is closed and so is an isomorphism between $\mathbb{F}_1$ and its range. But, in general, L can be an isomorphism from between $\mathbb{F}_1$ and its range does not satisfy the inequality (4.8) which is essential in the proof (cf. Example 4.18). The same observation is also true for $\mathcal{SH}(\mathbb{F}_1, \mathbb{F}_2)$ in $\mathcal{H}(\mathbb{F}_1, \mathbb{F}_2)$. This naturally implies that the set $\mathcal{GH}(\mathcal{F})$ of isomorphisms L of a Fréchet space $\mathbb{F}$ which belongs to $\mathcal{H}(\mathcal{F})$ is not open in $\mathcal{H}(\mathcal{F})$ in general.*

Example 4.18 *If $1 \leq p < \infty$, we define the Banach space*

$$\mathbb{F} = l^p(\mathbb{N}) = \left\{ f : \mathbb{N} \to \mathbb{R} \ : \ \|f\|_p := \left(\sum_{n \in \mathbb{N}} |f(n)|^p \right)^{1/p} < \infty \right\}.$$

We fix $1 \leq p_1 \leq p_2 < \infty$ and denote $\mathbb{F}_j := l^{p_j}(\mathbb{N})$ for $j \in \{1, 2\}$, so that we have a canonical continuous inclusion map $I_{p_1,p_2} \colon \mathbb{F}_1 \hookrightarrow \mathbb{F}_2$.

Then consider $\lambda_n := \frac{1}{n+1}$ for every $n \in \mathbb{N}$, so that $\lim\limits_{n \to \infty} \lambda_n = 0$, and define $\iota : \mathbb{F}_1 \to \mathbb{F}_2$ by $(\iota(f))(n) := \lambda_n f(n)$ for every $n \in \mathbb{N}$ and $f \in \mathbb{F}$. Then $(\iota - \lambda_n I_{p_1,p_2})_{n \in \mathbb{N}}$ is a sequence in $\mathcal{H}(\mathbb{F}_1, \mathbb{F}_2) \equiv \mathcal{L}(\mathbb{F}_1, \mathbb{F}_2)$ with the following properties:

(1) For every $n \in \mathbb{N}$, the operator $\lambda_n \mathbf{1}_{\mathbb{F}} - \iota$ fails to be injective. Indeed, we have $(\iota - \lambda_n I_{p_1,p_2})(e_n) = 0$, where $e_n := \chi_n \in \mathbb{F}_1$, that is, $e_n \colon \mathbb{N} \to \mathbb{R}$ is defined by $e_n(n) = 1$ and $e_n(m) = 0$ for every $m \in \mathbb{N} \setminus \{n\}$.

[8] i.e. T_n satisfies an inequality of type (4.9).

(2) The sequence $(\iota - \lambda_n I_{p_1,p_2})_{n\in\mathbb{N}}$ converges to ι for the norm topology of $\mathcal{H}(\mathbb{F}_1, \mathbb{F}_2)$.

(3) The operator ι is injective.

Then, according to the above properties, it follows that, for every neighbourhood U of the operator $\iota \in \mathcal{H}(\mathbb{F}_1, \mathbb{F}_2)$, there exists $n \in \mathbb{N}$ such that the non-injective operator $\iota - \lambda_n I_{p_1,p_2}$ belongs to U.

We are in situation to end this section by the following result:

Theorem 4.5

(1) The Banach space $\mathcal{H}_b(\mathbb{F})$ has a Banach-Lie algebra structure and the set $\mathcal{GH}_b(\mathbb{F})$ of uniformly bounded isomorphisms of $\mathbb{F}$ is open in $\mathcal{H}_b(\mathbb{F})$.

(2) $\mathcal{GH}_b(\mathbb{F})$ has a structure of Banach-Lie group whose Lie algebra is $\mathcal{H}_b(\mathbb{F})$.

(3) If $\mathbb{F}$ is identified with the projective $\varprojlim \mathbb{B}^n$ we denote by $\exp_n : \mathcal{L}(\mathbb{B}_n) \to \mathcal{GL}(\mathbb{B}_n)$, then we a have a well defined smooth map $\exp := \varprojlim \exp_n : \mathcal{H}_b(\mathcal{F}) \to \mathcal{GH}_b(\mathbb{F})$ which is a diffeomorphism from an open set of $0 \in \mathcal{H}_b(\mathcal{F})$ onto a a neighbourhood of $\mathrm{Id}_{\mathbb{F}}$.

Proof *Since $\mathcal{H}_b(\mathbb{F})$ is a unital Banach subalgebra of $\mathcal{H}_b(\mathbb{F})$, it follows that $\mathcal{H}_b(\mathbb{F})$ has a canonical structure of Banach-Lie algebra and the set $\mathcal{GH}_b(\mathbb{F})$ is exactly the intersection $\mathcal{IH}_b(\mathbb{F}_1\mathbb{F}_2) \cap \mathcal{IH}_b(\mathbb{F}_1\mathbb{F}_2)$. So, From Proposition 4.9, $\mathcal{GH}_b(\mathbb{F})$ is an open set of $\mathcal{H}_b(\mathbb{F})$. Note that $\mathcal{GH}_b(\mathbb{F})$ is the set of invertible elements of $\mathcal{H}_b(\mathbb{F})$. According to [Belt06], Example 2.21, it follows that $\mathcal{GH}_b(\mathbb{F})$ has a structure of Banach-Lie group[9]. Now From Theorem 2.42 (i) and [Belt06], Example 2.48, there exists an exponential map which is a local diffeomorphism from an open set of $0 \in \mathcal{H}_b(\mathcal{F})$ onto a a neighbourhood of $\mathrm{Id}_{\mathbb{F}}$. Finally, if $\mathbb{F}$ is identified with the projective limit $\varprojlim \mathbb{B}^n$, then the last property is obvious.*

4.12 Projective Limits of Generalized Frame Bundles

We extend the notion developed in [DGV16], 6.5, to the framework of generalized frame bundles defined over a projective sequence of Banach manifolds.

Let us consider a projective sequence of Banach vector bundles $\mathcal{E} = (E_n, \pi_n, M_n)_{\underleftarrow{n}}$ (cf. Definition 4.8) where the manifold M_n is modelled on the Banach space $\mathbb{M}_n$ and where $\mathbb{E}_n$ is the type fibre of E_n. The projective limit of $\mathcal{E}$ is a Fréchet vector bundle (E, π, M) (cf. Proposition 4.4).

We then define the space

$$\mathbf{P}(E_n) = \bigcup_{x_n \in M_n} \mathcal{H}_0^n\left(\mathbb{E}, (E_n)_{x_n}\right)$$

[9]From this Example 2.21 of [Belt06], it follows also that $\mathcal{GH}_b(\mathbb{F})$ is an open set of $\mathcal{H}_b(\mathbb{F})$.

where $\mathcal{H}_0^n\left(\mathbb{E},(E_n)_{x_n}\right)$ is the set of $(q_0,\dots,q_n)\in\prod_{i=0}^{n}\mathcal{L}is\left(\mathbb{E}_i,(E_i)_{\mu_i^n(x_n)}\right)$ such that

$$\forall(i,j)\in\{0,\dots,n\}^2,j\geq i,\ \lambda_i^j\circ q_j=q_i\circ\overline{\lambda_i^j}.$$

We denote by $\mathbf{p}_n:\mathbf{P}(E_n)\to M_n$ the natural projection. Then we have:

Proposition 4.10 *The quadruple $(\mathbf{P}(E_n),\mathbf{p}_n,M_n,\mathcal{H}_0^n(\mathbb{E}))$ is a Banach principal bundle over M_n with structural group $\mathcal{H}_0^n(\mathbb{E})$.*

Proof *Consider a Fréchet bundle atlas* $\left\{\left(U^\alpha=\varprojlim U_i^\alpha,\tau^\alpha=\varprojlim\tau_i^\alpha\right)\right\}_{\alpha\in A}$ *for the Fréchet bundle (E,π,M). For each $i\in\mathbb{N}$, this bundle atlas gives rise to a a bundle atlas $\{(U_i^\alpha,\Phi_i^\alpha)\}_{\alpha\in A}$ for the principal bundle $\ell(E_i)$ and the transition maps $T_i^{\alpha\beta}:U_i^\alpha\cap U_i^\beta\to\mathrm{GL}(\mathbb{E}_i)$ are the same for both these bundles (cf. Chapter 2, Proposition 2.10).*
We consider the map

$$\begin{array}{rccc}\mathbf{\Phi}_n^\alpha: & (\mathbf{p}_n)^{-1}(U_n^\alpha) & \to & U_n^\alpha\times\prod_{i=0}^{n}\mathrm{GL}(\mathbb{E}_n)\\ & (x_n;q_0,\dots,q_n) & \mapsto & (x_n;(\bar{\tau}_0^\alpha)_{\mu_0^n(x_n)}\circ q_0,\dots,(\bar{\tau}_0^\alpha)_{\mu_n^n(x_n)}\circ q_n)\end{array}$$

where $(\tau_i^\alpha)_{x_i}=pr_2\circ(\tau_i^\alpha)_{|(E_i)_{x_i}}$ (for more details see Chapter 2, § 2.2.2).

Since μ_i^j and τ_i^α are smooth maps, it follows that $\mathbf{\Phi}_n^\alpha$ is smooth. Now, according to property **(PSBVB5)** *and the definition of $\mathcal{H}_0^n\left(\mathbb{E},(E_n)_{x_n}\right)$, the map $\mathbf{\Phi}_n^\alpha$ takes values in $\mathcal{H}_0^n(\mathbb{E})$.*
Now the map

$$(x_n;g_0,\dots g_n)\mapsto\left(x_n;\{(\bar{\tau}_0^\alpha)\}_{\mu_0^n(x_n)}^{-1}(g_0),\dots\{(\bar{\tau}_n^\alpha)\}_{\mu_n^n(x_n)}^{-1}(g_n)\right)$$

is the inverse of $\mathbf{\Phi}_n^\alpha$ and this map is also smooth. It follows that τ_n^α is a trivialization.
The transition maps $\mathbf{T}_n^{\alpha\beta}$ are given by

$$\mathbf{T}_n^{\alpha\beta}(x_n)=\left(T_0^{\alpha\beta}(\mu_0^n(x_n)),\dots,T_n^{\alpha\beta}(\mu_n^n(x_n))\right).$$

Again, property **(PSBVB5)** *implies that, for all $0\leq i\leq j$,*

$$T_i^{\alpha\beta}(\mu_i^j(x_j);\overline{\lambda_i^j}(u))=\overline{\lambda_i^j}\circ T_j^{\alpha\beta}(x_j;u)$$

Thus $\mathbf{T}_n^{\alpha\beta}(x_n)$ takes values in $\mathcal{H}_0^n(\mathbb{E})$ which ends the proof.

Definition 4.24 *The quadruple $(\mathbf{P}(E_n),\mathbf{p}_n,M_n,\mathcal{H}_0^n(\mathbb{E}))$ is called the generalized frame bundle of E_n and is denoted by $\ell(E_n)$.*

For every $j \geq i$, we define the following projection

$$\begin{array}{cccc} \mathbf{r}_i^j : & \mathbf{P}(E_j) & \rightarrow & \mathbf{P}(E_i) . \\ & (q_1, \ldots, q_j) & \mapsto & (q_1, \ldots, q_i) \end{array}$$

Then, as for [DGV16], Lemma 6.5.3, from the previous results and definitions, we obtain:

Lemma 4.1 *The triple* $\left(\mathbf{r}_i^j, \mu_i^j, (h_0)_i^j\right)$ *is a principal bundle morphism from* $\left(\mathbf{P}(E_j), \mathbf{p}_j, M_j, \mathcal{H}_0^j(\mathbb{E})\right)$ *into* $\left(\mathbf{P}(E_j), \mathbf{p}_i, M_i, \mathcal{H}_0^i(\mathbb{E})\right)$.

According to the proof of Proposition 4.10 and Lemma 4.1 and adapting the proof of [DGV16], Proposition 6.5.4, we obtain:

Theorem 4.6 *The sequence* $(\ell(E_n))_{\underleftarrow{n}} = (\mathbf{P}(E_n), \mathbf{p}_n, M_n, \mathcal{H}_0^n(\mathbb{E}))_{\underleftarrow{n}}$ *is a projective sequence of principal bundles. The projective limit* $\underleftarrow{\lim}\mathbf{P}(E_n)$ *can be endowed with a structure of smooth Fréchet principal bundle over* $\underleftarrow{\lim}M_n$ *whose structural group is* $\mathcal{H}_0(\mathbb{E})$.

Remark 4.13 *Adapting [DGV16], Corollary 6.5.2 in the context of Theorem 4.6 and according to the proof of Proposition 4.10, we have a Fréchet principal bundle atlas* $\left\{\left(U^\alpha = \underleftarrow{\lim}U_n^\alpha, \mathbf{\Phi}^\alpha = \underleftarrow{\lim}\mathbf{\Phi}_n^\alpha\right)\right\}_{\alpha \in A}$ *such that, for each* $n \in \mathbb{N}$, *each* $\ell(E_n)$ *is trivializable over each* U_n^α *and the transition maps are*

$$\mathbf{T}_n^{\alpha\beta}(x_n) = \left(T_0^{\alpha\beta}(\mu_0^n(x_n)), \ldots, T_n^{\alpha\beta}(\mu_n^n(x_n))\right)$$

where $T_i^{\alpha\beta}$ *are the transition functions of the Banach bundle* (E_i, π_i, M_i) *for the atlas* $\{(U_i^\alpha, \tau_i^\alpha)\}_{\alpha \in A}$ *where* $0 \leq i \leq n$.

4.13 Projective Limits of G-Structures

Let $\left(\mathcal{G}_0^i(\mathbb{E}), \gamma_i^j\right)_{(i,j)\in\mathbb{N}^2,\ j \geq i}$ be a projective sequence of Banach-Lie groups associated with a sequence $(G_n)_{n\in\mathbb{N}}$ of weak Banach-Lie subgroup G_n of $\mathrm{GL}(\mathbb{E}_n)$.

Definition 4.25 *A sequence* $(F_n, \mathbf{p}_n|F_n, M_n, \mathcal{G}_0^n(\mathbb{E}))_{n\in\mathbb{N}}$ *is called a projective sequence of G-reductions of a projective sequence of Banach principal bundles* $(E_n, \mathbf{p}_n, M_n, \mathcal{H}_0^n(\mathbb{E}))_{\underleftarrow{n}}$ *if, for each* $n \in \mathbb{N}$, $\left(F_n, \mathbf{p}_{n|F_n}, M_n, \mathcal{G}_0^n(\mathbb{E})\right)$ *is a topological principal subbundle of* $(\mathbf{P}(E_n), \mathbf{p}_n, M_n, \mathcal{H}_0^n(\mathbb{E}))$ *such that* $\left(F, \pi_{|F}, M, G\right)$ *has its own smooth principal structure, and the inclusion is smooth.*

Then we have the following result:

Theorem 4.7 *Consider a sequence* $(\ell(E_n))_{n\in\mathbb{N}}=(\mathbf{P}(E_n),\mathbf{p}_n,M_n,\mathcal{H}_0^n(\mathbb{E}))_{n\in\mathbb{N}}$ *of generalized frame bundles over projective sequence of Banach manifolds. Let* $\left(F_n,\mathbf{p}_{n|F_n},M_n,\mathcal{G}_0^n(\mathbb{E})\right)_{\underleftarrow{n}}$ *be a projective sequence of G-reductions of* $(\ell(E_n))_{n\in\mathbb{N}}$.

Then $\underleftarrow{\lim}\mathbf{P}_n(E_n)$ *can be endowed with a structure of Fréchet principal bundle over* $\underleftarrow{\lim}M_n$ *whose structural group is* $\mathcal{G}_0(\mathbb{E})$.

Proof *(summarized proof) According to Remark 4.13 where we consider the atlas* $\left\{\left(U^\alpha=\underleftarrow{\lim}U_n^\alpha,\mathbf{\Phi}^\alpha=\underleftarrow{\lim}\mathbf{\Phi}_n^\alpha\right)\right\}_{\alpha\in A}$, *since each transition function* $T_i^{\alpha\beta}$ *belongs to* $\mathcal{G}_0^n(\mathbb{E})$, *thus the transition maps* $\mathbf{T}^{\alpha\beta}=\underleftarrow{\lim}\mathbf{T}_n^\alpha$ *associated with the atlas* $\{(U^\alpha,\mathbf{\Phi}^\alpha)\}_{\alpha\in A}$ *belong to* $\underleftarrow{\lim}\mathcal{G}_0^n(\mathbb{E})$.

4.14 Projective Limits of Tensor Structures

In this section, we only consider projective sequence $\left(E_i,\lambda_i^j\right)_{(i,j)\in\mathbb{N}^2,\ j\geq i}$ *of Banach bundles such that for* $\lambda_i^j:\pi_j^{-1}(x_j)\to\pi_i^{-1}(\delta_i^j(x_j))$ *is injective for all* $x_j\in M_j$ *and all* $j\geq i$ *and* $i\in\mathbb{N}$. *(cf. Remark 4.14)*

4.14.1 Projective Limits of Tensor Structures of Type (1, 1)

Recall that a tensor of type $(1,1)$ on a Banach space $\mathbb{E}$ is also an endomorphism of $\mathbb{E}$.

Definition 4.26

(1) Let $\left(\mathbb{E}_i,\overline{\lambda_i^j}\right)_{j\geq i}$ *be a projective sequence of Banach spaces. For any* $n\in\mathbb{N}$, *let* $S_n:\mathbb{E}_n\longrightarrow\mathbb{E}_n$ *be an endomorphism of the Banach space* $\mathbb{E}_n$. *A sequence* $(S_n)_{n\in\mathbb{N}}$ *is called a projective sequence of endomorphisms, or is coherent for short, if, for any integer* $j\geqslant i\geq 0$, *it fulfils the coherence condition:*

$$S_i\circ\overline{\lambda_i^j}=\overline{\lambda_i^j}\circ S_j.$$

(2) Let $(E_n,\pi_n,M_n)_{\underleftarrow{n}}$ *be a projective sequence of Banach vector bundles. A sequence* $(\mathscr{S}_n)_{n\in\mathbb{N}}$ *of endomorphisms* $\mathscr{S}_n$ *of* E_n *is called a projective sequence of endomorphisms, or a coherent sequence for short, if, for each* $x=\underleftarrow{\lim}x_n$, *the sequence* $\left((\mathscr{S}_n)_{x_n}\right)_{n\in\mathbb{N}}$ *is a coherent sequence of endomorphisms of* $(E_n)_{x_n}$.

We have the following properties:

Proposition 4.11 *Consider a coherent sequence $(\mathscr{S}_n)_{n\in\mathbb{N}}$ of endomorphisms $\mathscr{S}_n$ of E_n on a projective sequence of Banach vector bundles $(E_n, \pi_n, M_n)_{\underleftarrow{n}}$. Then the projective limit $\mathscr{S} = \varprojlim(\mathscr{S}_n)$ is well defined and is a smooth endomorphism of the Fréchet bundle $E = \varprojlim E_n$.*

Proof *By definition, a coherent sequence of endomorphisms is nothing but a projective sequence of linear maps; so the projective limit $S = \varprojlim S_n$ is a well-defined endomorphism of $\mathbb{E} = \varprojlim \mathbb{E}_n$. It follows that, for each $x = \varprojlim x_n \in M = \varprojlim M_n$, the projective limit $\mathscr{S}_x = \varprojlim(\mathscr{S}_n)_{x_n}$ is well defined. Now, from property* **(PSBVB5)***, there exists a local trivialization $\tau_n : \pi_n^{-1}(U_n) \longrightarrow U_n \times \mathbb{E}_n$ such that the following diagram is commutative:*

$$\begin{array}{ccc}
(\pi_i)^{-1}(U_i) & \xleftarrow{\lambda_i^j} & (\pi_j)^{-1}(U_j) \\
\Big\downarrow \tau_i & & \Big\downarrow \tau_j \\
U_i \times \mathbb{E}_i & \xleftarrow{\mu_i^j \times \overline{\lambda_i^j}} & U_j \times \mathbb{E}_j
\end{array}$$

This implies that the restriction of $\mathscr{S}_n$ to $(\pi_n)^{-1}(U_n)$ is a projective sequence of smooth maps and so, from [DGV16], Proposition 2.3.12, the restriction of $\mathscr{S}$ to $\pi^{-1}(U)$ is a smooth map where $U = \varprojlim U_n$, which ends the proof.

Let $(E_n, \pi_n, M_n)_{\underleftarrow{n}}$ be a projective sequence of Banach vector bundles. Consider a coherent sequence $(\mathscr{S}_n)_{n\in\mathbb{N}}$ of endomorphisms $\mathscr{S}_n$ defined on the Banach bundle $\pi_n : E_n \longrightarrow M_n$.
Fix any $x^0 = \varprojlim x_n^0 \in M = \varprojlim M_n$ and identify each typical fibre of E_n with $(E_n)_{x_n^0}$. For all $n \in \mathbb{N}$, we denote by G_n the isotropy group of $(\mathscr{S}_n)_{x_n^0}$ and by $\mathcal{G}_0^n$ the associated weak Banach-Lie subgroup of $\mathcal{H}_0^n$.

Then from according to Theorem 4.7, we have:

Theorem 4.8 *Let $(E_n, \pi_n, M_n)_{\underleftarrow{n}}$ be a projective sequence of Banach vector bundles. Consider a coherent sequence $(\mathscr{S}_n)_{n\in\mathbb{N}}$ of endomorphisms $\mathscr{S}_n$ defined on the Banach bundle $\pi_n : E_n \longrightarrow M_n$.*

Then $\varprojlim \mathbf{P}_n(E_n)$ can be endowed with a structure of Fréchet principal bundle over $\varprojlim M_n$ whose structural group is $\mathcal{G}_0(\mathbb{E})$ if and only if there exists a Fréchet atlas bundle $\left\{\left(U^\alpha = \varprojlim U_n^\alpha, \tau^\alpha = \tau_n^\alpha\right)\right\}_{n\in\mathbb{N},\alpha\in A}$ such that $\mathscr{S}_n^\alpha = \mathscr{S}_{n|U_n^\alpha}$ is locally modelled on $(\mathscr{S}_n)_{x_n^0}$ and the transition maps $T_n^{\alpha\beta}(x_n)$ take values into the isotropy group G_n, for all $n \in \mathbb{N}$ and all $\alpha, \beta \in A$ such that $U_n^\alpha \cap U_n^\beta \neq \emptyset$.

Proof *According to Chapter 2, Proposition 2.12, for each $n \in \mathbb{N}$, the Banach principal bundle $\ell(E_n)$ can be endowed with a structure of Banach principal bundle over M_n whose structural group is G_n if and only if* *there exists an atlas bundle* $\{(U_n^\alpha, \tau_n^\alpha)\}_{n\in\mathbb{N},\alpha\in A}$ *such that* $\mathscr{S}_n^\alpha$ *is locally modelled on* $(\mathscr{S}_n)_{x_n^0}$ *and the transition functions* $T_{\alpha\beta}(x)$ *belong to the isotropy group* G_n *for all* $x_n \in U_\alpha \cap U_\beta$*. Now, assume that we have Fréchet atlas* $\left\{U^\alpha = \varprojlim (U_n^\alpha)\ \tau^\alpha = \varprojlim (\tau_n^\alpha)\right\}_{n\in\mathbb{N},\alpha\in A}$ *for which such a property is true for all* $n \in \mathbb{N}$*. For all* $\alpha \in A$ *and for all* $j \geq i \geq 0$*, we have* $\tau_i^\alpha \circ \lambda_i^j = \mu_i^j \times \overline{\lambda_i^j} \circ \tau_j^\alpha$*. Thus we obtain*

$$T_i^{\alpha\beta} \circ \overline{\lambda_i^j} = \overline{\lambda_i^j} \circ T_j^{\alpha\beta}$$

and so $(T_1^{\alpha\beta}, \dots, T_n^{\alpha\beta})$ *belongs to* $\mathcal{G}_0^n(\mathbb{E})$*, which implies that* $\varprojlim \mathbf{P}_n(E_n)$ *can be endowed with a structure of Fréchet principal bundle over* $\varprojlim M_n$ *whose structural group is* $\mathcal{G}_0(\mathbb{E})$ *according to § 4.13.*

Conversely, under such an assumption, according to Theorem 4.7, consider a projective sequence $\left(F_n, \pi_{n|F_n}, M_n, \mathcal{G}_0^n(\mathbb{E})\right)_{n\in\mathbb{N}}$ *of G-reductions for* $(\ell(\mathbb{E}_n))_{n\in\mathbb{N}}$*. From the proof of this theorem and Remark 4.13, each transition map* $T_n^{\alpha\beta}$ *takes values in* G_n *and so the frame bundle of* (E_n, π_n, M_n) *can be endowed with a structure of Banach principal bundle over* M_n *whose structural group is* G_n*, (cf. § 2.2.1) which ends the proof.*

4.14.2 Projective Limits of Tensor Structures of Type $(2, 0)$

Definition 4.27

(1) Let $\left(\mathbb{E}_i, \overline{\lambda_i^j}\right)_{j\geq i}$ *be a projective sequence of Banach spaces.*
For any $n \in \mathbb{N}$*, let* $\omega_n : \mathbb{E}_n \times \mathbb{E}_n \longrightarrow \mathbb{R}$ *be a bilinear form of the Banach space* $\mathbb{E}_n$*.* $(\omega_n)_{n\in\mathbb{N}}$ *is called a projective sequence of 2-forms if it fulfills the coherence condition*

$$\forall (i,j) \in \mathbb{N}^2 : j \geq i \geq 0,\ \omega_j = \omega_i \circ \left(\overline{\lambda_i^j} \times \overline{\lambda_i^j}\right).$$

(2) Let $(E_n, \pi_n, M_n)_{\underleftarrow{n}}$ *be a projective sequence of Banach vector bundles. A sequence* $(\Omega_n)_{n\in\mathbb{N}}$ *of bilinear forms* Ω_n *on* E_n *is called a projective sequence of bilinear forms, or is a coherent sequence for short, if for each* $x = \varprojlim x_n$*, then we have*

$$(\Omega_j)_{x_j} = (\Omega_i)_{x_i} \circ \left(\lambda_i^j \times \lambda_i^j\right).$$

We have the following properties:

Proposition 4.12 *Consider a sequence* $(\Omega_n)_{n\in\mathbb{N}}$ *of coherent bilinear forms* Ω_n *of* E_n *on a projective sequence of Banach vector bundles* $(E_n, \pi_n, M_n)_{\underleftarrow{n}}$*. Then* $\Omega = \varprojlim(\Omega_n)$ *is well defined and is a smooth endomorphism of the Fréchet bundle* $E = \varprojlim E_n$*.*

Proof *Fix some point* $x = \varprojlim x_i$. *Then* $\left(\pi_i^{-1}(x_i), (\lambda_i^j)_{x_j}\right)_{j \geq i}$ *is a projective sequence of Banach spaces and so* $\left(\pi_i^{-1}(x_i) \times \pi^{-1}(x_i), (\lambda_i^j)_{x_j} \times (\lambda_i^j)_{x_j}\right)_{j \geq i}$ *is a projective sequence of Banach spaces. Now the coherence condition for the sequence* $(\Omega_i, x_i)_{i \in \mathbb{N}}$ *implies that, if* $u = \varprojlim u_i$ *and* $v = \varprojlim v_i$, *the projective limit* $\Omega_x(u,v) = \varprojlim (\Omega_i)_{x_i}(u_i, v_i)$ *is well defined. The end of the proof is similar to the analogue end of the proof of Proposition 4.11.*

Remark 4.14 *As mentioned at the beginning of this section, we only consider the case where* λ_i^j *is injective in restriction to each fibre. If it is not the case, since* $\Omega_j = (\lambda_i^j)^* \Omega_i$ *for* $j \geq i$ *the restriction of* Ω_j *to* $\ker \lambda_i^j$ *must be zero (cf. Remark 4.16). In particular, even if* Ω_i *is symplectic, and* $\ker \lambda_i^j \neq \{0\}$ *then* Ω_j *can not be symplectic.*

Let $(E_n, \pi_n, M_n)_{\underleftarrow{n}}$ be a projective sequence of Banach vector bundles. Consider a coherent sequence $(\Omega_n)_{n_i n \mathbb{N}}$ of bilinear forms Ω_n defined on the Banach bundle $\pi_n : E_n \longrightarrow M_n$.

Fix any $x^0 = \varprojlim x_n^0 \in M = \varprojlim M_n$ and identify each typical fibre of E_n with $(E_n)_{x_n^0}$. For all $n \in \mathbb{N}$, we denote by G_n the isotropy group of $(\Omega_n)_{x_n^0}$ and by $\mathcal{G}_0^n$ the associated weak Banach-Lie subgroup of $\mathcal{H}_0^n$.

Theorem 4.9 *Let* $(E_n, \pi_n, M_n)_{\underleftarrow{n}}$ *be a projective sequence of Banach vector bundles. Consider a coherent sequence* $(\Omega_n)_{n \in \mathbb{N}}$ *of bilinear forms* Ω_n *defined on the Banach bundle* $\pi_n : E_n \longrightarrow M_n$.
Then $\varprojlim \mathbf{P}_n(E_n)$ *can be endowed with a structure of Fréchet principal bundle over* $\varprojlim M_n$ *whose structural group is* $\mathcal{G}_0(\mathbb{E})$ *if and only if there exists a Fréchet bundle atlas*

$$\left\{\left(U^\alpha = \varprojlim U_n^\alpha, \tau^\alpha = \varprojlim \tau_n^\alpha\right)\right\}_{n \in \mathbb{N}, \alpha \in \boldsymbol{A}}$$

such that $\Omega_n^\alpha = \Omega_{n|U_n^\alpha}$ *is locally modelled on* $(\Omega_n)_{x_n^0}$ *and the transition maps* $T_n^{\alpha\beta}(x_n)$ *take values in the isotropy group* G_n, *for all* $n \in \mathbb{N}$ *and all* $\alpha, \beta \in A$ *such that* $U_n^\alpha \cap U_n^\beta \neq \emptyset$.

The proof of this theorem is formally the same as the proof of Theorem 4.8 and so is left to the reader.

4.14.3 Examples of Projective Limits of Tensor Structures

Again in this section we only consider projective sequence $\left(E_i, \overline{\lambda_i^j}\right)_{\underleftarrow{i}}$ of Banach bundles such that for $\lambda_i^j : \pi_j^{-1}(x_j) \to \pi_i^{-1}(\delta_i^j(x_j))$ is injective for all $x_j \in M_j$ and all integers i and j such that $j \geq i$

4.14.3.1 Projective Limits of Compatible Weak Almost Kähler and Almost Para-Kähler Structures

We consider the context and notations of Chapter 2, § 2.1.
If Ω (resp. g, resp. $\mathcal{I}$) is a weak symplectic form (resp. a weak Riemannian metric, resp. an almost complex structure) on a Fréchet bundle (E, π, M), the compatibility of any pair of the data $(\Omega, g, \mathcal{I})$ can be easily adapted to this Fréchet context. We can also easily transpose the notion of "local tensor $\mathscr{T}$ locally modelled on a linear tensor $\mathbb{T}$" (cf. Definition 2.31) in the Fréchet framework. We will see that such situations can be obtained by "projective limit" of a sequence $(E_n, \pi_n, M_n)_{\underleftarrow{n}}$ of Banach vector bundles.

Now, we introduce the following notations:

- Let $(\mathbb{T}_n)$ be a sequence of coherent tensors of type $(1,1)$ or $(2,0)$ on a projective sequence of Banach spaces $\left(\mathbb{E}_i, \overline{\lambda_i^j}\right)_{j \geq i}$.
  We denote by $G_n(\mathbb{T}_n)$ the isotropy group of $\mathbb{T}_n$. If $\mathbb{T} = \underleftarrow{\lim}\mathbb{T}_n$, then $\mathcal{G}_0^n(\mathbb{T})$ is the Banach-Lie group associated with the sequence $(G_n(\mathbb{T}_n))_{n\in\mathbb{N}}$. We set $\mathcal{G}_0(\mathbb{T}) = \underleftarrow{\lim}\mathcal{G}_0^n(\mathbb{T})$.
- Given a sequence $(E_n, \pi_n, M_n)_{\underleftarrow{n}}$ of Banach vector bundles, we provide each Banach bundle E_n with a symplectic form Ω_n and/or a weak Riemannian metric g_n and/or a complex structure $\mathcal{I}_n$.

Assume that we have a sequence of weak Riemaniann metrics $(g_n)_{n\in\mathbb{N}}$ on the sequence $(E_n)_{n\in\mathbb{N}}$. For each $x_n \in M_n$, we denote by $(\widehat{E}_n)_{x_n}$ the completion of normed space $(\check{E}_n)_{x_n} := \left((E_n)_{x_n}, || \; ||_{(g_n)_{x_n}}\right)$ and we set $\check{E}_n = \bigcup\limits_{x_n \in M_n} (\check{E}_n)_{x_n}$ and $\widehat{E} = \bigcup\limits_{x_n \in M_n} (\widehat{E}_n)_{x_n}$.

According to Chapter 2, Definition 2.33, let us consider the following assumptions:

Definition 4.28 *We say that the sequence $(E_n, \pi_n, M_n, g_n)_{n\in\mathbb{N}}$ with separable typical fibre satisfies the assumption of projective limit Hilbert extension (**(PLHE)** for short) if the following conditions are satisfied:*

(1) the sequence $(g_n)_{n\in\mathbb{N}}$ is coherent;

(3) each (E_n, π_n, M_n, g_n) satisfies the g_n-Hilbert extension property;

(4) there exists a covering $\{U^\alpha = \underleftarrow{\lim}U_n^\alpha\}_{\alpha\in A}$ of $M = \underleftarrow{\lim}M_n$ with trivialization $\tau^\alpha = \underleftarrow{\lim}\tau_n^\alpha : E_{|U} \to U \times \mathbb{E}$ such that the bundle atlas $\{(U_n^\alpha, \tau_n^\alpha)\}_{\alpha\in A}$ where $\tau_n^\alpha : (E_n)_{|U_n^\alpha} \to U_n^\alpha \times \mathbb{E}_n$ is a trivialization which satisfies the assumption (2) of the g_n-Hilbert extension for each n.

From Chapter 2, Theorem 2.4, we then obtain:

Theorem 4.10 *Consider a projective sequence* $(E_n, \pi_n, M_n, g_n)_{n\in\mathbb{N}}$ *which satisfies the assumption* **(PLHE)**. *Then we have:*

(1) The sequence $\left(\widehat{E}_n, \widehat{\pi}_n, M_n\right)_{\underleftarrow{n}}$ *(respectively* $\left(\check{E}_n, \pi_n, M_n\right)_{\underleftarrow{n}}$*) is a projective sequence of Hilbert bundles. the H-locally convex bundle* $\check{E} = \underleftarrow{\lim}\check{E}_n = E = \underleftarrow{\lim}E_n$ *as sets and is dense in the Fréchet Hilbert bundle* $\widehat{E} = \underleftarrow{\lim}\widehat{E}_n$.

(2) Denote by $\mathfrak{i}_n : \check{E}_n \to E_n$ *and* $\iota_n : \check{E}_n \to \widehat{E}_n$ *are the canonical injection morphisms, then* $\mathfrak{i} = \underleftarrow{\lim}\mathfrak{i}_n : \check{E} \to E$ *is the identity map and* $\iota = \underleftarrow{\lim}\iota_n : \check{E} \to \widehat{E}$ *is the natural inclusion morphism.*

(3) Let $\widehat{\Omega}_n$, *(resp.* $\widehat{g}_n$, *resp.* $\widehat{\mathcal{I}}_n$*) the extension of* Ω_n *(resp.* g_n, *resp.* $\mathcal{I}_n$*) to* $\widehat{E}_n$ *(cf. Chapter 2, Theorem 2.4). If the sequence* $(\Omega_n)_{n\in\mathbb{N}}$, *(resp.* $(g_n)_{n\in\mathbb{N}}$, *resp.* $(\mathcal{I}_n)_{n\in\mathbb{N}}$*) is coherent, so is the sequence* $(\widehat{\Omega}_n)_{n\in\mathbb{N}}$ *(resp.* $(\widehat{g}_n)_{n\in\mathbb{N}}$, *resp.*$(\widehat{\mathcal{I}}_n)_{n\in\mathbb{N}}$*). We set* $\widehat{\Omega} = \underleftarrow{\lim}\widehat{\Omega}_n$ *(resp.* $\widehat{g} = \underleftarrow{\lim}\widehat{g}_n$, *resp.* $\widehat{\mathcal{I}} = \underleftarrow{\lim}\widehat{\mathcal{I}}_n$*). We then have* $\Omega = \underleftarrow{\lim}\Omega_n = \iota^*\widehat{\Omega}$, $g = \underleftarrow{\lim}g_n = \iota^*\widehat{g}$ *and* $\mathcal{I} = \underleftarrow{\lim}\mathcal{I}_n = \iota^*\widehat{\mathcal{I}}$. *Moreover, if any pair of the data* $(\Omega_n, g_n, \mathcal{I}_n)$ *are compatible for all* $n \in \mathbb{N}$, *then* $(\widehat{\Omega}, \widehat{g}, \widehat{\mathcal{I}})$ *and* $(\Omega, g, \mathcal{I})$ *on the Fréchet vector bundle* $\widehat{E}$ *and* E *respectively, are compatible structures.*

(4) Let $(\ell(E_n))_{n\in\mathbb{N}} = (\mathbf{P}(E_n), \pi_n, M_n, \mathcal{H}_0^n(\mathbb{E}))_{n\in\mathbb{N}}$ *be the associated sequence of generalized frame bundles. Denote by* $\mathscr{T}_n$ *any tensor of the triple* $(\Omega_n, g_n, \mathcal{I}_n)$. *Assume that* $\mathscr{T}_n$ *defines a* $\mathbb{T}$*-tensor structure on* E_n *for each* $n \in \mathbb{N}$. *Then* $\underleftarrow{\lim}\mathbf{P}_n(E_n)$ *is a Fréchet principal bundle over* $\underleftarrow{\lim}M_n$ *whose structural group is* $\mathcal{G}_0(\mathbb{T})$.
Moreover, if the sequence $(\Omega_n, g_n, \mathcal{I}_n)_{n\in\mathbb{N}}$ *is coherent and each triple is a tensor structure* $(\Omega_n, g_n, \mathcal{I}_n)$ *on* E_n, *then* $(\Omega, g, \mathcal{I})$ *is also a tensor structure on* E *which compatible if* $(\Omega_n, g_n, \mathcal{I}_n)$ *fulfils this property.*

(5) Under the assumption **(PLHE)** , *all the properties of (4) are also valid for the sequence* $\left(\widehat{\Omega}_n, \widehat{g}_n, \widehat{\mathcal{I}}_n\right)_{n\in\mathbb{N}}$ *relatively to* $\widehat{E}$.

Proof

(1) and (2). Following the proof of Theorem 2.1 in Chapter 2, for each $x_n \in M_n$ *and each* $n \in \mathbb{N}$, *there exists a local trivialization* $\widehat{\tau}_n : \widehat{\pi}_n^{-1}(U_n) \longrightarrow U_n \times \widehat{\mathbb{E}}_n$ *and* $\tau_n : \pi_n^{-1}(U_n) \longrightarrow U_n \times \mathbb{E}_n$ *around* x_n *such that* $\widehat{\tau}_n \circ \iota_n = \tau_n$ *where* $\widehat{\tau}_n \circ \iota_n\left(\widehat{\pi}_n^{-1}(U_n)\right)$ *is dense in* $U_n \times \widehat{\mathbb{E}}_n$.
Now, we can choose such trivializations so that the property **(PLHE)**, *(2) is satisfied for the sequence* $(E_n, \pi_n, M_n)_{n\in\mathbb{N}}$. *For each* $x_n \in U_n$, *denote by* $(\check{E}_n)_{x_n}$ *the fibre* $(E_n)_{x_n}$ *provided with the norm associated with the inner product associated with* $(g_n)_{x_n}$. *Since the sequence* $(g_n)_{n\in\mathbb{N}}$ *is coherent, it follows that that each bounding map* f_i^j *gives rise to an isometry from* $(\check{E}_j)_{x_j}$ *into* $(\check{E}_i)_{x_i}$ *where* $x_i = \delta_i^j(x_j)$ *and so* f_i^j *must be injective. Thus, we can extend it to a bounding map* $(\widehat{f_i^j})_{x_j} : (\widehat{E}_j)_{x_j} \to (\widehat{E}_i)_{x_i}$, *which is also an*

isometry. The same arguments impliy that $\lambda_i^j : \mathbb{E}_j \to \mathbb{E}_i$ *gives rise to an isometry* $\check{\lambda}_i^j : \check{\mathbb{E}}_j \to \check{\mathbb{E}}_i$ *and so extends to an isometry* $\widehat{\lambda}_i^j : \widehat{\mathbb{E}}_j \to \widehat{\mathbb{E}}_i$.

Now by assumption **(PLBVB)** *for the projective sequence* $(E_n, \pi_n, M_n)_{\underleftarrow{n}}$, *we have*

$$\forall (i,j) \in \mathbb{N}^2 : j \geq i,\ \forall \alpha \in \mathbf{A},\ (\delta_i^j \times \lambda_i^j) \circ \tau_j^\alpha = \tau_i^\alpha \circ f_i^j.$$

It follows that the same relations are true for the projective sequence $(\check{E}_n, \pi_n, M_n)_{\underleftarrow{n}}$ *and from the assumption (2) of* **(PLHE)**, *we obtain*

$$(\delta_i^j \times \widehat{\lambda}_i^j) \circ \widehat{\tau}_j^\alpha = \widehat{\tau}_i^\alpha \circ \widehat{f}_i^j.$$

This implies that the assumption **(PLBVB)** *is satisfied for the projective sequences* $(\check{E}_n, \pi_n, M_n)_{\underleftarrow{n}}$ *and* $\left(\widehat{E}_n, \widehat{\pi}_n, M_n\right)_{\underleftarrow{n}}$. *Thus,* $\check{E} = \varprojlim \check{E}_n$ *is a Fréchet Hilbert bundle over* M *and* $\widehat{E} = \varprojlim \widehat{E}_n$ *is a* H*-locally convex bundle over* M. *Again from the assumption* **(HE)**, *since* $\iota_n : \check{E}_n \to \widehat{E}_n$ *is an injective bundle morphism with dense range, we have an injective vector bundle morphism* $\iota = \varprojlim \iota_n$ *from* $\check{E} \to \widehat{E}$, *if* $\bar{E}_n$ *with dense range. Note as sets, we have* $\check{E} = E$. *If* $\mathtt{i}$ *(resp.* $\mathtt{i}_n$ *) is the identity map from* $\check{E}$ *to* E *(resp. from* $\check{E}_n$ *to* E_n*), we have* $\mathtt{i} = \varprojlim \mathtt{i}_n$. *Moreover, we also have* $g = \varprojlim g_n$, $\widehat{g} = \varprojlim \widehat{g}_n$ *and finally* $g = \iota^* \hat{g}$ *which ends the proof of (1) and (2)*

(3) is a consequence of Chapter 2, Theorem 2.4 and the coherence of each sequence of these tensors.

(4) and (5). With the notations of (3), let $(\mathscr{T}_n)_{n\in\mathbb{N}}$ *anyone of the sequences* $\{\Omega_n, g_n, \mathcal{I}_n\}$ *and assume that the sequence* $(\mathscr{T}_n)_{n\in\mathbb{N}}$ *is coherent. Denote by* $\widehat{\mathscr{T}_n}$ *the extension of* $\mathscr{T}_n$ *to* $\widehat{E}_n$ *(which is well defined by Chapter 2, Theorem 2.4). Using the same arguments as the ones used for the bonding maps, it follows that* $(\widehat{\mathscr{T}_n})_{n\in\mathbb{N}}$ *is also a coherent sequence. Thus* $\mathscr{T} = \varprojlim \mathscr{T}_n$ *and* $\widehat{\mathscr{T}} = \varprojlim \widehat{\mathscr{T}_n}$ *are well defined and by construction we have* $\mathscr{T} = \iota^* \widehat{\mathscr{T}}$. *Moreover assume that each* $\mathscr{T}_n$ *defines a* $\mathbb{T}_n$*-structure on* E_n. *Then, from Chapter 2, Theorem 2.4 ,* $\widehat{\mathscr{T}_n}$ *is a* $\widehat{\mathbb{T}}$*-structure on* $\widehat{E}_n$. *Thus, from Theorem 4.8 or Theorem 4.9, each principal bundle* $\mathbf{P}_n(E_n)$ *(resp.* $\mathbf{P}_n\left(\widehat{E}_n\right)$*) is well defined and we can reduce its structural group* $\mathcal{G}_0^n(\mathbb{T})$ *(resp.* $\mathcal{G}_0^n\left(\widehat{\mathbb{T}}\right)$*). The last part of the proof is a direct consequence of the previous properties.*

Corollary 4.1 *Consider an almost Kähler (resp. para-Kähler) structure* $(\Omega_n, g_n, \mathcal{I}_n)_{n\in\mathbb{N}}$ *on a projective sequence* $(E_n, \pi_n, M_n)_{\underleftarrow{n}}$ *of Banach vector bundles with separable typical fibre. If* $(\Omega_n, g_n, \mathcal{I}_n)_{n\in\mathbb{N}}$ *is coherent, then* $(\Omega = \varprojlim \Omega_n, g = \varprojlim g_n, \mathcal{I} = \varprojlim \mathcal{I}_n)$ *is an almost Kähler (resp. para-Käler) structure on the Fréchet bundle* $E = \varprojlim E_n$.

4.14.3.2 Application to Sets of Smooth Maps

Let M be a connected manifold of dimension m and N a connected compact manifold of dimension n. We denote by $\mathsf{C}^k(N,M)$ the set of maps $f : N \longrightarrow M$ of class C^k. It is well known that $\mathsf{C}^k(N,M)$ is a Banach manifold modelled on the Banach space $\mathsf{C}^k(N,\mathbb{R}^m)$ which is separable. If $\lambda_i^j : \mathsf{C}^j(N,M) \longrightarrow \mathsf{C}^i(N,M)$ is the natural projection, the sequence $\left(\mathsf{C}^k(N,M)\right)_{k\in\mathbb{N}}$ is a projective sequence of Banach manifolds. By the way, the projective limit $\mathsf{C}^\infty(N,M) = \underleftarrow{\lim}\mathsf{C}^k(N,M)$ is provided with a Fréchet structure which is the usual Fréchet manifold structure on $\mathsf{C}^\infty(N,M)$.

Assume that M is a Kähler manifold. Let ω, $\mathfrak{g}$ and $\mathfrak{J}$ be the associated symplectic form, Riemannian metric and complex structure on M, respectively.

If ν is a volume measure on N, on the tangent bundle $T\mathsf{C}^k(N,M)$ we introduce (cf. [Kum13a] and [Kum15]):

- $(\Omega_k)_f(X,Y) = \displaystyle\int_N \omega\left(X_{f(t)}, Y_{f(t)}\right) d\nu(t);$
- $(g_k)_f(X,Y) = \displaystyle\int_N \mathfrak{g}\left(X_{f(t)}, Y_{f(t)}\right) d\nu(t);$
- $(\mathcal{I}_k)_f(X)(t) = \mathfrak{J}\left(X_{f(t)}\right).$

It is clear that Ω_k (resp. g_k) is a weak bilinear form (resp. a Riemaniann metric) on $\mathsf{C}^k(N,M)$ and $\mathcal{I}_k$ is an almost complex structure on $T\mathsf{C}^k(N,M)$. Moreover, the compatibility of the triple $(\omega,\mathfrak{g},\mathfrak{J})$ clearly implies the compatibility of $(\Omega_k, g_k, \mathcal{I}_k)$; in fact $(\Omega_k, g_k, \mathcal{I}_k)_{k\in\mathbb{N}^*}$ is a coherent sequence. Then, from Corollary 4.1, we get an almost Kähler structure on $\mathsf{C}^\infty(N,M)$. Note that this structure which can be directly defined on $\mathsf{C}^\infty(N,M)$ by the same formulae.

If M is a para-Kähler manifold, we can provide $\mathsf{C}^\infty(N,M)$ in the same way.

Remark 4.15 *If M is a Kähler manifold, then the almost complex structure is integrable and, in particular, $N_{\mathfrak{J}} \equiv 0$. It is clear that that we also have $N_{\mathcal{I}_k} \equiv 0$ for all $k \in \mathbb{N}$ and $N_{\mathcal{I}} \equiv 0$. Thus $\mathcal{I}$ is formally integrable but not integrable in general (cf. [Kum13a] in the case of sets of paths). On the opposite, if M is a para-Kähler manifold then again the Nijenhuis tensor of the almost para-complex structure $\mathcal{J}_k$ is zero and, by Theorem 2.10, this implies that the almost para-complex structure $\mathcal{J}$ on $\mathsf{C}^\infty(N,M)$ is also integrable.*

4.15 Darboux Charts on a Projective Limit of Sequences of Weak Symplectic Banach Manifolds

4.15.1 Projective Limit of a Sequence of Compatible Weak Symplectic Forms on a Projective Sequence of Banach Spaces

Let B be a symmetric or a skew-symmetric bilinear form on a Banach space $\mathbb{E}$ and $\mathbb{V}$ a Banach subspace of $\mathbb{E}$. Recall that the *B-orthogonal subspace* $\mathbb{V}_B^{\perp}$ is defined by

$$\mathbb{V}_B^{\perp} = \{x \in \mathbb{E}, : \forall y \in \mathbb{V}, B(x,y) = 0\}.$$

When there is no ambiguity, this set is simply denoted $\mathbb{V}^{\perp}$. Note that since B is symmetric or skew-symmetric, $\mathbb{V}_B^{\perp} = \{x \in \mathbb{E} : \forall y \in \mathbb{V}, B(y,x) = 0\}$ and so $(\mathbb{V}^{\perp})^{\perp} = \mathbb{V}$.

If $\mathbb{V}^0 = \{\xi \in \mathbb{E}^* : \forall u \in \mathbb{V}, \xi(u) = 0\}$ is the annihilator of $\mathbb{V}$, then $\mathbb{V}^{\perp} = (B^{\flat})^{-1}(\mathbb{V}^0)$.

Given two Banach subspaces $\mathbb{V}_1$ and $\mathbb{V}_2$ of $\mathbb{E}$, the following relations are classical:

- If $\mathbb{V}_1 \subset \mathbb{V}_2$ then $\mathbb{V}_2{}^{\perp} \subset \mathbb{V}_1^{\perp}$ and, in particular, for any subspace $\mathbb{V}$, $\mathbb{E}^{\perp} \subset \mathbb{V}^{\perp}$.
- $(\mathbb{V}_1 + \mathbb{V}_2)^{\perp} = \mathbb{V}_1^{\perp} \cap \mathbb{V}_2{}^{\perp}$.
- $(\mathbb{V}_1 \cap \mathbb{V}_2)^{\perp} = \mathbb{V}_1^{\perp} + \mathbb{V}_2{}^{\perp}$.

Let B_1 (resp. B_2) be a symmetric or a skew-symmetric bilinear form on a Banach space $\mathbb{E}_1$ (resp. $\mathbb{E}_2$) and $\ell : \mathbb{E}_1 \to \mathbb{E}_2$ a continuous map. Then we have:

Definition 4.29 *We say that ℓ is a weak isometry between B_1 and B_2 if $\ell(\mathbb{E}_1)$ is dense in $\mathbb{E}_2$ and we have:*

$$\ker \ell \cap (\ker \ell)^{\perp} = \{0\} \text{ and } \ell^* B_2 = B_1 \text{ on } (\ker \ell)^{\perp} \tag{4.10}$$

Note that the condition "$\ker \ell \cap (\ker \ell)^{\perp} = \{0\}$" is equivalent to the condition "the restriction of B_1 to $(\ker \ell)^{\perp}$ is non-degenerate".

Proposition 4.13 *Let B_1 (resp. B_2) be a weak symmetric or a skew-symmetric on a Banach space $\mathbb{E}_1$ (resp. $\mathbb{E}_2$) and let $\ell : \mathbb{E}_1 \to \mathbb{E}_2$ be a continuous map. We set $\mathbb{P} = (\ker \ell)$ and denote by $\overline{B}_1$ the restriction of B_1 to $\mathbb{P}^{\perp}$. We have the following properties:*

(1) If ℓ is a weak isometry between B_1 and B_2, then $\overline{B}_1$ and $B_{2|\ell(\mathbb{E}_1)}$ are non-degenerate, and $\ker \ell^* B_2 = \ker \ell$.

(2) *If B_1 and $\overline{B}_1$ are non-degenerate, ℓ is a weak isometry between B_1 and B_2 if and only if $\ell^* B_2 = B_1$ on $(\ker \ell)^\perp$ and, in this case, the restriction of B_2 to $\ell(\mathbb{E}_1)$ is non-degenerate.*

(3) *Let B_3 be a symmetric or skew-symmetric bilinear form on a Banach space $\mathbb{E}_3$ and $\ell' : \mathbb{E}_2 \to \mathbb{E}_3$ a continuous linear map. if ℓ_1 (resp. ℓ_2) is a weak isometry between B_1 and B_2 (resp. B_2 and B_3), then $\ell_2 \circ \ell_1$ is a weak isometric between B_1 and B_3.*

Note that if ℓ is a weak isometry between B_1 and B_2, the restriction $\overline{\ell}$ of ℓ to $(\ker \ell)^\perp$ is an isomorphism and

$$\overline{\ell}^* B_2 = \ell^* B_{2|(\ker \ell)^\perp} = B_{1|(\ker \ell)^\perp} \tag{4.11}$$

Proof *(1) We have $\overline{B}_1(u,v) = 0$ for all $v \in \mathbb{P}^\perp$ if and only if u belongs $\mathbb{P} \cap \mathbb{P}^\perp$, which implies that $\overline{B}_1$ is non-degenerate. Since $\overline{\ell}$ is an isomorphism, from (4.11), it follows that $B_{2|\ell(\mathbb{E}_1)}$ is non-degenerate. Now, v belongs to $\ker(\ell^*(B_2))^\flat$ if and only if $B_2(\ell(u), \ell(v))$, if and only if $\ell(u) = 0$.*

(2) Assume B_1 and $\overline{B}_1$ are non-degenerate. We must show that $\mathbb{E}_1 = \mathbb{P} \oplus \mathbb{P}^\perp$. But $(\mathbb{P} \oplus \mathbb{P}^\perp)^\perp = \mathbb{P}^\perp \cap \mathbb{P} = \{0\}$ and since B_1 is non-degenerate, it follows that $\{0\}^\perp = \mathbb{E}_1 = \mathbb{P} \oplus \mathbb{P}^\perp$, which ends the proof of (2) according to (4.11).

(3) Under the assumptions of (3), we have $\mathbb{E}_1 = \ker \ell_1 \oplus (\ker \ell_1)^\perp$ and $\mathbb{E}_2 = \ker \ell_2 \oplus (\ker \ell_2)^\perp$. Now, since the inclusion of $\ell_1(E_1)$ in $\mathbb{E}_2$ continuous, we set $\mathbb{P}_2 = \ker \ell_2 \cap \ell_1(\mathbb{E}_1)$; we then have $(\mathbb{P}_2)^\perp = (\ker \ell_2)^\perp \cap \ell_1(\mathbb{E}_1)$ and so $\ell_1(E_1) = \mathbb{P}_2 \oplus (\mathbb{P}_2)^\perp$. Let $\overline{\ell}_1$ be the restriction of ℓ_1 to $(\ker \ell_1)^\perp$; it is an isomorphism onto $(\mathbb{P}_2)^\perp$. If $\mathbb{P} = \overline{\ell}_1^{-1}(\mathbb{P}_2)$ and $\mathbb{X} = \overline{\ell_1}^{-1}((\mathbb{P}_2)^\perp)$, then $(\ker \ell_1)^\perp = \mathbb{P} \oplus \mathbb{X}$. By construction, $\ker \ell_2 \circ \ell_1 = \ker \ell_1 \oplus \mathbb{P}$ and we have $(\ker \ell_1 \oplus \mathbb{P})^\perp = (\ker \ell_1)^\perp \cap \mathbb{P}^\perp = \mathbb{X}$. Indeed $\mathbb{X}$ is contained in $(\ker \ell_1)^\perp$, and $\mathbb{X} = \overline{\ell_1}^{-1}((\mathbb{P}_2)^\perp)$ and $(\overline{\ell_1}^{-1})^*(B_{1|(\ker \ell_1)^\perp}) = B_{2|\ell_1(\mathbb{E}_1)}$, this implies that $\mathbb{X}$ is the orthogonal of $\mathbb{P}$ in $(\ker \ell_1)^\perp$. Now, the restriction of $\overline{\ell_2 \circ \ell_1}$ of $\ell_2 \circ \ell_1$ to $(\ker(\ell_2 \circ \ell_1))^\perp = \mathbb{X}$ is an isomorphism. For any $(u,v) \in \mathbb{X}^2$, we have:

$$B_1(u,v) = B_2(\overline{\ell_1}(u), \overline{\ell_1}(u)) = B_3(\overline{\ell_2 \circ \ell_1}(u), \overline{\ell_2 \circ \ell_1}(v)).$$

So the proof is completed.

Definition 4.30 *Let $\left(\mathbb{E}_i, \ell_i^j\right)_{j \geq i}$ be a reduced projective sequence Banach spaces and $(\omega_i)_{i \in \mathbb{N}}$ be a sequence of (linear) weak symplectic forms ω_i on $\mathbb{E}_i$. We say that $(\omega_i)_{i \in \mathbb{N}}$ is a compatible sequence of symplectic forms if each ℓ_i^{i+1} is a weak isometry between ω_{i+1} and ω_i, for all $i \in \mathbb{N}$.*

Remark 4.16

(1) *According to the properties of the sequence* $\left(\ell_i^j\right)_{j\geq i}$ *and Proposition 4.13 (3), if a sequence of weak symplectic 2-forms* ω_i *is compatible, then* ℓ_i^j *is a weak isometry between* ω_j *and* ω_i*, for all* $(i,j)\in\mathbb{N}^2$ *such that* $j\geq i$.

(2) *Note that if* $\ker\ell_i^j\neq\{0\}$*, the condition* $\omega_j=(\ell_i^j)^*\omega_i$ *implies that* $\mathbb{E}_j^{\perp}\neq\{0\}$ *and so* ω_j *cannot be symplectic (cf. Remark 4.14).*

We then have the following property:

Proposition 4.14 *Let* $(\omega_i)_{i\in\mathbb{N}}$ *be a compatible sequence of symplectic forms on a reduced projective sequence* $\left(\mathbb{E}_i,\ell_i^j\right)_{j\geq i}$*. Then if* $u=\varprojlim u_i$ *and* $v=\varprojlim v_i$ *in* $\mathbb{E}=\varprojlim\mathbb{E}_i$,

$$\omega:(u,v)\mapsto\varprojlim\omega_i(u_i,v_i)$$

defines a weak symplectic 2-form on $\mathbb{E}$.

The proof of this Proposition needs the following Lemma which will be also used later:

Lemma 4.2 *Let* $\mathbb{E}$ *be a projective limit of a reduced projective sequence* $\left(\mathbb{E}_i,\ell_i^j\right)_{j\geq i}$*. Assume that, for all* $(i,j)\in\mathbb{N}^2$ *such that* $j\geq i$*, the kernel of* ℓ_i^j *is supplemented.*

(1) *For each* $(i,j)\in\mathbb{N}^2$ *such that* $j\geq i$ *we have a decomposition*

$$\mathbb{E}_j=\mathbb{E}_i^j\oplus\mathbb{E}_{i+1}^j\oplus\cdots\oplus\mathbb{E}_{j-1}^j\oplus\ker\ell_{j-1}^j \tag{4.12}$$

with the following properties for all $j\geq l$

(a) $\ker\ell_l^j=\mathbb{E}_{l+1}^j\oplus\cdots\oplus\mathbb{E}_{j-2}^j\oplus\ker\ell_{j-1}^j$;

(b) *the restriction of* $(\ell')_l^j$ *of* ℓ_l^j *to* $(\mathbb{E}_i^j\oplus\cdots\oplus\mathbb{E}_l^j)$ *is injective with dense range in* $\mathbb{E}_l$;

(c) $\ell_l^j(\mathbb{E}_h^j)$ *is dense in* $\mathbb{E}_h^l$ *for all* $h\leq l\leq j$*, is dense in* $\ker\ell_{l-1}^l$ *for* $h=l$ *and* $\ell_l^j(\mathbb{E}_h^j)=\{0\}$ *for* $l<h\leq j-1$.

(2) *Let* $\ell_i:\mathbb{E}\to\mathbb{E}_i$ *the canonical projection. Then* $\mathbb{E}=\ker\ell_i\oplus\breve{\mathbb{F}}_i$ *and the restriction* ℓ_i' *of* ℓ *to* $\breve{\mathbb{F}}_i$ *is a continuous injective map into* $\mathbb{E}_i$ *with dense range. Moreover, if* $\|\ \|_i$ *is a norm on* $\mathbb{E}_i$*, then* $\nu_i=\|\ \|_i\circ\ell_i$ *is a semi-norm on* $\mathbb{E}$ *and the restriction of* ν_i *to* $\breve{\mathbb{F}}_i$ *is a norm and in this case,* ℓ_i' *is an isometry. In particular, the completion* $\overline{\breve{\mathbb{F}}_i}$ *of* $\breve{\mathbb{F}}_i$ *is isomorphic to* $\mathbb{E}_i$.

(3) *We set* $\mathbb{X}_i=\ker\ell_{i-1}^i$ *for* $i\geq 1$ *and* $\mathbb{X}_0=\mathbb{E}_0$*. If* $\breve{\mathbb{E}}_j=\prod_{l=0}^{j}\mathbb{X}_l$*, then there exist bounding maps* $\kappa_i^j:\breve{\mathbb{E}}_j\to\breve{\mathbb{E}}_i$ *with dense range, so that* $\left(\breve{\mathbb{E}}_i,\kappa_i^j\right)_{j\geq i}$ *is*

a reduced projective sequence. If $\breve{\mathbb{E}} = \varprojlim \breve{\mathbb{E}}_i$*, there exists an injective continuous linear map* $\varprojlim \theta_i : \mathbb{E} \to \breve{\mathbb{E}}$ *with dense range where* θ_i *is an injective linear map from* $\mathbb{E}_i$ *into* $\breve{\mathbb{E}}_i$ *with dense range. Moreover, if* $\left(\mathbb{E}_i, \ell_i^j\right)_{j \geq i}$ *is a surjective projective sequence, then each* θ_i *for* $i \in \mathbb{N}$ *and* θ *are isomorphisms.*

Remark 4.17

(1) From the properties of the sequence $\left(\ell_i^j\right)$ *and Proposition 4.13 (3), if* $(\omega_i)_{i \in \mathbb{N}}$ *is a sequence of compatible weak symplectic 2-forms, then* ℓ_i^j *is a weak isometry between* ω_j *and* ω_i*, for all* $i \in \mathbb{N}$ *and all* $j \geq i$*.*

(2) Consider the assumptions of Proposition 4.14. If $\left(\mathbb{E}_i, \ell_i^j\right)$ *is a* ILB *sequence, we have* $\ker \ell_i^j = \{0\}$ *for all* $j \geq i$ *and* $i \in \mathbb{N}$*. Thus* $(\omega_i)_{i \in \mathbb{N}}$ *is a sequence of compatible symplectic forms on this projective system if and only if, for all* $j \geq i$ *and* $i \in \mathbb{N}$*, then* $\omega_j = (\lambda_i^j)^* \omega_i$*, and* ω_0 *is symplectic. But in general, if for some pair* (i, j)*,* $\ker \ell_i^j \neq \{0\}$*, the condition* $\omega_j = (\ell_i^j)^* \omega_i$ *implies that* $\mathbb{E}_j^{\perp} \neq \{0\}$ *and so this does not implies that* ω_j *is symplectic.*

(3) In Proposition 4.14, when $\left(\mathbb{E}_i, \ell_i^j\right)$ *is a surjective*[10] *projective sequence, the symplectic form* ω *on* $\mathbb{E}$ *has the property that the induced form on* $\ker \ell_i$ *is symplectic and so we have* $\mathbb{E} = \ker \ell_i \oplus (\ker \ell_i)^{\perp}$ *where* $(\ker \ell_i)^{\perp}$ *is the orthogonal of* $\ker \ell_i$ *(relative to* ω*).*

Proof *(1) Fix some* $i \in \mathbb{N}$ *and assume that, for all* $i \leq l \leq j$*, we have a decomposition of type (4.12) with properties (a), (b) and (c).*
It is clear that this assumption is true for $l = i$ *and* $l = i + 1$*. At first, we have a decomposition*

$$\mathbb{E}_{j+1} = \ker \ell_j^{j+1} \oplus \mathbb{F}_{j+1}.$$

Therefore the restriction $(\ell')_j^{j+1}$ *of* ℓ_j^{j+1} *to* $\mathbb{F}_{j+1}$ *is an injective continuous map from* $\mathbb{F}_{j+1}$ *into* $\mathbb{E}_j$ *and* $(\ell')_j^{j+1}$ *and* ℓ_i^{j+1} *have the same range, so* $(\ell')_j^{j+1}(\mathbb{F}_{j+1})$ *is dense in* $\mathbb{E}_j$*. According to (4.12), each vector space* $\mathbb{E}_j^{j+1} = [(\ell')_j^{j+1}]^{-1}(\ker \ell_{j-1}^j)$ *,* $\mathbb{E}_l^{j+1} = [(\ell')_j^{j+1})]^{-1}(\mathbb{E}_l^j)$ *for all* $i \leq l < j$ *are Banach subspaces of* $\mathbb{F}_{j+1}$ *and we have the following decomposition:*

$$\mathbb{F}_{j+1} = \mathbb{E}_i^{j+1} \oplus \mathbb{E}_{i+1}^{j+1} \oplus \mathbb{E}_{i+2}^{j+1} \oplus \cdots \oplus \mathbb{E}_j^{j+1}.$$

It follows that, for $j > l > i$*, we have*

$$\begin{aligned}(\ell_j^{j+1})^{-1}(\ker \ell_l^j) &= (\ell_j^{j+1})^{-1}(\mathbb{E}_{l+1}^j \oplus \cdots \oplus \mathbb{E}_{j-1}^j \oplus \ker \ell_{j-1}^j) \\ &= \mathbb{E}_{l+1}^{j+1} \oplus \cdots \oplus \mathbb{E}_{j-1}^{j+1} \oplus \mathbb{E}_{j-1}^j \oplus \ker \ell_j^{j+1}\end{aligned}$$

[10] That is each ℓ_i^j is surjective.

and also:

$$\begin{aligned}\ell_l^{j+1}(\mathbb{E}_i^{j+1} \oplus \mathbb{E}_{i+1}^{j+1} \oplus \cdots \oplus \mathbb{E}_j^{j+1}) &= \ell_l^j(\ell_j^{j+1}(\mathbb{E}_i^{j+1} \oplus \mathbb{E}_{i+1}^{j+1} \oplus \cdots \oplus \mathbb{E}_j^{j+1}) \\ &= \ell_l^j(\mathbb{E}_i^j \oplus \mathbb{E}_{i+1}^j \oplus \cdots \oplus \mathbb{E}_{j-1}^j \oplus \ker \ell_{l-1}^l) \\ &= \ell_l^j(\mathbb{E}_j)\end{aligned}$$

which is dense in $\mathbb{E}_l$.
For $l = j$, *we have*

$$\ell_j^{j+1}(\mathbb{E}_i^{j+1} \oplus \mathbb{E}_{i+1}^{j+1} \oplus \mathbb{E}_{i+2}^{j+1} \oplus \cdots \oplus \mathbb{E}_j^{j+1}) = \ell_j^{j+1}(\mathbb{E}_{j+1})$$

which is dense in $\mathbb{E}_j$. *Finally, according to the definition of* $(\ell')_j^{j+1}$ *and the definition* $\mathbb{E}_h^{j+1}$ *it follows that* $(\ell')_j^{j+1}$ *is an injective continuous map from* $\mathbb{E}_h^{j+1}$ *into* $\mathbb{E}_h^j$ *with dense range for all* $i \leq h < j+1$. *Thus (1) is proved.*

(2) For i *fixed, on the one hand, from the property of bonding maps, the sequence* $\left(\ker \ell_i^j, (\ell_j^{j'})_{|\ker \ell_i^j}\right)_{j' \geq j}$ *is a projective sequence of Banach spaces. Thus if* $\ell_i = \varprojlim \ell_i^j$, *then* $\ker \ell_i = \varprojlim \ker \ell_i^j$ *which is a closed Fréchet subspace of* $\mathbb{E}$.

On the other hand, we have $\mathbb{E}_j = \ker \ell_i^j \oplus \mathbb{E}_i^j$ *and from (c), the sequence* $\left(\mathbb{E}_i^j, (\ell_i^j)_{|\mathbb{E}_i^j}\right)_{j \geq j}$ *is a projective sequence of Banach spaces. Thus if* $\breve{\mathbb{F}}_i = \varprojlim \mathbb{E}_i^j$, *we have* $E = \ker \ell_i \oplus \breve{\mathbb{F}}_i$. *It follows that the restriction* ℓ_i' *of* ℓ_i *to* $\breve{\mathbb{F}}_i$ *is an isomorphism onto* $\ell_i(\breve{\mathbb{F}}_i)$ *which is dense in* $\mathbb{E}_i$. *If* $|| \; ||_i$ *is a norm on* $\mathbb{E}_i$, *then* $\nu_i = || \; || \circ \ell_i$ *is a semi-norm on* $\mathbb{E}$ *whose kernel is precisely* $\ker \ell_i$. *Thus, the restriction of* ν_i *to* $\breve{\mathbb{F}}_i$ *is a norm and so* ℓ_i' *is an isometry which ends the proof of (2).*

(3) We set $\mathbb{X}_0 = \mathbb{E}_0$ *and* $\mathbb{X}_h = \ker \ell_{h-1}^h$ *and* $\breve{\mathbb{E}}_j = \prod_{h=0}^{j} \mathbb{X}_h$. *We consider the followings maps:*

$\theta_j : \mathbb{E}_j \to \breve{\mathbb{E}}_j$ *defined in the following way: for* $i \in \mathbb{N}$, *according to the decomposition (4.12), each* $x^j \in \mathbb{E}_j$ *can be written as a sequence* $(x_l^j)_{0 \leq h \leq j}$ *where* x_l^j *belongs to* $\mathbb{E}_l^j$ *and so* $\theta_j\left((x_l^j)\right) = \ell_l^j\left((x_l^j)\right)$ *belongs to* $\breve{\mathbb{E}}_i$ *from property (c)*[11] *and so is well defined.*

$\kappa_i^j : \breve{\mathbb{E}}_j \to \breve{\mathbb{E}}_i$ *defined in the following way: if* $\overline{x'}_i = (x_0', \dots, x_l' \dots, x_i')$ *and* $\overline{y'}_j = (y_0', \dots, y_l' \dots, y_j')$ *then* $\kappa_i^j\left(\overline{x'}_j = (x_l')\right) := \left((\ell_l^j(x_l')\right)$ *for* $0 \leq l \leq i$ *and for all* $i \in \mathbb{N}$ *and* $j \geq i$.

Note that θ_j *is a continuous linear map which is injective. Indeed, if* (x_l') *belongs to* $\theta_j(\mathbb{E}_j)$ *then* x_l' *belongs to* $\ell_l^j(\mathbb{E}_l^j) \subset \ker \ell_{l-1}^l$. *The restriction* ℓ_l^j *to* $\mathbb{E}_l^j$

[11] Take $h = l$ in property (c).

being injective according to property (b), we have a unique $x_l^j \in \mathbb{E}_l^j$ *such that* $\ell_l^j(x_l^j) = x_l'$ *for all* $l \in \{0, \dots, j\}$. *But, from property (c),* $\ell_l^j(\mathbb{E}_l^j)$ *is dense in* $\mathbb{X}_l$, *thus* θ_j *has a dense range in* $\mathbb{E}_j'$. *According to these notations, we can see that we have* $\theta_i \circ \ell_i^j = \kappa_i^j \circ \theta_j$.

It follows that we get an injective continuous linear map $\theta = \varprojlim \theta_i$ *from* $\mathbb{E} = \varprojlim \mathbb{E}_i$ *to* $\mathbb{E}' = \varprojlim \mathbb{E}_i'$ *But* $\mathbb{E}$ *can be identified with*

$$\left\{ (x_i) \in \prod_{i \in \mathbb{N}} \mathbb{E}_i, \, : \, x_l^i = \ell_i^j(x_l^j), \;\; 0 \leq l \leq i, \; j \geq i, \; i \in \mathbb{N} \right\}.$$

In the same way, $\breve{\mathbb{E}}$ *can be identified with*

$$\left\{ (\overline{x_i'}) \in \prod_{i \in \mathbb{N}} \breve{\mathbb{E}}_i, \, : \, x_l' = \kappa_i^j(x_l'), \; 0 \leq l \leq i, \; j \geq i, \; i \in \mathbb{N} \right\}.$$

This implies that θ *is a continuous injective map from* $\mathbb{E}$ *to* $\breve{\mathbb{E}}$ *with dense range.*

Now assume that ℓ_i^j *is surjective for all* $(i, j) \in \mathbb{N}^2$ *such that* $j \geq i$. *Then, clearly, this implies that* κ_i^j *is also surjective and so* θ_j *is an isomorphism for all* $j \geq i$. *In this way,* θ *is also an isomorphism.*

Proof of Proposition 4.14 *From our assumption, on the sequence* $(\omega_i)_{i \in \mathbb{N}}$ *and Proposition 4.13, we have a decomposition* $\mathbb{E}_i = \ker \ell_{i-1}^i \oplus \mathbb{F}_i$ *with* $\mathbb{F}_i = (\ker \ell_{i-1}^i)^\perp$ *for all* $i \geq 1$. *Thus we can apply Lemma 4.2. Thus, for all* $1 \leq j \leq n$, *we have a decomposition*

$$\mathbb{E}_j = \ker \ell_{j-1}^j \oplus \mathbb{F}_j \text{ with } \mathbb{F}_j = \mathbb{E}_0^j \oplus \mathbb{E}_1^j \oplus \mathbb{E}_2^j \oplus \cdots \oplus \mathbb{E}_{j-1}^j.$$

At first, ω_0 *is a symplectic form* $\omega_{\mathbb{X}_0}$ *on* $\mathbb{X}_0 = \mathbb{E}_0$. *From the assumption on the sequence* $(\omega_i)_{i \in \mathbb{N}}$, *by induction on* i, ω_i *induces a symplectic form* $\omega_{\mathbb{X}_i}$ *on the space* $\mathbb{X}_i (:= \ker \ell_{i-1}^i)$. *In this way, we obtain a symplectic form* ω_i' *on* $\breve{\mathbb{E}}_i$ *by*

$$\omega_i'(\overline{u'}_i, \overline{v'}_i) = \sum_{l=0}^{i} \omega_{\mathbb{X}_l}(u_l', v_l') \tag{4.13}$$

if $\overline{u'}_i = (u_0', \dots u_i')$ *and* $\overline{v'}_i = (v_0', \dots v_i')$. *According to the notations of the proof of Lemma 4.2, for all* $(i, j) \in \mathbb{N}^2$ *such that* $j \geq i$, *from the definition of* κ_i^j *and* θ_j, *it is easy to see that*

$$\omega_{j|\mathbb{X}_l} = \left((\kappa_i^j)^* \omega_i' \right)_{|\mathbb{X}_l} \;\; : \forall \;\; 0 \leq l \leq i \text{ and } \omega_i = \theta_i^* \omega_i'. \tag{4.14}$$

Now the sequence $\left(\breve{\mathbb{E}}_i \times \breve{\mathbb{E}}_i, \kappa_i^j \times \kappa_i^j\right)_{j \geq i}$ *is a reduced projective sequence. Thus according to (4.14), we can define a* 2*-form* ω' *on* $\breve{\mathbb{E}}$ *in the following way:*

if $\overline{u'} = (u'_l) \in \mathbb{E}$ *and* $\overline{v'} = (v'_l) \in \breve{\mathbb{E}}$ *then*

$$\omega'(\overline{u'}, \overline{v'}) = \varprojlim \omega'_i\left((u'_0, \ldots, u'_i), (v'_0, \ldots, v'_i)\right).$$

It remains to show that ω' *is symplectic. At first, note that by construction of* ω'_i *we have* $\omega'_i(u,v) = 0$ *for all* $u \in \mathbb{X}_j$ *and* $v \in \prod_{l \neq j, 0 \leq l \leq i} \mathbb{X}_l$*. Assume that* $\omega'(\overline{u'}, \overline{v'}) = 0$ *for all* $\overline{v'} \in \mathbb{E}'$*; this implies that, for all* $l \in \mathbb{N}$*, we have* $\omega_{\mathbb{X}_l}(u'_l, v'_l) = 0$*, for all* $v_l \in \mathbb{X}_l$ *and so we must have have* $u'_l = 0$ *for all* $l \in \mathbb{N}$*. Now, since* $\theta = \varprojlim \theta_i$ *is injective, and* $\omega_i = \theta_i^* \omega'_i$ *and the range of* θ *is dense this implies the results for* ω*.*

As in finite dimension, we introduce:

Definition 4.31 *Let* $\mathbb{E} = \varprojlim \mathbb{E}_i$ *be a projective limit of a surjective projective sequence* $\left(\mathbb{E}_i, \ell_i^j\right)_{j \geq i}$*. Consider a (weak) symplectic form* ω *on* $\mathbb{E}$ *such that* $\mathbb{E} = \ker \ell_i \oplus (\ker \ell_i)^\perp$*. We will say that* ℓ_i *is a symplectic submersion*[12]*.*

Remark 4.18 *In the context of Definition 4.31, the restriction of* ℓ_i *to* $(\ker \ell_i)^\perp$ *is an isomorphism onto* $\mathbb{E}_i$ *and so we have a well symplectic form* ω_i *on* $\mathbb{E}_i$ *such that* $\omega = \ell_i^* \omega_i$ *in restriction to* $(\ker \ell_i)^\perp$*.*

We have the following type of converse of Proposition 4.14:

Proposition 4.15 *Let* $\left(\mathbb{E}_i, \ell_i^j\right)_{j \geq i}$ *be a surjective projective sequence of Banach spaces and* $E = \varprojlim E_i$*. If* ω *is a symplectic form on* E *such that* $\ell_i : E \to E_i$ *is a symplectic submersion for all* $i \in \mathbb{N}$*, then* ω *induces a symplectic form* ω_i *on* E_i*. Moreover,* (ω_i) *is a sequence of compatible symplectic forms and the projective limit associated with this sequence is precisely* ω*.*

Proof *Since for* $j \geq i$*,* $\ell_i = \ell_i^j \circ \ell_j$ *this implies* $\ker \ell_i = \ker \ell_j \oplus (\ell')_j^{-1}(\ker \ell_i^j)$*. Thus we have* $\ker \ell_j \subset \ker \ell_i$ *and so* $(\ker \ell_i)^\perp \subset (\ker \ell_j)^\perp$*. As we have seen previously, there exists a (unique) symplectic form* ω_j *on* $\mathbb{E}_j$ *such that* $\omega = \ell_j^* \omega_j$ *on* $(\ker \ell_j)^\perp$*. Since for any* $j \in \mathbb{N}$*, the restriction* ℓ'_j *to* $(\ker \ell_j)^\perp$ *is an isomorphism onto* $\mathbb{E}_j$*, we have:*

$\omega_j(u_j, v_j) = \omega(\ell'_j(u'_j), \ell'_j(v'_j))$ *for all* $(u'_j, v'_j) \in \left((\ker \ell_j)^\perp\right)^2$ *with* $u_j = \ell'_j(u'_j)$ *and* $v_j = \ell'_j(v'_j)$*.*

[12]This terminology is classical in finite dimension symplectic geometry.

But, since $(\ker \ell_i)^\perp \subset (\ker \ell_j)^\perp$, *it follows that, for any* $(u'_i, v'_i) \in \left((\ker \ell_i)^\perp\right)^2$, *we have*

$$\begin{aligned}\omega(\ell'_i(u'_i), \ell'_i(v'_i)) &= \omega(\ell_i^j \circ \ell'_j(u'_i), \ell_i^j \circ \ell'_j(v'_i)) \\ &= (\ell_i^j)^* \omega(\ell'_j(u'_i), \ell'_j(v'_i)).\end{aligned}$$

Thus we obtain

$$(\omega_j) = (\ell_i^j)^* \omega_i \text{ on } \ell'_j\left((\ker \ell_i)^\perp\right).$$

The proof will be completed if we show that $\ell'_j\left((\ker \ell_i)^\perp\right) = (\ker \ell_i^j)^\perp$. *But this results follows from* $\ell'_j(\ker \ell_i) = \ker \ell_i^j$.

4.15.2 Projective limit of a sequence of compatible weak symplectic forms on a projective sequence of Banach bundles

Definition 4.32 *Let* $\left(E_i, \lambda_i^j\right)_{\underleftarrow{i}}$ *be a projective sequence of Banach bundles over a projective sequence* $\left(M_i, \delta_i^j\right)_{j \geq i}$ *of manifolds and let* $(\omega_i)_{i \in \mathbb{N}}$ *be a sequence of weak symplectic forms* ω_i *on* E_i. *If* $\mathbb{E}_i$ *is the typical fibre of* E_i, *assume that the following properties are satisfied:*

(RPSBS) *The sequence* $\left(\mathbb{E}_i, \overline{\lambda_i^j}\right)_{j \geq i}$ *is a reduced projective sequence of Banach spaces.*

We say that $(\omega_i)_{i \in \mathbb{N}}$ *is a sequence of compatible symplectic forms if the sequence* $((\omega_i)_{x_i})_{i \in \mathbb{N}}$ *is a sequence of compatible (linear) weak symplectic forms on the projective sequence* $\left(\pi_i^{-1}(x_i), (\lambda_i^j)_{x_j}\right)_{j \geq i}$ *of Banach spaces.*

Under the context of this Definition, we have

Theorem 4.11 *Consider a projective sequence* $\left(E_i, \lambda_i^j\right)_{\underleftarrow{i}}$ *of Banach bundles over a projective sequence of Banach manifolds* $\left(M_i, \delta_i^j\right)_{j \geq i}$ *which satisfies the assumption* **(RPSBS)**.

(1) Let $(\omega_i)_{i \in \mathbb{N}}$ *be a sequence of compatible weak symplectic forms on* $\left(E_i, \lambda_i^j\right)_{\underleftarrow{i}}$. *Then* $\omega = \varprojlim \omega_i$ *is a well-defined weak symplectic form on the Fréchet bundle* $E = \varprojlim E_i$ *over* $M = \varprojlim M_i$.

(2) Conversely, let ω *be a weak symplectic form on a projective limit bundles* $(E = \varprojlim E_i, \pi = \varprojlim \pi_i, M = \varprojlim M_i)$ *of a submersive projective sequence of Banach bundles* $(E_i, \pi_i, M_i)_{\underleftarrow{i}}$. *Assume that for each* $x = \varprojlim x_i$, *the map* $(\lambda_i)_x : \pi^{-1}(x) \to \pi_i^{-1}(x_i)$ *is a symplectic submersion. Then* ω *induces a weak symplectic 2-form* ω_i *on* E_i *which gives rise to a family of compatible*

weak symplectic forms. Moreover, the 2-form on E defined by this sequence (ω_i) is precisely the given 2-form ω.

We obtain directly the following Corollary:

Corollary 4.2 *A 2-form ω on a Fréchet bundle, projective limit ($E = \underleftarrow{\lim} E_i, \pi = \underleftarrow{\lim} \pi_i, M = \underleftarrow{\lim} M_i$) of a submersive sequence of Banach fibre bundles $\left(E_i, \lambda_i^j\right)_{\underleftarrow{i}}$ is a weak symplectic form if and only if there exists a sequence of compatible weak symplectic forms $(\omega_i)_{i\in\mathbb{N}}$ on E_i such that $\omega = \underleftarrow{\lim} \omega_i$.*

By taking $E_i = TM_i$, we obtain the following Corollary:

Corollary 4.3

(1) Let $\left(M_i, \lambda_i^j\right)_{j\geq i}$ be a reduced sequence of Banach manifolds and $(\omega_i)_{i\in\mathbb{N}}$ a sequence of compatible weak symplectic forms. Then $\omega = \underleftarrow{\lim} \omega_i$ is a weak symplectic form on $M = \underleftarrow{\lim} M_i$

(2) Let ω be a 2-form on a projective limit $M = \underleftarrow{\lim} M_i$ of a submersive sequence of manifolds $\left(M_i, \delta_i^j\right)_{j\geq i}$. Then ω is a weak symplectic form if and only if each $T_x\delta_i : T_xM \to T_{x_i}M_i$ is a symplectic submersion for each $i \in \mathbb{N}$.

Definition 4.33 *Let $M = \underleftarrow{\lim} M_i$ be a projective limit of a submersive sequence $\left(M_i, \delta_i^j\right)_{j\geq i}$ of manifolds and ω a weak symplectic form on M. We say that the canonical projection $\delta_i : M \to M_i$ is a symplectic submersion if the restriction of ω to each fibre $\delta^{-1}(x_i)$ is a symplectic form.*

Remark 4.19 *Under the context of this corollary, each map $\delta_i^j : M_j \to M_i$ is a symplectic submersion in the classical sense in finite-dimensional symplectic geometry, that is ω_j (resp. ω_i) is a symplectic form on M_j (resp. M_i) such that the restriction of ω_j is a symplectic form on each fibre of δ_i^j.*

Proof of Theorem 4.11 *From Proposition 4.14, we know that ω is well defined. Now ω is a smooth 2-form since it is a projective limit of smooth 2 forms, which ends the proof of (1).*
Now let ω be a weak symplectic form on a projective limit bundle $E = \underleftarrow{\lim} E_i, \pi = \underleftarrow{\lim} \pi_i, M = \underleftarrow{\lim} M_i$ which satisfies the assumptions of (2). Given some $x = \underleftarrow{\lim} x_i \in M$, since $(\lambda_i)_x : \pi^{-1}(x) \to \pi_i^{-1}(x_i)$ is a symplectic submersion, the restriction $(\lambda_i)'$ of λ_i to $\ker(\lambda_i)_x^{\perp}$ is an isomorphism on $\pi_i^{-1}(x_i)$ and so $(\omega_i)_{x_i} = \{[(\lambda_i)_x]^{-1}\}^*(\omega_x)_{|\ker(\lambda_i)_x^{\perp}}$ is a symplectic form on $\pi_i^{-1}(x_i)$. It remains to show that $x_i \mapsto \omega_{x_i}$ is smooth.

Fix some $x = \varprojlim x_i \in M$. There exists $\phi(U) \times \mathbb{E}$ with the following commutative diagram

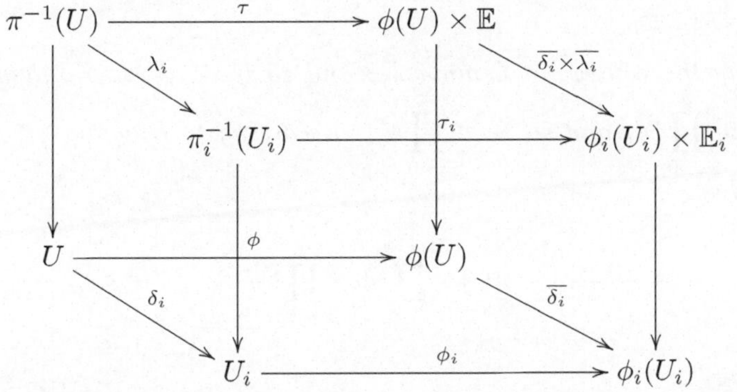

(4.15)

Let Ω be the symplectic form on $\phi(U) \times \mathbb{E}$ such that $\omega = \tau^*\Omega$. According to Proposition 4.7, $\ker \lambda_i$ *is a sub-bundle of E. Now, since ω is a smooth symplectic form and the orthogonal* $\ker(\lambda_i)_z^{\perp}$ *is a supplemented space of* $\ker(\lambda_i)_z$ *for all $z \in M$, it follows that* $\ker(\lambda_i)_z^{\perp}$ *is a Banach sub-bunlde of E and so the Diagram (4.15) has the more precise version, after shrinking U if necessary:*

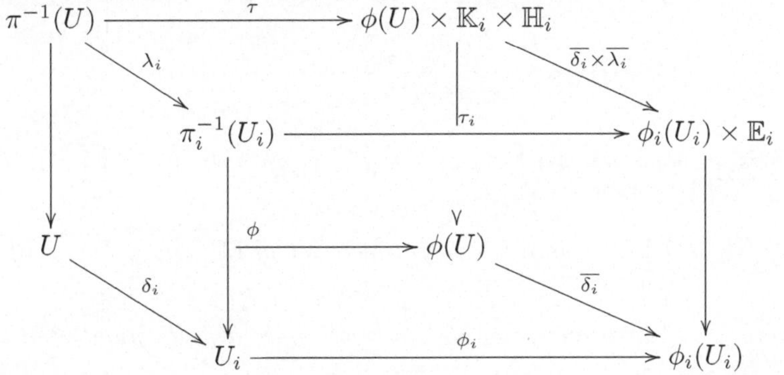

(4.16)

where $\mathbb{K}_i$ is the Kernel of $\overline{\lambda_i}$ and $\mathbb{H}_i$ is the orthogonal of $\mathbb{K}_i$ relative to $\Omega_{\phi(x)}$ over $\phi(x)$. Since the restriction $\overline{\lambda_i}'$ of $\overline{\lambda_i}$ to $\mathbb{H}_i$ is an isomorphism onto $\mathbb{E}_i$ and so $\overline{\delta_i} \times \overline{\lambda_i}'$ is an isomorphism from $\phi(U) \times \mathbb{H}_i$ onto $\phi(U) \times \mathbb{E}_i$. Thus, $(\Omega_i) = [\overline{\delta_i} \times \overline{\lambda_i}']^* \left((\Omega)_{|\phi(U)\times\mathbb{H}_i}\right)$ is a symplectic form on $\phi_i(U_i) \times \mathbb{E}_i)$ and so $\omega_i = \tau_i^*(\Omega)$ is a smooth symplectic form.
The end of the proof follows from Proposition 4.15.

Proof of Corollary 4.3 *According to the assumption of this Corollary, after applying Theorem 4.11, the proof will be completed if we prove that the 2-form ω defined by the closed 2-form ω_i is also closed and if ω is a closed 2-form on the projective limit $M = \varprojlim M_i$ (each induced 2 form ω_i induced on M_i is closed). Under the notations of the proof of Theorem 4.11, we have*

$\mathbb{E}_i = \mathbb{M}_i$ *and* $\mathbb{E} = \mathbb{M}$;

$\lambda_i^j = T\delta_i^j$, $\overline{\ell_i^j} = \overline{\delta_i^j}$, $\lambda_i = T\delta_i$;

$\tau_i = T\phi_i$, $\tau = T\phi$.

We can apply the context of Lemma 4.2 and so if $\breve{\mathbb{M}}_i = \ker \overline{\delta_i^j}$, *then* $\mathbb{M}_n$ *is isomorphic to* $\prod_{i=0}^{n} \breve{\mathbb{M}}_i$ *and so* $\mathbb{M} \equiv \prod_{i=0}^{\infty} \breve{\mathbb{M}}_i$. *According to Diagram 4.16 in our context, we have*

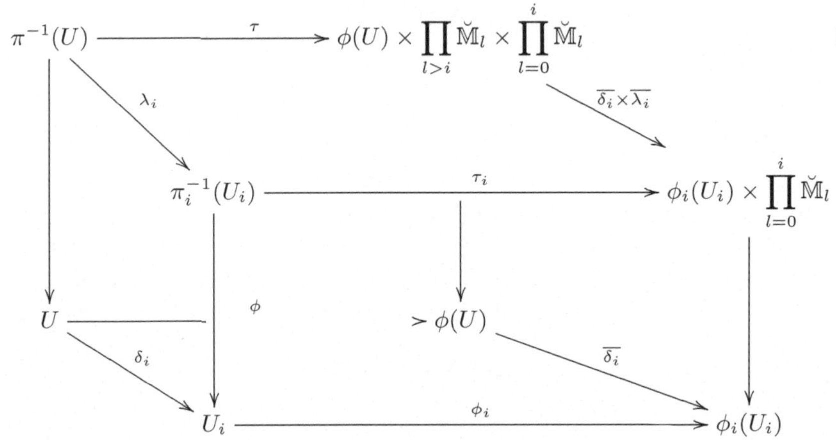

(4.17)

and $\phi(U)$ *is an open set in* $\prod_{i=0}^{\infty} \breve{\mathbb{M}}_i$ *and* $\phi_i(U_i)$ *is an open set of* $\prod_{l=0}^{i} \breve{\mathbb{M}}_l$. *Thus,* $\phi_i(U_i)$ *is of type* $\prod_{l=0}^{i} U_l'$ *where* U_l' *is an open set of* $\mathbb{M}_i'$ *and* $\phi(U)$ *is of type* $\prod_{l=0}^{\infty} U_l'$ *where* U_l' *is an open set of* $\breve{\mathbb{M}}_l$ *and with only a finite number of* $l \geq i$ *for which* $U_l \neq \breve{\mathbb{M}}_l$.

(1) Assume that ω *is a projective limit of the sequence* $(\omega_i)_{i\in\mathbb{N}}$. *As in the proof of Theorem 4.11 let* Ω_i *be the form on* $\phi_i(U_i)$ *induced by* ω_i *and we denote by* Ω *the symplectic form on* $\phi(U)$ *induced by* ω *according to the context of Diagram 4.17. If* ι_l *be the natural inclusion of* U_l' *in* $\prod_{l=0}^{i} \breve{\mathbb{M}}_l$, *we set* $\overline{\Omega_l} = \iota_i^* \Omega_i$ *for* $l \geq i$. *Note that* $\overline{\Omega_l}$ *does not depend on the choice of the integer* $i \geq l$. *As* Ω_i *is closed , it follows that* $\overline{\Omega_l}$ *is closed. Note that each subbundle* $U_l' \times \mathbb{E}_l'$ *is the tangent bundle of* U_l'. *But, from the construction of* ω *(and so,* Ω*), if* X_1 *and* X_2 *are vector fields on* $\phi(U)$ *which are tangent to* U_{l_1}' *and* U_{l_2}'*,respectively, we*

have $\Omega(X_1,X_2)=0$ *if* $l_1\neq l_2$ *and* $\Omega(X_1,X_2)=\overline{\Omega_l}(X_1,X_2)$ *if* $=l_2=l_1=l$. *This implies that* Ω *is closed.*

(2) Assume that ω *is a symplectic form such that* $T_x\delta_i:T_xM\to T_{x_i}M_i$ *is a symplectic submersion. Then from Theorem 4.14 (2),* ω *induces a non-degenerate 2-form* ω_i *on* M_i.

Again let Ω *(resp.* Ω_i*) be the 2-form on* $\phi(U)$ *(resp.* $\phi_i(U_i)$*) according to the context of Diagram 4.17. We must show that each* Ω_i *is closed. Since* Ω *is the projective limit of the sequence* $(\Omega_i)_{i\in\mathbb{N}}$*, according to Theorem 4.11 (2). Thus, as previously, if* X_1 *and* X_2 *are vector fields on* $\phi(U)$ *which are tangent to* U'_{l_1} *and* U'_{l_2} *respectively, we have* $\Omega(X_1,X_2)=0$ *if* $l_1\neq l_2$ *and* $\Omega(X_1,X_2)=\overline{\Omega_l}(X_1,X_2)$*, if* $=l_2=l_1=l$*. Thus* $\Omega_i=\iota_i^*\Omega$ *if* ι_i *is the natural inclusion of* $\phi_i(U_i)$ *in* $\phi(U)$*. It follows that each* ω_i *is closed.*

4.15.3 Weak Symplectic Forms on a Submersive Projective Sequence of Reflexive Banach Bundles and Darboux-Bambusi Assumption

Let $\mathbb{E}=\varprojlim\mathbb{E}_i$ be a projective limit of a projective sequence of reflexive Banach spaces $\left(\mathbb{E}_i,\overline{\lambda_i^j}\right)_{j\geq i}$. We can provide each Banach space $\mathbb{E}_i$ with a norm $||\ ||_i$ such that $||\overline{\lambda_i^{i+1}}||_i^{\mathrm{op}}\leq 1$ for $i\in\mathbb{N}$.

We consider a sequence $(\omega_i)_{i\in\mathbb{N}}$ of weak symplectic forms on $\mathbb{E}_i$ and let $\omega_i^\flat:\mathbb{E}_i\to\mathbb{E}_i^*$ be the associated bounded linear operator. Following the end of Chapter 2, § 2.1.1, we consider the norm $||u||_{\omega_i^\flat}=||\omega_i^\flat(u)||_i^*$ where $||\ ||_i^*$ is the canonical norm on $\mathbb{E}_i^*$ associated with $||\ ||_i$. We have seen that the inclusion of the Banach space $(\mathbb{E},||\ ||_i)$ in the normed space $(\mathbb{E}_i,||\ ||_{\omega_i^\flat})$ is continuous and we have denoted by $\widehat{\mathbb{E}}_i$ the Banach space which is the completion of $(\mathbb{E}_i,||\ ||_{\omega_i^\flat})$. Recall that from Remark 2.1, the Banach space $\widehat{\mathbb{E}}_i$ does not depend on the choice of the norm $||\ ||_i$ on $\mathbb{E}_i$. Again, according to the end of § 2.1.1, $\omega_i^\flat$ can be extended to an isometry between $\hat{\mathbb{E}}_i$ and $\mathbb{E}_i^*$. Moreover, $\omega_i^\flat$ is an isomorphism from $\mathbb{E}_i$ to $\widehat{\mathbb{E}}_i^*$.

Lemma 4.3

(1) The sequence $(\widehat{\mathbb{E}}_i^*)_{i\in\mathbb{N}}$ *is a projective sequence of Banach spaces and so* $\widehat{\mathbb{E}}^*=\varprojlim\widehat{\mathbb{E}}_i^*$ *is well defined. Moreover, if* $\overline{\lambda_i^j}$ *is surjective and its kernel is split, then the bonding map* $\widehat{\overline{\lambda_i^j}}=\widehat{\mathbb{E}}_j^*\to\widehat{\mathbb{E}}_i^*$ *also satisfies this assumption.*

(2) The projective limit $\omega^\flat=\varprojlim\omega_i^\flat$ *is well defined and is an isomorphism from* $\mathbb{E}$ *to* $\widehat{\mathbb{E}}^*$

Proof *(1) It is sufficient to show that* $\overline{\lambda_i^j}$ *and* $\omega_i^\flat$ *give rise to a map* $\widehat{\overline{\lambda_i^j}}$ *from* $\widehat{\mathbb{E}}_j^*$ *into* $\widehat{\mathbb{E}}_i^*$ *and if* $\overline{\lambda_i^j}$ *is surjective and with a split kernel so is* $\widehat{\overline{\lambda_i^j}}$. *Indeed since* $\omega_i^\flat$ *is an isomorphism from* $\mathbb{E}_i$ *to* $\widehat{\mathbb{E}}_i^*$, *the bonding map* $\widehat{\overline{\lambda_i^j}} = \omega_i^\flat \circ \overline{\lambda_i^j} \circ (\omega_i^\flat)^{-1}$ *satisfied the announced properties in (1).*

(2) is obvious.

Now we consider a reduced projective sequence $(E_i, \pi_i, M_i)_{\underleftarrow{i}}$ of Banach vector bundles where the typical fibre $\mathbb{E}_i$ is reflexive. The projective limit $E = \underleftarrow{\lim} E_i$ has a structure Fréchet bundle over $M = \underleftarrow{\lim} M_i$ with typical fibre $\mathbb{E} = \underleftarrow{\lim} \mathbb{E}_i$ (cf. Proposition 4.4).

Consider a sequence $(\omega_i)_{i\in\mathbb{N}}$ of compatible weak symplectic forms ω_i on E_i. According to the previous notations, since $\mathbb{E}_i$ is reflexive, we denote by $(\widehat{E_i})_{x_i}$ the Banach space which is the completion of $(E_i)_{x_i}$ provided with the norm $\|\ \|_{(\omega_i)^\flat_{x_i}}$. Then $(\omega_i)_{x_i}$ can be extended to a continuous bilinear map $(\hat{\omega}_i)_{x_i}$ on $(E_i)_{x_i} \times (\widehat{E_i})_{x_i}$ and $(\omega_i)^\flat_{x_i}$ becomes an isomorphism from $(E_i)_{x_i}$ to $(\widehat{E_i})^*_{x_i}$. We set

$$\widehat{E_i} = \bigcup_{x_i \in M_i} (\widehat{E_i})_{x_i}, \quad \widehat{E_i}^* = \bigcup_{x_i \in M_i} (\widehat{E_i}^*)_{x_i}$$

According to the assumption of Theorem 2.6 we introduce the following terminology:

Definition 4.34 *Let* $(E_i, \pi_i, M_i)_{\underleftarrow{i}}$ *be a reduced projective sequence of Banach bundles whose typical fibre* $\mathbb{E}_i$ *is reflexive. Consider a sequence* $(\omega_i)_{i\in\mathbb{N}}$ *of compatible weak symplectic forms* ω_i *on* E_i. *We say that the sequence* $(\omega_i)_{i\in\mathbb{N}}$ *satisfies the Bambusi-Darboux assumptions around* $x^0 \in M$ *if there exists a projective limit chart* $U = \underleftarrow{\lim} U_i$ *around* x^0 *such that:*

(i) *for each* $i \in \mathbb{N}$, $(\widehat{E}_i)_{|U_i}$ *is a trivial Banach bundle;*

(ii) *for each* $i \in \mathbb{N}$, ω_i *can be extended to a smooth field of continuous bilinear forms on* $(E_i)_{|U_i} \times (\widehat{E}_i)_{|U_i}$.

Remark 4.20 *The Bambusi-Darboux assumptions introduced in Definition 4.34 can be considered as a generalization of the necessary and sufficient conditions in the generalization of Darboux Theorem for reflexive Banach setting by Bambusi in [Bam99].*

Proposition 4.16 *Consider a sequence* $(\omega_i)_{i\in\mathbb{N}}$ *of compatible symplectic forms* ω_i *on* E_i *which satisfies the Bambusi-Darboux assumptions around* $x^0 \in M$. *Then we have the following properties:*

(1) The projective limit $(\widehat{E}^*)_{|U} = \underleftarrow{\lim}(\widehat{E_i}^*)|U_i$ is well defined and is a trivial Fréchet bundle with typical fibre $\widehat{\mathbb{E}} = \underleftarrow{\lim}\widehat{\mathbb{E}}_i$.

(2) The sequence $(\omega_i^\flat)$ of isomorphisms from $E_{i|U_i}$ to $\widehat{E}^*_{i\ |U_i}$ induces an isomorphism from $E_{|U}$ to $(\widehat{E}_i^*)_{|U}$.

Proof *(1) From our assumptions, for each i, we have a sequence of trivializations $\widehat{\tau}_i : (\widehat{E}_i)_{|U_i} \to U_i \times \widehat{\mathbb{E}}_i$. Thus we obtain a sequence $\widehat{\tau}_i^{-1} : U_i \times \widehat{\mathbb{E}}_i^* \to (\widehat{E}_i^*)_{|U_i}$ of isomorphisms of trivial bundles. Now, from the proof of Lemma 4.3, we have the bonding map $\widehat{\overline{\lambda_i^i}} : \widehat{\mathbb{E}}_j^* \to \widehat{\mathbb{E}}_i^*$ and by restriction to U_j we have a bonding map $\delta_i^j : U_j \to U_i$. So we get a bundle morphism $\delta_i^j \times \widehat{\overline{\lambda_i^i}}$ from $U_j \times \widehat{\mathbb{E}}_j^*$ to $U_i \times \widehat{\mathbb{E}}_i^*$. Now the map*

$$\widehat{\tau}_i^{-1} \circ (\delta_i^j \times \widehat{\overline{\lambda_i^i}}) \circ \widehat{\tau}_j$$

is a bonding map for the projective sequence of trivial bundles $\left((\widehat{E}_i^*)_{|U_i}\right)_{\underleftarrow{n}}$. Therefore the projective limits $\widehat{\tau} = \underleftarrow{\lim}\widehat{\tau}_i$ and $\widehat{E}^*_{|U} = \underleftarrow{\lim}(\widehat{E}_i^*)_{|U_i}$ are well defined and $\widehat{\tau}$ is a Fréchet isomorphism bundle from $U \times \widehat{\mathbb{E}^*}$ to $\widehat{E}^*_{|U}$, which ends the proof of (1).

(2) At first, from Proposition 4.14, then $\omega = \underleftarrow{\lim}\omega_i$ is a 2-form on E. From our assumption, since for each $i \in \mathbb{N}$ we can extend ω_i to a bilinear onto $(E_i)_{|U_i} \times (\widehat{E}_i)_{|U_i}$, this implies that $\omega_i^\flat$ is an isomorphism from $(E_i)_{|U_i}$ to $(\widehat{E}_i^*)_{|U_i}$. Consider the sequence of bonding maps $\left(\widehat{\overline{\lambda_i^i}}\right)_{i\in\mathbb{N}}$ for the projective sequence $\left((\widehat{E}^*)_{|U_i}\right)_{\underleftarrow{i}}$ previously defined. Then we have the following commutative diagram:

$$\begin{CD}
U_j \times \mathbb{E}_j @>{\tau_j^{-1}}>> (E_j)_{|U_j} @>{\omega_j^\flat}>> (\widehat{E}_j^*)_{|U_j} @>{\widehat{\tau}_j}>> U_j \times \widehat{\mathbb{E}}_j^* \\
@V{\delta_i^j \times \ell_i^j}VV @V{\ell_i^j}VV @V{\widehat{\ell}_i^j}VV @V{\delta_i^j \times \widehat{\ell}_i^j}VV \\
U_i \times \mathbb{E}_i @>>{\tau_i^{-1}}> (E_i)_{|U_i} @>>{\omega_i^\flat}> (\widehat{E}_i^*)_{|U_i} @>>{\widehat{\tau}_i}> U_i \times \widehat{\mathbb{E}}_i^*
\end{CD}$$

It follows that the projective limit $\omega^\flat = \underleftarrow{\lim}\omega_i^\flat$ is well defined and is an isomorphism from $E_{|U}$ to $\widehat{E}^*_{|U}$.

4.15.4 Problem of Existence of Darboux Charts on a Strong Reduced Projective Sequence of Banach Manifolds

4.15.4.1 Conditions of Existence of Darboux Charts

Let $\left(M_i, \delta_i^j\right)_{j \geq i}$ be a submersive or decreasing projective sequence of Banach manifolds where M_i is modelled on a Banach space $\mathbb{M}_i$. We first apply the previous results for $E_i = TM_i$.

Theorem 4.12

(1) *Consider a sequence $(\omega_i)_{i \in \mathbb{N}}$ of compatible weak symplectic forms ω_i on M_i. Then, for each $x \in M$, the projective limit $\omega_x^\flat = \underleftarrow{\lim}(\omega_i)_{x_i}^\flat$ is well defined and is an isomorphism from T_xM to $(\widehat{T_xM})^*$. Moreover $\omega_x(u,v) = \omega_x^\flat(u)(v)$ defines a smooth weak symplectic form on M.*

(2) *Let ω be a symplectic form on a submersive projective limit manifold $M = \underleftarrow{\lim} M_i$. For all $i \in \mathbb{N}$, assume that the canonical projection $\delta_i : M \to M_i$ is a symplectic submersion. Then there exists a symplectic form ω_i on M_i such that $\delta_i^* \omega_i = \omega$ in restriction to $(\ker \delta_i)^\perp$ and the sequence $(\omega_i)_{i \in \mathbb{N}}$ is a sequence of compatible weak symplectic forms such that the weak symplectic form which is the projective limit of $(\omega_i)_{i \in \mathbb{N}}$ on M is exactly ω.*

Proof *(1) Since $\omega(u,v) = \omega^\flat(u)(v)$, by application of Proposition 4.16 to $E_i = TM_i$, we obtain that ω is non-degenerate. The proof of that ω is closed is formally the same as in the proof of Corollary 4.3 (1).*

(2) is a direct consequence of Corollary 4.3 (2).

As in the Banach context, we introduce the notion of Darboux chart:

Definition 4.35 *Let ω be a weak symplectic form on the direct limit $M = \underleftarrow{\lim} M_i$. We say that a chart (V, ψ) around x_0 is a Darboux chart if $\psi^* \omega^0 = \omega$ where ω^0 is the constant form on $\psi(U)$ defined by $(\psi^{-1})^* \omega_{x_0}$.*

We have the following necessary and sufficient conditions of existence of Darboux charts on a submersive projective sequence of Banach manifolds:

Theorem 4.13 *Let $\left(M_i, \delta_i^j\right)_{j \geq i}$ be a submersive or decreasing projective sequence of Banach manifolds where M_i is modelled on a reflexive Banach space $\mathbb{M}_i$.*

(1) *Consider a sequence $(\omega_i)_{i \in \mathbb{N}}$ of compatible symplectic forms ω_i on M_i and let ω be the symplectic form which is the projective limit of $(\omega_i)_{i \in \mathbb{N}}$ on $M = \underrightarrow{\lim} M_i$. Assume that the following property is satisfied:*

 (D) *There exists a projective limit chart $(U = \underleftarrow{\lim} U_i, \phi = \underleftarrow{\lim} \phi_i)$ around x^0 such that, if $x_i^0 = \delta_i(x^0) \in M_i$, then (U_i, ϕ_i) is a Darboux chart around x_i^0 for ω_i.*

Then (U, ϕ) is a Darboux chart around x0 for ω.

(2) Let ω be a weak symplectic form on a submersive projective limit $M = \varprojlim M_i$ such that $\delta_i : M \to M_i$ is a symplectic submersion. Assume that there exists a Darboux chart(V, ϕ) around x^0 in M.
If ω_i is the symplectic form on M_i induced by ω, then there exists a projective limit chart $(U = \varprojlim U_i, \phi = \varinjlim \phi_i)$ around x^0 such that the property (D) is satisfied.

Proof *(1) Assume that the assumption **(D)** is true and that $(M_i)_{i\in\mathbb{N}}$ is a reduced projective sequence of Banach manifolds. We fix some $x^0 \in M$. We consider a projective limit chart $(U = \varprojlim U_i, \phi = \varprojlim \phi_i)$ around x^0 such that, if $x_i^0 = \delta_i(x^0) \in U_i$, then (U_i, ϕ_i) is a Darboux chart around x_i^0 for ω_i. Now we have the following commutative diagram:*

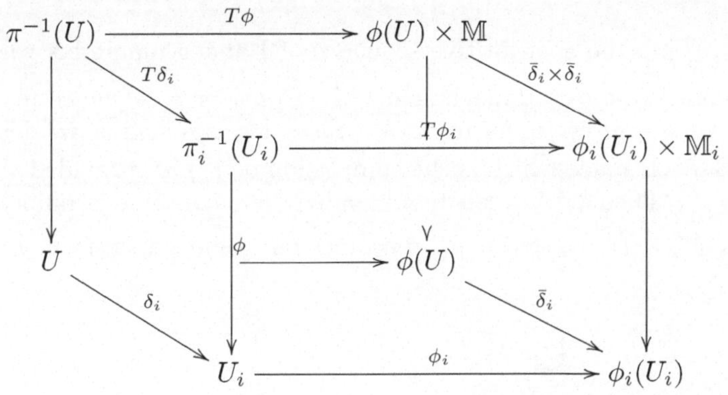

(4.18)

According to this diagram and modulo the diffeomorphisms ϕ and ϕ_i, we may assume that

- *U is an open neighbourhood of $x^0 \equiv 0 \in \mathbb{M}$, and U_i is a neighbourhood of $0 \in \mathbb{M}_i$;*
- *ω is a smooth 2-form on U and ω_i is a constant 2-form on U_i.*

Now if $x = \varprojlim x_i \in U$, $u = \varprojlim u_i$ and $v = \varprojlim v_i$, since ω_i is constant on U_i it follows that $(\omega_i)_{x_i}(u_i, v_i)$ is independent of $x_i \in U_i$; so the value

$$\omega_x(u, v) = \varprojlim (\omega_i)_{x_i}(u_i, v_i)$$

is independent of the point x, which ends the proof of (1).

(2) Let ω be a weak symplectic form on $M = \varprojlim M_i$ such that, for all $i \in \mathbb{N}$, $\delta_i : M \to M_i$ is a symplectic submersion. Assume that we have a Darboux chart $(U = \varprojlim U_i, \phi = \varinjlim \phi_i)$ around x^0 for ω. Fix some $i \in \mathbb{N}$. In the context of Diagram (4.18), we have $\mathbb{M} \equiv \mathbb{K}_i \times \mathbb{H}_i$ where $\mathbb{K}_i$ is the kernel of $T_0\delta_i$ and $\mathbb{H}_i$ is the orthogonal of $\mathbb{K}_i$ in $\mathbb{M}_i \equiv T_0 M$ (cf. Diagram (4.15) with, for all $i \in \mathbb{N}$, $E_i = TM_i$). Thus again, modulo the diffeomorphisms ϕ and ϕ_i, we may assume that

- $x^0 \equiv 0 \in U \subset \mathbb{K}_i \times \mathbb{H}_i$, $x_i^0 \equiv 0 \in U_i \subset \mathbb{M}_i$;
- ω *is a constant* 2*-form on* U *and* ω_i *is a smooth* 2*-form on* U_i.

Recall that the restriction of $\overline{\delta_i}$ *to* $\mathbb{H}_i$ *is an isomorphism onto* $\mathbb{M}_i$*, thus we may also assume that* $\mathbb{H}_i = \mathbb{M}_i$*. In this way, we have* $\delta_i^* \omega_i = \omega$ *in restriction to* $\mathbb{H}_i = \mathbb{M}_i$*. Thus, with our identification,* ω_i *is nothing but the restriction of* ω *to* $U_i \times \mathbb{M}_i$ *and so* ω_i *is a constant* 2*-form on* U_i *whose value is fixed by the restriction of* ω *to* $\mathbb{M}_i$.

4.15.4.2 Problem of Existence of Darboux Chart in General

In this subsection, we will explain why, even in the context of a submersive projective sequence of weak symplectic Banach manifolds which satisfies the assumption of Theorem 2.6, in general, there does not exist any Darboux chart for the induced symplectic form on the projective limit.

Let $\left(M_i, \delta_i^j\right)_{j \geq i}$ be a projective sequence of Banach manifolds where M_i is modelled on a *reflexive* Banach space $\mathbb{M}_i$. Consider a sequence $(\omega_i)_{i \in \mathbb{N}}$ of compatible weak symplectic forms on M_i. Since $\mathbb{M}_i$ is reflexive, we denote by $\widehat{T_{x_i} M_i}$ the Banach space which is the completion of $T_{x_n} M_i$ provided with the norm $|| \; ||_{(\omega_i)_{x_i}}$. Then $(\omega_i)_{x_i}$ can be extended to a continuous bilinear map $(\hat{\omega}_i)_{x_i}$ on $T_{x_i} M_i \times \widehat{T_{x_i} M_i}$ and $(\omega_i)_{x_i}^\flat$ becomes an isomorphism from $T_{x_i} M_i$ to $(\widehat{T_{x_i} M_i})^*$. We set

$$\widehat{TM_i} = \bigcup_{x_i \in M_i} \widehat{T_{x_i} M_i}, \quad \widehat{TM_i}^* = \bigcup_{x_i \in M_i} \widehat{T_{x_i} M_i}^*.$$

Then by application of Proposition 4.16, we have:

Proposition 4.17 *Let* $\left(M_i, \delta_i^j\right)_{j \geq i}$ *be a reduced projective sequence of Banach manifolds whose model is a reflexive Banach space* $\mathbb{M}_i$*. Consider a sequence* (ω_i) *of compatible weak symplectic forms* ω_i *on* M_i*. Assume that we have the following assumptions*[13] *at* $x^0 \in M = \varprojlim M_i$*:*

(i) there exists a limit chart $(U = \varprojlim U_i, \phi = \varprojlim \phi_i)$ *around* x^0 *such that* $(\widehat{TM}_i)_{|U_i}$ *is a trivial Banach bundle.*

(ii) ω_i *can be extended to a smooth field of continuous bilinear forms on* $(TM_i)_{|U_i} \times (\widehat{TM}_i)_{|U_i}$ *for all* $i \in \mathbb{N}$.

Then $\widehat{T^*M}_{|U}$ *is a trivial bundle. If* ω *is the symplectic form defined by the sequence* $(\omega_i)_{i \in \mathbb{N}}$*, then the morphism*

$$\omega^\flat : TM \to T^*M$$

induces an isomorphism from $TM_{|U}$ *to* $\widehat{TM}^*_{|U}$.

[13] These assumptions correspond to the Bambusi-Darboux assumptions in Definition 4.34.

Note that the context of Proposition 4.17 covers the particular framework of projective limit of strong symplectic Banach manifolds $(M_i, \omega_i)_{i\in\mathbb{N}}$.

We will expose which arguments are needed to prove a Darboux theorem in the context of reduced projective sequence of Banach manifolds under the assumptions of Proposition 4.17. In fact, we point out the problems that arise in establishing the existence of a Darboux chart by Moser's method.

Case 1.– Assume that $M = \varprojlim M_i$ is a reduced projective limit.
Fix some point $a = \varprojlim a_n \in M$. In the context on Proposition 4.17, on the projective limit chart (U, ϕ) around a, we can replace U by $\phi(U)$, ω by $\phi^*\omega$ on the open subset $\phi(U)$ of the Fréchet space $\mathbb{M}$. Thus, if ω^0 is the constant form on U defined by ω_a, we consider the 1-parameter family

$$\omega^t = \omega^0 + t\overline{\omega}, \text{ with } \overline{\omega} = \omega - \omega^0.$$

Since ω^t is closed and $\mathbb{M}$ is a Fréchet space, by [KrMi97], Lemma 33.20, there exists a neighbourhood $V \subset U$ of a and a 1-form α on V such that $d\alpha = \overline{\omega}$ which is given by

$$\alpha_x := \int_0^1 s.\overline{\omega}_{sx}(x, \ )ds. \tag{4.19}$$

Now, for all $0 \leq t \leq 1$, $\omega^t_{x_0}$ is an isomorphism from $T_aM \equiv \mathbb{M}$ onto $\widehat{T_aM} \equiv \widehat{\mathbb{M}}^*$. In the Banach context, using the fact that the set of invertible operators is open in the set of operators, after restricting V, we may assume that $(\omega^t)^\flat$ is a field of isomorphisms from $\mathbb{M}$ to $\widehat{\mathbb{M}}^*$. *Unfortunately, this result is not true in the Fréchet setting* (cf. § 3.19). Therefore, the classical proof does not work in this way in general.

Case 2.– Assume that M is a submersive projective limit.
According to Theorem 4.12, assume that the canonical projection $\delta_i : M \to M_i$ is a symplectic submersion, for all $i \in \mathbb{M}$. Then ω induces a symplectic form ω_i on M_i. Therefore, for each i, let α_i be the 1-form induced by α on $\phi_i(U_i \cap V)$. Then we have $\omega_i = d\alpha_i$ and also

$$(\alpha_i)_{x_i} = \int_0^1 s.(\overline{\omega}_i)_{sx_i}(x_i, \ )ds \tag{4.20}$$

where $\bar{\omega}_i = \omega_i - \omega_i^0$ is associated with the 1-parameter family $\omega_i^t = \omega^i + t\bar{\omega}_i$. We are exactly in the context of the proof of Theorem 2.6 and so the local flow $\mathrm{Fl}_t^{X_i}$ of $X_i^t = ((\omega_i^t)^\flat)^{-1}(\alpha_i)$ is a local diffeomorphism from a neighbourhood W_i of a_i in V_i and, in this way, we build a Darboux chart around a_i in M_i. Therefore, after restricting each W_i, if necessary, assume that:

(PLDC) (projective limit Darboux chart) We have a projective sequence of such open sets $(W_i)_{n\in\mathbb{N}}$, then on $W = \varprojlim W_i$, the family of local diffeomorphisms $F^t = \varprojlim F_i^t$ is well defined.

Recall that $\omega^\flat = \varprojlim \omega_i^\flat$ and $\omega^\flat$ is an isomorphism. Thus according to the previous notations, we have a time dependent vector field

$$X^t = ((\omega^t)^\flat)^{-1}(\alpha)$$

and again, we have $L_{X^t}\omega^t = 0$. Of course, if the **(PLDC)** assumption on $(W_n)_{n\in\mathbb{N}}$ is true, then $X^t = \varprojlim X_i^t$. So we obtain a Darboux chart as in the Banach context. Note that, in this case, we are in the context of Theorem 4.13.

Remark 4.21 *In fact, under the assumption* **(PLDC)**, *the flow* Fl_t *is the local flow (at time* $t \in [0,1]$*) of* $X^t = \varprojlim X_i^t$ *where* $X_i^t = ((\omega_n^t)^\flat)^{-1}(\alpha_i)$ *(with the previous notations). Unfortunately, according to Remark 4.5, outside particularity special cases, the "Darboux chart" assumption is not true in general, since, in general,*

$$\bigcap_{j \geq i_0} \delta_i^j(W_j)$$

is not an open neighbourhood of a_{i_0}.

4.15.5 A Generalization of Kumar-Darboux's Theorem

In this section, we will give an analog version of Kumar-Darboux's Theorem in [Kum13b] but with weaker assumptions in some sense (cf. Remark 4.22), but before the statement of this Theorem, we begin by some preliminaries.

4.15.5.1 Preliminaries and Results

Definition 4.36

(1) *Let* $\mathbb{E}$ *be a Banach space,* M *be a Banach manifold and* $E = M \times \mathbb{E}$ *be the trivial bundle over* M. *A function* $|| \; || : E \to \mathbb{R}$ *is a Finsler structure on* E *if*

   i) *for each* $x \in M$, *the map* $u \mapsto ||u||_x := ||(x,u)||$ *is a norm which defines the topology of* $\mathbb{E}$;

   ii) *given* $x_0 \in M$, *and* $k \geq 1$, *there exists a neighbourhood* U *of* x_0 *in* M *such that*

$$\forall x \in U, \forall u \in \mathbb{E}, \; \frac{1}{k}||u||_{x_0} \leq ||u||_x \leq k||u||_{x_0}.$$

(2) *Let* $\pi : E \to M$ *be a Banach bundle with typical fibre* $\mathbb{E}$. *A map* $|| \; || : E \to \mathbb{R}$ *is called a Finsler norm on* E, *if for any* $x \in M$, *there exists a trivilization* $\Phi : E_{|U} \to U \times \mathbb{E}$ *over an open neighbourhood* U *of* x *such that* $|| \; || \circ \Phi$ *is a Finsler structure on* $U \times \mathbb{E}$.

If M is a Banach manifold modelled on $\mathbb{M}$, given a chart (U, ϕ) on M, a trivialization $\Phi : E_{|U} \to U \times \mathbb{E}$, and a norm $|| \; ||$ on $\mathbb{E}$, then we get a canonical

Finsler norm on $U \times \mathbb{E}$, (again denoted $\|\ \|$), and so *via* Φ we can provide $E_{|U}$ with an associated Finsler norm.

Consider a submersive projective sequence $(E_i, \pi_i, M_i, \lambda_i^j, \delta_i^j)_{j \geq i}$ of Banach bundles. For each $x = \underleftarrow{\lim} x_i \in M = \underleftarrow{\lim} M_i$, there exist a projective chart bundle limit $(U = \underleftarrow{\lim} U_i, \phi = \underleftarrow{\lim} \phi_i)$ and a trivialisation $(U, \Phi = \underleftarrow{\lim} \Phi_i)$ of $E = \underleftarrow{\lim} E_i$ over U where Φ_i is a trivialization of E_i over U_i. If we identify $\mathbb{E}_i$ with the fibre $(E_i)_{x_i}$ then Φ_i is an isomorphism from $(E_i)_{|U_i}$. Given a sequence of norms $((\|\ \|_i)_{x_i})_{i \in \mathbb{N}}$ on $(E_i)_{x_i}$ by the previous argument, we obtain an associated Finsler norm $\|\ \|_i$ on $(E_i)_{|U_i}$.
Under the assumptions of Proposition 4.17, given a sequence of norms $((\|\ \|_i)_{x_i})_{i \in \mathbb{N}}$ on $T_{a_i} M_i$, we can consider the family of norms $(\|\ \|_{(\omega_i)^\flat_{a_i}})$ on $T_{a_i} M_i$ which gives rise a family of norm $\|\ \|^*_{a_i}$ on $\widehat{T_{a_i} M_i}^*$. Thus the bundle $(\widehat{TM_i}^*)_{|U}$ is provided with an associated Finsler norm $\|\ \|^*_i$. Recall that on the set of morphisms $\eta_i : TM_{i|U} \to (\widehat{TM_i}^*)_{|U}$ we have an operator norm

$$\|\eta_i\|_i^{\mathrm{op}} = \sup\{\|\eta_i(v)\|_i^* : \|v\|_i \leq 1\}.$$

The vector field $\mathbf{E}_{\mathbb{M}}$ characterized by[14]

$$\mathbf{E}_{\mathbb{M}}(f)(x) = \frac{d}{dt}_{|t=0} f(e^t x)$$

is the Euler vector field on $\mathbb{M}$ (in a classical sense). For any chart (U, ϕ) in M, $\mathbf{E} = \phi_*^{-1}(\mathbf{E}_{\mathbb{M}})$ is called an *an Euler vector field* on U. We will always consider such local Euler vector fields.

According to these preliminaries we have:

Theorem 4.14 *Let $M = \underleftarrow{\lim} M_i$ be a submersive projective limit of reflexive Banach manifolds M_i and $(\omega_i)_{i \in \mathbb{N}}$ a compatible sequence of symplectic forms (resp. ω a symplectic form on M such that $\delta_i : M \to M_i$ is a symplectic submersion).*
Assume that there exists a chart $(U = \underleftarrow{\lim} U_i, \phi = \underleftarrow{\lim} \phi_i)$ around a and sequence of norms $(\|\ \|_{a_i})_{i \in \mathbb{N}}$ on $T_{a_i} M_i$ such that, for all $i \in \mathbb{N}$, we have:

(DB) *the Darboux-Bambusi assumptions are satisfied on U_i;*

(SN) *if $(\|\ \|_i^{\mathrm{op}})$ is the sequence of Finsler norms operator associated with $(\|\ \|_i)_{i \in \mathbb{N}}$ on $TM_{i|U_i}$ and $\left((\|\ \widehat{\|_i^*})\right)_{i \in \mathbb{N}}$ on $(\widehat{TM_i})^*_{|U_i}$, there exists a continuous positive map K on U such that*

$$\forall x_i \in U_i, \ \ \|((\omega_i)^\flat_{x_i})\|_i^{\mathrm{op}} \leq K(x);$$

[14]The same is true for any locally convex topological vector space.

(EN) *consider the Euler vector field* $\mathbf{E}_i = \phi_*^{-1}(\mathbf{E}_{\mathbb{M}_i})$ *on* U_i *then there exists a continuous positive map* C *on* U *such*

$$\forall x_i \in U_i, \ \ ||(\omega_i - \phi_i^*\Omega_i)^\flat_{x_i}(\mathbf{E}_i)||_i^{\widehat{*}} \leq C(x)$$

where Ω_i *is the constant 2-form on* $\phi_i(U_i)$ *defined by* $(\phi_i^{-1})^*\left((\omega_i)_{a_i}\right)$.

Then there exists a Darboux chart around a.

4.15.5.2 Proof of Theorem 4.14

The context of Theorem 4.14 is exactly the **case 2** exposed in § 4.15.4.2 with the following assumptions:

(i) there exists a limit chart $(U = \varprojlim U_i, \phi = \varprojlim \phi_i)$ around a such that $(\widehat{TM_i})_{|U_i}$ is a trivial Banach bundle.

(ii) ω_i can be extended to a smooth field of continuous bilinear forms on $(TM_i)_{|U_i} \times (\widehat{TM_i})_{|U_i}$ for all $i \in \mathbb{N}$.

(iii)

$$\exists K(x) > 0 : \ \forall x_i = \delta_i(x) \in U_i, \ \forall i \in \mathbb{N}, \ ||((\omega_i)^\flat_{x_i})||_i^{\text{op}} \leq K(x); \tag{4.21}$$

$$\exists C(x) > 0 : \ \forall x_i = \delta_i(x) \in U_i, \ \forall i \in \mathbb{N}, \ ||(\omega_i - \phi_i^*\Omega_i)^\flat_{x_i}(\mathbf{E}_i)||_i^* \leq C(x) \tag{4.22}$$

Thus, according to Proposition 4.17, $(\widehat{TM}^*)_{|U} = \varprojlim (\widehat{TM_i}^*)_{|U_i}$ is a trivial bundle and $\omega = \varprojlim \omega_i$ and the morphism

$$\omega^\flat : TM \to T^*M$$

induces an isomorphism from $TM_{|U}$ to $(\widehat{TM}^*)_{|U}$.

Since the problem is local, without loss of generality, we can assume that $M_i = U_i$ (resp. $M = U$) is an open set of $\mathbb{M}_i$ (resp. $\mathbb{M}$) and $a = 0 \in \mathbb{M}$. By the way, we have $TM_i = U_i \times \mathbb{M}_i$ and $(\widehat{TM_i}^*)_{|U_i} = U_i \times \widehat{\mathbb{M}_i}^*$. In this context, $x_i \mapsto ((\omega_i)^\flat_{x_i})$ is a smooth map from U_i to $\mathcal{GL}(\mathbb{M}_i, \widehat{\mathbb{M}}_i)$ and $x_i \mapsto ((\omega_i)^\flat_{x_i})^{-1}$ is a smooth map from U_i to $\mathcal{GL}(\widehat{\mathbb{M}}_i, \mathbb{M}_i)$.

Now each Finsler norm $|| \ ||_i$ on $\mathbb{M}_i$ (resp. $|| \ ||_i^*$ on $\widehat{\mathbb{M}_i}^*$) induces a seminorm ν_i on $\mathbb{M}$ (resp. ν_i^* on $\widehat{\mathbb{M}}^*$). Since $(\omega_x^\flat) = \varprojlim (\omega_i)^\flat_{x_i})$ and $(\omega_x^\flat)^{-1} = \varprojlim (\omega_x^\flat)^{-1}$, according to our assumptions, it follows that $(\omega_i)^\flat_x$ (resp. $((\omega_i)^\flat_x)^{-1}$) belongs to $\mathcal{GH}_b(\mathbb{M}, \widehat{\mathbb{M}}^*)$ (resp. to $\mathcal{GH}_b(\widehat{\mathbb{M}}^*, \mathbb{M})$) (cf. § 4.11).

But, for all $t \in [0,1]$, we have $\omega_a^t = \omega_a$ so $(\omega_a^t)^\flat$ belongs to $\mathcal{GH}_b(\mathbb{M}, \widehat{\mathbb{M}}^*)$. By compactness of $[0,1]$, as $x \mapsto (\omega_x^t)^\flat$ is a smooth map from U to $\mathcal{H}_b(\mathbb{M}, \widehat{\mathbb{M}}^*)$ and by a continuity argument, from Proposition 4.9, we may assume that $(\omega_x^t)^\flat$

belongs to $\mathcal{GH}_b(\mathbb{M}, \widehat{\mathbb{M}}^*)$ for all $t \in [0,1]$, after shrinking U if necessary. Thus $((\omega_x^t)^\flat)^{-1}$ exists and belongs to $\mathcal{GH}_b(\widehat{\mathbb{M}}^*, \mathbb{M})$. Let denote by $|| \; ||_\infty^{\text{op}}$ the norm on $\mathcal{GH}_b(\mathbb{M}, \widehat{\mathbb{M}}^*)$ and $\mathcal{GH}_b(\widehat{\mathbb{M}}^*, \mathbb{M})$. Thus if $K = ||\omega_0^\flat||_\infty^{\text{op}}$ and $k = ||(\omega_0\flat)^{-1}||_\infty^{\text{op}}$, again from Proposition 4.9, by a continuity argument, there exists an open neighbourhood $W = \varprojlim W_i$ of $a \in U$ such that

$$||(\omega_x^t)^\flat||_\infty^{\text{op}} \leq 2K \;\; \forall x \in W \quad t \in [0,1] \tag{4.23}$$

$$(||((\omega_x^t)^\flat)^{-1}||_\infty^{\text{op}} \leq 2k \;\; \forall x \in W \quad t \in [0,1] \tag{4.24}$$

On the other hand, from its definition, it follows that the sequence $(\mathbf{E}_i)$ is a projective sequence of vector fields and its projective limit $\mathbf{E} = \varprojlim \mathbf{E}_i$ is the Euler vector field on $\mathbb{M}$ in the classical sense. Since in our context have $(\phi_i^*\Omega_i)_{x_i}$ is simply $(\Omega_i)_{x_i} = \omega_{a_i}$ for all $x_i \in U_i$, then $(\omega_i - \phi_i^*\Omega_i)_{x_i}^\flat(\mathbf{E}_i)$ is simply $(\omega_i - \Omega_i)_{x_i}^\flat(\mathbf{E}_i)$ and so we get

$$\varprojlim(\omega_i - \Omega_i)_{x_i}^\flat(\mathbf{E}_i) = (\omega - \Omega)_x^\flat(\mathbf{E}) \tag{4.25}$$

where obviously $\Omega_x = \varprojlim(\Omega_i)_{x_i} = \omega_a$ for all $x \in U$.
On the other hand by same arguments used to obtain (4.23), it is clear that from (4.22), after shrinking W if necessary, we get

$$||(\omega - \Omega_i)_x^\flat(\mathbf{E})||_\infty^* \leq 2C \quad \forall x \in W \tag{4.26}$$

Note that from the definition of $\mathbf{E}_i$, in our context, we have $\mathbf{E}_i(x_i) = x_i$ and so $\mathbf{E}(x) = x$. Thus the relations (4.20) and (4.19) are, respectively, exactly

$$(\alpha_i)_{x_i} = \int_0^1 (\omega_i - \Omega_i)_{sx_i}^\flat(\mathbf{E}_i)ds; \tag{4.27}$$

$$\alpha_x = \int_0^1 (\omega - \Omega)_{sx}^\flat(\mathbf{E})ds. \tag{4.28}$$

Thus, from (4.26), we have

$$||\alpha_x||_\infty^* \leq 2C, \; \forall x \in W \tag{4.29}$$

where $|| \; ||_\infty^*$ is the norm for uniformly bounded elements of $\widehat{\mathbb{M}}^*$.

According to Remark 4.21, the time dependent vector field $X^t = ((\omega^t)^\flat)^{-1}(\alpha)$ is defined on a star-shaped open set W_1 in W at 0. The following Lemma asserts that X^t satisfies the assumption of Theorem A.2 which ends the proof.

Lemma 4.4 *We can choose the star-shaped open set V of $0 \in U$ such that X^t is Lipschitz on V, that is: there exists a constant $\kappa >$ such that $\forall (x,x') \in V^2$, $\forall t \in [0,1]$, $\forall i \in \mathbb{N}$, $\nu_i(X_x^t - X_{x'}^t) \leq \kappa \nu_i(x - x')$.*

Proof *Note that from (4.25), (4.27) and (4.28) it follows that*

$$\alpha = \varprojlim \alpha_i.$$

Since from (4.27) the map $x_i \mapsto (\alpha_i)_{x_i}$ is smooth, this implies that $x \mapsto \alpha_x$ is a smooth map from W_1 to $\widehat{\mathbb{M}}^*$ which is uniformly bounded from (4.29). Therefore this map is κ_1-Lipschitz for some $\kappa_1 > 0$ on some star-shaped open W_2 of $0 \in W_1$ that is

$$\forall (x, x') \in W_2^2,\ \forall i \in \mathbb{N},\ \nu_i^*(\alpha_x - \alpha_{x'}) \leq \kappa_1 \nu_i(x - x')$$

From (4.24), by the same type of arguments for $x \mapsto ((\omega_x^t)^\flat)^{-1}$ we obtain that there exists a star-shaped open neighbourhood of W_3 of $0 \in W_1$ such that this map is κ_2-Lipschitz.

Taking in account (4.29), we obtain

$$\begin{array}{ll} & \nu_i(X_x^t - X_{x'}^t) \\ \leq & \nu_i\left(((\omega_x^t)^\flat)^{-1}(\alpha_x) - ((\omega_{x'}^t)^\flat)^{-1}(\alpha_x)\right) + \nu_i\left(((\omega_{x'}^t)^\flat)^{-1}(\alpha_x - \alpha_{x'})\right) \\ \leq & 2C\kappa_2 \nu_i(x - x') + 2k\kappa_1 \nu_i(x - x') \end{array}$$

for any $t \in [0,1]$ and any $i \in \mathbb{N}$. This ends the proof by taking $V = W_3$ and $\kappa = 2C\kappa_2 + 2k\kappa_1$.

4.16 Examples and Counter Example about the Existence of a Projective Limit of Darboux Charts

Example 4.19 *Existence of Darboux chart around any point.— According to [Pel18], 4, the set $\mathsf{L}_k^p(\mathbb{S}^1, M)$ of Sobolev loops of class L_p^k has a Banach structrue manifold and if where (M, ω) is a symplectic manifold, we can provide $\mathsf{L}_k^p(\mathbb{S}^1, M)$ with a weak symplectic form Ω_k and around any $\gamma \in \mathsf{L}_k^p(\mathbb{S}^1, M)$, we have a Darboux chart (cf. [Pel18] Theorem 32). Moreover, $\mathsf{L}_k^2(\mathbb{S}^1, M)$ is a Hilbert space and Ω_k is a strong symplectic form. If we denote by $\mathsf{L}^\infty(\mathbb{S}^1, M)$ the set of smooth loops in M, we have $\mathsf{L}^\infty(\mathbb{S}^1, M) = \varprojlim \mathsf{L}_k^2(\mathbb{S}^1, M)$ and this space is a* ILH*-manifold. It is easy to see that the sequence of forms $(\Omega_k)_{k \in \mathbb{N}}$ are compatible and since the projective sequence $\left(\mathsf{L}_k^2(\mathbb{S}^1, M)\right)_{k \in \mathbb{N}}$ is reduced, we get a weak symplectic form $\Omega = \varprojlim \Omega_k$ on $\mathsf{L}^\infty(\mathbb{S}^1, M)$. In fact, Ω can be defined directly in the same way as Ω_k on each $\mathsf{L}_k^p(\mathbb{S}^1, M)$.*
When $M = \mathbb{R}^{2m}$, consider the canonical (linear) Darboux form ω on $\mathbb{R}^{2m}$. Then we have a global Darboux chart for Ω on $\mathsf{L}^\infty(\mathbb{S}^1, \mathbb{R}^{2m})$ (cf. [Kum13b]). Of course, since we also have a global Darboux chart on each $\mathsf{L}_k^2(\mathbb{S}^1, \mathbb{R}^{2m})$, we then get an example of projective limit of Darboux charts.

Example 4.20 *Existence of a Darboux chart under Darboux-Bambusi assumptions.— Let $(\mathbb{M}_i)_{i\in\mathbb{N}}$ be a sequence of Banach spaces. Consider the submersive projective sequence of Banach spaces $\left(\overline{\mathbb{M}}_i = \prod_{k=1}^{i} \mathbb{M}_k\right)_{i\in\mathbb{N}^*}$ of Banach spaces where $\overline{\delta}_i^j : \overline{\mathbb{M}}_j \to \overline{\mathbb{M}}_i$ is the canonical projection. Then the projective limit $\overline{\mathbb{M}}$ is the product $\prod_{k=1}^{\infty} \mathbb{M}_k$. On $\overline{\mathbb{M}}$ the projective limit topology is the product topology and it is also the topology of Fréchet manifold.*
Now, assume that on each $\mathbb{M}_k$ we have a weak symplectic form ω_k such that, for some $\bar{a} = \varprojlim(a_1, \ldots, a_n) \in \overline{\mathbb{M}}$, each symplectic form ω_k satisfies the assumptions (i) and (ii) of Theorem 2.12 at a_k and for all $k \in \mathbb{N}^*$. Then from this Theorem, around the point $a_k \in \mathbb{M}_k$, we have a Darboux chart (V_k, F_k). For any $\bar{x}_n := (x_1, \ldots, x_n) \in \overline{\mathbb{M}}_n$ and $\bar{u}_n := (u_1, \ldots, u_n)$, $\bar{v}_n = (v_1, \ldots, v_n)$ in $T_{\bar{x}_n}\overline{\mathbb{M}}_n$ we define the 2 form

$$\bar{\omega}_n(\bar{u}_n, \bar{v}_n) := \sum_{k=1}^{n} \omega_k(u_k, v_k).$$

Then $\bar{\omega}_n$ is also a weak symplectic form on $\overline{\mathbb{M}}_n$ and it is easy to see that $(\overline{V}_n, \overline{F}_n)$ is a Darboux chart for $\bar{\omega}_n$ around $\bar{x}_n$. Now it is clear that the sequence $(\bar{\omega}_n)_{n\in\mathbb{N}}$ of weak symplectic forms are compatible and so give rise to a weak symplectic form $\bar{\omega}$ on $\overline{\mathbb{M}}$. Then $(\overline{V} = \varprojlim \overline{V}_n, \varprojlim \overline{F}_n)$ is a Darboux chart around $\bar{a} := \varprojlim \bar{a}_n$ if V is an open set that is $\overline{V}_n = \overline{M}_n$ for any $n \in \mathbb{N}$ execpt eventually for a finite subset $J \subset \mathbb{N}$. Such a situation occurs in the following contexts:

(1) ω_k is a linear Darboux form on the Banach space $\mathbb{M}_k$ for all $k \in \mathbb{N}$ eventually outside of finite set J.

(2) ω_k is a weak linear symplectic form on the reflexive Banach space $\mathbb{M}_k$ for all $k \in \mathbb{N}^*$ eventually outside of finite set J (cf. [CaPe20], Proposition B.3 (3))

(3) $\mathbb{H}$ is a separable infinite-dimensional real Hilbert space and we consider:

- $\mathbb{M}_k = \mathbb{H}$ *for each integer $k \in \mathbb{N}$;*
- $S_k : \mathbb{H} \to \mathbb{H}$ *is a compact operator with dense range, but proper subset of $\mathbb{H}$, which is self adjoint and positive*[15] *(such an operator is injective)*
- $\hat{\omega}$ *a linear Darboux form on $\mathbb{H}$ and $\omega_k = S_k^*\hat{\omega}$ for at most a finite number of integers and otherwise $S_k = Id_{\mathbb{H}}$.*

From the example produced in [Mars72], we can obtain the following example for which there is no Darboux chart on a submersive projective limit of symplectic Banach manifolds:

---

[15]Such operators S_k exist since the Hilbert space $\mathbb{H}$ is separable and infinite-dimensional.

Example 4.21 *Non-existence of Darboux chart.— Let $\mathbb{H}$ be a separable infinite-dimensional real Hilbert space endowed with its inner product $< \, , \, >$. If g is a weak Riemannian metric on $\mathbb{H}$, we may use the trivialization $T\mathbb{H} = \mathbb{H} \times \mathbb{H}$ to define a weak symplectic form ω in the following way ([Mars81]):*

$$2\,\omega_{(x,e)}((u,v),(u',v')) = D_x g_x(e,u).u' - D_x g_x(e,u').u + g_x(v',u) - g_x(v,u').$$

Then the operator $\omega^\flat_{(x,e)} : T_{(x,e)}\mathbb{H} \times \mathbb{H} \to T^*_{(x,e)}\mathbb{H} \times \mathbb{H}$ can be written as a matrix of type

$$\frac{1}{2}\begin{pmatrix} \Gamma_{(x,e)} & -g^\flat_x \\ g^\flat_x & 0 \end{pmatrix} \tag{4.30}$$

Since $g^\flat_x$ is always injective by assumption, it follows that $\omega^\flat_{(x,e)}$ is always injective and is surjective if and only $g^\flat_x$ is so. It follows that if Σ is the set of points $x \in \mathbb{H}$ where $g^\flat_x$ is not surjective, then the set of points $(x,e) \in T\mathbb{H}$ where $\omega_{(x,e)}$ is not a strong symplectic form is precisely $\Sigma \times \mathbb{H}$.

As at the end of the above Example, let $S : \mathbb{H} \to \mathbb{H}$ be a compact operator with dense range, but proper subset of $\mathbb{H}$, which is self adjoint and positive. Given a fixed $a \in \mathbb{H}$, then $A_x = ||x-a||^2 Id_{\mathbb{H}} + S$ is a smooth field of bounded operators of $\mathbb{H}$ which is an isomorphism for all $x \neq a$ and $A_a(\mathbb{H}) \neq \mathbb{H}$ but $A_a(\mathbb{H})$ is dense in $\mathbb{H}$ (cf. [Mars72]). Then $g_x(e,f) =< A_x(e), f >$ is a weak Riemaniann metric and the associated symplectic form $\omega_{(x,e)}$ is not a strong symplectic form if and only if (x,e) belongs to $\{a\} \times \mathbb{H}$ and, in this case, the range of $\omega^\flat_{(x,e)}$ is dense in $T^*_{(x,e)}(\mathbb{H} \times \mathbb{H}) \equiv \mathbb{H} \times \mathbb{H}$.

For each $k \in \mathbb{N}^*$ and any $x \in \mathbb{H}$ we set

$$(A_k)_x = ||x - \frac{a}{k}||^2 Id_{\mathbb{H}} + S.$$

We consider the Hilbert space $\overline{\mathbb{H}}_n = \prod_{k=1}^{n} \mathbb{H}_k$ where $\mathbb{H}_k = \mathbb{H}$ and provided with the inner product

$$< (x_1,\dots,x_n),(y_1,\dots,y_n) >_n = \sum_{k=1}^{n} < x_k, y_k >$$

As in the previous example, we identify $\overline{\mathbb{H}}_n$ with $\overline{\mathbb{H}}_n \times \{0\}$ in $\overline{\mathbb{H}}_{n+1}$. From now on, we will use the notations introduced in Example 4.20.
For any $\bar{x}_n = (x_1,\dots,x_n) \in \overline{\mathbb{H}}_n$, we set

$$(\ell_n)_{\bar{x}_n} = \left((A_1)_{x_1},\dots,(A_n)_{x_n}\right).$$

We denote by $\bar{g}_n$ the Riemannian metric on $\overline{\mathbb{H}}_n$ defined by

$$(\bar{g}_n)_{\bar{x}_n}(\bar{u}_n,\bar{v}_n) =< (\ell_n)_{(x_1,\dots,x_n)}(\bar{u}_n), \bar{v}_n >_n$$

for all $\bar{u}_n$ *and* $\bar{v}_n$ *in* $\overline{\mathbb{H}}_n$*. Thus we can consider the weak symplectic form* $\overline{\omega_n}$ *associated with* $\overline{g_n}$ *as above. Therefore the maximal open set on which* $\bar{\omega}_n$ *is a strong symplectic form is the open set*

$$\mathcal{U}_n = T\overline{\mathbb{H}}_n \setminus \bigcup_{k=1}^{n} (\{\frac{a}{k}\} \times \prod_{k=2}^{n} \mathbb{H}_k) \times \overline{\mathbb{H}}_n$$

By construction, for all $j \geq n$ *and* $n \in \mathbb{N}^*$*, we have*

$$\delta_n^j \circ (\ell_j)_{\delta_n^j(\bar{x}_j)} = (\ell_n)_{\bar{x}_n} \circ \delta_n^j.$$

We set $\overline{\mathbb{H}} = \varprojlim \overline{\mathbb{H}}_n$*. From all the above considerations, it follows that the sequence* $(\bar{\omega}_n)_{n\in\mathbb{N}^*}$ *is a family of compatible weak symplectic forms which induces a weak symplectic form* $\bar{\omega}$ *on the Fréchet manifold* $T\overline{\mathbb{H}} \equiv \overline{\mathbb{H}} \times \overline{\mathbb{H}}$ *since, as in the general case, the cotangent space* $T^*_{(\bar{x},\bar{u})}(\overline{\mathbb{H}} \times \overline{\mathbb{H}})$ *does not have a Fréchet structure, which implies that* $\bar{\omega}^\flat$ *can not be surjective.*
Now, for each $n \in \mathbb{N}^*$*, as* $(0,0)$ *belongs to the open set* $\mathcal{U}_n$*, we have a Darboux chart* $(\overline{V}_n, \overline{F}_n)$ *around* $(0,0) \in T\overline{\mathbb{H}}_n$ *from the classical Darboux Theorem for strong symplectic Banach manifold (cf. [Mars81] or [Wei71] for instance). Since* $\bar{\omega}_n$ *is a strong symplectic form on* $\overline{V}_n$ *we must have* $\overline{V}_n \subset \mathcal{U}_n$*. But from the definition of* $\mathcal{U}_n$*, it follows that*

$$\delta_1^n(\overline{V}_n) \cap T\overline{\mathbb{H}}_1 \subset \{(x,u) \in \mathbb{H} \times \mathbb{H} \ : \ ||x|| < \frac{1}{n}\}.$$

Therefore, according to Remark 4.5, the sequence $(\overline{V}_n, \overline{F}_n)$ *is not a projective sequence of charts and so there is* ***no Darboux chart*** *for* $\bar{\omega}$ *around* $(0,0) \in T\overline{\mathbb{H}}$*.*

.

Example 4.22 *Existence of Darboux chart under Theorem 4.14 assumptions.— We again consider an Hilbert space a separable infinite-dimensional real Hilbert space endowed with its inner product* $< \, , \, >$ *and for each* $k \in \mathbb{N}^*$ *and any* $x \in \mathbb{H}$ *we set*

$$(\Lambda_k)_x = \frac{1}{k^2}(||x||^2 S + Id_{\mathbb{H}})$$

where S *is a compact operator with dense range, which is self adjoint and positive. Since the set of* $\mathcal{GL}(\mathbb{H})$ *automorphisms of* $\mathbb{H}$ *is open in the Banach space* $\mathcal{L}(\mathbb{H})$ *of endomorphisms of* $\mathbb{H}$*, there exists a ball* $B(0, 2\eta_k)$ *such that* $\Lambda_k(x)$ *belongs to* $\mathcal{GL}(\mathbb{H})$ *and moreover such that*

$$||\Lambda_k(x)||^{\text{op}} \leq \frac{2}{k^2}$$

for all $x \in B(a, 2\eta_k)$*. Consider a smooth map* $\theta_k : \mathbb{R}^+ \to \mathbb{R}^+$ *such that* $\theta_k(t) = t$ *for* $0 \leq t \leq \eta_k$ *and* $\theta_k(t) = 0$ *for* $t > 2\eta_k$ *and we set*

$$(A_k)_x = \frac{1}{k^2}(\theta_k(||x||^2)S + Id_{\mathbb{H}})$$

Since $|\theta(t)| \leq 2\eta_k$ *for all* $t \in \mathbb{R}^+$ *we obtain*

$$||A_k(x)||^{\text{op}} \leq \frac{2}{k^2}. \tag{4.31}$$

for all $x \in \mathbb{H}$.

Note since θ_k *is smooth with compact support, there exist* $C_k > 0$ *such that* $||d\theta_k|| \leq C_k$ *and so*

$$||d\theta(||x||^2)||^{\text{op}} \leq 4C_k\eta_k \tag{4.32}$$

for all $x \in \mathbb{H}$.

Now, on $\mathbb{M} = \mathbb{H} \times \mathbb{H}$ *provided with the inner product*

$$< (x, x'), (y, y') >=< x, x' > + < y, y' >$$

we consider the 1*-form*

$$(\alpha_k)_{(x,x')}(u, u') = (A_k)_x(u) - (A_k)_{x'}(u)$$

and we set $\omega_k = d\alpha_k$. *Then we have*

$$\begin{aligned}
&(\omega_k)_{(x,x')}((u, u'), (v, v')) = \\
&\frac{1}{2k^2}((2d\theta_k(||x||^2) < x, u > S(v)+ < ((\theta(||x||^2)S + \text{Id}_{\mathbb{H}})(v), u > \\
&- < ((\theta(||x||^2)S + \text{Id}_{\mathbb{H}})(v), u >)) \\
&-\frac{1}{2k^2}(-2d\theta_k(||x'||^2) < x', u' > S(v')+ < ((\theta(||x'||^2)S + \text{Id}_{\mathbb{H}})(v'), u' > \\
&- < ((\theta(||x'||^2)S + Id_{\mathbb{H}})(u'), v' >).
\end{aligned}$$

It follows that $(\omega_k)^\flat_x(u, u')$ *can be written as the matrix* $\frac{1}{2k^2}\begin{pmatrix} X_k \\ Y_k \end{pmatrix}$. *where*

$$\begin{aligned}
X_k \;=\; &\big((2d\theta_k(||x||^2) < x, u > S(\ )+ < ((\theta(||x||^2)S + \text{Id}_{\mathbb{H}})(\ ), u > \\
&- < ((\theta(||x||^2)S + \text{Id}_{\mathbb{H}})(u), >)\big)
\end{aligned}$$

and

$$\begin{aligned}
Y_k \;=\; &\big(2d\theta_k(||x'||^2) < x', u' > S(\ )- < ((\theta(||x'||^2)S + \text{Id}_{\mathbb{H}})(\ ), u' > \\
&+ < ((\theta(||x'||^2)S + \text{Id}_{\mathbb{H}})(u'), > \big)
\end{aligned}$$

According to the expressions of X_k *and* Y_k *and the majorations (4.31) and (4.32), since* $\theta_k \equiv 0$ *for* $t > 2\eta_k$, *we then obtain*

$$||(\omega_k)^\flat_{(x,x')}||^{\text{op}} \leq \frac{1}{k^2}\,(2 + 4||S||^{\text{op}}\eta_k(C_k + 1)) \tag{4.33}$$

Now, from the construction of A_k*, for* $x = 0$*,* $(\omega_k)^\flat_{(0,0)}$ *is an isomorphism. Thus, for* η_k *small enough, it follows that* $(\omega_k)^\flat_{(x,x')}$ *is an isomorphism for all* $(x,x') \in \mathbb{M}$ *and so* ω_k *is a strong symplectic form on* $\mathbb{M}$*.*
Note that if we choose a decreasing sequence (η_k) *so is the sequence* (C_k) *and in this way for any* $k \geq 1$ *we get*

$$||(\omega_k)^\flat_{(x,x')}||^{\text{op}} \leq \frac{1}{k^2}\left(2 + 4||S||^{\text{op}}\eta_1(C_1+1)\right) \tag{4.34}$$

for all $(x,x') \in \mathbb{M}$*.*

Again using the expressions of X_k *and* Y_k *and the majorations (4.31), and (4.32), since* $\theta_k \equiv 0$ *for* $t > 2\eta_k$ *and and if* (η_k) *is decreasing, we obtain*

$$||(\omega_k)^\flat_{x,y)}(\mathbf{E}(x,x')) - (\omega_k)^\flat_{(00)}(\mathbf{E}(x,x'))||^{\text{op}} \leq \frac{1}{k^2}||S||^{\text{op}}(4\eta_1)^2(C_1+1) \tag{4.35}$$

for all $(x,x') \in \mathbb{M}^2$ *where* $\mathbf{E}$ *is the classical Euler vector fields on* $\mathbb{M}$*.*

By analogy with Example 4.20, we consider the Hilbert space $\overline{\mathbb{M}}_n = \prod_{k=1}^{n} \mathbb{M}_k$ *where* $\mathbb{M}_k = \mathbb{M}$ *for* $k = 1, \dots, n$ *and provided with the inner product*

$$< (x_1, \dots, x_n), (y_1, \dots, y_n) >_n = \sum_{k=1}^{n} < x_k, y_k >$$

From now on, we will use analogous notations introduced in the previous Example 4.21. For any $\bar{x}_n = (x_1, \dots, x_n) \in \overline{\mathbb{M}}_n$*, here we set*

$$(\ell_n)_{\bar{x}_n} = ((A_1)_{x_1}, \dots, (A_n)_{x_n}).$$

We can consider the symplectic form $\bar{\omega}_n = (\omega_1, \dots \omega_n)$ *on* $\overline{M}_n$ *where* ω_k *is the symplectic form on* $\mathbb{M}_k$ *associated with* (ℓ_k) *(as in Example 4.20) and we choose a decreasing sequence* (η_k)*. At point* $\bar{x}_n$ *the linear map* $(\bar{\omega}_n)^\flat_{\bar{x}_n}$ *is the matrix of diagonal blocks* $(\omega_k)^\flat_{x_k}$*. According to (4.33), there exists* $K > 0$ *(independent of* n*) such that*

$$||(\bar{\omega}_n)^\flat_{\bar{x}_n}||^{\text{op}} \leq \left(\sum_{k=1}^{n}\frac{1}{k^2}\right)K \leq \frac{\pi^2}{6}K \tag{4.36}$$

for all $\bar{x}_n \in \overline{\mathbb{M}}_n$*.*

On the other hand, denote by $\mathbf{E_k}$ *the Euler vector fields on* $\mathbb{M}_k$ *then the Euler vector field* $\overline{\mathbf{E}}_n$ *on* $\overline{M}_n$ *is* $(\mathbf{E}_1, \dots, \mathbf{E}_n)$*. Then by analog arguments, from (4.35), there exists a* $C > 0$ *such that*

$$||(\bar{\omega}_n)^\flat_{\bar{x}_n}(\bar{\mathbf{E}}_n(\bar{x}_n) - (\bar{\omega}_n)^\flat_{\bar{0}}(\bar{\mathbf{E}}_n(\bar{x}_n)||^{\text{op}} \leq \frac{\pi^2}{6}C \tag{4.37}$$

Now, as in the Example 4.20, if we set $\overline{\mathbb{M}} = \varprojlim \overline{\mathbb{M}}_n$ it follows that the sequence $(\bar{\omega}_n)_{n\in\mathbb{N}^*}$ is a family of compatible strong symplectic forms which induces a weak symplectic form $\bar{\omega}$ on the Fréchet manifold $\overline{M}$. In particular, the assumptions of Proposition 4.17 are satisfied around point $\bar{0} \in \overline{M}$ for the global chart $(\overline{M}, Id)$. Moreover, according to (4.36) and (4.37), on this chart, the assumptions **(SN)** *and* **(EN)** *are satisfied. It follows that there exists a Darboux chart around $\bar{0} \in \overline{M}$ which is a global chart.*

Remark 4.22 *In [Kum15] the required conditions for a sequence of compatible weak symplectic forms can be weakened. Precisely condition (2) is needed only for the sequence $(\alpha_i)_{i\in\mathbb{N}}$ defined by (4.20). So according to this Remark, it is easy to see that the Example 4.22 satisfies the other assumptions required in Theorem 5.1 in [Kum15]. Now consider any finite sequences of weak symplectic manifolds Banach manifold $(M_1, \sigma_1), \dots, (M_p, \sigma_p)$ which satisfies the Darboux-Bambusi Theorem assumptions around some points $a_i \in M_i$, $i \in \{1, \dots, p\}$ but not for a global chart (that is the associated vector field X_i^t given in the Moser's method is not defined on the whole manifold M_i for $i \in \{1, \dots, p\}$). For instance, one can take for M_i the Sobolev Banach manifold of loops $\mathsf{L}_k^p(\mathbf{S}, M)$ where M is a finite-dimensional symplectic manifold and use [Pel18], Theorem 5.9. Then by taking $M_n = \prod_{i=1}^{k} M_i \times \overline{\mathbf{M}}_n$ we can build a projective sequence of weak symplectic Banach manifolds for which the assumptions of Theorem 4.14 are satisfied but not the assumptions of [Kum15], Theorem 5.1 since on each M_i the vector fields X_i^t is not globally defined on M_i. Of course the condition (2) in for compatible sequences of weak symplectic form is not checked either.*

4.17 Projective Limits of Finsler-Banach Manifolds

Definition 4.37 *Let $(E_i, \pi_i, M_i, \rho_i)_{\underleftarrow{i}}$ be a projective sequence of Banach bundles. A sequence $(|| \; ||^i)_{i\in\mathbb{N}}$ of Finsler metrics on E_i is called a projective sequence of Finsler metrics if we have*

(PSFM) $\qquad \forall (i,j) \in \mathbb{N}^2 : \; j \geq i, \; || \; ||^j \circ \lambda_i^j = || \; ||^i.$

Under Definition 4.37, we set $|| \; || = \varprojlim || \; ||^i$ and $|| \; ||$ is called a *projective Finsler metric* on $E = \varprojlim E_i$. Note that if $\lambda_i : E \to E_i$ is the canonical map which is the direct limit $\varprojlim_{j\geq i} \lambda_i^j$ then, on each fibre E_x, we have a seminorm ν_x^i associated with $|| \; ||_{x_i}^i$ which is defined by $\nu_x^i(u) = ||u_i||_{x_i}^i$ if $\lambda_i(u) = u_i$. In this way, we get a continuous map $\nu^i : E \to \mathbb{R}$ which is continuous and called the *Finsler semi-metric* associated with $|| \; ||^i$. A projective Finsler metric on $M = \varprojlim M_i$ is a projective Finsler metric on TM.

Remark 4.23 *According to Definition 4.37, when M is a projective limit of a projective system $(M_i, \delta_i^j)_{j \geq i}$, for each $x \in M$, there exists a projective chart bundle limit $(U = \underleftarrow{\lim} U_i, \tau = \underleftarrow{\lim} \tau_i)$ a Finsler metric $|| \ ||^i$ on $(E_i)_{|U_i} \equiv U_i \times (E_i)_{x_i}$. It follows that we can associate a Finsler semi-metric ν^i on $(E_i)_{|U_i} \equiv U_i \times (E_i)_{x_i}$ and there exists a constant k_i such that (cf. Definition 4.36):*

$$\forall i \in \mathbb{N}, \ \forall (z, u) \in U_i \times (E_i)_{x_i}, \ \frac{1}{k_i} \nu_x^i(u) \leq \nu_z^i(u) \leq k_i \nu_x^i(u). \tag{4.38}$$

In [Eft15], K. Eftekharinasab defines a notion of Finsler metric on a Fréchet manifold M which is stronger than the previous Definition. More precisely, in Eftekharinasab's definition, when M is a projective limit of a sequence $(M_i, \delta_i^j)_{j \geq i}$, the constant k_i which appear in (4.38) must be independent of i.

Let $\Phi = \underleftarrow{\lim} \Phi^i$ be a Finsler metric on $M = \underleftarrow{\lim} M_i$. As in the Banach setting, for any C^1-piecewise curve $\gamma : [a, b] \to M$ its length

$$l(\gamma) = \int_a^b \Phi_{\gamma(t)}(\dot{\gamma}(t)) dt.$$

is well defined. If each M_i is connected, so is M, and we can again define a distance d_Φ on M in the same way and the topology defined d_Φ.

4.18 Projective Limits of Anchored Bundles and Lie Algebroids

In this section, $\left((\mathcal{A}_i, \pi_i, M_i), \left(\zeta_i^j, \delta_i^j \right) \right)_{(i,j) \in \mathbb{N}^2, j \geq i}$ is a projective sequence of Banach vector bundles with typical fibre $\mathbb{A}_i$.

4.18.1 Projective Limits of Banach Anchored Bundles

Definition 4.38 $\left((\mathcal{A}_i, \pi_i, M_i, \rho_i), \left(\zeta_i^j, \delta_i^j \right) \right)_{(i,j) \in \mathbb{N}^2, j \geq i}$ *is called a projective sequence of anchored bundles if*

(PSBAB1) $\left(\mathcal{A}_i, \zeta_i^j \right)_{j \geq i}$ *is a projective sequence of Banach vector bundles $(\pi_i : \mathcal{A}_i \to M_i)_{i \in \mathbb{N}}$ over the projective sequence of manifolds $\left(M_i, \delta_i^j \right)_{j \geq i}$;*

(PSBAB2) *For all $(i, j) \in \mathbb{N}^2$ such that $j \geq i$, one has*

$$\rho_i \circ \zeta_i^j = T\delta_i^j \circ \rho_j$$

We then have the essential result:

Theorem 4.15 *Let* $\left(\left(\mathcal{A}_i, \pi_i, M_i, \rho_i\right), \left(\zeta_i^j, \delta_i^j\right)\right)_{(i,j)\in\mathbb{N}^2, j\geq i}$ *be a projective sequence of Banach anchored bundles.*
Then $\left(\mathcal{A} = \underleftarrow{\lim}\mathcal{A}_i, \pi = \underleftarrow{\lim}\pi_i, M = \underleftarrow{\lim}M_i \rho = \underleftarrow{\lim}\rho_i\right)$ *is a Fréchet anchored bundle with typical fibre* $\mathbb{A} = \underleftarrow{\lim}\mathbb{A}_i$.

Proof *First remark that the projective limit* $\underleftarrow{\lim}M_i$ *is endowed with a differential manifold structure as defined in 3.40.*
$\left(\underleftarrow{\lim}\mathcal{A}_i, \underleftarrow{\lim}\pi_i, \underleftarrow{\lim}M_i\right)$ *is a Fréchet vector bundle whose structural group is* $H^0(\mathbb{A})$ *(cf. Proposition 4.4).*
The projective limit of the (vector) tangent bundles $\left(\underleftarrow{\lim}TM_i, \underleftarrow{\lim}p_i, \underleftarrow{\lim}M_i\right)$ *is equipped with a Fréchet vector bundle structure is the tangent vector bundle of* M *(cf. [Gal04], Theorem 2.1.)*
According to the condition $\rho_i \circ \zeta_i^j = T\delta_i^j \circ \rho_j$, *the projective limit* $\rho = \underleftarrow{\lim}\rho_i$ *is a Fréchet bundle morphism from* $\underleftarrow{\lim}\,\mathcal{A}_i$ *to* $\underleftarrow{\lim}TM_i$.

4.18.2 Projective Limits of Banach Lie Algebroids

Let $\left(\left(\mathcal{A}_i, \pi_i, M_i\right), \left(\zeta_i^j, \delta_i^j\right)\right)_{(i,j)\in\mathbb{N}^2, j\geq i}$ be a projective sequence of Banach bundles and consider $\zeta_i = \underleftarrow{\lim}\zeta_i^j : \mathcal{A} \to \mathcal{A}_i$ and $\delta_i = \underleftarrow{\lim}\delta_i^j : M = \underleftarrow{\lim}M_i \to M_i$. If U is an open set in M, then there exists a family $(U_i)_{i\in\mathbb{N}}$ of open sets U_i of M_i such that $U = \underleftarrow{\lim}U_i$. This means that $\delta_i(U) \subset U_i$ but, in general, $\delta_i(U)$ is not an open set of M_i.

Any (local) section $\mathfrak{a}$ of $\mathcal{A} = \underleftarrow{\lim}\mathcal{A}_i$ is not in general a projective limit of local sections $\mathfrak{a}_i$ of $\mathcal{A}_i$. If $\mathfrak{a}$ is a section of $\mathcal{A}_U$, in general the relation $\mathfrak{a}_i \circ \delta_i = \zeta_i \circ \mathfrak{a}$ does not define a section of $\mathcal{A}_i$ over $\delta_i(U)$ and, even if this is true, $\mathfrak{a}_i$ cannot be extended to a section defined on U_i.

Definition 4.39 $\left(\left(\mathcal{A}_i, \pi_i, M_i, \rho_i, [.,.]_{\mathcal{A}_i}\right), \left(\zeta_i^j, \delta_i^j\right)\right)_{(i,j)\in\mathbb{N}^2, j\geq i}$ *is called a projective sequence of Lie algebroids if* $\left(\left(\mathcal{A}_i, \pi_i, M_i, \rho_i\right), \left(\zeta_i^j, \delta_i^j\right)\right)_{(i,j)\in\mathbb{N}^2, j\geq i}$ *is a projective sequence of Banach anchored bundles and the sequence* $\left([.,.]_{\mathcal{A}_i}\right)_{i\in\mathbb{N}}$ *of Lie brackets satisfies the following property:*

(PSBLA) $\zeta_i^j : \mathcal{A}_j \to \mathcal{A}_i$ *is a Lie morphism from* $\left(\mathcal{A}_j, \pi_j, M_j, \rho_j, [.,.]_{\mathcal{A}_i}\right)$ *to* $\left(\mathcal{A}_i, \pi_i, M_j, \rho_i, [.,.]_{\mathcal{A}_i}\right)$ *over* δ_i^j.

We now fix some projective sequence of Lie algebroids

$$\left(\left(\mathcal{A}_i, \pi_i, M_i, \rho_i, [.,.]_{\mathcal{A}_i}\right), \left(\zeta_i^j, \delta_i^j\right)\right)_{(i,j)\in\mathbb{N}^2, j\geq i}.$$

Note that from Theorem 4.15, its projective limit is an anchored Fréchet bundle $(\mathcal{A}, \pi, M, \rho)$. Before stating the result on the structure of partial Fréchet Lie algebroid on this projective limit, we have

Proposition 4.18 *For any open set U of M, there exists a projective sequence $\left(U_i, \delta_{i\,|U_j}^j\right)_{(i,j)\in\mathbb{N}^2, j\geq i}$ such that $U = \varprojlim U_i$.*
Let $\mathfrak{P}(\mathcal{A}_U)$ be the $C^\infty(U)$ module generated by the set of sections $\mathfrak{a}$ of $\mathcal{A}$ defined on U such that $\mathfrak{a} = \varprojlim \mathfrak{a}_i$ where $\mathfrak{a}_i$ is a section of $\mathcal{A}_i$ over U_i. Then we have:

(1) For any open subset U in M, there exists a well-defined Lie bracket $[.,.]_{\mathcal{A}_U}$ on $\mathfrak{P}(\mathcal{A}_U)$ such that $(\mathfrak{P}(\mathcal{A}_U), [.,.]_{\mathcal{A}_U})$ has a structure of Lie algebra and ρ induces a Lie algebra morphism from $(\mathfrak{P}(\mathcal{A}_U), [.,.]_{\mathcal{A}_U})$ to the Lie algebra $(\mathfrak{X}(U), [.,.])$ of vector fields on U.

(2) For each $x \in M$, there exists a chart domain U around x in M such that $\mathcal{A}_U$ is trivializable over U and for each $v \in T_x M$, there exists a section $\mathfrak{a} \in \mathfrak{P}(\mathcal{A}_U)$ such that $\mathfrak{a}(x) = v$.

It is clear that the family $\mathfrak{P}_M(\mathcal{A}) := \{\mathfrak{P}(\mathcal{A}_U),\ U \text{ open set in } M\}$ defines a subsheaf of the sheaf $\Gamma_M(\mathcal{A}) := \{\Gamma(\mathcal{A}_U),\ U \text{ open set in } M\}$ of sections of $\mathcal{A}$. Note that from Remark 4.4, $\mathfrak{P}(\mathcal{A}_U)$ is strictly contained in $\Gamma(\mathcal{A}_U)$ for all open set U in M.
Now, from Proposition 4.18 (1), it follows that $[.,.]_{\mathcal{A}_U}$ is a sheaf of Lie brackets on $\mathfrak{P}_M(\mathcal{A})$ simply denoted $[.,.]_{\mathcal{A}}$. Thus, from Proposition 4.18, we obtain the following result:

Theorem 4.16 *Let $\left((\mathcal{A}_i, \pi_i, M_i, \rho_i, [.,.]_{\mathcal{A}_i}), \left(\zeta_i^j, \delta_i^j\right)\right)_{(i,j)\in\mathbb{N}^2, j\geq i}$ be a projective sequence of Banach-Lie algebroids.*
The Fréchet anchored bundle $(\mathcal{A}, \pi, M, \rho)$ has a structure of strong partial Lie algebroid $(\mathcal{A}, \pi, M, \mathfrak{P}_M(\mathcal{A}))$.

Proof of Proposition 4.18 *(1) According to the projective limit topology on M, if U is an open set in M, there exists an open set U_i in M_i such that $\delta_i(U) \subset U_i$ and so $U = \varprojlim U_i$. By the way, we have $\mathcal{A}_U = \varprojlim\{\mathcal{A}_i\}_{U_i}$. Let $\mathfrak{a} = \varprojlim \mathfrak{a}_i$ and $\mathfrak{a}' = \varprojlim \mathfrak{a}'_i$ be two sections of E_U which are projective limits of sections $\mathfrak{a}_i$ and $\mathfrak{a}'_i$ of $\mathcal{A}_i$ over U_i. Now, according to Definition 3.82, since ζ_i^j is a Lie morphism, over δ_i^j, the pairs of sections $(\mathfrak{a}_i, \mathfrak{a}_j)$ and $(\mathfrak{a}'_i, \mathfrak{a}'_j)$ are δ_i^j related and so we have*

$$\zeta_i^j([\mathfrak{a}_j, \mathfrak{a}'_j]_{\mathcal{A}_j})(x_j) = ([\mathfrak{a}_i, \mathfrak{a}'_i]_{\mathcal{A}_i})\,(\delta_i^j(x_j)) \tag{4.39}$$

It follows that $\varprojlim[\mathfrak{a}_i, \mathfrak{a}'_i]_{\mathcal{A}_i}$ belongs to $\mathfrak{P}(\mathcal{A}_U)$. We set $[\mathfrak{a}, \mathfrak{a}']_{\mathcal{A}_U} = \varprojlim[\mathfrak{a}_i, \mathfrak{a}'_i]_{\mathcal{A}_i}$.

For any $f \in C^\infty(U)$ we set

$$[\mathfrak{a}, f\mathfrak{a}']_{\mathcal{A}_U} = df\,(\rho(\mathfrak{a}))\,\mathfrak{a}' + f[\mathfrak{a}, \mathfrak{a}'] \tag{4.40}$$

By the way, $[\mathfrak{a}, f\mathfrak{a}']_{\mathcal{A}_U}$ belongs to $\mathfrak{P}(\mathcal{A}_U)$. Note that from the skew-symmetry of $[.,.]_{\mathcal{A}_i}$ for all $i \in \mathbb{N}$, it follows that

$$[f\mathfrak{a}, \mathfrak{a}'] = -df\,(\rho(\mathfrak{a}'))\,\mathfrak{a} + f[\mathfrak{a}, \mathfrak{a}']$$

for any $f \in C^\infty(U)$.

We will write the bracket $[\mathfrak{a}, \mathfrak{a}']_{\mathcal{A}_i}$ in local coordinates.

Fix some $x \in M$. According to Definition 4.16, there exists a chart $\left(U = \varprojlim U_i, \phi = \varprojlim \phi_i\right)$ where (U_i, ϕ_i) is a chart around $x_i = \delta_i(x)$ and such that we have also a trivialization $\tau_i : \{\mathcal{A}_i\}_{U_i} \to \phi_i(U_i) \times \mathbb{A}_i$ and $\tau = \varprojlim \tau_i : \{\mathcal{A}_i\}_U \to \phi(U) \times \mathbb{A}$ is a trivialization. According to § 3.15.3 and Remark 3.13, for each $i \in \mathbb{N}$, we have

$$[\mathfrak{a}_i, \mathfrak{a}_i']_{\mathcal{A}_i}(x_i) \equiv \mathsf{d}_{\mathsf{x}_i}\mathsf{a}_i'\left(\{\rho_i\}_{\mathsf{x}_i}(\mathsf{a}_i(\mathsf{x}_i))\right) - \mathsf{d}_{\mathsf{x}_i}\mathsf{a}_i\left(\{\rho_i\}_{\mathsf{x}_i}(\mathsf{a}_i'(\mathsf{x}_i))\right) + (\mathsf{C}_i)_{\mathsf{x}_i}\left(\mathsf{a}_i(\mathsf{x}_i), \mathsf{a}_i'(\mathsf{x}_i)\right)$$

From the compatibility with bonding maps for the sequence of sections (a_i) and (a_i'), of differentials $(\mathsf{d}_{\mathsf{x}_i})$, of anchors (ρ_i), and of brackets $([.,.]_{\mathcal{A}_i})$ it follows that we must have the compatibility with bonding maps between the operators $(\mathsf{C}_i)_{\mathsf{x}_i} \in L^2_{\text{alt}}(\mathbb{A}_i; \mathbb{A}_i)$ and $(\mathsf{C}_j)_{\mathsf{x}_j} \in L^2_{\text{alt}}(\mathbb{A}_j; \mathbb{A}_j)$ for all $(i,j) \in \mathbb{N}^2$ s.t. $j \geq i$. Therefore, for $[\mathfrak{a}, \mathfrak{a}']_{\mathcal{A}_U}$, we obtain

$$[\mathfrak{a}, \mathfrak{a}']_{\mathcal{A}_U}(x) \equiv \mathsf{d}_{\mathsf{x}}\mathsf{a}'(\rho_{\mathsf{x}}(\mathsf{a}(\mathsf{x})) - \mathsf{d}_{\mathsf{x}}\mathsf{a}(\rho_{\mathsf{x}}(\mathsf{a}'(\mathsf{x}))) + \mathsf{C}_{\mathsf{x}}(\mathsf{a}(\mathsf{x}), \mathsf{a}'(\mathsf{x})) \tag{4.41}$$

where $\mathsf{C}_{\mathsf{x}} = \varprojlim\,(\mathsf{C}_i)_{\mathsf{x}_i}$ is a smooth map from U to $L^2_{\text{alt}}(\mathbb{A}, \mathbb{A})$.
Fix some sections $\mathfrak{a}$ and $\mathfrak{a}'$ in $\mathfrak{P}(\mathcal{A}_U)$. There exists sections $\mathfrak{b}_1, \dots \mathfrak{b}_n$ and $\mathfrak{b}_1', \dots \mathfrak{b}_m'$, each one being a projective limit of sections of $\{\mathcal{A}_i\}_{U_i}$, and smooth functions $f_1, \dots, f_n$, $g_1', \dots, g_m'$ in $C^\infty(U)$ such that

$$\mathfrak{a} = \sum_{l=1}^{n} f_l \mathfrak{b}_l, \quad \mathfrak{a}' = \sum_{k=1}^{m} f_k' \mathfrak{b}_k' \tag{4.42}$$

Then we have

$$[\mathfrak{a}, \mathfrak{a}']_{\mathcal{A}_U} = \sum_{1 \leq l \leq n, 1 \leq k \leq m} [f_l \mathfrak{b}_l, f_k' \mathfrak{b}_k']_{\mathcal{A}_U} \tag{4.43}$$

We will show that the equation (4.43) defines an almost Lie bracket on $\mathfrak{P}(\mathcal{A}_U)$.

Indeed, according to (4.40) and (4.41), the expression of (4.43) in local coordinates is:

$$\begin{aligned}
[\mathfrak{a}, \mathfrak{a}']_U(x) &= \sum_{1 \leq l \leq n, 1 \leq k \leq m} [f_l \mathfrak{b}_l, f_k' \mathfrak{b}_k]_U \\
&\equiv \sum_{1 \leq l \leq n, 1 \leq k \leq m} \mathsf{C}(\mathsf{f}_l \mathsf{b}_l, \mathsf{f}_k' \mathsf{b}_k') + \mathsf{f}_l \mathsf{d}\mathsf{f}_k'(\rho_{\mathsf{x}}(\mathsf{b}_l))\mathsf{b}_k' + \mathsf{f}_l \mathsf{f}_k' \mathsf{d}\mathsf{b}_k'((\rho_{\mathsf{x}}(\mathsf{b}_l)) \\
&\quad - \mathsf{f}_k' \mathsf{d}\mathsf{f}_l(\rho_{\mathsf{x}}(\mathsf{b}_k'))\mathsf{b}_l + \mathsf{f}_l \mathsf{f}_k' \mathsf{d}\mathsf{b}_l((\rho_{\mathsf{x}}(\mathsf{b}_k')) \\
&= \mathsf{d}\mathsf{a}'(\rho(\mathsf{a})) - \mathsf{d}\mathsf{a}(\rho(\mathsf{a}')) + \mathsf{C}(\mathsf{a}, \mathsf{a}').
\end{aligned} \tag{4.44}$$

The last member of the equation (4.44) implies that $[\mathfrak{a},\mathfrak{a}']_{\mathcal{A}_U}$ *only depends on the* 1*-jets of* $\mathfrak{a}$ *and* $\mathfrak{a}'$*. Moreover, if the germ at* x *of* $\mathfrak{a}'$ *is identically zero, this last member implies that the germ at* x *of* $[\mathfrak{a},\mathfrak{a}']_{\mathcal{A}_U}$ *is also zero. In particular, the germ* $[\mathfrak{a},\mathfrak{a}']_{\mathcal{A}_U}(x)$ *does not depend on the decompositions (4.42). This means that, for any open set* U *and any* $\mathfrak{a}$ *and* $\mathfrak{a}'$ *in* $\mathfrak{P}(\mathcal{A}_U)$*, we have built a well-defined skew-symmetric bilinear form on* $\mathfrak{P}(\mathcal{A}_U)$ *with values in* $\mathfrak{P}(\mathcal{A}_U)$ *which depends on the* 1*-jets of sections in* $\mathfrak{P}(\mathcal{A}_U)$*. Moreover, again from (4.40) and (4.43), it follows that* $[.,.]_{\mathcal{A}_U}$ *satisfies the Leibniz property and so (4.43) defines an almost Lie bracket on* $\mathfrak{P}(\mathcal{A}_U)$*.*

We will show that the almost Lie bracket on $\mathfrak{P}(\mathcal{A}_U)$ is in fact a Lie bracket.

For all $i\in\mathbb{N}$*, we have* $[\rho_i(\mathfrak{a}_i),\rho_i(\mathfrak{a}_i')]=\rho_i([\mathfrak{a}_i,\mathfrak{a}_i']_{\mathcal{A}_i})$*. According to the compatibility with bonding maps of the Lie bracket* $[.,.]$ *on* M_i *and* $[.,.]_{\mathcal{A}_i}$ *on* $\mathcal{A}_i$ *and* ρ_i*, we have*

$$T\delta_i^j([\rho_j(\mathfrak{a}_j),\rho_j(\mathfrak{a}_j')]-\rho_j([\mathfrak{a}_j,\mathfrak{a}_j']_{\mathcal{A}_j})=[\rho_i(\mathfrak{a}_i),\rho_i(\mathfrak{a}_i')]-\rho_i([\mathfrak{a}_i,\mathfrak{a}_i']_{\mathcal{A}_i})\circ\delta_i^j$$

It follows that if $\mathfrak{a}=\varprojlim\mathfrak{a}_i$ *and* $\mathfrak{a}'=\varprojlim\mathfrak{a}_i'$*, then* $[\rho(\mathfrak{a}),\rho(\mathfrak{a}')]=\rho([\mathfrak{a},\mathfrak{a}']_{\mathcal{A}_U})$*. Now, for* $f\in C^\infty(U)$ *we also have*

$$[\rho(\mathfrak{a}),\rho(f\mathfrak{a}')]=\rho([\mathfrak{a},f\mathfrak{a}']_{\mathcal{A}_i})=f\rho([\mathfrak{a},\mathfrak{a}']_{\mathcal{A}_i})$$

since ρ *is a bundle morphism and the term* $df\,(\rho(\mathfrak{a})).\rho(\mathfrak{a}')$ *appears in each member. We then obtain* $[\rho(\mathfrak{a}),\rho(\mathfrak{a}')]=\rho([\mathfrak{a},\mathfrak{a}']_{\mathcal{A}_U})$*.*
Finally, we consider the alternating $\mathbb{R}$*-trilinear map* $J_{\mathcal{A}_U}:\mathfrak{P}(\mathcal{A}_U)^3\to\mathfrak{P}(\mathcal{A}_U)$ *given by*

$$J_{\mathcal{A}_U}(\mathfrak{a},\mathfrak{a}',\mathfrak{a}'')=[\mathfrak{a},[\mathfrak{a}',\mathfrak{a}'']_{\mathcal{A}_U}]_{\mathcal{A}_U}+[\mathfrak{a}',[\mathfrak{a}'',\mathfrak{a}]_{\mathcal{A}_U}]_{\mathcal{A}_U}+[\mathfrak{a}'',[\mathfrak{a},\mathfrak{a}']_{\mathcal{A}_U}]_{\mathcal{A}_U}.$$

By relation (3.18) in the proof of Proposition 3.19, we have

$$J_{\mathcal{A}_U}(\mathfrak{a},\mathfrak{a}',f\mathfrak{a}'')=fJ_{\mathcal{A}_U}(\mathfrak{a},\mathfrak{a}',\mathfrak{a}'').$$

Note that since $J_{\{\mathcal{A}_i\}_{U_i}}$ *are sheaf operators and the Lie brackets* $[.,.]_{\mathcal{A}_i}$ *are compatible with bonding maps, if* $\mathfrak{a}=\varprojlim\mathfrak{b}_i$*,* $\mathfrak{a}'=\varprojlim\mathfrak{b}_i'$ *and* $\mathfrak{a}''=\varprojlim\mathfrak{b}_i''$*, then we have*

$$J_{\mathcal{A}_U}(\mathfrak{a},\mathfrak{a}',\mathfrak{a}'')=\varprojlim(J_{\{\mathcal{A}_i\}_{U_i}}(\mathfrak{b}_i,\mathfrak{b}_i',\mathfrak{b}_i'').$$

Now as the Jacobiator $J_{\mathcal{A}_i}$ *is identically zero for each* $i\in\mathbb{N}$*, it follows that we have* $J_{\mathcal{A}_U}(\mathfrak{a},\mathfrak{a}',f\mathfrak{a}'')=fJ_{\mathcal{A}_U}(\mathfrak{a},\mathfrak{a}',\mathfrak{a}'')=0$*, for any* $f\in C^\infty(U)$*; clearly, this implies that* $J_{\mathcal{A}_U}$ *is identically zero on* $\mathfrak{P}(\mathcal{A}_U)$*, which ends the proof of (2).*

(2) Since Point (2) is a local problem, we can consider it in local coordinates. In this way, any section of $\mathfrak{P}(\mathcal{A}_U)$ *is a smooth map from an open set* U *in the Fréchet space* $\mathbb{M}$ *to the Fréchet space* $\mathbb{E}$*. Thus* $\mathsf{U}=\varprojlim\mathsf{U}_i$ *where each* U_i *is open in* $\mathbb{M}_i$*. Now if* (x,v) *belongs to* $\{\mathsf{x}\}\times\mathbb{E}$*, then* $\mathsf{v}=\varprojlim\mathsf{v}_\mathsf{i}$ *where each* $\mathsf{v}_\mathsf{i}\in\mathbb{E}_i$*. This implies that the constant map* $\mathsf{a}(\mathsf{z})=\mathsf{v}$ *is the projective limit of the sequence of constant maps* a_i *on* U_i*, that is* $\mathsf{a}_\mathsf{i}(\mathsf{z}_\mathsf{i})=\mathsf{v}_\mathsf{i}$*, which ends the proof.*

4.19 Notes

The references for the notion of projective limits of different Banach structures (manifolds, Lie groups, bundles) is the book [DGV16] written by C. T. J. Dodson, G. Galanis and E. Vassiliou and the papers [AbbMa99], [ADGS09], [AgSu10] and [DoGa04]. Note that the differential calculus *à la Leslie* is used in this framework.
One of the main issues one meets in the setting of projective limits is that $\mathrm{GL}(\mathbb{E})$ where $\mathbb{E} = \varprojlim \mathbb{E}_n$ cannot be endowed with a structure of Lie group; so, in order to obtain a structure of fibre bundles on a projective limit of Banach fibre bundles, the generalized Lie group $H^0(\mathbb{E})$ introduced in [Gal98] as a projective limit of the Banach-Lie groups must be considered.
For the projective limits of Banach-Lie groups, the reader is also referred to [Omo97], [Gal96] and [AbbMa99].
Our basic references for the theory of jets which appears as a fundamental example of projective limit are [Pal68], [Che91], [Sau89], [IVV04], [KrVe11] and [Vit17].
Essential results about projective limits of tensor structures can be found in [CaPe20].
A version of the Darboux theorem for projective limit of symplectic Darboux forms can be found in [Kum15] and in a Fréchet context in [Eft19]. A discussion about the existence of a Darboux theorem for projective limit of symplectic Banach manifolds can be found in [Pel21] where a stronger result than in [Kum15] is given.

5

Direct Limits

The notion of inductive limit introduced in 1931 by L. Pontryagine generalizes to structures the concept of limit used in analysis.
The notion of direct limits of Banach structures occur in various frameworks: finite-dimensional vector spaces ($\mathbb{R}^\infty = \underrightarrow{\lim}\mathbb{R}^n$ is the basic standard example), Sobolev spaces, manifolds (finite-dimensional such as $\mathbb{S}^\infty = \underrightarrow{\lim}\mathbb{S}^n$ or modeled on Silva spaces[1]), classical Lie groups such as the linear group $\mathrm{GL}(\infty, \mathbb{R}) = \underrightarrow{\lim}\, \mathrm{GL}(n, \mathbb{R})$, bundles, Lie algebroids and tensors.
In all these situations, the direct limit can be endowed, under conditions, with a convenient structure.
Similar to what was achieved in the previous chapter, in this one, we present results on direct limits of Banach geometrical and Lie structures under adequate assumptions.
In particular, a structure of the convenient Lie group $\mathrm{G}(\mathbb{E})$ is defined on the direct limit $\mathbb{E}$ of an ascending sequence of supplemented Banach spaces and can play the rôle of structural group.

5.1 Direct Limits of Categories

Definition 5.1 *Let* $(I, \preccurlyeq)$ *be a directed set and* C *a category. A directed system (or inductive system) is a family* $\mathcal{S} = \left\{ \left(Y_i, \varepsilon_i^j \right) \right\}_{(i,j)\in I^2,\ i \preccurlyeq j}$ *where* Y_i *is an object of the category* C *and* $\varepsilon_i^j : Y_i \to Y_j$ *is a morphism (bonding map) where:*

(DS1) $\forall i \in I,\ \varepsilon_i^i = \mathrm{Id}_{Y_i}$;

(DS2) $\forall (i,j,k) \in I^3 : i \preccurlyeq j \preccurlyeq k,\ \varepsilon_j^k \circ \varepsilon_i^j = \varepsilon_i^k$.

$$Y_i \xrightarrow{\varepsilon_i^j} Y_j \xrightarrow{\varepsilon_j^k} Y_k, \qquad Y_i \xrightarrow{\varepsilon_i^k} Y_k$$

[1]The group $\mathrm{Diff}^\omega(K)$ of real analytic diffeomorphisms on a real compact analytic manifold K is modelled on a Silva space.

DOI: 10.1201/9781003435587-5

The directed system $\left\{\left(Y_i, \varepsilon_i^j\right)\right\}_{(i,j)\in I^2,\ i\preccurlyeq j}$ is a (covariant) functor $I \to \mathsf{C}$ (cf. Appendix B).

When $I = \mathbb{N}$ with the usual order relation, countable direct systems are called *direct sequences*.

Definition 5.2 *An inductive cone (or co-cone) over $\mathcal{S}$ is a pair $\left(Y, \{\varepsilon_i\}_{i\in I}\right)$ where $Y \in \mathsf{ob}(\mathsf{C})$ and $\varepsilon_i : Y_i \to Y$ is such that $\varepsilon_j \circ \varepsilon_i^j = \varepsilon_i$ whenever $i \preccurlyeq j$.*

Definition 5.3 *An inductive cone $\left(Y, \{\varepsilon_i\}_{i\in I}\right)$ is a direct limit (or a colimit) of $\mathcal{S}$ if for every inductive cone $\left(Z, \{\varphi_i\}_{i\in I}\right)$ over $\mathcal{S}$ there exists a unique morphism $\eta : Y \to Z$ such that $\eta \circ \varepsilon_i = \varphi_i$. We then write $Y = \varinjlim \mathcal{S}$ or $Y = \varinjlim Y_i$.*

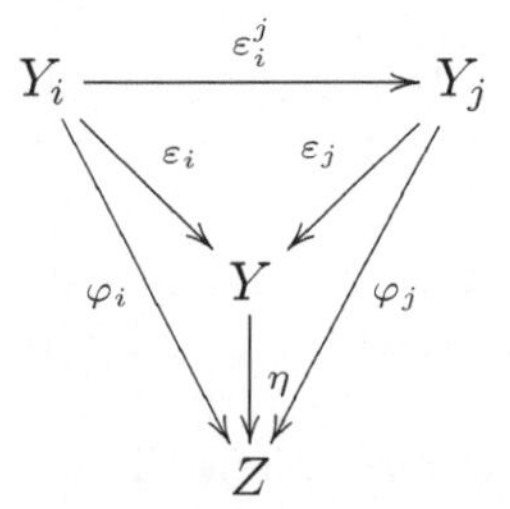

Direct limits in the categories of sets, groups and rings always exist and can be described as follows.
Let $\{Y_i\}_{i\in I}$ be a family of sets. The *disjoint union* of this family is the set

$$\bigsqcup_{i\in I} Y_i = \bigcup_{i\in I} \{(x,i) : x \in Y_i\}.$$

Consider the following equivalence relation $\sim$: for $x_i \in Y_i$ and $x_j \in Y_j$,

$$x_i \sim x_j \Leftrightarrow \exists k \in I :\ i \preccurlyeq k, j \preccurlyeq k,\ \varepsilon_i^k(x_i) = \varepsilon_j^k(x_j).$$

Then the direct limit is $\varinjlim Y_i = \bigsqcup_{i\in I} Y_i / \sim$.

Projective and direct limits can be linked *via* homomorphisms:

$$\mathbf{Hom}\left(\varinjlim X_i, Y\right) = \varprojlim \mathbf{Hom}(X_i, Y).$$

Definition 5.4 *Let $\left\{\left(X_i, \varepsilon_i^j\right)\right\}_{(i,j)\in I^2, i\preccurlyeq j}$ and $\left\{\left(Y_i, \zeta_i^j\right)\right\}_{(i,j)\in I^2, i\preccurlyeq j}$ two directed systems with respective limits X and Y. A family $\{f_i\}_{i\in I}$ of mappings $f_i : X_i \to Y_i$ satisfying the coherence condition:*

(CCDM)

$$\forall (i,j) \in I^2 : i \preccurlyeq j,\ f_j \circ \varepsilon_i^j = \zeta_i^j \circ f_i$$

is called a direct family of mappings (for $I = \mathbb{N}$, we have the notion of direct sequence of mappings).
The limit of the direct family of mappings $\{f_i\}_{i\in I}$ is the map

$$\begin{array}{rccc} f: & X = \varinjlim X_i & \to & Y = \varinjlim Y_i \\ & \{x_i\}_{i\in I} & \mapsto & \{f_i(x_i)\}_{i\in I} \end{array}$$

5.2 Direct Limits of Topological Spaces

Let $\left\{\left(X_i, \varepsilon_i^j\right)\right\}_{(i,j)\in I^2,\ i\preccurlyeq j}$ be a direct system of topological spaces and continuous maps. The direct limit $\{(X, \varepsilon_i)\}_{i\in I}$ of the sets becomes the direct limit in the category TOP of topological spaces if X is endowed with the direct limit topology (DL-topology for short), i.e. the final topology with respect to the inclusion maps $\varepsilon_i : X_i \to X$ which corresponds to the finest topology which makes the maps ε_i continuous. So $O \subset X$ is open if and only if $\varepsilon_i^{-1}(O)$ is open in X_i for each $i \in I$.

Definition 5.5 *Let $\mathcal{T} = ((X_n, \varepsilon_n^m))_{(m,n)\in\mathbb{N}^2,\ n\leq m}$ be a direct sequence of topological spaces such that each ε_n^m is injective. Without loss of generality, we may assume that we have*

$$X_0 \subset X_1 \subset \cdots \subset X_n \subset X_{n+1} \subset \cdots$$

and ε_n^{n+1} becomes the natural inclusion.

(ASTS) *$\mathcal{T}$ will be called an ascending sequence of topological spaces.*

(SASTS) *Moreover, if each ε_n^m is a topological embedding, then we will say that $\mathcal{T}$ is a* strict ascending sequence of topological spaces *(expanding sequence in the terminology of [Han71]).*

Notation 5.1 *The direct sequence $((X_n, \varepsilon_n^m))_{(n,m)\in\mathbb{N}^2,\ n\leq m}$ will be denoted $(X_n, \varepsilon_n^m)_{n\leq m}$.*

If $(X_n, \varepsilon_n^m)_{n\leq m}$ is a strict ascending sequence of topological spaces, each ε_n is a topological embedding from X_n into $X = \varinjlim X_n$.

Let us give some properties of ascending sequences of topological spaces ([Glo05], Lemma 1.7):

Proposition 5.1 *Let $(X_n, \varepsilon_n^m)_{n\leq m}$ be an ascending sequence of topological spaces. Equip $X = \bigcup\limits_{n\in\mathbb{N}} X_n$ with the* DL-*topology. Then we have:*

(1) If each X_n is T_1, then X is T_1.

(2) *If* $O_n \subset X_n$ *is open and* $O_1 \subset O_2 \subset \cdots$, *then* $O = \bigcup_{n\in\mathbb{N}} O_n$ *is open in* X *and the* DL-*topology on* $O = \varinjlim O_n$ *coincides with the topology induced by* X.

(3) *If each* X_n *is locally compact, then* X *is Hausdorff.*

(4) *If each* X_n *is* T_1 *and* $K \subset X$ *is compact, then* $K \subset X_n$ *for some* n.

Remark 5.1 *Unfortunately, a direct limit of Hausdorff topological spaces need not be Hausdorff (see [Herr69] for an example of such a situation). Sufficient conditions on* $(X_n)_{n\in\mathbb{N}}$ *under which the direct limit* $X = \bigcup_{n\in\mathbb{N}} X_n$ *is Hausdorff can be found in [HaSt72].*

However, we have the following result:

Proposition 5.2 *Let* $(X_n, \varepsilon_n^m)_{n\leq m}$ *be a strict ascending sequence of topological spaces; equip* $X = \bigcup_{n\in\mathbb{N}} X_n$ *with the* DL-*topology. Then we have:*

(1) *Assume that for each* n, X_n *is closed in* X_{n+1}.

   (a) *If each* X_n *is normal then* X *is normal;*

   (b) *If each* X_n *is Hausdorff and paracompact then* X *is normal.*
   In particular, in each previous situation, X *is Hausdorff.*

(2) *Assume that for each* n, X_n *is open in* X_{n+1} *and Hausdorff, then* X *is Hausdorff.*

Proof *(1) (a) see [Han71] Proposition 4.3.*
(1) (b) It is well known that any Hausdorff and paracompact topological space is normal and then such a topological space is Hausdorff.
(2) see [Han71] Proposition 4.2.

Remark 5.2 *The direct limit of an ascending sequence* $(X_n)_{n\in\mathbb{N}}$ *is equal to the direct limit of* $(X_n)_{n\in\mathbb{N},\ n_0\leq n}$

Proposition 5.3 *(Continuity criterion, [HSTH01]). Let* $(X_n, i_n^m)_{n\leq m}$ *and* $(Y_n, j_n^m)_{n\leq m}$ *be two ascending sequences of topological spaces and consider a sequence of consistent maps* $f_n : X_n \to Y_n$. *If every map* f_n *is continuous, then the direct limit* $f = \varinjlim f_n$ *is continuous with respect to the* DL-*topologies on* X *and* Y.

Remark 5.3 *Let* $(X_n, i_n^m)_{n\leq m}$ *and* $(Y_n, j_n^m)_{n\leq m}$ *be two ascending sequences of topological spaces and consider a sequence of consistent maps* $f_n : X_n \to Y_n$. *If every map* f_n *is injective (resp. surjective) so is the direct limit* $f = \varinjlim f_n$.

5.3 Ascending Sequences of Normed Spaces

5.3.1 Locally Convex Topology on an Ascending Sequence of Normed Spaces

Let $\mathbb{E}_0 \subset \mathbb{E}_1 \subset \cdots$ an increasing sequence of normed spaces where all the bonding maps $i_n^m : \mathbb{E}_n \to \mathbb{E}_m$ are linear and continuous (i.e. bounded operators).
The *locally convex direct limit* of this ascending sequence is the union $\mathbb{E} = \bigcup\limits_{n\in\mathbb{N}} \mathbb{E}_n$ endowed with the finest locally convex topology under which all the maps $i_n : \mathbb{E}_n \to \mathbb{E}$ are continuous; this final topology is called *locally convex direct limit topology* and denoted LDLC-topology for short.
For this topology, a convex subset U is a neighborhood of 0 if and only if $U \cap \mathbb{E}_n$ is a neighborhood of $0 \in \mathbb{E}_n$ for each $n \in \mathbb{N}$.

The following result given [Dah11], Proposition 1.2.1 is the adaptation of [RoRo64], Chapter V, Proposition 4 for the setting of inductive limits of l.c.t.v.s.

Proposition 5.4 *Let $\underline{\varepsilon} = (\varepsilon_n)_{n\in\mathbb{N}}$ be any sequence of positive numbers. Then the set*

$$V(\underline{\varepsilon}) = \bigcup_{n\in\mathbb{N}} \left(B_1(\varepsilon_1) + \cdots + B_n(\varepsilon_n)\right)$$

is a neighbourhood of $0 \in \mathbb{E}$, where, for $i \in \mathbb{N}$, $B_i(\varepsilon_i)$ is the ball of center 0 in $\mathbb{E}_n$ and of radius ε_i.
Moreover, the set of such neighborhoods is a basis of neighborhoods of $0 \in \mathbb{E}$.

The LDLC-topology satisfies the universal property:
a linear map f from $\mathbb{E}$ to a l.c.t.v.s. $\mathbb{F}$ is continuous if and only if each restriction $f|_{\mathbb{E}_n}$ is continuous with respect to the topology of $\mathbb{E}_n$.

For each $n \in \mathbb{N}$, we fix a norm $\| \ \|_n$ on $\mathbb{E}_n$ such that:

$$\forall n \in \mathbb{N}, \ \| \ \|_{n+1} \leq \| \ \|_n \text{ on } \mathbb{E}_n.$$

Then, for $m \geq n$, for any $r \geq 0$, we have $B_n(r) \subset B_m(r)$.

Theorem 5.1 *Consider an ascending sequence $(\mathbb{E}_n, i_n^m)_{n\leq m}$ of normed spaces and denote by $\mathbb{E} = \bigcup\limits_{n\in\mathbb{N}} \mathbb{E}_n$ its direct limit.*

(1) The set $B(r) = \bigcup_{n\in\mathbb{N}} B_n(r)$ is an absolutely convex absorbent open set in $\mathbb{E}$ for the LDLC*-topology.*

(2) The set

$$\mathcal{B} = \{x + B(r), \ x \in \mathbb{E}, \ r > 0\}$$

is a basis for the LDLC*-topology on $\mathbb{E}$.*

(3) Consider an ascending sequence of open sets $(U_n)_{n\in\mathbb{N}}$ *in* $\mathbb{E}_n$*. Then* $U = \bigcup_{n\in\mathbb{N}} U_n$ *is an open set for the* LDLC*-topology on* $\mathbb{E}$.

Proof

(1) Since in a normed space, any ball $B(\varepsilon)$ *is an absolutely convex absorbent set, this implies the same property for* $B(r)$ *in* $\mathbb{E}$*. Now, for any* $x \in B(r)$ *let* p *be the smallest integer such that* x *belongs to* $B_p(r)$*. For each* $n \geq p$*, there exists* $\varepsilon_n > 0$ *such that* $B_n(x, \varepsilon_n) \subset B_n(r)$*. We set* $\epsilon_n = 0$ *for* $1 \leq n \leq p$ *and* $\underline{\varepsilon} = (\varepsilon_n)$*. According to Proposition 5.4, the set* $V(\underline{\varepsilon})$ *is a neighbourhood of* $0 \in \mathbb{E}$ *and so* $x + V(\underline{\varepsilon})$ *is a neighbourhood of* x *contained in* $B(r)$ *which implies that* $B(r)$ *is an open set for the locally convex space topology on* $\mathbb{E}$.

(2) is a consequence of Proposition 5.4 and the definition of a basis neighborhoods.

(3) Consider an ascending sequence of open sets U_n *in* $\mathbb{E}_n$*. Fix some* $x \in U$ *and denote by* p *the smallest integer such that* $x \in U_p$*. For* $n \geq p$*, there exists a ball* $B_n(\epsilon_n)$ *such that* $B_n(x, \varepsilon_n) = x + B_n(\varepsilon_n)$ *which is contained in* U_n*. Thus if again we set* $\varepsilon_n = 0$ *for* $1 \leq n \leq p$ *and* $\underline{\varepsilon} = (\varepsilon_n)$*. Then* $x + V(\underline{\varepsilon})$ *is a neighbourhood of* x *in* U *and so* U *is an open set for the* LDLC*-topology.*

Note that the LDLC-topology does not coincides with the DL-topology. Moreover, in general, the locally convex topology on the direct limit of an ascending sequence of normed space need not be Hausdorff (for more details the reader can see [Dah11] or [Flo79]).

Definition 5.6 *An ascending sequence* $(\mathbb{E}_n, i_n^m)_{n \leq m}$ *of normed spaces is called a* strict ascending sequence a normed spaces *if* $\mathbb{E}_n$ *is a closed normed subspace of* $\mathbb{E}_{n+1}$.

5.3.2 Direct Sequences of Smooth Maps

We consider strict ascending sequences $(\mathbb{E}_n, i_n^m)_{n \leq m}$ and $(\mathbb{F}_n, j_n^m)_{n \leq m}$. We denote by $\mathbb{E}$ and $\mathbb{F}$ the respective direct limit of $(\mathbb{E}_n, i_n^m)_{n \leq m}$ and $(\mathbb{F}_n, j_n^m)_{n \leq m}$.

The following proposition gives results about continuity and differentiability of direct sequence of mappings, i.e. sequence of mappings fulfilling the compatibility condition **(CCDM)** of Definition 5.4 between $\mathbb{E}$ and $\mathbb{F}$.

Proposition 5.5 *Let* $U_1 \subset \cdots \subset U_n \subset \cdots$ *be an ascending sequence* $(U_n)_{n\in\mathbb{N}}$ *of open sets of* $(\mathbb{E}_n)_{n\in\mathbb{N}}$ *and set* $U = \bigcup\limits_{n\in\mathbb{N}} U_n$*. Assume that we have a direct sequence of mappings* $f_n : U_n \to \mathbb{F}_n$ *of class* C^k *for* $k \in \mathbb{N} \cup \{0, \infty\}$ *and set* $f = \varinjlim f_n$*. Then we have:*

(1) If each f_n *is continuous so is* f.

(2) If each f_n *is of class* C^1 *so is* f *and*

$$\forall x \in U_n,\ d_x f = \varinjlim d_x f_n.$$

(3) If each f_n is of class C^k so is f.

(4) If each f_n is a C^k diffeomorphism from U_n into an open set V_n of $\mathbb{F}$, then f is a C^k diffeomorphism from U to $V = \bigcup_{n\in\mathbb{N}} V_n$

Proof *At first from Theorem 5.1, $U = \bigcup\limits_{n\in\mathbb{N}} U_n$ is an open set of $\mathbb{E}$ for its* LDLC *topology. Moreover, we may assume that $U_n = U_{n+1} \cap \mathbb{E}_n$ for all $n \in \mathbb{N}$. We denote by $i_n = \varinjlim_{m\geq n} i_m : \mathbb{E}_n \to \mathbb{E}$ and $j_n = \varinjlim_{m\geq n} j_m : \mathbb{F}_n \to \mathbb{F}$ the natural inclusions. By the way, we can identify $\mathbb{E}_n$ and $\mathbb{F}_n$ as subspaces of $\mathbb{E}$ and $\mathbb{F}$, respectively. Thus we will consider each f_n as a map from $U_n = \mathbb{E}_n \cap U$ to $\mathbb{F}$.*

(1) If each f_n is continuous, then, for any open set V in $\mathbb{F}$, $V_n = V \cap \mathbb{F}_n$ is an open set of $\mathbb{F}_n$ and $f_n^{-1}(V_n)$ is open in U_n; so $\bigcup\limits_{n\in\mathbb{N}} f_n^{-1}(V_n)$ is open of $\mathbb{E}$. It follows that $f = \varinjlim f_n$ must be continuous.

(2) Assume that each f_n is of class C^1. Fix some $x \in U$ and let p be the smallest integer such that x belongs to U_p. Then since $\mathbb{E}_p$ is closed in any $\mathbb{E}_n$ for $n \geq p$, it follows that if $p \geq 2$, for $1 \leq n < p$ then $U \cap \mathbb{E}_n = \emptyset$. Fix some $n \geq p$ and choose some $u \in \mathbb{E}_n$. There exists some interval $[0, \epsilon]$ such that $x + tu$ belongs to U_n for $t \in [0, \epsilon]$. Consider the map.

$$\begin{cases} R_n(t) = \dfrac{f_n(x+tu) - f_n(x) - d_x f_n(tu)}{t}, & \text{for } t \in]0, \epsilon] \\ R_n(0) = 0 \end{cases}$$

This map is continuous at 0 if f_n is of class C^1. Since for $m \geq n$, we have $U_m \cap \mathbb{E}_n = U_n$, it follows that $x + [0, \epsilon]u$ belongs to U_m On the other hand, the restriction of f_m to U_n is f_n and as the ascending sequence $\mathbb{E}_n$ is strict, so the restriction of $d_x f_m$ to $\mathbb{E}_n$ must be $d_x f_n$. From all these properties, it follows that the restriction of R_m to $[0, \epsilon]$ is R_n. So the map $R = \varinjlim R_n$ is well defined and we have

$$\begin{cases} R(t) = \dfrac{f(x+tu) - f_n(x) - d_x f(tu)}{t}, & \text{for } t \in]0, \epsilon] \\ R(0) = 0 \end{cases}$$

and so R is continuous at 0 and we have

$$d_x f(u) := \lim_{t\to 0} \frac{(f(x+tu) - f(x)}{t}.$$

Since this relation is true for any $u \in \mathbb{E}$, this means that $d_x f$ is well defined and we have

$$d_x f = \varinjlim d_x f_n. \tag{5.1}$$

The relation (5.1) implies that $d_x f$ is linear on $\mathbb{E}$ and continuous from (1).

(3) Now the relation(5.1) being true for any $x \in \mathbb{E}$, *we have* $df = \varinjlim df_n$. *But since* $df_n : U_n \times \mathbb{E}_n \to \mathbb{F}_n, df : U \times \mathbb{E} \to \mathbb{F}$, $U \times \mathbb{E} = \varinjlim U_n \times \mathbb{E}_n$ *and* $\mathbb{F} = \varinjlim \mathbb{F}_n$, *from (1), it follows that df is continuous.*

Finally, using analogous arguments and Chapter 3, Definition 3.21, by induction on k, *we can show that* f *is of class* C^k *if each* f_n *is so.*

(4) Assume that each f_n *is a* C^k *diffeomorphism from* U_n *to* V_n, *then* f *is a bijective map from* U *to* $V = \bigcup_{n\in\mathbb{N}} V_n$ *of class* C^k. *Let* $g_n = f_n^{-1}$. *The previous argument shows that* $g = \varinjlim g_n$ *is a* f *is a bijective map from* V *to* $U = \bigcup_{n\in\mathbb{N}} U_n$ *of class* C^k *and we have* $g = f^{-1}$, *which ends the proof.*

Proposition 5.6 *Consider a* c^∞*-open set* U *in* $\mathbb{E}$ *and a smooth map* $f : U \to \mathbb{F}$ *and set* $U_n = U \cap \mathbb{E}_n$. *Given any* $x = \varinjlim x_n \in U$, *we consider an ascending sequence* (B_n) *of open balls* B_n *in* U_n *centred in* x_n. *There exists an increasing map* $n \mapsto j_n$ *in* $\mathbb{N}$ *such that* $f(B_n) \subset \mathbb{F}_{j_n}$ *and, if* $B = \varinjlim B_n$ *and* $f_n = f_{|B_n}$, *then* $f_{|B} = \varinjlim f_n$.

Proof *We denote by* $i_n = \varinjlim_{m\geq n} i_m : \mathbb{E}_n \to \mathbb{E}$ *and* $j_n = \varinjlim_{m\geq n} j_m : \mathbb{F}_n \to \mathbb{F}$ *the natural inclusions. By the way, we can identify* $\mathbb{E}_n$ *and* $\mathbb{F}_n$ *with subspaces of* $\mathbb{E}$ *and* $\mathbb{F}$, *respectively.*
Since U *is a* c^∞*-open set, from [CaPe19] Proposition 3.7,* U *is an open set for the DL topology. We set* $U_n = U \cap \mathbb{E}_n$. *Then* U_n *is an open set of* $\mathbb{E}_n$ *which gives rise to an ascending sequence of open sets such that* $U = \varinjlim U_n$. *According to the assumptions, since each* B_n *is a bounded set and* f *is bounded over each bounded set in* $\mathbb{E}$, *it follows that there exists* $j_n \in \mathbb{N}$ *such that* $f(B_n) \subset \mathbb{F}_{j_n}$ *(cf. [CaPe19], Lemma 3.6). Since the sequence* $(\mathbb{F}_n)_{n\in\mathbb{N}}$ *is ascending, without loss of generality, we may assume that* $j_n \geq n$. *Since* $B_n \subset B_{n+1}$, *we may assume that* $j_{n+1} > j_n$. *By the way, the set* $\{j_n, n \in \mathbb{N}\} \subset \mathbb{N}$ *is countable. By [CaPe19], Proposition 3.5, we have* $\mathbb{F} = \varinjlim \mathbb{F}_{j_n}$, *which implies the results.*

Remark 5.4 *From Proposition 5.6, each smooth map* $f : \mathbb{E} \to \mathbb{F}$ *can be written locally as a direct limit of smooth maps from* $\mathbb{E}_n$ *to some* $\mathbb{F}_{j_n}$. *But in general we will have* $j_n > n$ *and so* f *cannot be written as a direct limit of maps from* $\mathbb{E}_n$ *to* $\mathbb{F}_n$.
For example, consider $\mathbb{E} = \mathbb{F} = \mathbb{R}^\infty$ *corresponding to the direct sum of the spaces*[2] $\mathbb{R}^n$ *for* $n \in \mathbb{N}$, *provided with the seminorms* $||(x_n)_{n\in\mathbb{N}}||_k = |x_k|$. *Consider the map* $\begin{array}{cccc} f: & \mathbb{R}^\infty & \to & \mathbb{R}^\infty \\ & (x_n) & \mapsto & (x_{2n}) \end{array}$. $f_n = f_{|\mathbb{R}^n}$ *is then a smooth map and since* $f = \varinjlim f_n$, f *is smooth. Now, given any point* $(a_n)_{n\in\mathbb{N}} \in \mathbb{R}^\infty$ *and any sequence of balls* $B_n = \{(x_1, \dots, x_n) \in \mathbb{R}^n : |x_k - a_k| < \varepsilon_k\}$, *then* $f(B_n) \subset \mathbb{R}^{2n}$. *Assume that* f *can be written as a direct limit of smooth maps* $\phi_n : B_n \to \mathbb{R}^n$.

[2] The convenient space $\mathbb{R}^\infty$ is defined in Chapter 3, Example 3.10.

There would imply that $\phi_n = f_{|B_n}$ *and so* $f(B_n) \subset \mathbb{R}^n$*, which is a contradiction with the definition of* f*.*

5.4 Ascending Sequences of Banach Spaces

Definition 5.7 *An ascending sequence* $(\mathbb{E}_n, i_n^m)_{n \leq m}$ *of Banach spaces is a* strict ascending sequence of Banach spaces *if* $\mathbb{E}_n$ *is a Banach subspace of* $\mathbb{E}_{n+1}$ *for each* $n \in \mathbb{N}$*.*

Definition 5.8 *A locally convex limit of ascending sequence of Banach spaces is called an (LB)-space.*
If the sequence is strict we speak of LB-space or strict (LB)-space.

Definition 5.9 *A locally convex limit of ascending sequence of Banach spaces where each inclusion is compact is called a Silva space or a* (DFS)*-space (for dual of Fréchet-Schwartz space).*

As indicated in [Glo11], Silva spaces are very well-behaved direct limits: if $\mathbb{E} = \bigcup\limits_{n \in \mathbb{N}} \mathbb{E}_n$ is a Silva space, it is Hausdorff, complete, boundedly regular[3] and the LCDL-topology coincides with the DL-topology (results of [Flo71] recalled in op. cit., Proposition 4.4).
Moreover, some interesting infinite-dimensional Lie groups are modelled on Silva spaces[4].

An (LB)-space does not have a structure of convenient space in general. However, since every Banach space is convenient, then each LB-space has a structure of convenient space (see [KrMi97], Theorem 2.15). In particular, the direct limit of an ascending sequence of finite-dimensional Banach spaces has a structure of convenient space.
Note that if $\mathbb{E} = \underrightarrow{\lim}\mathbb{E}_n$ and $\mathbb{F} = \underrightarrow{\lim}\mathbb{F}_n$ are LB-spaces, then we can identify $\mathbb{E} \times \mathbb{F} = \underrightarrow{\lim}(\mathbb{E}_n \times \mathbb{F}_n)$ with $\underrightarrow{\lim}\mathbb{E}_n \times \underrightarrow{\lim}\mathbb{F}_n$ is a l.c.t.v.s. (cf. [HSTH01], Theorem 4.3) and so $\mathbb{E} \times \mathbb{F}$ is a convenient vector space (cf.[KrMi97]).

We now consider a general situation which gives rise to a convenient structure on a direct limit of an ascending sequence of Banach spaces.

[3] $\mathbb{E} = \bigcup\limits_{n \in \mathbb{N}} \mathbb{E}_n$ is *boundedly regular* if every bounded subset of $\mathbb{E}$ is contained in some $\mathbb{E}_n$.

[4] The infinite-dimensional Lie group $\mathrm{Diff}^{\omega}(K)$ of real analytic diffeomorphisms of a compact real analytic manifold K is modelled on a Silva space (cf. [Lesl82] and [KrMi97], Theorem 43.4).

Proposition 5.7 *Let $(\mathbb{E}_n, i_n^m)_{n \leq m}$ be an ascending sequence of Banach spaces. Assume that there exists an infinite subset $I \subset \mathbb{N}$ such that $\mathbb{E}_I = \bigcup_{i \in I} \mathbb{E}_i = \underrightarrow{\lim} \mathbb{E}_i$ is an LB-space. Then $\mathbb{E} = \bigcup_{n \in \mathbb{N}} \mathbb{E}_n = \mathbb{E}_I$ and $\mathbb{E}$ is an LB-space.*

Lemma 5.1 *If $\mathbb{E} = \underrightarrow{\lim} \mathbb{E}_n$ is an LB-space, then for the* LDLC*-topology we have:*

(1) $\mathbb{E}$ is Hausdorff and boundedly regular.

(2) Let $f : \mathbb{E} \longrightarrow \mathbb{F}$ be a linear map where $\mathbb{F}$ is a Banach space. The following properties are equivalent:

(i) f is bounded;

(ii) each restriction f_n of f to $\mathbb{E}_n$ is continuous;

(iii) f is continuous.

Proof *(1) As $\mathbb{E}$ is endowed with a convenient structure, it must be Hausdorff. On the other hand, since $\mathbb{E}$ is an LB-space, it must be boundedly regular (see for example [PeBo87]).*

(2) (i) $\Longrightarrow$ (ii): if f is bounded, then its restriction f_n of f to $\mathbb{E}_n$ is bounded, so $f_n : \mathbb{E}_n \longrightarrow \mathbb{F}$ is continuous.

(ii) $\Longrightarrow$ (iii): assume that each restriction f_n of f to $\mathbb{E}_n$ is continuous. To prove that f is continuous, it is sufficient to show that for any ball $B(0,r)$ in $\mathbb{F}$, then $f^{-1}(B(0,r))$ is an open convex set of $\mathbb{E}$. But we have

$$f^{-1}(B(0,r)) = \bigcup_{n \in \mathbb{N}} f_n^{-1}(B(0,r)).$$

As each f_n is continuous, $U_n = f_n^{-1}(B(0,r))$ is an open set of $\mathbb{E}_n$. Moreover, $U_n \subset U_{n+1}$. On the other hand, as $B(0,r)$ is convex, so is U_n. There $U = f^{-1}(B(0,r)) = \bigcup_{n \in \mathbb{N}} U_n$ is also convex. So U is an open set of $\mathbb{E}$ (relative to the LCDL-topology).

(iii) $\Longrightarrow$ (i): assume now that f is continuous and consider a bounded set B of E. From (1), B is contained in some $\mathbb{E}_n$. But as the inclusion of $\mathbb{E}_n$ in $\mathbb{E}$ is continuous, f_n is continuous and so $f(B) = f_n(B)$ is bounded in $\mathbb{F}$.

The following result corresponds to [CaPe19], Proposition 3.7.

Proposition 5.8 *On an LB-space $\mathbb{E} = \underrightarrow{\lim} \mathbb{E}_n$, the* DL*-topology coincides with the c^∞-topology.*

Proof *Let B be an absolutely convex bounded set of $\mathbb{E}$. From Lemma 5.1, B is contained in some $\mathbb{E}_n$ and then $\mathbb{E}_B$ is a vector subspace of $\mathbb{E}_n$ equipped with the Minkowski norm p_B. The closed unit ball B_n in $\mathbb{E}_n$ is also an absolutely convex bounded set and there exists $\alpha \geq 1$ such that $B \subset \alpha.B_n$. Remark that if p_n is the given norm on $\mathbb{E}_n$, then p_n is the Minkowski functional associed to B_n.*

If $v \in \mathbb{E}_B$*, then* $\dfrac{v}{p_B(v)}$ *belongs to* $\alpha.B_n$*; so we get*

$$p_n \leq \alpha.p_B.$$

Therefore the inclusion of $\mathbb{E}_B$ *in* $\mathbb{E}_n$ *is continuous.*
Let U *be an open set of the DL-topology on* $\mathbb{E}$*. Then for any* $n \in \mathbb{N}$*,* $U \cap \mathbb{E}_n$ *is open in* $\mathbb{E}_n$*. Thus, given any absolutely convex bounded set* B *of* $\mathbb{E}$*, if* $\mathbb{E}_B$ *is contained in* $\mathbb{E}_n$*, then* $U \cap \mathbb{E}_B = U \cap \mathbb{E}_n \cap \mathbb{E}_B$ *is open in* $\mathbb{E}_B$*. It follows from Proposition 3.6 that* U *is a* c^∞*-open.*
Conversely, if U *is a* c^∞*-open, as* $\mathbb{E}_n = \mathbb{E}_{B_n}$ *and the norm* p_{B_n} *is the given norm on* $\mathbb{E}_n$*, again from Proposition 3.6, it follows that* U *is an open set of the DL-topology on* $\mathbb{E}$*.*

Remark 5.5 *In Proposition 5.8, the fact that an LB-space is regular is essential. More generally, if an (LB)-space is convenient and boundedly regular, such a result is also true. However, for the sake of simplicity, we limit ourselves to the LB-space context.*

Proposition 5.9 *The direct limit* $\mathbb{E}$ *of a strict ascending sequence* $(\mathbb{E}_n, i_n^m)_{n \leq m}$ *of normed spaces provided with the* LDLC*-topology is Hausdorff.*

Proof *Consider a strict ascending sequence* $(\mathbb{E}_n, i_n^m)_{n \leq m}$ *of normed spaces. For each* $n \in \mathbb{N}$ *we denote by* $\overline{\mathbb{E}}_n$ *the Banach space which is the completion of* $\mathbb{E}_n$*. If* $\bar{i}_n^m$ *is the inclusion of* $\overline{\mathbb{E}}_n$ *in* $\overline{\mathbb{E}}_m$*, then the sequence* $(\overline{\mathbb{E}}_n, \bar{i}_n^m)_{n \leq m}$ *is a strict ascending sequence of Banach spaces, so its direct limit* $\overline{\mathbb{E}}$ *can be endowed with a complete locally convex topological vector space which is Hausdorff. Moreover, the direct limit* $\mathbb{E}$ *of the sequence* $(\mathbb{E}_n, i_n^m)_{n \leq m}$ *is locally convex topological contained in* $\overline{\mathbb{E}}$ *and its* LDLC*-topology is induced by the* LDLC*-topology of* $\overline{\mathbb{E}}$*. Since this topology is Hausdorff the same is true for* $\mathbb{E}$*.*

We have the following link between C^∞-smoothness on each member $\mathbb{E}_n$ (cf. Chapter 3, Definition 3.21) and conveniently smoothness on $\mathbb{E}$ for an LB-space:

Lemma 5.2 *Let* $\mathbb{E} = \underrightarrow{\lim} \mathbb{E}_n$ *be an LB-space,* $U \subset \mathbb{E} = \underrightarrow{\lim} \mathbb{E}_n$ *an open set of* E *(for the* DL*-topology) and* $U_n = U \cap \mathbb{E}_n$ *the associated open set in* $\mathbb{E}_n$*. Given a map* $f : U \longrightarrow \mathbb{F}$ *where* $\mathbb{F}$ *is a convenient space,* f *is conveniently smooth if and only if* $f_n = f_{|U \cap \mathbb{E}_n}$ *is* C^∞ *for each* $n \in \mathbb{N}$*.*

Proof *This proof is an adaption of the proof of [Glo05], Lemma 1.9.*
First, note that on an open set of a Banach space, we have equivalence between conveniently smoothness and C^∞*-differentiability (cf. [Bom67]).*
Assume that f *is conveniently smooth on* U*. Given any smooth curve* $\gamma : \mathbb{R} \longrightarrow U \cap \mathbb{E}_n$*, then* γ *is a smooth curve in* U*, so* $f \circ \gamma$ *is smooth. As* $U \cap \mathbb{E}_n$ *is an open set of a Banach space, it follows that* $f_n = f_{|U \cap \mathbb{E}_n}$ *is* C^∞*.*

Conversely, assume that $f_n = f_{|U\cap\mathbb{E}_n}$ *is* C^∞ *and let* $\gamma : \mathbb{R} \longrightarrow U$ *be a smooth curve. Fix* $a < b$ *and* $k \in \mathbb{N}$. *From the bounded regularity of* E, *it follows that there exists* $N \in \mathbb{N}$, $N \geq n$, *such that all the sets*

$$\left\{ \frac{(\gamma^{(j)}(t) - \gamma^{(j)}(s))}{s-t} : (s,t) \in]a,b[\times]a,b[, s \neq t \right\} \quad \text{and } \gamma^{(j)}(]a,b[)$$

where $\gamma^{(j)} : \mathbb{R} \to \mathbb{E}$ *is the* j^{th} *derivative of* γ, *are contained and bounded in* $\mathbb{E}_N$ *for all* $j \in \{0, \dots, k\}$. *Since* $\gamma^{(j)}_{|]a,b]}$ *is a Lipschitz curve in the Banach space* $\mathbb{E}_N$, *so there exists a primitive* η_j *on* $]a,b[$. *Thus* η_j *and* $\gamma^{(j-1)}_{|]a,b]}$ *have the same derivative* $\gamma^{(j)}_{|]a,b]}$ *and so these curves differ by a constant for all* $j \in \{1, \dots, k\}$. *We conclude that* $\gamma^{(j-1)} :]a,b[\to E_N$ *is a* C^1*-curve whose derivative is* $\gamma^{(j)}_{|]a,b]}$. *As a consequence,* $\gamma_{|]a,b]}$ *is a* Lip^k*-curve in* $\mathbb{E}_N$. *We can choose* $b - a$ *small enough such that* $\gamma(]a,b[) \subset U \cap \mathbb{E}_n \subset U \cap \mathbb{E}_N$. *Since the restriction of* f *to* $U \cap \mathbb{E}_n$ *is smooth, it follows that, for any* $k \in \mathbb{N}$, $f \circ \gamma :]a,b[\to \mathbb{R}$ *is* Lip^k *for any* $a < b$ *where* $b - a$ *is small enough. From Proposition 3.5, we get that* $f \circ \gamma$ *is smooth in* E *and so* f *is conveniently smooth (cf. Definition 3.29).*

Proposition 5.10 *Let* $\mathbb{E} = \varinjlim \mathbb{E}_n$ *and* $\mathbb{F} = \varinjlim \mathbb{F}_n$ *be LB-spaces and* $U_1 \subset \cdots \subset U_n \subset \cdots$ *an ascending sequence* $(U_n)_{n\in\mathbb{N}}$ *of open sets of* $(E_n)_{n\in\mathbb{N}}$ *and set* $U = \bigcup\limits_{n\in\mathbb{N}} U_n$. *Assume that we have a sequence of* $f_n : U_n = U \cap \mathbb{E}_n \longrightarrow \mathbb{F}_n$ *which are* C^∞.

(1) Then $f = \varinjlim f_n$ *is a conveniently smooth map from* U *to* F.

(2) Let $f_n : \mathbb{E}_n \longrightarrow \mathbb{F}_n$ *be a sequence of continuous linear maps; then* f *is a linear map from* $\mathbb{E}$ *to* $\mathbb{F}$ *which is conveniently smooth and continuous for the* DL*-topologies.*

Proof *(1) According to Proposition 5.1,* U *is an open set of* $\mathbb{E}$ *and* $U = \varinjlim U_n$. *So* $f = \varinjlim f_n : U \longrightarrow \mathbb{F}$ *is a well-defined continuous map (for the DL-topology). From Lemma 5.2,* f *is then a conveniently smooth map from* U *to* F.

Under the assumption of (2), the associated map $f = \varinjlim f_n$ *is linear and continuous. Now, as* f_n *is continuous linear between Banach spaces,* f_n *must be* C^∞ *and by (1),* f *must be conveniently smooth.*

Proposition 5.11 *Let* $(\mathbb{E}_n)_{n\in\mathbb{N}}$ *be a strict ascending sequence of Banach spaces,* U *a* c^∞ *open set in* $\mathbb{E} = \varinjlim \mathbb{E}_n$[5] *and we set* $U_n = U \cap \mathbb{E}_n$. *Let* $(\mathbb{F}_n)_{n\in\mathbb{N}}$ *be a strict ascending sequence of normed spaces. If* $(f_n : U_n \to \mathbb{F}_n)_{n\in\mathbb{N}}$ *is a direct sequence of maps of class* C^∞, *then* $f = \varinjlim f_n : U \to \mathbb{F} = \varinjlim \mathbb{E}_n$ *is convenient smooth.*

[5] $\mathbb{E}$ has a structure of convenient space (cf. Proposition 5.8).

Proof *At first a curve* $c : \mathbb{R} \to U$ *is smooth if and only if it is* Lip^k *for any* $k \in \mathbb{N}$ *(cf. Chapter 3, Proposition 3.5). Then using this fact and the bounded regularity of* $\mathbb{E}$ *, the proof is formally the same as the proof of Lemma 5.2.*

Consider a sequence $(\mathbb{E}_n)_{n\in\mathbb{N}}$ of supplemented Banach spaces and denote $\breve{\mathbb{E}}_0$, $\breve{\mathbb{E}}_1$, ... the Banach spaces such that:

$$\begin{cases} \mathbb{E}_0 = \breve{\mathbb{E}}_0 \\ \forall n \in \mathbb{N},\ \mathbb{E}_{n+1} \simeq \mathbb{E}_n \times \breve{\mathbb{E}}_{n+1} \end{cases}$$

For $(i,j) \in \mathbb{N}^2$, $i < j$, consider the injections

$$\begin{array}{rccc} \iota_i^j : & \mathbb{E}_i \simeq \breve{\mathbb{E}}_0 \times \cdots \times \breve{\mathbb{E}}_i & \to & \mathbb{E}_j \simeq \breve{\mathbb{E}}_0 \times \cdots \times \breve{\mathbb{E}}_j \\ & (x'_0, \ldots, x'_i) & \mapsto & (x'_0, \ldots, x'_i, 0, \ldots, 0) \end{array}$$

So $\left(\mathbb{E}_i, \iota_i^j\right)_{i\leq j}$ is a direct sequence.

Definition 5.10 $\left(\mathbb{E}_n, \iota_n^{n+1}\right)_{n\in\mathbb{N}}$ *is called an ascending sequence of supplemented Banach spaces.*

Example 5.1 *Let* $(\mathbb{E}_n)_{n\in\mathbb{N}}$ *be a sequence of Banach spaces. Consider the ascending sequence of Banach spaces* $\left(\mathbb{F}_n = \bigoplus_{k=0}^{n} \mathbb{E}_k\right)_{n\in\mathbb{N}}$ *of Banach spaces. It is an ascending sequence of supplemented Banach spaces. If we provide each* $\mathbb{E}_n$ *with a norm* $\|\ \|_n$ *we have a canonical norm* $|||\ |||_n = \sum_{k=0}^{n} \|\ \|_k$ *on* $\mathbb{F}_n$ *which is isomorphic to* $\prod_{k=0}^{n} \mathbb{E}_k$*. Thus, the direct limit* $\mathbb{F}$ *of the ascending sequence* $(\mathbb{F}_n)_{n\in\mathbb{N}}$ *is isomorphic to the vector space*

$$\left\{ (x_n)_{n\in\mathbb{N}} \in \prod_{k=0}^{\infty} \mathbb{E}_k \ :\ x_n = 0 \ \text{except for a finite number of indices } n \right\}.$$

The direct-limit $\mathbb{F}$ *is a topological locally convex vector space and this topology is in fact the "box topology" (cf. [Glo05]). A basis for this topology is*

$$\left\{ \bigoplus_{k=0}^{\infty} U_k := \mathbb{F} \cap \prod_{k=0}^{\infty} U_k \ :\ U_k \ \text{open set in } \mathbb{E}_k \right\}.$$

Remark 5.6 *Note that for any ascending sequence* $\left(\mathbb{E}_n, \iota_n^{n+1}\right)_{n\in\mathbb{N}}$ *of supplemented Banach spaces, for a fixed choice of each supplemented subspace* $\breve{\mathbb{E}}_{n+1}$ *of* $\mathbb{E}_n$ *in* $\mathbb{E}_{n+1}$*, we obtain an ascending sequence of supplemented Banach spaces which is isomorphic to the situation of Example 5.1.*

5.5 Differential Equations on Direct Limits of Banach Spaces

In Appendix, § A.2, following [DGV16], we have recalled the result obtained by these authors about the problem of existence and unicity of solutions for ordinary differential equations on Fréchet spaces. Here, we are interested in the analogous problem on direct limits of increasing sequences of supplemented Banach spaces.

5.5.1 Sufficient Conditions of Existence of Solutions for Direct Limit of ODEs

The existence and unicity, with fixed initial conditions, of solutions for a time depending differential equation on direct limit of an ascending sequence of Banach spaces is not true in general (cf. Examples 5.2-6). However, we will exhibit *sufficient conditions* under which such existence and unicity are true on a direct limit of an ascending sequence of supplemented Banach spaces.

Consider an ascending sequence $\left(\mathbb{E}_n, \iota_n^{n+1}\right)_{n\in\mathbb{N}}$ of supplemented Banach spaces as defined in Definition 5.10. We provide this sequence $(\mathbb{E}_n)_{n\in\mathbb{N}}$ with norms $\|\ \|_n$ by induction in the following way:

1. We choose any norm $\|\ \|_0$ on $\mathbb{E}_0$;
2. Assume that we have built a norm $\|\ \|_n$ on $\mathbb{E}_n$, we then choose any norm $\|\ \|'_n$ on $\breve{\mathbb{E}}_n$ and provide $\mathbb{E}_{n+1}$ with the norm $\|\ \|_{n+1} = \|\ \|_n + |\ \|\ \|'_n$.

Since, by construction, we have an increasing sequence of norms $(\|\ \|_n)_{n\in\mathbb{N}}$ such that the restriction of $\|\ \|_{n+1}$ to $\mathbb{E}_n$ is $\|\ \|_n$, the direct limit $\|\ \| = \underrightarrow{\lim}\|\ \|_n$ is well defined and is a continuous map on $\mathbb{E} = \underrightarrow{\lim}\mathbb{E}_n$.

Definition 5.11 *We will say that a sequence of norms $(\|\ \|_n)_{n\in\mathbb{N}}$ is coherent if the restriction of $\|\ \|_{n+1}$ to $\mathbb{E}_n$ is $\|\ \|_n$.*

For example, the sequence of norms $|||\ |||_n$ in Example 5.1 is coherent.

Proposition 5.12 *Let $(\|\ \|_n)_{n\in\mathbb{N}}$ be a sequence of coherent norms. Then $\|\ \| = \underrightarrow{\lim}\|\ \|_n$ is well defined. Moreover, the topology defined by $\|\ \|$ is coarser than the c^∞-topology of $\mathbb{E}$.*

Proof *At first, note that, for any $x \in \mathbb{E}$, from the coherence of the sequence of norms, we have $\|x\|_n = \|x\|_{n'}$ if $x \in \mathbb{E}_n \subset \mathbb{E}_{n'}$; so $\|x\|$ is well defined and it follows that $\|x\| = \underrightarrow{\lim}\|x\|_n$. To prove that $\|\ \|$ is a norm, from its definition, we only have to show the triangular inequality because the other properties are obvious from the previous argument. Fix some x and y in $\mathbb{E}$; let n_0 be the smallest integer n such that x and y belong to E_n. Therefore we have*

$$\|x+y\| = \|x+y\|_{n_0} \leq \|x\|_{n_0} + \|y\|_{n_0} = \|x\| + \|y\|.$$

Consider the open balls $B(a,r) = \{x \in \mathbb{E} \, : ||x-a|| < r\}$ and set $B_n(a,r) = B(a,r) \cap E_n$. From the coherence of the sequence of norms, we have $B_n(a,r) = \{x \in \mathbb{E}_n \, : ||x-a||_n < r\}$ and so $B_n(a,r)$ is open in $\mathbb{E}_n$. It follows that $B(a,r)$ is an open set for the direct limit topology. This implies that the topology defined by $|| \; ||$ is coarser than the direct limit topology. But the direct limit topology is the same as the the c^∞-topology of $\mathbb{E}$ (cf. [CaPe19], Proposition 3.7).

Note that in the Example 5.1 the topology defined by the norm $||| \; ||| = \underrightarrow{\lim} ||| \; |||_n$ coincides with the c^∞-topology of $\mathbb{E}$. The same is true for the coherent sequence of norms previously built on an ascending sequence of supplemented Banach spaces according to Remark 5.6.
Assume that $(|| \; ||_n)_{n \in \mathbb{N}}$ is a sequence of coherent norms on $\mathbb{E}_n$. We denote by $\mathcal{L}(\mathbb{E}_n)$ the set of continuous linear endomorphisms of $\mathbb{E}_n$ provided by the norm $|| \; ||_n^{\text{op}}$ associated with $|| \; ||_n$. If $(T_n)_{n \in \mathbb{N}}$ is a sequence of endomorphisms of $\mathbb{E}_n$, such that the restriction of T_{n+1} to $\mathbb{E}_n$ is T_n, we have

$$||T_{n+1}||_{n+1}^{\text{op}} \leq ||T_n||_n^{\text{op}} \text{ and } ||(T_{n+1})_{|E_n}||_{n+1}^{\text{op}} = ||T_n||_n^{\text{op}}.$$

From now on, and in this section, we assume that the ascending sequence $(\mathbb{E}_n)_{n \in \mathbb{N}}$ of Banach spaces is provided with a coherent sequence $(|| \; ||_n)_{n \in \mathbb{N}}$ of norms.

Given a compact interval I, we fix the map

$$\begin{array}{rccc} f = \underrightarrow{\lim} f_n : & I \times U & \to & \mathbb{E} \\ & (t,x) & \mapsto & \underrightarrow{\lim} f_n\left(t, x_n\right) \end{array}$$

On $\mathbb{E} = \underrightarrow{\lim} \mathbb{E}_n$, we consider the time depending differential equation

$$\dot{x} = f\left(t, x\right) \tag{5.2}$$

Our purpose is to find sufficient local conditions on f under which, around some point $a \in \mathbb{E}$, there exists a neighbourhood V of a and some closed interval $J = [t_0 - \tau, t_0 + \tau] \subset I$ such that for any $x \in V$ we have a solution γ^x of (5.2) defined on J with initial condition $\gamma^x(t_0) = x$.

For any $n \geq 0$, we consider the differential equation in the Banach space $\mathbb{E}_n$

$$x_n' = f_n\left(t, x_n\right). \tag{5.3}$$

Now fix some $a = \underrightarrow{\lim} a_n \in U$. Fix some $t_0 \in I$ and let $T > 0$ be such that $J =]t_0 - T, t_0 + T[\subset I$. If n_0 is the smallest integer such that $a \in \mathbb{E}_n$, then $a_n = a_{n_0}$ for $n \geq n_0$. For each $n \geq n_0$, we consider a closed ball $\bar{B}_n(a_n, r_n) \subset U_n$ centered at a_n (relative to the norm $|| \; ||_n$).

Assume that we have the following assumption:

($\mathbf{A}_n$) There exists a constant $K_n > 0$ such that:

(i) $\forall (t, x_n) \in I \times U_n$, $\|f_n(t, x_n)\|_n \leq K_n$;

(ii) $\forall (t, x_n) \in I \times U_n$, $\|D_2 f_n(t, x_n)\|_n^{\text{op}} \leq K_n$
where $\| \ \|_n^{\text{op}}$ is the operator norm on the set $\mathcal{L}(\mathbb{E}_n)$ of continuous endomorphisms of $\mathbb{E}_n$ and D_2 is the differential relative to the second variable.

From the classical Theorem of existence of solution of an ordinary differential equation (cf. [Cart71], Corollary 1.7.2, for instance), we can conclude that there exists $0 < \tau < \min\{r_n/K_n, T\}$ such that each differential equations (5.3) has a unique solution $\gamma_n^{x_n}$ defined on $J = [t_0 - \tau, t_0 + \tau] \subset I$ with initial condition $\gamma_n^{x_n}(t_0) = x_n$, for any x_n in the open ball $B_n(a_n, r_n - \tau K_n)$ (relatively to the norm $\| \ \|_n$).

Assume that we have following complementary assumptions:

(B) The sequence (r_n/K_n) is lower bounded[6].

(C) The sequence $(r_n - \tau K_n)$ is lower bounded.

Theorem 5.2 *(Existence of solutions) With the previous notations, under the assumptions* **($\mathbf{A}_n$)** *for all $n \geq n_0$,* **(B)** *and* **(C)***, the sequence of open neighbourhoods of a_n*

$$(V_n = B_n(a_n, r_n - \tau K_n) \cap B_n(a_{n_0}, r_{n_0} - \tau K_{n_0}))$$

is an ascending sequence of open sets and we have:
for each $x = \underrightarrow{\lim} x_n \in V = \underrightarrow{\lim} V_n$, there exists a unique solution $\gamma^x : [t_0 - \tau, t_0 + \tau] \to \mathbb{E}$ of the ODE

$$\dot{x} = f(t, x) \tag{5.4}$$

with initial condition $\gamma^x(t_0) = x$.

Remark 5.7 *We have another simpler criterion of existence of an integral curve for the ODE (5.2). Assume that for some $a \in \mathbb{E}$, for each n, we have an integral curve $\gamma_n : [t_0 - \tau, t_0 + \tau] \to \mathbb{E}_n$ which is a solution of (5.3) with initial conditions $\gamma_n(t_0) = a_n$, for all $n \geq n_0$. The assumption that the restriction of f_{n+1} to $I \times \mathbb{E}_n$ is f_n, implies that $\gamma = \underrightarrow{\lim} \gamma_n$ is well defined on $[t_0 - \tau, t_0 + \tau]$ and is an integral curve of (5.2) with initial condition $\gamma(t_0) = a$. This condition is of course necessary but it is not a useful condition. Note that if we take $\mathbb{E}_n = \mathbb{E}$ for each $n \in \mathbb{N}$, then the assumptions of Theorem 5.2 reduces to the assumption of the classical Theorem of existence of solution of an ODE in the Banach framework. However, Theorem 5.2 need very restrictive conditions which are of course not necessary to have the same conclusion (cf. Examples 5.2-3).*

[6]Then we can choose τ independent of n.

5.5.2 Comments and Examples

The validity of the assumptions $(\mathbf{A}_n)$ for all $n \geq n_0$, $(\mathbf{B})$ and $(\mathbf{C})$ suggests the following comments:

Comments 5.1

(1) According to the previous notations, we set:

$$\begin{cases} K_n^0(a) = \sup_{t\in I} ||f_n(t,a)||_n \\ K_n^1(a) = \sup_{t\in I} ||D_2 f_n(t,a)||_n^{\mathrm{op}} \end{cases}$$

Assume that we have

$(\mathbf{A}'_n)$ $K = \sup_{n\in\mathbb{N}^*} (K_n^0(a), K_n^1(a)) < +\infty$, *and there exists an ascending sequence* (U_n) *of neighbourhoods of* a_n *such that* $(\mathbf{A}_n)$ *is true for* $K_n = 2\sup(K_n^0(a), K_n^1(a))$*;*

$(\mathbf{B}')$ $r = \inf_{n\in\mathbb{N}^*} (r_n) > 0$.

Then, for $0 < \tau \leq \inf(r/2K, T)$*, all the conditions* $(\mathbf{A}_n)$, $(\mathbf{B})$ *and* $(\mathbf{C})$ *are satisfied, and so we can apply Theorem 5.2 (cf. Example 5.2-3.)*

(2) Note that for each n*, the assumption* $(\mathbf{A}_n)$ *is always satisfied at least on some neighbourhood* W_n *of* $a_n \in U_n$*, but since the restriction of* f_{n+1} *to* U_n *is* f_n *then* $K_{n+1} \leq K_n$ *and* W_{n+1} *can eventually not contain* W_n*. So even after reduction of* U_n *to* $W_n \cap W_{n_0}$*, the sequence* $W_n \cap W_{n_0}$ *need not be an ascending sequence of open sets and again we can have*

$$\bigcap_{n \geqslant n_0} W_n \cap W_{n_0} = \{a\}.$$

For instance, under the assumption $(\mathbf{A}'_n)$*, since each* f_n *is a Lipschitz map and the restriction of* f_{n+1} *to* U_n *is* f_n*, the sequence of open sets*

$$W_n = \{x_n \in U_n \ : \forall t \in I,\ ||f_n(t,x_n)||_n \leq K\ \}$$

will be a decreasing sequence of open sets and so the previous situation is generally what happens (cf. Example 5.2-4).

(3) However, assume that we can choose an ascending sequence (U_n) *such that* $(\mathbf{A}_n)$ *is true for all* $n \geq n_0$*. Since* (K_n) *is an increasing sequence, the assumption* $(\mathbf{B})$ *is a condition on the comparison between the growth of* K_n *and of the diameter of* U_n *and there is no general reason for which* $(\mathbf{B})$ *is satisfied (cf. Example 5.2-4).*

(4) Assume that $(\mathbf{A}_n)$ *and* $(\mathbf{B})$ *are satisfied, it is clear that there can exist some opposition between* $(\mathbf{B})$ *and* $(\mathbf{C})$.

In conclusion, according to the fact that we impose the existence of a sequence of coherent norms, these conditions are too much strong and so Theorem 5.2 seems to be of weak interest **in the general case**.

Example 5.2 *In all the following examples, unless otherwise specified, we assume that each ascending sequence* $(\mathbb{E}_n)_{n\in\mathbb{N}}$ *of Banach spaces is provided with a sequence* $(\|\ \|_n)_{n\in\mathbb{N}}$ *of coherent norms and we will use the previous notations.*

(1) *Assume that* f_n *is a linear map in the second variable. Since the interval* I *is compact, the assumption* $(\mathbf{A}_n)$ *is true for each* $n \geq n_0$ *and for any neighbourhood* U_n *of* a*. Note that* $K_n^1(a)$ *does not depend on* a *and on* $U_n = B(a_n, R_n)$ *we have*

$$\forall (t, x_n) \in I \times U_n,\ ||f_n(t, x_n)||_n \leq K_n^1 R_n.$$

If the sequence $(K_n)_{n\in\mathbb{N}}$ *is bounded, this implies that we can always find a sequence* $(R_n)_{n\in\mathbb{N}}$ *such that* $(\mathbf{A}_n)$, $(\mathbf{B})$ *and* $(\mathbf{C})$ *are satisfied and Theorem 5.2 can be applied. Note that, since for each* n*, we have a linear ODE, we can directly show the existence of solutions for the ODE (5.2) simply by using the general result about the existence of solutions of a linear ODE in a Banach space and Remark 5.7. Of course, the first description must be considered as an illustration of Theorem 5.2. This also justifies the conclusions at the end of Comments 5.1.*

(2) *Assume that* $f_n = f_N$*, for all* $n \geq N$*. This situation can occur when* $\mathbb{E}_n = \mathbb{E}_N$ *for* $n \geq N$ *or also when* $\mathbb{E}_n$ *is supplemented in* $\mathbb{E}_{n+1}$*, for* $n \geq N$*. Then, again in this case, clearly all the assumptions* $(\mathbf{A}_n)$, $(\mathbf{B})$ *and* $(\mathbf{C})$ *are satisfied but, of course, we can get the result directly without using such conditions.*

(3) *According to example 5.1, let* $(\mathbb{E})_{n\in\mathbb{N}}$ *be a sequence of Banach spaces with* $\mathbb{E}_n = \mathbb{E}$ *for all* $n \in \mathbb{N}$ *and again we set* $\mathbb{F}_n = \bigoplus_{k=0}^{n} \mathbb{E}_n$*. On* $\mathbb{F}_n \equiv \mathbb{E}_0 \times \cdots \times \mathbb{E}_n$*, we denote by* $\bar{x}_n \in \mathbb{F}_n$ *the* $(n+1)$*-uple* $(x_0, \ldots, x_n)$*. Let* $\dot{x} = f(t, x)$ *be a differential equation on* $\mathbb{E}$ *where* $f : I \times U \to \mathbb{E}$ *is a smooth map where* U *is a neighbourhood of* 0*. We consider the differential equation on* $\mathbb{F}_n$

$$\dot{\bar{x}}_n = \bar{f}_n(t, \bar{x}_n)$$

where $\bar{f}_n = (f, \ldots, f)$*.*
If $||\ ||$ *is some fixed norm on* $\mathbb{E}$*, we assume that there exists a constant* $K > 0$ *such that:*

(i) $\forall (t, x) \in I \times U,\ ||f(t, x|| \leq K;$
(ii) $\forall (t, x) \in I \times U,\ ||D_2 f(t, x)||^{\mathrm{op}} \leq K.$

Note since I *is a compact interval, after shrinking* U *if necessary, such conditions are always satisfied. Then, by applying the classical result about* *the existence of a solution of a differential equation in $\mathbb{E}$, there exists a neighbourhood V of 0 in U such that, for each $x \in V$, we have a solution γ^x of $\dot{x} = f(t,x)$, defined on some interval $[t_0-\tau, t_0+\tau]$, such that $\gamma^x(t_0) = x$. Thus by applying Remark 5.7, the differential equation $\dot{x} = \bar{f}(t,x)$ where $\bar{f} = \varinjlim \bar{f}_n$ has a solution $\gamma^{\bar{x}} : [t_0 - \tau, t_0 + \tau] \to \mathbb{F}$ for all $\bar{x} \in \prod_{k=0}^{\infty} V_k \cap \mathbb{F}$ with $\gamma^{\bar{x}}(t_0) = \bar{x}$ and $V_k = V$ for all $k \in \mathbb{N}$. It is clear that the condition* **(C)** *is not satisfied in this context. Thus the assumptions of Theorem are sufficient but not necessary.*

(4) We provide $\mathbb{R}^n$ with the norm $||(x_1, \ldots, x_n)||_n = \sum_{i=1}^{n} |x_i|$. Then $(|| \, ||_n)_{n \in \mathbb{N}^*}$ is a coherent sequence of norms. Note that this context is a particular case of the context of the previous Example 3 with $\mathbb{E} = \mathbb{R}$.
Now, on $\mathbb{R}$, we consider the function $\theta : u \mapsto u^2 + 1$. Then the family of solutions of the differential equation $u' = \theta(u)$ is given by $u : t \mapsto \tan(t - C)$ for any constant C. For $t \in [-1,1]$ and $y_i \in]-\frac{\pi}{2}, \frac{\pi}{2}[$, we set

$$\varphi_i(t, y_i) = \frac{t}{i^2}(y_i^2 + 1).$$

Let U_n be the open set $\left(\left]-\frac{\pi}{2}, \frac{\pi}{2}\right[\right)^n \subset \mathbb{R}^n$. On U_n, we consider the differential equation $x_n' = f_n(t, x_n)$ where $f_n(t, x_n) = (\varphi_1(t, y_1), \ldots, \varphi_n(t, y_n))$ and $x_n = (y_1, \ldots, y_n)$. Then, on U_n, we have

$$\forall (t, x_n) \in [-1,1] \times U_n, \ ||f_n(t, x_n)||_n \leq \left(\frac{\pi^2}{4} + 1\right)\left(\sum_{i=1}^{n} \frac{1}{i^2}\right)$$

$$\forall (t, x_n) \in [-1,1] \times U_n, \ ||D_2 f_n(t, x_n)||_n^{\mathrm{op}} \leq \left(\frac{\pi^2}{4} + 1\right)\left(\sum_{i=1}^{n} \frac{1}{i^2}\right)$$

So the assumption **($\mathbf{A}_n$)** *is satisfied for all $n > 0$. Note that we have*

$$U = \varinjlim U_n = \left(\left]-\frac{\pi}{2}, \frac{\pi}{2}\right[\right)^{\infty} \cap \mathbb{R}^{\infty}.$$

If $K_n = \left(\frac{\pi^2}{4} + 1\right)\left(\sum_{i=1}^{n} \frac{1}{i^2}\right)$, then we have $K_n \leq \left(\frac{\pi^2}{4} + 1\right) \times \frac{\pi^2}{6}$. Now the closed ball $\bar{B}_n(0, \frac{3}{2}))$ is contained in U_n for each $n > 0$ and we have

$$\frac{3}{2K_n} \geq \left(\frac{\pi^2}{4} + 1\right) \times \frac{\pi^2}{6} > 1$$

Thus the assumption **(B)** *is satisfied.*

We set $K = \left(\frac{\pi^2}{4}+1\right) \times \frac{\pi^2}{6} > 0$ *and we choose* $\tau = \frac{1}{K}$. *Then the assumption* **(C)** *is satisfied. Therefore, we can apply Theorem 5.2 and its conclusion is valid on the ball* $B(0, \frac{1}{2})$.

(5) *With the same notations as in the previous example, we consider*

$$\varphi_i(t, y_i) = \frac{t}{i}(y_i^2 + 1).$$

If $f_n(t, x_n) = (\varphi_1(t, y_1), \dots, \varphi_n(t, y_n))$, *then, in this case, on* $U_n = \left(\left]-\frac{\pi}{2}, \frac{\pi}{2}\right[\right)^n$, *we have*

$$\forall (t, x_n) \in [-1, 1] \times U_n,\ ||f_n(t, x_n)||_n \leq \left(\frac{\pi^2}{4}+1\right)\left(\sum_{i=1}^{n} \frac{1}{i}\right)$$

$$\forall (t, x_n) \in [-1, 1] \times U_n,\ ||D_2 f_n(t, x_n)||_n^{\mathrm{op}} \leq \left(\frac{\pi^2}{4}+1\right)\left(\sum_{i=1}^{n} \frac{1}{i}\right)$$

Of course, the condition $(\mathbf{A}_n)$ *will be satisfied. But, in this situation, for any closed ball* $\bar{B}(0, r_n)$ *contained in* U_n, *we have* $r_n \leq \pi/2$. *Since the sequence* $\left(\sum_{i=1}^{n} \frac{1}{i}\right)_{n \in \mathbb{N}^*}$ *is not bounded, the condition* **(B)** *is not satisfied.*
Now, we can impose that, for $n \geq N$, *we have a constant* $K > 0$ *such that:*

(i) $\forall (t, x) \in [-1, 1] \times U_n,\ ||f_n(t, x)||_n \leq K$;
(ii) $\forall (t, x) \in [-1, 1] \times U_n,\ ||D_2 f_n(t, x)||_n^{\mathrm{op}} \leq K$.

Then **(B)** *and* **(C)** *will be satisfied. But in order to get such a relation on* U_n, *it is necessary to shrink* U_n *for* $n > N$ *and it is easy to see that, in this case, we will be in the situation of Comments 5.1 (2), and so the conditions* $(\mathbf{A}_n)$ *cannot be satisfied.*

(6) *With the same notations as in the previous example, we consider*

$$\varphi_i(t, y_i) = (iy_i)^2 + 1.$$

Again, we set $f_n(t, x_n) = (\varphi_1(t, y_1), \dots, \varphi_n(t, y_n))$ *with* $x_n = (y_1, \dots, y_n)$. *Thus the* maximal solution *of the differential equation* $\dot{x}_n = f_n(t, x_n)$ *with initial condition* 0 *for* $t = 0$ *is given by*

$$t \mapsto (\tan t, \tan 2t, \dots, \tan nt) \text{ for } t \in [-\frac{\pi}{2n}, \frac{\pi}{2n}].$$

Thus from Remark 5.7, on $\mathbb{R}^\infty$, *the differential equation* $\dot{x} = f(t, x)$, *where* $f = \varinjlim f_n$, *has no solutions through* 0.

5.6 Direct Limits of Ascending Sequences of Banach Manifolds

Definition 5.12 *$\mathcal{M} = (M_n, \varepsilon_n^{n+1})_{n\in\mathbb{N}}$ is called an ascending sequence of Banach manifolds if, for any $n \in \mathbb{N}$, $(M_n, \varepsilon_n^{n+1})$ is a submanifold*[7] *of M_{n+1}.*

Proposition 5.13 *Let $\mathcal{M} = (M_n, \varepsilon_n^{n+1})_{n\in\mathbb{N}}$ be an ascending sequence of Banach manifolds.*
Assume that for $x \in M = \underrightarrow{\lim} M_n$, there exists a sequence of charts $((U_n, \phi_n))_{n\in\mathbb{N}}$ of $(M_n)_{n\in\mathbb{N}}$, such that:

(ASC1) *$(U_n)_{n\in\mathbb{N}}$ is an ascending sequence of chart domains;*

(ASC2) $$\forall n \in \mathbb{N},\ \phi_{n+1} \circ \varepsilon_n^{n+1} = \iota_n^{n+1} \circ \phi_n.$$

Then $U = \underrightarrow{\lim} U_n$ is an open set of M endowed with the DL*-topology and $\phi = \underrightarrow{\lim}\phi_n$ is a well-defined map from U to $\mathbb{M} = \underrightarrow{\lim}\mathbb{M}_n$.*
Moreover, ϕ is a continuous homeomorphism from U onto the open set $\phi(U)$ of $\mathbb{M}$.

Definition 5.13 *We say that an ascending sequence $\mathcal{M} = (M_n, \varepsilon_n^{n+1})_{n\in\mathbb{N}}$ of Banach manifolds has the direct limit chart property* **(DLCP)** *at x if it satisfies both* **(ASC1)** *and* **(ASC2)**.

We then have the fundamental result (cf. [CaPe19]).

Theorem 5.3 *Let $\mathcal{M} = (M_n)_{n\in\mathbb{N}}$ be an ascending sequence of Banach C^∞-manifolds, modelled on the Banach spaces $\mathbb{M}_n$. Assume that*

(ASBM1) *$(M_n)_{n\in\mathbb{N}}$ has the direct limit chart property* **(DLCP)** *at each point $x \in M = \underrightarrow{\lim} M_n$;*

(ASBM2) *$\mathbb{M} = \underrightarrow{\lim}\mathbb{M}_n$ is an LB-space.*

Then there is a unique n.n.H. convenient manifold structure on $M = \underrightarrow{\lim} M_n$ modelled on the convenient space $\mathbb{M}$ such that the topology associated with this structure is the DL-topology on M.
In particular, for each $n \in \mathbb{N}$, the canonical injection $\varepsilon_n : M_n \longrightarrow M$ is an injective conveniently smooth map and (M_n, ε_n) is a closed submanifold of M. Moreover, if each M_n is locally compact or is open in M_{n+1} or is a paracompact Banach manifold closed in M_{n+1}, then $M = \underrightarrow{\lim} M_n$ is provided with a Hausdorff convenient manifold structure.

Proof *Let $\mathscr{A}$ be the set of all sequences of charts $\{(U_n^\alpha, \phi_n^\alpha)\}_{\alpha\in\mathbf{A},\, n\in\mathbb{N}}$ of $(M_n)_{n\in\mathbb{N}}$ such that $(U_n)_{n\in\mathbb{N}}$ is an ascending sequence of chart domains. We set $V_n^\alpha = \phi_n^\alpha(U_n^\alpha)$, $U^\alpha = \underrightarrow{\lim} U_n^\alpha$ and $\phi^\alpha = \underrightarrow{\lim}\phi_n^\alpha$. From Proposition 5.13, ϕ^α is a homeomorphism from U^α to the open set $V^\alpha = \phi^\alpha(U^\alpha) = \underrightarrow{\lim} V_n^\alpha$ of*

[7] cf. Definition 1.9.

$\mathbb{M}$. Then, according to Proposition 5.1 (2), U^α and V^α are open sets of the DL-topology on M and $\mathbb{M}$ respectively. From our assumption, $\mathscr{A}$ is then a topological atlas of M modelled on the convenient space $\mathbb{M}$. Note that, from Proposition 5.8, each V^α is also a c^∞-open set.
Let us prove that the change of charts are conveniently smooth diffeomorphisms. Consider two charts (U^α, ϕ^α) and (U^β, ϕ^β) around $x \in M$. We consider

$$\theta_n^{\beta\alpha} = \phi_n^\beta \circ (\phi_n^\alpha)^{-1} : \phi_n^\alpha\left(U_n^\alpha \cap U_n^\beta\right) \longrightarrow \phi_n^\beta\left(U_n^\alpha \cap U_n^\beta\right).$$

For each $n \in \mathbb{N}$, the pairs $(U_n^\alpha, \phi_n^\alpha)$ and $(U_n^\beta, \phi_n^\beta)$ are charts of M_n and the intersection of their domains is not empty. It follows that the map $\theta_n^{\beta\alpha}$ is a C^∞ local diffeomorphism of $\mathbb{M}_n$. But the construction of U^α and U^β implies that $U^\alpha \cap U^\beta$ is the direct limit of $(U_n^\alpha \cap U_n^\beta)_{n\in\mathbb{N}}$ in M. It follows that $\phi^\alpha\left(U^\alpha \cap U^\beta\right)$ is the direct limit of $\phi_n^\alpha\left(U_n^\alpha \cap U_n^\beta\right)_{n\in\mathbb{N}}$. In the same way, we have $\phi^\alpha\left(U^\alpha \cap U^\beta\right) = \underrightarrow{\lim}\phi_n^\alpha\left(U_n^\alpha \cap U_n^\beta\right)$. Therefore we get a direct limit map $\theta^{\beta\alpha} = \underrightarrow{\lim}\theta_n^{\beta\alpha}$ from the open set $\phi^\alpha\left(U^\alpha \cap U^\beta\right)$ onto the open set $\phi^\beta\left(U^\alpha \cap U^\beta\right)$ of $\mathbb{M}$. Again the sets $\phi^\alpha\left(U^\alpha \cap U^\beta\right)$ and $\phi^\beta\left(U^\alpha \cap U^\beta\right)$ are c^∞-open sets of $\mathbb{M}$. As each $\theta_n^{\beta\alpha}$ is a C^∞ diffeomorphism of E_n, Lemma 5.2 implies that $\theta^{\alpha\beta}$ is a conveniently smooth diffeomorphism from $\phi^\alpha\left(U^\alpha \cap U^\beta\right)$ onto $\phi^\beta\left(U^\alpha \cap U^\beta\right)$. Therefore we obtain that $\mathscr{A}$ is a convenient atlas on M. Note that the topology of M defined by such an atlas is exactly the DL-topology on M.
Therefore, if M_n is locally compact, then M is Hausdorff from Proposition 5.1. In the same way, from Proposition 5.2, if M_n is open in M_{n+1} or is paracompact and closed in M_{n+1} for each n, then M is Hausdorff. Thus in each of the previous particular cases, M is provided with a Hausdorff convenient manifold structure.

Now we prove the uniqueness of this convenient structure. Assume that Y is a convenient manifold structure modelled on the convenient vector space $\mathbb{M}$ and $h_n : M_n \to Y$ a C^∞ map for each $n \in \mathbb{N}$ s.t. $\left(Y, (h_n)_{n\in\mathbb{N}}\right)$ is a cone over $\mathcal{M}$. Then there is a uniquely determined continuous map $h : M \longrightarrow Y$ s.t. $h_{|M_n} = h_n$. Let $x \in M$; we can find a chart (U, φ) around x in the atlas $\mathscr{A}$ where $f = \underrightarrow{\lim} f_n U$ for charts $\varphi_n : U_n \longrightarrow E_n$. Let $\psi : W \subset Y \longrightarrow V$ be a chart for Y (W is an open set of Y). Then $O = \left(h \circ \varphi^{-1}\right)^{-1}(W)$ is an open set of $U \subset \mathbb{M}$ and $O_n = O \cap \mathbb{M}_n$ is open in $\mathbb{M}_n$ for each n. Consider $g = \psi \circ h \circ \varphi^{-1}|_O^V : O \to V$. Then $g_{|O_n} = \psi \circ h_n \circ \varphi_n^{-1}|_{O_n}^V : O_n \to V$ is C^∞ for each $n \in \mathbb{N}$. Hence g is c^∞ (cf. Proposition 5.10), so is h on the open neighborhood U of x and hence on all of M because x is arbitrary. Thus $\left(M, (\varepsilon_n)_{n\in N}\right) = \underrightarrow{\lim}\mathcal{M}$ in the category of c^∞-manifolds. The uniqueness of a convenient structure of manifold on M follows from the universal property of direct limits.

Moreover, if each M_n is paracompact, then $M = \underrightarrow{\lim} M_n$ is provided with a Hausdorff convenient manifold structure.

Example 5.3

(1) Let $(M_n, \varepsilon_n^{n+1})_{n\in\mathbb{N}}$ be an ascending sequence of finite-dimensional manifolds. $M = \underrightarrow{\lim} M_n$ is a Hausdorff convenient manifold modelled on $\mathbb{R}^\infty$ (cf. [Glo05]).

(2) If $(M_n, \varepsilon_n^{n+1})_{n\in\mathbb{N}}$ is an ascending sequence of finite-dimensional manifolds then $\mathrm{L}_k^p\left(\mathbb{S}^1, \underrightarrow{\lim} M_n\right) = \underrightarrow{\lim} \mathrm{L}_k^p\left(\mathbb{S}^1, M_n\right)$ is a Hausdorff convenient manifold modelled on $\mathrm{L}_k^p\left(\mathbb{S}^1, \mathbb{R}^\infty\right)$ (cf. [Pel18]).

(3) More generally, let N be a compact finite-dimensional manifold and let $\left(M_n, \varepsilon_n^{n+1}\right)_{n\in\mathbb{N}}$ be an ascending sequence of finite-dimensional manifolds. By analogous arguments as for the ascending sequence $\left(\mathrm{L}_k^p\left(\mathbb{S}^1, M_n\right)\right)_{n\in\mathbb{N}}$ we can show that $\mathrm{L}_k^p\left(N, \underrightarrow{\lim} M_n\right) = \underrightarrow{\lim} \mathrm{L}_k^p\left(N, M_n\right)$ is a Hausdorff convenient manifold modelled on $\mathrm{L}_k^p\left(N, \mathbb{R}^\infty\right)$.

One can find in [CaPe19], Proposition 4.5, the following result:

Theorem 5.4 *Let $(M_n)_{n\in\mathbb{N}}$ be an ascending sequence of C^∞-Banach manifolds modelled on the ascending sequence of Banach spaces $(\mathbb{M}_n)_{n\in\mathbb{N}}$ where for all $n \in \mathbb{N}$, M_n is closed in M_{n+1}. If the sequence $(M_n)_{n\in\mathbb{N}}$ satisfies the following assumptions*

(i) $\mathbb{M}_n$ is a supplemented subspace in $\mathbb{M}_{n+1}$ for each $n \in \mathbb{N}$;

(ii) there exists a linear connection on TM_n for each $n \in \mathbb{N}$,

then $(M_n)_{n\in\mathbb{N}}$ has the direct chart limit property and $M = \underrightarrow{\lim} M_n$ provided with the DL*-topology, has a unique structure of n.n.H. convenient manifold modelled on $\mathbb{M} = \underrightarrow{\lim} \mathbb{M}_n$. If, moreover, each M_n is paracompact, then M has a Hausdorff convenient manifold structure.*

Sketch of the proof. *This Theorem is a consequence of Proposition 5.13 if we can show that the assumptions implies the direct limit chart property around each point of the direct limit. Now this property is obtained by induction using the following Lemma whose proof is a generalization of the arguments used in the proof of Theorem 5.1 in Chapter IV of [Lan95] over contractible open sets:*

Lemma 5.3 *Let N_1 be a complemented immersed submanifold, modelled on F_1, of a Banach manifold N, modeled on F with $F_1 \subset F$. Assume that there exists a linear connection on TN. Given any chart (U_1, ϕ_1) of x in N_1 such that U_1 is a contractible set, there exists a chart (U, Φ) of $x \in N_1$ such that U is contractible, $U \cap N_1 = U_1$ and $\Phi_{|U_1} = \phi_1$.*

(for details see Lemma 4.7 in [CaPe19])

Remark 5.8 *Note that under the assumption (i) of Theorem 5.4, if each M_n is regular, then M_n admits a smooth partition of unity and so condition (ii) is always satisfied. In this case, the direct limit $M = \underrightarrow{\lim} M_n$ is a Hausdorff convenient manifold.*

5.7 Direct Limits of Ascending Sequences of Topological Groups

Let us consider an ascending sequence $G_0 \subset G_1 \subset \cdots$ of topological group where all inclusion maps $\iota_n^{n+1} : G_n \to G_{n+1}$ are continuous group homomorphisms. We endow the direct set $G = \bigcup_{n \in \mathbb{N}} G_n$ with a structure of group and the DLG-topology which is the finest group topology making each inclusion $\iota_n : G_n \to G$ continuous. Then G is a (not necessarily Hausdorff) topological group (cf. [Glo11], 3.). The DLG-topology is the so-called Bamboo-Shoot topology described in [TSH98] and [HSTH01]. It is obvious that the DL-topology is finer than the DLG-topology.

Example 5.4 *Let $G_n := \mathbb{Q} \times \mathbb{R}^{n-1}$ be equipped with the addition and the topology induced by $\mathbb{R}^n$. It is proved in [TSH98], Example 1.2, that the* DL-*topology does not make the group operation continuous; so, in this case, the* DL-*topology is strictly finer than the* DLG-*topology.*

The reader can find in [Yam98] situations where both topologies coincide.

5.8 Direct Limits of Ascending Sequences of Banach-Lie Groups

The direct limits of finite-dimensional Lie-groups appear in the structure and representation theory of Lie algebras.
Interesting Lie groups often appear as direct limits $G = \bigcup_{n \in \mathbb{N}} G_n$ of increasing sequence of Lie groups $G_0 \subset G_1 \subset \cdots$ where the inclusions $\varepsilon_n^{n+1} : G_n \to G_{n+1}$ are smooth homomorphisms.
In [Glo05], Glöckner shows that direct limits of finite-dimensional Lie groups can be provided with a structure of Lie group and he obtains an analogue of Lie's third theorem.

The class of infinite-dimensional Lie groups has a very rich representation theory (cf. [Nee98], [Wol05], ...).

Example 5.5 *The group* $\mathrm{Diff}_c(M)$ *of compactly supported diffeomorphisms*[8] *of a smooth finite dimension manifold $\sigma-$compact M is the direct limit of the increasing sequence of the groups* $\mathrm{Diff}_{K_n}(M)$ *of diffeomorphisms with support contained in K_n where $K_1 \subset K_2 \subset \cdots$ is an exhaustive sequence of compact*

[8] For such a diffeomorphism φ, the set $\{x \in M : \varphi(x) = x\}$ has compact closure.

sets of M such that $K_n \subset \overset{\circ}{K}_{n+1}$. The manifold structure of $\text{Diff}_c(M)$ *is modelled on the* LF*-space* $\mathfrak{X}_c(M)$ *of compactly supported smooth fields.*

In order to obtain an interesting structure on the direct limit $G = \bigcup_{n\in\mathbb{N}} G_n$ on an increasing sequence of Banach-Lie groups G_n, an essential hypothesis is the existence of a direct limit chart on G which is built from charts of the Lie groups G_n.

Definition 5.14 *Let $G = \bigcup_{n\in\mathbb{N}} G_n$ be the union of an increasing sequence of Banach-Lie groups where the inclusions $\varepsilon_n^m : G_n \to G_m$ are C^∞ homomorphisms and where G_n is a subgroup of G.*
We say that G has the direct limit chart property if

(DLCPBLG) *there exists local charts $\phi_n : G_n \supset U_n \to V_n \subset \mathfrak{g}_n$ of G_n centered at 1_n (neutral element) for $n \in \mathbb{N}$ (where $\mathfrak{g}_n$ stands for the Lie algebra of G_n) such that, for any $n \leq m$, $U_n \subset U_m$, $\phi_m|_{U_n} = \mathbf{L}(i_n^m) \circ \phi_n$ and $V = \bigcup_{n\in\mathbb{N}} V_n$ is an open set in the direct limit $\varinjlim \mathfrak{g}_n$ assumed as Hausdorff.*

This hypothesis is sufficient to get a structure of C^∞-Lie group on a direct limit of C^∞-Lie groups which have a structure of $k_\omega-$space (cf. [Glo07], Proposition 1.4.3)[9].

Proposition 5.14 *Let $G = \bigcup_{n\in\mathbb{N}} G_n$ be a group which is the union of an ascending sequence of C^∞-Lie groups. Assume that*
(i) G has a candidate $\phi : U \longrightarrow V \subset \varinjlim \mathfrak{g}_n$ for a direct limit chart;
(ii) $\mathfrak{g}_n$ is a $k_\omega-$space for each $n \in \mathbb{N}$.
Then on $G = \varinjlim G_n$ there exists a unique C^∞-Lie group structure making $\phi_{|W}$ a direct chart limit for G around 1, where W is an open neighbourhood of 1 contained in U.

We have the same conclusion if we replace (ii) by
G_n is a Banach Lie group for each $n \in \mathbb{N}$ and the inclusion map $\mathfrak{g}_n \longrightarrow \mathfrak{g}_m$ is a compact linear operator for all $n < m$.

One can find in a paper of Glöckner ([Glo05], version arXiv:math/0606078, Remark 14.8) the following general result.

Proposition 5.15 *Let $G = \bigcup_{n\in\mathbb{N}} G_n$ be a group which is the union of an ascending sequence of (Hausdorff) convenient Lie groups. Equip the vector space*

---

[9] A Hausdorff topological space X is called a k_ω-space if there exists an ascending sequence $K_0 \subset K_1 \subset \cdots$ of compact subsets of X s.t. $X = \bigcup_{n\in\mathbb{N}} K_n$ and $U \subset X$ is open if and only if $U \cap K_n$ is open in K_n for each $n \in \mathbb{N}$.

$\varinjlim \mathfrak{g}_n$ with the locally convex vector topology associated with the direct limit bornology which is assumed to be Hausdorff. We require that G admits a candidate for a direct limit chart in the convenient sense and that each bounded subset in $\mathfrak{g}$ is a bounded subset of some $\mathfrak{g}_n$.
Then G may be endowed with a structure n.n.H. convenient Lie group.

We then have the fundamental result about direct limits of Banach Lie groups (cf. [CaPe19], Theorem 4.13).

Theorem 5.5 *Let $G = \bigcup_{n\in\mathbb{N}} G_n$ be a group which is the union of an ascending sequence of Banach Lie groups. Assume that the direct limit $\mathfrak{g} = \varinjlim \mathfrak{g}_n$ of the ascending sequence $(\mathfrak{g}_n)_{n\in\mathbb{N}}$ of associated Lie algebras is an LB-space. If G admits a candidate for a direct limit chart, then G can be endowed with a structure of n.n.H. convenient Lie group modelled on the LB-space $\mathfrak{g}$.*

Examples 5.1

(1) Let $(G_n)_{n\in\mathbb{N}}$ be a sequence of Banach-Lie groups. The weak direct product $\prod^*_{n\in\mathbb{N}} G_n$ of this sequence is the set of elements $(g_n) \in \prod_{n\in\mathbb{N}} G_n$ such that $g_n = 1_n$ except for a finite number of integers n. It can be provided with a structure of convenient Lie group whose Lie algebra $\prod^*_{n\in\mathbb{N}^*} \mathbf{L}(G_n)$ is the set of elements $(u_n) \in \prod_{n\in\mathbb{N}} \mathbf{L}(G_n)$ such that $u_n = 0$ except for a finite number of integers n.

(2) Let $(G_n)_{n\in\mathbb{N}}$ be an ascending sequence of Banach-Lie groups. For each n, $\mathrm{L}^p_k(\mathbb{S}^1, G_n)$ can be provided with a group structure with pointwise multiplication in G_n. According to the structure of Banach manifold on $\mathrm{L}^p_k(\mathbb{S}^1, G_n)$, it is easy to see that we get a Banach-Lie group structure on $\mathrm{L}^p_k(\mathbb{S}^1, G_n)$ whose Lie algebra is $\mathrm{L}^p_k(\mathbb{S}^1, \mathbf{L}(G_n))$ provided with the pointwise Lie Bracket. Then $\mathrm{L}^p_k\left(\mathbb{S}^1, \varinjlim G_n\right) = \varinjlim \mathrm{L}^p_k(\mathbb{S}^1, G_n)$ can be provided with a structure of convenient Lie group whose Lie algebra is $\mathrm{L}^p_k(\mathbb{S}^1, \mathbf{L}(G))$.

On the other hand, the reader can find the following criterion in [Dah11]:

Theorem 5.6 *Let $\mathcal{G} = (G_n)_{n\in\mathbb{N}}$ be an ascending sequence of Banach-Lie groups such that all inclusion maps $j_n : G_n \to G_{n+1}$ are analytic group morphisms and assume that we have the following properties:*

(i) For each $n \in \mathbb{N}$, there exists a norm $|| \; ||_n$ on the Lie algebra $\mathfrak{g}_n$ defining its Banach space structure, such that its Lie bracket satisfies the inequality

$$\forall (x,y) \in \mathfrak{g}_n^2,\ ||[x,y]||_n \leq ||x||_n ||y||_n$$

and such that the bounded linear operator $\mathbb{L}(j_n) : \mathfrak{g}_n \to \mathfrak{g}_{n+1}$ has a norm operator bounded by 1;

(ii) *The locally convex structure of vector space* $\mathfrak{g} = \varinjlim \mathfrak{g}_n$ *is Hausdorff;*

(iii) *The map* $\exp_G = \bigcup_{n\in\mathbb{N}^*} \exp_{G_n} : \bigcup_{n\in\mathbb{N}^*} \mathfrak{g}_n \longrightarrow \bigcup_{n\in\mathbb{N}^*} G_n$ *is injective on some neighborhood of* 0.

Then $G = \varinjlim G_n = \bigcup_{n\in\mathbb{N}} G_n$ *has an analytic structure of Lie group modelled on* $\mathfrak{g}$ *and* $\exp_G$ *is an analytic diffeomorphism from some neighborhood of* 0 *to a neighborhood of* $1 \in G$.

One can find in [CaPe19] the following results about direct limits of linear groups.

Theorem 5.7 *Let* $(\mathbb{E}_n)_{n\in\mathbb{N}}$ *be an ascending sequence of Banach spaces such that* $\mathbb{E}_n$ *is a supplemented Banach subspace of* $\mathbb{E}_{n+1}$.
Then $\mathbb{E} = \bigcup_{n\in\mathbb{N}} \mathbb{E}_n$ *is an LB-space and* $\mathcal{L}(\mathbb{E}) = \bigcup_{n\in\mathbb{N}} \mathcal{L}(\mathbb{E}_n)$ *is also an LB-space, where* $\mathcal{L}(\mathbb{E}_n)$ *is the Banach space of continuous linear operators of* $\mathbb{E}_n$. *Moreover,* $\mathcal{GL}(\mathbb{E}) = \bigcup_{n\in\mathbb{N}} \mathcal{GL}(\mathbb{E}_n)$ *has a structure of convenient Lie group modelled on* $\mathcal{L}(\mathbb{E})$, *where* $\mathcal{GL}(\mathbb{E}_n)$ *is the Banach-Lie group of linear continuous automorphisms of* $\mathbb{E}_n$.

We end this section with some results about completness of direct limits given in [Glo19].

Definition 5.15 *A net* $\{g_\alpha\}$ *in a topological group* G *is called a left Cauchy net if, for each identity neighborhood* $U \subset G$

$$\exists \gamma \in A,\ \forall (\alpha,\beta) \in A^2 : \gamma \preccurlyeq \alpha, \gamma \preccurlyeq \beta, g_\beta^{-1}.g_\alpha \in U.$$

If every left Cauchy net in G *is convergent, then* G *is called complete.*

We then have a result of completness for the weak direct product

$$G := \prod_{\alpha\in A}^{*} G_\alpha := \left\{ (g_\alpha) \in \prod_{\alpha\in A} G_\alpha : g_\alpha = 1_\alpha \text{ for almost all } \alpha \right\}$$

Proposition 5.16 $\prod_{\alpha\in A}^{*} G_\alpha$ *is complete for each family* $\{G_\alpha\}_{\alpha\in A}$ *of complete Lie groups* G_α.

Examples 5.2

(1) *Every strict direct limit of finite-dimensional group is complete.*

(2) *The diffeomorphism group* $\mathrm{Diff}_c(M)$ *defined in Example 5.5 is complete.*

5.9 The Fréchet Topological Group G (𝔼)

Let $(\mathbb{E}_n, \iota_n^{n+1})_{n\in\mathbb{N}}$ be an ascending sequence of supplemented Banach spaces where $\mathbb{E} = \underrightarrow{\lim}\mathbb{E}_n$ can be endowed with a structure of convenient vector space. The group $\mathrm{GL}(\mathbb{E})$ does not admit any reasonable topological structure. So we have to replace it by the convenient Lie group $\mathrm{G}(\mathbb{E})$ defined as follows.

Because $\mathbb{E}_0 \subset \mathbb{E}_1 \subset \cdots$ is a direct sequence of supplemented Banach spaces, there exists Banach subspaces $\breve{\mathbb{E}}_0, \breve{\mathbb{E}}_1, \ldots$ such that:

$$\begin{cases} \mathbb{E}_0 = \breve{\mathbb{E}}_0, \\ \forall i \in \mathbb{N}, \mathbb{E}_{i+1} \backsimeq \mathbb{E}_i \times \breve{\mathbb{E}}_{i+1} \end{cases}$$

For $(i,j) \in \mathbb{N}^2$, $i \leq j$, we have the injection

$$\begin{array}{rccc} \iota_i^j : & \mathbb{E}_i \backsimeq \breve{\mathbb{E}}_0 \times \cdots \times \breve{\mathbb{E}}_i & \to & \mathbb{E}_j \backsimeq \breve{\mathbb{E}}_0 \times \cdots \times \breve{\mathbb{E}}_j \\ & (x'_0, \ldots, x'_i) & \mapsto & (x'_0, \ldots, x'_i, 0, \ldots, 0) \end{array}$$

Any $A_{n+1} \in GL(\mathbb{E}_{n+1})$ is represented by $\begin{pmatrix} A_n & B_{n+1} \\ A'_n & B'_{n+1} \end{pmatrix}$ where $A_n \in \mathcal{L}(\mathbb{E}_n, \mathbb{E}_n)$, $A'_n \in \mathcal{L}\left(\mathbb{E}_n, \breve{\mathbb{E}}_{n+1}\right)$, $B_{n+1} \in \mathcal{L}\left(\breve{\mathbb{E}}_{n+1}, \mathbb{E}_n\right)$ and $B'_{n+1} \in \mathcal{L}\left(\breve{\mathbb{E}}_{n+1}, \breve{\mathbb{E}}_{n+1}\right)$.
The group

$$GL_0(\mathbb{E}_{n+1}|\mathbb{E}_n) = \{A \in GL(\mathbb{E}_{n+1}) : A(\mathbb{E}_n) = \mathbb{E}_n\}$$

can be identified with the Banach-Lie sub-group of operators of type $\begin{pmatrix} A_n & B_{n+1} \\ 0 & B'_{n+1} \end{pmatrix}$ (cf. [ChSt76]). The set

$$G_n = \{A_n \in GL(\mathbb{E}_n) : \forall k \in \{0, \ldots, n-1\}, A_n(\mathbb{E}_k) = \mathbb{E}_k\}$$

can be endowed with a structure of Banach-Lie subgroup.
An element A_n of G_n can be seen as

$A_n =$ 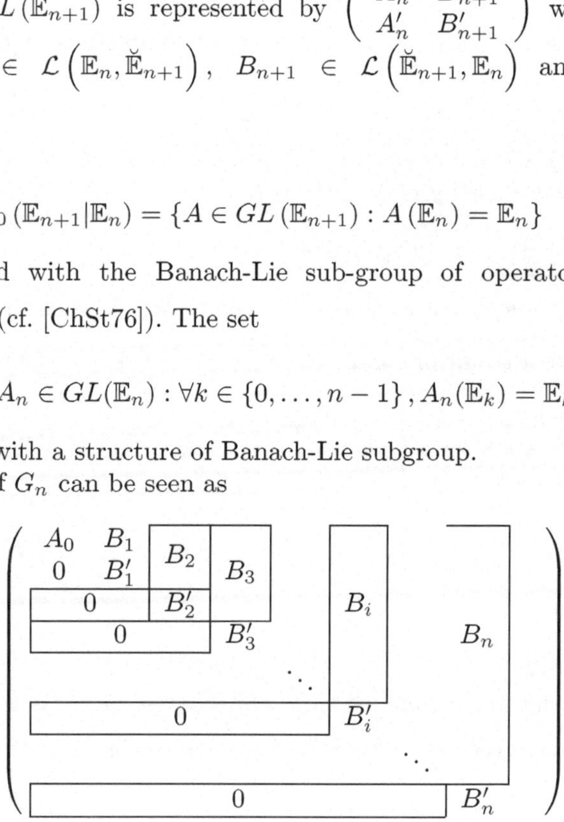

For $0 \leq i \leq j \leq k$, we consider the following diagram

$$\begin{array}{ccc}
\mathbb{E}_k & \xrightarrow{A_k} & \mathbb{E}_k \\
\uparrow \iota_j^k & & \downarrow P_j^k \\
\mathbb{E}_j & \xrightarrow{A_j} & \mathbb{E}_j \\
\uparrow \iota_i^j & & \downarrow P_i^j \\
\mathbb{E}_i & \xrightarrow{A_i} & \mathbb{E}_i
\end{array}$$

where $P_i^j : \mathbb{E}_j \to \mathbb{E}_i$ is the projection along the direction $\breve{\mathbb{E}}_{i+1} \oplus \cdots \oplus \breve{\mathbb{E}}_j$. The map

$$\begin{array}{cccc} \theta_i^j : & G_j & \to & G_i \\ & A_j & \mapsto & P_i^j \circ A_j \circ \iota_i^j \end{array}$$

is perfectly defined and we have

$$\left(\theta_i^j \circ \theta_j^k\right)(A_k) = \theta_i^j \left[\theta_j^k (A_k)\right] = \theta_i^j \left(P_j^k \circ A_j \circ \iota_j^k\right) = P_i^j \circ P_j^k \circ A_j \circ \iota_j^k \circ \iota_i^j.$$

Because $P_i^j \circ P_j^k = P_i^k$ (projective sequence) and $\iota_j^k \circ \iota_i^j = \iota_i^k$ (inductive sequence), we have

$$\left(\theta_i^j \circ \theta_j^k\right)(A_k) = P_i^k \circ A_j \circ \iota_i^k = \theta_i^k (A_k).$$

So $\left(G_i, \theta_i^j\right)_{j \geq i}$ is a projective sequence of Banach-Lie groups and the projective limit $\mathrm{G}(\mathbb{E}) = \underleftarrow{\lim} G_i$ can be endowed with a structure of Fréchet topological group.

5.10 Lie Subgroups of G (𝔼)

Let $(H_i)_{i \in \mathbb{N}}$ be a sequence where each H_i is a Banach-Lie subgroup of G_i where

$$\forall (i,j) \in I^2, i \leq j, \ \theta_i^j (H_j) = H_i.$$

Then $(H_i, \theta_i^j)_{(i,j) \in I^2, i \leq j}$ is a projective sequence and the space $H(\mathbb{M}) = \underleftarrow{\lim} H_i$ can be endowed with a structure of Fréchet topological group.

5.11 Direct Limits of Banach and Normed Vector Bundles

Definition 5.16 *$\left((E_n,\pi_n,M_n),\left(\lambda_n^{n+1},\varepsilon_n^{n+1}\right)\right)_{n\in\mathbb{N}}$ is called a ascending sequence of Banach or normed vector bundles if the following assumptions are satisfied:*

(ASBVB1) *$\mathcal{M}=(M_n)_{n\in\mathbb{N}}$ is an ascending sequence of Banach C^∞-manifolds, where M_n is modelled on the Banach space $\mathbb{M}_n$ such that $\mathbb{M}_n$ is a supplemented Banach subspace of $\mathbb{M}_{n+1}$ and the inclusion $\varepsilon_n^{n+1}:M_n\to M_{n+1}$ is a C^∞ injective map such that $(M_n,\varepsilon_n^{n+1})$ is a closed submanifold of M_{n+1};*

(ASBVB2) *The sequence $(E_n)_{n\in\mathbb{N}}$ is an ascending sequence such that the sequence of typical fibre $(\mathbb{E}_n)_{n\in\mathbb{N}}$ of $(E_n)_{n\in\mathbb{N}}$ is an ascending sequence of Banach or normed spaces.*

(ASBVB3) *For each $n\in\mathbb{N}$, $\pi_{n+1}\circ\lambda_n^{n+1}=\varepsilon_n^{n+1}\circ\pi_n$ where $\lambda_n^{n+1}:E_n\to E_{n+1}$ is the natural inclusion;*

(ASBVB4) *Any $x\in M=\varinjlim M_n$ has the direct limit chart property* **(DLCP)** *for $(U=\varinjlim U_n,\phi=\varinjlim\phi_n)$;*

(ASBVB5) *For each $n\in\mathbb{N}$, there exists a trivialization $\tau_n:(\pi_n)^{-1}(U_n)\to U_n\times\mathbb{E}_n$ such that, for any $i\leq j$, the following diagram is commutative:*

$$\begin{array}{ccc}
(\pi_i)^{-1}(U_i) & \xrightarrow{\lambda_i^j} & (\pi_j)^{-1}(U_j) \\
{\scriptstyle \tau_i}\downarrow & & \downarrow{\scriptstyle \tau_j} \\
U_i\times\mathbb{E}_i & \xrightarrow{\varepsilon_i^j\times\iota_i^j} & U_j\times\mathbb{E}_j
\end{array}$$

Moreover, if in assumption **(ASBVB1)** *each space $\mathbb{E}_n$ is a supplemented subspace of $\mathbb{E}_{n+1}$, we say that $\left((E_n,\pi_n,M_n),\left(\lambda_n^{n+1},\varepsilon_n^{n+1}\right)\right)_{n\in\mathbb{N}}$ is a strong ascending sequence of Banach or normed vector bundles.*

For example, the sequence $\left((TM_n,\pi_n,M_n),\left(T\varepsilon_n^{n+1},\varepsilon_n^{n+1}\right)\right)_{n\in\mathbb{N}}$ is a strong ascending sequence of Banach vector bundles whenever $(M_n)_{n\in\mathbb{N}}$ is an ascending sequence which has the direct limit chart property at each point of $x\in M=\varinjlim M_n$ whose model $\mathbb{M}_n$ is supplemented in $\mathbb{M}_{n+1}$.

Notation 5.2 *From now on and for the sake of simplicity, the ascending sequence of vector bundles $\left((E_n,\pi_n,M_n),\left(\lambda_n^{n+1},\varepsilon_n^{n+1}\right)\right)_{n\in\mathbb{N}}$ will be denoted $(E_n,\pi_n,M_n)_{\underrightarrow{n}}$.*

We then have the following result which is some generalization of Proposition 4.18. in [CaPe19]

Proposition 5.17

(1) Let $(E_n, \pi_n, M_n)_{\underrightarrow{n}}$ be a ascending sequence of Banach vector bundles. We have:

(i) $\underrightarrow{\lim} E_n$ has a structure of n.n.H. convenient manifold modelled on the LB-space $\underrightarrow{\lim}\mathbb{M}_n \times \underrightarrow{\lim}\mathbb{E}_n$ which has a Hausdorff convenient structure if and only if M is Hausdorff.

(ii) $\left(\underrightarrow{\lim} E_n, \underrightarrow{\lim}\pi_n, \underrightarrow{\lim} M_n\right)$ can be endowed with a structure of convenient vector bundle whose typical fibre is $\underrightarrow{\lim}\mathbb{E}_n$. If this ascending sequence is strong, $\left(\underrightarrow{\lim} E_n, \underrightarrow{\lim}\pi_n, \underrightarrow{\lim} M_n\right)$ is a convenient bundle whose structural group is the Fréchet topological group $\mathrm{G}(\mathbb{E})$.

(2) Let $(E_n, \pi_n, M_n)_{\underrightarrow{n}}$ be a ascending sequence of normed vector bundles. Then $\left(\underrightarrow{\lim} E_n, \underrightarrow{\lim}\pi_n, \underrightarrow{\lim} M_n\right)$ is a locally convex bundle whose fibre is isomorphic to $\mathbb{E} = \underrightarrow{\lim}\mathbb{E}_n$.[10]

Proof *Proof of Point (1),(i).- Consider (x, v) in some E_n. Then x belongs to M_n. According to assumptions* **(ASBVB3)** *and* **(ASBVB4)**, *there exists a chart (U, ϕ) around x in $M = \underrightarrow{\lim} M_i$ where $\phi : \bigcup_{i \geq n} U_i \to V = \bigcup_{i \geq n} \phi_i(U_i)$, (U_i, ϕ_i) being a chart around x in M_i and $V_i = \phi_i(U)_i \subset \mathbb{M}_i$. We also have a trivialization $\tau_i : \pi_i^{-1}(U_i) \to U_i \times \mathbb{E}_i$ which gives rise, via ϕ_i to a trivialization $\psi_i : \pi_i^{-1}(U_i) \to V_i \times \mathbb{E}_i \subset \mathbb{M}_i \times \mathbb{E}_i$.*
From the assumption **(ASBVB4)**, *we get the following commutative diagram*

$$\begin{array}{ccc}
(\pi_i)^{-1}(U_i) & \xrightarrow{\lambda_i^j} & (\pi_j)^{-1}(U_j) \\
\downarrow{\scriptstyle \tau_i} & & \downarrow{\scriptstyle \tau_j} \\
U_i \times \mathbb{E}_i & \xrightarrow{\varepsilon_i^j \times \iota_i^j} & U_j \times \mathbb{E}_j
\end{array}$$

The previous arguments imply that the sequence of Banach manifolds $\{E_n\}_{n \in \mathbb{N}^*}$ has the direct limit chart property at any point $(x, v) \in \underrightarrow{\lim} E_n$. Therefore, from Theorem 5.3, there exists a unique structure of n.n.H. convenient manifold on $E = \underrightarrow{\lim} E_n$ whose topology coincides with the DL-topology on E. In particular this structure is Hausdorff if and only if M is so.
The map $\tau_i : \pi_i^{-1}(U_i) \longrightarrow U_i \times \mathbb{E}_i$ can be written $\tau_i(y_i, u_i) = (y_i, \tilde{\tau}_i(y_i)(u_i))$ where $u_i \mapsto \tilde{\tau}_i(y_i)(u_i)$ is an isomorphism of Banach spaces from $\pi_i^{-1}(y_i)$ to $\mathbb{E}_i$.

Consider an atlas $\mathcal{A} = \{(U^\alpha = \underrightarrow{\lim} U_i^\alpha, \phi^\alpha = \underrightarrow{\lim}\phi_i^\alpha)\}_{\alpha \in \mathbf{A}}$ on M. From the proof of the first part, the set $\hat{\mathcal{A}} = \{(\pi^{-1}(U^\alpha) = \underrightarrow{\lim}\pi_i^{-1}(U_i^\alpha), \tau^\alpha = \underrightarrow{\lim}\tau_i^\alpha)\}_{\alpha \in \mathbf{A}}$ is an atlas for the manifold E.

---

[10] cf. Appendix D, Definition D.1.

Now, if $U_i^\alpha \cap U_i^\beta \neq \emptyset$,

$$\tau_i^\alpha \circ (\tau_i^\beta)^{-1} : \phi_i^\beta(U_i^\alpha \cap U_i^\beta) \times \mathbb{E}_i \to \phi_i^\alpha(U_i^\alpha \cap U_i^\beta) \times \mathbb{E}_i \tag{5.5}$$

can be written

$$(\overline{y}_i^\beta, \overline{u'}_i^\beta \mapsto (\phi_i^\alpha \circ (\phi_i^\beta)^{-1}(\overline{y}_i^\beta), [\tilde{\tau}_i^\alpha(y_i)] \circ [\tilde{\tau}_i^\beta(y_i)]^{-1}(\overline{u_i}^\beta) \tag{5.6}$$

where $\overline{y}_i^\beta = \phi_i^\beta(y_i)$. *With these notations,* $\overline{y}_i^\beta \mapsto \Theta_i^{\alpha\beta}(\overline{y}_i^\beta) = [\tilde{\tau}_i^\alpha(y_i)] \circ [\tilde{\tau}_i^\beta(y_i)]^{-1}$ *is a conveniently smooth map.*
From assumption **(ASBVB3)** *and assumption* **(ASBVB5)** *written over the open sets* U_i^α *and* U_j^α *(resp.* U_i^β *and* U_j^β*), we have:*

$$\left(\tilde{\tau}_j^\alpha\left(y_j\right)\right)^{-1} \circ \iota_i^j = \lambda_i^j \circ \left(\tilde{\tau}_i^\alpha\left(y_i\right)\right)^{-1}$$

Finally we get:

$$\Theta_j^{\alpha\beta}(\bar{y}_j) \circ \iota_i^j = \iota_i^j \circ \Theta_i^{\alpha\beta}(\bar{y}_i)$$

So if $\bar{y} = \underrightarrow{\lim}\bar{y}_i$, *from the above relation, one can define the transition function* $\Theta^{\alpha\beta} = \underrightarrow{\lim}\Theta_j^{\alpha\beta}$ *for the atlas* $\hat{\mathcal{A}}$ *and since each* $\Theta_j^{\alpha\beta}$ *is a smooth field of isomorphisms of* $\mathbb{E}_i$, *it follows the* $\Theta^{\alpha\beta}$ *is a smooth of isomorphisms of* $\mathbb{E}$.
From assumptions **(ASBVB3)**, *it follows that we have a well-defined conveniently smooth projection* $\pi = \underrightarrow{\lim}\pi_i : \underrightarrow{\lim}E_i \longrightarrow \underrightarrow{\lim}M_i$ *given by* $\pi(x, v) = x$ *and, with the previous notations, we also have* $\underrightarrow{\lim}\left(\pi_i\right)^{-1}\left(U_i\right) = \pi^{-1}\left(\underrightarrow{\lim}U_i\right)$.
This ends the proof of Point (1), (i)
Proof of Point (1)-(ii).– The main difficulty is to define the structural group $G\left(\mathbb{E}\right)$ *where* $\mathbb{E} = \underrightarrow{\lim}\mathbb{E}_n$. *In fact,* $G\left(\mathbb{E}\right)$ *is much larger than the direct limit of the linear groups* $GL\left(\mathbb{E}_n\right)$.
Let $\mathbb{E}_1 \subset \mathbb{E}_2 \subset \cdots$ *be the direct sequence of supplemented Banach spaces associated with the direct sequence* $E_1 \subset E_2 \subset \cdots$*; so there exists Banach subspaces* $\mathbb{E}'_1, \mathbb{E}'_2, \ldots$ *such that:*

$$\begin{cases} \mathbb{E}_1 = \mathbb{E}'_1, \\ \forall i \in \mathbb{N}^*, \mathbb{E}_{i+1} \backsimeq \mathbb{E}_i \times \mathbb{E}'_{i+1} \end{cases}$$

For $(i, j) \in \mathbb{N}^* \times \mathbb{N}^*$, $i \leq j$, *we have the injection*

$$\begin{array}{rccc} \iota_i^j : & \mathbb{E}_i \backsimeq \mathbb{E}'_1 \times \cdots \times \mathbb{E}'_i & \to & \mathbb{E}_j \backsimeq \mathbb{E}'_1 \times \cdots \times \mathbb{E}'_j \\ & (x'_1, \ldots, x'_i) & \mapsto & (x'_1, \ldots, x'_i, 0, \ldots, 0) \end{array}$$

Any $A_{n+1} \in GL\left(\mathbb{E}_{n+1}\right)$ *is represented by the matrix* $\begin{pmatrix} A_n & B_{n+1} \\ A'_n & B'_{n+1} \end{pmatrix}$
where $A_n \in \mathcal{L}\left(\mathbb{E}_n, \mathbb{E}_n\right)$, $A'_n \in \mathcal{L}\left(\mathbb{E}_n, \mathbb{E}'_{n+1}\right)$, $B_{n+1} \in \mathcal{L}\left(\mathbb{E}'_{n+1}, \mathbb{E}_n\right)$ *and* $B'_{n+1} \in \mathcal{L}\left(\mathbb{E}'_{n+1}, \mathbb{E}'_{n+1}\right)$.
The group

$$GL_0\left(\mathbb{E}_{n+1} | \mathbb{E}_n\right) = \left\{A \in GL\left(\mathbb{E}_{n+1}\right) : A\left(\mathbb{E}_n\right) = \mathbb{E}_n\right\}$$

can be identified with the Banach-Lie sub-group of operators of type $\begin{pmatrix} A_n & B_{n+1} \\ 0 & B'_{n+1} \end{pmatrix}$ *(cf. [ChSt76]). The set*

$$G_n = \{A_n \in GL(\mathbb{E}_n) : \forall k \in \{1, \ldots, n-1\}, A_n(\mathbb{E}_k) = \mathbb{E}_k\}$$

can be endowed with a structure of Banach-Lie subgroup.

For $1 \leq i \leq j \leq k$*, we consider the following diagram:*

$$\begin{array}{ccc} \mathbb{E}_k & \underset{\longrightarrow}{A_k} & \mathbb{E}_k \\ \iota_j^k \uparrow & & \downarrow P_j^k \\ \mathbb{E}_j & \underset{\longrightarrow}{A_j} & \mathbb{E}_j \\ \iota_i^j \uparrow & & \downarrow P_i^j \\ \mathbb{E}_i & \underset{\longrightarrow}{A_i} & \mathbb{E}_i \end{array}$$

where $P_i^j : \mathbb{E}_j \longrightarrow \mathbb{E}_i$ *is the projection along the direction* $\mathbb{E}'_{i+1} \oplus \cdots \oplus \mathbb{E}'_j$*.*
The map

$$\begin{array}{cccc} \theta_i^j : & G_j & \longrightarrow & G_i \\ & A_j & \mapsto & P_i^j \circ A_j \circ \iota_i^j \end{array}$$

is perfectly defined and we have:

$$\left(\theta_i^j \circ \theta_j^k\right)(A_k) = \theta_i^j\left[\theta_j^k(A_k)\right] = \theta_i^j\left(P_j^k \circ A_j \circ \iota_j^k\right) = P_i^j \circ P_j^k \circ A_j \circ \iota_j^k \circ \iota_i^j$$

Because $P_i^j \circ P_j^k = P_i^k$ *(projective system) and* $\iota_j^k \circ \iota_i^j = \iota_i^k$ *(inductive system), we have:*

$$\left(\theta_i^j \circ \theta_j^k\right)(A_k) = P_i^k \circ A_j \circ \iota_i^k = \theta_i^k(A_k)$$

So $\left(G_i, \theta_i^j\right)_{i \leq j}$ *is a projective system of Banach-Lie groups and the projective limit* $G(\mathbb{E}) = \varprojlim G_n$ *can be endowed with a structure of metrizable complete topological group.*

Now in the relation (5.5), replacing $\mathbb{E}_i$ *by* $\mathbb{E}'_1 \times \cdots \times \mathbb{E}'_i$ *and, in relation (5.6), replacing* $\overline{u_i}^\beta \in \mathbb{E}_i \simeq \mathbb{E}'_1 \times \cdots \times \mathbb{E}'_i$ *by its decomposition* $(\overline{u'}_1^\beta, \ldots, \overline{u'}_i^\beta) \in \mathbb{E}'_1 \times \cdots \times \mathbb{E}'_i$*, it follows that transition function* $\Theta^{\alpha\beta}(\bar{y})$ *as an element of the metrizable complete topological group* $G(\mathbb{E})$ *which ends the proof of ends Point (1)-(ii).*

Proof of Point (2)

This proof is an adaptation, step by step[11] *, of the proof of Point (1)-(i) by using the context and results of Appendix D.*

---

[11]Without the sentence "Therefore, from Theorem 5.3, there exists a unique structure of n.n.H. convenient manifold on $E = \varinjlim E_n$ whose topology coincides with the DL-topology on E".

5.12 Direct Limits of Banach Connections

Definition 5.17 *Let* $(E_i, \pi_i, M_i)_{\underrightarrow{i}}$ *be an ascending sequence of Banach vector bundles where* $\varepsilon_i^j : M_i \to M_j$ *and* $\lambda_i^j : E_i \to E_j$ *are the compatible bonding maps.*
A sequence of connections $D_i : TE_i \to E_i$ *is called a ascending sequence of Banach connections if*

$$\lambda_i^{i+1} \circ D_i = D_{i+1} \circ T\lambda_i^{i+1}.$$

Then we have the following result (cf. [CaPe19], Theorem 5.6).

Theorem 5.8 *Let* $(D_i)_{i\in\mathbb{N}}$ *be a strong ascending sequence of Banach connections on an ascending sequence* $(E_i, \pi, M_i)_{\underrightarrow{i}}$ *of Banach bundles and assume that* $(M_i)_{i\in\mathbb{N}}$ *has the direct limit chart property at each point of* $x \in M = \underrightarrow{\lim} M_i$.
Then the direct limit $D = \underrightarrow{\lim} D_i$ *is a connection on the convenient vector bundle* $\left(\underrightarrow{\lim} E_i, \underrightarrow{\lim} \pi_i, \underrightarrow{\lim} M_i\right)$.

Using the intrinsic link between a connection D and a Koszul connection ∇, we get the following result which proof is left to the reader:

Corollary 5.1 *Let* $(D_n)_{n\in\mathbb{N}}$ *be a strong ascending sequence of Banach connections on a direct sequence* $(E_n, \pi_n, M_n)_{\underrightarrow{n}}$ *of Banach bundles and consider the associated sequence* $(\nabla_n)_{n\in\mathbb{N}}$ *of Koszul connections.*
The direct limit $\nabla = \underrightarrow{\lim} \nabla_n$ *is a Koszul connection on the convenient vector bundle* $\left(\underrightarrow{\lim} E_n, \underrightarrow{\lim} \pi_n, \underrightarrow{\lim} M_n\right)$.

Example 5.6 *Denote by* $L^p_{\text{loc}}(\mathbb{R}^m)$ *the space of locally* L^p *functions on* $\mathbb{R}^m$ $(1 \leq p < +\infty)$. *A function belongs to* $L^p_{\text{loc}}(\mathbb{R}^m)$ *if and only if its restriction to any compact set* K *of* $\mathbb{R}^m$ *belongs to* $L^p(K)$. *Since* $\mathbb{R}^m$ *is an ascending sequence of compact sets* K_n^m *(where* $K_n^m \subset \overset{\circ}{K}{}_{n+1}^m$*), we have*

$$L^p_{\text{loc}}(\mathbb{R}^m) = \underrightarrow{\lim} L^p(K_n^m).$$

Moreover, the closure $\widehat{K_n^m}$ *of the open set* $K_{n+1}^m \backslash K_n^m$ *is also compact and we have* $L^p(K_{n+1}^m) = L^p(K_n^m) \oplus L^p\left(\widehat{K_n^m}\right)$. *Therefore the sequence of Banach spaces* $(L^p(K_n^m))_{n\in\mathbb{N}^*}$ *is an ascending sequence of supplemented Banach spaces. Since the tangent bundle to each* $L^p(K_n^m)$ *is trivial, there exists a (trivial) Koszul connection on this Banach bundle. Therefore we get a Koszul connection on* $L^p_{\text{loc}}(\mathbb{R}^m)$.

Example 5.7 *Let $(H_n)_{n\in\mathbb{N}}$ be a sequence of Banach-Lie groups and consider the Banach-Lie group of cartesian products $G_n = \prod_{k=1}^{n} H_k$. The weak direct product $\prod^*_{k\in N^*} H_k$ is the set of all sequences $(h_n)_{n\in\mathbb{N}^*}$ such that $h_n = 1$ for all but finitely many n. The weak direct product is a topological group for the box topology (see [Glo05], 4.). In fact, this weak direct product has a structure of Lie group modelled on the locally convex topological space $\bigoplus_{k\in\mathbb{N}^*} \mathfrak{H}_k$ where $\mathfrak{H}_k$ is the Lie algebra of H_k. The tangent space TG_n is the vector bundle $G_n \times \bigoplus_{k=1}^{n} \mathfrak{H}_k$. Moreover, TG_n is a supplemented subbundle of TG_{n+1} and is naturally endowed with the (trivial) Koszul connection.*

5.13 Direct Limits of Banach Principal Bundles

Definition 5.18 *A sequence $\mathcal{P} = \left((P_n, \pi_n, M_n, G_n), (\lambda_n^{n+1}, \mu_n^{n+1}, \gamma_n^{n+1})\right)_{n\in\mathbb{N}}$ of Banach principal bundles is called a strong ascending sequence of Banach principal bundles if the following assumptions are satisfied:*

- **(ASBPB1)** *$\mathcal{P} = (P_n, \lambda_n)_{n\in\mathbb{N}}$ is an ascending sequence of Banach manifolds which has the direct limit chart property* **(DLCP)** *at any point;*
- **(ASBPB2)** *$(G_n, \gamma_n)_{n\in\mathbb{N}}$ is an ascending sequence of C^∞ Banach-Lie groups where the inclusions $\varphi_n^m : G_n \to G_m$ are C^∞ homomorphisms which has the direct limit chart property* **(DLCPBLG)***;*
- **(ASBPB3)** *$\mathcal{M} = (M_n, \varepsilon_n)_{n\in\mathbb{N}}$ is an increasing sequence of Banach manifold which has the direct limit chart property* **(DLCP)** *at any point relatively to a chart domain $U = \varinjlim U_n$;*
- **(ASBPB4)** *for any $n \in \mathbb{N}$, $\left(\lambda_n^{n+1}, \varepsilon_n^{n+1}, \varphi_n^{n+1}\right)$ is a principal bundle morphism from (P_n, π_n, M_n, G_n) to $(P_{n+1}, \pi_{n+1}, M_{n+1}, G_{n+1})$;*
- **(ASBPB5)** *for any $x = \varinjlim x_n$ in M and for any $n \in \mathbb{N}$, the bundle P_n is trivializable over U_n.*

Notation 5.3 *For the sake of simplicity, in the sequel, the strong ascending sequence of principal bundles $\left((P_n, \pi_n, M_n, G_n), (\lambda_n^{n+1}, \mu_n^{n+1}, \gamma_n^{n+1})\right)_{n\in\mathbb{N}}$ will be denoted $(P_n, \pi_n, M_n, G_n)_{\underrightarrow{n}}$.*

Theorem 5.9 *If $\mathcal{P} = (P_n, \pi_n, M_n, G_n)_{\underrightarrow{n}}$ is a strong ascending sequence of Banach principal bundles, then $\left(\varinjlim P_n, \varinjlim \pi_n, \varinjlim M_n, \varinjlim G_n\right)$ can be endowed with a structure of convenient principal bundle.*

Proof *According to* **(ASBPB1)**, $P = \varinjlim P_n$ *can be endowed with a structure of convenient manifold.*
Thanks to **(ASBPB2)**, $G = \varinjlim G_n$ *can be endowed with a structure of convenient Lie group. According to* **(ASBPB3)**, $M = \varinjlim M_n$ *can be endowed with a structure of convenient manifold. The projection* $\pi = \varinjlim \pi_i : \varinjlim P_i \to \varinjlim M_i$ *is well defined and is smooth.*

Let p *be an element of* $\varinjlim P_n$. *We can assume that* p *belongs to some* P_n; *in particular,* $x = \pi(p)$ *belongs to* M_n. *So, according to* **(ASBPB5)**, *there exists a local chart* (U, ϕ) *centered at* x *where* $U = \varinjlim U_n$ *is a* c^∞*-open set of* M, $\phi = \varinjlim \phi_n$ *and where* $P_{n|U_n}$ *is trivializable.*
For $n \leq i$, *the local trivialiation map* $\Psi_i : \pi^{-1}(U_i) \to U_i \times G_i$ *can be defined. Because, for* $i \leq j$, *we have* $\pi_j \circ \lambda_i^j = \left(\varepsilon_i^j \times \varphi_i^j\right) \circ \pi_i$, *we can define a map* $\Psi : \pi^{-1}(U) \to U \times G$ *which is a bijection from* $\pi^{-1}(U) \subset P = \varinjlim P_n$ *onto the open set* $U \times G$ *of* $M \times G$.
Now, let us study the transition mappings. Because P_i *is a principal bundle, the transition functions are linked with right action on the Lie group* G_i:

$$\Psi_i^\alpha \circ \left(\Psi_i^\beta\right)^{-1}(x_i, g_i) = \left(x_i, \Psi_i^{\alpha\beta}(x_i) . g_i\right).$$

On the other side, the equivariance of the right action of the Lie group G_i

$$\lambda_i^j(p_i \cdot g_i) = \lambda_i^j(p_i) \cdot \varphi_i^j(g_i)$$

gives rise to a right action for which the following diagram is commutative:

$$\begin{array}{ccc}
\left(U_i^\alpha \cap U_i^\beta\right) \times G_i & \xrightarrow{\varepsilon_i^j \times \varphi_i^j} & \left(U_i^\alpha \cap U_i^\beta\right) \times G_i \\
\Psi_i^\alpha \circ \left(\Psi_i^\beta\right)^{-1} \downarrow & & \downarrow \Psi_j^\alpha \circ \left(\Psi_j^\beta\right)^{-1} \\
\left(U_i^\alpha \cap U_i^\beta\right) \times G_i & \xrightarrow{\varepsilon_i^j \times \varphi_i^j} & \left(U_j^\alpha \cap U_j^\beta\right) \times G_j
\end{array}$$

So the transition functions $\Psi^\alpha \circ \left(\Psi^\beta\right)^{-1}$ *are diffeomorphisms of* $\left(U^\alpha \cap U^\beta\right) \times G$ *because they are of the form* $(x, g) \longmapsto \left(x, \Psi^{\alpha\beta}(x) . g\right)$ *where* $\Psi^{\alpha\beta}(x) . g$ *is perfectly defined because* $L_{\Psi_j^{\alpha\beta}(x_j)} \circ \varphi_i^j = \varphi_i^j \circ L_{\Psi_i^{\alpha\beta}(x_i)}$ *where* L *is the left translation of the Lie group and where* $x_j = \varepsilon_i^j(x_i)$.

We then have a family of local trivializations $\{(U_\alpha, \tau_\alpha)\}_{\alpha \in A}$ *of* P *where the projection* $\pi : P \to M$ *is smooth.*
This structure is equipped with a right action $\begin{array}{ccc} P \times G & \to & P \\ (p, g) & \longmapsto & p \cdot g \end{array}$ *which is perfectly defined thanks to* **(ASBPB3)**.
So (P, π, M, G) *is endowed with a convenient principal bundle structure.*

5.14 Direct Limits of Frame Bundles

Definition 5.19 *A sequence* $(\ell(E_n))_{n\in\mathbb{N}} = (P(E_n), \pi_n, M_n, G_n)_{n\in\mathbb{N}}$ *of frame bundles is called an ascending sequence of frame bundles if the following assumptions are satisfied:*

(ASFB1) $\mathcal{M} = (M_n, \varepsilon_n^{n+1})_{n\in\mathbb{N}}$ *is an ascending sequence of Banach manifolds;*

(ASFB2) *for any* $n \in \mathbb{N}$, $(\lambda_n^{n+1}, \varepsilon_n^{n+1}, \gamma_n^{n+1})$ *is a morphism of principal bundles from* $(P(E_n), \pi_n, M_n, G_n)$ *to* $(P(E_{n+1}), \pi_{n+1}, M_{n+1}, G_{n+1})$ *where* λ_n^{n+1} *is defined via local sections* s_α^n *of* $P(E_n)$ *and fulfil the condition*

$$\lambda_n^{n+1} \circ s_\alpha^n = s_\alpha^{n+1} \circ \varepsilon_n^{n+1}$$

(ASFB3) *any* $x \in M = \underrightarrow{\lim} M_n$ *has the direct limit chart property for* $\left(U = \underrightarrow{\lim} U_n, \phi = \underrightarrow{\lim} \phi_n\right)$*;*

(ASFB4) *For any* $x \in M = \underrightarrow{\lim} M_n$, *there exists a trivialization* $\tau_n : (\pi_n)^{-1}(U_n) \to U_n \times \mathbb{E}_n$ *such that the following diagram is commutative:*

$$\begin{array}{ccc} (\pi_n)^{-1}(U_n) & \xrightarrow{\lambda_n^{n+1}} & (\pi_{n+1})^{-1}(U_{n+1}) \\ {\scriptstyle \tau_n}\downarrow & & \downarrow{\scriptstyle \tau_{n+1}} \\ U_n \times \mathbb{E}_n & \xrightarrow{\varepsilon_n^{n+1} \times \iota_n^{n+1}} & U_{n+1} \times \mathbb{E}_{n+1} \end{array}$$

for each $n \in \mathbb{N}$.

Theorem 5.10 *If* $(\ell(E))_{\underrightarrow{n}} = (P(E_n), \pi_n, M_n, G_n)_{\underrightarrow{n}}$ *is an ascending sequence of frame bundles, then the direct limit*

$$\left(\underrightarrow{\lim} P(E_n), \underrightarrow{\lim} \pi_n, \underrightarrow{\lim} M_n, \mathrm{G}(\mathbb{E})\right)$$

can be endowed with a structure of convenient principal bundle.

Proof *According to* **(ASFB1)** *and* **(ASFB3)** *corresponding to the limit chart property* **(DLCP)**, $M = \underrightarrow{\lim} M_n$ *can be endowed with a structure of n.n.H. convenient manifold (cf. [CaPe19], 4.1).*
From **(ASFB2)** *and* **(PB1)**, *we deduce* $\pi_{n+1} \circ \lambda_n^{n+1} = \varepsilon_n^{n+1} \circ \pi_n$. *So the projection* $\pi = \underrightarrow{\lim} \pi_n : \underrightarrow{\lim} P(E_n) \to \underrightarrow{\lim} M_n$ *is well defined. Let* u *be an element of* $P = \underrightarrow{\lim} P(E_n)$*; so* u *belongs to some* $P(E_n)$. *In particular,* $x = \pi(u) \in M_n$ *where* $\pi = \underrightarrow{\lim} \pi_n$. *According to* **(ASFB3)** *and* **(ASFB4)**, *there exists a local chart* (U, ϕ) *of the convenient manifold* M *whose domain contains* x. *Moreover, there exists a local trivialization* $\Psi_n : \pi_n^{-1}(U_n) \to U_n \times G_n$ *of* $P(E_n)$.

For $i \geq n$, one can define the local trivilization $\Psi_i : \pi_i^{-1}(U_i) \to U_i \times G_i$ of $P(E_i)$. According to **(ASFB2)**, *the map $\Psi : \pi^{-1}(U) \to U \times \mathrm{G}(\mathbb{E})$ can be defined and it is one to one.*
Let us study the transition functions. Using the commutativity of the diagram

$$\begin{array}{ccc} \left(U_i^\alpha \cap U_i^\beta\right) \times \mathrm{G}_i & \xrightarrow{\varepsilon_i^j \times J_i^j} & \left(U_j^\alpha \cap U_j^\beta\right) \times \mathrm{G}_j \\ {\scriptstyle \Psi_i^\alpha \circ \left(\Psi_i^\beta\right)^{-1}} \downarrow & & \downarrow {\scriptstyle \Psi_j^\alpha \circ \left(\Psi_j^\beta\right)^{-1}} \\ \left(U_i^\alpha \cap U_i^\beta\right) \times \mathrm{G}_i & \xrightarrow{\varepsilon_i^j \times J_i^j} & \left(U_j^\alpha \cap U_j^\beta\right) \times \mathrm{G}_j \end{array}$$

we obtain transition functions $\Psi^\alpha \circ \left(\Psi^\beta\right)^{-1}$ which are diffeomorphisms of $\left(U^\alpha \cap U^\beta\right) \times \mathrm{G}(\mathbb{E})$. We then have a set of local trivializations $\{(U^\alpha, \Psi^\alpha)\}_{\alpha \in A}$ of P where the projection $\pi : P \to M$ is smooth. This structure is endowed with a right action well defined according to **(ASFB 2)**.

5.15 Direct Limits of G-Structures

Considering the reductions of the frame bundles, we then get the following result whose proof is rather analogous to the previous one: we have to check the different compatibility conditions and replace the structure group $\mathrm{G}(\mathbb{E})$ by $H(\mathbb{E})$ (cf. § 5.10).

Theorem 5.11 *Let $(\ell(E_n))_{\underrightarrow{n}} = (P_n(E_n), \mathbf{p}_n, M_n, G_n)_{\underrightarrow{n}}$ be an ascending sequence of frame bundles.*
Let $\left(F_n, \mathbf{p}_{n|F_n}, M_n, H_n\right)_{n \in \mathbb{N}}$ be a sequence of associated G-structures where, for any $n \in \mathbb{N}$, $\left(\left(\lambda_n^{n+1}\right)_{|F_n}, \varepsilon_n^{n+1}, \theta_n^{n+1}\right)$ is a morphism of principal bundles.
Then $\left(\underrightarrow{\lim} F_n, \underrightarrow{\lim} \mathbf{p}_{n|F_n}, \underrightarrow{\lim} M_n, H(\mathbb{E})\right)$ can be endowed with a convenient G-structure.

5.16 Direct Limits of Tensor Structures

5.16.1 Direct Limits of Tensor Structures of Type $(1,1)$

In a similar way to the projective framework of $(1,1)$-tensors, we introduce the following definition and results.

Definition 5.20

(1) Let $\left(\mathbb{E}_i, \overline{\lambda_i^j}\right)_{i \leq j}$ *be an ascending sequence of supplemented Banach spaces. For any* $n \in \mathbb{N}$, *let* $S_n : \mathbb{E}_n \to \mathbb{E}_n$ *be an endomorphism of the Banach space* $\mathbb{E}_n$.
$(S_n)_{n\in\mathbb{N}}$ *is called an ascending sequence of endomorphisms if it fulfils the coherence conditions*

$$\forall (i,j) \in \mathbb{N}^2 : i \leq j,\ \overline{\lambda_i^j} \circ S_i = S_j \circ \overline{\lambda_i^j}.$$

(2) Let $(E_i, \pi_i, M_i)_{\underrightarrow{i}}$ *be an ascending sequence of supplemented Banach vector bundles. A sequence* $(\mathscr{S}_n)_{n\in\mathbb{N}}$ *of endomorphisms* $\mathscr{S}_n$ *of* E_n *is called a coherent sequence of endomorphisms if, for each* $x = \underrightarrow{\lim} x_n$, *the sequence* $\left((\mathscr{S}_n)_{x_n}\right)_{n\in\mathbb{N}}$ *is a coherent sequence of endomorphisms of* $\left((E_n)_{x_n}\right)_{n\in\mathbb{N}}$, *that is:*

$$\forall (i,j) \in \mathbb{N}^2 : i \leq j,\ \lambda_i^j \circ (\mathscr{S}_i)_{x_i} = (\mathscr{S}_j)_{x_j} \circ \lambda_i^j.$$

Proposition 5.18

(1) Consider a sequence $(S_n)_{n\in\mathbb{N}}$ *of coherent endomorphisms* S_n *of* E_n *on an ascending sequence of supplemented Banach spaces* $(\mathbb{E}_n)_{n\in\mathbb{N}}$. *If* $G(S_n)$ *is the isotropy group of* S_n, *then* $G(S_n)$ *is a convenient subgroup of* $\mathrm{G}(\mathbb{E})$.

(2) Consider a sequence $(\mathscr{S}_n)_{n\in\mathbb{N}}$ *of coherent endomorphisms* $\mathscr{S}_n$ *of* E_n *on an ascending sequence of supplemented Banach vector bundles* $(E_i, \pi_i, M_i)_{\underrightarrow{i}}$. *Then the direct limit* $\mathscr{S} = \underrightarrow{\lim}(\mathscr{S}_n)$ *is well defined and is a smooth endomorphism of the convenient bundle* $E = \underrightarrow{\lim} E_n$.

Proof *(1) Under our assumption, we have* $\mathbb{E}_0 \subset \mathbb{E}_1 \subset \cdots$ *and Banach spaces* $\mathbb{E}'_0, \mathbb{E}'_1, \ldots$ *such that:*

$$\begin{cases} \mathbb{E}_0 = \breve{\mathbb{E}}_0, \\ \forall i \in \mathbb{N}, \mathbb{E}_{i+1} \backsimeq \mathbb{E}_i \times \breve{\mathbb{E}}_{i+1} \end{cases}$$

Of course, $G(S_0) \subset \mathrm{GL}(\mathbb{E}) = \mathrm{G}(\mathbb{E})$. *Assume that, for all* $0 \leq i \leq n$, *we have* $G(S_i) \subset \mathrm{G}(\mathbb{E}_i)$. *Since* $\mathbb{E}_{n+1} \backsimeq \mathbb{E}_n \times \breve{\mathbb{E}}_{n+1}$ *and the sequence* $(S_n)_{n\in\mathbb{N}}$ *is coherent, if* T_{n+1} *belongs to* $G(A_{n+1})$ *it follows that* $T_{n+1} \circ S_n = T_{n+1} \circ S_n$. *Thus, if* T_n *is the restriction of* T_{n+1} *to* E_n, *then* T_n *belongs to* $\mathrm{GL}(\mathbb{E}_n)$ *and so* T_n *can be written as a matrix of type*

$$\begin{pmatrix} T_n & T_n^1 \\ 0 & T_{n+1}^2 \end{pmatrix}$$

Thus T_{n+1} *belongs to* $\mathrm{G}(\mathbb{E}_{n+1})$ *which ends the proof of (1).*

(2) By definition, a coherent sequence of endomorphisms is nothing but an ascending system of linear maps; so the direct limit $S = \underrightarrow{\lim} S_n$ *is a well*

endomorphism of $\mathbb{E} = \underrightarrow{\lim}\mathbb{E}_n$. *It follows that, for each* $x = \underrightarrow{\lim} x_n \in M = \underrightarrow{\lim} M_n$, *the direct limit* $\mathscr{S}_x = \underrightarrow{\lim}(\mathscr{S}_n)_{x_n}$ *is well defined. Now from property* **(ASBVB5)**, *for each* $n \in \mathbb{N}$, *there exists a trivialization* $\tau_n : (\pi_n)^{-1}(U_n) \to U_n \times \mathbb{E}_n$ *such that, for any* $i \leq j$, *the following diagram is commutative:*

$$\begin{array}{ccc}
(\pi_i)^{-1}(U_i) & \xrightarrow{\lambda_i^j} & (\pi_j)^{-1}(U_j) \\
\tau_i \downarrow & & \downarrow \tau_j \\
U_i \times \mathbb{E}_i & \xrightarrow{\varepsilon_i^j \times \iota_i^j} & U_j \times \mathbb{E}_j
\end{array}$$

This implies that the restriction of $\mathscr{S}_n$ *to* $(\pi_i)^{-1}(U_n)$ *is an ascending system of smooth maps and so, from [CaPe19], Lemma 3.9, the restriction of* $\mathscr{S}$ *to* $\pi^{-1}(U)$ *is a smooth map where* $U = \underrightarrow{\lim} U_n$ *and so the proof is completed.*

Let $(E_i, \pi_i, M_i)_{\underrightarrow{i}}$ be an ascending sequence of supplemented Banach vector bundles. Consider a coherent sequence $(\mathscr{S}_n)_{n \in \mathbb{N}}$ of endomorphisms $\mathscr{S}_n$ defined on the Banach bundle $\pi_n : E_n \to M_n$.
Fix any $x^0 = \underrightarrow{\lim} x_n^0 \in M = \underrightarrow{\lim} M_n$ and identify each typical fibre of E_n with $(E_n)_{x_n^0}$. For all $n \in \mathbb{N}$, we denote by $G((\mathscr{S}_n)_{x_n^0})$ the isotropy group of $(\mathscr{S}_n)_{x_n^0}$.

Now, by the same type of arguments as the ones used in the proof of Theorem 4.8, but replacing "projective sequence" by "ascending sequence" and Fréchet atlas by convenient atlas, we obtain:

Theorem 5.12 *Let* $(E_i, \pi_i, M_i)_{\underrightarrow{i}}$ *be an ascending sequence of supplemented Banach vector bundles. Consider a coherent sequence* $(\mathscr{S}_n)_{n \in \mathbb{N}}$ *of endomorphisms* $\mathscr{S}_n$ *defined on the Banach bundle* $\pi_n : E_n \to M_n$.

Then $\underrightarrow{\lim}\mathbf{p}_i(E_i)$ *can be endowed with a structure of convenient principal bundle over* $\underrightarrow{\lim} M_i$ *whose structural group is* $G(\mathscr{S}_{x^0}) = \underrightarrow{\lim} G(\mathscr{S}_i)_{x_i^0}$ *if and only if there exists a convenient atlas bundle* $\left\{\left(U^\alpha = \underrightarrow{\lim} U_n^\alpha, \Psi^\alpha = \Psi_n^\alpha\right)\right\}_{n \in \mathbb{N}, \alpha \in A}$ *such that* $\mathscr{S}_n^\alpha = \mathscr{S}_{n|U_n^\alpha}$ *is locally modelled on* $(\mathscr{S}_n)_{x_n^0}$ *and the transition maps* $\Psi_n^{\alpha\beta}(x_n)$ *take values in the isotropy group* $G((\mathscr{S}_n)_{x_n^0})$, *for all* $n \in \mathbb{N}$ *and all* $(\alpha, \beta) \in A^2$ *such that* $U_n^\alpha \cap U_n^\beta \neq \emptyset$.

5.16.2 Direct Limits of Tensor Structures of Type $(2, 0)$

We adapt the results obtained for coherent sequences of $(1, 1)$-tensors to such sequences of $(2, 0)$-tensors.

Definition 5.21

(1) Let $\left(\mathbb{E}_i, \overline{\lambda_i^j}\right)_{i \leq j}$ *be an ascending sequence of supplemented of Banach spaces.*

For any $n \in \mathbb{N}$, let $\omega_n : \mathbb{E}_n \times \mathbb{E}_n \to \mathbb{R}$ be a bilinear form of the Banach space $\mathbb{E}_n$.
$(\omega_n)_{n\in\mathbb{N}}$ is called a projective sequence of 2-forms if it fulfils the coherence condition,

$$\forall (i,j) \in \mathbb{N}^2 : i \leq j,\ \omega_i = \omega_j \circ \left(\overline{\lambda_i^j} \times \overline{\lambda_i^j} \right).$$

(2) *Let $(E_n, \pi_n, M_n)_{\underrightarrow{n}}$ be a direct sequence of Banach vector bundles. A sequence $(\omega_n)_{n\in\mathbb{N}}$ of bilinear forms ω_n on E_n is called a coherent sequence of bilinear forms if, for each $x = \underrightarrow{\lim} x_n$, the sequence $\left((\Omega_n)_{x_n} \right)$ is a coherent sequence of bilinear forms of $(E_n)_{x_n}$, that is*

$$\forall (i,j) \in \mathbb{N}^2 : i \leq j,\ (\omega_i)_{x_i} = (\omega_j)_{x_j} \circ (\lambda_i^j \times \lambda_i^j)$$

We have the following properties:

Proposition 5.19

(1) *Consider a sequence $(\omega_n)_{n\in\mathbb{N}}$ of coherent bilinear forms ω_n of E_n on an ascending sequence of supplemented Banach spaces $(\mathbb{E}_n)_{n\in\mathbb{N}}$. If $G(\omega_n)$ is the isotropy group of ω_n, then $G(\omega) = \underrightarrow{\lim} G(\omega_n)$ is a convenient subgroup of $\mathrm{G}(\mathbb{E})$.*

(2) *Consider a sequence $(\Omega_n)_{n\in\mathbb{N}}$ of coherent bilinear forms Ω_n of E_n on an ascending sequence of supplemented Banach vector bundles $(E_n, \pi_n, M_n)_{\underrightarrow{n}}$. Then the projective limit $\Omega = \underrightarrow{\lim}(\Omega_n)$ is well defined and is a smooth bilinear form of the convenient bundle $E = \underrightarrow{\lim} E_n$.*

Let $(E_n, \pi_n, M_n)_{\underrightarrow{n}}$ be an ascending sequence of supplemented Banach vector bundles. Consider a coherent sequence $(\Omega_n)_{n\in\mathbb{N}}$ of bilinear forms Ω_n defined on the Banach bundle $\pi_n : E_n \to M_n$.
Fix any $x^0 = \underrightarrow{\lim} x_n^0 \in M = \underrightarrow{\lim} M_n$ and identify each typical fibre of E_n with $(E_n)_{x_n^0}$. For all $n \in \mathbb{N}$, we denote by $G((\Omega_n)_{x_n^0})$ the isotropy group of $(\Omega_n)_{x_n^0}$.

Theorem 5.13 *Let $(E_i, \pi_i, M_i)_{\underrightarrow{i}}$ be an ascending sequence of supplemented Banach vector bundles. Consider a coherent sequence $(\Omega_n)_{n\in\mathbb{N}}$ of bilinear forms Ω_n defined on the Banach bundle $\pi_n : E_n \to M_n$.*

Then $\underrightarrow{\lim} \mathbf{p}_n(E_n)$ can be endowed with a structure of Fréchet principal bundle over $\underrightarrow{\lim} M_n$ whose structural group can be reduced to $G((\Omega)_{x^0}) = \underrightarrow{\lim} G((\Omega_n)_{x_n^0})$ if and only if there exists a convenient atlas bundle $\left\{ \left(U^\alpha = \underrightarrow{\lim} U_n^\alpha, \Psi^\alpha = \Psi_n^\alpha \right) \right\}_{n\in\mathbb{N}, \alpha\in A}$ such that $\Omega_n^\alpha = \Omega_{n|U_n^\alpha}$ is locally modelled $(\Omega_n)_{x_n^0}$ and the transition maps $\Psi_n^{\alpha\beta}(x_n)$ take values in the isotropy group $G((\Omega_n)_{x_n^0})$, for all $n \in \mathbb{N}$ and all $(\alpha, \beta) \in A^2$ such that $U_n^\alpha \cap U_n^\beta \neq \emptyset$.

5.17 Examples of Direct Limits of Tensor Structures

5.17.1 Direct Limits of Compatible Weak Almost Kähler and Almost Para-Kähler Structures

We consider the context and notations of Chapter 2, § 2.1.8.
If Ω (resp. g, resp. $\mathcal{I}$) is a weak symplectic form (resp. a weak Riemannian metric, resp. an almost complex structure) on a convenient bundle (E, π, M), the compatibility of any pair of the data $(\Omega, g, \mathcal{I})$ can be easily adapted to this convenient context. We can also easily transpose the notion of "local tensor $\mathscr{T}$ locally modelled on a linear tensor $\mathbb{T}$" to the convenient framework. We will see that such situations can be obtained by some "direct limit" of a sequence $(E_n, \pi_n, M_n)_{n\in\mathbb{N}}$ of Banach vector bundles. Once again, this context is very similar to the context of § 4.14.3 and we will only write down the details without any proof.

First, we introduce the following corresponding notations:

- Let $(\mathbb{T}_n)_{n\in\mathbb{N}}$ be a sequence of coherent tensor of type $(1,1)$ or $(2,0)$ on an ascending sequence of supplemented Banach spaces $\left(\mathbb{E}_i, \overline{\lambda_i^j}\right)_{i\leq j}$. We denote by $G(\mathbb{T}_n)$ the isotropy group of $\mathbb{T}_n$. If $\mathbb{T} = \underrightarrow{\lim}\mathbb{T}_n$, then $G(\mathbb{T}) = \underrightarrow{\lim}G(\mathbb{T}_n)$ is a convenient sub-group of $\mathrm{G}(\mathbb{E})$
- Given an ascending sequence $(E_n, \pi_n, M_n)_{\underrightarrow{n}}$ of supplemented Banach vector bundles, we provide each Banach bundle E_n with a symplectic form Ω_n and/or a weak Riemannian metric g_n and/or an almost complex structure $\mathcal{I}_n$.

Assume that we have a sequence of weak Riemaniann metrics $(g_n)_{n\in\mathbb{N}}$ on the ascending sequence $(E_n)_{n\in\mathbb{N}}$. For each $x_n \in M_n$, we denote by $(\widehat{E}_n)_{x_n}$ the completion of normed space $(\check{E}_n)_{x_n} := \left((E_n)_{x_n}, \|\ \|_{(g_n)_{x_n}}\right)$ and we set $\check{E}_n = \displaystyle\bigcup_{x_n\in M_n} (\check{E}_n)_{x_n}$ and $\widehat{E} = \displaystyle\bigcup_{x_n\in M_n} (\widehat{E}_n)_{x_n}$.

According to Definition 2.33, Chapter 2, and as in Chapter 4 Definition 4.28, let us consider the following assumptions:

Definition 5.22 *We say that the ascending sequence $(E_n, \pi_n, M_n, g_n)_{n\in\mathbb{N}}$ with separable typical fibre satisfies the assumption of direct limit Hilbert extension (* **(DLHE)** *for short) if the following conditions are satisfied:*

(1) the sequence $(g_n)_{n\in\mathbb{N}}$ is coherent;

(3) each (E_n, π_n, M_n, g_n) satisfies the g_n-Hilbert extension property;

(4) there exists a covering $\{U^\alpha = \underrightarrow{\lim}U_n^\alpha\}_{\alpha\in\mathbf{A}}$ of $M = \underrightarrow{\lim}M_n$ with trivialization $\tau^\alpha = \underrightarrow{\lim}\tau_n^\alpha : E_{|U} \to U \times \mathbb{E}$ such that the bundle atlas $\{(U_n^\alpha, \tau_n^\alpha)\}_{\alpha\in A}$

where $\tau_n^\alpha : (E_n)_{|U_n^\alpha} \to U_n^\alpha \times \mathbb{E}_n$ is a trivialization which satisfies the assumption (2) of the g_n-Hilbert extension for each n;

From Theorem 2.4, we then obtain

Theorem 5.14 *Consider an ascending sequence $(E_n, \pi_n, M_n, g_n)_{n\in\mathbb{N}}$ which satisfies the assumption* **(DLHE)**. *Then we have:*

(1) The sequence $\left(\widehat{E}_n, \widehat{\pi}_n, M_n\right)_{\underrightarrow{n}}$ (respectively $(\check{E}_n, \pi_n, M_n)_{\underrightarrow{n}}$ is an ascending sequence of Hilbert bundles. The locally convex bundle $\check{E} = \varinjlim \check{E}_n = E = \varinjlim E_n$ (as sets and) is dense in the convenient Hilbert bundle $\widehat{E} = \varinjlim \widehat{E}_n$.

(2) If $\mathfrak{i}_n : \check{E}_n \to E_n$ and $\iota_n : \check{E}_n \to \widehat{E}_n$ are the canonical injection morphisms, then $\mathfrak{i} = \varinjlim \mathfrak{i}_n : \check{E} \to E$ is the identity map and $\iota = \varinjlim \iota_n : \check{E} \to \widehat{E}$ is the natural inclusion morphism.

(3) Let $\widehat{\Omega}_n$, (resp. $\widehat{g}_n$, resp. $\widehat{\mathcal{I}}_n$) the extension of Ω_n (resp. g_n, resp. $\mathcal{I}_n$) to $\widehat{E}_n$ (cf. Chapter 2, Theorem 2.4). If the sequence $(\Omega_n)_{n\in\mathbb{N}}$, (resp. $(g_n)_{n\in\mathbb{N}}$, resp. $(\mathcal{I}_n)_{n\in\mathbb{N}}$) is coherent, so is the sequence $(\widehat{\Omega}_n)_{n\in\mathbb{N}}$ (resp. $(\widehat{g}_n)_{n\in\mathbb{N}}$, resp.$(\widehat{\mathcal{I}}_n)_{n\in\mathbb{N}}$). We set $\widehat{\Omega} = \varinjlim \widehat{\Omega}_n$ (resp. $\widehat{g} = \varinjlim \widehat{g}_n$, resp. $\widehat{\mathcal{I}} = \varinjlim \widehat{\mathcal{I}}_n$). We then have $\Omega = \varinjlim \Omega_n = \iota^* \widehat{\Omega}$, $g = \varinjlim g_n = \iota^* \widehat{g}$ and $\mathcal{I} = \varinjlim \mathcal{I}_n = \iota^* \widehat{\mathcal{I}}$. Moreover, if any pair of the data $(\Omega_n, g_n, \mathcal{I}_n)$ are compatible for all $n \in \mathbb{N}$, then $(\widehat{\Omega}, \widehat{g}, \widehat{\mathcal{I}})$ and $(\Omega, g, \mathcal{I})$ on the Fréchet vector bundle $\widehat{E}$ and E respectively are compatible structures.

(4) Let $(\ell(E_n))_{n\in\mathbb{N}} = (\mathbf{P}(E_n), \pi_n, M_n, G(\mathbb{E}))_{n\in\mathbb{N}}$ be the associated sequence of generalized frame bundles. Denote by $\mathscr{T}_n$ any tensor of the triple $(\Omega_n, g_n, \mathcal{I}_n)$. Assume that $\mathscr{T}_n$ defines a $\mathbb{T}$-tensor structure on E_n for each $n \in \mathbb{N}$. Then $\varinjlim \mathbf{P}_n(E_n)$ is a Fréchet principal bundle over $\varinjlim M_n$ whose structural group is $G(\mathscr{T})$.
Moreover, if the sequence $(\Omega_n, g_n, \mathcal{I}_n)_{n\in\mathbb{N}}$ is coherent and each triple is a tensor structure $(\Omega_n, g_n, \mathcal{I}_n)$ on E_n, then $(\Omega, g, \mathcal{I})$ is also a tensor structure on E. Moreover, $(\Omega, g, \mathcal{I})$ are compatible if $(\Omega_n, g_n, \mathcal{I}_n)$ are compatible.

(5) Under the assumption **(DLHE)** *, all the properties of (4) are also valid for the sequence $\left(\widehat{\Omega}_n, \widehat{g}_n, \widehat{\mathcal{I}}_n\right)_{n\in\mathbb{N}}$ relatively to $\widehat{E}$.*

The proof of this Theorem is step by step is formally the same of the proof of Theorem 4.10 Chapter 4 only by replacing projective limit by direct limit and so is left to the reader as an exercise.

Corollary 5.2 *Consider an almost Kähler (resp. para-Kähler) structure $(\Omega_n, g_n, \mathcal{I}_n)_{n\in\mathbb{N}}$ on a projective sequence $(E_n, \pi_n, M_n)_{\underrightarrow{n}}$ of Banach vector bundles with a separable typical fibre. If $(\Omega_n, g_n, \mathcal{I}_n)_{n\in\mathbb{N}}$ is coherent, then $(\Omega = \varinjlim \Omega_n, g = \varinjlim g_n, \mathcal{I} = \varinjlim \mathcal{I}_n)$ is an almost Kähler (resp. para-Käler) structure on the Fréchet bundle $E = \varinjlim E_n$.*

5.17.2 Application to Sobolev Loop Spaces

It is well known that if M is a connected manifold of dimension m, the set $\mathsf{L}^p_k(\mathbb{S}^1, M)$ of loops $\gamma : \mathbb{S}^1 \to M$ of Sobolev class L^p_k is a Banach manifold modelled on the Sobolev space $\mathsf{L}^p_k(\mathbb{S}^1, \mathbb{R}^m)$ (see Chapter 1, § 1.8.1 or [Pel16]). Note that the space $\mathsf{L}^p_k(\mathbb{S}^1, \mathbb{R}^m)$ is separable (cf. [Vis13]).

Let us now consider an ascending sequence $(M_n)_{n\in\mathbb{N}}$ of finite-dimensional manifolds. Then the direct limit $M = \underrightarrow{\lim} M_n$ is modelled on the convenient space $\mathbb{R}^\infty$. According to [Pel16], Proposition 34, the direct limit $\mathsf{L}^p_k(\mathbb{S}^1, M) = \underrightarrow{\lim}(\mathsf{L}^p_k(\mathbb{S}^1, M_n))$ has a Hausdorff convenient manifold structure modelled on the convenient space $\mathsf{L}^p_k(\mathbb{S}^1, \mathbb{R}^\infty) = \underrightarrow{\lim}(\mathsf{L}^p_k(\mathbb{S}^1, \mathbb{R}^m))$.

Assume now that each M_n is a Kähler manifold and let ω_n, $\mathfrak{g}_n$ and $\mathfrak{J}_n$ be the associated symplectic form, Riemannian metric and complex structure on M_n respectively. Moreover, we suppose that each sequence $(\omega_n)_{n\in\mathbb{N}}$, $(\mathfrak{g}_n)_{n\in\mathbb{N}}$ and $(\mathfrak{J}_n)_{n\in\mathbb{N}}$ is a coherent sequence on the ascending sequence $(M_n)_{n\in\mathbb{N}}$.

For p and k fixed, on the tangent bundle $T\mathsf{L}^p_k(\mathbb{S}^1, M_n)$, we introduce (cf. [Kum13a] and [Kum15]):

– $(\Omega_n)_\gamma(X,Y) = \displaystyle\int_{\mathbb{S}^1} \omega_n \left(X_{\gamma(t)}, Y_{\gamma(t)}\right) dt;$

– $(g_n)_\gamma(X,Y) = \displaystyle\int_{\mathbb{S}^1} \mathfrak{g}_n \left(X_{\gamma(t)}, Y_{\gamma(t)}\right) d\nu(t);$

– $(\mathcal{I}_n)_\gamma(X)(t) = \mathfrak{J}_n \left(X_{\gamma(t)}\right).$

It is clear that Ω_n (resp. g_n) is a weak bilinear form (resp. a Riemaniann metric) on $\mathsf{L}^p_k(\mathbb{S}^1, M)$ and $\mathcal{I}_n$ is an almost complex structure on $T\mathsf{L}^p_k(\mathbb{S}^1, M)$. The compatibility of the triple $(\omega, \mathfrak{g}, \mathfrak{J})$ implies clearly the compatibility of any $(\Omega_n, g_n, \mathcal{I}_n)$. Moreover, $(\Omega_n, g_n, \mathcal{I}_n)_{n\in\mathbb{N}}$ is a coherent sequence. Then, from Corollary 5.2, we get an almost Kähler structure on $\mathsf{L}^p_k(\mathbb{S}^1, M)$.

If each M_n is a para-Kähler manifold, we can provide $\mathsf{L}^p_k(\mathbb{S}^1, M)$ with an almost para-Kähler structure in the same way.

Remark 5.9 *As in the projective limit context, if each M_n is a Kähler manifold, then the almost complex structure is integrable and, in particular, the Nijenhuis tensor $N_{\mathfrak{J}_n}$ is zero. It is clear that we also have $N_{\mathcal{I}_n} \equiv 0$ for all $n \in \mathbb{N}$ and also $N_{\mathcal{I}} \equiv 0$, by classical direct limit argument; thus $\mathcal{I}$ is formally integrable but not integrable in general. On the opposite, if each M_n is a para-Kähler manifold, then, again, the Nijenhuis tensor of the almost para-complex structure $\mathcal{J}_n$ is zero and by Chapter 2, Theorem 2.10, this implies that the almost para-complex structure $\mathcal{J}$ on $\mathsf{L}^p_k(\mathbb{S}^1, M)$ is also integrable.*

5.18 Symplectic Forms on Direct Limits of Ascending Sequences of Banach Manifolds

5.18.1 Symplectic Forms on Direct Limits of Ascending Sequences of Reflexive Banach Spaces

Let $\mathbb{E} = \underrightarrow{\lim}\mathbb{E}_n$ be a direct limit of an ascending sequence of reflexive Banach spaces $(\mathbb{E}_n)_{n\in\mathbb{N}}$. Let us choice a norm $\| \ \|_n$ on the Banach space $\mathbb{E}_n$ such that the restriction of $\| \ \|_{n+1} \leq \| \ \|_n$ in restriction to $\mathbb{E}_n$ for all $n \in \mathbb{N}$.

We consider a sequence $(\omega_n)_{n\in\mathbb{N}}$ of symplectic forms on $\mathbb{E}_n$ such that ω_n is the restriction of ω_{n+1} to $\mathbb{E}_n$ and let $\omega_n^\flat : \mathbb{E}_n \to \mathbb{E}_n^*$ be the associated bounded linear operator. In Chapter 2, § 2.1.4, we have considered the norm $||u||_{\omega_n^\flat} = ||\omega_n^\flat(u)||_n^*$ where $|| \ ||_n^*$ is the canonical norm on $\mathbb{E}_n^*$ associated with $|| \ ||_n$. We have seen that the inclusion of the Banach space $(\mathbb{E}, || \ ||_n)$ in the normed space $(\mathbb{E}_n, || \ ||_{\omega_n^\flat})$ is continuous and we have denoted by $\widehat{\mathbb{E}}_n$ the Banach space which is the completion of $(\mathbb{E}_n, || \ ||_{\omega_n})$. Recall that the Banach space $\widehat{\mathbb{E}}_n$ does not depend on the choice of the norm $|| \ ||_n$ on $\mathbb{E}_n$.
Again, according to Chapter 2, § 2.1.4, $\omega_n^\flat$ can be extended to an isometry between $\hat{\mathbb{E}}_n$ and $\mathbb{E}_n^*$. Moreover, $\omega_n^\flat$ is an isomorphism from $\mathbb{E}_n$ to $\widehat{\mathbb{E}}_n^*$.

Lemma 5.4 *$\widehat{\mathbb{E}}_n^*$ is an ascending sequence of Banach spaces and so $\widehat{\mathbb{E}}^* = \underrightarrow{\lim}\widehat{\mathbb{E}}_n^*$ is well defined.*

Recall that $\mathbb{E} = \underrightarrow{\lim}\mathbb{E}_n$ is a convenient space (see [CaPe19], Proposition 3.7). Given a bounded skew-symmetric bilinear form ω on $\mathbb{E}$, we again denote $\omega^\flat$ the bounded linear operator from $\mathbb{E}$ to its dual $\mathbb{E}^*$ defined as usually by $\omega^\flat(u) = \omega(u, \ )$. Let ι_n be the natural inclusion of $\mathbb{E}_n$ in $\mathbb{E}$ and set $\omega_n = \iota_n^*\omega$. If u and v belong to $\mathbb{E}$, then u and v belong to some $\mathbb{E}_n$ and so we have $\omega(u, v) = \omega_n(u, v)$.

Definition 5.23 *A weak symplectic form on $\mathbb{E}$ is a bounded skew-symmetric bilinear form ω on $\mathbb{E}$ which is non-degenerate where $\omega^\flat$ is injective.*

Proposition 5.20 *With the previous notations, we have*

(1) $\omega_n = \iota_n^*\omega$ is a weak symplectic form on $\mathbb{E}_n$ and $\omega_n^\flat = \iota_n^* \circ \omega^\flat \circ \iota_n$. Moreover, the sequence $(\omega_n)_{n\in\mathbb{N}}$ is coherent, and if $u = \underrightarrow{\lim}u_n$ and $v = \underrightarrow{\lim}v_n$, then we have $\omega(u, v) = \underrightarrow{\lim}\omega_n(u_n, v_n)$, and we have $\omega(u, v) = \omega^\flat(u)(v)$ where $\omega^\flat = \underrightarrow{\lim}(\omega_n^\flat)$.

(2) Conversely, let $(\omega_n)_{n\in\mathbb{N}}$ be a coherent sequence of symplectic forms. If $u = \underrightarrow{\lim}u_n$ and $v = \underrightarrow{\lim}v_n$, then $\omega(u, v) = \underrightarrow{\lim}\omega_n(u_n, v_n)$ is well defined and is asymplectic form on $\mathbb{E}$.

Proof *(1) Recall that a set B in $\mathbb{E}$ is bounded if and only if B is a bounded subset of some $\mathbb{E}_n$ ([CaPe19], Lemma 3.6). Consider the subset*

$$B_n = \{u_n \in \mathbb{E}_n \ : \ ||u_n|| \leq 1\}.$$

Then $\hat{B} = \iota_n(B_n)$ is a bounded subset of $\mathbb{E}$ and so

$$\exists K_n > 0 : \ \forall\,(u,v) \in B^2, \ |\omega(u,v)| \leq K_n.$$

This implies that

$$\exists K_n > 0 : \ \forall\,(u,v) \in B_n^2, \ |\omega(u,v)| \leq K_n.$$

and so ω_n is a continuous skew-symmetric bilinear form on $\mathbb{E}_n$. Now we have $\omega_n^\flat = \iota_n^* \circ \omega^\flat \circ \iota_n$. Indeed fix some $v \in \mathbb{E}_n$. For any $u \in \mathbb{E}_n$, we have

$$\omega_n^\flat(u)(v) = \omega(\iota(u), \iota(v)) = \iota_n^* \circ \omega^\flat(\iota_n(u))(v).$$

This implies that $\omega_n^\flat = \iota_n^* \circ \omega^\flat \circ \iota_n$ and so ω_n is non-degenerate.
Clearly, by construction, the sequence $(\omega_n)_{n\in\mathbb{N}}$ is coherent. Fix some u, v in $\mathbb{E}$. Let n_0 be the smallest integer such that u and v belong to $\mathbb{E}_n$. Then, for $n \geq n_0$, we have $u_n = u_{n_0}$ and $v_n = v_{n_0}$; so, for all $n \geq n_0$, $\omega(u,v) = \omega_n(u_n, v_n)$. It follows that, for all $n \geq n_0$, $\omega^\flat(u)(v) = \omega_n^\flat(u_n)(v_n)$, which implies the result. Finally, we have already seen that $\omega^\flat = \varinjlim(\omega_n^\flat)$ is well defined. Now, if as previously, n_0 is the smallest integer such that u and v belong to $\mathbb{E}_n$ then, for all $n \geq n_0$, we have $\omega_n^\flat(u)(v) = \omega_n(u_n, v_n)$. Since $\omega^\flat(u)(v) = \varinjlim(\omega_n^\flat(u_n)(v_n)$, this implies the last property.

(2) The same arguments as in the last part of the previous proof permits to show that ω is well defined and, by construction, it is a skew-symmetric bilinear form. We have to show that ω is bounded. If B is a bounded set of $\mathbb{E}$, then B is contained in some $\mathbb{E}_n$ and so $\omega(u,v) = \omega_n(u,v)$ when u and v belong to $\mathbb{E}_n$. Since ω_n is continuous,

$$\exists K > 0 : \ \forall\,(u,v) \in B^2, \ |\omega_n(u,v)| \leq K$$

and so ω is a bounded bilinear form. It remains to show that $\omega^\flat$ is injective. Assume that $\omega(u,v) = 0$ for all $v \in \mathbb{E}$. Let n_0 be the smallest integer such that u belongs to E_{n_0}. Our assumption implies that $\omega(u,v) = 0$ for all $v \in \mathbb{E}_n$ and all $n \geq n_0$. But $\omega(u,v) = \omega_n(u,v)$ for all $v \in \mathbb{E}_n$. Since each ω_n is symplectic, this implies that $u = 0$.

5.18.2 Symplectic Forms on Direct Limits of Ascending Sequences of Banach Bundles

We consider an ascending sequence $(E_i, \pi_i, M_i)_{\underrightarrow{i}}$ of Banach vector bundles where the typical fibre $\mathbb{E}_i$ is reflexive. The direct limit $E = \varinjlim E_i$ has a structure of n.n.H. convenient bundle over $M = \varinjlim M_i$ with typical fibre $\mathbb{E} = \varinjlim \mathbb{E}_i$ (cf. [CaPe19]).

Definition 5.24 *A sequence $(\omega_n)_{n\in\mathbb{N}}$ of symplectic forms ω_n on E_n for $n \in \mathbb{N}$ is called an ascending sequence of symplectic forms or is coherent for short if we have*

$$\forall n \in \mathbb{N},\ (\lambda_n^{n+1})^* \omega_{n+1} \circ \varepsilon_n^{n+1} = \omega_n.$$

where λ_n^{n+1} (resp. ε_n^{n+1}) is the natural inclusion of E_n (resp. M_n) in E_{n+1} (resp. M_{n+1}).

Consider a coherent sequence $(\omega_n)_{n\in\mathbb{N}}$ of symplectic forms on E_n. According to the notations of Chapter 2, § 2.1.4, since $\mathbb{E}_n$ is reflexive, we denote by $\widehat{(E_n)}_{x_n}$ the Banach space which is the completion of $(E_n)_{x_n}$ provided with the norm $|| \ ||_{(\omega_n)^\flat_{x_n}}$. Then $(\omega_n)_{x_n}$ can be extended to a continuous bilinear map $(\hat{\omega}_n)_{x_n}$ on $(E_n)_{x_n} \times \widehat{(E_n)}_{x_n}$ and $(\omega_n)^\flat_{x_n}$ becomes an isomorphism from $(E_n)_{x_n}$ to $\widehat{(E_n)}^*_{x_n}$. We set

$$\widehat{E_n}^* = \bigcup_{x_n \in M_n} \widehat{(E_n)}^*_{x_n}$$

Proposition 5.21

(1) In the previous context, consider a coherent sequence $(\omega_n)_{n\in\mathbb{N}}$ of symplectic forms ω_n on E_n. Then, for each $x = \varinjlim x_n \in M$, if $u = \varinjlim u_n$ and $v = \varinjlim v_n$ belong to E_x, then we have $\omega_x(u,v) = \varinjlim (\omega_n)_{x_n}(u_n, v_n)$ and we get a smooth symplectic form on E. Moreover, we have $\omega_x(u,v) = \omega_x^\flat(u)(v)$ where $\omega_x^\flat = \varinjlim (\omega_n^\flat)_{x_n}$.

(2) Let ω be a symplectic form on M. The inclusion $\varepsilon_n^{n+1} : M_n \to M_{n+1}$ (resp. the map $\lambda_n^{n+1} : E_n \to E_{n+1}$) defines a smooth injective map $\varepsilon_n : M_n \to M$ (resp. an injective convenient bundle morphism $\lambda_n : E_n \to E$). If we set $\omega_n = \lambda_n^*(\omega) \circ \varepsilon_n$, then ω_n is a symplectic form on E_n and $(\omega_n)_{n\in\mathbb{N}}$ is a sequence of coherent symplectic forms. Moreover, the symplectic form on M associated with this family is exactly ω.

Proof *(1) Consider $x = \varinjlim x_n \in M$ and $u = \varinjlim u_n$ and $v = \varinjlim v_n$ in E_x. Let n_0 be the smallest integer such that x belongs to M_n and u, v belong to E_{x_n}. Then, for all $n \geq n_0$, we have $x_n = x_{n_0}$, $u_n = u_{n_0}$ and $v_n = v_{n_0}$. This implies that $\omega_x(u,v) = (\omega_n)_{x_n}(u_n, v_n) = (\omega_{n_0})_{x_{n_0}}(u_{n_0}, v_{n_0})$. From Proposition 5.20, it follows that ω_x is a symplectic form on E_x and $\omega_x(u,v) = \omega_x^\flat(u)(v)$. Consider any bundle chart $\left(U = \varinjlim U_n, \tau = \varinjlim \tau_n\right)$, where τ_n is a trivialization of $(E_n)_{|U_n}$ (cf. § 5.11). Since $(x_n, u_n, v_n) \mapsto (\omega_n)_{x_n}(u_n, v_n)$ is a smooth map from $(E_n)_{|U_n}$ to $\mathbb{R}$, it follows that $(x,u,v) \mapsto \omega_x(u,v)$ is a smooth map from $E_{|U}$ to $\mathbb{R}$ (cf. [CaPe19], Lemma 3.9). This implies that ω is a smooth symplectic form.*

(2) Let ω be a symplectic form on E. From the universal properties of a direct limit and because $\varepsilon_n^{n+1} : M_n \to M_{n+1}$ is an injective smooth map (resp. $\lambda_n^{n+1} : E_n \to E_{n+1}$ is an injective Banach bundle morphism), we obtain a

smooth injective map $\varepsilon_n : M_n \to M$ (resp. an injective convenient bundle morphism $\lambda_n : E_n \to E$) (cf. [CaPe19], Lemma 3.9 again). This implies that ω_n is a non-degenerate 2-form on E_n. But, from the definition of ω_n, it is clear that $(\omega_n)_{n\in\mathbb{N}}$ is a sequence of coherent symplectic forms. Now, by application of the proof of (1) to the sequence $(\omega_n)_{n\in\mathbb{N}}$, the symplectic form defined by the sequence $(\omega_n)_{n\in\mathbb{N}}$ is clearly the given symplectic form ω.

According to the assumption of Theorem 2.12, we introduce the following terminology:

Definition 5.25 *Let $(E_i, \pi_i, M_i)_{\underrightarrow{i}}$ be a sequence of strong ascending Banach bundles whose typical fibre $\mathbb{E}_i$ is reflexive. Consider a coherent sequence $(\omega_n)_{n\in\mathbb{N}}$ of symplectic forms ω_n on E_n. We say that the sequence $(\omega_n)_{n\in\mathbb{N}}$* satisfies the Bambusi-Darboux assumption *around $x^0 \in M$ if*

(B-D) *there exists a domain of a direct limit chart $U = \underrightarrow{\lim} U_n$ around x^0 such that:*

- *(i) for each $n \in \mathbb{N}$, $(\widehat{E}_n)_{|U_n}$ is a trivial Banach bundle;*
- *(ii) for each $n \in \mathbb{N}$, ω_n can be extended to a smooth field of continuous bilinear forms on $(E_n)_{|U_n} \times (\widehat{E}_n)_{|U_n}$.*

Under these assumptions we have:

Proposition 5.22 *Consider a coherent sequence $(\omega_n)_{n\in\mathbb{N}}$ of symplectic forms ω_n on E_n which satisfies the Bambusi-Darboux assumption* **(B-D)** *around $x^0 \in M$. Then we have the following properties:*

(1) The direct limit $\widehat{E}^*_{|U} = \underrightarrow{\lim} \widehat{E_n}^*_{|U_n}$ is well defined and is a trivial convenient bundle with typical fibre $\widehat{\mathbb{E}}^* = \underrightarrow{\lim} \widehat{\mathbb{E}}^*_n$.

(2) The sequence $(\omega_n^\flat)_{n\in\mathbb{N}}$ of isomorphisms from E_n to $\widehat{E}^*_n$ induces an isomorphism from $E_{|U}$ to $\widehat{E}^*_{|U}$.

Proof *(1) From our assumptions, for each $\in \mathbb{N}$, we have a sequence of trivializations $\widehat{\tau}_n : (\widehat{E}_n)_{|U_n} \to U_n \times \widehat{\mathbb{E}}_n$. Thus we obtain a sequence $\widehat{\tau}_n^{-1} : U_n \times \widehat{\mathbb{E}}^*_n \to (\widehat{E}_n)_{|U_n}$ of isomorphisms of trivial bundles. Now, we have natural inclusions $\widehat{\iota}_n^{n+1} : \widehat{\mathbb{E}}^*_n \to \widehat{\mathbb{E}}^*_{n+1}$ and $\varepsilon_n^{n+1} : U_n \to U_{n+1}$. So we get an injective bundle morphism δ_n^{n+1} from $U_n \times \widehat{\mathbb{E}}^*_n$ to $U_{n+1} \times \widehat{\mathbb{E}}^*_{n+1}$. Therefore, the map defined by $\widehat{\lambda}_n^{n+1}(x,u) = (\tau^*_{n+1})_x \circ \delta_n^{n+1} \circ ((\tau_n)^*_n)^{-1})_x(u)$, for all $(x,u) \in (\widehat{E}_n)_{|U_n}$, is a bonding map for the ascending sequence of trivial bundles $\left((\widehat{E}^*_n)_{|U_n}\right)_{n\in\mathbb{N}}$. Therefore, the direct limits $\widehat{\tau}^* = \underrightarrow{\lim}\widehat{\tau}^*_n$ and $\widehat{E}_{|U} = \underrightarrow{\lim}(\widehat{E}^*_n)_{|U_n}$ are well defined and $\widehat{\tau}$ is a convenient isomorphism bundle from $U \times \widehat{\mathbb{E}}$ to $\widehat{E}_{|U}$, which ends the proof of Point 1.*

(2) At first, from Proposition 5.21, 1., the sequence $(\omega_n)_{n\in\mathbb{N}}$ defines a symplectic form on E. From our assumption, since for each $n \in \mathbb{N}$, we can

extend ω_n *to a bilinear onto* $(E_n)_{|U_n} \times (\widehat{E}_n)_{|U_n}$*, this implies that* $\omega_n^\flat$ *is an isomorphism from* $(E_n)_{|U_n}$ *to* $(\widehat{E}_n^*)_{|U_n}$*. Consider the sequence of bonding maps* $\widehat{\lambda}_n^{n+1}$ *for the ascending sequence of* $\left((\widehat{E}_n^*)_{|U_n}\right)_{n\in\mathbb{N}}$ *previously defined. Then we have the following commutative diagram:*

$$\begin{array}{ccccccc}
U\times\widehat{E}_{n+1}^* & \xleftarrow{\widehat{\tau}_{n+1}} & (\widehat{E}_{n+1}^*)_{|U} & \xleftarrow{\omega_{n+1}^\flat} & (E_{n+1})_{|U} & \xleftarrow{\tau_{n+1}^{-1}} & U\times\mathbb{E}_{n+1} \\
\uparrow & & \uparrow \widehat{\lambda_n^{n+1}} & & \uparrow \lambda_n^{n+1} & & \uparrow \\
U\times\widehat{E}_n^* & \xleftarrow{\widehat{\tau}_n} & (\widehat{E}_n^*)_{|U} & \xleftarrow{\omega_n^\flat} & (E_n)_{|U} & \xleftarrow{\tau_n^{-1}} & U\times\mathbb{E}_n
\end{array}$$

It follows that the direct limit $\omega^\flat = \underrightarrow{\lim}\,\omega_n^\flat$ *is well defined and is an isomorphism from* $E_{|U}$ *to* $\widehat{E}_{|U}^*$.

5.18.3 Darboux Charts for Symplectic Forms on Direct Limits of Symplectic Banach Manifolds

In Chapter 2, § 2.5, we have established a Darboux theorem for reflexive Banach manifolds.

In a wider context like Fréchet or convenient manifolds, a symplectic form is always weak. Unfortunately, the Moser method does not work anymore in this framework since the flow of a vector filed need not exist in general (cf. [KrMi97], 32). In the situation of the projective limit approach for the Fréchet context as described in the book [DGV16], P. Kumar proves a version of the Darboux Theorem, using Moser's method, under very strong assumptions (see [Kum15]).

Here, we are interested in the problem of the validity of the Darboux Theorem on a direct limit of a sequence of Banach manifolds, under the assumption of existence of a limit chart (cf. Definition 5.13), endowed with a sequence of coherent symplectic forms.

In this situation, we give simple *necessary and sufficient conditions for the existence of a Darboux chart around some point of the direct limit* (Theorem 5.16). Unfortunately, these conditions are not satisfied in general. For instance, we can look for applying the Moser's method to the direct limit of Banach symplectic manifolds which satisfy the Darboux-Bambusi Theorem assumptions. But, as in the Fréchet framework, *apart from trivial cases or very strong assumptions*, a direct limit X of a sequence $(X_n)_{n\in\mathbb{N}}$ of vector fields X_n on M_n does not have a local flow in general.

This implies that the *Moser's method does not work in general in this context.*

However, under very strong assumptions, we could prove a Darboux Theorem. But it seems that these conditions are so restrictive that they are not satisfied in a large context. Thus we believe that such a result would not be relevant.

5.18.3.1 Conditions of Existence of Darboux Charts

Let $(M_n, \varepsilon_n^{n+1})_{n\in\mathbb{N}}$ be an ascending sequence of Banach manifolds where M_n is modelled on a Banach space $\mathbb{M}_n$.

Moreover, the sequence $(TM_n, \pi_n, M_n)_{\underrightarrow{n}}$ associated with the ascending sequence $\left(M_n, \varepsilon_n^{n+1}\right)_{n\in\mathbb{N}}$ has a direct limit $TM = \underrightarrow{\lim} TM_n$ which has a structure of convenient vector bundle whose typical fibre is $\underrightarrow{\lim}\mathbb{M}_n$ over $M = \underrightarrow{\lim} M_n$.

The following definition is classic (cf. [KrMi97], 48).

Definition 5.26 *A symplectic form on a convenient manifold is a differential 2-form ω on M which is non-degenerate and such that $d\omega = 0$*

Theorem 5.15

(1) In the previous context, consider a coherent sequence $(\omega_n)_{n\in\mathbb{N}}$ of symplectic forms ω_n on M_n. Then, for each $x \in M$, the direct limit $\omega_x^\flat = \underrightarrow{\lim}(\omega_n)_{x_n}^\flat$ is well defined and is an isomorphism from T_xM to $(\widehat{T_xM})^*$. Moreover, $\omega_x(u,v) = \omega_x^\flat(u)(v)$ defines a smooth symplectic form on M.

(2) Let ω be a symplectic form on M. The inclusion $\varepsilon_n : M_n \to M$ is a smooth injective map. If we set $\omega_n = \varepsilon_n^*(\omega)$, then ω_n is a symplectic form on M_n and the sequence $(\omega_n)_{n\in\mathbb{N}}$ is a sequence of coherent symplectic forms. Moreover the symplectic form on M associated with this sequence is exactly ω.

Proof *(1) By application Proposition 5.21, 1. to $E_n = TM_n$, we obtain the first part. It only remains to prove that $d\omega = 0$. Let ε_n be the inclusion of M_n in M. Since each inclusion $\varepsilon_n^j : M_n \to M_j$ is a smooth injective map, so is ε_n. Therefore we have*

$$\varepsilon_n^*(d\omega) = d_n\varepsilon_n^*(\omega) = d_n\omega_n$$

where d_n is the exterior differential on M_n (cf. [KrMi97], 33). Consider $u = \underrightarrow{\lim} u_n$ and $v = \underrightarrow{\lim} v_n$ in T_xM where $x = \underrightarrow{\lim} x_n$. Since as a set, we have $T_xM = \bigcup_{n\in\mathbb{N}} T_{x_n}M_n$, let n_0 be the smallest integer n such that u and v belong to $T_{x_n}M_n$. Then, for all $n \geq n_0$, we have

$$\begin{aligned} d\omega_x(u,v) &= d\omega_x(T\varepsilon_n(u), T\varepsilon_n(v)) = \varepsilon_n^*(d\omega_x)(u,v) \\ &= d_{n_0}\varepsilon_n^*(\omega_x)(u,v) = d_n(\omega_n)_{x_n}(u_n, v_n) \\ &= 0. \end{aligned}$$

As $u = \underrightarrow{\lim} u_n$ and $v = \underrightarrow{\lim} v_n$, we have $(\omega_n)_x(u_n, v_n) = (\omega_{n_0})_{x_{n_0}}(u_{n_0}, v_{n_0})$ for all $n \geq n_0$, which ends the proof of (1).

(2) Again the first part is a consequence of Proposition 5.21. Now since $d\omega = 0$, we obtain $d_n\omega_n = 0$.

But from the definition of ω_n, it is clear that $(\omega_n)_{\mathbb{N}}$ is a sequence of coherent *symplectic forms. Now, by application of the proof of (1) to the sequence $(\omega_n)_{\mathbb{N}}$, the symplectic form defined by (ω_n) is clearly the given symplectic form ω.*

As in the Banach context, we introduce the notion of Darboux chart:

Definition 5.27 *Let ω be a weak symplectic form on the direct limit $M = \underrightarrow{\lim} M_n$. We say that a chart (V, ψ) around x_0 is a Darboux chart if $\psi^*\omega^0 = \omega$ where ω^0 is the constant form on $\psi(U)$ defined by $(\psi^{-1})^*\omega_{x_0}$.*

We have the following necessary and sufficient conditions of existence of Darboux charts on a direct limit of ascending sequence of Banach manifolds:

Theorem 5.16 *Let $\left(M_n, \varepsilon_n^{n+1}\right)_{n\in\mathbb{N}}$ be an ascending sequence of Banach manifolds where M_n is modelled on a reflexive Banach space $\mathbb{M}_n$.*

(1) Consider a coherent sequence $(\omega_n)_{n\in\mathbb{N}}$ of symplectic forms ω_n on M_n and let ω be the symplectic form generated by (ω_n) on $M = \underrightarrow{\lim} M_n$. Assume that the following property is satisfied:

(H) *There exists a limit chart $(U = \underrightarrow{\lim} U_n, \phi = \underrightarrow{\lim} \phi_n)$ around x_0 such that if x_0 belongs to M_n, then (U_n, ϕ_n) is a Darboux chart around x_0 for ω_n.*

Then (U, ϕ) is a Darboux chart around x_0 for ω.

(2) Let ω be a weak symplectic form on the direct limit $M = \underrightarrow{\lim} M_n$. Assume that there exists a Darboux chart (V, ψ) around x_0 in M. If ω_n is the symplectic form on M_n induced by ω, then there exists a limit chart $(U = \underrightarrow{\lim} U_n, \phi = \underrightarrow{\lim} \phi_n)$ around x_0 such that the property **(H)** *is satisfied.*

Proof *(1) Assume that the assumption* **(H)** *is true. Note that, by construction of ω, we have $\omega_n = \varepsilon_n^*\omega$. Fix some $x \in U$ and (u, v) in $(T_xM)^2$. Denote by n_0 the small integer such x belongs to M_n and (u, v) are in $(T_xM_n)^2$. Then, for any $n \geq n_0$, we have $\omega_x(u, v) = (\omega_n)_x(u, v)$. For the same reason, if ω_n^0 is the "model" in the Darboux chart for ω_n, we also have $\omega^0(u, v) = \omega_n^0(u, v)$. From the properties of compatibility of ϕ and ϕ_n, the induced form on $\phi_n(U_n)$ by $\phi^*\omega$ is precisely $\phi_n^*(\omega_n)$ and, moreover, the constant 2-form on $\phi_n(U_n)$ induced by ω^0 is ω_n^0. But according to our assumption, we have $\omega_n^0(T_x\phi_n(u), T_x\phi_n(v) = \omega^0((T_x\phi_n(u), T_x\phi_n(v))$. Thanks to the compatibility conditions between $T_x\phi$ and $T_x\phi_n$, we then obtain:*

$$\begin{aligned}\omega(T_x\phi(u), T_x\phi(v) &= \omega_n^0((T_x\phi_n(u), T_x\phi_n(v)) \\ &= \omega^0(T_x\phi(u), T_x\phi(v)).\end{aligned}$$

Since $T_x\phi_n$ is an isomorphism onto $\mathbb{M}_n$, this ends the proof of (1).

(2) Given a symplectic form ω, we can consider the induced form $\omega_n = \varepsilon_n^*\omega$ on M_n. Consider a Darboux chart (V, ψ) for ω around x_0. We set $V_n = V \cap M_n$

and $\psi_n = \psi \circ \varepsilon_n$. *From the condition of compatibility of* ψ *and the inclusion of* $\mathbb{M}_n$ *in* $\mathbb{M}$, *it follows that* $\psi(V \cap M_n) = \psi_n(V_n)$ *and* ψ_n *is a diffeomorphism from* V_n *to* $\psi(V_n)$. *Therefore we have* $V = \underrightarrow{\lim} V_n$ *and* $\psi = \underrightarrow{\lim} \psi_n$. *Using the same argument as in the first part, we show that* (V_n, ψ_n) *is a Darboux chart for* ω_n.

5.18.3.2 The Problem of Existence of a Darboux Chart in General

In this subsection, we will explain why, even in the context of an ascending sequence of symplectic Banach manifolds which satisfies the assumption of Theorem 2.6, in general, there does not exist Darboux charts for the induced symplectic form on the direct limit.

Let $\left(M_n, \varepsilon_n^{n+1}\right)_{n\in\mathbb{N}}$ be an ascending sequence of Banach manifolds where M_n is modelled on a *reflexive* Banach space $\mathbb{M}_n$. Consider a coherent sequence $(\omega_n)_{n\in\mathbb{N}}$ of symplectic forms on M_n. According to the notations of Chapter 2, § 2.1.1, since $\mathbb{M}_n$ is reflexive, we denote by $\widehat{T_{x_n}M_n}$ the Banach space which is the completion of $T_{x_n}M_n$ provided with the norm $|| \ ||_{(\omega_n)_{x_n}}$. Then $(\omega_n)_{x_n}$ can be extended to a continuous bilinear map $(\hat{\omega}_n)_{x_n}$ on $T_{x_n}M_n \times \widehat{T_{x_n}M_n}$ and $(\omega_n)^\flat_{x_n}$ becomes an isomorphism from $T_{x_n}M_n$ to $(\widehat{T_{x_n}M_n})^*$. We set

$$(\widehat{TM_n})^* = \bigcup_{x_n\in M_n} (\widehat{T_{x_n}M_n})^*.$$

Then, by application of Proposition 5.22, we have:

Proposition 5.23 *Let* $\left(M_n, \varepsilon_n^{n+1}\right)_{n\in\mathbb{N}}$ *be an ascending sequence of Banach manifolds whose model is a reflexive Banach space* $\mathbb{M}_n$. *Consider a coherent sequence* $(\omega_n)_{n\in\mathbb{N}}$ *of symplectic forms* ω_n *on* E_n. *Assume that we have the following assumptions:*

(i) There exists a limit chart $(U = \underrightarrow{\lim} U_n, \phi = \underrightarrow{\lim} \phi_n)$ *around* x_0 *such that* $(\widehat{TM_n})_{|U_n}$ *is a trivial Banach bundle.*

(ii) ω_n *can be extended to a smooth field of continuous bilinear forms on* $(TM_n)_{|U_n} \times (\widehat{TM}_n)_{|U_n}$.

Then $\widehat{T^*M}_{|U}$ *is a trivial bundle. If* ω *is the symplectic form defined by the sequence* $(\omega_n)_{n\in\mathbb{N}}$, *then the morphism*

$$\omega^\flat : TM \to T^*M$$

induces an isomorphism from $TM_{|U}$ *to* $\widehat{T^*M}_{|U}$.

Note that the context of Proposition 5.23 covers the particular framework of sequences $(M_n, \omega_n)_{n\in\mathbb{N}}$ of strong symplectic Banach manifolds.

We will expose which arguments are needed to prove a Darboux theorem in the context of direct limit of Banach manifolds under the assumption of

Proposition 5.23. In fact, we point out the problems that arise in establishing the existence of a Darboux chart by Moser's method: too much strong assumptions are needed to obtain solutions for ODE on direct limit of Banach spaces (cf. [Pel18]), and so *we are not sure that this result can have non-trivial applications.* So we omit to give such a precise Darboux Theorem. Fix some point $a = \underrightarrow{\lim} a_n \in M$. In the context on Proposition 5.23, as in the proof of Theorem 2.6, on the direct limit chart (U, ϕ) around a, we can replace U by $\phi(U)$, ω by $(\phi)^*\omega$ on $\phi(U)$ in the convenient space $\mathbb{M}$. Thus, if ω^0 is the constant form on U defined by ω_a, we consider the 1-parameter family

$$\omega^t = \omega^0 + t\overline{\omega}, \text{ with } \overline{\omega} = \omega - \omega^0.$$

Since ω^t is closed and $\mathbb{M}$ is a convenient space, by [KrMi97], Lemma 33.20, there exists a neighbourhood $V \subset U$ of a and a 1-form α on V such that $\alpha = d\overline{\omega}$ which is given by

$$\alpha_x = \int_0^1 s.\overline{\omega}_{sx}(x, \ )ds.$$

Now, for all $0 \leq t \leq 1$, $\omega^t_{x_0}$ is an isomorphism from $T_aM \equiv \mathbb{M}$ onto $\widehat{T_aM} \equiv \widehat{\mathbb{M}}^*$. In the Banach context, using the fact that the set of invertible operators is open in the set of operators, after restricting V, we may assume that $(\omega^t)^\flat$ is a field of isomorphisms from $\mathbb{M}$ to $\widehat{\mathbb{M}}^*$.

Unfortunately, *this result is not true in the convenient setting* (cf. [KrMi97]). Therefore, the classical proof does not work in this way.

However, let ω_n be the symplectic form induced by ω on M_n. Therefore, for each n, let α_n be the 1-form α_n induced by α on $\phi_n(U_n \cap V)$. Then we have $\omega_n = d\alpha_n$ and also

$$(\alpha_n)_{x_n} = \int_0^1 s.(\overline{\omega}_n)_{sx_n}(x_n, \ )ds$$

where $\bar{\omega}_n = \omega_n - \omega_n^0$ is associated with the 1-parameter family $\omega_n^t = \omega^n + t\bar{\omega}_n$. We are exactly in the context of the proof of Theorem 2.6 and so the local flow $\mathrm{Fl}_t^{X_n}$ of $X_n^t = ((\omega_n^t)^\flat)^{-1}(\alpha_n)$ is a local diffeomorphism from a neighbourhood W_n of a in V_n and, in this way, we build a Darboux chart around a_n in M_n. Therefore, after shrinking each W_n, assume that:

(DCASBM) [Direct limit Darboux chart] We have an ascending sequence of such open sets $(W_n)_{n\in\mathbb{N}}$ then on $W = \underrightarrow{\lim} W_n$, the family of local diffeomorphisms $\mathrm{Fl}_t^X = \underrightarrow{\lim} F_t^{X_n}$ is defined on W.

Recall that $\omega^\flat = \underrightarrow{\lim}\omega_n^\flat$ and $\omega^\flat$ is an isomorphism. Thus according to the previous notations, we have a time dependent vector field

$$X^t = ((\omega^t)^\flat)^{-1}(\alpha)$$

and again, we have $L_{X^t}\omega^t = 0$. Of course, if the direct limit Darboux chart assumption **(DCASBM)** on $(W_n)_{n\in\mathbb{N}}$ is true, then $X^t = \underrightarrow{\lim} X_n^t$. So we obtain a Darboux chart as in the Banach context. Note that, in this case, we are in the context of Theorem 5.16.

Remark 5.10 *In fact, under the direct limit Darboux chart assumption* **(DCASBM)**, *the flow* Fl_t *is the local flow (at time* $t \in [0,1]$*) of* $X^t = \underrightarrow{\lim} X_n^t$ *where* $X_n^t = ((\omega_n^t)^\flat)^{-1}(\alpha_n)$ *(with the previous notations). Unfortunately, the direct limit Darboux chart assumption* **(DCASBM)** *is not true since we have* $\bigcap_{n\in\mathbb{N}} W_n = \{a\}$ *in general.*

For the problems of solutions of ODE on direct limit of Banach manifolds linked to existence of a Darboux chart, see [Pel18], Remark 3.5.2.

From the example of [Mars81], we can obtain the following example for which *there is no Darboux chart* on a direct limit of symplectic Banach manifolds:

Example 5.8 *Let* $\mathbb{H}$ *be a Hilbert space and endowed with its inner product* $<\,,\,>$. *If* g *is a Riemannian metric on* $T\mathbb{H} = \mathbb{H}\times\mathbb{H}$, *we can define a symplectic form* ω *in the following way:*

$$2\,\omega_{(u,e)}((e_1,e_2),(e_3,e_4)) = D_u g_u(e,e_2).e_3 - D_u g_u(e,e_3).e_1 + g_u(e_4,e_1) - g_u(e_2,e_3).$$

Let $S : \mathbb{H} \to \mathbb{H}$ *be a compact operator with dense range, but proper subset of* $\mathbb{H}$, *which is selfadjoint and positive. Given a fixed* $e \in \mathbb{H}$, *then* $A_x = ||x-e||^2 Id_{\mathbb{H}} + S$ *is a smooth field of bounded operators of* $\mathbb{H}$ *which is an isomorphism for all* $x \neq e$ *and* $A_e(\mathbb{H}) \neq \mathbb{H}$ *but* $A_e(\mathbb{H})$ *is dense in* $\mathbb{H}$. *Then* $g_x(e,f) =< A_x(e), f >$ *is a weak Riemanian metric and the associated symplectic form* ω *is such that* $\omega_u^\flat$ *is an isomorphism for* $x \neq e$ *and the range of* $\omega_e^\flat$ *is only dense in* $T_e\mathbb{H}$ *(cf. [Mars81]).*
Let $(e_k) \in \mathbb{H}^{\mathbb{N}^*}$ *be a sequence which converges to* 0. *For each* $n > 0$, *we provide* $\mathbb{H}_n = \mathbb{H} \oplus \mathbb{R}^n$ *of the the inner product* $<\,,\,>_n$ *obtained from the inner product* $<\,,\,>$ *on* $\mathbb{H}$ *and the canonical one on* $\mathbb{R}^n$. *Now, we consider the continuous operator* $S_n = S \oplus Id_{\mathbb{R}^n}$ *and, for any* $x_n \in \mathbb{H}_n$, *we set*

$$(A_n)_{x_n} = ||x_n - e_n||^2 Id_{\mathbb{H}_n} + S_n.$$

We denote by g_n *the Riemannian metric on* $\mathbb{H}_n$ *defined by*

$$(g_n)_{x_n}(e_n, f_n) =< (A_n)_x(e_n), f_n >$$

and we consider the symplectic form ω_n *associated with* g_n *as previously. We set* $\mathcal{H} = \underrightarrow{\lim}\mathbb{H}_n$ *and* $M_n = \left(\mathcal{H} \setminus \bigcup_{n\in\mathbb{N}^*}\{e_n\}\right) \bigcap \mathbb{H}_n$. *Now, the sequence* $(\omega_n)_{n\in\mathbb{N}^*}$ *induces a family of coherent strong symplectic forms on* M_n *which induces a weak symplectic form* ω *on* $M = \underrightarrow{\lim} M_n$ *since the cotangent space* $T_x^*\mathcal{H}$ *is the projective limit of* $T_{x_n}^* M_n$ *and so is a Fréchet space, which implies that*

$\omega^\flat$ cannot be surjective. Now, for each n, we have a Darboux chart (V_n, F_n) around 0. But from the previous construction, we must have $\bigcap\limits_{n\in\mathbb{N}^*} V_n = \{0\}$. Therefore there is no Darboux chart for ω around $0 \in M$ according to Theorem 5.16. Note that since each ω_n is a strong symplectic form, the assumptions of Proposition 5.23 are satisfied.

5.19 Direct Limits of Anchored Bundles and Lie Algebroids

In this section, $\left((\mathcal{A}_i, \pi_i, M_i), \left(\eta_i^j, \varepsilon_i^j \right) \right)_{(i,j)\in\mathbb{N}^2, i\leq j}$ is a strong ascending sequence of Banach vector bundles with typical fibre $\mathbb{A}_i$ and we denote by

$$\left(\mathcal{A} = \underrightarrow{\lim} \mathcal{A}_i, \pi = \underrightarrow{\lim} \pi_i, M = \underrightarrow{\lim} M_i \right)$$

its direct limit which is a convenient vector bundle whose typical fibre is $\mathbb{A} = \underrightarrow{\lim} \mathbb{A}_i$ and whose structural group is $\mathrm{G}(\mathbb{E})$ (cf. Proposition 5.17).

5.19.1 Direct Limits of Anchored Bundles

Definition 5.28 $\left((\mathcal{A}_i, \pi_i, M_i, \rho_i) \left(\eta_i^j, \varepsilon_i^j \right) \right)_{(i,j)\in\mathbb{N}^2, i\leq j}$ *is a strong ascending sequence of anchored bundles if*

(SASBVB1) $\left(\mathcal{A}_i, \eta_i^j \right)_{i\leq j}$ *is a strong ascending sequence of Banach vector bundles* $(\pi_i : \mathcal{A}_i \to M_i)_{i\in\mathbb{N}}$ *over the strong ascending sequence of manifolds* $\left(M_i, \delta_i^j \right)_{i\leq j}$;

(SASBVB2) *For all* $(i,j) \in \mathbb{N}^2$ *such that* $i \leq j$, *one has*

$$\rho_j \circ \eta_i^j = T\varepsilon_i^j \circ \rho_i$$

We then have the essential result:

Theorem 5.17 *Let* $\left((\mathcal{A}_i, \pi_i, M_i, \rho_i), \left(\eta_i^j, \varepsilon_i^j \right) \right)_{(i,j)\in\mathbb{N}^2, i\leq j}$ *be a strong sequence of Banach anchored bundle.*
Then $\left(\mathcal{A}, \pi, M, \rho = \underrightarrow{\lim} \rho_i \right)$ *is a convenient anchored bundle.*

Proof *The direct limit* $\left(\underrightarrow{\lim} TM_i, \underrightarrow{\lim} p_{M_i}, \underrightarrow{\lim} M_i \right)$ *is exactly the tangent vector bundle of* M *(cf. [CaPe19]).*
According to the condition $\rho_j \circ \eta_i^j = T\varepsilon_i^j \circ \rho_i$ *for all* $(i,j) \in \mathbb{N}^2$ *s.t.* $i \leq j$, *the direct limit* $\rho = \underrightarrow{\lim} \rho_i$ *is a convenient morphism from* $\mathcal{A} = \underrightarrow{\lim} \mathcal{A}_i$ *to* $TM = \underrightarrow{\lim} TM_i$.

5.19.2 Strong ascending sequence of Banach Lie Algebroids

For analog reasons as in the context of projective sequences of Banach bundles, any section of the direct limit of a strong ascending sequence of Banach Lie algebroids is not a direct limit of sections. Thus again, in general, the direct limit of a strong ascending sequence of Banach Lie algebroids will only have a structure of partial Lie algebroid.

Definition 5.29 $\left((\mathcal{A}_i,\pi_i,M_i,\rho_i,[.,.]_{\mathcal{A}_i}),\left(\eta_i^j,\varepsilon_i^j\right)\right)_{(i,j)\in\mathbb{N}^2,i\leq j}$ *is called a strong ascending sequence of Lie algebroids if the sequence* $\left((\mathcal{A}_i,\pi_i,M_i,\rho_i),\left(\eta_i^j,\varepsilon_i^j\right)\right)_{(i,j)\in\mathbb{N}^2,i\leq j}$ *is a strong ascending sequence of anchored bundles and if, for all* $(i,j)\in\mathbb{N}^2$ *s.t.* $i\leq j$, *we have:*

(SASLA) $\eta_i^j:\mathcal{A}_i\to\mathcal{A}_j$ *is a Lie morphism from* $(\mathcal{A}_i,\pi_i,M_i,\rho_i,[.,.]_{\mathcal{A}_i})$ *to* $(\mathcal{A}_j,\pi_j,M_j,\rho_j,[.,.]_{\mathcal{A}_j})$ *over* ε_i^j.

From now on, the ascending sequence of Banach Lie algebroids $\left((\mathcal{A}_i,\pi_i,M_i,\rho_i,[.,.]_{\mathcal{A}_i}),\left(\eta_i^j,\varepsilon_i^j\right)\right)_{(i,j)\in\mathbb{N}^2,i\leq j}$ is fixed. Note that from Theorem 5.17, we have an anchored convenient bundle $(\mathcal{A},\pi,M,\rho)$ on $\mathcal{A}$.
As in the projective limit context, we have the analog following Proposition:

Proposition 5.24 *For any open set* U *of* M, *there exists an ascending sequence* $\left(U_i,\varepsilon_{i\,|U_j}^j\right)_{(i,j)\in\mathbb{N}^2,i\leq j}$ *such that* $U=\underrightarrow{\lim}U_i$. *Let* $\mathfrak{P}(\mathcal{A}_U)$ *be the* $C^\infty(U)$ *module generated by the set of sections* $\mathfrak{a}$ *of* $\mathcal{A}$ *defined on* U *and such that* $\mathfrak{a}=\underrightarrow{\lim}\mathfrak{a}_i$ *where* $\mathfrak{a}_i$ *is a section of* $\mathcal{A}_i$ *over* U_i. *Then we have:*

(1) For any open subset U *in* M, *there exists a well-defined Lie bracket* $[.,.]_{\mathcal{A}_U}$ *on* $\mathfrak{P}(\mathcal{A}_U)$ *such that* $(\mathfrak{P}(\mathcal{A}_U),[.,.]_{\mathcal{A}_U})$ *has a structure of Lie algebra and* ρ *induces a Lie algebra morphism from* $(\mathfrak{P}(\mathcal{A}_U),[.,.]_{\mathcal{A}_U})$ *to the Lie algebra* $\mathfrak{X}(U),[.,.])$ *of vector fields on* U.

(2) For each $x\in M$, *there exists a chart domain* U *around* x *in* M *such that* $\mathcal{A}_U$ *is trivializable over* U *and for each* $v\in T_xM$, *there a section* $\mathfrak{a}\in\mathfrak{P}(\mathcal{A}_U)$ *such that* $\mathfrak{a}(x)=v$.

As in the projective context, the family

$$\mathfrak{P}_M(\mathcal{A}):=\{\mathfrak{P}(\mathcal{A}_U),\ U\text{ open set in }M\}$$

also defines a subsheaf of the sheaf $\Gamma_M(\mathcal{A}):=\{\Gamma(\mathcal{A}_U),\ U\text{ open set in }M\}$ of sections of $\mathcal{A}$. Note that from Remark 5.4, $\mathfrak{P}(\mathcal{A}_U)$ is strictly contained in $\Gamma(\mathcal{A}_U)$ for all open sets U in M.
Now, from Proposition 5.24 (1), it follows that $[.,.]_{\mathcal{A}_U}$ is a sheaf of Lie brackets on $\mathfrak{P}_M(\mathcal{A})$ which will be denoted also $[.,.]_{\mathcal{A}}$. Thus, from Proposition 5.24 we obtain:

Theorem 5.18 *Let* $\left((\mathcal{A}_i, \pi_i, M_i, \rho_i, [.,.]_{\mathcal{A}_i}), \left(\eta_i^j, \varepsilon_i^j\right)\right)_{(i,j)\in\mathbb{N}^2, i\leq j}$ *be a strong ascending sequence of Banach-Lie algebroids. The convenient anchored bundle* $(\mathcal{A}, \pi, M, \rho)$ *has a structure of strong partial Lie algebroid* $(\mathcal{A}, \pi, M, \mathfrak{P}_M(\mathcal{A}))$.

Proof of Proposition 5.24

(1) According to the direct limit topology on M*, if* U *is an open set in* M*, there exists an open set* U_i *in* M_i *such that* $U_i = \varepsilon_i(U)$ *is open in* M_i *and so* $U = \underrightarrow{\lim} U_i$ *(cf. [CaPe19]). In this way, we have* $\mathcal{A}_U = \underrightarrow{\lim}\{\mathcal{A}_i\}_U$*. Let* $\mathfrak{a} = \underrightarrow{\lim}\mathfrak{a}_i$ *and* $\mathfrak{a}' = \underrightarrow{\lim}\mathfrak{a}'_i$ *be two sections of* E_U *which are direct limits of sections* $\mathfrak{a}_i$ *and* $\mathfrak{a}'_i$ *of* $\mathcal{A}_i$ *over* U_i*. Now, according to Definition 3.82, since* η_i^j *is a Lie morphism over* ε_i^j *and the pairs of sections* $(\mathfrak{a}_i, \mathfrak{a}_j)$ *and* $(\mathfrak{a}'_i, \mathfrak{a}'_j)$ *are* ε_i^j *related, we have*

$$\eta_i^j\left([\mathfrak{a}_i, \mathfrak{a}'_i]_{\mathcal{A}_i}\right) = [\mathfrak{a}_j, \mathfrak{a}'_j]_{\mathcal{A}_j} \circ \varepsilon_i^j \tag{5.7}$$

It follows that $\underrightarrow{\lim}[\mathfrak{a}_i, \mathfrak{a}'_i]_{\mathcal{A}_i}$ *belongs to* $\mathfrak{P}(\mathcal{A}_U)$*. We set* $[\mathfrak{a}, \mathfrak{a}']_{\mathcal{A}_U} = \underrightarrow{\lim}[\mathfrak{a}_i, \mathfrak{a}'_i]_{\mathcal{A}_i}$.

For any $f \in C^\infty(U)$, we set:

$$[\mathfrak{a}, f\mathfrak{a}']_{\mathcal{A}_U} = df\,(\rho(\mathfrak{a}))\,\mathfrak{a}' + f[\mathfrak{a}, \mathfrak{a}']_{\mathcal{A}_U} \tag{5.8}$$

In this way, $[\mathfrak{a}, f\mathfrak{a}']_{\mathcal{A}_U}$ *belongs to* $\mathfrak{P}(\mathcal{A}_U)$*. Note that from the skew-symmetry of* $[.,.]_{\mathcal{A}_i}$ *for all* $i \in \mathbb{N}$*, it follows that*

$$[f\mathfrak{a}, \mathfrak{a}'] = -df\,(\rho(\mathfrak{a}'))\,\mathfrak{a} + f[\mathfrak{a}, \mathfrak{a}']_{\mathcal{A}_U}$$

for any $g \in C^\infty(U)$.

We will write the bracket $[\mathfrak{a}, \mathfrak{a}']_{\mathcal{A}_U}$ in local coordinates.

Fix some $x \in M$*. According to Definition 5.16, there exists a chart* $(U = \underrightarrow{\lim} U_i, \phi = \underrightarrow{\lim}\phi_i)$ *where* (U_i, ϕ_i) *is a chart around* $x_i = \varepsilon_i(x)$ *and such that we also have a trivialization* $\tau_i : \{\mathcal{A}_i\}_{U_i} \to \phi_i(U_i) \times \mathbb{A}_i$*; w and then* $\tau = \underrightarrow{\lim}\tau_i : \{\mathcal{A}_i\}_U \to \phi(U) \times \mathbb{A}$ *is a trivialization. According to §3.15.3 and (5.9) for each* $i \in \mathbb{N}$*, we have*

$$[\mathfrak{a}_i, \mathfrak{a}'_i]_{\mathcal{A}_i}(x_i) \equiv \mathsf{d}_{\mathsf{x}_i}\mathsf{a}'_i\left(\{\rho_i\}_{\mathsf{x}_i}(\mathsf{a}_i(\mathsf{x}_i))\right) \mathsf{d}_{\mathsf{x}_i}\mathsf{a}_i(\{\rho_i\}_{\mathsf{x}_i}(\mathsf{a}'_i(\mathsf{x}_i))) + \{\mathsf{C}_i\}_{\mathsf{x}_i}(\mathsf{a}_i(\mathsf{x}_i), \mathsf{a}'_i(\mathsf{x}_i)).$$

and so for $[\mathfrak{a}, \mathfrak{a}']_{\mathcal{A}_U}$ *we obtain*

$$[\mathfrak{a}, \mathfrak{a}']_{\mathcal{A}_U}(x) \equiv \mathsf{d}_{\mathsf{x}}\mathsf{a}'(\rho_{\mathsf{x}}(\mathsf{a}(\mathsf{x})) - \mathsf{d}_{\mathsf{x}}\mathsf{a}(\rho_{\mathsf{x}}(\mathsf{a}'(\mathsf{x}))) + \mathsf{C}_{\mathsf{x}}(\mathsf{a}(\mathsf{x}), \mathsf{a}'(\mathsf{x})) \tag{5.9}$$

where $\mathsf{C}_{\mathsf{x}} = \underrightarrow{\lim}\{\mathsf{C}_i\}_{\mathsf{x}_i}$ *is a smooth map from* U *to* $L^2_{alt}(\mathbb{A}; \mathbb{A})$.

Fix some sections $\mathfrak{a}$ *and* $\mathfrak{a}'$ *in* $\mathfrak{P}(\mathcal{A}_U)$. *There exists sections* $\mathfrak{b}_1, \dots \mathfrak{b}_n$ *and* $\mathfrak{b}'_1, \dots \mathfrak{b}'_m$*, each one being a direct limit of sections of* $\{\mathcal{A}_i\}_{U_i}$ *and smooth functions* $f_1, \dots, f_n$, $g'_1, \dots, g'_m$ *in* $C^\infty(U)$ *such that*

$$\mathfrak{a} = \sum_{l=1}^{n} f_l \mathfrak{b}_l, \quad \mathfrak{a}' = \sum_{k=1}^{m} f'_k \mathfrak{b}'_k \tag{5.10}$$

Then we set

$$[\mathfrak{a}, \mathfrak{a}']_{\mathcal{A}_U} = \sum_{1 \leq l \leq n, 1 \leq k \leq m} [f_l \mathfrak{b}_l, f'_k \mathfrak{b}'_k]_{\mathcal{A}_U} \tag{5.11}$$

We will show that the equation (5.11) defines an almost Lie bracket on $\mathfrak{P}(\mathcal{A}_U)$.

Indeed, according to (4.40) and (5.9), the expression of (5.11) in local coordinates is:

$$\begin{aligned}
([\mathfrak{a}, \mathfrak{a}']_{\mathcal{A}_U})(x) &= \sum_{1 \leq l \leq n, 1 \leq k \leq m} ([f_l \mathfrak{b}_l, f'_k \mathfrak{b}_k]_{\mathcal{A}_U})(x) \\
&\equiv \sum_{1 \leq l \leq n, 1 \leq k \leq m} \mathsf{C}(\mathsf{f}_l \mathsf{b}_l, \mathsf{f}'_k \mathsf{b}'_k) + \mathsf{f}_l \mathsf{d}\mathsf{f}'_k(\rho_{\mathsf{x}}(\mathsf{b}_l))\mathsf{b}'_k + \mathsf{f}_l \mathsf{f}_k \mathsf{d}\mathsf{b}'_k((\rho_{\mathsf{x}})(\mathsf{b}_l)) \\
&\quad - \mathsf{f}'_k \mathsf{d}\mathsf{f}_l(\rho_{\mathsf{x}}(\mathsf{b}'_k))\mathsf{b}_l + \mathsf{f}_l \mathsf{f}'_k \mathsf{d}\mathsf{b}_l((\rho_{\mathsf{x}}(\mathsf{b}'_k)) \\
&= \left(\mathsf{d}\mathsf{a}'(\rho(\mathsf{a})) - \mathsf{d}\mathsf{a}(\rho(\mathsf{a}')) + \mathsf{C}(\mathsf{a}, \mathsf{a}')\right)(x).
\end{aligned} \tag{5.12}$$

The last member of the equation (5.12) implies that $[\mathfrak{a}, \mathfrak{a}']_{\mathcal{A}_U}$ *only depends on the* 1*-jets of* $\mathfrak{a}$ *and* $\mathfrak{a}'$. *Moreover, if the germ at* x *of* $\mathfrak{a}'$ *is identically zero, this last member implies that the germ at* x *of* $[\mathfrak{a}, \mathfrak{a}']_{\mathcal{A}_U}$ *is also zero. In particular, the germ* $[\mathfrak{a}, \mathfrak{a}']_{\mathcal{A}_U}(x)$ *does not depend on the decompositions (5.10). This means that, for any open set* U *and any* $\mathfrak{a}$ *and* $\mathfrak{a}'$ *in* $\mathfrak{P}(\mathcal{A}_U)$, *we have built a well-defined skew-symmetric bilinear form on* $\mathfrak{P}(\mathcal{A}_U)$ *with values in* $\mathfrak{P}(\mathcal{A}_U)$ *which depends on the* 1*-jets of sections in* $\mathfrak{P}(\mathcal{A}_U)$. *Moreover, again from (4.40) and (5.11), it follows that* $[.,.]_{\mathcal{A}_U}$ *satisfies the Leibniz property and so (5.11) defines an almost Lie bracket on* $\mathfrak{P}(\mathcal{A}_U)$.

We will show that the almost Lie bracket on $\mathfrak{P}(\mathcal{A}_U)$ is in fact a Lie bracket.

For all $i \in \mathbb{N}$, *we have* $[\rho_i(\mathfrak{a}_i), \rho_i(\mathfrak{a}'_i)] = \rho_i([\mathfrak{a}_i, \mathfrak{a}'_i]_{\mathcal{A}_i})$. *According to the compatibility with bonding maps of the Lie brackets* $[.,.]$ *on* M_i, $[.,.]_{\mathcal{A}_i}$ *on* $\mathcal{A}_i$ *and* ρ_i *we have*

$T\varepsilon_i^j([\rho_i(\mathfrak{a}_i), \rho_i(\mathfrak{a}'_i)]) - \rho_i([\mathfrak{a}_i, \mathfrak{a}'_i]_{\mathcal{A}_i}) = ([\rho_j(\mathfrak{a}_j), \rho_j(\mathfrak{a}'_j)] - \rho_j([\mathfrak{a}_j, \mathfrak{a}'_j]_{\mathcal{A}_j}) \circ \varepsilon_i^j$.
It follows that if $\mathfrak{a} = \varinjlim \mathfrak{a}_i$ *and* $\mathfrak{a}' = \varinjlim \mathfrak{a}'_i$ *then* $\rho(\mathfrak{a}), \rho(\mathfrak{a}') = \rho([\mathfrak{a}, \mathfrak{a}']_{\mathcal{A}_U})$. *Now, for* $f \in C^\infty(U)$, *we also have* $[\rho(\mathfrak{a}), \rho(f\mathfrak{a}')] = \rho([\mathfrak{a}, f\mathfrak{a}']_{\mathcal{A}_i}) = f\rho([s, s']_{\mathcal{A}_i})$ *since* ρ *is a bundle morphism. We then get*

$[\rho(\mathfrak{a}), \rho(\mathfrak{a}')] = \rho([\mathfrak{a}, \mathfrak{a}']_{\mathcal{A}_U})$.

Finally, we consider the alternating $\mathbb{A}$*-trilinear map* $J_{\mathcal{A}_U} : \mathfrak{P}(\mathcal{A}_U)^3 \to \mathfrak{P}(\mathcal{A}_U)$ *given by*

$$J_{\mathcal{A}_U}(\mathfrak{a}, \mathfrak{a}', \mathfrak{a}") = [\mathfrak{a}, [\mathfrak{a}', \mathfrak{a}"]_{\mathcal{A}_U}]_{\mathcal{A}_U} + [\mathfrak{a}', [\mathfrak{a}", \mathfrak{a}]_{\mathcal{A}_U}]_{\mathcal{A}_U} + [\mathfrak{a}", [\mathfrak{a}, \mathfrak{a}']_{\mathcal{A}_U}]_{\mathcal{A}_U}.$$

By relation (3.18) in the proof of Proposition 3.19, we have $J_{\mathcal{A}_U}(\mathfrak{a}, \mathfrak{a}', f\mathfrak{a}'') = f J_{\mathcal{A}_U}(\mathfrak{a}, \mathfrak{a}', \mathfrak{a}'')$*. Now since the Jacobiator* $J_{\mathcal{A}_i}$ *is identically zero for each* $i \in \mathbb{N}$*, it follows that if* $\mathfrak{a} = \varprojlim \mathfrak{b}_i$*,* $\mathfrak{a}' = \varprojlim \mathfrak{b}'_i$ *and* $\mathfrak{a}'' = \varprojlim \mathfrak{b}''_i$*, we have* $J_{\mathcal{A}_U}(\mathfrak{a}, \mathfrak{a}', f\mathfrak{a}'') = f J_{\mathcal{A}_U}(\mathfrak{a}, \mathfrak{a}', \mathfrak{a}'') = 0$ *for any* $f \in C^\infty(U)$*. This clearly implies that* $J_{\mathcal{A}_U}$ *is identically zero on* $\mathfrak{P}(\mathcal{A}_U)$*, which ends the proof of Point (1).*

(2) Since (2) is a local problem, we can consider it in local coordinates. In this way, any section of $\mathfrak{P}(\mathcal{A}_U)$ *is a smooth map from an open set* U *in the Fréchet space* $\mathbb{M}$ *to the Fréchet space* $\mathbb{E}$*. Thus,* $\mathsf{U} = \varinjlim \mathsf{U}_i$ *where each* U_i *is open in* $\mathbb{M}_i$*. Now, if* (x, v) *belongs to* $\{\mathsf{x}\} \times \mathbb{E}$*, then* $\mathsf{v} = \varinjlim \mathsf{v}_\mathsf{i}$ *where each* v_i *belongs to* $\mathbb{E}_i$*. This implies that the constant map* $\mathsf{a}(\mathsf{z}) = \mathsf{v}$ *is the projective limit of the sequence of constant maps* a_i *on* U_i*, that is* $\mathsf{a}_i(\mathsf{z}_i) = \mathsf{v}_i$*, which ends the proof.*

5.20 Notes

The notion of direct limit of categories is exposed in [McL98] (for sets see [Bou70]).
For direct limits of topological spaces the references are [RoRo64], [Bou71], [Han71], [HSTH01] and [Glo05].
For the setting of l.c.t.v.s. the reader is referred to [RoRo64], Chapter V where the finest locally convex topology on the direct limit of such spaces is described. In the particular case of direct limit of normed spaces, the reference is [Dah11].
Interesting results about direct limits of Lie groups were obtained by Glöckner in various papers: directs limits of finite-dimensional Lie groups in [Glo03] and [Glo05] and infinite dimensional in [Glo11] (see also [Nee05], [Dah11] and [Nee06] where one can find an interesting historical panorama about the Lie theory of locally convex groups). The property of completeness of infinite-dimensional Lie groups is studied in [Glo18] and the regularity is developed in [Glo19].
In [CaPe19], Theorem 4.8 gives an essential result about the structure of not necessarily Hausdorff convenient manifold for a direct limit of an ascending sequence of C^∞ Banach manifolds having the direct limit chart property and for which the direct limit of the modelled Banach spaces is LB. In this paper, considering the direct limit $\mathbb{E} = \varinjlim \mathbb{E}_n$ of Banach vector spaces for which the

group $\mathrm{GL}(\mathbb{E})$ does not admit any interesting topological structure, is introduced the convenient Lie group $\mathrm{G}(\mathbb{E})$. So using $\mathrm{G}(\mathbb{E})$ as structural group, the direct limit of a strong ascending sequence of Banach vector bundles can be endowed with a convenient structure. Some analogous results are obtained for direct limits of various geometrical Banach structures (principal bundles, frame bundles, G-structures, tensor structures and Lie algebroids).
Essential results about direct limits of tensor structures can be found in [CaPe20].
The main reference for the study of symplectic forms on direct limits of ascending sequences of Banach manifolds is the article [Pel18] where one can find a discussion about the conditions of existence of Darboux charts.

6

Some Properties of Convenient Lie Algebroids and Prolongations

In classical mechanics, the configuration space is a finite-dimensional smooth manifold M where the tangent bundle $p_M : TM \to M$ corresponds to the velocity space (cf. [CaMa01]). This geometrical object plays a relevant rôle in the Lagrangian formalism between tangent and cotangent bundles (cf. [Kle62]). In [Wei71], Weinstein develops a generalized theory of Lagrangian Mechanics on Lie algebroids. In [Lib55], Libermann shows that such a formalism is not possible, in general, if we consider the tangent bundle of a Lie algebroid. The notion of prolongation of a Lie algebroid introduced by Higgins and Mackenzie [HiMa90] offers a nice context in which such a formalism was generalized by Martínez (cf. [Mar01a] and [Mar01b]).
In this chapter, the processus of prolongation, along a convenient fibred manifold $\mathbf{p} : \mathcal{M} \to M$, for an anchored convenient bundle $(\mathcal{A}, \pi, M, \rho)$ is generalized and one obtains a convenient anchored bundle $(\mathbf{T}\mathcal{M}, \hat{\mathbf{p}}, \mathcal{M}, \hat{\rho})$.
Unfortunately, even in the Banach setting, if $(\mathcal{A}, \pi, M, \rho, [.,.]_{\mathcal{A}})$ is a convenient Lie algebroid, the prolongation of the Lie bracket can be defined only on "projectable sections".
However, Lie derivative, exterior differential and Nijenhuis tensor are well defined on $\mathbf{T}\mathcal{M}$.

6.1 $\mathcal{A}$-Connections on a Convenient Bundle

By analogy with a Koszul connection, we define the notion of $\mathcal{A}$-connection.

Definition 6.1 *Let $(\mathcal{A}, \pi, M, \rho)$ be an anchored convenient bundle. An $\mathcal{A}$-connection on a convenient bundle $\mathbf{p} : \mathcal{E} \to M$ is a $\mathbb{R}$-bilinear map $\nabla : \Gamma(\mathcal{A}) \times \Gamma(\mathcal{E}) \to \Gamma(\mathcal{E})$ which fulfils, for any function f on M, $\mathfrak{a} \in \Gamma(\mathcal{A})$ and $\mathfrak{b} \in \Gamma(\mathcal{E})$, the following properties:*

(AC1) $\nabla_{\mathfrak{a}}(f\mathfrak{b}) = df(\rho(\mathfrak{a}))\mathfrak{b} + f\nabla_{\mathfrak{a}}\mathfrak{b}$;

(AC2) $\nabla_{f\mathfrak{a}}\mathfrak{b} = f\nabla_{\mathfrak{a}}\mathfrak{b}$.

DOI: 10.1201/9781003435587-6

As for a Koszul connection, *we also must assume that such operator* ∇ *is localizable* (cf. [CaPe19], 5.2). Therefore, locally, we can associate to a $\mathcal{A}$-connection on $\mathcal{A}$ its Christoffel components whose expression is left to the reader.
Now, if we consider two $\mathcal{A}$-connections ∇ and ∇' on $\mathcal{A}$ then $\nabla - \nabla'$ is a tensor of type $(2,1)$ on $\mathcal{A}$.
Assume that there exists a Koszul connection ∇ on $\mathcal{A}$, then we get a $\mathcal{A}$-connection $\bar{\nabla}$ defined by

$$\overline{\nabla}_{\mathfrak{a}_1}\mathfrak{a}_2 = \nabla_{\rho(\mathfrak{a}_1)}\mathfrak{a}_2.$$

Therefore if there exists a Koszul connection on $\mathcal{A}$ (cf. [CaPe19], 5.2) there also exists a $\mathcal{A}$-connection on $\mathcal{A}$.

Remark 6.1

(1) If ∇ *is a Koszul connection on* TM*, we can associate an* $\mathcal{A}$*-connection on* $\mathcal{A}$*:*

$$\nabla'_{\mathfrak{a}_1}\mathfrak{a}_2 = \nabla_{\rho(\mathfrak{a}_2)}\mathfrak{a}_1 + [\mathfrak{a}_1, \mathfrak{a}_2]_{\mathcal{A}}$$

which, of course, is different from the $\mathcal{A}$*-connection* $\overline{\nabla}_{\mathfrak{a}_1}\mathfrak{a}_2 = \nabla_{\rho(\mathfrak{a}_1)}\mathfrak{a}_2$.

(2) If ∇ *is a Koszul connection on* $\mathcal{A}$ *we can associate an* $\mathcal{A}$*-connection on* TM*:*

$$\nabla''_{\mathfrak{a}}X = \rho(\nabla_X \mathfrak{a}) + [\rho(\mathfrak{a}), X].$$

(3) If ∇ *is a Koszul connection on a Banach-Lie algebroid* $(\mathcal{A}, \pi, M, \rho, [.,.]_{\mathcal{A}})$*, then we have*

$$\nabla''_{\mathfrak{a}_1}\rho(\mathfrak{a}_2) = \rho(\nabla'_{\mathfrak{a}_1}\mathfrak{a}_2).$$

(4) Let ∇ *and* ∇' *be two* $\mathcal{A}$*-connections on* $\mathcal{A}$*. Then* $\nabla - \nabla'$ *is a tensor of type* $(1,1)$ *on* $\mathcal{A}$*.*
Conversely, given any tensor $\mathfrak{T}$ *of type* $(1,1)$ *on* $\mathcal{A}$*, for any* $\mathcal{A}$*-connection* ∇*,* $\nabla' = \nabla + \mathfrak{T}$ *is a* $\mathcal{A}$*-connection.*

As in finite dimension, if $(\mathcal{A}, \pi, M, \rho, [.,.]_{\mathcal{A}})$ is a Banach-Lie algebroid, the torsion T_∇ of the $\mathcal{A}$-connection ∇ on $\mathcal{A}$ is defined by

$$T_\nabla(\mathfrak{a}_1, \mathfrak{a}_2) = \nabla_{\mathfrak{a}_1}\mathfrak{a}_2 - \nabla_{\mathfrak{a}_2}\mathfrak{a}_1 - [\mathfrak{a}_1, \mathfrak{a}_2]_{\mathcal{A}}.$$

Proposition 6.1 *Assume that there exists an* $\mathcal{A}$*-connection on a Banach-Lie algebroid* $(\mathcal{A}, \pi, M, \rho, [.,.]_{\mathcal{A}})$*. Then there exists an* $\mathcal{A}$*-connection* ∇ *on* $\mathcal{A}$ *with zero torsion:*

$$\nabla_{\mathfrak{a}_1}\mathfrak{a}_2 - \nabla_{\mathfrak{a}_2}\mathfrak{a}_1 = [\mathfrak{a}_1, \mathfrak{a}_2]_{\mathcal{A}}.$$

Proof *Let* $\bar{\nabla}$ *be an* $\mathcal{A}$*-connection and* $\bar{T}$ *its torsion. Then* $\nabla = \bar{\nabla} + \dfrac{1}{2}\bar{T}$ *is an* $\mathcal{A}$*-connection on* $\mathcal{A}$ *which has the required property.*

Remark 6.2 *Proposition 6.1 implies that if there exists an* $\mathcal{A}$*-connection on Banach-Lie algebroid* $(\mathcal{A}, \pi, M, \rho, [.,.]_{\mathcal{A}})$*, then the Lie bracket* $[.,.]_{\mathcal{A}}$ *can be defined by such an* $\mathcal{A}$*-connection. The converse is not true: for instance the Lie bracket of vector fields* $[.,.]$ *on a Banach manifold* M *provides a Banach-Lie algebroid structure* $(TM, \pi, M, \mathrm{Id}, [.,.])$ *on* TM *but there can exist no linear connection on* TM*.*

6.2 Foliations and Banach-Lie Algebroids

We first recall the classical notion of integrability of a distribution on a Banach manifold M (cf. [Pel12]; see also Chapter 8).

(1) A *distribution* Δ on the Banach manifold M is an assignment $\Delta : x \mapsto \Delta_x \subset T_x M$ on M where Δ_x is a subspace of $T_x M$. The distribution Δ is called closed if Δ_x is closed in $T_x M$ for all $x \in M$.

(2) A vector field X on M, defined on an open set $\mathrm{Dom}(X)$, is called *tangent to a distribution* Δ if $X(x)$ belongs to Δ_x for all $x \in \mathrm{Dom}(X)$.

(3) Let X be a vector field tangent to a distribution Δ and Fl_t^X its flow. We say that Δ is X-invariant if $T_x \,\mathrm{Fl}_t^X(\Delta_x) = \Delta_{\mathrm{Fl}_t^X(x)}$ for all t for which $\mathrm{Fl}_t^X(x)$ is defined.

(4) A distribution Δ on M is called integrable if, for all $x_0 \in M$, there exists a weak submanifold (N, ϕ) of M such that $\phi(y_0) = x_0$ for some $y_0 \in N$ and $T\phi(T_y N) = \Delta_{\phi(y)}$ for all $y \in N$. In this case (N, ϕ) is called an integral manifold of Δ through x. A leaf L is a weak submanifold which is a maximal integral manifold.

(5) A distribution Δ is called involutive if for any vector fields X and Y on M tangent to Δ the Lie bracket $[X, Y]$ defined on $\mathrm{Dom}(X) \cap \mathrm{Dom}(Y)$ is tangent to Δ.

Classically, in the Banach context, when Δ is a supplemented subbundle of TM, according to the Frobenius Theorem, involutivity implies integrability.

In finite dimension, the famous results of H. Sussman and P. Stefan give necessary and sufficient conditions for the integrability of smooth distributions.

Few generalizations of these results in the framework of Banach manifolds can be found in [ChSt76]. In the context of this section, we have (cf. Chapter 8, Theorem 8.5):

Theorem 6.1 *Let* $(\mathcal{A}, \pi, M, \rho, [.,.]_{\mathcal{A}})$ *be a split Banach-Lie algebroid. If* $\rho(\mathcal{A})$ *is a closed distribution, then this distribution is integrable.*

Note that if ρ is a Fredholm morphism, the assumptions of Theorem 6.1 are always satisfied. In the Hilbert framework, only the closeness of ρ is required.

6.3 Pseudo-Riemannian and Riemannian Structures on a Banach-Lie Algebroid

In this section, $(\mathcal{A}, \pi, M, \rho, [.,.]_{\mathcal{A}})$ is a **fixed Banach-Lie algebroid** and let $\pi^* : \mathcal{A}^* \to M$ be the dual bundle of $\pi : \mathcal{A} \to M$.

6.3.1 Pseudo-Riemannian structures and connections

A field g of symmetric bilinear forms g_x over $\mathcal{A}_x$ is a *pseudo-Riemannian metric* if the map $g^\flat : \mathcal{A} \to \mathcal{A}^*$ given by $g^\flat(\mathfrak{u}) = g(\mathfrak{u}, \ )$ is an injective morphism of bundles. In general, if $\mathbb{A}$ is not a Banach space, then $g^\flat$ need not be an isomorphism. When $g^\flat$ is an isomorphism, we say that g is a *strong pseudo-Riemannian metric* on $\mathcal{A}$. In this case, $\mathbb{A}$ must be reflexive. When $g^\flat$ is not surjective, we say that g is a *weak pseudo-Riemannian metric*. If, moreover, each g_x defines an inner product on each fibre $\mathcal{A}_x$, we will say that g is a *strong (resp. weak) Riemaniann metric* if $g^\flat$ is surjective (resp. non-surjective). In this situation, $\mathbb{A}$ can be provided with a Hilbert norm (resp. pre-Hilbert norm).

Definition 6.2 *We say that an $\mathcal{A}$-connection ∇ on $\mathcal{A}$ is a metric connection if there exists a weak pseudo-Riemannian metric g on $\mathcal{A}$ such that*

$$L^\rho_{\mathfrak{a}} g(\mathfrak{u},\mathfrak{v}) = g(\nabla_{\mathfrak{a}}\mathfrak{u},\mathfrak{v}) + g(\nabla_{\mathfrak{a}}\mathfrak{v},\mathfrak{u}) \tag{6.1}$$

for all (local) sections $\mathfrak{a}$, $\mathfrak{u}$ and $\mathfrak{v}$ of $\mathcal{A}$. We also say that ∇ is a g-metric connection.

Remark 6.3

(1) In fact a Koszul connection ∇ on $\mathcal{A}$, gives rise to an $\mathcal{A}$-connection $\bar{\nabla}$ defined by
$\bar{\nabla}_{\mathfrak{u}}\mathfrak{v} = \nabla_{\rho(\mathfrak{u})}\mathfrak{v}$. If the relation (6.1) applied to $\bar{\nabla}$ is satisfied, we say that ∇ is a g-metric Koszul connection.

(2) As classically in finite dimension, we can extend an $\mathcal{A}$ connection to the set of all tensors on $\mathcal{A}$. In particular, ∇g is given by:

$$\nabla_{\mathfrak{a}}\left(g(\mathfrak{u},\mathfrak{v})\right) = L^\rho_{\mathfrak{a}} g(\mathfrak{u},\mathfrak{v}) - g(\nabla_{\mathfrak{a}}\mathfrak{u},\mathfrak{v}) - g(\nabla_{\mathfrak{a}}\mathfrak{v},\mathfrak{u}).$$

Thus ∇ is g-metric if and only if $\nabla g \equiv 0$.

(3) Recall that, as in finite dimension, given any curve $\gamma : [0,T] \to M$, the parallel transport along γ is defined in the following way: a vertical section $\mathfrak{u}$ along γ is parallel if

$$\forall s \in [0,T],\ \nabla_{\dot{\gamma}}(\mathfrak{u})(\gamma(s)) = 0.$$

Now, for any u in the fibre $\mathcal{A}_{\gamma(0)}$, there exists a unique parallel section $\mathfrak{u}$ along γ such that $\mathfrak{u}(\gamma(0)) = u$. We then get an isomorphism τ_s from $\mathcal{A}_{\gamma(0)}$ to $\mathcal{A}_{\gamma(s)}$.
Then, the property $\nabla g \equiv 0$ means that τ_s is an isometry for any $s \in [0,T]$ and any curve γ in $\mathcal{M}$.

Assume that g is a strong pseudo-Riemannian metric on $\mathcal{A}$, then $\mathcal{A}^*$ is also provided with a strong pseudo-Riemannian metric g^* *via* the isomorphism $g^\flat$. Moreover, as in finite dimension, the following relation:

$$2g(\nabla_{\mathfrak{a}}\mathfrak{b},\mathfrak{c}) = L^\rho_{\mathfrak{a}}g(\mathfrak{b},\mathfrak{c}) + L^\rho_{\mathfrak{b}}g(\mathfrak{a},\mathfrak{c}) - L^\rho_{\mathfrak{c}}g(\mathfrak{a},\mathfrak{b}) \\ +g([\mathfrak{c},\mathfrak{a}]_{\mathcal{A}},\mathfrak{b}) + g([\mathfrak{c},\mathfrak{b}]_{\mathcal{A}},\mathfrak{a}) + g([\mathfrak{a},\mathfrak{b}]_{\mathcal{A}},\mathfrak{c}). \tag{6.2}$$

defines an $\mathcal{A}$-connection on $\mathcal{A}$ which is g-metric and without torsion. As usually this connection is called the *Levi-Civita connection.*

Assume now that g is a weak pseudo-Riemannian metric on a Banach-Lie algebroid. Then of course $g^\flat(\mathcal{A})$ is a Banach bundle over M which is (strictly) contained in $\mathcal{A}^*$. Unfortunately, in general, there is no Levi-Cevita connection associated with g. However, in many examples, such a connection exists even if g is only weak. For a complete panorama of such Examples, the reader can see [Bil04] and references inside.

6.3.2 Strong Riemannian Metrics

In the particular case where g is a strong Riemannian metric on $\mathcal{A}$, we have the following complementary result:

Theorem 6.2 *Assume that there exists a strong Riemannian metric g on the Banach-Lie algebroid $(\mathcal{A},\pi,M,\rho,[.,.]_{\mathcal{A}})$. Then we have the following properties:*

(1) For each $x\in M$, the g-orthogonal $(\ker\rho_L)^\perp_x$ of $\ker\rho_x$ is a supplemented Banach space of $\ker\rho_x$ and so $(\mathcal{A},\pi,M,\rho,[.,.]_{\mathcal{A}})$ is split.

(2) If, moreover, the distribution $\rho(\mathcal{A})$ is closed, then $(\mathcal{A},\pi,M,\rho,[.,.]_{\mathcal{A}})$ is integrable. The pull-back $\mathcal{A}_L$ of $\mathcal{A}$ to a leaf L has a natural Lie algebroid structure $(\mathcal{A}_L,\pi_{|L},L,\rho_L,[.,.]_{\mathcal{A}_L})$ induced by $(\mathcal{A},\pi,M,\rho,[.,.]_{\mathcal{A}})$ and we have a decomposition $\mathcal{A}_L$ in a Whitney sum of Banach bundles

$$\mathcal{A}_L = \ker\rho_L \oplus (\ker\rho_L)^\perp$$

with the following properties:

(a) the bracket of $[.,.]_{\mathcal{A}_L}$ induces a Lie bracket $[.,.]_{\ker\rho_L}$ on $\ker\rho_L$ and then $(\ker\rho_L,L,[.,.]_{\ker\rho_L})$ is a Lie algebra bundle over L;

(b) the orthogonal projection on $(\ker\rho_L)^\perp$ of the bracket of (local) sections of $(\ker\rho_L)^\perp$ defines a Lie bracket $[.,.]_L$ on $(\ker\rho_L)^\perp$ and then $((\ker\rho_L)^\perp,L,(\rho_L)_{|(\ker\rho_L)^\perp},[.,.]_L)$ is a Banach-Lie algebroid;

(c) $(\rho_L)_{|(\ker\rho_L)^\perp}$ is an algebroid isomorphism. In particular, g induces a canonical strong Riemannian metric on L.

Definition 6.3 *Under the assumption of Theorem 6.2 (2), the Banach-Lie algebroid* $(\mathcal{A}_L, \pi_{|L}, L, \rho_L, [.,.]_{\mathcal{A}_L})$ *will be called the restriction of* $(\mathcal{A}, M, \rho, [.,.]_{\mathcal{A}})$ *to* L.

Proof *(1) Consider a fibre* $\mathcal{A}_x$ *and choose any norm* $|| \ ||_x$ *on it. Now since* g_x *is a bounded bilinear form, there exists a constant* C *such that*

$$g_x(a,a) \leq C||a||_x^2.$$

Therefore, we have a pre-Hilbert structure on $\mathcal{A}_x$ *and we denote by* $\widetilde{\mathcal{A}}_x$ *the associated Hilbert space. Since* $\ker \rho_x$ *is closed in the Banach space* $(\mathcal{A}_x, || \ ||_x)$*, it is also a closed subspace of the pre-Hilbert space* $\mathcal{A}_x$*. Let* H_x *be the orthogonal of* $\ker \rho_x$ *in* $\widetilde{\mathcal{A}}_x$ *and set* $(\ker \rho_x)^{\perp} = H_x \cap \mathcal{A}_x$*. In fact,* $(\ker \rho_x)^{\perp}$ *is closed in* $(\mathcal{A}_x, || \ ||_x)$ *and we have* $\mathcal{A}_x = \ker \rho_x \oplus (\ker \rho_x)^{\perp}$*. Of course,* $(\ker \rho_x)^{\perp}$ *is the orthogonal of* $\ker \rho_x$ *according to* g_x*, which ends the proof of (1).*

(2) From Theorem 6.1, the distribution $\rho(\mathcal{A})$ *is integrable. Let* L *be a leaf of this foliation. Since* L *is a closed submanifold of* M*, if* $\iota : L \to M$ *is the inclusion, the pull-back* $\mathcal{A}_L$ *of* $\mathcal{A}$ *is a Banach bundle over* L*. Now, if* $\bar{\iota} : \mathcal{A}_L \to \mathcal{A}$ *is the canonical morphism over* ι*, we set* $\rho_L = \rho \circ \bar{\iota}$*. Since* $\rho(\mathcal{A}_x) = T_xL$ *for any* $x \in L$*, we get an anchored bundle* $(\mathcal{A}_L, L, \rho_L)$*. Finally, since each (local) section* $\mathfrak{u}$ *of* $\mathcal{A}_L$ *is of type* $\tilde{\mathfrak{u}} \circ \iota$ *where* $\tilde{\mathfrak{u}}$ *is a (local) section of* $\mathcal{A}$*, it follows that* $[.,.]_{\mathcal{A}}$ *induces a Lie bracket* $[.,.]_{\mathcal{A}_L}$ *on* $\mathcal{A}_L$*. Now the anchor* ρ_L *is surjective and the Banach-Lie algebroid* $(\mathcal{A}_L, \pi_{|L}L, \rho_L, [.,.]_{\mathcal{A}_L})$ *is split. It follows that this algebroid is transitive according to [BGJP19], 6.3, and so we have the announced Whitney decomposition and property (a).*

Clearly, the composition of $[.,.]_{\mathcal{A}_L}$ *with the orthogonal projection on* $(\ker \rho_L)^{\perp}$ *defines a Lie bracket on the Banach anchored bundle* $((\ker \rho_L)^{\perp}, L, (\rho_L)_{|(\ker \rho_L)^{\perp}})$*. This ends the proof of affirmation (b) since* $(\rho_L)_{|(\ker \rho_L)^{\perp}}$ *is an isomorphism.*

6.3.3 A Characteristic Subbundle of $T^*\mathcal{A}^*$

Since this result will be used in the next chapter, in this section we consider the general situation of a convenient bundle $\mathcal{A}$ and s, as usual, its dual bundle will be denoted $\pi' : \mathcal{A}' \to M$[1]. Fix some open U set in M and denote by $C^\infty(\mathcal{A}'_U)$ the set of smooth functions on $\mathcal{A}'_U$. Any $f \in C^\infty(\mathcal{A}'_U)$ is called *linear* if the restriction of f to each fibre $\mathcal{A}'_x = \pi'^{-1}(x)$ is a linear map.

First of all, for any section $\mathfrak{a} \in \Gamma(\mathcal{A}_U)$, we associate the linear function $\Phi_{\mathfrak{a}}$ on $\mathcal{A}'_U$ defined by

$$\Phi_{\mathfrak{a}}(\sigma) = < \eta, \mathfrak{a} \circ \pi'(\sigma) > .$$

We then have the following properties ([CaPe12]):

[1]By convention, in the convenient setting, $\mathcal{A}'$ is the (bounded) dual bundle of $\mathcal{A}$ and in the Banach setting $\mathcal{A}'$ and $\mathcal{A}^*$ are equal. The same is true for $T'\mathcal{A}'$ and $T^*\mathcal{A}^*$.

Lemma 6.1

(1) for each local section $\mathfrak{a}$ of $\mathcal{A}_U$, the map $\Phi_{\mathfrak{a}}$ is a smooth map on $\mathcal{A}'_U$ and the map $\mathfrak{a} \mapsto \Phi_{\mathfrak{a}}$ is linear and injective; we also have

$$\forall \mathfrak{a} \in \Gamma(\mathcal{A}_U), \forall f \in C^\infty(U),\ \Phi_{f\mathfrak{a}} = (f \circ \pi')\Phi_{\mathfrak{a}}$$

(2) In a local trivialization (cf. Chapter 3, § 3.15.3), we have

$$< d\Phi_{\mathfrak{a}}(\sigma), \eta_2 >=< \mathfrak{a} \circ \pi'(\sigma), \eta_2 >$$

for any $\eta_2 \in \mathcal{A}'$, considered as vertical fibre $T_\sigma \mathcal{A}'$;.

(3) If for some $f \in C^\infty(U)$, we have $d_\sigma(\Phi_{\mathfrak{a}} + f \circ \pi') = 0$ then $\mathfrak{a} \circ \pi'(\eta) = 0$

Proposition 6.2 *The set*

$$T^\flat \mathcal{A}' = \bigcup_{\sigma \in \mathcal{A}'} \{d_\sigma(\phi_{\mathfrak{a}} + f \circ \pi'),\ \ \mathfrak{a} \in \Gamma(\mathcal{A}_U),\ f \in C^\infty(U)\}$$

(where U is any chart domain around x) is a well-defined closed submanifold of $T'\mathcal{A}'$ and if $q^\flat$ is the restriction of $q_{\mathcal{A}'}$ to $T^\flat \mathcal{A}'$ then

$$q^\flat : T^\flat \mathcal{A}' \to \mathcal{A}'$$

is a closed convenient subbundle of typical fibre $\mathbb{M}' \times \mathbb{A}$. In particular, $T^\flat \mathcal{A}' = T'\mathcal{A}'$ if and only if $\mathbb{A}$ is reflexive.

6.4 Prolongation of a Convenient Lie Algebroid along a Fibration

6.4.1 Prolongation of a Convenient Anchored Bundle

Let $\mathbf{p} : \mathcal{E} \to M$ be a convenient vector bundle with typical fibre $\mathbb{E}$ and $\mathcal{M}$ an open submanifold such that the restriction of $\mathbf{p}$ to $\mathcal{M}$ is a surjective fibration over M of typical fibre $\mathbb{O}$ (open subset of $\mathbb{E}$). We consider some anchored convenient bundle $(\mathcal{A}, \pi, M, \rho)$.

Notation 6.1 *If U is an open set in M, we denote by $\mathcal{E}_U$, $\mathcal{M}_U$ and $\mathcal{A}_U$ the restriction of $\mathcal{E}$, $\mathcal{M}$ and $\mathcal{A}$ to U respectively. In particular, if (U, ϕ) is a chart such that $\mathcal{E}_U$ and $\mathcal{A}_U$ are trivializable, then $T M_U = T\mathcal{M}_U$ and $T\mathcal{M}_U$ are also trivializable. In this case, as in Chapter 3, § 3.15.3, we have the following trivializations and local coordinates:*

- $\mathcal{E}_U \equiv U \times \mathbb{E}$ *and* $\mathcal{M}_U \equiv U \times \mathbb{O}$ *with local coordinates* $\mathsf{m} = (\mathsf{x}, \mathsf{e})$;
- $T\mathcal{M}_U \equiv (U \times \mathbb{O}) \times \mathbb{M} \times \mathbb{E}$ *with local coordinates* $(\mathsf{m}, \mathsf{v}, \mathsf{z})$.

For $m \in \mathbf{p}^{-1}(x)$ we set

$$\mathbf{T}^{\mathcal{A}}_m \mathcal{M} = \{(a, \mu) \in \mathcal{A}_x \times T_m\mathcal{M} : \ \rho(a) = T\mathbf{p}(\mu)\}.$$

An element of $\mathbf{T}^{\mathcal{A}}_m \mathcal{M}$ will be denoted (m, a, μ).
We set $\mathbf{T}^{\mathcal{A}}\mathcal{M} = \bigcup\limits_{m \in \mathcal{M}} \mathbf{T}^{\mathcal{A}}_m \mathcal{M}$ and we consider the projection $\hat{\mathbf{p}} : \mathbf{T}^{\mathcal{A}}\mathcal{M} \to \mathcal{M}$ defined by $\hat{\mathbf{p}}(m, a, \mu) = m$.
We introduce the following context:

(1) Let $\widetilde{\pi} : \widetilde{\mathcal{A}} \to \mathcal{M}$ be the pull-back of the bundle $\pi : \mathcal{A} \to M$ by $\mathbf{p} : \mathcal{M} \to M$. We denote by $\bar{\mathbf{p}}$ the canonical vector bundle such that the following diagram is commutative:

$$\begin{array}{ccc} \widetilde{\mathcal{A}} & \xrightarrow{\bar{\mathbf{p}}} & \mathcal{A} \\ {\scriptstyle \widetilde{\pi}}\downarrow & & \downarrow{\scriptstyle \pi} \\ \mathcal{M} & \xrightarrow{\mathbf{p}} & M \end{array} \tag{6.3}$$

(2) Consider the map

$$\begin{array}{cccc} \hat{\rho} : & \mathbf{T}^{\mathcal{A}}\mathcal{M} & \to & T\mathcal{M} \\ & (m, a, \mu) & \mapsto & (m, \mu) \end{array}$$

and let $\mathbf{p}_{\widetilde{\mathcal{A}}} : \mathbf{T}^{\mathcal{A}}\mathcal{M} \mapsto \widetilde{\mathcal{A}}$ be the map defined by $\mathbf{p}_{\widetilde{\mathcal{A}}}(m, a, \mu) = (m, a)$. Then the following diagrams are commutative

$$\begin{array}{ccc} \mathbf{T}^{\mathcal{A}}\mathcal{M} & \xrightarrow{\mathbf{p}_{\widetilde{\mathcal{A}}}} & \widetilde{\mathcal{A}} \\ {\scriptstyle \hat{\mathbf{p}}}\downarrow & & \downarrow{\scriptstyle \widetilde{\pi}} \\ \mathcal{M} & \xrightarrow{\mathrm{Id}} & \mathcal{M} \end{array} \qquad \begin{array}{ccc} \mathbf{T}^{\mathcal{A}}\mathcal{M} & \xrightarrow{\hat{\rho}} & T\mathcal{M} \\ {\scriptstyle \mathbf{p}_{\mathcal{A}}}\downarrow & & \downarrow{\scriptstyle T\mathbf{p}} \\ \mathcal{A} & \xrightarrow{\rho} & TM \end{array}$$

(3) If $\mathbf{p}_{\mathcal{M}} : T\mathcal{M} \mapsto \mathcal{M}$ is the tangent bundle, consider the associated vertical bundle $\mathbf{p}^{\mathbf{V}}_{\mathcal{M}} : \mathbf{T}\mathcal{M} \to \mathcal{M}$. Then there exists a canonical isomorphism bundle ν from the pull-back $\widetilde{\mathbf{p}} : \widetilde{\mathcal{E}} \to \mathcal{M}$ of the the bundle $\mathbf{p} : \mathcal{E} \to M$ over $\mathbf{p} : \mathcal{M} \to M$ to $\mathbf{p}^{V}_{\mathcal{M}} : \mathbf{V}\mathcal{M} \to \mathcal{M}$ so that the following diagram is commutative:

$$\begin{array}{ccc} \widetilde{\mathcal{E}} & \xrightarrow{\nu} & \mathbf{V}\mathcal{M} \\ {\scriptstyle \widetilde{\mathbf{p}}}\downarrow & & \downarrow{\scriptstyle \mathbf{p}^{V}_{\mathcal{M}}} \\ \mathcal{M} & \xrightarrow{\mathrm{Id}} & \mathcal{M} \end{array} \tag{6.4}$$

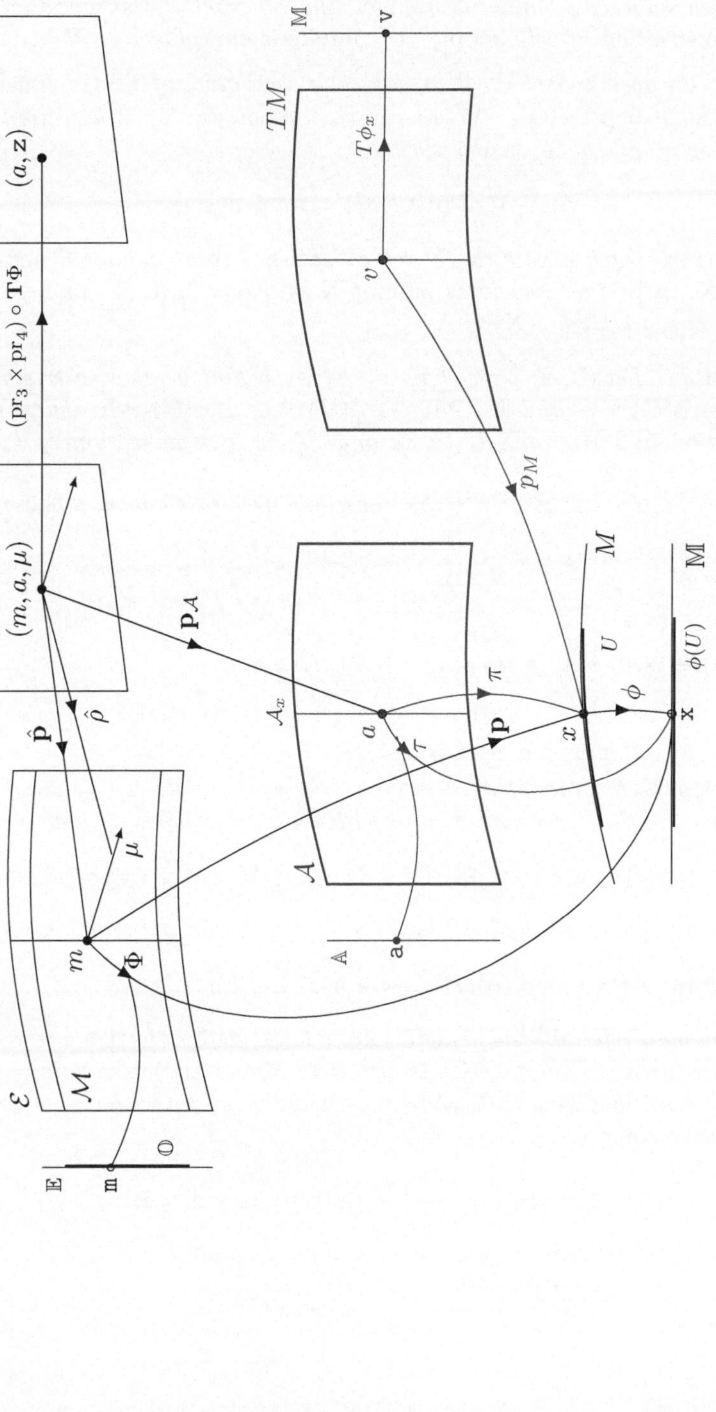

FIGURE 6.1
Prolongation of a Lie algebroid along a fibration

Theorem 6.3

(1) $\hat{\mathbf{p}} : \mathbf{T}^{\mathcal{A}}\mathcal{M} \to \mathcal{M}$ *is a convenient bundle with typical fibre* $\mathbb{A} \times \mathbb{E}$ *and* $(\mathbf{T}^{\mathcal{A}}\mathcal{M}, \hat{\mathbf{p}}, \mathcal{M}, \hat{\rho})$ *is an anchored bundle.*

(2) $\mathbf{p}_{\widetilde{\mathcal{A}}}$ *is a surjective bundle morphism whose kernel is a subbundle of* $\mathbf{T}\mathcal{M}$*. The restriction of* $\hat{\rho}$ *to* $\ker \mathbf{p}_{\widetilde{\mathcal{A}}}$ *is a bundle isomorphism on* $\mathbf{V}\mathcal{M}$*.*

(3) Given an open subset V*, then, for each section* $\mathbf{X}$ *of* $\mathbf{T}\mathcal{M}$ *defined on the open set* $\mathcal{V} = \hat{\mathbf{p}}^{-1}(V) \subset \mathcal{M}$*, there exists a pair* $(\mathfrak{a}, X)$ *of a section* $\mathfrak{a}$ *of* $\widetilde{\mathcal{A}}$ *and a vector field* X *on* $\mathcal{V}$ *such that*

$$\forall m \in \mathcal{V},\ T\mathbf{p}(X(m)) = \rho \circ \widetilde{\mathbf{p}}(\mathfrak{a}(m)). \tag{6.5}$$

Conversely such a pair $(\mathfrak{a}, X)$ *which satisfies (6.5) defines a unique section* $\mathbf{X}$ *on* $\mathcal{V}$*, the associated pair of* $\mathbf{X}$ *is precisely* $(\mathfrak{a}, X)$ *and with these notations, we have* $\hat{\rho}(\mathbf{X}) = X$*.*

Proof *(1) Let* (U, ϕ) *be a chart on* M *such that we have a trivialization* $\tau : \mathcal{A}_U \to \phi(U) \times \mathbb{A}$ *and* $\Phi : \mathcal{M}_U \to \phi(U) \times \mathbb{O} \subset \phi(U) \times \mathbb{E}$*. Then* $T\phi$ *is a trivialization of* TM_U *on* $\phi(U) \times \mathbb{M}$ *and* $T\Phi$ *is a trivialization of* $T\mathcal{M}_U$ *on* $\phi(U) \times \mathbb{O} \times \mathbb{M} \times \mathbb{E}$*.*
To be very precise, according to the notations used in Chapter 3, § 3.15.3, we have

$\phi(x) \equiv \mathsf{x}$;

$\tau(x, a) \equiv (\mathsf{x}, \mathsf{a})$;

$\Phi(x, e) \equiv (\mathsf{x}, \mathsf{e})$ *and for* $m = (x, e)$, $\Phi(m) \equiv \mathsf{m}$;

$T\phi(x, v) \equiv (\mathsf{x}, \mathsf{v})$;

$T\Phi(m, \mu) = T\Phi(x, e, v, z) \equiv (\mathsf{x}, \mathsf{e}, \mathsf{v}, \mathsf{z})$

where $\equiv$ *stands for "denoted".*
By the way, in this local context and with these notations, we have

$$\mathbf{T}^{\mathcal{A}}\mathcal{M}_U \equiv \{(\mathsf{x}, \mathsf{e}, \mathsf{a}, \mathsf{v}, \mathsf{z}) \in \phi(U) \times \mathbb{O} \times \mathbb{A} \times \mathbb{M} \times \mathbb{E} \ : \ \mathsf{v} = \rho_{\mathsf{x}}(\mathsf{a})\} \tag{6.6}$$

where ρ *corresponds to the local expression of the anchor. It follows that:*

$$\mathbf{T}^{\mathcal{A}}\mathcal{M}_U \equiv \{(\mathsf{x}, \mathsf{e}, \mathsf{a}, \rho_{\mathsf{x}}(\mathsf{b}), \mathsf{z}) \ : (\mathsf{x}, \mathsf{e}, \mathsf{a}, \mathsf{z}) \in \phi(U) \times \mathbb{O} \times \mathbb{A} \times \mathbb{E}\}, \tag{6.7}$$

and so the map $\mathbf{T}\Phi : \mathbf{T}^{\mathcal{A}}\mathcal{M}_U \to \phi(U) \times \mathbb{O} \times \mathbb{A} \times \mathbb{E}$ *defined by* $\mathbf{T}\Phi(x, e, a, v, z) \equiv (\mathsf{x}, \mathsf{e}, \mathsf{a}, \mathsf{z})$ *is a smooth map which is bijective. Moreover, for each* $m = (x, e) \in \mathcal{M}_U$ *the restriction* $\mathbf{T}\Phi_m$ *to* $\mathbf{T}_m\mathcal{M}_U$ *is clearly linear and we have the following commutative diagram*

$$\begin{array}{ccc} \mathbf{T}^{\mathcal{A}}\mathcal{M}_U & \xrightarrow{\ \mathbf{T}\Phi\ } & \phi(U) \times \mathbb{O} \times \mathbb{A} \times \mathbb{E} \\ \Big\downarrow{\scriptstyle \hat{\mathbf{p}}} & & \Big\downarrow{\scriptstyle \hat{\pi}_1} \\ \mathcal{M}_U & \xrightarrow{\ \ \Phi\ \ } & \phi(U) \times \mathbb{O} \end{array}$$

This shows that $\mathbf{T}\Phi$ *is a local trivialization of* $\mathbf{T}^{\mathcal{A}}\mathcal{M}_U$ *modelled on* $\mathbb{A}\times\mathbb{E}$. *Now, in this local context,* $\hat{\rho}(x,e,a,v,z)\equiv(\mathsf{x},\mathsf{e},\mathsf{a},\mathsf{z})$.

Consider two such chart domains U *and* U' *in* M *such that* $U\cap U'\neq\emptyset$. *Then we have:*

- *the transition maps associated with the trivializations* τ *and* τ' *in* $\mathcal{A}$ *are of type*
$$(\mathsf{x},\mathsf{a})\mapsto(\mathsf{t}(\mathsf{x}),\mathsf{G}_{\mathsf{x}}(\mathsf{a}))$$
where $\mathsf{x}\mapsto\mathsf{G}_{\mathsf{x}}$ *takes value in* $\mathrm{GL}(\mathbb{A})$ *and is a smooth map from* $U\cap U'$ *to* $\mathrm{L}(\mathbb{A})$*;*
- *the transition maps associated with the trivializations* Φ *and* Φ' *in* $\mathcal{M}$ *are of type*
$$(\mathsf{x},\mathsf{e},\mathsf{a},\mathsf{v})\mapsto(\mathsf{t}(\mathsf{x}),\mathsf{F}_{\mathsf{x}}(\mathsf{e}))$$
where $\mathsf{x}\mapsto\mathsf{F}_{\mathsf{x}}$ *takes value in* $\mathrm{GL}(\mathbb{E})$ *and is a smooth map from* $U\cap U'$ *to* $\mathrm{L}(\mathbb{E})$*;*
- *if* $\widetilde{\Phi}$ *and* $\widetilde{\Phi}'$ *are the trivializations of* $\widetilde{\mathcal{A}}$ *associated with* Φ *and* Φ', *the transition maps associated with trivializations* $\widetilde{\Phi}$ *and* $\widetilde{\Phi}'$ *in* $\widetilde{\mathcal{A}}$ *are of type*
$$(\mathsf{x},\mathsf{e},\mathsf{a})\mapsto(\mathsf{t}(\mathsf{x}),\mathsf{F}_{\mathsf{x}}(\mathsf{e}),\mathsf{G}_{\mathsf{x}}(\mathsf{a}));$$
- *the transition maps associated with trivializations* $T\Phi$ *and* $T\Phi'$ *in* $T\mathcal{M}$ *are of type*
$$(\mathsf{x},\mathsf{e},\mathsf{v},\mathsf{z})\mapsto\big(\mathsf{t}(\mathsf{x}),\mathsf{F}_{\mathsf{x}}(\mathsf{e}),D_{\mathsf{x}}\mathsf{t}(\mathsf{y}),\mathsf{H}_{(\mathsf{x},\mathsf{e})}(\mathsf{v})\big)$$
where $(\mathsf{x},\mathsf{e})\mapsto\mathsf{H}_{(\mathsf{x},\mathsf{e})}$ *takes value in* $GL(\mathbb{E})$ *and is a smooth smooth map from* U *to* $\mathrm{L}(\mathbb{E}$*;*
- *the transition maps associated with trivializations* $\mathbf{T}\Phi$ *and* $\mathbf{T}\Phi'$ *in* $\mathbf{T}^{\mathcal{A}}\mathcal{M}$ *are of type*
$$(\mathsf{x},\mathsf{e},\mathsf{a},\mathsf{z})\mapsto\big(\mathsf{t}(\mathsf{x}),\mathsf{F}_{\mathsf{x}}(\mathsf{e}),\mathsf{G}_{\mathsf{a}}(\mathsf{a}),\mathsf{H}_{(\mathsf{x},\mathsf{e})}(\mathsf{z}))\big).$$

Clearly, this implies that $\hat{\mathbf{p}}:\mathbf{T}\mathcal{M}\to\mathcal{M}$ *is a convenient bundle.*

Now, in a trivialization τ *and* $T\phi$, *we write* $\rho(x,a)\equiv(\mathsf{x},\mathsf{a})\mapsto(\mathsf{x},\rho_{\mathsf{x}}(\mathsf{a}))$. *If, in another trivialization* τ' *and* $T\phi'$, *we write* $\rho(x,a)\equiv(\mathsf{x}',\mathsf{a}')\mapsto(\mathsf{x}',\rho'_{\mathsf{x}'}(\mathsf{a}'))$, *then, for the associated transition maps, we have*
$$\rho'_{x'}=D_xt\circ\rho_x\circ G_x.$$
It follows easily that $\hat{\rho}$ *is a bundle convenient morphism.*

(2) By construction, the following diagram is commutative:

$$\begin{array}{ccc}
\mathbf{T}^{\mathcal{A}}\mathcal{M} & \xrightarrow{\hat{\rho}} & T\mathcal{M} \\
{\scriptstyle \mathbf{p}_{\widetilde{\mathcal{A}}}}\downarrow & & \downarrow{\scriptstyle p_{\mathcal{M}}} \\
\widetilde{\mathcal{A}} & \xrightarrow{\widetilde{\pi}} & \mathcal{M}
\end{array}$$

In the trivialization $\widetilde{\Phi} : \widetilde{A}_{\mathcal{M}_U} \to \phi(U) \times \mathbb{O} \times \mathbb{A}$*, using the same convention as previously, we have*

$$\{(x, e, a, v, z) \mapsto \mathbf{p}_{\widetilde{\mathcal{A}}}(x, e, a, v)\} \equiv \{(\mathsf{x}, \mathsf{e}, \mathsf{a}, \mathsf{v}, \mathsf{z}) \mapsto (\mathsf{x}, \mathsf{e}, \mathsf{a})\}.$$

Thus, by analog arguments as in the proof of (1), it is clear that $\mathbf{p}_{\widetilde{\mathcal{A}}}$ *is compatible with the transition maps associated with the trivializations over the chart domains* U *and* U' *of* M *for* $\mathbf{T}^{\mathcal{A}}\mathcal{M}$ *and for* $\widetilde{\mathcal{A}}$*. Thus* $\mathbf{p}_{\widetilde{\mathcal{A}}} : \mathbf{T}^{\mathcal{A}}\mathcal{M} \to \widetilde{\mathcal{A}}$ *is a Banach bundle morphism which is surjective.*
From the construction of $\mathbf{T}^{\mathcal{A}}\mathcal{M}$ *we have*

$$\ker \mathbf{p}_{\widetilde{\mathcal{A}}} = \{(m, a, v) \in \mathbf{T}^{\mathcal{A}}_m\mathcal{M} \quad : \rho(a) = 0\}.$$

The definition of $\hat{\rho}$ *implies that its restriction to* $\ker \mathbf{p}_{\widetilde{\mathcal{A}}}$ *is an isomorphism onto* $\mathbf{V}\mathcal{M}$*, which ends the proof of (2).*

(3) Let $\mathbf{X}$ *be a section of* $\mathbf{T}\mathcal{M}$ *defined on* $\mathcal{V} = \hat{\mathbf{p}}^{-1}(V)$*. According to (2), if we set* $\mathfrak{a} = \mathbf{p}_{\widetilde{\mathcal{A}}} \circ \mathbf{X}$ *and* $X = \hat{\rho}(\mathbf{X})$*, the pair* $(\mathfrak{a}, X)$ *is well defined and from the definition of* $\mathbf{T}^{\mathcal{A}}\mathcal{M}$ *the relation (6.5) is satisfied. Conversely, if* $\mathfrak{a}$ *is a section* $\mathfrak{a}$ *of* $\widetilde{\mathcal{A}}$ *and* X *a vector field on* $\mathcal{V}$*, the relation (6.5) means exactly that* $\mathbf{X}(m) = (\widetilde{\mathbf{p}}(\mathfrak{a}(m)), X(m))$ *belongs to* $\mathbf{T}^{\mathcal{A}}_m\mathcal{M}$*; so we get a section* $\mathbf{X}$ *of* $\mathbf{T}^{\mathcal{A}}\mathcal{M}$*. Now it is clear that* $\mathfrak{a} = \mathbf{p}_{\widetilde{\mathcal{A}}} \circ \mathbf{X}$ *and* $X = \hat{\rho}(\mathbf{X})$*.*

Definition 6.4 *The anchored bundle* $(\mathbf{T}^{\mathcal{A}}\mathcal{M}, \hat{\mathbf{p}}, \mathcal{M}, \hat{\rho})$ *is called the prolongation of* $(\mathcal{A}, \pi, M, \rho)$ *over* $\mathcal{M}$*. The subbundle* $\ker \mathbf{p}_{\widetilde{\mathcal{A}}}$ *will be denoted* $\mathbf{V}^{\mathcal{A}}\mathcal{M}$ *and is called the vertical subbundle.*

Remark 6.4 *According to the proof of Theorem 6.3, if* $\mathbf{V}^{\mathcal{A}}\mathcal{M}_U$ *is the restriction of* $\mathbf{V}^{\mathcal{A}}\mathcal{M}$ *to* $\mathcal{M}_U$*, we have*

$$\mathbf{V}^{\mathcal{A}}\mathcal{M}_U \equiv \{(\mathsf{x}, \mathsf{e}, \mathsf{a}, 0, \mathsf{z}) \quad : (\mathsf{x}, \mathsf{e}, \mathsf{a}, \mathsf{z}) \in \phi(U) \times \mathbb{O} \times \mathbb{A} \times \mathbb{E}\}.$$

Examples 6.1 *For simplicity in these examples we assume that the model space* $\mathbb{M}$ *of* M *and the typical fibre* $\mathbb{A}$ *of* $\mathcal{A}$ *are Banach spaces.*

(1) If $\mathcal{A} = \mathcal{M} = TM$ *then we have* $\mathbf{T}\mathcal{A} = TTM$ *and* $\mathbf{T}\mathcal{A}^* = TT^*M$ *and the anchor* $\hat{\rho}$ *is the identity.*

(2) If $\mathcal{A} = \mathcal{M}$ *then* $\mathbf{T}^{\mathcal{A}}\mathcal{A}$ *is simply denoted* $\mathbf{T}\mathcal{A}$ *and the anchor* $\hat{\rho}$ *is the map* $(x, a, b, c) \mapsto (x, a, \rho(b), \nu(c))$ *from* $\mathbf{T}\mathcal{A}$ *to* $T\mathcal{A}$*.*

(3) If $\mathcal{M} = \mathcal{A}^*$ *then* $\mathbf{T}^{\mathcal{A}}\mathcal{A}^*$ *is simply denoted* $\mathbf{T}\mathcal{A}^*$ *and the anchor* $\hat{\rho}$ *is the map* $(x, \xi, a, \eta) \mapsto (x, \xi, \rho(a), \nu(\eta)$ *from* $\mathbf{T}\mathcal{A}^*$ *to* $T\mathcal{A}^*$*.*

(4) If $\mathcal{M} = \mathcal{A} \times_M \mathcal{A}^*$ *then* $\mathbf{T}^{\mathcal{A}}\mathcal{M}$ *is simply denoted* $\mathbf{T}(\mathcal{A} \times \mathcal{A}^*)$ *and the anchor* $\hat{\rho}$ *is the map* $(x, a, \xi, b, c, \eta) \mapsto (x, a, \xi, \rho(b), \nu(v), \eta))$ *from* $\mathbf{T}(\mathcal{A} \times \mathcal{A}^*)$ *to* $T(\mathcal{A} \times \mathcal{A}^*)$*.*

(5) If $\mathcal{M}$ *is a conic submanifold of* $\mathcal{A}$ *(cf [Pel19]), then* $\mathbf{T}^{\mathcal{A}}\mathcal{M}$ *is simply denoted* $\mathbf{T}\mathcal{M}$ *and the anchor* $\hat{\rho}$ *is the same expression as in (2).*

Remark 6.5 *We come back to the previous general context: $(\mathcal{A}, \pi, M, [.,.]_{\mathcal{A}})$ is a convenient Lie algebroid, $\mathbf{p} : \mathcal{E} \to M$ is a convenient vector bundle and $\mathcal{M}$ is an open submanifold of $\mathcal{E}$ which is fibred on M.*
Let (U, ϕ) be a chart of M such that $\mathcal{A}_U$ (resp. $\mathcal{E}_U$) is trivial and denote

$$\tau : \mathcal{A}_U \to \phi(U) \times \mathbb{A} \text{ (resp. } \Phi : \mathcal{E}_U \to \phi(U) \times \mathbb{E})$$

the associated trivialization (cf. notations in the proof of Theorem 6.3). Then we have a canonical anchored bundle

$$(\phi(U) \times \mathbb{A}, \pi_1, \phi(U), \mathsf{r} = T\phi \circ \rho \circ \phi^{-1})$$

where $\pi_1 : \phi(U) \times \mathbb{A} \to \phi(U)$.
Therefore, the prolongation $\phi(U) \times \mathbb{A}$ over $\phi(U) \times \mathbb{E}$ is then

$$\mathbf{T}^{\phi(U)\times\mathbb{A}}(\phi(U) \times \mathbb{E}) = \phi(U) \times \mathbb{E} \times \mathbb{A} \times \mathbb{E}$$

and the anchor $\hat{\mathsf{r}}$ is the map $(\mathsf{x}, \mathsf{e}, \mathsf{a}, \mathsf{z}) \mapsto (\mathsf{x}, \mathsf{e}, \mathsf{r}(\mathsf{a}), \mathsf{z})$.
Note that the bundle $\mathbf{T}^{\phi(U)\times\mathbb{A}}(\phi(U) \times \mathbb{E}) = \phi(U) \times \mathbb{E} \times \mathbb{A} \times \mathbb{E}$ can be identified with the subbundle

$$\{(\mathsf{x}, \mathsf{e}, \mathsf{a}, 0, \mathsf{v}) \in U \times \mathbb{E} \times \mathbb{A} \times \mathbb{M} \times \mathbb{E}\}$$

of $T(\phi(U) \times \mathbb{E}) = \phi(U) \times \mathbb{E} \times \mathbb{A} \times \mathbb{M} \times \mathbb{E}$.
An analog description is true for any open set $\mathcal{U}$ in $\mathcal{E}$ such that $\mathbf{p}(\mathcal{U}) = U$ since $\mathcal{U}$ is contained in $\mathcal{M}_U$.

Given a fibred morphism $\Psi : \mathcal{M}_1 \to \mathcal{M}_2$ between the fibred manifolds $\mathbf{p}_1 : \mathcal{M}_1 \to M_1$ and $\mathbf{p}_2 : \mathcal{M}_2 \to M_2$ over $\psi : M_1 \to M_2$ and a morphism of anchored bundles φ between $(\mathcal{A}_1, \pi_1, M_1, \rho_1)$ and $(\mathcal{A}_2, \pi_2, M_2, \rho_2)$, over $\psi : M_1 \to M_2$, we get a map

$$\begin{array}{llcl} \mathbf{T}\Psi : & \mathbf{T}^{\mathcal{A}_1}\mathcal{M}_1 & \to & \mathbf{T}^{\mathcal{A}_2}\mathcal{M}_2 \\ & (m_1, a_1, \mu_1) & \mapsto & (\Psi(m_1), \varphi(a_1), T_{m_1}\Psi(\mu_1)) \end{array}$$

Remark 6.6 *As in the proof of Theorem 6.3, consider a chart $(\mathcal{U}, \Phi)$ of $\mathcal{M}$ where $\mathbf{p}(\mathcal{U}) = U$ and let $\tau : \mathcal{A}_U \to U \times \mathbb{A}$ be an associated trivialization of $\mathcal{A}_U$. According to Example 6.1 (6), since Φ is a smooth diffeomorphism from $\mathcal{U}$ onto its range $\phi(U) \times \mathbb{O}$ in $\phi(U) \times \mathbb{E}$, then $\mathbf{T}\Phi : \mathbf{T}^{\mathcal{A}}\mathcal{M}_{|\mathcal{U}} \to \mathbf{T}^{\phi(U)\times\mathbb{A}}\mathcal{U} = \Phi(\mathcal{U}) \times \mathbb{A} \times \mathbb{E}$ is a bundle isomorphism and so $(\mathbf{T}^{\mathcal{A}}\mathcal{M}_{|\mathcal{U}}, \mathbf{T}\Phi)$ is a chart for $\mathbf{T}^{\mathcal{A}}\mathcal{M}$.*

Notation 6.2 *From now on, the anchored bundle $(\mathcal{A}, \pi, M, \rho)$ is fixed and, if no confusion is possible, we simply denote by $\mathbf{T}\mathcal{M}$ and $\mathbf{V}\mathcal{M}$ the sets $\mathbf{T}^{\mathcal{A}}\mathcal{M}$ and $\mathbf{V}^{\mathcal{A}}\mathcal{M}$ respectively. The bundle $\mathbf{V}\mathcal{M}$ will be considered as a subbundle of $\mathbf{T}\mathcal{M}$ as well as of $T\mathcal{M}$.*

6.4.2 Basic Forms and Tensors on a Prolongation

Let $\mathbf{T}^*\mathcal{M}$ the dual bundle of $\mathbf{T}\mathcal{M}$. A k-form ω on $\mathbf{T}\mathcal{M}$ (i.e. a smooth section of the bundle $\text{Ł}^k\mathbf{T}^*\mathcal{M}$) is called *semi-basic* if $i_{\mathbf{Z}}\omega = 0$ for any vertical section $\mathbf{Z}$. In the same way, a vector valued k-form ω on $\mathbf{T}\mathcal{M}$ (i.e. a smooth section of the bundle $\text{Ł}^k\mathbf{T}^*\mathcal{M} \otimes \mathsf{L}^k\mathbf{T}\mathcal{M}$) is called *semi-basic* if ω takes its values in $\mathbf{V}\mathcal{M}$ and if $i_{\mathbf{Z}}\omega = 0$ when $\mathbf{Z}$ is vertical. More generally, any tensor Θ of type (k,l) on $\mathbf{T}\mathcal{M}$ (i.e. a smooth section of the bundle $\bigotimes^k \mathbf{T}^*\mathcal{M} \bigotimes^l \mathbf{T}\mathcal{M}$) is called semi-basic if Θ is a section of $\bigotimes^k \mathbf{T}^*\mathcal{M} \bigotimes^l \mathbf{V}\mathcal{M}$ and $\Theta(\mathbf{X}_1, \cdots, \mathbf{X}_k)$ belongs to $\bigotimes^l(\mathbf{V}\mathcal{M})$, for any section $\mathbf{X}_1, \cdots, \mathbf{X}_k$ of $\mathbf{T}\mathcal{M}$ and this value is 0 as soon as one of these sections is vertical.

6.4.3 Connections on a Prolongation

Classically (cf. [KrMi97], 37), a *connection*[2] on a convenient vector bundle $\mathbf{p} : \mathcal{E} \to M$ is a Whitney decomposition $T\mathcal{E} = H\mathcal{E} \oplus V\mathcal{E}$. Now, as in finite dimension (see for instance [Gri72]), we introduce this notion on $\mathbf{T}\mathcal{M}$.

Definition 6.5 *A connection on* $\mathbf{T}\mathcal{M}$ *is a decomposition of this bundle in a Whitney sum* $\mathbf{T}\mathcal{M} = \mathbf{H}\mathcal{M} \oplus \mathbf{V}\mathcal{M}$.

Such a decomposition is equivalent to the datum of an endomorphism $\mathbf{N}$ of $\mathbf{T}\mathcal{M}$ such that $\mathbf{N}^2 = \text{Id}$ with $\mathbf{T}\mathcal{M} = \ker(\text{Id} + \mathbf{N})$ and $\mathbf{H}\mathcal{M} = \ker(\text{Id} - \mathbf{N})$ where Id is the identity morphism of $\mathbf{T}\mathcal{M}$. We naturally get two projections:

$$h_{\mathbf{N}} = \frac{1}{2}(\text{Id} + \mathbf{N}) : \mathbf{T}\mathcal{M} \to \mathbf{H}\mathcal{M}$$

$$v_{\mathbf{N}} = \frac{1}{2}(\text{Id} - \mathbf{N}) : \mathbf{T}\mathcal{M} \to \mathbf{V}\mathcal{M}.$$

$v_{\mathbf{N}}$ and $h_{\mathbf{N}}$ are called respectively the *vertical* and *horizontal* projector of $\mathbf{N}$. Using again the context of Remark 6.6, we have charts

$$\Phi : \mathcal{M}_U \to \phi(U) \times \mathbb{O} \quad \text{and} \quad \mathbf{T}\Phi : \mathbf{T}\mathcal{M}_U \to \phi(U) \times \mathbb{O} \times \mathbb{A} \times \mathbb{E}.$$

If $\mathbf{N}$ is a connection on $\mathbf{T}\mathcal{M}$, then $\mathsf{N} = \mathbf{T}\Psi \circ \mathbf{N} \circ \Psi^{-1}$ is a non-linear connection on the trivial bundle $\phi(U) \times \mathbb{O} \times \mathbb{A} \times \mathbb{E}$. Thus N can be written as a matrix field of endomorphisms of $\mathbb{A} \times \mathbb{E}$ of type

$$\begin{pmatrix} \text{Id}_{\mathbb{A}} & 0 \\ -2\mathsf{F} & -\text{Id}_{\mathbb{E}} \end{pmatrix} \tag{6.8}$$

and so the associated horizontal (resp. vertical) projector is given by

$$\mathsf{h}_{\mathsf{N}_\mathsf{m}}(\mathsf{a}, \mathsf{z}) = \frac{1}{2}(\mathsf{a}, \mathsf{z}) - \mathsf{F}(\mathsf{a});$$

$$\mathsf{v}_{\mathsf{N}_\mathsf{m}}(\mathsf{a}, \mathsf{z}) = \frac{1}{2}(\mathsf{a}, \mathsf{v}) + \mathsf{F}(\mathsf{a}).$$

[2]In the finite-dimensional context such a connection is sometimes called a *non-linear connection*.

The associated horizontal space in $\{\mathsf{m}\} \times \mathbb{A} \times \mathbb{E}$ is

$$\left\{\frac{1}{2}(\mathsf{a},\mathsf{z}) - \mathsf{F}(\mathsf{a}), (\mathsf{a},\mathsf{z}) \in \{\mathsf{m}\} \times \mathbb{A} \times \mathbb{E}\right\}$$

and the associated vertical space in $\{\mathsf{m}\} \times \mathbb{A} \times \mathbb{E}$ is $\{\mathsf{m}\} \times \{0\} \times \mathbb{E}$.
F is called the *(local) Christoffel symbol of* $\mathbf{N}$.
Let $\widetilde{\mathbf{p}} : \widetilde{\mathcal{A} \times \mathcal{E}} \to \mathcal{M}$ the fibered product bundle over $\mathcal{M}$ of $(\pi, \mathbf{p}) : \mathcal{A} \times \mathcal{E} \to M$. We have natural inclusions $\iota_1 : \widetilde{\mathcal{A}} \to \widetilde{\mathcal{A} \times \mathcal{E}}$ and $\iota_2 : \widetilde{\mathcal{E}} \to \widetilde{\mathcal{A} \times \mathcal{E}}$, given respectively by $\iota_1(m,a) = (m,a,0)$ and $\iota_1(m,z) = (m,0,z)$, such that

$$\widetilde{\mathcal{A} \times \mathcal{E}} = \iota_1(\widetilde{\mathcal{A}}) \oplus \iota_2(\widetilde{\mathcal{E}}). \tag{6.9}$$

With these notations, we have

Proposition 6.3

(1) There exists a non-linear connection $\mathbf{N}$ *on* $\mathbf{T}\mathcal{M}$ *if and only if there exists a convenient bundle morphism* $\mathbf{H}$ *from* $\widetilde{\mathcal{A}}$ *to* $\mathbf{T}\mathcal{M}$ *such that* $\mathbf{T}\mathcal{M} = \mathbf{H}(\widetilde{\mathcal{A}}) \oplus \mathbf{V}\mathcal{M}$*. In this case* $\mathbf{T}\mathcal{M}$ *is isomorphic to* $\widetilde{\mathcal{A} \times \mathcal{E}}$*.*

(2) Assume that $\mathbf{N}$ *is a connection on* $\mathbf{T}\mathcal{M}$*. Let* Υ *be a morphism from* $\mathbf{T}\mathcal{M}$ *to* $\mathbf{V}\mathcal{M}$ *such that* $\Upsilon(\mathbf{Z}) = 0$ *for any local vertical section* $\mathbf{Z}$*, i.e. semi-basic* $\mathbf{T}\mathcal{M}$*-tensor* Υ *of type* $(1,1)$*, then* $\mathbf{N} + \Upsilon$ *is a connection on* $\mathbf{T}\mathcal{M}$*. Conversely, given any non-linear connection* $\mathbf{N}'$ *on* $\mathbf{T}\mathcal{M}$*, there exists a unique semi-basic vector valued* Υ *such that* $\mathbf{N}' = \mathbf{N} + \Upsilon$*.*

According to this Proposition we introduce:

Definition 6.6 *A prolongation of* $\mathcal{A}$ *over* $\mathbf{p} : \mathcal{M} \to M$ *is called a split prolongation if there exists a Withney decomposition* $\mathbf{T}\mathcal{M} = \mathbf{K}\mathcal{M} \oplus \mathbf{V}\mathcal{M}$*.*

The following result is a clear consequence of Proposition 6.3.

Corollary 6.1 *Let* $\mathbf{T}\mathcal{M}$ *be a split prolongation. Then there exists a connection on* $\mathbf{T}\mathcal{M}$ *and* $\mathbf{T}\mathcal{M}$ *isomorphic to* $\widetilde{\mathcal{A} \times \mathcal{E}}$*.*

Proof of Proposition 6.3
(1) Assume that we have a connection $\mathbf{N}$ *on* $\mathbf{T}\mathcal{M}$ *and let* $\mathbf{T}\mathcal{M} = \mathbf{H}\mathcal{M} \oplus \mathbf{V}\mathcal{M}$ *be the associated Whitney decomposition.*
Let $\mathbf{H}$ *be the restriction of* $\mathbf{p}_{\widetilde{\mathcal{A}}}$ *to* $\mathbf{H}\mathcal{M}$*. Since* $\mathbf{V}\mathcal{M}$ *is the kernel of the surjective morphism* $\mathbf{p}_{\widetilde{\mathcal{A}}}$*, it follows that* $\mathbf{H}$ *is an isomorphism. Since we have an isomorphism* $\nu : \widetilde{\mathcal{E}} \to \mathbf{V}\mathcal{M}$*, according to (6.9), it follows easily that* $\widetilde{\mathcal{A} \times \mathcal{E}}$ *is isomorphic to* $\mathbf{T}\mathcal{M}$*. The converse is clear.*

(2) At first, if Υ *is semi basic, then* $\ker(\mathrm{Id} + \mathbf{N} + \Upsilon) = \mathbf{V}\mathcal{M}$ *and clearly the range of* $\mathrm{Id} + \mathbf{N} + \Upsilon$ *is a supplemented subbundle of* $\mathbf{V}\mathcal{M}$*.*

On the one hand, if $\mathbf{N}'$ *is a connection, we set* $\Upsilon = \mathbf{N}' - \mathbf{N}$. *Then* $\Upsilon(\mathbf{Z}) = 0$, *for all local vertical sections. On the other hand* $\Upsilon(\mathbf{X}) = (\mathrm{Id} + \mathbf{N}')(\mathbf{X}) - (\mathrm{Id} + \mathbf{N}(\mathbf{X})$ *which belongs to* $\mathbf{V}\mathcal{M}$.

A sufficient condition for the existence of a connection on $\mathbf{T}\mathcal{M}$ is given by the following result:

Theorem 6.4 *Assume that there exists a linear connection on the bundle* $\mathbf{p} : \mathcal{E} \to M$. *Then there exists a connection* $\mathbf{N}$ *on* $\mathbf{T}\mathcal{M}$.

Proof *let* $\mathbf{p}^*TM$ *(resp.* $\mathbf{p}^*\mathcal{E}$*) is the pull-back over* $\mathcal{M}$ *of* $TM \to M$ *(resp.* $\mathcal{E} \to M$*).*
If there exists a linear connection N *on the bundle* $\mathcal{E} \to M$, *from Theorem 3.6, there exists a convenient isomorphism bundle*

$$\kappa = (\kappa_1, \kappa_2) : T\mathcal{E} \to \mathbf{p}^*TM \oplus \mathbf{p}^*\mathcal{E}.$$

Therefore, for an open fibered submanifold $\mathcal{M}$ *of* $\mathcal{E}$, *by restriction, we obtain an isomorphism (again denoted* κ*).*

$$\kappa : T\mathcal{M} \to \mathbf{p}^*TM \oplus \mathbf{p}^*\mathcal{E}$$

Without loss of generality, we can identify $T\mathcal{M}$ *with* $\mathbf{p}^*TM \oplus \mathbf{p}^*\mathcal{E}$. *For* $m = (x, e)$, *the fibre* $\mathbf{T}_m\mathcal{M}$ *is then*

$$\mathbf{T}_m\mathcal{M} = \{(a, \mu) \in \mathcal{A}_x \times T_xM \times \mathbf{V}_m\mathcal{M} \ \ : \rho(a) = T\mathbf{p}(\mu)\}.$$

But, under our identification of $T\mathcal{M}$ *with* $\mathbf{p}^*TM \oplus \mathbf{p}^*\mathcal{E}$, *if* $m = (x, e)$, *this implies that* $\mu \in T_m\mathcal{M}$, *can be written as a pair* $(v, z) \in T_xM \times \mathcal{E}_x$ *and so we replace the condition* $\rho(a) = T\mathbf{p}(\mu)$ *by* $\rho_x(a) = v$.
Recall that, in the one hand, we have a chart (cf. proof of Theorem 6.3):

$$\mathbf{T}\Phi : \mathbf{T}\mathcal{M}_U \to \Phi(\mathcal{M}_U) \times \mathbb{A} \times \mathbb{E}.$$

The value of $\mathbf{T}\Phi(m, a, \mu)$ *can be written* $(\phi(x), \Phi_x(e), \tau_x(a), T_m\Phi(\mu))$ *(value denoted* $(\mathsf{x}, \mathsf{e}, \mathsf{a}, \mathsf{z})$*) with* $T_m\Psi(T\mathbf{p}(\mu)) = T_x\phi(\rho_x(a))$. *But, under our assumption, we have* $T_m\Phi(\mu)) = (T_x\phi(\rho_x(a)), \mathsf{z}) \in \{(\mathsf{x}, \mathsf{e})\} \times \mathbb{M} \times \mathbb{E}$ *and so we obtain*

$$\mathbf{T}\Phi(m, a, \mu) \equiv (\mathsf{x}, \mathsf{e}, \mathsf{a}, \mathsf{z}). \tag{6.10}$$

On the other hand, we have a trivialization $\widetilde{\tau \times \Phi}$ *from* $\widetilde{\mathcal{A} \times \mathcal{E}}_{\mathcal{M}_U}$ *to* $\Phi(\mathcal{M}_U) \times \mathbb{A} \times \mathbb{E}$ *over* Φ. *In fact, we have* $(\widetilde{\tau \times \Phi})(x, e, a, z) = (\phi(x), \Phi_x(e), \tau_x(a), \Phi_x(z))$. *According to our assumption, the map* $\widetilde{\Psi} : \widetilde{\mathcal{A} \times \mathcal{E}} \to \mathbf{T}\mathcal{M}$ *is given by*

$$\widetilde{\Psi}(m, a, z) = (m, a, \rho(a), z)$$

is well defined. In local coordinates, we have

$$\widetilde{\Psi}(e, x, a, \rho(a), z) \equiv (\mathsf{x}, \mathsf{e}, \mathsf{a}, \mathsf{z}).$$

Thus $\widetilde{\Psi}$ is the identity in local coordinates and so is a local bundle isomorphism. To complete the proof, we only have to show that under our assumption $\widetilde{\Psi}$ is a convenient bundle morphism.
By analogy with the notations used in the proof of Theorem 6.3 (1), let (U', ϕ') be another chart on M and consider all the corresponding trivializations $T\phi'$, Φ', and Φ'. We set $(\mathsf{x}', \mathsf{e}', \mathsf{v}', \mathsf{z}') = T\Phi'(\mathfrak{e}, \mathfrak{z})$ (i.e. $(\mathfrak{e}, \mathfrak{z}) \equiv (\mathsf{x}', \mathsf{e}', \mathsf{v}', \mathsf{z}')$ following our convention), in these new coordinates, we have $\widetilde{\Psi}(\mathfrak{e}, \mathfrak{z}) \equiv (\mathsf{x}', \mathsf{e}', \mathsf{a}', \mathsf{z}')$. Assume that $U \cap U' \neq \emptyset$. For the change of coordinates, we set $\theta_{\mathsf{x}} = \phi' \circ \phi^{-1}(\mathsf{x})$, and each associated transition map gives rise to a smooth field of isomorphisms of convenient spaces as follows:

$$\begin{aligned} T_{\mathsf{x}}\theta(\mathsf{v}) &= T_{\mathsf{x}}(\phi' \circ \phi^{-1})(\mathsf{v}) \\ \mathfrak{T}_{\mathsf{x}}(\mathsf{a}) &= \left(\tau' \circ \tau^{-1}\right)_{\mathsf{x}}(\mathsf{a}) \\ \Theta_{\mathsf{x}}(\mathsf{e}) &= (\Phi' \circ \Phi^{-1})_{\mathsf{x}}(\mathsf{e}). \end{aligned}$$

Thus, under our assumption, according to (6.10), in fact we have

$$\mathbf{T}\Phi' \circ \mathbf{T}\Phi^{-1}(\mathsf{x}, \mathsf{e}, \mathsf{a}, \mathsf{z}) = (\theta(\mathsf{x}), \Theta_{\mathsf{x}}(\mathsf{e}), \mathfrak{T}_{\mathsf{x}}(\mathsf{a}), \Theta_{\mathsf{x}}(\mathsf{z})).$$

Now with the previous notations we have

$$\begin{aligned} & \left(\widetilde{\tau' \times \Phi'}\right) \circ \left(\widetilde{\tau \times \Phi}\right)^{-1} (\mathsf{x}, \mathsf{e}, \mathsf{a}, \mathsf{z}) \\ = \; & (\theta(\mathsf{x}), \Theta_{\mathsf{x}}(\mathsf{e}), \mathfrak{T}_{\mathsf{x}}(\mathsf{a}), \Theta_{\mathsf{x}}(\mathsf{z})) \\ = \; & \mathbf{T}\Phi' \circ \mathbf{T}\Phi^{-1}(\mathsf{x}, \mathsf{e}, \mathsf{a}, \mathsf{z}). \end{aligned}$$

Since, in such local coordinates, $\widetilde{\Psi}$ is the identity map, so under our assumption $\widetilde{\Psi}$ is a convenient bundle isomorphism.

6.4.4 Vertical Lift and Liouville Section on a Prolongation

The canonical isomorphism ν, defined by (6.4), composed with $\hat{\rho}^{-1}_{|\mathbf{V}\mathcal{M}}$, is precisely

$$\begin{array}{rccl} vl_{\mathcal{M}}: & \widetilde{\mathcal{E}} = \mathcal{M} \times_M \mathcal{E} & \to & \mathbf{V}\mathcal{M} \\ & (x, e, z) & \mapsto & \dot{\gamma}(0) \end{array}$$

where $\gamma(t) = e + tz$.
The *vertical endomorphism* $\mathbf{J}_{\mathcal{M}}$ of $\mathbf{T}\mathcal{M}$ is given by

$$\mathbf{J}_{\mathcal{M}}(x, e, a, z) = (x, a, 0, vl_{\mathcal{M}}(x, e, z)).$$

As in finite dimension, (cf. [CLMM09]) if $\mathfrak{e}$ is a local section of $\widetilde{\mathcal{E}}$, then the *vertical lift* $\mathfrak{e}^v$ is defined by

$$\mathfrak{e}^v(x, e, z) = (x, e, 0, vl_{\mathcal{M}}(x, e, z)).$$

Remark 6.7 *Let* $\mathbf{Z}$ *be a vertical section, i.e. a section of* $\mathbf{V}\mathcal{M}$ *defined around some point* $m_0 \in \mathcal{U}$. *In the context of Examples 6.1 (6), Remark 6.6 and Remark 6.4, in the chart* $(\mathbf{T}\mathcal{M}_{|\mathcal{U}}, \mathbf{T}\Phi)$, $\mathbf{T}\Phi(\mathbf{Z})$ *is represented by some map* $(\mathsf{x}, \mathsf{e}) \mapsto (0, \mathsf{z}(\mathsf{x}, \mathsf{e}))$ *from* $\Phi(\mathcal{U})$ *to* $\mathbb{A} \times \mathbb{E}$ *and conversely. Thus a vertical section is always the lift of a section of* $\widetilde{\mathcal{E}}$.

The *Liouville section* $\mathbf{C}$ on $\mathbf{T}\mathcal{M}$ is the section defined by

$$\mathbf{C}(x, e) = (x, e, 0, vl_{\mathcal{M}}(x, e, e)).$$

We have the following characterization of $\mathbf{C}$:

Proposition 6.4 *A vertical section* $\mathbf{Z}$ *is the Liouville section if and only if, for any* $m_0 \in \mathcal{M}$, *there exists a chart* $(\mathcal{U}, \Phi)$ *around* m_0 *such that, in the associated chart* $(\mathbf{T}\mathcal{M}_{|\mathcal{U}}, \mathbf{T}\Phi)$, *we have* $\mathbf{Z}(m) \equiv (\mathsf{x}, \mathsf{e}, 0, \mathsf{e})$.

Proof *We use the context of Remark 6.7. Assume that* $\mathsf{Z}(\mathsf{x}, \mathsf{e}) = \mathsf{e}$. *Now since* $\Phi(\mathcal{U})$ *is an open subset of* $\mathbb{M} \times \mathbb{E}$, *around the point* $(\mathsf{x}, \mathsf{e}) = \Phi(x, e)$, *for* t *small enough, the curve* $t \mapsto \mathsf{e} + t\mathsf{e}$ *is contained in* $\Psi(\mathcal{U})$ *and so* $\mathbf{T}\Psi(\mathbf{Z})$ *is the Liouville section* $\mathbf{C}$ *of* $\mathbf{T}^{\phi(U)\times\mathbb{A}}\mathcal{U} = \Phi(\mathcal{U}) \times \mathbb{A} \times \mathbb{E}$. *It follows that*

$$\mathbf{Z}(x, e) = \mathbf{T}\Phi^{-1}(\mathbf{C}(\Phi(x, e)).$$

Since $\mathbf{C}(\mathsf{x}, \mathsf{e}) = (\mathsf{x}, \mathsf{e}, 0, \mathsf{e})$, *and* $\mathbf{T}\Phi^{-1}(\mathsf{x}, \mathsf{e}, 0, \mathsf{e}) = (x, e, 0, e)$, *this implies that* $\mathbf{Z}(x, e) = \mathbf{C}(x, e)$.
Now, looking for the vertical section $\mathbf{C}$, *in the local chart of type* $\mathcal{U} = \mathcal{M}_U$ *and trivialization* $\Phi : \mathcal{M}_U \to \phi(U) \times \mathbb{O}$, *we have* $\mathbf{C}(x, e) \equiv (\mathsf{x}, \mathsf{e}, 0, \mathsf{e})$, *which ends the proof.*

6.4.5 Prolongation of the Lie Bracket

As in the finite-dimensional framework, we will define a Lie bracket for smooth sections of $\hat{\mathbf{p}} : \mathbf{T}\mathcal{M} \to \mathcal{M}$.

According to the notations 6.1, for each open set U of M, we denote by $\Gamma(\mathbf{T}\mathcal{M}_U)$, $\Gamma(\mathbf{V}\mathcal{M}_U)$ and $\Gamma(\widetilde{\mathcal{A}}_U)$ the $C^\infty(\mathcal{M}_U)$-modules of sections of $\mathbf{T}\mathcal{M}_U$, $\mathbf{V}\mathcal{M}_U$ and $\widetilde{\mathcal{A}}_U$ respectively. We also denote $\Gamma(\mathcal{A}_U)$ the $C^\infty(U)$-module of sections $\pi : \mathcal{A}_U \to U$.

Definition 6.7 *Let* U *be an open subset of* M. *A section* $\mathbf{X}$ *in* $\Gamma(\mathbf{T}\mathcal{M}_U)$ *is called projectable if there exists* $\mathfrak{a} \in \Gamma(\mathcal{A}_U)$ *such that*

$$\mathbf{p}_{\mathcal{A}} \circ \mathbf{X} = \mathfrak{a}.$$

Therefore $\mathbf{X}$ is projectable if and only if there exists a vector field X on $\mathcal{M}_U$ and $\mathfrak{a} \in \Gamma(\mathcal{A}_U)$ such that (see Theorem 6.3)

$$\mathbf{X} = (\mathfrak{a} \circ \mathbf{p}, X) \text{ with } T\mathbf{p}(X) = \rho \circ \mathfrak{a}.$$

Assume now that $(\mathcal{A}, \pi, M, \rho, [.,.]_{\mathcal{A}})$ is a convenient Lie algebroid.

Let $\mathbf{X}_i = (\mathfrak{a}_i \circ \mathbf{p}_i, X_i)$, $i \in \{1, 2\}$ be two projectable sections defined on $\mathcal{M}_U$. We set

$$[\mathbf{X}_1, \mathbf{X}_2]_{\mathbf{T}\mathcal{M}} = ([\mathfrak{a}_1, \mathfrak{a}_2]_{\mathcal{A}} \circ \mathbf{p}, [X_1, X_2]) \tag{6.11}$$

Since $\rho([\mathfrak{a}_1, \mathfrak{a}_2]_{\mathcal{A}}) = [\rho(\mathfrak{a}_1), \rho(\mathfrak{a}_2)]$ and $T\mathbf{p}([X_1, X_2]) = [T\mathbf{p}(X_1), T\mathbf{p}(X_2)]$, it follows that $[\mathbf{X}_1, \mathbf{X}_2]_{\mathbf{T}\mathcal{M}}$ is a projectable section which is well defined. Moreover, we have

$$\hat{\rho}([\mathbf{X}_1, \mathbf{X}_2]_{\mathbf{T}\mathcal{M}}) = [\hat{\rho}(\mathbf{X}_1), \hat{\rho}(\mathbf{X}_2)]. \tag{6.12}$$

Comments 6.1 *Now, **in finite dimension**, it is well known that the module of sections of $\widetilde{\mathcal{A}}_U \to \mathcal{M}_U$ is the $C^\infty(\mathcal{M}_U)$-module generated by the set of sections $\mathfrak{a} \circ \mathbf{p}$ where $\mathfrak{a}$ is any section of $\mathcal{A}_U \to U$. Therefore, according to Theorem 6.3, the module $\Gamma(\mathbf{T}\mathcal{M}_U)$ is generated, as $C^\infty(\mathcal{M}_U)$-module, by the set of all projectable sections of $\Gamma(\mathbf{T}\mathcal{M}_U)$. This result is essentially a consequence of the fact that, in the local context used in the proof of Theorem 6.3, over such a chart domain U, the bundle $\widetilde{\mathcal{A}}_U$ is a finite-dimensional bundle and so the module of sections of $\widetilde{\mathcal{A}}_U$ over $\mathcal{M}_U$ is finitely generated as $C^\infty(\mathcal{M}_U)$. Thus, if $\mathcal{A}$ is a finite rank bundle over M, then the module $\Gamma(\mathbf{T}\mathcal{M}_U)$ is generated, as $C^\infty(\mathcal{M}_U)$-module, by the set of all projectable sections of $\Gamma(\mathbf{T}\mathcal{M}_U)$.*
Unfortunately, this is no longer true in the convenient context and even in the Banach setting in general.
Note that, under some types of approximation properties of $\mathbb{A}$, we can show that the module generated by the set of all projectable sections of $\Gamma(\mathcal{A}_U)$ as $C^\infty(\mathcal{M}_U)$-module, is dense in $\Gamma(\widetilde{\mathcal{A}}_U)$ as a convenient space (see for instance [KrMi97]). In this case, the $C^\infty(\mathcal{M}_U)$-module, generated by the set of all projectable sections of $\Gamma(\mathbf{T}\mathcal{M}_U)$ will be dense in $\Gamma(\mathbf{T}\mathcal{M}_U)$ (as a convenient space).

Definition 6.8 *We denote by $\mathfrak{P}(\mathbf{T}\mathcal{M}_U)$ the $C^\infty(\mathcal{M}_U)$-module $\Gamma(\mathbf{T}\mathcal{M}_U)$ generated by the set of projectable sections defined on U in the $C^\infty(\mathcal{M}_U)$-module. Each section of $\mathfrak{P}(\mathbf{T}\mathcal{M}_U)$ will be called a (local) quasi-projectable section of $\mathbf{T}\mathcal{M}$.*

Each module $\mathfrak{P}(\mathbf{T}\mathcal{M}_U)$ has the following properties:

Lemma 6.2

(1) For any open subset U in M, there exists a well-defined Lie bracket $[.,.]_{\mathbf{T}\mathcal{M}_U}$ on $\mathfrak{P}(\mathbf{T}\mathcal{M}_U)$ which satisfies the assumption of Definition 3.76, whose restriction to projectable sections is given by the relation (6.11) which provides $\mathfrak{P}(\mathbf{T}\mathcal{M}_U)$ with a Lie algebra structure on $\mathfrak{P}(\mathbf{T}\mathcal{M}_U)$ and the restriction of rĥo to $\mathfrak{P}(\mathbf{T}\mathcal{M}_U)$ is a Lie algebra morphism into $\mathcal{X}(\mathcal{M}_U)$.

(2) For each $x \in M$, there exists a chart domain U around x in M such that $\mathbf{T}\mathcal{M}_U$ is trivializable over $\mathcal{M}_U$ and for each $(m, v) \in \mathbf{T}_m\mathcal{M}$ where $m = (x, e) \in \mathcal{M}_U$, there exists a projectable section $\mathbf{X}$ defined on $\mathcal{M}_U$ such that $\mathbf{X}(m) = (a, v)$.

(3) Assume that we have a Whitney decomposition $\mathbf{T}\mathcal{M} = \mathbf{K}\mathcal{M} \oplus \mathbf{V}\mathcal{M}$ *and let* $p_{\mathbf{K}}$ *be the associated projection on* $\mathbf{K}\mathcal{M}$. *Then, for any section* $\mathbf{X} \in \mathfrak{P}(\mathbf{T}\mathcal{M}_U)$, *the induced section* $\mathbf{X}^K = p_{\mathbf{K}} \circ \mathbf{X}$ *belongs to* $\mathfrak{P}(\mathbf{K}\mathcal{M}_U) = \Gamma(\mathbf{K}\mathcal{M}_U) \cap \mathfrak{P}(\mathbf{T}\mathcal{M}_U)$. *In particular,* $\mathbf{X}^K$ *is projectable if and only if* $\mathbf{X}$ *is so.*

Proof *(1) According to the definition of and the Bracket of projectable sections, the results in Point 1 is a direct application of Theorem 3.7 in Chapter 3.*

(2) Choose $x_0 \in M$. *According to the proof of Theorem 6.3, there exists a chart domain* U *around* x *such that*

- $\mathcal{M}_U \equiv U \times \mathbb{O}$;
- $\mathcal{A}_U \equiv U \times \mathbb{A}$;
- $T\mathcal{M}_U \equiv U \times \mathbb{O} \times \mathbb{M} \times \mathbb{E}$;
- $\widetilde{\mathcal{A}}_U = U \times \mathbb{O} \times \mathbb{A}$;
- $\mathbf{T}\mathcal{M}_U \equiv U \times \mathbb{O} \times \mathbb{A} \times \mathbb{E}$.

Consider $(a_0, v_0) \in \mathbf{T}_{m_0}\mathcal{M}$ *where* $m = (x_0, e_0) \in \mathcal{M}_U$. *Using local coordinates, if* $m_0 \equiv (\mathsf{x}_0, \mathsf{e}_0)$ *and* $(a_0, v_0) \equiv (\mathsf{a}_0, \mathsf{v}_0)$, *we consider the section*

$$\mathsf{X} : U \times \mathbb{O} \to U \times \mathbb{O} \times \mathbb{A} \times \mathbb{E}$$

given by $\mathsf{X}(\mathsf{x}, \mathsf{e}) = (\mathsf{x}, \mathsf{e}, \mathsf{a}_0, \mathsf{v}_0)$. *Then, by construction, the corresponding local section* $\mathbf{X} \equiv X$ *is a projectable section defined on* $\mathcal{M}_U$.

(3) Under the assumptions of (3), note that the restriction $\mathbf{p}_{\mathbf{K}\mathcal{M}}$ *of* $\mathbf{p}_{\widetilde{\mathcal{A}}}$ *to* $\mathbf{K}\mathcal{M}$ *is an isomorphism onto* $\widetilde{\mathcal{A}}$. *Let* $\mathbf{X}$ *be a section of* $\mathbf{T}\mathcal{M}_U$. *Thus the difference* $\mathbf{X} - \mathbf{X}^K$ *is a vertical section. Since the sum of two projectable sections is a projectable section, it follows that* $\mathbf{X}$ *is projectable if and only if* $\mathbf{X}^K$ *is projectable. This implies that the same result is true for quasi-projectable sections.*

According to Chapter 3, Definition 3.80, in this context, we also have:

Notation 6.3 *When no confusion is possible, we simply denote by* $[.,.]_{\mathbf{T}\mathcal{M}}$ *the sheaf of brackets* $\{[.,.]_{\mathbf{T}\mathcal{M}_U}, U$ *open set in* $M\}$.

Remark 6.8 *According to Comments 6.1, when* $\mathcal{A}$ *is a finite rank vector bundle, then the bracket is a localizable Lie bracket on* $\Gamma(\mathbf{T}\mathcal{M})$.

Notation 6.4 *Sheaf* $\mathfrak{P}_{\mathcal{M}}$ *of sections of* $\mathbf{T}\mathcal{M}$.– *Since for any open set* $\mathcal{U}$ *of* $\mathcal{M}$, *if* $U = \mathbf{p}(\mathcal{U})$, *then* $\mathcal{U} \subset \mathcal{M}_U$ *and so we set* $\mathfrak{P}(\mathcal{U}) := \{(\mathbf{X}_{|\mathcal{U}}, \mathbf{X} \in \mathfrak{P}(\mathcal{M}_U\}$. *On the other hand, since the set of smooth sections of a Banach bundle defines a sheaf over its basis (cf. Appendix C), it follows that the set* $\{\mathfrak{P}(\mathbf{T}\mathcal{U}), \mathcal{U}$ *open set in* $\mathcal{M}\}$ *defines a sub-sheaf of modules* $\mathfrak{P}_{\mathcal{M}}$ *of the sheaf* *of modules $\Gamma_{\mathcal{M}}$ of sections of $\mathbf{T}\mathcal{M}$. According to Chapter 3, Definition 3.80, in this context, when no confusion is possible, we simply denote by $[.,.]_{\mathbf{T}\mathcal{M}}$ the sheaf of brackets generated by $\{[.,.]_{\mathbf{T}\mathcal{M}_U}, U$ open set in $M\}$.*

Notation 6.5 *Local version of the Lie bracket $[.,.]_{\mathbf{T}\mathcal{M}}$.– Each section $\mathbf{X}$ in $\mathfrak{P}(\mathbf{T}\mathcal{M}_U)$ has a decomposition:*

$$\mathbf{X} = \sum_{i=1}^{p} f_i \mathbf{X}_i$$

with $f_i \in C^\infty(\mathcal{M}_U)$ and $\mathbf{X}_i$ are projectable sections for all $i \in \{1, \dots, p\}$.

We consider a chart domain U in M for which the situation considered in the proof of Lemma 6.2 (2) is valid. In the associated local coordinates, for any section $\mathbf{X}$ of $\mathbf{T}\mathcal{M}_U$, we have $\mathbf{X}(x, e) \equiv (\mathsf{x}, \mathsf{e}, \mathsf{a}(\mathsf{x}, \mathsf{e}), \mathsf{z}(\mathsf{x}, \mathsf{e}))$. Now, $\mathbf{X}$ is projectable if and only if a only depends on x. Under these notations, if $\mathbf{X}' \equiv (\mathsf{x}, \mathsf{e}, \mathsf{a}'(\mathsf{e}), \mathsf{z}'(\mathsf{x}, \mathsf{e}))$ is another projectable section, we have (cf (6.11) and Notations 6.11):

$$\begin{aligned}[\mathbf{X}, \mathbf{X}']_{\mathbf{T}\mathcal{M}}(x, e) \equiv &(\mathsf{x}, \mathsf{e}, \mathsf{C}_\mathsf{x}(\mathsf{a}, \mathsf{a}') + d\mathsf{a}'(\rho_\mathsf{x}(\mathsf{a})) - d\mathsf{a}(\rho_\mathsf{x}(\mathsf{a}')),\\ &d\mathsf{z}'(\rho_\mathsf{x}(\mathsf{a}), \mathsf{z}) - d\mathsf{z}((\rho_\mathsf{x}(\mathsf{a}'), \mathsf{z}'))).\end{aligned} \tag{6.13}$$

Now consider two sections $\mathbf{X}$ and $\mathbf{X}'$ in $\mathfrak{P}(\mathbf{T}\mathcal{M}_U)$. We can write $\mathbf{X} = \sum_{i=1}^{p} f_i \mathbf{X}_i$ and also $\mathbf{X}' = \sum_{j=1}^{q} f'_j \mathbf{X}'_j$ where $(f_i, f'_i) \in C^\infty(\mathcal{M}_U)^2$ and $\mathbf{X}_i$ and $\mathbf{X}'_j$ are projectable sections for all $i \in \{1, \dots, p\}$ and $j \in \{1, \dots, q\}$. Since the value of the Lie bracket $[\mathbf{X}, \mathbf{X}']_{\mathbf{T}\mathcal{M}}$ at m only depends on the 1-jets of $\mathbf{X}$ and $\mathbf{X}'$ and m, so this value does not depend on the previous decompositions. Now $\mathbf{X}$ and $\mathbf{X}'$ can be also written as a pair $(\mathfrak{a}, X)$ and $(\mathfrak{a}', X')$ respectively. Of course, if $m = (x, e)$, we have:

$$\mathfrak{a}(m) = \sum_{i=1}^{p} f_i(m) a_i(x) \text{ and } \mathfrak{a}'(m) = \sum_{j=1}^{q} f'_i(m) a'_i(x)$$

$$X = \sum_{i=1}^{p} f_i X_i \text{ and } X = \sum_{i=1}^{p} f_i X_i.$$

In local coordinates, we then have:

$$\mathbf{X} \equiv (\mathsf{a}, \mathsf{z}) = \left(\sum_{i=1}^{p} \mathsf{f}_i \mathsf{a}_i, \sum_{i=1}^{p} \mathsf{f}_i \mathsf{z}_i \right)$$

$$\mathbf{X}' \equiv (\mathsf{a}', \mathsf{z}') = \left(\sum_{j=1}^{q} \mathsf{f}'_j \mathsf{a}'_j, \sum_{j=1}^{q} \mathsf{f}_j \mathsf{z}_j \right).$$

Thus the Lie bracket $[\mathbf{X},\mathbf{X}']_{\mathbf{T}\mathcal{M}}$ has the following expression in local coordinates:

$$\begin{aligned}[\mathbf{X},\mathbf{X}']_{\mathbf{T}\mathcal{M}} \equiv (\mathsf{C}(\mathsf{a},\mathsf{a}') + d\mathsf{a}'\,(\rho_{\mathsf{x}}(\mathsf{a}')) \\ - d\mathsf{a}(\rho_{\mathsf{x}}(\mathsf{a}')), d\mathsf{z}'\,(\rho_{\mathsf{x}}(\mathsf{a}),\mathsf{z})) - d\mathsf{z}(\rho_{\mathsf{x}}(\mathsf{a}'),\mathsf{z}'))\end{aligned} \tag{6.14}$$

where $\mathsf{C} : U \to L^2_{\mathrm{alt}}(\mathbb{A})$ is defined in local coordinates by the Lie bracket $[.,.]_{\mathcal{A}}$ and where

$$\mathsf{C}(\mathsf{a},\mathsf{a}')(\mathsf{m}) = \sum_{i=1}^{p}\sum_{j=1}^{q} \mathsf{f}_i(\mathsf{m})\mathsf{f}'_j(\mathsf{m})\mathsf{C}_{\mathsf{x}}(\mathsf{a}_i(x),\mathsf{a}'_j(\mathsf{x})).$$

Now, from Lemma 6.2, $(\mathfrak{P}(\mathbf{T}\mathcal{M}_U),[.,.]_{\mathbf{T}\mathcal{M}})$ is a Lie algebra and $\hat{\rho}$ induces a Lie algebra morphism from $(\mathfrak{P}(\mathbf{T}\mathcal{M}_U),[.,.]_{\mathbf{T}\mathcal{M}})$ to the Lie algebra of vector fields on $\mathcal{M}_U$. In this way, so we get:

Theorem 6.5 *The set of modules $\{\mathfrak{P}(\mathbf{T}\mathcal{M}_U) : U$ open set in $M\}$ defines a sheaf $\mathfrak{P}_{\mathcal{M}}$ on $\mathcal{M}$ which gives rise to a strong partial convenient Lie algebroid on the anchored bundle $(\mathbf{T}\mathcal{M},\hat{p},\mathcal{M},\hat{\rho})$. Moreover, the restriction of the bracket $[.,.]_{\mathbf{T}\mathcal{M}}$ to the module of vertical sections induces a convenient Lie algebroid structure on the anchored subbundle $(\mathbf{V}\mathcal{M},\hat{p}_{|\mathbf{V}\mathcal{M}},\mathcal{M},\hat{\rho})$ which is independent on the bracket $[.,.]_{\mathcal{A}}$.*

From Remark 6.8, we obtain

Corollary 6.2 *If $\mathcal{A}$ is a finite-dimensional fibre Banach bundle, then $(\mathbf{T}\mathcal{M},\hat{p},\mathcal{M},\hat{\rho},[.,.]_{\mathbf{T}\mathcal{M}})$ is a convenient Lie algebroid .*

Proof of Theorem 6.5 *According to Notations 6.4,*

$$\{\mathfrak{P}(\mathbf{T}\mathcal{M}_U) : U \text{ open set in } M\}$$

defines a sheaf $\mathfrak{P}_{\mathcal{M}}$ on $\mathcal{M}$.

From Lemma 6.2, $(\mathfrak{P}(\mathbf{T}\mathcal{M}_U),[.,.]_{\mathbf{T}\mathcal{M}})$ is a Lie algebra and $\hat{\rho}$ induces a Lie algebra morphism from $(\mathfrak{P}(\mathbf{T}\mathcal{M}_U),[.,.]_{\mathbf{T}\mathcal{M}})$ to the Lie algebra of vector fields on $\mathcal{M}_U$. According to the definition of $\mathfrak{P}(\mathbf{T}\mathcal{U})$, the same properties is true for $(\mathfrak{P}(\mathbf{T}\mathcal{U}),[.,.]_{\mathbf{T}\mathcal{M}})$ for any open set $\mathcal{U}$ in $\mathcal{M}$. Thus we obtain a sheaf $\mathfrak{P}_{\mathcal{M}}$ of Lie algebras and $C^\infty_{\mathcal{M}}$ modules on $\mathcal{M}$, which implies that $(\mathbf{T}\mathcal{M},\hat{p},\mathcal{M},\hat{\rho},[.,.]_{\mathbf{T}\mathcal{M}})$ is a convenient partial Lie algebroid which is strong, according to Lemma 6.2 (2).

It remains to prove the last property. From its definition, the restriction of $\hat{\rho}$ to the vertical bundle $\mathbf{V}\mathcal{M} \to \mathcal{M}$ is an isomorphism onto the vertical bundle of the tangent bundle $T\mathcal{M} \to \mathcal{M}$. Now, from the definition of the bracket of projectable sections, it is clear that the bracket of two (local) vertical sections of $\mathbf{T}\mathcal{M} \to \mathcal{M}$ is also a (local) vertical section which is independent on the choice of $[.,.]_{\mathcal{A}}$.

Remark 6.9 *Given a fibred morphism* $\Psi : \mathcal{M}_1 \to \mathcal{M}_2$ *between fibred manifold* $\mathbf{p}_1 : \mathcal{M}_1 \to M_1$ *and* $\mathbf{p}_2 : \mathcal{M}_2 \to M_2$ *over* $\psi : M_1 \to M_2$ *and a morphism of anchored bundles* φ *between two Lie algebroids* $(\mathcal{A}_1, \pi_1, M_1, \rho_1, [.,.]_{\mathcal{A}_1})$ *and* $(\mathcal{A}_2, \pi_2, M_2, \rho_2, [.,.]_{\mathcal{A}_2})$*, over* $\psi : M_1 \to M_2$*, then the prolongation* $\mathbf{T}\Psi : \mathbf{T}^{\mathcal{A}_1}\mathcal{M}_1 \to \mathbf{T}^{\mathcal{A}_2}\mathcal{M}_2$ *is a Lie morphism of partial convenient Lie algebroids from* $(\mathbf{T}^{\mathcal{A}_1}\mathcal{M}_1, \mathcal{M}_1, \hat{\rho}_1, \mathfrak{P}_{\mathcal{M}_1})$ *to* $(\mathbf{T}^{\mathcal{A}_2}\mathcal{M}_2, \mathcal{M}_2, \hat{\rho}_2, \mathfrak{P}_{\mathcal{M}_2})$ *(cf. Remark 3.17).*

6.4.6 Prolongations and Foliations

We assume that $(\mathcal{A}, M, \rho, [.,.]_{\mathcal{A}})$ *is a split Banach Lie algebroid.*

Under the upper assumptions, by application of Theorem 6.3 and Chapter 8, Theorem 8.3, we obtain the following link between the foliation on M and the foliation on $T\mathcal{M}$:

Theorem 6.6 *Assume that* $(\mathcal{A}, \pi, M, \rho, [.,.]_{\mathcal{A}})$ *is split and the distribution* $\rho(\mathcal{A})$ *is closed.*

(1) The distribution $\hat{\rho}(\mathbf{T}\mathcal{A})$ *is integrable on* $\mathbf{T}\mathcal{M}$*.*

(2) Assume that $\mathcal{M} = \mathcal{A}$*. Let* L *be a leaf of* $\rho(\mathcal{A})$ *and* $(\mathcal{A}_L, \pi_{|L}, L, \rho_L, [.,.]_{\mathcal{A}_L})$ *be the Banach-Lie algebroid which is the restriction of* $(\mathcal{A}, \pi, M, \rho, [.,.]_{\mathcal{A}})$*. Then* $\hat{L} = \mathcal{A}_L = \pi^{-1}(L)$ *is a leaf of* $\hat{\rho}(\mathbf{T}\mathcal{A})$ *such that* $\hat{\mathbf{p}}(\hat{L}) = L$*.*

(3) In the previous context, the partial Banach-Lie algebroid which is the prolongation of $(\mathcal{A}_L, \pi_{|L}, L, \rho_L, \mathfrak{P}_{\mathcal{A}_L})$ *over* $\hat{L}$ *is exactly the restriction to* $\hat{L}$ *of the partial Banach-Lie algebroid* $(\mathbf{T}\mathcal{A}, \hat{p}, \mathcal{A}, \hat{\rho}, \mathfrak{P}_{\mathcal{A}})$*.*

Remark 6.10

(1) If there exists a weak Riemannian metric on $\mathcal{A}$*, the distribution* $\rho(\mathcal{A})$ *is closed and the assumptions of Theorem 6.6 are satisfied. These assumptions are always satisfied if* ρ *is a Fredholm morphism.*

(2) The set of leaves of the foliations defined by $\hat{\rho}(\mathbf{T}\mathcal{A})$ *is*

$$\{\mathcal{A}_L,\ L \text{ leaf of } \rho(\mathcal{A})\ \}.$$

Proof *From Theorem 6.3, if* $m = (x,e) \in \mathcal{A}$*, the fibre* $\mathbf{T}_m\mathcal{M}$ *can be identified with* $\mathcal{A}_x \times \widetilde{\mathcal{E}}_m \equiv \mathbb{A} \times \mathbb{E}$*, so we have* $\hat{\rho}_m(a,v) \equiv (\rho_{\mathsf{x}}(\mathsf{a}), \mathsf{v})$*. It follows that* $\ker(\hat{\rho}_m)$ *can be identified with* $\ker \rho_x \times \mathbf{V}_m\mathcal{M} \subset \mathcal{A}_x \times T_m\mathcal{A}$*. Thus,* $\ker \hat{\rho}_m$ *is split if and only if* $\ker \rho_x$ *is split. Moreover* $\hat{\rho}_m(\mathbf{T}_m\mathcal{A}) = \rho_x(\mathcal{A}_x) \times \mathbf{V}_m\mathcal{M}$ *is closed in* $\mathbf{T}_m\mathcal{M} \equiv \mathbb{M} \times \mathbb{E}$ *if and only if* $\rho_x(\mathcal{A}_x)$ *is closed in* T_xM*. Then (1) will be a consequence of Proposition 8.1 if we show that, for any* $(x,e) \in \mathcal{M}$*, there exists an open neighbourhood* U *of* x *in* M *such that the set* $\mathfrak{P}(\mathbf{T}\mathcal{M}_U)$ *is a generating upper set for* $\hat{\rho}(\mathbf{T}\mathcal{M})$ *around* (x,e) *(cf. Definition 8.16) which satisfies the condition* **(LB)**.

The point $m = (x, e) \in \mathcal{M}$ *is fixed and we choose a chart domain* U *of* $x \in M$ *such that* $\mathcal{M}_U$ *and* $\mathcal{A}_U$ *are trivializable. Without loss of generality, according to the notations in Chapter 3, § 3.15.3, we may assume that* $U \subset \mathbb{M}$, $\mathcal{M}_U = U \times \mathbb{O} \subset \mathbb{M} \times \mathbb{E}$, $\mathcal{A}_U = U \times \mathbb{A}$. *Thus, according to the proof of Theorem 6.3, we have* $\mathbf{T}\mathcal{M}_U = U \times \mathbb{O} \times \mathbb{A} \times \mathbb{E}$. *In this context, if* $\{\mathsf{x}\} \times \mathbb{A} = \ker \rho_{\mathsf{x}} \oplus \mathbb{S}$, *then*

$$\{(\mathsf{x}, \mathsf{e})\} \times \mathbb{A} \times \mathbb{E} = \{(\mathsf{x}, \mathsf{e})\} \times (\ker \rho_{\mathsf{x}} \oplus \mathbb{S}) \times \mathbb{E}.$$

Now consider the upper trivialization $\rho : U \times \mathbb{A} \to U \times \mathbb{M}(= TM_U)$ *and the associated upper trivialization:*

$$\hat{\rho} : U \times \mathbb{O} \times \mathbb{A} \times \mathbb{E} \to U \times \mathbb{O} \times \mathbb{A} \times \mathbb{M}(= T\mathcal{M}_U).$$

Then, from the definition of $\hat{\rho}$, *any upper vector field is of type*

$$\mathbf{X}_{(\mathsf{a},\mathsf{v})} = \hat{\rho}(\mathsf{a}, \mathsf{v}) = (\rho(\mathsf{a}), \mathsf{v})$$

for any $(\mathsf{a}, \mathsf{v}) \in \mathbb{A} \times \mathbb{E}$. *From the proof of Lemma 6.2 (2), it follows that such a vector field is the range, under* $\hat{\rho}$, *of a projectable section. Moreover, the stability of projectable sections under the Lie bracket* $[.,.]_{\mathbf{T}\mathcal{M}_U}$ *and the fact that* $\hat{\rho}$ *induces a Lie algebra morphism from* $\mathfrak{P}(\mathbf{T}\mathcal{M}_U)$ *into the Lie algebra of vector fields on* $\mathcal{M}_U$ *implies that condition* **(LB)** *is satisfied for the set* $\mathfrak{P}(\mathbf{T}\mathcal{M}_U)$.
Assume that $\mathcal{M} = \mathcal{A}$. *Fix some leaf* L *in* M. *If* $(\mathcal{A}_L, \pi_{|L}, L, \rho_L, [.,.]_{\mathcal{A}_L})$ *is the Banach-Lie algebroid which is the restriction of* $(\mathcal{A}, \pi, M, \rho, [.,.]_{\mathcal{A}})$ *then* $\pi^{-1}(L) = \mathcal{A}_L$ *and* $\rho_L(\mathcal{A}_L) = TL$. *Now by construction, the prolongation* $\mathbf{T}\mathcal{A}_L$ *of* $\mathcal{A}_L$ *relative to the Banach-Lie algebroid* $(\mathcal{A}_L, \pi_{|L}, L, \rho_L, [.,.]_{\mathcal{A}_L})$ *is characterized by*

$$\mathbf{T}_{(x,a)}\mathcal{A}_L = \{(b, (v, y)) \in \mathcal{A}_x \times T_{(x,a)}\mathcal{A}_L \ : \ \rho_x(b) = y\}.$$

It follows that $\mathbf{T}\mathcal{A}_L$ *is the restriction of* $\mathbf{T}\mathcal{A}$ *to* $\mathcal{A}_L$ *and also* $\hat{\rho}(\mathbf{T}\mathcal{A}_L) = T\mathcal{A}_L$. *Since* L *is connected, so is* $\mathcal{A}_L$ *and then* $\mathcal{A}_L$ *is a leaf of* $\hat{\rho}(\mathbf{T}\mathcal{M})$.

6.4.7 Derivative Operator on $\mathbf{T}\mathcal{M}$

The purpose of this section is to show that, although $\mathbf{T}\mathcal{M}$ *has only a convenient partial Lie algebroid structure, all the differential operators defined in Chapter 3, § 3.18.6 can also be defined on* $\mathbf{T}\mathcal{M}$.

Consider an open set $\mathcal{U}$ in $\mathcal{M}$. We simply denote $\mathbf{T}\mathcal{U}$ the restriction of $\mathbf{T}\mathcal{M}$ to $\mathcal{U}$. Note that $U = \mathbf{p}(\mathcal{U})$ is an open set in M and of course $\mathcal{U}$ is contained in the open set $\mathcal{M}_U$; so we have a natural restriction map from $\Gamma(\mathbf{T}\mathcal{M}_U)$ (resp. $\mathfrak{P}(\mathbf{T}\mathcal{M}_U)$) into $\Gamma(\mathbf{T}\mathcal{U})$ (resp. $\mathfrak{P}(\mathbf{T}\mathcal{U})$).

By application of results of Chapter 3, § 3.18.6, we then have the announced following result:

Theorem 6.7 *Fix some open set $\mathcal{U}$ in $\mathcal{M}$.*

(1) Fix some projectable section $\mathfrak{u} \in \Gamma(\mathbf{T}\mathcal{U})$. For any k-form ω on $\mathcal{U}$ (resp. any tensor of Q of type$(r,0)$ on $\mathbf{T}\mathcal{U}$) the Lie derivative $L^{\hat{\rho}}_{\mathfrak{u}}\omega$ (resp. $L^{\hat{\rho}}_{\mathfrak{u}}A$) is a well-defined k-form on $\mathcal{U}$ (resp. a tensor of type $(r,0)$ on $\mathbf{T}\mathcal{U}$).

(2) Fix some k-form ω on $\mathcal{U}$. The exterior derivative $d_{\hat{\rho}}\omega$ of ω is well defined.

6.5 Projective Limits of Prolongations of Banach-Lie Algebroids

Definition 6.9 *Consider a projective sequence of Banach-Lie algebroid bundles*

$$\left((\mathcal{A}_i, \pi_i, M_i, \rho_i, [.,.]_i), \left(\zeta_i^j, \xi_i^j, \delta_i^j\right)\right)_{(i,j)\in\mathbb{N}^2, j\geq i}$$

(resp. of Banach bundles $\left((\mathcal{E}_i, \pi_i, M_i, \rho_i, [.,.]_i), \left(\xi_i^j, \delta_i^j\right)\right)_{(i,j)\in\mathbb{N}^2, j\geq i}$).
A sequence of open fibred manifolds $\mathcal{M}_i$ of $\mathcal{E}_i$ is called compatible with algebroid prolongations, if, for all $(i,j) \in \mathbb{N}^2$ such that $j \geq i$, we have

(PSPBLAB1) $\xi_i^j(\mathcal{M}_j) \subset \mathcal{M}_i$;

(PSPBLAB2) $\mathbf{p}_i \circ \xi_i^j = \delta_i^j \circ \mathbf{p}_j$.

Under the assumptions of Definition 6.9, for each $i \in \mathbb{N}$, we denote by $(\mathbf{T}\mathcal{M}_i, \hat{\mathbf{p}}_i, \mathcal{M}_i, \hat{\rho}_i)$ the prolongation of $\mathcal{A}_i$ over $\mathcal{M}_i$ and $[.,.]_{\mathbf{T}\mathcal{M}_i}$ the prolongation of the Lie bracket $[.,.]_{\mathcal{A}_i}$ on projectable sections of $\mathbf{T}\mathcal{M}_i$.

We then have the following result.

Proposition 6.5 *Consider a sequence a projective sequence of Banach-Lie algebroid bundles*

$$\left((\mathcal{A}_i, \pi_i, M_i, \rho_i, [.,.]_i), \left(\zeta_i^j, \xi_i^j, \delta_i^j\right)\right)_{(i,j)\in\mathbb{N}^2, j\geq i}$$

(resp. Banach bundles $\left((\mathcal{E}_i, \pi_i, M_i, \rho_i, [.,.]_i), \left(\xi_i^j, \delta_i^j\right)\right)_{(i,j)\in\mathbb{N}^2, j\geq i}$)
and a sequence of open fibred manifolds $\mathcal{M}_i$ of $\mathcal{E}_i$ compatible with algebroid prolongations. Then

(1) $\left(\mathbf{T}\mathcal{M} = \varprojlim \mathbf{T}\mathcal{M}_i, \hat{\mathbf{p}} = \varprojlim \hat{\mathbf{p}}_i, M = \varprojlim \mathcal{M}_i, \hat{\rho} = \varprojlim \hat{\rho}_i\right)$ is a Fréchet anchored bundle which is the prolongation of the projective limit $\left(\mathcal{A} = \varprojlim \mathcal{A}_i, \pi = \varprojlim \pi_i, M = \varprojlim M_i\right)$ over M.

(2) Consider any open U in M and a sequence of open set U_i in M_i such that $U = \varprojlim U_i$. We denote by $\mathfrak{P}^{pl}(\mathbf{T}\mathcal{M}_U)$ the $C^\infty(U)$ module generated by all projective limits $\mathbf{X} = \varprojlim \mathbf{X}_i$ of projectable sections $\mathbf{X}_i$ of $\mathbf{T}\mathcal{M}_i$ over

$\{\mathcal{M}_i\}_{U_i}$.
Then there exists a Lie bracket $[.,.]_{\mathbf{T}\mathcal{M}}$ *defined on* $\mathfrak{P}^{pl}(\mathbf{T}\mathcal{M}_U)$ *which satisfies the assumptions of Definition 3.76 characterized by*

$$[\mathbf{X},\mathbf{X}']_{\mathbf{T}\mathcal{M}_U} = \underleftarrow{\lim}[\mathbf{X}_i,\mathbf{X}'_i]_{\mathbf{T}\mathcal{M}_i}$$

where $\mathbf{X} = \underleftarrow{\lim}\mathbf{X}_i$ *and* $\mathbf{X}' = \underleftarrow{\lim}\mathbf{X}'_i$.

Note that $\mathfrak{P}^{pl}(\mathbf{T}\mathcal{M}_U)$ is a submodule of the module $\mathfrak{P}(\mathbf{T}\mathcal{M}_U)$ generated by projectable sections $\mathbf{T}\mathcal{M}_U$. Therefore, by analog argument as used in the proof of Theorem 6.5, the set $\{\mathfrak{P}^{pl}(\mathbf{T}\mathcal{M}_U), U \text{ open set in } M\}$ generates a sheaf of $\mathfrak{P}^{pl}_{\mathcal{M}}$ over $\mathcal{M}$. Moreover, for any open set $\mathcal{U}$ in $\mathcal{M}$, according to Proposition 6.5, the restriction of $\hat{\rho}$ to each $\mathfrak{P}^{pl}(\mathbf{T}\mathcal{U})$ is a Lie algebra morphism into the Lie algebra morphism into the Lie algebra of vector fields on $\mathcal{U}$.. Thus we obtain:

Theorem 6.8 *Consider a sequence a projective sequence of Banach-Lie algebroid bundles*

$$\left((\mathcal{A}_i,\pi_i,M_i,\rho_i,[.,.]_i),\left(\zeta_i^j,\xi_i^j,\delta_i^j\right)\right)_{(i,j)\in\mathbb{N}^2,j\geq i}$$

(resp. Banach bundles $\left((\mathcal{E}_i,\pi_i,M_i,\rho_i,[.,.]_i),\left(\xi_i^j,\delta_i^j\right)\right)_{(i,j)\in\mathbb{N}^2,j\geq i}$*) and a sequence of open fibered manifolds* $\mathcal{M}_i$ *of* $\mathcal{E}_i$ *compatible with algebroid prolongations. Then* $\left(\mathbf{T}\mathcal{M} = \underleftarrow{\lim}\mathbf{T}\mathcal{M}_i, \hat{\mathbf{p}} = \underleftarrow{\lim}\hat{\mathbf{p}}_i, M = \underleftarrow{\lim}\mathcal{M}_i, \hat{\rho} = \underleftarrow{\lim}\hat{\rho}_i\right)$ *is a Fréchet anchored bundle which is the prolongation over* M *of* $\left(\mathcal{A} = \underleftarrow{\lim}\mathcal{A}_i, \pi = \underleftarrow{\lim}\pi_i, M = \underleftarrow{\lim}M_i\right)$. *Moreover* $\left(\mathbf{T}\mathcal{M},\hat{\mathbf{p}},\mathcal{M},\hat{\rho},\mathfrak{P}^{pl}_{\mathcal{M}}\right)$ *is a strong partial Fréchet Lie algebroid.*

Remark 6.11 *In general since the projective limit of a projective sequence of Banach algebroids has only a structure of partial Fréchet Lie algebroid, it follows that* $\mathfrak{P}^{pl}_{\mathcal{M}}$ *is a subsheaf of modules of* $\mathfrak{P}_{\mathcal{M}}$ *and the inclusion is strict in the sense that for each open* $\mathcal{U}$ *in* $\mathcal{M}$, *the inclusion of* $\mathfrak{P}^{pl}(\mathcal{U})$ *in* $\mathfrak{P}(\mathcal{U})$ *is strict. Thus we do not have a structure of strong partial Fréchet Lie algebroid defined on* $\mathfrak{P}_{\mathcal{M}}$ *as in Theorem 6.5.*

Proof of Proposition 6.5
(1) According to **(PSPBLAB1)** *and Chapter 4, Theorem 4.15, we obtain that* $\left(\underleftarrow{\lim}\mathcal{A}_i, \underleftarrow{\lim}\pi_i, \underleftarrow{\lim}M_i, \underleftarrow{\lim}\rho_i\right)$ *is a Fréchet anchored bundle.*
From **(PSPBLAB2)** *and Chapter 4, Proposition 4.4, we obtain a structure of Fréchet vector bundle on* $\left(\underleftarrow{\lim}\mathcal{E}_i, \underleftarrow{\lim}\mathbf{p}_i, \underleftarrow{\lim}M_i\right)$. *Since each* $\mathcal{M}_i$ *is an open manifold of* $\mathcal{E}_i$ *such that the restriction of* $\mathbf{p}_i$ *is a surjective fibration of* $\mathcal{M}_i$ *over* M_i *and we have* $\xi_i^j(\mathcal{M}_j) \subset \mathcal{M}_i$, *it follows that* $(\mathcal{M}_i,\xi_i^j)_{(i,j)\in\mathbb{N}^2,j\geq i}$ *is a projective sequence of Banach manifolds and so the restriction of* $\mathbf{p} = \underleftarrow{\lim}\mathbf{p}_i$

to $\mathcal{M} = \varprojlim \mathcal{M}_i$ is a surjective fibration onto M

Recall that

$$\mathbf{T}\mathcal{M}_j = \{(a_j, \mu_j) \in \mathcal{A}_{x_j} \times T_{m_j}\mathcal{M}_j : \ \rho_j(a_j) = T_{m_j}\mathbf{p}_j(\mu_j)\}.$$

Let (a_j, μ_j) be in $\mathbf{T}\mathcal{M}_j$ and consider $a_i = \xi_i^j(a_j)$ and $\mu_i = T\xi_i^j(\mu_j)$. We then have:

$$\rho_i(a_i) = \rho_i \circ \xi_i^j(a_j) = T\delta_i^j \circ \rho_j(a_j)$$

and also

$$T_{m_i}\mathbf{p}_i(\mu_i) = T_{m_i}\mathbf{p}_i \circ T\xi_i^j(\mu_j) = T\delta_i^j \circ T_{m_j}\mathbf{p}_j(\mu_j).$$

Since $\rho_j(a_j) = T_{m_j}\mathbf{p}_j(\mu_j)$, we then obtain $\rho_i(a_i) = T_{m_i}\mathbf{p}_i(\mu_i)$.

So $\mathbf{T}\xi_i^j : \mathbf{T}\mathcal{M}_j \to \mathbf{T}\mathcal{M}_i$ is a morphism of Banach bundles and we have the following commutative diagram

$$\begin{array}{ccc}
\mathbf{T}\mathcal{M}_i & \xleftarrow{\mathbf{T}\xi_i^j} & \mathbf{T}\mathcal{M}_j \\
\downarrow{\scriptstyle \hat{\rho}_i} & & \downarrow{\scriptstyle \hat{\rho}_j} \\
T\mathcal{M}_i & \xleftarrow{T\xi_i^j} & T\mathcal{M}_j
\end{array}$$

We deduce that $\left(\left(\mathbf{T}^{\mathcal{A}_i}\mathcal{M}_i, \hat{\mathbf{p}}_i, \mathcal{M}_i, \hat{\rho}_i\right), \left(\mathbf{T}\xi_i^j, \xi_i^j\right)\right)_{(i,j)\in\mathbb{N}^2, j\geq i}$ is a projective sequence of Banach anchored bundles. Applying again Theorem 4.15, we get a Fréchet anchored bundle structure on $\left(\varprojlim \mathbf{T}\mathcal{M}_\mathbf{i}, \varprojlim \hat{\mathbf{p}}_\mathbf{i}, \varprojlim \mathcal{M}_\mathbf{i}\right)$ over $\varprojlim \mathcal{M}_i$ and which appears as the prolongation of $\left(\varprojlim \mathcal{A}_i, \varprojlim \pi_i, \varprojlim M_i, \varprojlim \rho_i\right)$ over $\varprojlim \mathcal{M}_i$.

(2) Let U be an open set in M. There exists U_i in M_i such that $\delta_i(U) \subset U_i$ for each $i \in \mathbb{N}$ and so that $U = \varprojlim U_i$. Now, from the definition of $\{\mathcal{M}_i\}_{U_i}$, we must have $\mathcal{M}_U = \varprojlim \{\mathcal{M}_i\}_{U_i}$.

Recall that a projectable section $\mathbf{X}_i$ on $\{\mathcal{M}_i\}_{U_i}$ is characterized by a pair $(\mathfrak{a}_i, X_i)$ where $\mathfrak{a}_i$ is a section of $\{\mathcal{A}_i\}_{U_i}$ and X_i is a vector field on $\{\mathcal{M}_i\}_{U_i}$ such that $\rho_i \circ \mathfrak{a}_i = T\mathbf{p}_i(X_i)$. Assume that $\mathfrak{a} = \varprojlim \mathfrak{a}_i$ and $X = \varprojlim X_i$, from the compatibility with bonding maps for sequences of sections $(\mathfrak{a}_i)$, anchors (ρ_i) and vector fields (X_i) we must have then have $T\mathbf{p}(X) = \rho \circ \mathfrak{a}$. But from Theorem 6.3, it follows that $(\mathfrak{a} \circ \mathbf{p}, X)$ defines a projectable section $\mathbf{X}$ over $\mathcal{M}_U$ and so $\mathbf{X} = \varprojlim \mathbf{X}_i$.

On the other hand, recall that from (6.11), for each $i \in \mathbb{N}$, we have

$$[\mathbf{X}_i, \mathbf{X}'_i]_{\mathbf{T}\mathcal{M}_i} = ([\mathfrak{a}_i, \mathfrak{a}'_i]_{\mathcal{A}_i} \circ \mathbf{p}, [X_i, X'_i]) \tag{6.15}$$

Now since ξ_i^j *is a Lie algebroid morphism, over* δ_i^j*, according to Definition 3.82 we have*

$$\xi_i^j([\mathfrak{a}_j, \mathfrak{a}'_j]_{\mathcal{A}_j})(x_j) = ([\mathfrak{a}_i, \mathfrak{a}_i]_{\mathcal{A}_i})\left(\delta_i^j(x_j)\right). \tag{6.16}$$

Since $\delta_i^j \circ \mathbf{p}_j = \mathbf{p}_i \circ \lambda_i^j$ *this implies that:*

$$\xi_i^j([\mathfrak{a}_j, \mathfrak{a}'_j]_{\mathcal{A}_j}) \circ \mathbf{p}_j(m_j) = ([\mathfrak{a}_i, \mathfrak{a}'_i]_{\mathcal{A}_i}) \circ \mathbf{p}_i \circ \xi_i^j(m_j). \tag{6.17}$$

Naturally, since X_i *(resp.* X'_i*) and* X_j *(resp.* X'_j*) are* ξ_i^j *related, we also have*

$$[X_i, X'_i)]\left(\xi_i^j(m_j)\right) = T\xi_i^j\left([X_j, X_j]\right)(m_j). \tag{6.18}$$

From (6.15) and (6.4.1) we then obtain

$$\begin{aligned}
\mathbf{T}\xi_i^j\left([\mathbf{X}_i, \mathbf{X}'_j]_{\mathbf{T}\mathcal{M}_i}\right)(m_j) &= \left(\xi_i^j([\mathfrak{a}_j, \mathfrak{a}'_j]_{\mathcal{A}_j}) \circ \mathbf{p}_j(m_j), T_{m_j}\xi_i^j([X_i, X'_j])\right) \\
&= \left([\mathfrak{a}_i, \mathfrak{a}'_i]_{\mathcal{A}_i} \circ \mathbf{p}_i \circ \xi_i^j(m_j), [X_i, X'_i]) \circ \xi_i^j(m_j)\right) \\
&= ([\mathbf{X}_i, \mathbf{X}'_i]_{\mathbf{T}\mathcal{M}_i}) \circ \xi_i^j(m_j).
\end{aligned} \tag{6.19}$$

It follows that we can define:

$$[\mathbf{X}, \mathbf{X}']_{\mathbf{T}\mathcal{M}_U} = \varprojlim [\mathbf{X}_i, \mathbf{X}'_i]_{\mathbf{T}\mathcal{M}_i}. \tag{6.20}$$

Now, since each bracket $[.,.]_{\mathbf{T}\mathcal{M}_i}$ *satisfies the Jacobi identity, from* (6.20)*, the same is true for* $[.,.]_{\mathbf{T}\mathcal{M}}$ *on projective limit* $\mathbf{X}$ *and* $\mathbf{X}'$ *of projective sections* $(\mathbf{X}_i)$ *and* $(\mathbf{X}'_i)$*. Finally, as*

$$[\hat{\rho}_i(\mathbf{X}_i), \hat{\rho}_i\mathbf{X}'_i)]_{\mathbf{T}\mathcal{M}_i} = \hat{\rho}_i\left([\mathbf{X}_i, \mathbf{X}_i]_{\mathbf{T}\mathcal{M}_i}\right). \tag{6.21}$$

from from the compatibility with bonding maps for sequences of sections $(\mathfrak{a}_i)$*, anchors* (ρ_i) *vector fields* (X_i) *and Lie brackets* $[.,.]_{\mathbf{T}\mathcal{M}_i}$ *on projective sequence* $(\mathbf{T}\mathcal{M}_i)$*, it follows that* $\hat{\rho}$ *satisfies the same type of relation as (6.21).*

Now using by same arguments as used in the proof of Lemma 6.2 we show that we can extend this bracket to a Lie bracket on the module $\mathfrak{P}^{pl}(\mathbf{T}\mathcal{M}_U)$ *so that we have a Lie algebra and the restriction of* $\hat{\rho}$ *to this Lie algebra is a morphism of Lie algebra into the Lie algebra of vector fields on* $\mathcal{M}_U$*.*

6.6 Direct Limits of Prolongations of Banach-Lie Algebroids

As in the previous section we introduce

Definition 6.10 *Consider a direct sequence of Banach-Lie algebroid bundles*

$$\left((\mathcal{A}_i,\pi_i,M_i,\rho_i,[.,.]_{\mathcal{A}_i}),\left(\eta_i^j,\chi_i^j,\varepsilon_i^j\right)\right)_{(i,j)\in\mathbb{N}^2,i\leq j}$$

(resp. of Banach bundles $\left((\mathcal{E}_i,\pi_i,M_i),\left(\chi_i^j,\varepsilon_i^j\right)\right)_{(i,j)\in\mathbb{N}^2,i\leq j}$*).*
A sequence of open fibred manifolds $\mathcal{M}_i$ *of* $\mathcal{E}_i$ *is called compatible with algebroid prolongations, if for all* $(i,j)\in\mathbb{N}^2$ *such that* $i\leq j$*, we have*

(DSPBLAB1) $\chi_i^j(\mathcal{M}_j)\subset\mathcal{M}_i;$

(DSPBLAB2) $\varepsilon_i^j\circ\mathbf{p}_i=\mathbf{p}_j\circ\chi_i^j.$

Remark 6.12 *The context of direct limit in which we work concerns ascending sequences of Banach manifolds* $(M_i)_{i\in\mathbb{N}}$ *where* M_i *is a closed submanifold of* M_{i+1}*. The reason of this assumption is essentially that the direct limit has a natural structure of (n.n.H) convenient manifold.*
Although each manifold $\mathcal{M}_i$ *is open in* $\mathcal{E}_i$*, since* $\mathcal{E}_i$ *is a closed subbundle of* $\mathcal{E}_j$*, it follows that* $\left(\mathcal{M}_i,\chi_i^j\right)_{(i,j)\in\mathbb{N}^2,i\leq j}$ *is an ascending sequence of convenient manifolds.*

As in the previous section, for each $i\in\mathbb{N}$, we denote by $(\mathbf{T}\mathcal{A}_i,\hat{\mathbf{p}}_i,\mathcal{A}_i,\hat{\rho}_i)$ the prolongation of $\mathcal{A}_i$ over $\mathcal{M}_i=\mathcal{A}_i$ and $[.,.]_{\mathbf{T}\mathcal{A}_i}$ the prolongation of the Lie bracket $[.,.]_{\mathcal{A}_i}$ on projectable sections of $\mathbf{T}\mathcal{A}_i$.

Adapting the argument used in proof of Proposition 6.5 to this setting of strong ascending sequences and direct limits, we have the result below. Note that, in this context, the prolongation is not Hausdorff in general. However, all the arguments used in the proofs are local and so they still work in this context.

Proposition 6.6 *Consider a sequence a projective sequence of Banach-Lie algebroid bundles*

$$\left((\mathcal{A}_i,\pi_i,M_i,\rho_i,[.,.]_{\mathcal{A}_i}),\left(\eta_i^j,\chi_i^j,\varepsilon_i^j\right)\right)_{(i,j)\in\mathbb{N}^2,i\leq j}$$

(resp. Banach bundles $\left((\mathcal{E}_i,\pi_i,M_i,\rho_i,[.,.]_i),\left(\xi_i^j,\varepsilon_i^j\right)\right)_{(i,j)\in\mathbb{N}^2,i\leq j}$*) and a sequence* $(\mathcal{M}_i)_{i\in\mathbb{N}}$ *where, for any* $i\in\mathbb{N}$*,* $\mathcal{M}_i$ *is an open fibered manifolds of* $\mathcal{E}_i$ *compatible with algebroid prolongations. Then*

(1) $\left(\varinjlim\mathbf{T}\mathcal{M}_i,\varinjlim\hat{\mathbf{p}}_i,\varinjlim\mathcal{M}_i,\varinjlim\hat{\rho}_i\right)$ *is a convenient anchored bundle which is the prolongation of* $\left(\varinjlim\mathcal{A}_i,\varinjlim\pi_i,M_i\right)$ *over* $\varinjlim\mathcal{M}_i$*.*

(2) *Consider any of open U in M and a sequence $(U_i)_{i\in\mathbb{N}}$ where, for any $i \in \mathbb{N}$, U_i is open in M_i and such that $U = \underrightarrow{\lim} U_i$. We denote by $\mathfrak{P}^{dl}(\mathbf{T}\mathcal{M}_U)$ the $C^\infty(U)$ module generated by all direct limits $\mathbf{X} = \underrightarrow{\lim}\mathbf{X}_i$ of projectable sections $\mathbf{X}_i$ of $\mathbf{T}\mathcal{M}_i$ over $\{\mathcal{M}_i\}_{U_i}$.*
Then there exists a Lie bracket $[.,.]_{\mathbf{T}\mathcal{M}_\mathcal{U}}$ defined on $\mathfrak{P}^{pl}(\mathbf{T}\mathcal{M}_U)$ which satisfies the assumptions of Definition 3.76 characterized by

$$[\mathbf{X},\mathbf{X}']_{\mathbf{T}\mathcal{M}_U} = \underrightarrow{\lim}[\mathbf{X}_i,\mathbf{X}'_i]_{\mathbf{T}\mathcal{M}_i}$$

where $\mathbf{X} = \underrightarrow{\lim}\mathbf{X}_i$ and $\mathbf{X}' = \underrightarrow{\lim}\mathbf{X}'_i$.

As in the context of Projective sequences, $\mathfrak{P}^{dl}(\mathbf{T}\mathcal{M}_U)$ is a submodule of the module $\mathfrak{P}(\mathbf{T}\mathcal{M}_U)$ generated by the projectable sections $\mathbf{T}\mathcal{M}_U$. Therefore, again by analog argument as used in the proof of Theorem 6.5, the set $\{\mathfrak{P}^{dl}(\mathbf{T}\mathcal{M}_U), U \text{ open set in } M\}$ generates a sheaf of $\mathfrak{P}^{dl}_{\mathcal{M}}$ over $\mathcal{M}$. Moreover, for any open set $\mathcal{U}$ in $\mathcal{M}$, according to Proposition 6.6, the restriction of $\hat{\rho}$ to each $\mathfrak{P}^{dl}(\mathbf{T}\mathcal{U})$ is a Lie algebra morphism into the Lie algebra of vector fields on $\mathcal{U}$. Thus we obtain:

Theorem 6.9 *Consider a direct sequence of Banach-Lie algebroid bundles*

$$\left((\mathcal{A}_i,\pi_i,M_i,\rho_i,[.,.]_{\mathcal{A}_i}),\left(\eta_i^j,\chi_i^j,\varepsilon_i^j\right)\right)_{(i,j)\in\mathbb{N}^2,i\leq j}$$

(resp. Banach bundles $\left((\mathcal{E}_i,\pi_i,M_i,\rho_i,[.,.]_i),\left(\xi_i^j,\varepsilon_i^j\right)\right)_{(i,j)\in\mathbb{N}^2,i\leq j}$) and a sequence of open fibered manifolds $\mathcal{M}_i$ of $\mathcal{E}_i$ compatible with algebroid prolongations.
Then $\left(\underrightarrow{\lim}\mathbf{T}\mathcal{M}_i,\underrightarrow{\lim}\hat{\mathbf{p}}_i,\underrightarrow{\lim}\mathcal{M}_i,\underrightarrow{\lim}\hat{\rho}_i\right)$ is a convenient anchored bundle which is the prolongation of $\left(\underrightarrow{\lim}\mathcal{A}_i,\underrightarrow{\lim}\pi_i,M_i\right)$ over $\underrightarrow{\lim}\mathcal{M}_i$. Moreover $(\mathbf{T}\mathcal{M},\hat{\mathbf{p}},\mathcal{M},\hat{\rho},\mathfrak{P}^{dl}_{\mathcal{M}})$ is a strong partial convenient Lie algebroid.

6.7 Notes

On finite-dimensional Lie algebroids, the notion of $\mathcal{A}$-connections was firstly introduced in [Fer02] and also plays an important rôle in [CrFe11].
In [Bouc11], on a Lie algebroid provided with a Riemannian metric, in particular $\mathcal{A}$-connections which respect the Riemannian metric, there exists such a unique connection with free torsion, again called the Levi-Civita connection. As in classic Riemannian geometry, this $\mathcal{A}$-connection plays a fundamental rôle. The reader will find in [Bouc11] a complete description of geometrical properties in such a situation. The generalization to these notions in the convenient or even in the Banach setting, as exposed in § 6.1 and 6.3, points out

many problems. Notice, for example, that, in finite dimension, there always exists a Riemannian metric on an algebroid, which is no longer true in the Banach setting. Let us mention some other problems:

- The classical result of the existence of an isomorphism between the cotangent bundle of the dual of an algebroid and its tangent space (cf. [GGU06]) is not true in general.
- The Levi-Civita connection of a Riemannian metric in the general setting of convenient manifold does not exist without strong assumptions (cf. [MMM12]).
- The distribution of the anchor defined on a convenient Lie algebroid is not integrable in general. Even in the Banach setting, we need some assumptions for the validity of such result (cf. § 6.2).

The prolongation of a finite-dimensional Lie algebroid over a fibered manifold, introduced by Higgins and Mackenzie in [HiMa90], offers a nice context in which a Lagrangian formalism can be adapted as it was done by Martínez in [Mar01a]; moreover, such a prolongation has again a Lie algebroid structure. This last result is not true in the convenient setting and even in the Banach or Hilbert one.

7

Partial Poisson Structures

The concept of Poisson structure is a fundamental mathematical tool in classical mechanics, mathematical physics, hydrodynamics and quantum mechanics.
In this chapter, following the works of A. Odzijewicz, T. Ratiu and their collaborators in [OdRa03], [OdRa08] and [Rat11] in the Banach framework, the notion of partial Poisson structure on a convenient manifold M is introduced. This is done by specifying a (weak) subbundle $T^\flat M$ of $T'M$ and an antisymmetric morphism $P : T^\flat M \to TM$ such that the bracket $\{f, g\}_P = - < df, P(dg) >$ defines a Poisson bracket on the algebra $\mathcal{A}$ of smooth functions f on M whose differential df induces a section of $T^\flat M$. In particular, to each such function $f \in \mathcal{A}$ is associated a Hamiltonian vector field $P(df)$. This notion takes naturally place in the framework of infinite-dimensional weak symplectic manifolds and Lie algebroids. After having defined this concept, we will illustrate it by a lot of natural examples. We will also consider the particular situations of projective (resp. direct) limits of such Banach structures.
Finally, we will also give some results on the existence of (weak) symplectic foliations naturally associated with some particular partial Poisson structures.

7.1 Partial Poisson Manifolds

7.1.1 General Properties of a Partial Poisson Manifold

Let M be a convenient manifold modelled on a convenient space $\mathbb{M}$. We denote by : $p_M : TM \to M$ its kinematic tangent bundle ([KrMi97], 28.12) and by $p'_M : T'M \to M$ its kinematic cotangent bundle ([KrMi97], 33.1).

Definition 7.1 *A vector subbundle $p^\flat : T^\flat M \to M$ of $p'_M : T'M \to M$ where $p^\flat : T^\flat M \to M$ is a convenient bundle is a weak subbundle of $p'_M : T'M \to M$ if the canonical injection $\iota : T^\flat M \to T'M$ is a convenient bundle morphism.*

Following [KrMi97], § 48, for any open set U in M we introduce:

Definition 7.2 *Let $\mathfrak{A}(U)$ be the set of smooth functions $f \in C^\infty(U)$ such*

DOI: 10.1201/9781003435587-7

that each iterated derivative $d^k f(x) \in L^k_{\mathrm{sym}}(T_xM, \mathbb{R})$ ($k \in \mathbb{N}^*$) satisfies[1]*:*

$$\forall (u_2, \dots, u_k) \in (T_xM)^{k-1},\ d^k_x f(., u_2, \dots, u_k) \in T^\flat_x M. \tag{7.1}$$

Remark 7.1

1. *If $T^\flat M = T'M$, then for any open set U in M, the algebra $C^\infty(U)$ satisfies the assumption of Definition 7.2 and so in this case $\mathfrak{A}(U) = C^\infty(U)$.*
2. *Consider $f \in \mathfrak{A}(U)$, then for any $x \in U$, any $k \in \mathbb{N}$ any $u_2, \dots, u_k$ in T_xM, if A is an endomorphism of T_xM then the linear map $u \mapsto d^k_x f(A(u), u_2, \dots, u_k)$ belongs to $T^\flat_x M$*

Proposition 7.1 *Fix any open set U in M,*

1. *The set $\mathfrak{A}(U)$ is a subalgebra of $C^\infty(U)$.*
2. *For each $k \in \mathbb{N}$ and local vector fields $X_1, \dots, X_k$ on U, the map $x \mapsto d^k f(X_1, \dots, X_k)(x)$ belongs to $\mathfrak{A}(U)$.*

Proof *1. It is clear that $\mathfrak{A}(U)$ is a real vector subspace of $C^\infty(U)$. We must show that for any $f, g \in \mathfrak{A}(U)$ then the product fg belongs to $\mathfrak{A}(U)$. We fix such f and g in $\mathfrak{A}(U)$. Since df and dg are sections of $T^\flat M$ on U, it follows $d_x(fg)$ belongs to $T^\flat_x M$ for all $x \in U$. More generally for any $k \in \mathbb{N}$, any $x \in U$ and any $u_2, \dots, u_k$ in T_xM we have the following formulae for the derivatives*

$$\begin{aligned} & d^k_x(fg)(., u_2, \dots, u_k) \\ = & \sum_{l=1}^{k} \Big(\sum_{\sigma} d^l_x f(u_{\sigma_1}, \dots, u_{\sigma_l}) d^{k-l}_x g(., u_{\sigma_{l+1}}, \dots, u_{\sigma_{k-1}}) \\ & + d^l_x g(u_{\sigma_1}, \dots, u_{\sigma_l}) df^{k-l}_x(., u_{\sigma_{l+1}}, \dots, u_{\sigma_{k-1}}) \Big) \end{aligned}$$

where the summation on σ is for all $(l, k-l-1)$ shuffles σ of the set $\{1, \dots, l, \dots, k-1\}$.
Since f and g belong to $\mathfrak{A}(U)$, it follows that the second member of the previous relation is a section of $T^\flat_x M$, which ends the proof of the first part.

2. Let us consider the domain U of a local chart (U, ϕ) around x for which $T^\flat M$ is trivializable; so, the inclusion $T^\flat M \to T'M$ corresponds to the inclusion from $U \times \mathbb{M}^\flat$ into $U \times \mathbb{M}'$.
Fix k vector fields $X_1, \dots X_k$ on U; each vector field can be seen as a linear map from $\mathbb{M}$ to $\mathbb{M}$. We then set $\varphi = d^k f(X_1, \dots, X_k)$.
The differential $d^l_x \varphi(u_1, \dots, u_l)$ is a sum of terms of type

$$d^{k+h}_x f(u^{\sigma^0}, X_1^{h_1}(u^{\sigma^1}), \dots X_j^{h_j}(u^{\sigma^j}), \dots X_k^{h_k}(u^{\sigma^k})) \tag{7.2}$$

with the following notations and constraints:

---

[1] This condition in particular implies that for any bounded linear functional λ on $T^\flat_x M$ then $u \mapsto \lambda(d^k_x f(u, u_2, \dots, u_k))$ is bounded on bounded sets of T_xM.

- *$h_0 + h_1 + \cdots + h_k = k + h$ where $h \in \{0, \ldots, l-k\}$;*
- *for $j \in \{0, \ldots, k\}$,*
  - *if $h_j = 0$ then σ^j is an empty set,*
  - *if $h_j > 0$ then σ^j is a strictly increasing sequence $\sigma^j = \left(\sigma^j_1, \ldots, \sigma^j_{h_j}\right)$ and $u^{\sigma^j} = (u_{\sigma^j_1}, \ldots u_{\sigma^j_{h_j}})$;*
- *for $j \in \{1, \ldots, k\}$,*
  - *if $h_j = 0$ then $X^0_j(u^{\sigma^j}) = X_j(x)$ (i.e. σ^j is an empty set),*
  - *if $h_j > 0$ then $X^{h_j}_j(u^{\sigma^j}) = T^{h_j}_x X_j(u_{\sigma^j_1}, \ldots, u_{\sigma^j_{h_j}})$;*
- *$\left\{\{\sigma^j_1, \ldots \sigma^j_{h_j}\},\ j \in \{0, \ldots k\},\ \sigma^j \neq \emptyset\right\}$ is a partition of $\{1, \ldots, l\}$.*

Note that, according to the previous conditions, a term of type (7.2) is uniquely defined by the partition $\left\{\{\sigma^j_1, \ldots \sigma^j_{h_j}\},\ j \in \{0, \ldots, k\}, \sigma^j \neq \emptyset\right\}$ and $d^l_x\varphi(u_1, \ldots, u_l)$ is the sum of terms of type (7.2) for all such partitions

Now, **if we delete** u_1**, in** *$d^l_x\varphi(u_1, \ldots, u_l)$, then, for each term of type (7.2) will give rise to a term in one and only one of the following situations, (according to the associated partition):*

(i) there exists $i \in \{1, \ldots, h_0\}$ with $\sigma^0_i = 1$. Then in all terms associated with such a partition, after having deleted u_1, we obtain a term of type

$$d^{k+h}_x f(u_{\sigma^0_1}, \ldots, \widehat{u_{\sigma^0_i}}, \ldots, u_{\sigma^0_{h_0}}, X^{h_1}_1(u^{\sigma^1}), \ldots X^{h_j}_j(u^{\sigma^j}), \ldots X^{h_k}_k(u^{\sigma^k}))$$

which belongs to $T^\flat_x M$, according to characterization of $\mathfrak{A}(U)$ and the symmetry of $d_x f^{k+l}$;

(ii) there exists $j \in \{1, \ldots, k\}$ such that some $\sigma^j_i = 1$. Therefore, after deleting u_1 in $X^{\sigma^j}_j(u^{\sigma^j})$, we obtain $T^{h_j}_x X_j(u_{\sigma^j_1}, \ldots, \widehat{u_{\sigma^j_i}}, \ldots, u_{\sigma^j_{h_j}})$ which defines an endomorphism of T_xM and all the other terms $X^{h_i}_i(u^{\sigma^i})$, for $i \neq j$ belongs to T_xM. Again, by same arguments in (i) and Remark 7.1, 2., in all terms associated with such a partition, the value of $d^{k+h}_x f(u^{\sigma^0}, X^{h_1}_1(u^{\sigma^1}), \ldots X^{h_j}_j(u^{\sigma^j}), \ldots, X^{h_k}(u^{\sigma^k}))$ after deleting $u_{\sigma^j_i}$, belongs to $T^\flat_x M$.

This implies that $d_x\varphi(., u_2, \ldots, u_l)$ belongs to $T^\flat_x M$ for any $u_2, \ldots, u_l$ in T_xM.

Since $d^l_x\varphi(u_1, u_2, \ldots, u_l)$ is symmetric in $(u_1, \ldots, u_l)$, it follows that

$$\forall i \in \{1, \ldots, l\},\ d^l_x\varphi(u_1, \ldots, \widehat{u_i}, \ldots, u_l) \in T^\flat_x M$$

Now, as $d_x^{k+l} f$ *and* $T_x^{h_j} X_j$ *for* $j \in \{1, \ldots, k\}$*, are bounded symmetric maps, it follows that* $d_x^l \varphi$ *is also a bounded symmetric map.*
Finally, such a result is true for any integer k *and* l*, any* $x \in U$ *and any local vector fields* $X_1, \ldots, X_k$ *on* U *and so the proof is completed.*

Consider the canonical bilinear crossing $< \, , \, >$ between $T'M$ and TM.

Definition 7.3 *A morphism* $P : T^\flat M \to TM$ *is called skew-symmetric if it satisfies the relation*

$$< \xi, P(\eta) >= - < \eta, P(\xi) > \tag{7.3}$$

for ξ *and* η *of* $T_x^\flat M$*.*
We say that P *is an almost Poisson anchor.*

Given such a morphism P, on $\mathfrak{A}(U)$ we define:

$$\{f, g\}_P = - < df, P(dg) > \tag{7.4}$$

In these conditions, the relation (7.4) defines a skew-symmetric bilinear map $\{.,.\}_P : \mathfrak{A}(U) \times \mathfrak{A}(U) \to C^\infty(U)$.

Lemma 7.1 *The bilinear map* $\{.,.\}_P$ *takes values in* $\mathfrak{A}(U)$ *and satisfies the Leibniz property:*

$$\{f, gh\}_P = g\{f, h\}_P + h\{f, g\}_P$$

Proof *Since* $\{f, g\}_P = -df(P(dg))$ *for all* f *and* g *in* $\mathfrak{A}(U)$*, from Proposition 7.1, 2., it follows that* $df(P(dg))$ *belongs to* $\mathfrak{A}(U)$*.*
Finally, from the definition of the bracket, we have

$$\begin{aligned} \{f, gh\}_P &= - < df, Pd(gh) >= -g < df, P(dh) > -h < df, P(dg) > \\ &= g\{f, h\}_P + h\{f, g\}_P \end{aligned}$$

which ends the proof.

Definition 7.4 *Let* $p^\flat : T^\flat M \to M$ *be a weak subbundle of* $p'_M : T'M \to M$ *and* $P : T^\flat M \to TM$ *an almost Poisson anchor.*

(1) We say that $(T^\flat M, M, P, \{.,.\}_P)$ *is a partial Poisson structure on* M *if the bracket* $\{.,.\}_P$ *satisfies the Jacobi identity*

$$\{f, \{g, h\}_P\}_P + \{g, \{h, f\}_P\}_P + \{h, \{f, g\}_P\}_P = 0.$$

In this case, P *is called a Poisson anchor*

(2) Let $\mathfrak{A}$ *be a subalgebra of* $\mathfrak{A}(M)$ *such that the restriction of* $\{.,.\}_P$ *to* $\mathfrak{A} \times \mathfrak{A}$ *takes values in* $\mathfrak{A}$*. We say that* $(M, \mathfrak{A}, \{.,.\}_P)$ *is a partial Poisson manifold.*

Note that, in particular, if $(T^\flat M, M, P, \{.,.\}_P)$ is a partial Poisson structure, then $(M, \mathfrak{A}(M), \{.,.\}_P)$ is always a partial Poisson manifold.

Definition 7.5 *Let* $(M_1, \mathfrak{A}_1, \{.,.\}_{P_1})$ *and* $(M_2, \mathfrak{A}_2, \{.,.\}_{P_2})$ *two partial Poisson manifolds. A smooth map* $\varphi : M_1 \to M_2$ *is called a Poisson morphism if the induced map* $\varphi^* : \mathcal{C}^\infty(M_2) \to \mathcal{C}^\infty(M_1)$*, defined by* $\varphi^*(f) := f \circ \varphi$*, is such that*

$$\begin{cases} \varphi^*(\mathfrak{A}_2) \subset \mathfrak{A}_1 \\ \{\varphi^*(f), \varphi^*(g)\}_{P_1} = \varphi^*(\{f,g\}_{P_2}) \end{cases}$$

If M is a Hilbert (resp. Banach, resp. Fréchet) manifold and if the weak subbundle $T^\flat M$ is a Hilbert (resp. Banach, resp. Fréchet) bundle, the partial Poisson manifold $(M, \mathfrak{A}, \{.,.\}_P)$ will be called a partial Poisson Hilbert (resp. Banach, resp Fréchet) manifold.

As classically, given a partial Poisson manifold $(M, \mathfrak{A}, \{.,.\}_P)$, any function $f \in \mathfrak{A}$ is called a *Hamiltonian* and the associated vector field $X_f = P(df)$ is called a *Hamiltonian vector field.*

We then have $\{f, g\} = X_f(g)$ and also $[X_f, X_g] = X_{\{f,g\}}$ (see [NST14]), which is equivalent to

$$P(d\{f,g\}) = [P(df), P(dg)] \tag{7.5}$$

7.1.2 Examples

Example 7.1 *Finite-dimensional Poisson manifold.— A finite-dimensional Poisson manifold* $(M, \mathcal{C}^\infty(M), \{.,.\})$ *(cf. [Marl83]) is a particular case of partial Poisson manifold. Indeed, to each function* f *on* M *is associated a Hamiltonian vector field* X_f*. Since* $T^\flat M = T^*M$ *is locally generated by differential of functions, therefore the map* $df \mapsto X_f$ *extends to a unique skew-symmetric morphism of bundles* $P : T^*M \to TM$ *such that* $P(df) = X_f$*.*

Example 7.2 *Banach-Poisson manifold.— Let* M *be a Banach manifold. The notion of Banach-Poisson manifold was defined and developed in [OdRa08] and [Rat11]. These authors assume that there exists a Poisson bracket* $\{.,.\}$ *on* $C^\infty(M)$ *such that to each linear functional* ξ *on* M *is associated a section* ξ' *of the bidual* $T''M$ *which, in fact, belongs to* $TM \subset T''M$*. This gives rise to a skew-symmetric morphism* $P : T'M \to TM$*. Conversely, given such a morphism, we get a bracket* $\{.,.\}$ *on* $C^\infty(M)$ *as given in (7.4). Therefore, if this bracket satisfies the Jacobi identity, we get the previous notion of Banach-Lie Poisson manifold (see [Pel12]). Therefore* M *is endowed with a partial Poisson structure.*

Example 7.3 *Weak symplectic convenient manifold.— A weak symplectic manifold is a convenient manifold* M *endowed with a closed 2-form* ω *such that the associated morphism*

$$\begin{array}{rccc} \omega^\flat : & TM & \to & T'M \\ & X & \mapsto & \omega(X,.) \end{array}$$

is injective (see [KrMi97], 48). Therefore if the range $T^\flat M = \omega^\flat(TM)$ *is a weak subbundle of* $T'M$ *we have a skew-symmetric morphism* $P = (\omega^\flat)^{-1} : T^\flat M \to TM$*. According to [KrMi97], 48, we have a bracket which satisfies the Leibniz property on* $\mathfrak{A}(M)$*. Moreover, since* ω *is closed, this bracket satisfies the Jacobi identity (see [KrMi97], Theorem 48.8); so we obtain a partial Poisson structure on* M.

Remark 7.2 *If* M *is a weak symplectic Banach manifold endowed with a weak symplectic 2-form* ω*, then the range* $T'M = \omega^\flat(TM)$ *is not always a weak subbundle of* T^*M *(cf. Marsden example in [Mars72]). However, on* M *we have a subalgebra* $\mathfrak{A}(M)$ *of* $C^\infty(M)$ *which has the same characterization as in Definition 7.2. Then in [KrMi97], 48, the authors shows that it can be provided with a Lie Poisson bracket. The reader will find a generalization of such a situation in the Banach setting in [Tum20].*

Remark 7.3 *One can also find in [Kole07] how* $C^\infty(\mathbb{S}^1)$ *can be endowed with a weaker structure than a partial Poisson one: since the product of two functionals is no well defined (cf. [Olv93], p. 357), the Leibniz rule has no counterpart in this situation.*

7.1.3 Poisson Bracket Associated with Convenient Lie Algebroid

In this section we adapt to the convenient setting the link between Lie algebroid and partial Poisson bracket as exposed in [CaPe12].

Let $\pi : \mathcal{A} \to M$ be a convenient vector bundle and $\pi_* : \mathcal{A}' \to M$ its associated dual convenient bundle. The arguments used in Lemma 4.6 and Proposition 4.7 in [CaPe12] can be easily adapted to the convenient setting and so we can define a characteristic closed subbundle $q^\flat : T^\flat\mathcal{A}' \to \mathcal{A}'$ of $q_{\mathcal{A}'} : T'\mathcal{A}' \to \mathcal{A}'$.

Definition 7.6 *Let* U *be an open set in* M *and denote by* $\mathcal{A}_U$ *(resp.* $\mathcal{A}'_U$*) the restriction of* $\mathcal{A}$ *(resp.* $\mathcal{A}'$ *) to* U*. A function* $\Phi \in C^\infty(\mathcal{A}'_U)$ *is called linear, if the restriction of* Φ *to each fibre* $\mathcal{A}'_x$ *is a linear map into* $\mathbb{R}$[2]

According to Chapter 6, § 6.3.3, to a $\mathfrak{a}$ be a section of $\mathcal{A}_U$ we have associate a map $\Phi_{\mathfrak{a}} : \mathcal{A}' \to \mathbb{R}$ p defined by

$$\Phi_{\mathfrak{a}}(\sigma) = <\sigma, \mathfrak{a} \circ \pi'(\sigma)> . \tag{7.6}$$

and $\mathfrak{a} \mapsto \Phi_{\mathfrak{a}}$ is linear and injective (cf. Chapter 6, Lemma 6.1).

[2]Since Φ is smooth, its restriction to $\mathcal{A}'_x$ is linear and smooth, so must be bounded (cf. [KrMi97]).

We denote by $\mathfrak{A}(\mathcal{A}'_U)$ the set of smooth functions on $\mathcal{A}'_U$ which satisfy the properties of Definition 7.2.

Recall that (cf. Chapter 6, § 6.3.3), we have

$$T^\flat_\sigma \mathcal{A}' = \{d_\sigma(\Phi_\mathfrak{a} + f \circ \pi'); \mathfrak{a} \in \Gamma(\mathcal{A}_U),\ f \in C^\infty(U)\} \tag{7.7}$$

for any small enough open set U around x.

Each function of type $\Phi_\mathfrak{a}$ and $f \circ \pi'$ are such that $d\Phi_\mathfrak{a}$ and $df \circ \pi'$ belong to $\Gamma(T^\flat \mathcal{A}'_U)$ for any section $\mathfrak{a}$ of $\Gamma(\mathcal{A}'_U)$ and any function f of $C^\infty(U)$.

Remark 7.4 *Any $\varphi \in C^\flat(\mathcal{A}'_U)$ whose restriction to each fibre $\mathcal{A}'_x$ is constant naturally defines a function $\bar{\varphi}$ on U such that $\varphi = \bar{\varphi} \circ \pi'$.*
Therefore, we can identify the subalgebra $\pi'(C^\infty(U)) = \{f \circ \pi',\ f \in C^\infty(U)\}$ of $C^\flat(\mathcal{A}'_U)$ with $C^\infty(U)$.

Now, it is clear that the set $C^\flat(\mathcal{A}'_U)$ of functions $F \in C^\infty(\mathcal{A}'_U)$ whose differential belongs to $\Gamma(T^\flat \mathcal{A}'_U)$ has a structure of algebra. Note that from Definition 7.2, if F belongs to $\mathfrak{A}(\mathcal{A}'_U)$, this clearly implies that dF must be a section of $T^\flat \mathcal{A}'$. Since dF is a smooth section of $T'\mathcal{A}'$ and that $T^\flat \mathcal{A}'$ is closed convenient subbundle of $T'\mathcal{A}'$, it follows that dF is a smooth section of $T^\flat \mathcal{A}'_U$ and so $\mathfrak{A}(\mathcal{A}'_U)$ is contained in $C^\flat(\mathcal{A}'_U)$. In fact we have

Proposition 7.2 *Let U be an open set which is a chart domain. The algebra $\mathfrak{A}(\mathcal{A}'_U)$ is equal to $C^\flat(\mathcal{A}'_U)$. This algebra is generated by functions of type $f \circ \pi'$ and $\Phi_\mathfrak{a}$, for any $f \in C^\infty(U)$ and any $\mathfrak{a} \in \Gamma(\mathcal{A}_U)$.*

Proof *Since it is a local problem, according to notations in Chapter 6, § 6.3.3, we may assume that*

- $\mathcal{A} = U \times \mathbb{A}$,
- $\mathcal{A}' = U \times \mathbb{A}'$,

and so

- $T\mathcal{A}' = U \times \mathbb{M} \times \mathbb{A}' \times \mathbb{A}'$
- $T^\flat \mathcal{A}' = U \times \mathbb{M} \times \mathbb{M}' \times \mathbb{A} \subset U \times \mathbb{M} \times \mathbb{M}' \times \mathbb{A}''$

where $\mathbb{A}$ is considered as a subspace of $\mathbb{A}''$.
Taking into account these equivalences, we adopt in this proof the following notations

- $\mathfrak{a} = (x, a)$ *on* $\mathcal{A} = U \times \mathbb{A}$
- $\sigma = (x, \xi)$ *on* $\mathcal{A}' = U \times \mathbb{A}'$
- (σ, X, η) *on* $T\mathcal{A}' = U \times \mathbb{A}' \times \mathbb{M} \times \mathbb{A}'$
- (σ, η, A) *on* $T'\mathcal{A}' = U \times \mathbb{A}' \times \mathbb{M}' \times \mathbb{A}''$

Thus, over such an open set U, from relation (7.7), the module $\Gamma(T^\flat \mathcal{A}'_U)$ of smooth sections of $T^\flat \mathcal{A}'_U$ is generated by the differentials of functions of type $f \circ \pi_*$ and $\Phi_\mathfrak{a}$ with $f \in C^\infty(U)$ and $\mathfrak{a} \in \Gamma(\mathcal{A}_U)$.
Therefore, $C^\flat(\mathcal{A}'_U)$ is generated by functions of type $f \circ \pi_*$ and $\Phi_\mathfrak{a}$ for any $f \in C^\infty(U)$ and any $\mathfrak{a} \in \Gamma(\mathcal{A}_U)$.
According to the previous notations, we have

$$T^\flat_\sigma \mathcal{A}' = \{(x,\xi\} \times \mathbb{M}' \times \mathbb{A} \tag{7.8}$$

Now, it is clear that, for each $f \circ \pi_*$ with $f \in C^\infty(U)$ then, for any vector $(X_2,\eta_2),\dots,(X_k,\eta_k)$ in $T_\sigma \mathcal{A}'$, we have

$$d^k_\sigma(f \circ \pi')(.,(X_2,\eta_2),\dots(X_k,\eta_k)) = d^k_\sigma f(.,X_2,\dots,X_k)$$

and so belongs to $\{\sigma\} \times \mathbb{M} \times \{0\} \subset T_\sigma \mathcal{A}'$ from (7.8). It follows that $f \circ \pi'$ belongs to $\mathfrak{A}(\mathcal{A}'_U)$.
Therefore, the proof will be complete if we show that each function of type $\Phi_\mathfrak{a}$ also belongs to $\mathfrak{A}(\mathcal{A}'_U)$.
We have

$$d_\sigma \Phi_\mathfrak{a}(X,\eta) = <\xi, d_x\mathfrak{a}(X)> + <\eta,\mathfrak{a}>$$

By induction on $k \geq 1$, we can show:

$$\begin{aligned} d^k_\sigma \Phi_\mathfrak{a}((X_1,\eta_1),\dots,(X_k,\eta_k)) &= <\xi, d^k_x\mathfrak{a}(X_1,\dots,X_k)> \\ &+ \sum_{i=1}^{k} <\eta_i, d^{k-1}_x\mathfrak{a}(X_1,\dots,\widehat{X_i},\dots,X_k)> \end{aligned} \tag{7.9}$$

For ξ fixed, the map $f : x \mapsto \Phi_\mathfrak{a}(x,\xi)$ is a smooth map from U to $\mathbb{R}$ and the first term in the RHS of (7.9) is $d^k f$; so $d^k f(.,X_2,\dots X_k)$ belongs to $T^\flat_\sigma \mathcal{A}'$ as we have already seen.
On the other hand, for x fixed and $1 \leq i \leq k$, we consider the map

$$f_i : \eta \mapsto <\eta, d^{k-1}_x\mathfrak{a}(X_1,\dots,\widehat{X_i},\dots,X_k)>$$

defined on $\mathbb{A}'$. Since $d^{k-1}_x\mathfrak{a}$ belongs to the space $L_S(\mathbb{M}^{k-1};\mathbb{A})$ of continuous symmetric $(k-1)$-linear forms, so $d^{k-1}_x\mathfrak{a}(X_1,\dots,\widehat{X_i},\dots,X_k)$ belongs to $\mathbb{A}$ and then df_i is a continuous linear form $<.,d^{k-1}_x\mathfrak{a}(X_1,\dots,\widehat{X_i},\dots,X_k)>$ on $\mathbb{A}'$ which belongs to $\mathbb{A} \subset \mathbb{A}''$; so it belongs to $T^\flat_\sigma \mathcal{A}'$.

In consequence, if in the relation (7.9), we omit (X_1,η_1), it follows that $d^k_\sigma \Phi_\mathfrak{a}(.,(X_2,\eta_2)\dots,(X_k,\eta_k))$ belongs to $T^\flat_\sigma \mathcal{A}'$, which ends the proof.

Definition 7.7 *Consider a partial Poisson structure $(T^\flat \mathcal{A}', \mathcal{A}', P, \{.,.\}_P)$ on $\mathcal{A}'$. We say that P is a linear Poisson anchor or that the Poisson bracket $\{.,.\}_P$ is linear if, for any open set U and any linear function Φ_1 and Φ_2 in $\mathfrak{A}(\mathcal{A}'_U)$, the Poisson bracket $\{\Phi_1,\Phi_2\}_P$ is linear.*

Now, as in [CaPe12], we give an adaptation to the convenient context of the classical result about equivalence between partial Poisson structure on $\mathcal{A}'$ and structure of convenient Lie algebroid on $\mathcal{A}$ (see for example [Marl08]).

Theorem 7.1 *Let $P : T^\flat\mathcal{A}' \to T\mathcal{A}'$ be a Poisson anchor on $\mathcal{A}'$. Then there exists a unique structure of convenient Lie algebroid $(\mathcal{A}, \pi, M, \rho, [.,.]_P)$ characterized in the following way, for any open set U in M:*

$$\forall (\mathfrak{a}_1, \mathfrak{a}_2) \in \Gamma(\mathcal{A}_U)^2, \ \Phi_{[\mathfrak{a}_1,\mathfrak{a}_2]_P} = \{\Phi_{\mathfrak{a}_1}, \Phi_{\mathfrak{a}_2}\}_P \tag{7.10}$$

$$\forall f \in C^\infty(U), \forall \mathfrak{a} \in \Gamma(\mathcal{A}_U), \ \{\Phi_\mathfrak{a}, f \circ \pi'\}_P = df(\rho(\mathfrak{a})) \circ \pi'. \tag{7.11}$$

Conversely, a structure of convenient Lie algebroid $(\mathcal{A}, \pi, M, \rho, [.,.]_\rho)$ defines a unique linear Poisson anchor $P : T'\mathcal{A}' \to T\mathcal{A}'$ which gives rise to a linear Poisson bracket $\{.,.\}_\rho$ on $\mathfrak{A}(\mathcal{A}'_U)$, for any open set U in M, which is characterized by relations (7.10) and (7.11). Moreover, the unique structure of convenient Lie algebroid $(\mathcal{A}, \pi, M, \rho, [.,.]_P)$ to P is the original one.

From Proposition 7.2 and Theorem 7.1, we obtain:

Corollary 7.1 *Let $(\mathcal{A}, \pi, M, \rho, [.,.]_P)$ be a convenient Lie algebroid and $\{.,.\}_P$ the Poisson Lie bracket associated with this structure.*
Then $\left(M, C^\flat(\mathcal{A}'), \{.,.\}_P\right)$ is a partial Poisson manifold.

For the first part of the proof of Theorem 7.1, we need the following Lemma:

Lemma 7.2 *Consider a linear Poisson anchor $P : T^\flat\mathcal{A}' \to T\mathcal{A}'$ on $\mathcal{A}'$ and $\{.,.\}_P$ its associated Poisson bracket. For any open set U we have:*

(1) For any section $\mathfrak{a} \in \Gamma(\mathcal{A}_U)$ and any $f \in C^\infty(U)$, the bracket $\{\Phi_\mathfrak{a}, f \circ \pi'\}_P$ belongs to $C^\infty(U)$[3]

(2) If f and g belong to $\mathfrak{A}(\mathcal{A}'_U)$, then $\{f \circ \pi', g \circ \pi'\}_P = 0$.

Proof of Lemma 7.2

Consider any $\mathfrak{a}$ and $\mathfrak{a}'$ in $\Gamma(\mathcal{A}_U)$, and any $f \in C^\infty(U)$. From Leibniz rule for the bracket $\{.,.\}_P$, we get:

$$\{\Phi_\mathfrak{a}, (f \circ \pi')\Phi_{\mathfrak{a}'}\}_P = (f \circ \pi')\{\Phi_\mathfrak{a}, \Phi_{\mathfrak{a}'}\}_P + \Phi_{\mathfrak{a}'}\{\Phi_\mathfrak{a}, f \circ \pi'\}_P \tag{7.12}$$

Moreover $\{\Phi_\mathfrak{a}, (f \circ \pi')\Phi_{\mathfrak{a}'}\}_P$ is a linear function on $\mathcal{A}'_U$. Since $(f \circ \pi')\{\Phi_\mathfrak{a}, \Phi_{\mathfrak{a}'}\}_P$ is also a linear map, it follows from (7.12) that $\Phi_{\mathfrak{a}'}\{\Phi_\mathfrak{a}, f \circ \pi'\}_P$ is a linear function on $\mathcal{A}'_U$ for any $\mathfrak{a}' \in \Gamma(\mathcal{A}_U)$. It follows that $\{\Phi_\mathfrak{a}, f \circ \pi'\}_P$ must be constant on each fibre and then we get (1).
On the other hand, by same argument, we have

$$\{(f \circ \pi')\Phi_{\mathfrak{a}'}, g \circ \pi'\}_P = (f \circ \pi')\{\Phi_{\mathfrak{a}'}, g \circ \pi'\}_P + \Phi_{\mathfrak{a}'}\{f \circ \pi', g \circ \pi'\}_P$$

[3] cf. Remark 7.4.

So from part (1), we deduce that , $\Phi_{\mathfrak{a}'}\{f\circ\pi', g\circ\pi'\}_P$ *belongs to* $\pi'(C^\infty(U))$ *for any* $\mathfrak{a}'\in\Gamma(\mathcal{A}_U)$*. If follows that we must have* $\{f\circ\pi', g\circ\pi'\}_P=0$ *and we get (2).*

Proof of Theorem 7.1 *Consider a linear Poisson anchor* $P: T^\flat\mathcal{A}'\to T\mathcal{A}'$ *and* $\{.,.\}_P$ *its associated bracket. We fix some open set* U *in* M*. As* $\Phi:\mathfrak{a}\to\Phi_\mathfrak{a}$ *is injective, the pairing* $[.,.]_P$ *given by (7.10) is well defined and is* $\mathbb{R}$*-bilinear and skew-symmetric. It follows from Lemma 7.2, 1. and the Leibniz property of* $\{.,.\}_P$ *that, for some fixed* $\mathfrak{a}\in\Gamma(\mathcal{A}_U)$*, the map* $f\mapsto\{\Phi_s, f\circ\pi'\}_P$ *defines a derivation* $\rho(\mathfrak{a}$ *on* $C^\infty(U)$ *which is characterized by (taking in account Remark 7.4):*

$$\{\Phi_\mathfrak{a}, f\circ\pi'\}_P=< d(f\circ\pi'), P(d\Phi_\mathfrak{a})>=< df, T\pi'(P(\Phi_\mathfrak{a}))>$$

But since $P(d\Phi_\mathfrak{a})$ *is a vector field on* $\mathcal{A}'_U$*, it follows that* $T\pi' P(d\Phi_\mathfrak{a})$ *is a vector field on* U*. We set* $\rho(\mathfrak{a})=T\pi'(P(\Phi_\mathfrak{a}))$*.*
Assume that $d_\sigma\Phi_{\mathfrak{a}_1}=d_\sigma\Phi_{\mathfrak{a}_2}$*. Then, if* $\pi'(\sigma)=x$*, from Chapter 6, Lemma 6.1 3., we must have* $\mathfrak{a}_1(x)=\mathfrak{a}_2(x)$*, which implies that* $\rho(\mathfrak{a})$ *only depends on the value of* $\mathfrak{a}$ *at* $\pi*(\sigma)$*.*
Now, from (7.6), for any $g\in C^\infty(U)$*, we have*

$$\Phi_{g\mathfrak{a}}=(g\circ\pi')\Phi_\mathfrak{a} \tag{7.13}$$

and so, from Leibniz property of $\{.,.\}_P$ *and Lemma 7.2, 2., we have*

$$\{\Phi_{g\mathfrak{a}}, f\circ\pi'\}_P=g\circ\pi'\{\Phi_\mathfrak{a}, f\circ\pi'\}_P+\Phi_\mathfrak{a}\{g\circ\pi', f\circ\pi'\}_P=g\circ\pi'\{\Phi_\mathfrak{a}, f\circ\pi'\}_P$$

which implies that $\rho(g\mathfrak{a})=g\rho(\mathfrak{a})$*.*
Since P *is a bundle morphism, from its definition, it follows that* ρ *is a well-defined bundle morphism from* $T^\flat\mathcal{A}'$ *to* $T\mathcal{A}'$*.*
From (7.10), (7.11), and (7.13), for any $\mathfrak{a}_1$ *and* $\mathfrak{a}_2$ *in* $\Gamma(\mathcal{A}_U)$ *and any* $f\in C^\infty(U)$*, we have:*

$$\Phi_{[\mathfrak{a}_1, f\mathfrak{a}_2]}=\{\Phi_{\mathfrak{a}_1},(f\circ\pi')\Phi_{\mathfrak{a}_2}\}_P=(f\circ\pi')\{\Phi_{\mathfrak{a}_1},\Phi_{\mathfrak{a}_2}\}_P+\Phi_{\mathfrak{a}_2}df(\rho(\mathfrak{a}_1))$$

So we get the following relation:

$$[\mathfrak{a}_1, f\mathfrak{a}_2]_P=f[\mathfrak{a}_1,\mathfrak{a}_2]_P+df(\rho(\mathfrak{a}_1))\mathfrak{a}_2$$

Finally, since $\{.,.\}_P$ *satisfies the Jacobi identity, for any* $\mathfrak{a}_i\in\Gamma(\mathcal{A}_U)$ *where* $i\in\{1,2,3\}$*, we have*

$$\begin{aligned} 0 &= \{\{\Phi_{\mathfrak{a}_1},\Phi_{\mathfrak{a}_2}\}_P,\Phi_{\mathfrak{a}_3}\}_P+\{\{\Phi_{\mathfrak{a}_2},\Phi_{\mathfrak{a}_3}\}_P,\Phi_{\mathfrak{a}_1}\}_P+\{\{\Phi_{\mathfrak{a}_3},\Phi_{\mathfrak{a}_2}\}_P,\Phi_{\mathfrak{a}_1}\}_P \\ &= \Phi_{[[\mathfrak{a}_1,\mathfrak{a}_2]_P,\mathfrak{a}_3]_P+[[\mathfrak{a}_2,\mathfrak{a}_3]_P,\mathfrak{a}_1]_P+[[\mathfrak{a}_3,\mathfrak{a}_2]_P,\mathfrak{a}_1]_P} \end{aligned} \tag{7.14}$$

The injectivity of Φ *implies the Jacobi identity for* $[.,.]_P$*.*

Conversely, let $(\mathcal{A}, \pi, M, \rho, [.,.]_\rho)$ *be a convenient Lie algebroid. We want to associate a linear Poisson anchor* $P : T^\flat\mathcal{A}' \to T\mathcal{A}'$ *such that, according to the first part, the induced Lie algebroid structure on* $(\mathcal{A}', M, \rho)$ *is exactly* $(\mathcal{A}, \pi, M\rho, [.,.]_\rho)$. *For this, for any open set* U *in* M, *we must have:*

$$\{\Phi_{\mathfrak{a}}, f\circ\pi'\}_P = df(\rho(\mathfrak{a}))\{f\circ\pi', g\circ\pi'\}_P = 0$$

for all $\mathfrak{a} \in \Gamma(\mathcal{A}_U)$ *and any* f *and* g *in* $C^\infty(U)$.

We define $P : T^\flat_\sigma\mathcal{A}' \to T_\sigma\mathcal{A}'$ *as follows.*
According to the local notations used at the end of the proof of Proposition 7.2, for any $(\eta_1, \mathsf{w}_2) \in T^\flat_\sigma\mathcal{A}'$, *there exists a germ at* x *of function* f *on* M *and a germ* $\mathfrak{a}$ *at* x *of section of* $\mathcal{A}$ *such that* $\mathsf{w}_2 = d_\sigma\Phi_{\mathfrak{a}|V_\sigma\mathcal{A}'}$ *and a function* f *on* M *such that* $d_\sigma(f\circ\pi')) = \eta_1 - \dfrac{\partial\Phi_{\mathfrak{a}}}{\partial \mathsf{x}}(\sigma)$

By the way, we set

$$P_\sigma(\eta_1, \mathsf{w}_2) = \Phi_{[\mathfrak{a},.]_\rho}(\sigma) + L_{\rho(\mathfrak{a}}(.)\circ\pi'(\sigma) - L_{\rho(.)}(f)\circ\pi'(\sigma) \tag{7.15}$$

For σ *fixed, we must show that the value of* $P_\sigma(\eta, \mathfrak{a})$ *belongs to* $T_\sigma\mathcal{A}'$ *and does not depends of the choice of* $\mathfrak{a}(x)$ *and* f *such that such that the value of* $d_\sigma(\Phi_{\mathfrak{a}} + f\circ\pi')$ *is fixed.*

For simplicity we write $\sigma := (x, \xi)$. *In our notations context, each local sections* $\mathfrak{a}$ *and* $\mathfrak{b}$ *of* $\mathcal{A}$ *are smooth map from* U *to* $\mathbb{A}$. *From relation (3.15) § 3.18.3 Chapter 3, we have*

$$\begin{aligned}
\Phi_{[\mathfrak{a},\mathfrak{b}]_\rho}(\sigma) &= <\xi, d_x\mathfrak{b}(\rho(\mathfrak{a})) - d_x\mathfrak{a}(\rho(\mathfrak{b})) + C_x(\mathfrak{a}(x), \mathfrak{b}(x))> \\
&= \frac{\partial\Phi_{\mathfrak{b}}}{\partial \mathsf{x}}(\rho(\mathfrak{a}(x)) - \frac{\partial\Phi_{\mathfrak{a}}}{\partial \mathsf{x}}(\rho(\mathfrak{b}(x)) + <\xi, C_x(\mathfrak{a}(x), \mathfrak{b}(x))> \\
&= T_\sigma\pi'(d_\sigma\Phi_{\mathfrak{b}})(\rho(\mathfrak{a})) - T_\sigma\pi'(d_\sigma\Phi_{\mathfrak{a}})(\rho(\mathfrak{b})) + <\xi, C_x(\mathfrak{a}(x), \mathfrak{b}(x))>
\end{aligned} \tag{7.16}$$

Thus since (η_1, w_2) *is represented by* $d_\sigma(\Phi_{\mathfrak{a}} + f\circ\pi')$, *we obtain*

$$P_\sigma(\eta_1, \mathsf{w}_2) = -\frac{\partial\Phi_{\mathfrak{a}}}{\partial \mathsf{x}}\circ\rho_x(.) + <\xi, C_x(\mathfrak{a}(x), .)> - d_x f\circ\rho_x(.) + \rho(\mathfrak{a}(x)) \tag{7.17}$$

Therefore we can write $P_\sigma(\eta_1, \mathsf{w}_2)$ *as a pair* $(X, \eta) \in T_\sigma\mathcal{A}'$ *with* $X = \rho(\mathfrak{a}(x))$ *and* $\eta = -T_\sigma\pi'(d_\sigma(\Phi_{\mathfrak{a}} + f\circ\pi')\circ\rho_x(.) + <\xi, C_x(\mathfrak{a}(x), .>$.

Now, if $\mathfrak{b}$ *and* g *are such that* $d_\sigma(\Phi_{\mathfrak{a}} + f\circ\pi') = d_\sigma(\Phi_{\mathfrak{b}} + g\circ\pi')$ *that is* $\mathsf{v}_2 = \mathfrak{b}(x)$ *and then we have* $d_\sigma(\Phi_{(\mathfrak{a}-\mathfrak{b})} + (f-g)\circ\pi') = 0$ *and so from Chapter 6, Lemma 6.1, we have* $\mathfrak{a}(x) = \mathfrak{b}(x)$. *Thus* $\rho(\mathfrak{a}(x)) = \rho(\mathfrak{b}(x))$ *and*

$$\begin{aligned}
& T_\sigma\pi'(d_\sigma(\Phi_{\mathfrak{a}} + f\circ\pi')\circ\rho_x(.) + <\xi, C_x(\mathfrak{a}(x), .> \\
= \; & T_\sigma\pi'(d_\sigma(\Phi_{\mathfrak{b}} + g\circ\pi')\circ\rho_x(.) + <\xi, C_x(\mathfrak{b}(x), .>.
\end{aligned}$$

which implies that the value of $P_\sigma((\eta_1, \mathsf{w}_2)$ *does not depend on the choice of* $\mathfrak{a}$ *and* f *such that* (η_1, w_2) *is represented by* $d_\sigma(\Phi_\mathfrak{a} + f \circ \pi')$. *In particular, from (7.17), we also have*

$$P_\sigma(\eta_1, \mathsf{w}_2) = (\rho_x(\mathsf{w}_2), -\eta_1 \circ \rho_x(.) + < \xi, C_x(\mathsf{w}_2, .) >) \qquad (7.18)$$

Since ρ_x *is a bounded linear map and* C_x *a bounded bilinear map, from (7.18), it follows that* P_σ *is a bounded linear map, and also in such local coordinates* $\sigma \mapsto P_\sigma$ *is a smooth maps since* $x \mapsto \rho_x$ *and* $x \mapsto C_x$ *are so. Finally, it follows that* P *is a convenient vector bundle morphism.*

Under the previous notations, consider any $(\theta_1, \mathsf{v}_2) \in T^\flat_\sigma \mathcal{A}'$ *represented by* $d_\sigma(\Phi_\mathfrak{b} + g \circ \pi')$, *that is* $\mathsf{v}_2 = \mathfrak{b}(x)$ *and* $\theta_1 = \dfrac{\partial \Phi_\mathfrak{a}}{\partial \mathsf{x}}(\sigma) + d_\sigma(f \circ \pi'))$. *Therefore, from (7.16) and (7.17) we obtain:*

$$\begin{aligned} < (\theta_1, \mathsf{v}_2), P(\eta_1, \mathsf{w}_2) > &= \Phi_{[\mathfrak{a},\mathfrak{b}]}(\sigma) + d_x g(\rho(\mathfrak{a})) - d_x f(\rho(\mathfrak{b})) \\ &= \theta_1(\rho(\mathsf{w}_2)) - \eta_1(\rho(\mathsf{v}_2)) + < \xi, C_x(\mathsf{w}_2, \mathsf{v}_2) > \end{aligned} \qquad (7.19)$$

From relation (7.19), we see that P *is skew-ymmetric relative to the pairing* $< .,. >$ *and therefore* P *is an almost Poisson anchor.*

As usual, over any open set U *in* M, *to* P, *we can associate a pairing* $\{.,.\}_P$ *on* $\mathfrak{A}(\mathcal{A}'_U)$ *which can be defined only on functions of type* $\Phi_\mathfrak{a}$ *and* $f \circ \pi'$ *from Proposition 7.2 that is, from (7.19):*

(1) $\{\Phi_{\mathfrak{a}_1}, \Phi_{\mathfrak{a}_2}\}_P =< \Phi_{\mathfrak{a}_2}, P(\Phi_{\mathfrak{a}_1}) >= \Phi_{[\mathfrak{a}_2, \mathfrak{a}_1]_\rho}$,
which is equivalent to (7.10).

(2) $\{\Phi_\mathfrak{a}, f \circ \pi'\}_P = - < df \circ \pi', P(\Phi_\mathfrak{a}) >= df(\rho(\mathfrak{a})) \circ \pi'$,
which is equivalent to (7.11).

(3) $\{f \circ \pi', g \circ \pi'\}_P = - < df \circ \pi', P(dg \circ \pi') >= 0$,
which is equivalent to Lemma 7.2, 2.

The Leibniz property of $\{.,\}_P$ *is a consequence of the Leibniz property of* $[.,.]_\rho$ *previous properties (1), (2) and (3) and the fact that* $\mathfrak{A}(\mathcal{A}'_U)$ *is generated by functions of type* $f \circ \pi'$ *and* $\Phi_\mathfrak{a}$.

It remains to show that P *satisfies the Jacobi identity. From (7.14), the Jacobi identity is satisfied for a triple* $(\Phi_{\mathfrak{a}_1}, \Phi_{\mathfrak{a}_2}, \Phi_{\mathfrak{a}_3})$. *For a a triple* $(\Phi_{\mathfrak{a}_1}, \Phi_{\mathfrak{a}_2}, f_3 \circ \pi')$, *the Jacobi identity is a consequence of the fact that* $[X, Y] = L_X \circ L_Y - L_Y \circ L_X$ *for any vector fields* X *and* Y *on* U. *For a a triple* $(\Phi_{\mathfrak{a}_1}, f_2 \circ \pi', f_3 \circ \pi')$, *each term in the Jacobi identity is identically zero. The Jacobi identity for any triple* (F_1, F_2, F_3) *in* $\mathfrak{A}(\mathcal{A}'_U)$ *is then a consequence of the fact that* $\mathfrak{A}(\mathcal{A}'_U)$ *is generated by functions of type* $f \circ \pi'$ *and* $\Phi_\mathfrak{a}$ *and the Leibniz property of* $\{.,.\}_P$.

Finally, the Lie bracket $[.,.]_P$ *on sections of* $\mathcal{A}_U$ *associated with* P *is exactly the original one* $[.,.]_\rho$ *from properties (1) and (2), which ends the proof.*

Remark 7.5 *In the proof of the last Part of Theorem 7.1, we have shown that* $P(d_\sigma(\Phi_{\mathfrak{a}} + f \circ \pi'))$ *depends only on the value of* $d_\sigma(\Phi_{\mathfrak{a}} + f \circ \pi')$*. Thus, from (7.15), it follows that the vertical part of* $P(d_\sigma(\Phi_{\mathfrak{a}}+f\circ\pi'))$ *is* $\Phi_{[\mathfrak{a},.]_\rho}(\sigma)+ d_\sigma(f \circ \pi') \circ \rho$ *and so the Hamiltonian vector field of* $\Phi_{\mathfrak{a}} + f \circ \pi'$ *at* σ *is*

$$\left(\rho(\mathfrak{a} \circ \pi'(\sigma), \Phi_{[\mathfrak{a},.]_\rho}(\sigma) + d_\sigma(f \circ \pi') \circ \rho\right) \in T_\sigma \mathcal{A}'.$$

7.2 Partial Lie Algebroids and Partial Poisson Manifolds

We refer to Chapter 3, § 3.18.4, for the notion of partial convenient Lie algebroid and we will use the same notations. The first section states how a partial convenient Lie algebroid structure is defined on the anchored convenient Lie algebroid $(T^\flat M, M, P)$ for a partial convenient Poisson structure $(T^\flat M, M, P, \{.,.\}_P)$.

7.2.1 Partial Lie Algebroid

It is well known that in finite dimension, the cotangent space of a Poisson manifold can be provided with a Lie algebroid structure. We will show that we have a comparable weaker result in our context. In this purpose, the following Lemma is basic:

Lemma 7.3 *Let* $(T^\flat M, M, P, \{.,.\}_P)$ *be a partial Poisson structure.*
For each $x \in M$*, there exists an open neighbourhood* U *of* x *such that the vector space* $\mathfrak{P}_x(U) = \{d_x f,\ f \in \mathfrak{A}(U)\}$ *is equal to the fibre* $T^\flat_x M$*.*

Proof *Fix some* $x \in M$*. Since it is a local problem, according to the assumptions of the definition of a partial Poisson manifold, there exists an open neighbourhood* U *of* $x \in M$ *such that the bundles* TM*,* $T'M$ *and* $T^\flat M$ *are trivializable over* U*. Thus, without loss of generality, we may assume that* U *is an open set of* $\mathbb{M}$*, so that* $TM_{|U} = U \times \mathbb{M}$*,* $T'M_{|U} = U \times \mathbb{M}'$*,* $T^\flat M_{|U} = U \times \mathbb{E}$ *and if* $i : \mathbb{E} \to \mathbb{M}'$ *is the natural inclusion then* $\iota_{|U}(x,u) = (x, i(u))$*. In this situation, any section of* $T^\flat M$ *over* U *is characterized by a smooth map* $\alpha : U \to \mathbb{E} \subset \mathbb{M}'$*.*
Now, choose any $\alpha \in T^\flat_x M \subset T'_x M$ *and consider* $\begin{array}{rccc} f_\alpha : & U & \to & \mathbb{R} \\ & z & \mapsto & <\alpha, z> \end{array}$.
Then $d_x f_\alpha(u) =<\alpha, u>$ *for all* $u \in U$*, so* f_α *belongs to* $\mathfrak{A}(U)$ *since* $d^k_x f_\alpha = 0$ *for* $k \geq 2$*, which ends the proof.*

Now we are situation to prove the following proposition[4].

[4] This result is a more precise version of [PeCa19].

Proposition 7.3 *Let $(M, \mathfrak{A}(M), \{.,.\}_P)$ be a partial Poisson structure and $P : T^\flat M \to TM$ the associated Poisson anchor. Then we have the following properties:*

(1) We can define a sheaf of almost bracket $[.,.]_P$ on the anchored bundle $(T^\flat M, M, P)$ by:

$$[\alpha, \beta]_P = L_{P(\alpha)}\beta - L_{P(\beta)}\alpha - d < \alpha, P(\beta) > \tag{7.20}$$

for any open set U in M and section α and β in $\mathfrak{P}(U)$ where L_X is the Lie derivative.
Moreover $[.,.]_P$ satisfies:

$$\forall (f, g) \in (\mathfrak{A}(U))^2 \,,\; [df, dg]_P = d\{f, g\}_P \tag{7.21}$$

(2) Let $\mathfrak{P}_M$ be the sheaf of $\mathcal{C}^\infty(U)$-modules generated by the set $\{df, f \in \mathfrak{A}(U)\}$. Then $(\mathfrak{P}_M, [.,.]_P)$ is a sheaf of Poisson-Lie algebra. In particular, $(T^\flat M, p^\flat_M, M, P, \mathfrak{P}_M)$ is a strong partial convenient Lie algebroid and $(T^\flat M, M, P, [.,.]_P)$ is an almost Lie algebroid.

Proof

(1) At first we must show that $[.,.]_P$ takes values in $T^\flat M$. Since it is a local problem, we may assume that M is an open set of $\mathbb{M}$ so that $TM_U = U \times \mathbb{M}$, $T'M_U = U \times \mathbb{M}'$, $T^\flat M_U = U \times \mathbb{E}$ and if $i : \mathbb{E} \to \mathbb{M}'$ is the natural inclusion, then $\iota(x, u) = (x, i(u))$. In this way, any section α can be considered as a smooth map from U to $\mathbb{E}$ and any vector field Y on U can be also considered as a smooth map $Y : U \to \mathbb{M}$. It follows that, for each $x \in U$, the differential $d_x\alpha$ of the map α belongs to the space $L(\mathbb{M}, \mathbb{E})$ of bounded linear maps from $\mathbb{M}$ to $\mathbb{E}$ and d_xY belongs to $L(\mathbb{M}, \mathbb{M})$. Thus $d_x\alpha(Y_x, \;)$ belongs to $\mathbb{E}$ and $\alpha_x \circ d_xY$ belongs also to $\mathbb{E}$. Now, the Lie derivative $L_Y\alpha$ is defined by:

$$L_Y\alpha(Z) = d(< \alpha, Z >)(Y) - \alpha \circ dZ(Y) + \alpha \circ dY(Z) = d\alpha(Y, Z) - \alpha \circ dY(Z).$$

Under these notations, any section α and β of $T^\flat M_U$ are smooth maps from U to $\mathbb{E}$. We set $X = P(\alpha)$ and $Y = P(\beta)$. Since

$$L_Y\alpha = d\alpha(Y, \;) - \alpha \circ dY. \tag{7.22}$$

the previous arguments imply that $L_Y\alpha$ is a smooth map from U to $\mathbb{E}$. Of course, the same is true for $L_X\beta$. Finally, also from the previous argument, $d(< \beta, X >)$ is a smooth map from U to $\mathbb{E}$ and so $[.,.]_P$ takes values in $T^\flat M$. More precisely, using the relation

$$d < \alpha, P\beta >= \frac{1}{2}\left(d < \alpha, P\beta > -d < \beta, P\alpha >\right).$$

In our context, we finally obtain

$$[\alpha, \beta]_P = d\beta(P(\alpha), .) - d\alpha(P\beta, .) - \frac{1}{2}(< \beta, dP(\alpha, .) > - < \alpha, dP(\beta, .) >). \tag{7.23}$$

This implies that $[.,.]_P$ *satisfies the Property* **(AL2)** *in Chapter 3, Definition 3.76. The Leibniz property of* $[.,.]_P$ *(i.e. Property* **(AL1)** *in Definition 3.76) is a direct consequence of (7.23).*
For a smooth function f*, we have* $L_Y(df) = d < df, Y >$*. So from the definition of* $[.,.]_P$*, we have*

$$\begin{aligned}[df, dg]_P &= d < df, Pdg > -d < dg, Pdf > -d < df, Pdg > \\ &= -d < df, Pdg > \\ &= d\{f, g\}\end{aligned}$$

which ends the proof of (1).
(2) As any $\alpha \in \mathfrak{P}(U)$ *can be written* $\alpha = \sum_{i \in I} f_i dg_i$ *where* I *is a finite set of indexes and where each* f_i *(resp.* g_i*) belongs to* $C^\infty(U)$ *(resp.* $\mathcal{A}(U)$*), using the Leibniz property of the bracket and the property (7.21), it follows that the restriction of the bracket (7.21) to* $\mathfrak{P}(U)$ *takes value in* $\mathfrak{P}(U)$*. Now we have*

$$[df, [dg, dh]_P]_P = d\{f, \{g, h\}_P\}_P$$

Since the Poisson bracket satisfies the Jacobi identity, the previous relation implies

$$J_P(df_1, df_2, df_3) = 0$$

for all f_2, f_2, f_3 *in* $\mathfrak{A}(U)$*. On the other hand, from (7.21) and (7.5) it follows that*

$$P([df_1, df_2]_P) = [P(df_1), P(df_2)] \tag{7.24}$$

for any f_1, f_2 *in* $\mathfrak{P}(U)$*.*

According to Proposition 3.19 point (2), for any functional linear combination $\sum_{i=1}^{n} g_i df_i$*, we have*

$$J_P(\sum_{i=1}^{n} g_i df_i, dh, dg) = \sum_{i=1}^{n} g_i J_P(df_i, dg, dh) = 0$$

For any $f_i, g, h \in \mathfrak{A}(U)$ *and* $g_i \in C^\infty(U)$ *for* $i = 1, \ldots, n$*.*
Therefore J_P *vanishes identically in restriction to the module* $\mathfrak{P}(U)$*. The Leibntiz property of the almost bracket* $[.,.]_P$ *on local sections of* $T^\flat M$ *and relation (7.21) imply that the restriction of* P *to* $\mathfrak{P}(U)$ *is a Lie algebra morphism, which, according to Lemma 7.3, complete the proof of (2).*

Remark 7.6 *In finite dimension, in fact, we have a structure of Lie algebroid on* $T^*M = T'M$*. In infinite dimension, according to Chapter 3, Proposition 3.19 (1), Lemma 7.3 and relation (7.24), for any germs of sections* α_1 *and* α_2*, we have*

$$P([\alpha_1, \alpha_2]_P) = [P(\alpha_1), P(\alpha_2)] \tag{7.25}$$

Thus P is a morphism of Lie algebras. Unfortunately, from Chapter 3, Proposition 3.19 (2), this does not imply in general that the Jacabiator is identically zero on germs of sections of $T^{\flat}M$. Thus, in general, $T^{\flat}M$ need not be provided with a convenient Lie algebroid structure in this way. We only have a "pre-Lie algebroid structure" (cf. Remark 3.12)

According to this remark, from Proposition 3.19 (3), if P is injective this implies that the Jacobiator is identically zero and so we get the following result:

Corollary 7.2 *Let $(T^{\flat}M, M, P, \{.,.\}_P)$ be a partial Poisson structure. If P is an injective morphism then $(T^{\flat}M, M, P, [.,.]_P)$ is a Lie algebroid. This situation occurs in particular for some weak symplectic structures (cf. Example 7.3).*

It is well known that, in finite dimension, if $P : T^*M \to TM$ is an almost Poisson anchor, then we have a skew-symmetric contravariant tensor Λ of type $(2,0)$ on T^*M defined by

$$\Lambda(\alpha, \beta) =< \beta, P\alpha > . \tag{7.26}$$

Then P defines a Poisson structure on M if and only if the Schouten bracket $[\Lambda, \Lambda]$ vanishes identically on M (cf. [Lic77]). In fact, the property $[\Lambda, \Lambda] = 0$ is equivalent to the Jacobi identity of the bracket $\{.,.\}_P$ naturally associated with P (cf. Lemma 7.1).

Now consider an almost Poisson anchor $P : T^{\flat}M \to TM$. On any open set U in M, and any σ_1 and σ_2 in $\mathfrak{P}(U)$, the map $[P, P] : \mathfrak{P}(U) \times \mathfrak{P}(U) \to \mathfrak{X}(U)$ given by

$$[P, P](\sigma_1, \sigma_2) = P([\sigma_1, \sigma_2]_P) - [P\sigma_1, P\sigma_2] \tag{7.27}$$

is well defined and is bilinear; moreover, its value in T_xM only depends on the values of σ_1 and σ_2 at x (cf. arguments in the proof of Proposition 3.19, 1.). Thus $[P, P]$ is a tensor of type $(2,1)$ with takes values in TM. On the other hand, as in finite dimension, to P is associated a skew-symmetric tensor Λ on $T^{\flat}M$ of type $(2,0)$ also defined by (7.26). Thus we can consider the tensor $[\Lambda, \Lambda]$ of type $(3,0)$ on $T^{\flat}M$ defined by

$$[\Lambda, \Lambda](\sigma_1, \sigma_2, \sigma_3) =< \sigma_3, [P, P](\sigma_2, \sigma_1) > . \tag{7.28}$$

for any $\sigma_i \in \mathfrak{P}(U)$ for $i \in \{1, 2, 3\}$.

The following Lemma is an adaptation to our context of [MaMo84], Appendix B.1.

Lemma 7.4 *Using the previous notations,*

(1) for any $\sigma_i \in \mathfrak{P}(U)$*,* $i \in \{1,2,3\}$*, we have:*

$$\begin{aligned}&[\Lambda,\Lambda](\sigma_1,\sigma_2,\sigma_3)\\ =\ &< L_{P\sigma_1}\sigma_2, P\sigma_3 > + < L_{P\sigma_2}\sigma_3, P\sigma_1 > + < L_{P\sigma_3}\sigma_1, P\sigma_2 >\end{aligned}$$

(2) If $\sigma_i = df_i$ *with* $f_i \in \mathfrak{A}(U)$*,* $i \in \{1,2,3\}$*,*

$$[\Lambda,\Lambda](df_1,df_2,df_3) = \{f_1,\{f_2,f_3\}_P\}_P + \{f_2,\{f_3,f_1\}_P\}_P + \{f_3,\{f_1,f_2\}_P\}_P.$$

Proof *For each* $x \in U$*, consider the adjoint* $P_x^* : T_x^*M \to (T_x^\flat M)^*$ *of* P_x*. From the skew-symmetry of* P *it follows that* $P^*_{|T_x^\flat M} = -P$ *and so*

$$P^*(T_x^\flat M) \subset T_xM \subset T_x''M.$$

From this remark and the properties of the Lie derivative, we obtain:

$$\begin{aligned}&< L_{P\sigma_1}\sigma_2, P\sigma_3 > + < L_{P\sigma_2}\sigma_3, P\sigma_1 > + < L_{P\sigma_3}\sigma_1, P\sigma_2 >\\ &= - < \sigma_3, P(L_{P\sigma_1}\sigma_2) > + L_{P\sigma_2} < \sigma_3, P\sigma_1 > - < \sigma_3, [P\sigma_2, P\sigma_1] >\\ &+ L_{P\sigma_3} < \sigma_1, P\sigma_2 > - < \sigma_1, [P\sigma_3, P\sigma_2] >\end{aligned}$$

Since we have $L_{P\sigma_2} < \sigma_3, P\sigma_1 >= -L_{P\sigma_2} < \sigma_1, P\sigma_3 >$ *and*

$$-L_{P\sigma_2} < \sigma_1, P\sigma_3 >= - < L_{P\sigma_2}\sigma_1, P\sigma_3 > + < \sigma_1, [P\sigma_2, P\sigma_3] >,$$

the last member of the previous equality becomes:

$$\begin{aligned}&- < \sigma_3, P(L_{P\sigma_1}\sigma_2) > - < L_{P\sigma_2}\sigma_1, P\sigma_3 > - < \sigma_3, [P\sigma_2, P\sigma_1] >\\ &- < d < \sigma_2, P\sigma_1 >, P\sigma_3 >\\ &=< \sigma_3, P(L_{P\sigma_2}\sigma_1) - P(L_{P\sigma_1}\sigma_2) + P(d < \sigma_2, P\sigma_1 >) - < \sigma_3, [P\sigma_2, P\sigma_1] >\\ &=< \sigma_3, P[\sigma_2,\sigma_1] - [P\sigma_2, P\sigma_1] >\end{aligned}$$

which ends the proof of (1).

Recall that an expression of the Poisson bracket of $\{.,.\}_P$ *is*

$$\{f_1, f_2\}_P =< df_1, Pf_2 >= L_{Pdf_2} df_1$$

and so

$$\{f_1,\{f_2,f_3\}_P\}_P =< L_{Pdf_2}(df_1), Pdf_3 > .$$

Then (2) is a consequence of (1).

The following definition will be justified in the next section:

Definition 7.8 *Let* $P : T^\flat M \to TM$ *be an almost Poisson anchor and* Λ *the associated contravariant tensor of type* $(3,0)$*. Then* $2[\Lambda,\Lambda]$ *is called the Schouten bracket of* Λ*.*

Thus according to Lemma 7.4, as in finite dimension, we have:

Theorem 7.2 *Let* $P : T^\flat M \to TM$ *be an almost Poisson anchor and* Λ *the associated contravariant tensor of type* $(3,0)$*. Then* P *defines a partial Poisson structure on* M *if and only if* $[\Lambda,\Lambda] = 0$*.*

7.2.2 Schouten Bracket on $T^\flat M$

In this section, we fix an almost Poisson anchor $P : T^\flat M \to TM$. Then we have a structure of strong partial convenient Lie algebroid $(T^\flat M, p^\flat_M, M, P, \mathfrak{P}_M)$ on $T^\flat M$. We will use the definition of the Schouten bracket on a Poisson manifold given in [FeMa14], 1.4, to propose a generalization of the Schouten bracket on $T^\flat M$.

As a generalization of the notion of derivation on $C^\infty(U)$ we introduce:

Definition 7.9 *Let U be an open set in M.*

1. *For $k \geq 1$, a k-alternating derivation on $\mathfrak{A}(U)$ is a k-alternating bounded linear map $D : (\mathfrak{A}(U))^k \to \mathfrak{A}(U)$ such that*

$$\begin{aligned} & D(f_1, \dots, f_{i-1}, gh, f_{i+1} \dots, f_k) \\ = \;& gD(f_1, \dots, f_{i-1}, h, f_{i+1} \dots, f_k) + D(f_1, \dots, f_{i-1}, g, f_{i+1} \dots, f_k)h \end{aligned}$$

for all $i \in \{1, \dots, k\}$ and all $f_1, \dots, f_1, \dots, f_{i-1}, g, h, f_{i+1} \dots, f_k$ in $\mathfrak{A}(U)$.

2. *A k-alternating derivation D on $\mathfrak{A}(U)$ is called of 1st-order if $D(f_1, \dots, f_k)$ only depends on the 1-jets of each f_i for $i \in \{1, \dots, k\}$.*

Notation 7.1 *We denote by*

$\mathrm{Der}_k(\mathfrak{A}(U))$ *the set of k-alternating derivations on $\mathfrak{A}(U)$;*

$\mathrm{Der}^1_k(\mathfrak{A}(U))$ *the subset of k-alternating derivations of first order.*

Example 7.4 *For $k = 1$, from the definition, a 1-alternating derivation is simply a derivation on $\mathfrak{A}(U)$ in the usual sense. Since TM is a closed subbundle of $T''M$, each local vector field induces a derivation of $C^\infty(U)$ by*

$$X(f) =< df, X >= i_X(df).$$

But from Proposition 7.1, if f belongs to $\mathfrak{A}(U)$, then $X(f) \in \mathfrak{A}(U)$. Therefore X defines a derivation of $\mathfrak{A}(U)$ of 1st-order.

Example 7.5 *Let $(X_1, \dots X_k)$ be a finite sequence of vector fields on an open set U in M. Then we can consider the decomposable non zero multi-vector $X_1 \wedge \cdots \wedge X_k$. Now, for each $(f_1, \dots, f_k) \in (\mathfrak{A}(U))^k$, consider the map:*

$$(f_1, \dots, f_k) \mapsto \det\left(i_{X_i}(df_j)\right). \tag{7.29}$$

If each f_i belongs to $\mathfrak{A}(U)$, then $i_{X_i}(df_j)$ also belongs to $\mathfrak{A}(U)$ (cf. Example 7.4). Thus the map (7.29) takes values in $\mathfrak{A}(U)$. In fact, this maps gives rise to a k-alternating derivation on $\mathfrak{A}(U)$ associated with $X_1 \wedge \cdots \wedge X_k$ and which only depends on the value of this multi-vector and not of such a decomposition. It is a derivation of 1st-order.

Example 7.6 *More generally, consider a finite sequence $(X_1, \dots, X_m)$ of decomposable multi-vectors of dimension k on U. For any function $f_1, \dots, f_m \in \mathfrak{A}(U)$, then $D = \sum_{i=1}^{m} f_i X_i$ is also a derivation of first order.*

All these examples motivate the following definition.

Definition 7.10 *A derivation $D \in \mathrm{Der}_k(\mathfrak{A}(U))$ of first order is called a k-alternating kinematic derivation of $\mathfrak{A}(U)$ if, for any fixed $f_2, \dots, f_k$ in $\mathfrak{A}(U)$, there exists a vector field X on U such that:*

$$D(f, f_2, \dots, f_k) = df(X) \tag{7.30}$$

for any $f \in \mathfrak{A}(U)$.

Notation 7.2 *The set of k-alternating kinematic derivations of $\mathfrak{A}(U)$ is denoted $\mathbf{D}_k(\mathfrak{A}(U))$.*

Remark 7.7 *If D is a k-alternating kinematic derivation, it follows that, for any $i \in \{1 \dots, k\}$, the map $f_i \mapsto D(f_1, \dots, f_k)$ for $\{f_1, \dots, \widehat{f_i}, \dots, f_k\}$ fixed , there exists a vector field X_i on U such that $df_i(X_i) = D(f_1, \dots, f_k)$. Note that such a vector field may be not unique (cf. Remark 7.8).*

Of course, Example 7.4, Example 7.5 and Example 7.6 are alternating kinematic derivations. In finite dimension, all derivations of $C^\infty(U)$ are kinematic (cf. [FeMa14]). This is not true in the infinite setting, even in the Banach setting as we can see in the following Example.

Example 7.7 *Consider a non-reflexive Banach space $\mathbb{M}$ and set $T^\flat M = T'M$. Then for any c^∞ open U in $\mathbb{M}$, we consider $\mathfrak{A}(U) = C^\infty(U)$. So any operational vector field X of order r on $\mathbb{M}$[5] defines a 1-alternating derivation of order r on $\mathfrak{A}(U)$ which is not kinematic. More generally, if $X_1, \dots, X_r$ are operational vector fields of order r on U, then, in the same way as in Example 7.5, $X_1 \wedge \cdots \wedge X_r$ gives rise to a k-alternating derivation of order r on $\mathfrak{A}(U)$ which is not kinematic even for $r = 1$.*

As in [FeMa14], 1.4, we introduce:

Definition 7.11 *If D belongs to $\mathrm{Der}_k(\mathfrak{A}(U))$ and D' to $\mathrm{Der}_{k'}(\mathfrak{A}(U))$*

(1) we have

$$\begin{aligned} & D \circ D'(f_1, \dots, f_{k'}, f_{k'+1}, \dots, f_{k+k'-1}) \\ = & \sum_{\sigma} (-1)^{\mathrm{sign}(\sigma)} D\left(D'(f_{\sigma(1)}, \dots, f_{\sigma(k')}), f_{\sigma(k'+1)}, \dots, f_{\sigma_{(k+k'-1)}})\right) \end{aligned}$$

for all $f_i \in \mathfrak{A}(U)$, $i \in \{1, \dots k + k' - 1\}$ and where the sum is for all $(k', k-1)$-shuffles σ.

(2) We set

$$[D, D'] = D \circ D' - (-1)^{(k-1)(k'-1)} D' \circ D. \tag{7.31}$$

[5] cf. [KrMi97], 32.

(3) The wedge product of $D \wedge D'$ of D and D' is defined by:

$$\begin{aligned} & D \wedge D'(f_1, \dots, f_{k+k'}) \qquad (7.32) \\ = & \frac{1}{k!}\frac{1}{k'!}\sum_{\sigma}(-1)^{\mathrm{sign}\sigma} D(f_{\sigma(1)}, \dots, f_{\sigma(k)}) D'(f_{\sigma(k+1)}, \dots, f_{\sigma(k+k')}) \end{aligned}$$

where the summation is over all (k, k')-shuffles σ of $\{1, \dots, k+k'\}$.

Then as in [FeMa14], p. 17, Problem 1.3, Proposition 2.2, we can show:

Proposition 7.4 *Let U be an open set in M. Then we have the following properties:*

(1) $Der_k(\mathfrak{A}(U))$ has a structure of $\mathfrak{A}(U)$-module and $Der_k^1(\mathfrak{A}(U))$ is a submodule.

(2) If D belongs to $\mathrm{Der}_k(\mathfrak{A}(U))$ and D' to $\mathrm{Der}_{k'}(\mathfrak{A}(U))$ then $[D, D']$ belongs to $\mathrm{Der}_{(k+k'-1)}(\mathfrak{A}(U))$.

(3) The bracket $[.,.]$ is $\mathbb{R}$ bilinear on $\mathrm{Der}(\mathfrak{A}(U))$ and has the following properties:

(i) $$[D, D'] = -(-1)^{(k-1)(k'-1)}[D, D'].$$

(ii) (Generalized Jacobi identity)
For all $D \in \mathrm{Der}_k(\mathfrak{A}(U))$, $D' \in \mathrm{Der}_{k'}(\mathfrak{A}(U))$ and $D'' \in \mathrm{Der}_{k''}(\mathfrak{A}(U))$,

$$\begin{aligned} &(-1)^{(k-1)(k''-1)}[[D, D'], D''] + (-1)^{(k'-1)(k-1)}[[D', D''], D] \\ &+(-1)^{(k''-1)(k'-1)}[[D'', D], D'] = 0. \end{aligned}$$

Let $\pi : F \to M$ be a convenient bundle with typical fibre $\mathbb{F}$. As in Chapter 3, § 3.1.7.4, we consider the set $L^k_{\mathrm{alt}}(\mathbb{E}, \mathbb{F})$ of all bounded k-linear alternating mappings from $\mathbb{E}^k$ to $\mathbb{F}$ which is a convenient space. Using common atlas for the bundle structure of $p^\flat : T^\flat M \to M$ and $\pi : F \to M$ we can show that

$$p^\flat_k : L^k_{\mathrm{alt}}(T^\flat M; F) = \bigcup_{x \in M} L^k_{\mathrm{alt}}(T^\flat_x M; F) \to M$$

is a convenient vector bundle.
The vector space of local sections of $L^k_{\mathrm{alt}}(T^\flat M; F)$ over U is denoted by $\bigwedge^k \Gamma^*(T^\flat M_U; F_U)$ and is called the set of vectorial k-differential forms on $T^\flat M$.
The set

$$\{\bigwedge^k \Gamma^*(T^\flat M_U; F_U),\ U \text{ open set in } M\}$$

gives rise to a sheaf of modules over the sheaf C^∞_M of smooth functions on M. For $k \geq 1$, the values of a section $P \in \bigwedge^k \Gamma^*(T^\flat M_U; F_U)$ is characterized by its values on each k-uple $(df_1, \dots, df_k)$ for any $(f_1, \dots, f_k)$ in $(\mathfrak{A}(U))^k$.

When $F = M \times \mathbb{R}$ we simply denote $\bigwedge^k \Gamma^*(T^\flat M_U)$ the set $\bigwedge^k \Gamma^*(T^\flat M_U; U \times \mathbb{R})$.

Let $\iota : T^\flat M \to T\prime M$ be the inclusion morphism. Then ι^* is a bundle morphism from the bidual bundle $T''M$ to the dual bundle $T^\flat M'$ of $T^\flat M$. Note that TM is a closed subbundle of $T''M$ and so $\iota^*(TM)$ is contained in $T^\flat M'$ but in general it is not a subbundle

Definition 7.12 *Let U be an open set in M.*

(i) For $k = 1$, an element $\Lambda \in \bigwedge^1 \Gamma^*(T^\flat M_U) = \Gamma^*(T^\flat M_U)$ is admissible if there exists a vector field X on U such that $\Lambda = \iota^* X$.

(ii) For $k \geq 2$, a section $\Lambda \in \bigwedge^k \Gamma^*(T^\flat M_U)$ will be called admissible if there exists $P \in \bigwedge^{k-1} \Gamma^*(T^\flat M_U; TM_U)$ such that we have

$$\Lambda_x(\alpha_1, \alpha_2, \ldots, \alpha_k) =< \alpha_1, P_x(\alpha_2, \ldots, \alpha_k) >$$

for all $\alpha_1, \ldots, \alpha_k \in T_x^\flat M$.

Notation 7.3 *We denote by $\Gamma_k^*(\mathfrak{A}(U))$ the set of admissible elements of $\bigwedge^k \Gamma^*(T^\flat M_U)$ and by $\mathbf{D}_k^*(\mathfrak{A}(U))$ the set of admissible kinematic derivations of $\mathbf{D}_k(\mathfrak{A}(U))$.*

- For $k \geq 2$, the element P of $\bigwedge^{k-1} \Gamma^*(T^\flat M_U; TM_U)$ associated with the admissible section $\Lambda \in \bigwedge^k \Gamma^*(T^\flat M_U)$ will be denoted $\Lambda^\sharp$; we then have

$$\Lambda_x(\alpha_1, \alpha_2, \ldots, \alpha_k) = \langle \alpha_1, \Lambda_x^\sharp(\alpha_2, \ldots, \alpha_k) \rangle$$

  for all $\alpha_1, \ldots, \alpha_k \in T_x^\flat M$ and so $\iota^* \circ \Lambda^\sharp(\alpha_2, \ldots, \alpha_k) \in T^\flat M'$.
- For $k = 1$, we set $\Lambda^\sharp = X$ where X is a fixed vector field such that $\Lambda = \iota^* X$.

Example 7.8 *To any almost partial Poisson anchor $P : T^\flat M \to TM$ is associated a tensor $\Lambda \in \bigwedge^2 \Gamma^*(T^\flat M)$ which is admissible.*

Proposition 7.5

(1) $\Gamma_k^*(\mathfrak{A}(U))$ is a $\mathfrak{A}(U)$-module and the set $\{\bigwedge^k \Gamma^*(\mathfrak{A}(U), U$ open in $M\}$ is a $\mathfrak{A}_M$ sheaf of $\mathfrak{A}_M$ modules, if $\mathfrak{A}_M$ denotes the sheaf of rings defined by $\{\mathfrak{A}(U), U$ open set in $M\}$.

(2) For any $\Lambda \in \Gamma_k^*(\mathfrak{A}(U))$, $\Lambda(df_1, \ldots, df_k)$ takes values in $\mathfrak{A}(U)$ for all $f_1, \ldots, f_k$ in $\mathfrak{A}(U)$.

Proof

(1) The proof of Point (1) is elementary and is left to the reader as an exercise.

(2) This point is a consequence that $\Lambda^\sharp$ takes values in TM and Proposition 7.1.

Now, we have the following link between $\Gamma^*_k(\mathfrak{A}(U))$ and the set $\mathbf{D}_k(\mathfrak{A}(U))$ of k-alternating kinematic derivations on $\mathfrak{A}(U)$:

Proposition 7.6 *To any $\Lambda \in \Gamma^*_k(\mathfrak{A}(U))$ is associated a unique k-alternating kinematic derivation $D_\Omega \in \mathbf{D}_k(\mathfrak{A}(U))$ defined by*

$$D_\Lambda(f_1, \ldots, f_k) = \Lambda(df_1, \ldots, df_k). \tag{7.33}$$

for any $(f_1, \ldots, f_k) \in \mathfrak{A}(U)^k$. Moreover, this correspondence is injective.

Proof

Existence.

The map $f \mapsto df$ is a bounded linear derivation and Λ belongs to $\bigwedge^k \Gamma^*(T^\flat M_U)$, and so is bounded over bounded sets. If Λ is admissible then $\Lambda(df_1, \ldots, df_k)$ belongs to $\mathfrak{A}(U)$. Now the relation (7.33) gives rise to a well-defined k-alternating linear bounded map D^U_Λ which is a derivation of $\mathfrak{A}(U)$ and such that the vector field associated with the derivation $f \mapsto D(f, f_2; \ldots, f_k)$ is precisely $\Lambda^\sharp(df_2, \ldots, f_k)$ and so is a kinematic derivation.

Unicity.

At first note that for any open subset V of U, the restriction of Λ to V gives also a derivation D^V_Λ of $\mathfrak{A}(V)$ which is the restriction of D^U_Λ to $\mathfrak{A}(V)$. Assume that $D^U_\Lambda = D^U_\Omega$. Fix any $x \in U$, and $(\alpha_1, \ldots, \alpha_k) \in (T^\flat_x M)^k$. From Lemma 7.3, there exists an open set V around x in U and functions $f_1, \ldots, f_k$ in $\mathfrak{A}(V)$ such that $d_x f_i = \alpha_i$ for $i \in \{1, \ldots, k\}$. From the previous observation, we must have $D^V_\Lambda = D^V_\Omega$ which implies $\Lambda_x(\alpha_1, \ldots, \alpha_k) = \Omega_x(\alpha_1, \ldots, \alpha_k)$. This complete the proof.

Remark 7.8 *According to Definition 7.12, it follows that Exemples 7.4, 7.5 and 7.6 give also examples of admissible k-forms on $T^\flat M_U$ via the morphism ι^*.*

More generally, let $\bigwedge^k TM$ be the convenient bundle whose typical fibre is $\bigwedge^k \mathbb{M}$. Then $\bigwedge^k TM$ is a subbundle of $\bigwedge^k T''M$. Therefore, for any local section Ξ of $\bigwedge^k TM$ over U, $\iota^*\Xi$ belongs to $\bigwedge^k \Gamma^*(\mathfrak{A}(U))$. For $k \geq 2$, if Ξ is decomposable we can associate a morphism $\Xi^\flat : \bigwedge^{k-1} T'M_U \to T''M_U$ defined by

$$\Xi^\flat(\alpha_2, \ldots, \alpha_k) = \Xi(., \alpha_2, \ldots, \alpha_k)$$

whose range is contained in TM_U. More generally, if Ξ belongs to the $\mathfrak{A}(U)$-module generated by decomposable multi-vectors on U then $\iota^*\Xi$ is admissible and we have $\Xi^\flat = (\iota^*\Xi)^\sharp$. On the other hand, from Proposition 7.6, such a section induces a kinematic derivation D_Ξ on $\mathfrak{A}(U)$ and if Ξ' is another such section then $\mathcal{D}_\Xi = \mathcal{D}_{\Xi'}$ if and only if $\iota^*(\Xi^\flat - (\Xi')^\flat) = 0$.

In particular, if D is a kinematic derivation, two vector fields X and X' induce the same derivation $f \mapsto D(f, f_2, \dots, f_k)$ on $\mathfrak{A}(U)$ (for fixed $f_2, \dots, f_k$ in $\mathfrak{A}(U)$) if and only if $X - X'$ belongs to $\ker \iota^*_{|T''M_U}$.
For a given kinematic derivation D and for $f_2, \dots, f_k$ fixed, a vector field $X_{f_2 \dots, f_k}$ such that $df(X_{f_1,\dots,f_k}) = D(f, f_2, \dots, f_k)$ is unique if and only if the set

$$\{x \in U : \ \ker \iota^*_{|T''_x M_U} \cap T_x M_U = \{0\}\}$$

is dense in U. In this case the same is true for any previous section Ξ.

Remark 7.9 *The map $\Lambda \mapsto D_\Lambda$ from $\Gamma^*_k(\mathfrak{A}(U))$ to $\mathbf{D}_k(\mathfrak{A}(U))$ is not surjective in general in infinite dimension for two reasons:*

(i) If M is not a regular manifold, in general $\mathfrak{A}(U)$ could be strictly smaller that $\mathfrak{A}(V)$ for $V \subset U$ and so the same is true for $\mathbf{D}_k(\mathfrak{A}(U))$. Unfortunately we have no examples of such a situation.

(ii) If $T^\flat M \neq T'M$, from Remark 7.8, if the set

$$\{x \in U : \ \ker \iota^*_{|T''_x M_U} \cap T_x M_U = \{0\}\}$$

is not dense in U, then there exists an open set V in U such that

$$\dim \left(\ker \iota^*_{|T''_x M_V} \cap T_x M_V \right) \geq 1.$$

After restricting V if necessary, there exists a vector field Z which belongs to $\ker \iota^*_{|T''M_V} \cap TM_V$. Thus for instance, any kinematic derivation $D \in \mathbf{D}_2(\mathfrak{A}(V))$, for any $f \in \mathfrak{A}(V)$, and $g \in \mathfrak{A}(V)$ fixed, the set of vector fields X_g on V such that

$$df(X_g) = D(f, g)$$

contains at least the 1-dimensional affine subspace $X_g + \phi Z$ where $\phi \in \mathfrak{A}(V)$. So, if we had $\Lambda \in \Gamma^*_2(\mathfrak{A}(V))$ such that $D_\Lambda = D$, then $\Lambda^\sharp_x(dg)$ should belong to

$$X_g(x) + \ker \iota^*_{|T''_x M_V} \cap T_x M_V$$

and, in particular, could take any value in $X_g(x) + \phi(x)Z(x)$ and so could not be intrinsically defined.

Now according to Proposition 7.6, as in finite dimension (cf. [FeMa14]), we have

Proposition 7.7 *Let U be any open set in M. Then, for any $\Lambda \in \Gamma^*_k(\mathfrak{A}_U)$ and $\Omega \in \Gamma^*_l(\mathfrak{A}_U)$, the bracket $[D_\Lambda, D_\Omega]$ is a kinematic $(k+l-1)$-alternating derivation of $\mathfrak{A}(U)$ and there exists a unique element $[\Lambda, \Omega]_S \in \Gamma^*_{k+l-1}(\mathfrak{A}_U)$ such that*

$$D_{[\Lambda,\Omega]_S} = [D_\Lambda, D_\Omega].$$

Proof *According to Definition 7.11 and Proposition 7.6, we have*

$$\begin{aligned}&(D_\Lambda \circ D_\Omega(f_1,\dots,f_l,f_{l+1},\dots,f_{k+l-1}))(x) \qquad (7.34)\\ =& \sum_\sigma (-1)^{\mathrm{sign}(\sigma)} \Lambda_x \left(d_x\left(\Omega(df_{\sigma(1)},\dots,df_{\sigma(l)})\right), df_{\sigma(l+1)},\dots,df_{\sigma_{(k+l-1)}}\right)\end{aligned}$$

for all $f_i \in \mathfrak{A}(U)$, $i \in \{1,\dots,k+l-1\}$ and where the sum is for all $(l,k-1)$-shuffles σ.

But since $d^2 f_i = 0$, we have

$$d_x\left(\Omega(df_{\sigma(1)},\dots,df_{\sigma(l)})\right) = T_x\Omega((df_{\sigma(1)},\dots,df_{\sigma(l)}). \qquad (7.35)$$

where Ω is considered as a smooth map from U to $\bigwedge^l T^{\flat\prime}M$.

According to the definition of the bracket of derivations as defined in the relation (7.31), it follows that

$$([D_\Lambda, D_\Omega](f_1,\dots,f_l,f_{l+1},\dots,f_{k+l-1}))(x)$$

only depends on the 1-jets of Λ and Ω at x and the 1-jets of $f_1,\dots f_{k+l-1}$ at x. Thus $[D_\Lambda, D_\Omega]$ is a derivation of 1st-order. The proof will be completed by application of the following Lemma

Lemma 7.5 *There exists a unique element $[\Lambda,\Omega]_S \in \Gamma^*_{k+l-1}(\mathfrak{A}(U))$ such that*

$$[D_\Lambda, D_\Omega] = D_{[\Lambda,\Omega]_S}.$$

Proof of Lemma 7.5

Consider $\Lambda \in \Gamma^*_k(\mathfrak{A}_U)$ and $\Omega \in \Gamma^*_l(\mathfrak{A}_U)$. From Proposition 7.6, $D = D_\Lambda$ and $D' = D_\Omega$ are kinematic derivations.

As in the proof of Proposition 7.1, 2., let us consider the domain U of a local chart (U,ϕ) around x for which $T^\flat M$ is trivializable; so, the inclusion $T^\flat M \to T'M$ corresponds to the inclusion from $U \times \mathbb{M}^\flat$ into $U \times \mathbb{M}'$.
In this chart, $\Omega^\sharp : \left(T^\flat M\right)^{l-1} \to TM$ can be seen as a map from $U \subset \mathbb{M}$ to $L^{l-1}_{\mathrm{alt}}\left(\mathbb{M}^\flat;\mathbb{M}\right)$. So $T_x\Omega^\sharp$ is a bounded linear map from $\mathbb{M}$ to $L^{l-1}_{\mathrm{alt}}\left(\mathbb{M}^\flat;\mathbb{M}\right)$ which will be seen as an element of $L\left(\mathbb{M}\times\left(\mathbb{M}^\flat\right)^{l-1};\mathbb{M}\right)$.

Fix some functions $f_1,\dots,, f_k, f_{k+1}\dots, f_{k+l-1}$ in $\mathfrak{A}(U)$. According to relation (7.34), we have one and only one of the two situations:
case 1: *$\sigma_i = 1$ for some $1 \le i \le l$.*
Then we have

$$\Omega\left(df_{\sigma(1)},\dots,df_{\sigma(l)}\right) = (-1)^i \left\langle df_1, \Omega^\sharp\left(df_{\sigma(1)},\dots,\widehat{df_{\sigma(i)}},\dots,df_{\sigma(l)}\right)\right\rangle$$

Thus we get

$$d_x\Big(\Omega\left(df_{\sigma(1)},\dots,df_{\sigma(l)}\right)\Big) = (-1)^i \left\langle df_1, T_x\Omega^\sharp\left(df_{\sigma(1)},\dots,\widehat{df_{\sigma(i)}},\dots,df_{\sigma(l)}\right)(.)\right\rangle$$

This gives rise to:

$$\begin{aligned}&\Lambda_x\Big(d_x\Big(\Omega\left(df_{\sigma(1)},\dots,df_{\sigma(l)}\right)\Big),df_{\sigma(l+1)},\dots,df_{\sigma(k+l-1)}\Big)\\ =\;&(-1)^i\langle df_1,T_x\Omega^\sharp\left(\Lambda^\sharp\left(df_{\sigma(l+1)},\dots,df_{\sigma(k+l-1)}\right),\right.\\ &\left.df_{\sigma(1)},\dots,\widehat{df_{\sigma(i)}},\dots,df_{\sigma(l)}\right)\rangle\end{aligned}$$

where in the expression

$$T_x\Omega^\sharp\Big(\Lambda^\sharp\left(df_{\sigma(l+1)},\dots,df_{\sigma(k+l-1)}\right),df_{\sigma(1)},\dots,\widehat{df_{\sigma(i)}},\dots,df_{\sigma(l)}\Big)$$

$\Lambda^\sharp\left(df_{\sigma(l+1)},\dots,df_{\sigma(k+l-1)}\right)$ *is seen as an element of* $\mathbb{M}$.
case 2: $\sigma_i=1$ *for some* $l<i\leq k+l-1$.
Then we have

$$\begin{aligned}&\Lambda_x\Big(d_x\Big(\Omega\left(df_{\sigma(1)},\dots,df_{\sigma(l)}\right)\Big),df_{\sigma(l+1)},\dots,df_{\sigma(k+l-1)}\Big)\\ =\;&(-1)^{i-l+1}\Big\langle df_1,\Lambda^\sharp\Big(d_x\Big(\Omega\left(df_{\sigma(1)},\dots,df_{\sigma(l)}\right)\Big),\\ &df_{\sigma(l+1)},\dots,\widehat{df_{\sigma(i)}},\dots,df_{\sigma(k+l-1)}\Big)\Big\rangle\end{aligned}$$

Finally, we obtain

$$\begin{aligned}&[D_\Lambda,D_\Omega](f_1,\dots,,f_k,f_{k+1}\dots,f_{k+l-1})(x)\\ &=\langle df_1,P_+\left(df_2,\dots,,df_k,df_{k+1}\dots,df_{k+l-1}\right)\\ &-(-1)^{(k-1)(l-1)}P_-\left(df_2,\dots,,df_k,df_{k+1}\dots,df_{k+l-1}\right)\Big\rangle\end{aligned}\tag{7.36}$$

where

$$\begin{aligned}&P_+\left(df_2,\dots,,df_k,df_{k+1}\dots,df_{k+l-1}\right)\\ &=\sum_{i=1}^{l}(-1)^i\Big(\sum_{\sigma,\sigma_i=1}(-1)^{\operatorname{sign}(\sigma)}T\Omega^\sharp\left(df_{\sigma(1)},\dots,\widehat{df_{\sigma(i)}},\dots,df_{\sigma(l)}\right)\\ &\Big(\Lambda^\sharp\left(df_{\sigma(l+1)},\dots,df_{\sigma(k+l-1)}\right)\Big)\Big)\\ &+\sum_{i=l+1}^{k+l-1}(-1)^{i-l+1}\Big(\sum_{\sigma,\sigma_i=1}(-1)^{\operatorname{sign}(\sigma)}\Lambda^\sharp\Big(d\Big(\Omega\left(df_{\sigma(1)},\dots,df_{\sigma(l)}\right)\Big),\\ &df_{\sigma(l+1)},\dots,\widehat{df_{\sigma(i)}},\dots df_{\sigma(k+l-1)}\Big)\Big).\end{aligned}$$

and $P_-(df_2,\dots,df_k,df_{k+1}\dots,df_{k+l-1})$ *has the same type of expression by permutation of* Ω *and* Λ *and* k *and* l *respectively.*

It is clear that $P := P_+ - (-1)^{(k-1)(l-1)}P_-$ *takes values in* TM. *The skew-symmetry of the bracket* $[D_\Lambda, D_\Omega]$ *implies the skew-symmetry of* P *in* $df_2, \dots, df_k, df_{k+1} \dots, df_{k+l-1}$ *and the skew-symmetry of* $< df_1, P(df_2, \dots, df_{k+l-1}) >$ *in* $df_1, df_2, \dots, df_k, df_{k+1} \dots, df_{k+l-1}$. *Now from Lemma 7.3,* P_x *is map from* $\bigwedge^{k+l-2} T_x^\flat M$ *to* T_xM *and from its definition it follows easily that the map* $x \mapsto P_x$ *is smooth. Finally, using the same arguments as in the first part of the proof of Proposition 7.3 we can show that* P *is bounded on bounded sets. This implies that the relation (7.36) defines an element* $[\Lambda, \Omega]_S$ *of* $\Gamma^*_{k+l-1}(\mathfrak{A}(U))$.

According to Proposition 7.7, we can introduce the core notion of Schouten bracket:

Definition 7.13 *For any* $\Lambda \in \Gamma^*_k(\mathfrak{A}_U)$ *and* $\Omega \in \Gamma^*_l(\mathfrak{A}_U)$. *Then the Schouten bracket* $[\Lambda, \Omega]_S$ *is the unique element* $[\Lambda, \Omega]_S \in \Gamma^*_{k+l-1}(\mathfrak{A}(U))$ *such that*

$$D_{[\Lambda,\Omega]_S} = [D_\Lambda, D_\Omega].$$

Before listing the properties of the Schouten bracket, from Proposition 7.5, we can note that, we have:

$$\begin{aligned}[\Lambda, \Lambda]_S\,(df_1, df_2, df_3) &= [D_\Lambda, D_\Lambda]\,(df_1, df_2, df_3)\\ &= 2\left(\{f_1, \{f_2, f_3\}_P\}_P + \{f_2, \{f_3, f_1\}_P\}_P + \{f_3, \{f_1, f_2\}_P\}_P\right).\end{aligned}$$

which justifies *a posteriori* the Definition 7.8.
According to Remark 7.8, we have the following result:

Theorem 7.3 *The Schouten bracket has the following properties:*

(1) For any vector fields X *and* Y *on* U, ι^*X *and* ι^*Y *belong to* $\Gamma^*_1(\mathfrak{A}(U))$ *and we have*

$$\iota^*[X, Y] = [\iota^*X, \iota^*Y]_S.$$

(2) For any $\Omega \in \Gamma^*_k(\mathfrak{A}(U))$ *and* $\Phi \in \Gamma^*_h(\mathfrak{A}(U))$,

$$[\Omega, \Phi]_S = -(-1)^{(k-1)(h-1)}[\Phi, \Omega]_S.$$

(3) For any $\Omega \in \Gamma^*_k(\mathfrak{A}(U))$, $\Phi \in \Gamma^*_h(\mathfrak{A}(U))$ *and* $\Psi \in \Gamma^*_l(\mathfrak{A}(U))$,

$$[\Omega, \Phi \wedge \Psi]_S = [\Omega, \Phi]_S \wedge \Psi + (-1)^{(k-1)h}\Omega \wedge [\Phi, \Psi]_S.$$

(4) For any $\Omega \in \Gamma^*_k(\mathfrak{A}(U))$, $\Phi \in \Gamma^*_h(\mathfrak{A}(U))$ *and* $\Psi \in \Gamma^*_l(\mathfrak{A}(U))$,

$$\begin{aligned}&(-1)^{(k-1)(l-1)}[\Omega, [\Phi, \Psi]_S]_S + (-1)^{(h-1)(k-1)}[\Phi, [\Psi, \Omega]_S]_S\\ &+(-1)^{(l-1)(h-1)}[\Psi, [\Omega, \Phi]_S]_S = 0.\end{aligned}$$

that is the generalized Jacobi identity.

(5) Let $X_1 \wedge \cdots \wedge X_k$ and $Y_1 \wedge \cdots \wedge Y_h$ be decomposable multivectors; then $\iota^*(X_1 \wedge \cdots \wedge X_k)$ and $\iota^*(Y_1 \wedge \cdots \wedge Y_h)$ belongs to $\Gamma_k^*(\mathcal{A}(U))$ and $\Gamma_h^*(\mathcal{A}(U))$ respectively and we have

$$\begin{aligned} & [\iota^*(X_1 \wedge \cdots \wedge X_k), \iota^*(Y_1 \wedge \cdots \wedge Y_h)]_S \\ = \; & \iota^*\Big(\sum_{i,j} (-1)^{i+j} [X_i, Y_j] X_1 \wedge \cdots \wedge \widehat{X_i} \wedge \cdots \wedge X_k \\ & \wedge Y_1 \wedge \cdots \wedge \widehat{Y_j} \wedge \cdots \wedge Y_h \Big) \end{aligned}$$

Proof

Considering X and Y as 1-alternating derivations (cf. Example 7.4); by application of the definition of the bracket of derivations, on a chart domain, we have:

$$[\iota^* X, \iota^* Y]_S(df) = (X \circ Y - Y \circ X)(f).$$

According to Definition 7.13, on a chart domain, the Properties (2) , (3) and (4) are exactly the properties (3.i), (3.ii) and (3.iii) of Proposition 7.4.
As in finite dimension, Property (5) is obtain from Property (1) and Property (3) by induction.

7.3 Cohomology Associated with a Partial Poisson Manifold

7.3.1 Exterior Derivative Associated with a Lie Algebroid of a Partial Poisson Manifold

Consider a partial Poisson structure $(T^\flat M, M, P, \{\ ,\ \}_P)$. We have seen that $(T^\flat M, M, P, \mathfrak{P}_M)$ is a strong partial convenient Lie algebroid and we denote by d_P the associated exterior derivative operator (cf. Chapter 3, § 3.18.6.4). Then, if X is a vector field defined on an open set U in M, $\iota^* X$ is an element of $\bigwedge^1 \Gamma^*(T^\flat M_U)$ which will be denoted simply by X. By the way we have:

$$d_P X(\alpha, \beta) = L_\alpha^P < \beta, X > - L_\beta^P < \alpha, X > - < [\alpha, \beta]_P, X > \qquad (7.37)$$

for any sections α and β of $T^\flat M_U$.

Despite the fact that the Lie derivative of a k-form Ω on a strong partial Lie algebroid is defined only for sections in $\mathfrak{P}_M$ in our previous context, for any k-form in $\Gamma_k^*(\mathfrak{A}(U))$ we have:

Lemma 7.6 *Let $\Omega \in \Gamma_k^*(\mathfrak{A}(U))$. Then for any vector field X on U the Lie derivative*

$$L_X\Omega(\alpha_1,\dots,\alpha_k) = d\Big(\Omega(\alpha_1,\dots,\alpha_k)\Big)(X) - \sum_{i=1}^{k}\Omega(\alpha_1,\dots,L_X\alpha_i,\dots,\alpha_k) \tag{7.38}$$

is a well-defined k-form which belongs to $\Gamma_k^*(\mathfrak{A}(U))$.
Moreover, we have $L_X\Omega = [X,\Omega]_S$.

Proof *As in the proof of Proposition 3.21 in Chapter 3, we may assume that U is a c^∞-open set of $\mathbb{M}$ so that $T^\flat M$ is $U\times\mathbb{E}$. In this way, X, $\Omega^\sharp$, α_i are smooth maps from U to $\mathbb{M}$, $L_{\text{alt}}^{k-1}(\mathbb{E},\mathbb{M})$ and $\mathbb{E}'$ respectively. Moreover, for any $x\in U$, there exists a germ of function $f_i\in\mathfrak{A}(U)$ such that $(\alpha_i)_x = d_xf_i$ for $i\in\{1,\dots,k\}$. Thus it is sufficient to prove that $L_X\Omega(d_xf_1,\dots,d_xf_k)$ only depends on the 1-jets at x of X and Ω and on the value of df_i at x for $i\in\{1,\dots,k\}$. At first, note that as $(L_Xdf_i)_x = d_xf_i\circ T_xX_i$ is a section of $T_x^\flat M$ from Remark 7.1, 2.*

Now, we easily obtain:

$$\begin{aligned}
& L_X\Omega(d_xf_1,\dots d_xf_k)(x) \qquad (7.39)\\
= \quad & T_x\Omega(X)(d_xf_1,\dots d_xf_k) - \sum_{i=2}^{k}\Omega(d_xf_1,\dots,d_xf_i\circ T_xX,\dots,d_xf_k)\\
= \quad & < d_xf_1, T_x\Omega^\sharp(X)(d_xf_2\dots,d_xf_k) > - < d_xf_1, T_xX\left(\Omega^\sharp(d_xf_2\dots,d_xf_k)\right)\\
& > - \sum_{i=2}^{k} < d_xf_1, \Omega_x^\sharp(d_xf_2,\dots,d_xf_i\circ T_xX,\dots d_xf_k) > .
\end{aligned}$$

It follows that the last member of (7.39) only depends on the 1-jet at x of $\Omega^\sharp$ and X and on the value of d_xf_i for $i=1,\dots,k$.
Now, it is an easy exercise to see that the second member of (7.39) is nothing but the bracket $[D_{\iota^*X}, D_\Omega]$ which ends the proof from Lemma 7.5.

Recall that for any differential k-form Ω on $T^\flat M$, $d_P\Omega$ is a well-defined $(k+1)$-form on $T^\flat M$ (cf. Chapter 3, Proposition 3.22) whose value $d\Omega_x(\alpha_0,\dots\alpha_k)$ only depends on the value of α_i at x for $i\in\{0,\dots,k\}$. More precisely, for $\alpha_i = d_xf_i$, where f_i is a germ of function $f_i\in\mathfrak{A}(V)$, for some open set V around $x\in U$, we have:

$$\begin{aligned}
& (d_P\Omega)_x(d_xf_0,\dots,d_xf_k) \qquad (7.40)\\
= & \sum_{i=0}^{k}(-1)^i T_x\Omega\Big(P_x(d_xf_i)\Big)\left(d_xf_0,\dots,\widehat{d_xf_i},\dots,d_xf_k\right)\\
& + \sum_{0\le i<j\le k}(-1)^{i+j}\Omega_x\left([df_i,df_j]_P(x), d_xf_0,\dots,\widehat{d_xf_i},\dots,\widehat{d_xf_j},\dots,d_xf_k\right)).
\end{aligned}$$

In fact we have

Lemma 7.7 *For any $k \in \mathbb{N}$ and any $\Omega \in \Gamma^*_k(\mathfrak{A}(U))$, $d\Omega$ belongs to $\Gamma^*_{k+1}(\mathfrak{A}(U))$*

Proof *At first, from Proposition 7.3 and the definition of $\{.,.\}_P$, we have*

$$[df_i, df_j]_P(x) = d_x\{f_i, f_j\}_p = d_x(< df_j, P(df_i) >= d_x f_j \circ T_x P(., d_x f_i) \quad (7.41)$$

Since T_xP belongs to $\mathrm{L}\left(T_xM, \mathrm{L}\left(T_x^\flat M, T_xM\right) \equiv \mathrm{L}\left(T_xM \times T_x^\flat M, T_xM\right)\right.$ then $T_xP(., df_i)$ belongs to $\mathrm{L}\left(T_xM, T_xM\right)$. Now, using relation (7.41), in the second member of this relation we can write:

$$\begin{aligned}
&\sum_{i=0}^{k}(-1)^i T_x\Omega\left(P\left(df_i\right)\right)\left(d_xf_0,\ldots,\widehat{d_xf_i},\ldots,d_xf_k\right)\\
=\;&\sum_{i=1}^{k}(-1)^i < d_xf_0, T_x\Omega^\sharp\left(P(d_xf_i)\right)\left(d_xf_1,\ldots,\widehat{d_xf_i},\ldots,d_xf_k\right) >\\
&+T_x\Omega\left(Pd_xf_0\right)\left(d_xf_1,\ldots,d_xf_k\right)
\end{aligned}$$

and

$$\begin{aligned}
&\sum_{0\leq i<j\leq k}(-1)^{i+j}\left(\Omega_x\left(\left[df_i, df_j\right]_P, d_xf_0,\ldots,\widehat{d_xf_i},\ldots,\widehat{d_xf_j},\ldots,d_xf_k\right)\right)\\
=\;&\sum_{1\leq i<j\leq k}(-1)^{i+j}\Big(< d_xf_0, \Omega_x^\sharp(d_xf_j\circ T_xP(., d_xf_i),\\
&\qquad d_xf_1,\ldots,\widehat{d_xf_i},\ldots,\widehat{d_xf_j},\ldots,d_xf_k) > \Big)\\
&+\sum_{j=1}^{k}\Omega\left(d_xf_1,\ldots,df_j\circ T_xP(., df_0),\ldots,d_xf_k\right)
\end{aligned}$$

Now, according to the second member of relation (7.39) the term

$$T_x\Omega(Pd_xf_0)\left(d_xf_1,\ldots,d_xf_k\right)-\sum_{j=1}^{k}\Omega\left(d_xf_1,\ldots,df_j\circ T_xP(., df_0),\ldots,d_xf_k\right)$$

is nothing but $L_{P(df_0)}\Omega(df_1,\ldots,df_k)(x)$.
The summation

$$\begin{aligned}
&\sum_{i=1}^{k}(-1)^i < d_xf_0, T_x\Omega^\sharp\left(P(d_xf_i)\right)\left(d_xf_1,\ldots,\widehat{d_xf_i},\ldots,d_xf_k\right) >\\
&-\sum_{1\leq i<j\leq k}(-1)^{i+j} < d_xf_0, \Omega_x^\sharp\Big(d_xf_j\circ T_xP(., d_xf_i),\\
&\qquad d_xf_1,\ldots,\widehat{d_xf_i},\ldots,\widehat{d_xf_j},\ldots,d_xf_k\Big) >
\end{aligned}$$

can be written

$$< d_xf_0, \Pi_x(d_xf_1,\ldots,d_xf_k) >$$

where

$$\begin{aligned}
&\Pi_x(d_xf_1,\ldots,d_xf_k)\\
&= \sum_{i=1}^{k}(-1)^iT_x\Omega^\sharp\left(P(d_xf_i)\right)\left(d_xf_1,\ldots,\widehat{d_xf_i},\ldots,d_xf_k\right)\\
&\quad - \sum_{1\leq i<j\leq k}(-1)^{i+j}\Omega^\sharp_x(d_xf_j\circ T_xP(.,d_xf_i),\\
&\qquad d_xf_1,\ldots,\widehat{d_xf_i},\ldots,\widehat{d_xf_j},\ldots,d_xf_k)
\end{aligned}$$

By similar arguments as at the end of the proof of Lemma 7.5, we can show that the definition of Π *implies that* Π *belongs to* $\bigwedge^k\Gamma^*(T^\flat M_U;TM_U)$*. Since from Lemma 7.6,* $L_{P(df_0)}\Omega=[P(df_0),\Omega]_S$*, this complete the proof.*

Now we are in situation to generalize to our context, a classical result of [Lic77] in finite dimension:

Proposition 7.8 *Let* Λ *be the Poisson tensor associated with* P*. Then with the previous notations, for any vector fields* X *and* Y *on an open set* U*, we have:*

(1) $d_PX=-L_X\Lambda$

(2) $d_P[X,Y]=L_Xd_PY-L_Yd_PX$.

Remark 7.10

1. *According to Proposition 7.8, 1. and Lemma 7.6, we also have*

$$d_PX=[\Lambda,X]_S$$

   which, in finite dimension, is the usual characterization of the exterior derivative on the dual Lie algebroid $(T^*M,M,P,[.,.]_P)$ *(cf. [DuZu05]).*

2. *In finite dimension, recall that a Lie bi-algebroid is a pair of Lie algebroids* $(\mathcal{A},\pi,M,\rho,[.,.]_{\mathcal{A}})$ *and* $(\mathcal{A}^*,\pi',M,\rho^*,[.,.]_{\mathcal{A}^*})$ *where* $\pi':\mathcal{A}^*\to M$ *is the dual bundle of* $\pi:\mathcal{A}\to M$ *with the following compatibility condition*

$$d_{\rho^*}[X,Y]_{\mathcal{A}}=L_Xd_{\rho^*}Y-L_Yd_{\rho^*}X$$

   for all sections X *and* Y *of* $\mathcal{A}$ *(cf. [MaXu94]).*
   In particular, if M *is provided with a Poisson structure defined by a bivector* Λ *and a Poisson anchor* $P:T^*M\to M$*, then the pair* $(TM,p_{TM},M,\mathrm{Id},[.,.])$ *and* $(T^*M,p_{T^*M},P,[.,.]_P)$ *is a Lie bi-algebroid (cf. [MaXu94]). According to our notations, in this finite-dimensional context, the compatibility relation is exactly Proposition 7.8 (2).*
   Unfortunately, in our infinite-dimensional context, even if $T^\flat M=T'M$*, the Poisson structure, in general, does not provide* $T'M$ *with a Lie algebroid structure.*

Proof of Proposition 7.8

(1) On the one hand, note that Λ *belongs to* $\bigwedge^2 \Gamma^*(T^\flat M)$ *and so from the formulae of a Lie derivative of a 2-form on a partial strong convenient algebroid (cf. Chapter 3, § 3.18.6.3) and (7.26), we have*

$$L_X \Lambda(\alpha, \beta) = - < \beta, [P(\alpha,)X] > + < \alpha, [P(\beta), X] > -X(< \beta, P(\alpha) >).$$

Using the expression of the Lie derivative for the admissible 2-form Λ *given in Lemma 7.6 we obtain:*

$$L_X \Lambda(\alpha, \beta) = X(< \beta, P(\alpha) >) - < L_X \alpha, P(\beta) > + < L_X \beta, P(\alpha) > .$$

On the other hand, formulae of the exterior derivative of a 2-form on a partial strong convenient algebroid (cf. Chapter 3, § 3.18.6.3), we get:

$$d_P X(\alpha, \beta) = P(\alpha)(< \beta, X >) - P(\beta)(< \alpha, X >) - < [\alpha, \beta]_P, \alpha > .$$

Using the value of $[\alpha, \beta]_P$ *given in (7.20), we get:*

$$d_P X(\alpha, \beta) = < \beta, [P(\alpha), X] > - < \alpha, [P(\beta), X] > +X(< \beta, P(\alpha) >)$$

which ends the proof of (i).

(2) From the value of a Lie derivative given in Lemma 7.6 and (1), we have:

$$\begin{aligned} L_X d_P Y(\alpha, \beta) &= X\big(d_P Y(\alpha, \beta)\big) - d_P Y(L_X \alpha, \beta) - d_P Y(\alpha, L_X \beta) \\ &= X\big(Y(< \beta, P(\alpha) >)\big) - X\big(< L_Y \alpha, P(\beta) > \big) \\ &+X\big(< L_Y \beta, P(\alpha) > \big) - Y\big(< L_X \beta, P(\alpha) > \big) \\ &+ < L_X L_Y \alpha, P(\beta) > - < L_Y \beta, P(L_X \alpha) > \\ &-Y\big(< \beta, P(L_X \alpha) > \big) + < L_Y \alpha, P(L_X \beta) > \\ &- < L_X L_Y \beta, P(\alpha) > \end{aligned}$$

By permutation of X *and* Y*, we obtain the corresponding expression of* $L_Y d_P X$*. By a direct calculation we obtain easily that*

$$\begin{aligned} L_X d_P Y - L_Y d_P X = [X, Y]\{< \alpha, P(\beta) > - < L_{[X,Y]} \alpha, P\beta > \\ + < L_{[X,Y]} \beta, P(\alpha) \end{aligned}$$

which is exactly $d_P[X, Y](\alpha, \beta)$*.*

7.3.2 Lichnerowicz Cohomology Groups of a Partial Poisson Manifold

Consider a partial Poisson structure $(T^\flat M, M, P, \{\ ,\ \}_P)$. We have seen that $(T^\flat M, M, P, [\cdot, \cdot]_P)$ is a strong partial convenient Lie algebroid. According to

Lemma 7.7, the graded algebra $\bigwedge \Gamma^*(\mathfrak{A}(U)) = \bigoplus_{k\in\mathbb{N}} \Gamma_k^*(\mathfrak{A}(U))$ and for any open set U provided with d_P is an exterior derivative graded algebra.

So for $U = M$, we have the cohomology groups $\left(H^k(\mathfrak{A}(M))\right)_{k\geq 1}$, with coefficients in $\mathfrak{A}(M)$. We also have a cohomolgy sheaves of groups $\{H^k(\mathfrak{A}(U)),\ U \text{ open in } M\}$ with coefficients in the sheaf $\mathfrak{A}_M$.

Definition 7.14 *The cohomology groups $\left(H^k(\mathfrak{A}(M))\right)_{k\geq 0}$ are called the Lichnerowicz-Poisson cohomology groups of the partial Poisson manifold $(T^\flat M, M, P, \{\ ,\ \}_P)$ and $\{H^k(\mathcal{A}_U),\ U \text{ open in } M\}_{k\geq 0}$ are called cohomology sheaves of groups of such a structure.*

For any open set U in M, we denote by $\mathfrak{X}_P(U) = \{P(df), f \in \mathfrak{A}(U)\}$; we then have a long exact sequence

$$0 \longrightarrow H^0(\mathfrak{A}(U)) \longrightarrow \mathfrak{A}(U) \xrightarrow{\text{grad}_P} \mathfrak{X}_P(U) \longrightarrow H^1(\mathfrak{A}(U)) \longrightarrow 0 \quad (7.42)$$

where $\text{grad}_P(f) = P(df)$. More precisely, $H^0(\mathfrak{A}(U)) = \{f \in \mathfrak{A}(U),\ d_P f := df = 0\}$ and $H^1(\mathfrak{A}(U))$ is the first cohomology group of the algebroid $(T^\flat M_U, U, P_U, [\cdot,\cdot]_{P_U})$ which is exactly

$$H^1(\mathfrak{A}(U)) = \frac{\mathcal{X}^1(U)}{\mathcal{X}_P(U)}$$

where $\mathcal{X}^1(U) = \{X \in \mathcal{X}(U)\ :\ L_X\Lambda = -d_P X = 0\}$ (cf Proposition 7.8).

Of course, we have the same long exact sequence when $U = M$.

Remark 7.11 *The previous description of the cohomology associated with each graded differential algebra $(\bigwedge \Gamma^*(\mathfrak{A}(U)), d_P)$ is a generalization to the context of convenient partial Poisson manifold of the classical properties of Lichnerowicz cohomology associated with a finite-dimensional Poisson manifold (cf. for instance [DuZu05], Chapter 2).*

7.4 Convenient Poisson Morphisms

Let $\varepsilon : M_1 \to M_2$ be a smooth map between two convenient manifolds M_1 and M_2 modelled on $\mathbb{M}_1$ and $\mathbb{M}_2$ respectively.
We consider subbundles $T^\flat M_1$ and $T^\flat M_2$ of $T'M_1$ and $T'M_2$ respectively, whose respective inclusion are convenient bundle morphisms and we denote by $\mathbb{F}_1$ and $\mathbb{F}_2$ respectively their typical fibres.

Let $T\varepsilon : TM_1 \to TM_2$ be the tangent map of ε. The adjoint of the bounded operator $T_x\varepsilon : T_xM_1 \to T_{\varepsilon(x)}M_2$ is denoted by $T^*_{\varepsilon(x)}\varepsilon : T'_{\varepsilon(x)}M_2 \to T'_xM_1$ for all $x \in M$. If ε is an injective weak immersion, we can define an adjoint bundle morphism, denoted $T^*\varepsilon$, from $\{T'M_2\}_{|\varepsilon(M_2)}$ to $T'M_1$ by:

$$\left(T^*_{y'}\varepsilon\right)(\alpha') = \alpha' \circ T_x\varepsilon \text{ if } \varepsilon(x) = y' \tag{7.43}$$

More generally, such a morphism does not exist, but, for each point $\varepsilon(x) \in M_2$, the adjoint linear operator $T^*_x\varepsilon : T'_{\varepsilon(x)}M_2 \to T'_xM_1$ is characterized by the equation (7.43) and so is a bounded linear operator. Moreover, in our context, the restriction of $T^*_x\varepsilon$ to $T^\flat_{\epsilon(x)}M_2$ is also a bounded linear operator from $T^\flat_{\epsilon(x)}M_2$ into T'_xM since the inclusion of $T^\flat_{\epsilon(x)}M_2$ in $T'_{\epsilon(x)}M_2$ is bounded.

By abuse of notation, we will simply denote by $T^*\varepsilon$ the set of operators $T^*_{\varepsilon(x)}\varepsilon$ for $x \in M_1$.

Remark 7.12 *Assume that*

$$T^*_x\varepsilon(T^\flat_{\varepsilon(x)}M_2) \subset T^\flat_xM_1. \tag{7.44}$$

Then the linear operator $T^*_x\varepsilon_{|T^\flat_{\varepsilon(x)}M_2} : T^\flat_{\varepsilon(x)}M_2 \to T^\flat_xM_1$ is not bounded in general if $T^\flat_xM_1$ is provided with its own topology of convenient space. However, this operator is bounded if $T^\flat_xM_1$ is provided with the topology induced from the topology on T'_xM_1. In particular, if $T^\flat_xM_1$ is closed in T'_xM_1, then $T^*_x\varepsilon_{|T^\flat_{\varepsilon(x)}M_2} : T^\flat_{\varepsilon(x)}M_2 \to T^\flat_xM_1$ is bounded.

Definition 7.15 *Consider two partial Poisson manifolds $(M_1, \mathfrak{A}(M_1), \{.,.\}_{P_1})$ and $(M_2, \mathfrak{A}(M_2), \{.,.\}_{P_2})$. Assume that $T^\flat M_1$ is a closed subbundle of $T'M_1$. We will say that a smooth map $\varepsilon : M_1 \to M_2$ is compatible with these partial structures if*

(CPS) $\forall x \in M,\ T^*_x\varepsilon(T^\flat_{\varepsilon(x)}M_2) \subset T^\flat_xM_1$.

Then this property will be often cut down by

$$\left(T^*\varepsilon\right)\left(T^\flat M_2\right) \subset T^\flat M_1. \tag{7.45}$$

Consider a chart (U, ϕ) of a convenient manifold M and $T\phi : TM_{|U} = p_M^{-1}(U) \to \phi(U) \times \mathbb{M}$ the associated trivialization of TM. Then $T^*\phi : \phi(U) \times \mathbb{M}' \to T'M_{|U} = (p'_M)^{-1}(U)$ is a bundle isomorphism, and so its inverse morphism $T^*\phi^{-1}$ is a trivialization of $T'M$ canonically associated with the chart (U, ϕ).
Assume that we have the property (7.45). For any $x \in M_1$, there exist charts (U, ϕ) around x in M_1 and (U_2, ϕ_2) around $\varepsilon(x)$ in M_2 respectively such that:

(1) The restriction $T^*\phi^{-1}$ and $T^*\phi'^{-1}$ to $\{T^\flat M_1\}_{|U_1}$ and $\{T^\flat M_2\}_{|U_2}$ are trivializations onto $\phi(U_1) \times \mathbb{F}_1$ and $\phi_2(U_2) \times \mathbb{F}_2$ respectively;

(2) For all $z \in \phi(U)$, we have:

$$\{(T^*(\phi_2) \circ \varepsilon \circ \phi_1^{-1}) (\phi_2) \circ \varepsilon \circ \phi_1^{-1}(z)\} \times \mathbb{F}_2) \subset \{z\} \times \mathbb{F}_1$$

Such charts (U_1, ϕ_1) and (U_2, ϕ_2) will be called *compatible with this Property.*

Proposition 7.9 *Consider two partial Poisson manifolds $(M_1, \mathfrak{A}(M_1), \{.,.\}_{P_1})$ and $(M_2, \mathfrak{A}(M_2), \{.,.\}_{P_2})$ such that $T^\flat M_1$ is a closed subbundle of $T'M_1$.*
Given a smooth map $\varepsilon : M_1 \to M_2$, we have
$\varepsilon^*(\mathfrak{A}(U_2)) \subset \mathfrak{A}(U_1)$ for any open set U_1 of M_1 such that $\varepsilon(U_1)$ is contained in an open set U_2 of M_2 if and only if the property **(CPS)** *is satisfied.*

Sketch of the proof *According to Lemma 7.3, the assumptions of Proposition 7.9 implies clearly property* **(CPS)**. *We have to prove the converse.*

Suppose that property **(CPS)** *is satisfied. Since the problem is local, and according to compatible charts as previously, we may assume that, for $i \in \{1,2\}$, U_i is a c^∞ open set in $\mathbb{M}_i$ such that $TM_i = U_i \times \mathbb{M}_i$, $T^\flat M_i = U_i \times \mathbb{F}_i$ and ε is a smooth map from $U_1 \to U_2$ where*
– $T_x\varepsilon\left(\{x\} \times \mathbb{M}_1\right) \subset \{\varepsilon(x)\} \times \mathbb{M}_2$
– $T_x^*\varepsilon\left(\{\varepsilon(x)\} \times \mathbb{F}_2\right) \subset \{x\} \times \mathbb{F}_1$.
Moreover, by assumption, $\mathbb{F}_1$ is closed in $\mathbb{M}_1'$.

Since $\varepsilon^*(f) = f \circ \varepsilon$, we have

$$d_x(f \circ \varepsilon) = d_{\varepsilon(x)} f \circ T_x\varepsilon = T_x^*\varepsilon\left(d_{\varepsilon(x)} f\right)$$

Since f belongs to $\mathfrak{A}(U_2)$, the map $d_x f$ belongs to $T_y^\flat M_2$. Thus, according to relation (7.45), the previous relation implies that $d_x(f \circ \varepsilon)$ belongs to $T_x^\flat M_1$ and is bounded since $\mathbb{F}_1$ is closed in $\mathbb{M}_1'$.
As in the proof of Proposition 7.1, 2., by induction, we can show that

$$d_x^k(f \circ \varepsilon)(u_1, \ldots, u_k)$$

is a summation of terms of type

$$d_{\varepsilon(x)}^l f(T_x\varepsilon\left(u^{\sigma^1}\right), T_x^2\varepsilon\left(u^{\sigma^2}\right), \ldots, T_x^i\varepsilon\left(u^{\sigma^i}\right), \ldots, T_x^h\varepsilon\left(u^{\sigma^h}\right)) \qquad (7.46)$$

where

- *the set $\{\sigma^1, \ldots, \sigma^h\}$ is a partition of the set $\{1, \ldots, k\}$;*
- *for σ^i, one has:*
  – *each σ^i is a disjoint union of ν_i strictly increasing sequence of length i, that is $\sigma_j^i := \left((s_j^i)_1, \ldots, (s_j^i)_i \right)$ if $\nu_i \geq 1$;*

– *otherwise* σ^i *is an empty set*

- *for all* $1 \leq l \leq k$ *and* $h + l = k$, *we have*

$$\begin{cases} \nu_1 + 2\nu_2 + \cdots + i\nu_i + \cdots + h\nu_h &= k \\ \nu_1 + \cdots + \nu_i + \cdots + \nu_h &= l \end{cases}$$

- *If* $\nu_i \geq 1$ *and* $\sigma^i_j = \left(\left(s^i_j\right)_1, \ldots, \left(s^i_j\right)_i \right)$, *for* $j \in \{1, \ldots, \nu_i\}$, *we set*

$$\begin{aligned} u_{\sigma^i_j} &= \left(u_{(s^i_j)_1}, \ldots, u_{(s^i_j)_i}\right) \\ T^i_x \varepsilon \left(u^{\sigma^i}\right) &= \left(T^i_x \varepsilon \left(u_{\sigma^i_1}\right) \ldots, T^i_x \varepsilon \left(u_{\sigma^i_{\nu_i}}\right)\right) \end{aligned}$$

Note that the term (7.46) is completely defined by such a partition $\{\sigma^1, \ldots, \sigma^h\}$ *and the summation in the expression of* $d^k_x(f \circ \varepsilon)(u_1, \ldots, u_k)$ *is for all such partitions.*
For the end of the proof we need the following Lemma:

Lemma 7.8 *Let* ε *be a smooth map from a* c^∞ *open set* U *around* 0 *in convenient space* E_1 *into a convenient space* E_2.
Assume that there exists a convenient space F_2 *contained in* E_2' *with bounded inclusion and a closed convenient subspace* F_1 *of* E_1'. *If the restriction of* $(T_x\varepsilon)^*$ *to* F_2 *is a bounded linear map from* F_2 *to* F_1 *for any* $x \in U$ *any* $u_2, \ldots, u_n \in E_1$, *then* $((T^n_x\varepsilon)(., u_2, \ldots, u_n))^*$ *is a bounded linear map from* F_2 *to* F_1, *for integer* n *and* $x \in U$.

According to this Lemma, we fix such a partition $\{\sigma^1, \ldots, \sigma^h\}$. *Then* 1 *belongs to one and only one* $\sigma^{i_0}_j := ((s^{i_0}_j)_1, \ldots, (s^{i_0}_j)_{i_0})$ *and since this sequence is strictly increasing, we* $(s^{i_0}_j)_1 = 1$. *The corresponding term is*

$$T^{i_0}_x \varepsilon \left(u^{\sigma^{i_0}_j}\right) = T^l_x \varepsilon \left(u_{(s^{i_0}_j)_1}, \ldots, u_{(s^{i_0}_j)_{i_0}}\right)$$

After deleting $u_{(s^l_j)_1}$ *in this term we obtain the bounded linear map*

$$\widehat{\varepsilon_1} : u_1 \mapsto T^{i_0}_x \varepsilon(u_1, u_{(s^{i_0}_j)_2}, \ldots, u_{(s^{i_0}_j)_{i_0}})$$

By application of Lemma 7.8, the restriction of $(\widehat{\varepsilon_1})^*$ *to* $\mathbb{F}_2$ *is a bounded linear map into* $\mathbb{F}_1$.

Associated with this partition, we consider again the corresponding term (7.46)

$$d^l_{\varepsilon(x)} f\left(T_x\varepsilon\left(u^{\sigma^1}\right), T^2_x\varepsilon\left(u^{\sigma^2}\right), \ldots, T^i_x\varepsilon\left(u^{\sigma^i}\right), \ldots, T^h_x\varepsilon\left(u^{\sigma^h}\right)\right) \tag{7.47}$$

In this expression, all vectors $T_x\varepsilon\left(u^{\sigma^1}\right),\dots,T^i_x\varepsilon\left(u^{\sigma^i}\right),\dots,T^h_x\varepsilon\left(u^{\sigma^h}\right)$ *belong to* $T_{\varepsilon(x)}M_1\equiv\mathbb{F}_2$ *which we denote by* $v_1,\dots,v_h$*. Among these vectors, we have one and only one which is equal to* $\widehat{\varepsilon_1}(u_1)$ *and we may assume that* $v_1=\widehat{\varepsilon_1}(u_1)$*;*

Finally, after having deleted u_1 *in (7.47), we obtain a term of type*

$$\widehat{\varepsilon_1}^*\left(d^l_{\varepsilon(x)}f\left(.,v_2,\dots,\dots v_h\right)\right).$$

which belongs to $\mathbb{F}_1$ *from the construction of* $(\varepsilon_1)^*$*. But such a property is true for any term of the decomposition of* $d^k_x(f\circ\varepsilon)(.,u_2,\dots,u_k)$ *for any* $k\in\mathbb{N}$ *and any* $u_2,\dots,u_k$ *in* T_xM_1*. This implies that* $d^k_x(f\circ\varepsilon)(.,u_2,\dots,u_k)$ *belongs to* T_xM_1 *for any* $k\in\mathbb{N}$ *and any* $u_2,\dots,u_k$ *in* T_xM_1 *and so the proof is complete, according to Proposition 7.1.*

Proof of Lemma 7.8 *Fix some* $x\in U$*. By assumption the result is true for* $n=1$*. Assume that the result is true for all* $1\leq i<n$*. Then according to the definition of* $d^n_x\varepsilon$ *(cf. [KrMi97], 5.11), we have*

$$T^n_x\varepsilon(u,u_2,\dots,u_n)=\lim_{t\to 0}\left(\frac{1}{t}\left(T^{n-1}_{x+tu}\varepsilon\left(u_2,\dots,u_n\right)-T^{n-1}_x\varepsilon\left(u_2,\dots,u_n\right)\right)\right)$$

On the other hand, for any $\alpha\in F_2$*, we obtain*

$$(T^n_x\varepsilon(.,u_2,\dots,u_n))^*(\alpha)=\alpha\circ(T^n_x\varepsilon(.,u_2,\dots,u_n))$$

Now the map

$$t\mapsto\alpha\circ T^{n-1}_{x+tu}\varepsilon(.,u_2,\dots,u_n)\qquad(7.48)$$

is a smooth map from E_2' *to* E_1'*.*
Thus we have

$$\frac{d}{dt}_{|t=0}\left(\alpha\circ\left(T^{n-1}_{x+tu}\varepsilon(.,u_2,\dots,u_n)\right)\right)=\alpha\circ(T^n_x\varepsilon(.,u,u_2\dots,u_n))$$

But the inclusion of F_2 *in* E_2 *is bounded,* F_1 *is closed in* E_1' *and the map (7.48) takes values in* F_1 *from the inductive assumption. This implies that* $\alpha\circ(T^n_x\varepsilon(.,u,u_2\dots,u_n))$ *belongs to* F_1 *for any* $\alpha\in F_2$*. But the map* $\alpha\mapsto\alpha\circ(T^n_x\varepsilon(.,u,u_2\dots,u_n))$ *is a bounded map from* F_2 *to* E_1' *which takes values in* F_1*. Since* F_1 *is closed it follows that this map is a linear bounded map from* F_2 *to* F_1*.*

Remark 7.13 *Let* $P_1:T^\flat M_1\to TM_1$ *(resp.* $P_2:T^\flat M_2\to TM_2$*) be a bundle morphism over* M_1 *(resp.* M_2*). Assume that we have*

$$P_2(\varepsilon(x),\xi)=T\varepsilon\circ P_1(x,T^*_{\varepsilon(x)}\varepsilon(\xi))$$

for all $x\in M$ *and* $\xi\in T^*_{\varepsilon(x)}M_2$*. This relation will be simply denoted by*

$$P_2=T\varepsilon\circ P_1\circ T^*\varepsilon.$$

If the condition **(CPS)** *is fulfilled, from the definition of the Poisson bracket, we have*

$$\{\varepsilon^*(f), \varepsilon^*(g)\}_{P_1} = \varepsilon^*(\{f, g\}_{P_2})$$

if

$$P_2 = T\varepsilon \circ P_1 \circ T^*\varepsilon.$$

When ε *is a Poisson morphism between two partial Poisson manifolds* $(M_1, \mathfrak{A}_1, \{.,.\}_{P_1})$ *and* $(M_2, \mathfrak{A}_2, \{.,.\}_{P_2})$*, then we have the following relation between Hamiltonian fields:*

$$X_{\varepsilon^* f} = T\varepsilon(X_f)$$

7.5 Projective Limits of Partial Poisson-Banach Manifolds

In this section, as usually in the Banach setting, for any Banach manifold X, for the cotangent bundle we use the notation T^*X instead of $T'X$, as previously; in this context, T^*M is nothing but than $T'X$.

Let $(M_i, \mathfrak{A}(M_i), \{.,.\}_{P_i})_{i \in \mathbb{N}}$ be a sequence of partial Poisson-Banach manifolds where $p_i^\flat : T^\flat M_i \to M_i$ is a Banach subbundle of $p_{M_i}^* : T^*M_i \to M_i$ and $P_i : T^\flat M_i \to TM_i$ is a skew-symmetric morphism. We denote by $\mathbb{M}_i$ the Banach space on which M_i is modelled, and by $\mathbb{F}_i$ the model of the typical fibre of $p_i^\flat : T^\flat M_i \to M_i$ and we assume that $\mathbb{F}_i$ is a Banach subspace of the dual $\mathbb{M}_i^*$ of $\mathbb{M}_i$.

Definition 7.16 *The sequence* $(M_i, \mathfrak{A}(M_i), \{.,.\}_{P_i})_{i \in \mathbb{N}}$ *is called a projective sequence of partial Poisson-Banach manifolds if there exists submersions* $\delta_i^j : M_j \to M_i$ *such that* $\left(M_i, \delta_i^j\right)_{i \in \mathbb{N}}$ *is a submersive projective sequence of Banach manifolds fulfilling, for all* $i \in \mathbb{N}$*, the following properties:*

(PSPPBM1) $T^*\delta_i^{i+1}(T^\flat M_i) \subset T^\flat M_{i+1}$*;*

(PSPPBM2) $P_i = T\delta_i^{i+1} \circ P_{i+1} \circ T^*\delta_i^{i+1}$*;*

(PSPPBM3) *Around each* $x \in M$ *there exists a sequence of charts* $(U_i, \phi_i)_{i \in \mathbb{N}}$ *such that* $(U = \varprojlim(U_i), \phi = \varprojlim(\phi_i))$ *is a chart of* x *in* M*, so that the charts* (U_{i+1}, ϕ_{i+1}) *and* (U_i, ϕ_i) *are compatible with Property* **(PSPPBM1)**.

Remark 7.14 *Note that since* $\delta_i^j = \delta_i^{i+1} \circ \cdots \circ \delta_{j-1}^j$*, the sequence* $(M_i, \delta_i^j)_{j \geq i}$ *is a submersive projective sequence of Banach manifolds if and only if* $\delta_i^{i+1} : M_{i+1} \to M_i$ *is surjective. In this way, there exists a surjective operator* $\overline{\delta}_i^{i+1}$

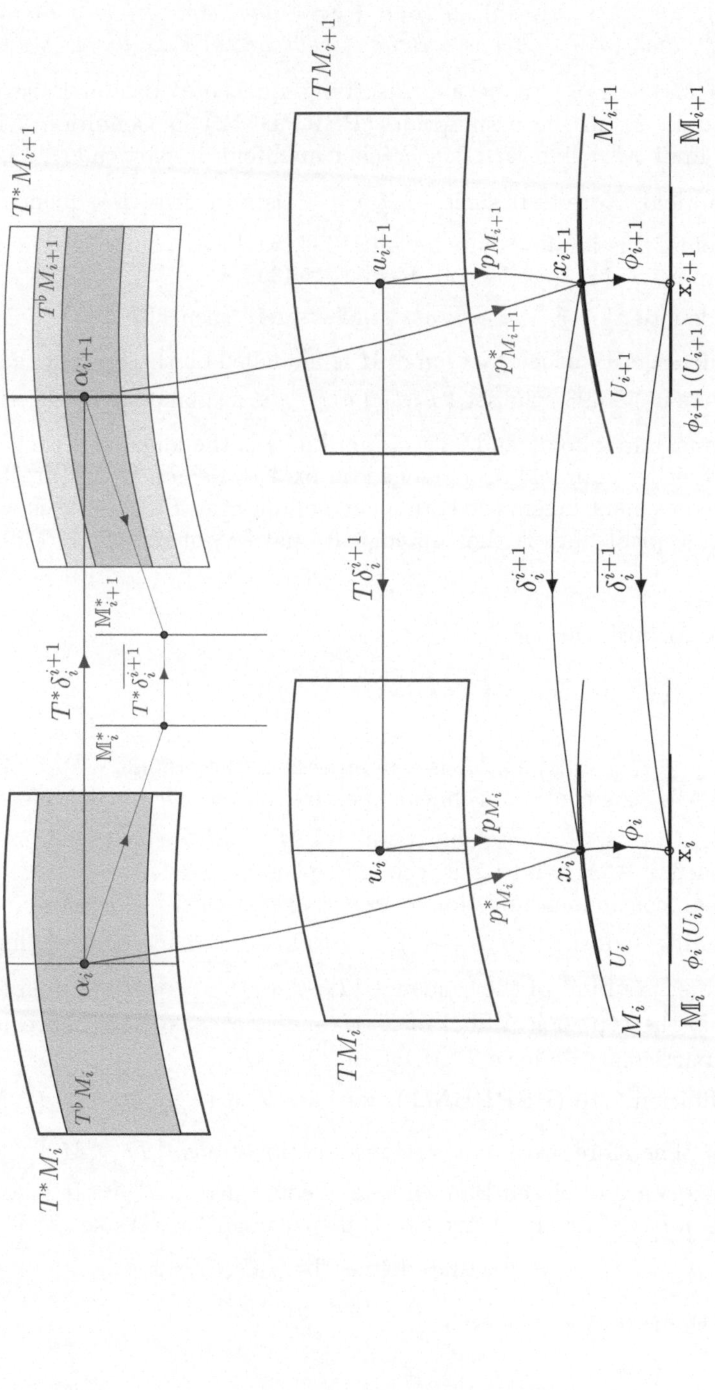

FIGURE 7.1
Projective sequence of partial Poisson structures

from $\mathbb{M}_{i+1}$ onto $\mathbb{M}_i$. Thus, the adjoint operator $(\bar{\delta}_i^{i+1})^* : \mathbb{M}_i^* \to \mathbb{M}_{i+1}^*$ is injective. Moreover $(\bar{\delta}_i^{i+1})^*(\mathbb{M}_i^*)$ is a Banach subspace of $\mathbb{M}_{i+1}^*$. Again, since $\mathbb{F}_i$ is the typical fibre of $T^\flat M_i$, according to assumptions **(PSPPBM1)** *and* **(PSPPBM3)**, *we must have $(\bar{\delta}_i^{i+1})^*(\mathbb{F}_i) \subset \mathbb{F}_{i+1}$ and, since $\mathbb{F}_i$ is a Banach subspace of $\mathbb{M}_i^*$, then $(\bar{\delta}_i^{i+1})^*(\mathbb{F}_i)$ is a Banach subspace of $\mathbb{F}_{i+1}$ for all $i \in \mathbb{N}$.*

Let $(M_i, \mathfrak{A}(M_i), \{.,.\}_{P_i})_{i\in\mathbb{N}}$ be a projective sequence of partial Poisson-Banach manifolds. From the assumption **(PSPPBM2)** in Definition 7.16, the projective limit $M = \varprojlim(M_i)$, is a Fréchet manifold; in particular M is a convenient manifold. Note that since $\left(M_i, \delta_i^j\right)_{j\geq i}$ is a projective sequence of Banach manifolds, this implies that, for all $i \in \mathbb{N}$, we have a surjective linear continuous map $\bar{\delta}_i^{i+1} : \mathbb{M}_{i+1} \to \mathbb{M}_i$ (cf. Remark 7.14).
For $j > i$, we denote $\bar{\delta}_i^j = \bar{\delta}_i^{i+1} \circ \cdots \circ \bar{\delta}_{j-1}^j : \mathbb{M}_j \to \mathbb{M}_i$; then $\left(\mathbb{M}_i, \bar{\delta}_i^j)\right)_{j\geq i}$ is a projective sequence of Banach spaces and M is modelled on the Fréchet space $\mathbb{M} = \varprojlim(\mathbb{M}_i)$. As in [Gal98], the set $\left(TM_i, T\delta_i^j\right)_{j\geq i}$ is a submersive projective sequence of Banach manifolds and $TM = \varprojlim(TM_i)$ is the kinematic tangent bundle of the Fréchet manifold M modelled on $\mathbb{M}$. Now, since M is a Fréchet manifold, the convenient kinematic cotangent bundle $p_M^* : T^*M \to M$ is well defined and its typical fibre is the topological dual $\mathbb{M}^*$ of $\mathbb{M}$ (cf. [KrMi97], 33.1).

We identify M with the set

$$\{x = (x_i)_{i\in\mathbb{N}} \in \prod_{i\in\mathbb{N}} M_i : x_i = \delta_i^j(x_j), j \geq i\}.$$

Since for each $j \geq i$, $\delta_i^j : M_j \to M_i$ is a submersion, the adjoint map $T^*\delta_i^j : T^*_{\delta_i^j(x_j)}M_i \to T^*_{x_j}M_j$ is a continuous linear injective map whose range is closed, for all $x_j \in M_j$. Now, we have a submersion $\delta_i : M \to M_i$ defined by $\delta_i(x) = x_i \in M_i$ for each $x \in M$, and again the transpose map $T^*\delta_i : T^*_{\delta_i(x)}M_i \to T^*_xM$ is a linear continuous injection whose range is closed. Therefore, we have an ascending sequence $\left(T^*_{\delta_i(x)}M_i\right)_{i\in\mathbb{N}}$ of closed Banach spaces. Since T_xM is the projective limit of the sequence $\left(T_{\delta_i(x)}M_i)\right)_{i\in\mathbb{N}}$, each vector space $\varinjlim(T^*_{\delta_i(x)}M_i)$ is the topological dual of T_xM and is a convenient space (cf. [CaPe19]). In particular, we have $T^*_xM = \varinjlim(T^*_{\delta_i(x)}M_i)$.
Now from Definition 7.16 **(PSPPBM1)**, we have $T^*\delta_i^j(T^*_{\delta_i^j(x_j)M_i}) \subset T^*_{x_j}M_j$ for all $x_j \in M_j$. Therefore, with our previous identifications, $\left(T^\flat_{\delta_i(x)}M_i\right)_{i\in\mathbb{N}}$ is an ascending sequence of closed Banach spaces contained in T^*_xM. It follows that $T^\flat_xM = \varinjlim(T^\flat_{\delta_i(x)}M_i)$ is a convenient subspace of T^*_xM. We set $T^\flat M = \bigcup_{x\in M} T^\flat_xM$ and $p^\flat : T^\flat M \to M$ the map defined by $p^\flat(x,\xi) = x$.
We then have the following proposition.

Proposition 7.10 *Let $(M_i, \mathfrak{A}(M_i), \{.,.\}_{P_i})_{i\in\mathbb{N}}$ be a projective sequence of partial Poisson-Banach manifolds where $p_i^\flat : T^\flat M_i \to M_i$ is a Banach subbundle of $p_{M_i}^* : T^*M_i \to M_i$ and $P_i : T^\flat M_i \to TM_i$ is a skew-symmetric morphism. Let $T_x^\flat M = \varinjlim(T_{\delta_i(x)}^\flat M_i)$ and $p^\flat : T^\flat M \to M$ the map defined by $p^\flat(x,\xi) = x$. We then have*

(1) $p^\flat : T^\flat M \to M$ is a convenient subbundle of the cotangent bundle $p_M^* : T^*M \to M$.

(2) For each $i \in \mathbb{N}$, there exists a canonical bundle morphism $T^\flat\delta_i : T^\flat M \to T^\flat M_i$ over δ_i such that $T^\flat\delta_i(x,\xi) = (x_i,\xi_i)$ if $x = \varprojlim(x_i)$ and $\xi = \varinjlim(\xi_i)$ where

$$P_i \circ T^\flat\delta_i(x,\xi) = T\delta_i^j \circ P_j \circ T^\flat\delta_j(x,\xi)$$

for all $(x,\xi) \in T^\flat M$ and $j > i$. Then $P = \varprojlim(P_i \circ T^\flat\delta_i)$ is a bundle morphism from $T^\flat M$ to TM which is skew-symmetric (relatively to the canonical dual pairing between T^*M and TM in restriction to $T^\flat M \times TM$).

Proof of Proposition 7.10 *In this proof, we will consider a projective sequence $(U_i, \phi_i)_{i\in\mathbb{N}}$ of charts such that $(U = \varprojlim(U_i), \phi = \varprojlim(\phi_i))$ is a chart of M.*

(1) At first we will show that $T_M^* M_l := \bigcup\limits_{x\in M} T^*\delta_l(T_{\delta_l(x)}^* M_l)$ is the total space of a convenient subbundle over M of $p_M^* : T^*M \to M$ and that $T_M^\flat M_l = \bigcup\limits_{x\in M} T^*\delta_l(T_{\delta_l(x)}^\flat M_l)$ is the total space of a convenient subbundle of $p^\flat : T^\flat M \to M$ which is also a closed subbundle of $p_M^* : T^*M \to M$.

Fix some chart $(U = \varprojlim(U_l), \phi = \varprojlim(\phi_l))$ as previously, and for each $l \in \mathbb{N}$, consider

$$T_M^* M_{l|U} := \bigcup_{x\in U} T^*\delta_l(T_{\delta_l(x)}^* M_l) \subset T^*M_{|U}.$$

Recall that we have $\delta_l(U) = U_l$ and $\bar{\delta}_l \circ \phi = \phi_l \circ \delta_l$ where $\bar{\delta}_l$ is the natural linear projection $\mathbb{M} = \varprojlim(\mathbb{M}_l)$ on $\mathbb{M}_l$ which is the projective limit of $\bar{\delta}_l^j$.

Therefore, if $\bar{\delta}_l^*$ is the adjoint of $\bar{\delta}_l$ then $\bar{\delta}_l^*$ is injective and so $\mathbb{M}^*$ can be identified with the direct limit of the ascending sequence $\left(\mathbb{M}_l^*, \bar{\delta}_l^*\right)_{l\in\mathbb{N}}$ of Banach spaces. It follows that $T^*\phi : \phi(U) \times \mathbb{M}^* \to T^*M_{|U}$ is a bundle isomorphism which is the inverse of the trivialization $T^*\phi^{-1}$ of the cotangent bundle over U. Moreover, on $TM_{|U}$, we also have

$$T\bar{\delta}_l \circ T\phi = T\phi_l \circ T\delta_l.$$

We then obtain, for all $x \in U$, on $\{\bar{\delta}_l(\phi(x))\} \times \mathbb{M}_l^*$

$$T^*\phi \circ T^*\bar{\delta}_l = T^*\delta_l \circ T^*\phi_l. \tag{7.49}$$

In particular

$$T^*\phi \circ T^*\bar{\delta}_l : \{\bar{\delta}_l(\phi(x))\} \times \mathbb{M}_l^* \to T^*\delta_l(T^*_{\delta_l(x)}M_l) \subset T^*_x M$$

is a linear map for all $x \in U$.
But $T^*\bar{\delta}_l \equiv \bar{\delta}_l^* : \{\bar{\delta}_l(t)\} \times \mathbb{M}_l^* \to \{t\} \times \mathbb{M}^*$ is an injective closed linear map for all $t \in \phi(U)$. Therefore, on the one hand, if we denote $\mathbb{E}_l = \bar{\delta}_l^*(\mathbb{M}_l^*)$, then $\bar{\delta}_l^*$ is an isomorphism from $\mathbb{M}_l^*$ onto the closed Banach subspace $\mathbb{E}_l$ of $\mathbb{M}^*$; so we can assume that $\mathbb{M}_l^*$ is contained in $\mathbb{M}^*$. With these identifications, the map $T^*\phi \circ T^*\bar{\delta}_l$ is nothing but the natural inclusion of $\phi(U) \times \mathbb{M}_l^*$ into $\phi(U) \times \mathbb{M}^*$.

In this way, we obtain $T^*\phi(\phi(U) \times \mathbb{M}_l^*) = T^*_M M_{l|U} \subset T^*M_{|U}$. This implies that $(T^*_M M_l)_{|U}$ is the total space of a closed trivial subbundle of $T^*M_{|U} \to U$ with typical fibre $\mathbb{M}_l^*$.

Now, recall that $\mathbb{F}_l$ is a Banach subspace of $\mathbb{M}_l^*$. From Equation (7.49), the assumptions of Proposition 7.10 and the previous arguments, we have

$$T^*\phi(\phi(U) \times \mathbb{F}_l) = \bigcup_{x \in U} T^*\delta_l(T^\flat_{\delta_l(x)}M_l) = T^\flat_M M_{l|U} \subset T^*M_{|U}.$$

Therefore $(T^\flat_M M_l)_{|U}$ is also the total space of a closed trivial subbundle of $(T^*M)_{|U} \to U$ with typical fibre $\mathbb{F}$.

Since $\bar{\delta}_i^{j*}$ is a linear continuous closed inclusion of $\mathbb{M}_i^*$ into $\mathbb{M}_j^*$, via any chart $(U = \varprojlim(U_l), \phi = \varprojlim(\phi_l))$ with the properties required at the beginning of the proof, we can build an injective morphism $(T_i^j)_U : T^*_M(M_i)_{|U} \to T^*_M(M_j)_{|U}$ given by

$$\forall (x, \xi_i) \in T^*_M M_{i|U}, (T_i^j)_U(x, \xi_i) = T^*\phi \circ \left(\bar{\delta}_i^j \times (\bar{\delta}_i^j)^*\right) \circ T^*\phi^{-1}(x, \xi_i).$$

Moreover, for any other chart $(U' = \varprojlim(U_i'), \phi' = \varprojlim(\phi_l'))$ of this type with $U \cap U' \neq \emptyset$, for all $(x, \xi_i) \in T^*_M M_{i|U \cap U'}$, we have

$$(T_i^j)_{U'}(x, \xi_i) = T^*\phi' \circ T^*\phi^{-1} \circ (T_i^j)_U \circ T^*\phi \circ T^*\phi'^{-1}(x, \xi_i). \tag{7.50}$$

On the one hand, since for $j = i$ we have $(\bar{\delta}_i^j)^* = \mathrm{Id}$, the relation (7.50) implies that we also have $(T_i^i)_U = (T_i^i)'_U = \mathrm{Id}$, which implies that $T^*_M(M_i)$ is a Banach bundle over M. On the other hand, for any $j \geq i$, the relation (7.50) implies that we get an injective bundle morphism $T_i^j : T^*_M M_i \to T^*_M M_j$ which is nothing but the inclusion of $T^*_M M_i$ into $T^*_M M_j$. In other words, $T^*_M M_i \to M$ is a Banach subbundle of $T^*_M M_j$ for $j \geq i$. Finally, from this construction, it follows that $\left(T^*_M M_i, T_i^j\right)_{i \leq j}$ is an ascending sequence of Fréchet manifolds which has the direct limit chart property at every point of $\varinjlim(T^*_M M_i)$. It follows that $\varinjlim(T^*_M M_i)$ is a convenient manifold modelled on $\mathbb{M} \times \mathbb{M}^*$; in particular, we have $T^*M = \varinjlim(T^*_M M_i)$. Moreover, around each point (x, ξ_i) in $T^*_M M_i$, there exists a chart $(T^*_M M_{i|U}, T^*\phi_{|T^*_M M_{i|U}})$ where (U, ϕ) is a chart *around x which has the required properties.*

Therefore $(T^*M_{|U} = \varinjlim(T^*_M M_{i|U}), (T\phi)^* = T^*\phi^{-1} = \varinjlim(T^*\phi_{|T^*_M M_{i|U}}))$ *is a chart around* $(x, \xi = \varinjlim(\xi_i))$ *in* T^*M*. This implies that* $T^*M = \varinjlim(T^*_M M_i)$ *is a convenient vector bundle over* M *whose typical fibre is* $\mathbb{M}^* = \varinjlim(\mathbb{M}^*_i)$*.*

Clearly the same arguments can be applied to $\left(T^\flat_M M_i, T^j_{i\,|(T^\flat_M M_i)}\right)_{i \leq j}$ *which leads to a proof of the fact that* $T^\flat_M M_i$ *is a Banach bundle over* M *and that* $T^\flat M = \varinjlim(T^\flat_M M_i)$*; we then get a convenient vector bundle* $p^\flat : T^\flat M \to M$ *with typical fibre* $\mathbb{F} = \varinjlim(\mathbb{F}_i)$*.*

(2) For each $i \in \mathbb{N}$*, consider the projection bundle* $\hat{p}^*_i : T^*_M M_i \to M$ *and* $\hat{p}^\flat_i : T^\flat_M M_i \to M$*. Obviously, these bundles are nothing but the pull-back over* $\delta_i : M \to M_i$ *of the bundles* $p^*_{M_i} : T^*M_i \to M_i$ *and* $p^\flat_i : T^\flat M_i \to M_i$*. Therefore, we have a bundle morphism* $\widehat{T}^*\delta_i$ *(resp.* $\widehat{T}^\flat\delta_i$*)over* δ_i *from* $T^*_M M_i$ *to* T^*M_i *(resp. from* $T^\flat_M M_i$ *to* $T^\flat M_i$*) such that its restriction to any fibre is an isomorphism defined by*

$$(x, \xi_i) \mapsto (\delta_i(x), \xi_i). \tag{7.51}$$

Recall that for all $i \leq j$*,* $T^j_i : T^*_M M_i \to T^*_M M_j$ *is the natural inclusion and* $T^*M = \varinjlim(T^*_M M_i)$*. Thus we have a bundle morphism* $T_i = \varinjlim_{j \geq i} T^j_i : T^*_M M_i \to T^*M_i$ *which is the natural inclusion of* $T^*_M M_i$ *into* T^*M*. But the restriction of* T^j_i *to* $T^\flat_M M_i$ *is the natural inclusion of* $T^\flat_M M_i$ *into* $T^*_M M_j$ *(cf. first part of the proof). Thus the restriction of* T_i *to* $T^\flat_M M_i$ *is the natural inclusion of* $T^\flat_M M_i$ *into* $T^\flat M = \varinjlim(T^\flat_M M_i)$

Therefore, from the assumptions **(PSPPBM1)** *and* **(PSPPBM2)** *in Definition 7.16, by induction on* j *for* $i < j$*, we get the following commutative diagram:*

$$\begin{array}{ccccccc}
T^\flat M & \xrightarrow{T_i} & T^\flat_M M_i & \xrightarrow{T^\flat\delta_i} & T^\flat M_i & \xrightarrow{P_i} & TM_i \\
\downarrow{\scriptstyle \mathrm{Id}} & & \downarrow{\scriptstyle T^j_i} & & \downarrow{\scriptstyle T^*\delta^j_i} & & \uparrow{\scriptstyle T\delta^j_i} \\
T^\flat M & \xrightarrow{T_j} & T^\flat_M M_j & \xrightarrow{T^\flat\delta_j} & T^\flat M_j & \xrightarrow{P_j} & TM_j
\end{array}$$

We set $\hat{P}_i(x, \xi) = P_i(\delta_i(x), T^\flat\delta_i \circ T_i(\xi))$*. Note that* $\hat{P}_i$ *is a bundle morphism from* $T^\flat M$ *into* TM_i *over* δ_i*. According to the previous commutative diagram, we obtain*

$$\hat{P}_j(x, \xi) = T\delta^j_i \circ \hat{P}_i(x, \xi).$$

Since $TM = \varprojlim(TM_i)$*, we get bundle morphism* $P = \varprojlim(\hat{P}_i) : T^\flat M \to TM$*. In particular, we have* $T\delta_i \circ P = \hat{P}_i$*. Recall that that* $T^\flat\delta_i$ *a bundle morphism over* δ_i *such that* $T^\flat\delta_i(x, \xi) = (x_i, \xi_i)$ *if* $x = \varprojlim(x_i)$ *and* $\xi = \varinjlim(\xi_i)$ *and also*

$\hat{P}_i = P_i \circ T^\flat \delta_i$.
It remains to prove that P *is skew-symmetric. Since* P_i *is skew-symmetric relatively the canonical pairing* $<\,,\,>_i$ *between* T^*M_i *and* TM_i *we have*

$$< T^\flat \delta_i(x,\eta), \hat{P}_i(x,\xi) >_i = - < T^\flat \delta_i(x,\xi), \hat{P}_i(x,\eta) >_i . \tag{7.52}$$

If $<\,,\,>$ *denotes the canonical pairing between* T^*M *and* TM, *for any kinematic differential forms* $\eta = \varinjlim(\eta_i)$ *and* $\xi = \varinjlim(\xi_i)$ *on* M *which are sections of* $p^\flat : T^\flat M \to M$. *Note that, in fact,* $\eta = \delta_i^*\eta_i$ *and* $\xi = \delta_i^*\xi_i$, *for any* $i \in \mathbb{N}$. *There exists an integer* i *such that* η *and* ξ *are sections of* $T_M^\flat M_i$. *Therefore*

$$\begin{aligned} <\eta, P(\xi)> &= <\delta_i^*\eta_i(P(x,\xi))> = \eta_i(T\delta_i \circ P(x,\xi)) = \eta_i(\hat{P}_i(x,\xi)) \\ &= < T^\flat \delta_i(x,\eta), \hat{P}_i(x,\xi) >_i . \end{aligned}$$

(7.52) implies that P *is skew-symmetric.*

Theorem 7.4 *Let* $(M_i, \mathfrak{A}(M_i), \{.,.\}_{P_i})_{i\in\mathbb{N}}$ *be a projective sequence of partial Poisson-Banach manifolds and* $M = \varprojlim(M_i)$. *There exists a weak subbundle* $p^\flat : T^\flat M \to M$ *of* $p_M^* : T^*M \to M$ *and a skew-symmetric morphism* $P : T^\flat M \to TM$ *such that* $(M, \mathfrak{A}_P(M), \{.,.\}_P)$ *is a partial Poisson manifold with the following characterization:*

If $\delta_i : M \to M_i$ *is the canonical projection, then* δ_i *is a Poisson morphism from* $(M, \mathfrak{A}_P(M), \{.,.\}_P)$ *to* $(M_i, \mathfrak{A}(M_i), \{.,.\}_{P_i})$ *for all* $i \in \mathbb{N}$, *and where we have*

$$\mathfrak{A}_P(M) = \varprojlim \mathfrak{A}(M_i)$$

$$\{.,.\}_P = \varprojlim \{.,.\}_{P_i}.$$

Proof *According to Proposition 7.10, to the skew-symmetric morphism* $P : T^\flat M \to TM$ *is associated the algebra* $\mathfrak{A}_P(M) = \varprojlim(\mathfrak{A}(M_i)$ *is a set of of smooth functions* $f : M \to \mathbb{R}$ *whose differential* df *is a section of* $T^\flat M \to M$ *which satisfies the assumptions of Definition 7.2.*
Consider the bilinear form $\{.,.\}_P = \varprojlim\{.,.\}_{P_i}$. *Of course, if* $f = \varprojlim(f_i)$ *and* $g = \varprojlim(g_i)$ *then* $f = f_i \circ \delta_i$ *and* $g = g_i \circ \delta_i$ *and so* $df = T^*\delta_i(df_i)$ *and* $dg = T^*\delta_i(dg_i)$. *Using the relation of compatibility*

$$P_i = T\delta_i^j \circ P_j \circ T^*\delta_i^j$$

and the relations

$$df_j = T^*\delta_i^j(df_i), \quad dg_j = T^*\delta_i^j(dg_i), \quad \{f_i, g_i\}_{P_i} = df_i(P_i(dg_i)),$$

from Remark 7.13, we obtain:

$$\{f_j \circ \delta_i^j, g_j \circ \delta_i^j\}_{P_j} = \{f_i, g_i\}_{P_i} \circ \delta_i^j.$$

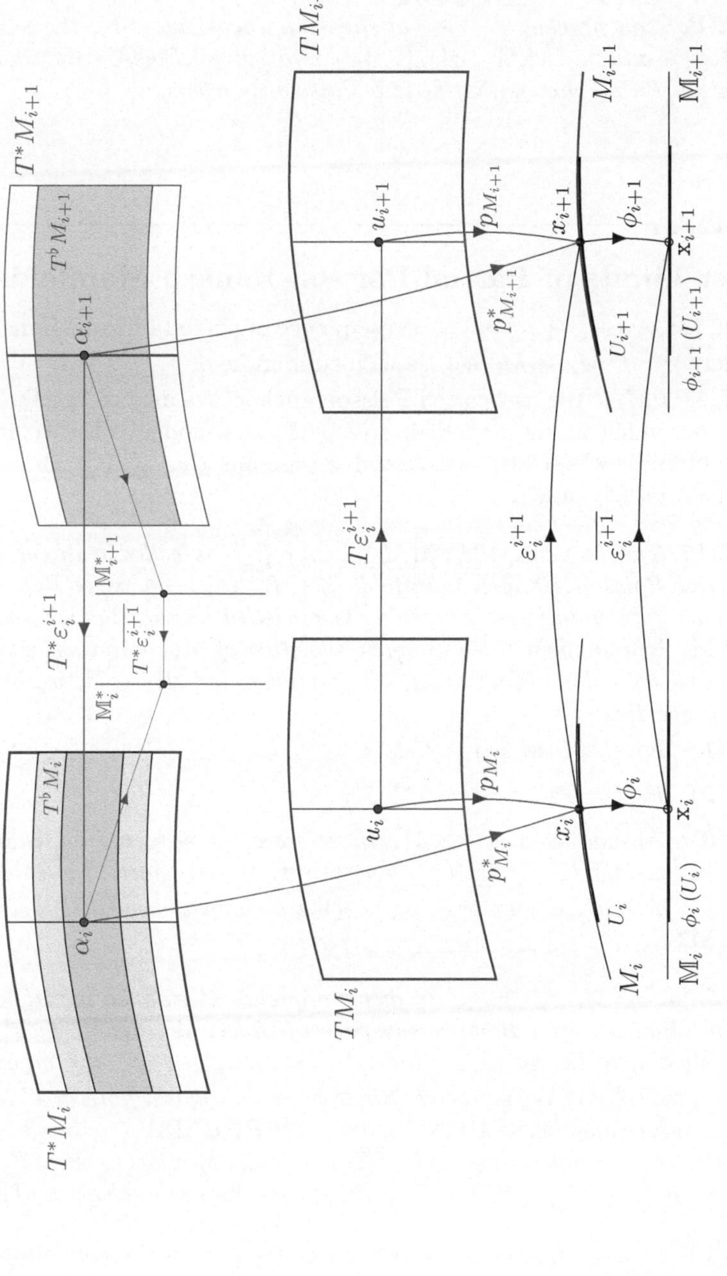

FIGURE 7.2
Direct sequence of partial Poisson structures

Thus we also have:

$$\{f_i \circ \delta_i, g_i \circ \delta_i\}_P = \{f_i, g_i\}_{P_i} \circ \delta_i. \tag{7.53}$$

Finally we get $\{f,g\}_P = \varprojlim \{f_i, g_i\}_{P_i}$.
Now, as each Poisson bracket $\{.,.\}_{P_i}$ *satisfies the Jacobi identity, the same is true for* $\{f,g\}_P$ *and so* $(M, \mathfrak{A}_P(M), \{.,.\}_P)$ *is a partial Poisson manifold. Finally, Equation (7.53) means that* δ_i *is a Poisson morphism.*

7.6 Direct Limits of Partial Poisson-Banach Manifolds

In this section, $(M_i, \mathfrak{A}(M_i), \{.,.\}_{P_i})_{i\in\mathbb{N}}$ is a sequence of partial Poisson-Banach manifolds where $p_i^\flat : T^\flat M_i \to M_i$ is a Banach subbundle of $p_{M_i}^* : T^*M_i \to M_i$ and $P_i : T^\flat M_i \to TM_i$ is the associated Poisson anchor. We denote by $\mathbb{M}_i$ the Banach space on which M_i is modelled, and by $\mathbb{F}_i$ the model of the typical fibre of the subbundle $p_i^\flat : T^\flat M_i \to M_i$ and we assume that $\mathbb{F}_i$ is a Banach subspace of the dual $\mathbb{M}_i^*$ of $\mathbb{M}_i$.

Definition 7.17 *The sequence* $(M_i, \mathfrak{A}(M_i), \{.,.\}_{P_i})_{i\in\mathbb{N}}$ *is called a direct sequence of partial Poisson Banach manifolds if* $(M_i)_{i\in\mathbb{N}}$ *is an ascending sequence of Banach* C^∞*-manifolds, where* M_i *is modelled on the Banach space* $\mathbb{M}_i$ *such that* $\mathbb{M}_i$ *is a supplemented Banach subspace of* $\mathbb{M}_{i+1}$ *and such that* $(M_i, \varepsilon_i^{i+1})$ *is a closed submanifold of* M_{i+1}*; moreover, for all* $i \in \mathbb{N}$*, we have the following properties:*

(DSPPBM1) $T^*\varepsilon_i^{i+1}(T^\flat M_{i+1}) \subset T^\flat M_i$*;*

(DSPPBM2) $P_{i+1} = T\varepsilon_i^{i+1} \circ P_i \circ T^*\varepsilon_i^{i+1}$*;*

(DSPPBM3) *Around each* $x \in M$*, there exists a sequence of charts* $(U_i, \phi_i)_{i\in\mathbb{N}}$ *such that* $(U = \varinjlim(U_i), \phi = \varinjlim(\phi_i))$ *is a chart of* x *in* M*, so that the charts* (U_i, ϕ_i) *and* (U_{i+1}, ϕ_{i+1}) *are compatible with Property* **(DSPPBM1)***.*

Remark 7.15 *Since* $(M_i, \varepsilon_i^{i+1})_{i\in\mathbb{N}}$ *is a direct sequence of Banach manifolds, we may assume that* $\mathbb{M}_i$ *is a Banach subspace of* $\mathbb{M}_{i+1}$*. We denote by* $\bar{\varepsilon}_i^{i+1}$ *the natural inclusion of* $\mathbb{M}_i$ *in* $\mathbb{M}_{i+1}$ *and* $(\bar{\varepsilon}_i^{i+1})* : \mathbb{M}_{i+1}^* \to \mathbb{M}_i^*$ *the adjoint operator. Note that* $(\bar{\varepsilon}_i^{i+1})^*$ *is surjective. Since* $\mathbb{F}_i$ *is the typical fibre of* $T^\flat M_i$*, according to assumptions* **(DSPPBM1)** *and* **(DSPPBM3)** *(compatibility with trivializations), we must have* $(\bar{\varepsilon}_i^{i+1})^*(\mathbb{F}_{i+1}) \subset \mathbb{F}_i$*. Moreover, since* $\mathbb{F}_{i+1}$ *is a Banach subspace of* $\mathbb{M}_{i+1}^*$*, then* $(\bar{\varepsilon}_i^{i+1})^*(\mathbb{F}_{i+1})$ *is a Banach subspace of* $\mathbb{F}_i$*.*

Let $(M_i, \mathfrak{A}(M_i), \{.,.\}_{P_i})_{i\in\mathbb{N}}$ be a direct sequence of partial Poisson-Banach manifolds.

From the assumption in Definition 7.17, **(DSPPBM3)** and according to Chapter 5, Theorem 5.3, the direct limit $M = \varinjlim(M_i)$, is a convenient manifold.
Without loss of generality, we can assume that $M_i \subset M_{i+1}$; so $M = \bigcup\limits_{i \in \mathbb{N}} M_i$ and ε_i^{i+1} is the natural inclusion of M_i into M_{i+1}. For $i < j$, we denote by $\varepsilon_i^j = \varepsilon_{j-1}^j \circ \cdots \circ \varepsilon_i^{i+1} : M_i \to M_j$ the natural inclusion.
Given a point $x \in M$, there exists $i \in \mathbb{N}$ such that x belongs to M_i; let n be the smallest of such integers i.
$T_x\varepsilon_n^k : T_xM_n \to T_xM_k$ is an injective continuous linear map for all $n < k$. Since each T_xM_k is isomorphic to the Banach space $\mathbb{M}_k$ for $n \leq k$, $(T_xM_k, T_x\varepsilon_n^k)_{n \leq k}$ is an ascending sequence of Banach spaces whose direct limit $T_xM = \varinjlim(T_xM_k) = \bigcup\limits_{k \geq n} T_xM_k$ is a convenient space. We set $TM = \bigcup\limits_{x \in M} T_xM$. Let $p_M : TM \to M$ be the canonical projection.
Now, from Definition 7.17 **(DSPPBM1)**, it follows that $T_x^*\varepsilon_n^k(T_x^\flat M_k)$ is a subset of $T_x^\flat M_n$ for all $n < k$. Therefore, $\left(T_x^\flat M_k, T_x^*\varepsilon_k^n\right)_{n<k}$ is a projective sequence of Banach spaces. We set $T_x^\flat M = \varprojlim(T_x^\flat M_k)$. In the same way, we can define the "projective dual" $T_x^*M = \varprojlim(T_x^*M_k)$ of T_xM. Of course, we have $T_x^\flat M \subset T_x^*M$ and these vector spaces provided with the inverse limit topology are Fréchet vector spaces. We set $T^\flat M = \bigcup\limits_{x \in M} T_x^\flat M$ and $T^*M = \bigcup\limits_{x \in M} T_x^*M$.
We have canonical projections $p^\flat : T^\flat M \to M$ and $p_M^* : T^*M \to M$. We then have the following proposition.

Proposition 7.11 *Let $(M_i, \mathfrak{A}(M_i), \{.,.\}_{P_i})_{i \in \mathbb{N}}$ be a direct sequence of partial Poisson-Banach manifolds.*

(1) $p_M : TM \to M$ is a convenient bundle which is the kinematic bundle of M.

(2) $p_M^* : T^*M \to M$ and $p^\flat : T^\flat M \to M$ are locally trivial Fréchet bundles over M. Moreover $p_M^* : T^*M \to M$ is the kinematic cotangent bundle of M.

(3) There exists a canonical bundle morphism $P : T^\flat M \to TM$ characterized, for $x_i \in M_i$, by

$$P(x_i, \xi) = P(x_i, \varprojlim_{k \geq i}(\xi_k)) = \varinjlim_{i \leq k}(P_k(x_i, \xi_k)).$$

Moreover, P is skew-symmetric, relatively to the canonical dual pairing between T^*M and TM in restriction to $T^\flat M \times TM$.

Proof *Let us fix some $x \in M$ and assume that x belongs to M_n where n is the smallest integer for which this property is true.*
Since $\left(M_i, \varepsilon_n^k\right)_{n \leq k}$ is an ascending sequence of Banach manifolds which has the limit chart property, there exists a chart $(U = \varinjlim_{n<k}(U_k), \phi = \varinjlim_{n<k}(\phi_k))$ around x such that (U_k, ϕ_k) is a chart around $\varepsilon_n^k(x_n)$ in M_k. After shrinking

U_n if necessary, we may assume that $U \cap M_i = \emptyset$ for $i < n$.

(1) cf. [CaPe19], proof of Proposition 4.18. (2)

(2) We begin by considering

$$T^*_{M_i}M = \bigcup_{x \in M_i} T^*_x M \text{ and } T^\flat_{M_i}M = \bigcup_{x \in M_i} T^\flat_x M$$

and we denote by $p_i^* : T^*_{M_i}M \to M_i$ and $p_i^\flat : T^\flat_{M_i}M \to M_i$ the associate canonical projections respectively. For each $(i,j) \in \mathbb{N}^2$ such that $i \leq j$, we set $T_{M_i}M_j = (TM_j)_{|M_i}$, $T^*_{M_i}M_j = (T^*M_j)_{|M_i}$. Then $T\varepsilon_i^j$ is a morphism from TM_i to TM_j and so we get a surjective morphism $T^*\varepsilon_i^j$ from $T^*_{M_i}M_j$ into T^*M_i given by $T^*_{x_i}\varepsilon_i^j(\xi) = \xi_{|T_{x_i}M_i}$. Thanks to the compatibilty relations of local charts, on $\bar{\varepsilon}_i^j \circ \phi_i(U_i) \times \mathbb{M}_j^*$, we have

$$T^*\varepsilon_i^j \circ T^*\phi_j = T^*\phi_i \circ T^*\bar{\varepsilon}_i^j. \tag{7.54}$$

But $T^*\bar{\varepsilon}_i^j$ from $\mathbb{M}_i \times \mathbb{M}_j^*$ onto $\mathbb{M}_i \times \mathbb{M}_i^*$ is the map $(y,\xi) \mapsto (y, \xi_{|\mathbb{M}_i})$. It follows that $\left(T^*_{M_i}M_j, T^*\varepsilon_i^j\right)_{j \geq i}$ is a submersive projective sequence of bundles over M_i. From [Gal98], it follows that

$$T^*_{M_i}M = \varprojlim(T^*_{M_i}M_j) \to M_i$$

is a Fréchet bundle. Note that $T^*_{M_i}M$ is a Fréchet manifold modelled on $\mathbb{M}_i \times \varprojlim_{j \geq i}(\mathbb{M}_j^*)$. According to Properties **(DSPPBM1)** *and* **(DSPPBM3)** *of Definition 7.17, the same arguments implies that $\left(T^\flat_{M_i}M_j, T^*\varepsilon^j_{i\,|(T^\flat{}_{M_i}M_j)}\right)_{j \geq i}$ is also a submersive projective sequence of bundles over M_i. Therefore $T^\flat_{M_i}M = \varprojlim_{j \geq i}\left(T^\flat_{M_i}M_j\right) \to M_i$ is then a Fréchet bundle and a Fréchet manifold modelled on $\mathbb{M}_i \times \varprojlim_{j \geq i}(\mathbb{F}_j)$.*

On the one hand, $\varepsilon_i^j : M_i \to M_j$ is the natural inclusion, which induces the natural inclusion $\hat{\varepsilon}_i^j : T^*_{M_i}M \to T^*_{M_j}M$, namely $\hat{\varepsilon}_i^j(\omega)(x_i) = \omega(\varepsilon_i^j(x_i))$ for any $x_i \in M_i$. Therefore $\left(T^*_{M_j}M, M_j, \hat{\varepsilon}_i^j\right)_{i \leq j}$ is a direct sequence of Fréchet bundles and so we have $T^*M = \varinjlim(T^*_{M_i}M)$.
If we set $T^*_{U_i}M = T^*_{M_i}M_{|U_i}$, in the same way, we also get

$$T^*_U M = \varinjlim T^*_{M_i}M_{|U_i} = \varinjlim(T^*_{U_i}M_i)$$

We have $\phi_i^j = \phi_{j|U_i} = \phi_i$ and so $\phi_i^j(U_i) = \phi_i(U_i)$. Therefore $T^*{\phi_i^j}^{-1}$ is a trivialization of $T^*_{U_i}M_j$ onto $\phi_i(U_i) \times \mathbb{M}_j^*$ which is the restriction of $T^*\phi_j$ to $T_{U_i}M_j$. According to (7.54) and the results of [Gal98], we get a trivialization

$\underrightarrow{\lim}(T^*\phi_i^{j^{-1}})$ *of* $T_{U_i}M$ *onto* $\phi_i(U_i)\times\mathbb{M}^*$. *Note that* $\underrightarrow{\lim}(T^*\phi_i^{j^{-1}})$ *is in fact the adjoint operator of* $\left(T\phi_j^{-1}\right)_{|\{\phi_i(x_i)\}\times\mathbb{M}^*}$ *for all* $x_i\in U_i$ *where* $\phi=\underrightarrow{\lim}(\phi_j)$ *and so* $T^*\phi_{|U_i}^{-1}=\underrightarrow{\lim}(T^*\phi_i^{j^{-1}})$.
But since $(U=\underrightarrow{\lim}(U_j),\phi=\underrightarrow{\lim}(\phi_j))$ *is a limit chart, for such fixed* i, *we have* $U=\bigcup_{i\leq j}U_j$ *and and then* $T^*\phi_j{}^{-1}|U_i=\underleftarrow{\lim}_{j\geq i}(T^*\phi_i^{j^{-1}})$ *is a trivialization of* T^*M *onto* $\phi(U)\times\mathbb{M}^*$.
Thus we get a direct sequence of charts for the direct limit

$$(T^*_{U_i}M=\underrightarrow{\lim}(T^*_{U_i}M),T^*\phi^{-1}{}_{|U_i}=\underrightarrow{\lim}(T^*\phi_i^{j^{-1}})$$

for the sequence $\left(T^*_{M_i}M,\hat{\varepsilon}_i^j\right)_{i\leq j}$ *around any point* $x\in M$ *which belongs to* M_i. *Note that each manifold* $T^*_{M_i}M$ *is a closed submanifold of* $T^*_{M_j}M$ *modelled on the Fréchet spaces* $\mathbb{M}_i\times\underleftarrow{\lim}_{l\geq i}(\mathbb{M}_l^*)$ *and* $\mathbb{M}_j\times\underleftarrow{\lim}_{l\geq j}(\mathbb{M}_l^*)$ *respectively and the first one is a closed Fréchet subspace of the second one. Note also that* $\underrightarrow{\lim}_{i\leq j}(\mathbb{M}_j)\times\underleftarrow{\lim}_{l\geq i}(\mathbb{M}_l^*)$ *is a convenient space which is diffeomorphic to* $\mathbb{M}\times\mathbb{M}^*$ *where* $\mathbb{M}=\underrightarrow{\lim}(\mathbb{M}_j)$. *By same arguments as in [CaPe19], Proposition 4.18, we can prove that* T^*M *is a convenient manifold whose structural group is a metrizable complete topological group.*
According to Definition 7.17, the arguments used to prove that $T^*M=\underrightarrow{\lim}(T^*_{M_i}M)$ *is a convenient bundle over* M *still work for* $T^\flat M=\underrightarrow{\lim}(T^\flat_{M_i}M)$ *(with typical fibre* $\mathbb{F}=\underleftarrow{\lim}(\mathbb{F}_l)$*)* .

(3) Fix some $i\in\mathbb{N}$. *According to assumption* **(DSPPBM2)** *of Definition 7.17, for* $i\leq j\leq l$, *by composition, we have over* M_l

$$P_l=T\varepsilon_j^l\circ P_j\circ T^*\varepsilon_j^l:T^\flat_{M_j}M_l\to T_{M_j}M_l$$

Therefore, this relation is also true in restriction to $T_{M_i}M_j$ *and, in this case, over* M_i, *we have the composition*

$$P_l=T\varepsilon_i^l\circ P_j\circ T^*\varepsilon_i^l:T^\flat_{M_i}M_j\to T_{M_i}M_l.$$

We set $P_{ji}^l=T\varepsilon_i^l\circ P_{j|M_i}$. *Therefore we have a morphism* $P_{ji}^l:T^\flat_{M_i}M_j\to T_{M_i}M_l$. *From the arguments developed in the proof of (1), it is easy to see that* $\left(T_{M_i}M_l,T\varepsilon_i^l\right)_{i\leq l}$ *is a direct sequence of Banach bundles. Since we have* $P_{ji}^l=T\varepsilon_i^l\circ P_j$, *we get a morphism* $\hat{P}_i^j=\underrightarrow{\lim}_{j\leq l}P_{ji}^l$ *from* $T^\flat_{M_i}M_j$ *to* $\underrightarrow{\lim}_{j\leq l}T_{M_i}M_j=\bigcup_{j\leq l}T_{M_i}M_l$. *Note that since we have* $TM_i\subset T_{M_i}M_{i+1}\subset\cdots\subset T_{M_i}M_j$, *then we also have*

$$\underrightarrow{\lim}_{j\leq l}T_{M_i}M_j=\bigcup_{l\geq i}T_{M_i}M_l=TM_{|M_i}=T_{M_i}M.$$

On the one hand, recall that $\left(T^\flat_{M_i}M_j, T^*\varepsilon^j_{i\,|T^\flat_{M_i}M_j)}\right)_{j\geq i}$ *is a submersive projective sequence of bundles over* M_i *(see the second part of the proof of Point (2)). On the other hand, we have a sequence of morphisms* $\hat{P}^j_i : T^\flat_{M_i}M_j \to T_{M_i}M$ *over* M_i *such that* $\hat{P}_{hi} = \hat{P}^j_i \circ T^*\varepsilon_{hj}$. *This implies that we get a morphism*

$$\hat{P}_i = \varprojlim_{j\geq i}(\hat{P}^j_i) : T^\flat_{M_i}M = \varprojlim_{j\geq i}T^\flat_{M_i}M_j \to T_{M_i}M$$

By construction, if $(x, \xi = \varprojlim_{k\geq i}(\xi_k)) \in T^\flat_{M_i}M$ *then*

$$\hat{P}_i(x,\xi) = (x, \varinjlim_{i\leq k}(P_k(x,\xi_k)).$$

Now recall that each P_k *is antisymmetric, i.e., in each fibre over* x*, we have the relation*

$$<\eta_k, P_k(x,\xi_k)> = - <\xi_k, P_k(x,\eta_k)> .$$

Thus given two covectors $\xi = \varprojlim_{k\geq i}(\xi_k)$ *and* $\eta = \varprojlim_{k\geq i}(\eta_k)$ *in* $T^\flat_x M$*, we obtain*

$$<\eta, \hat{P}_i(x,\xi)> = - <\xi, \hat{P}_i(x,\eta)> .$$

Finally, since we have ascending sequences $\left(T^\flat_{M_i}M\right)_{i\in\mathbb{N}}$ *and* $(T_{M_i}M)_{i\in\mathbb{N}}$*, from the construction of the sequence* $\left(\hat{P}_i\right)_{i\in\mathbb{N}}$ *of morphisms, we have* $\hat{P}_j(x,\xi) = \hat{P}_i(x,\xi)$ *for all* $j \geq i$*; we then obtain a morphism* $P : T^\flat M \to TM$ *which is antisymmetric relatively to the canonical duality pairing between* T^*M *and* TM *in restriction to* $T^\flat M \times TM$.

Theorem 7.5 *Let* $(M_i, \mathfrak{A}(M_i), \{.,.\}_{P_i})_{i\in\mathbb{N}}$ *be a direct sequence of partial Poisson-Banach manifolds and* $M = \varinjlim(M_i)$.
There exists a weak subbundle $p^\flat : T^\flat M \to M$ *of* $p^*_M : T^*M \to M$ *and a skew-symmetric morphism* $P : T^\flat M \to TM$ *such that* $(M, \mathfrak{A}_P(M), \{.,.\}_P)$ *is a partial Poisson manifold with the following characterization:*
If $\varepsilon_i : M_i \to M$ *is the canonical injection then* ε_i *is a Poisson morphism from* $(M_i, \mathfrak{A}(M_i), \{.,.\}_{P_i})$ *to* $(M, \mathfrak{A}(M), \{.,.\}_P)$ *for all* $i \in \mathbb{N}$*, and we have*

$$\mathfrak{A}_P(M) = \varinjlim \mathfrak{A}(M_i)$$

$$\{.,.\}_P = \varinjlim \{.,.\}_{P_i}.$$

Proof *According to Proposition 7.11, we have a vector subbundle* $T^\flat M \to M$ *of* $T^*M \to M$ *and a skew-symmetric morphism* $P : T^\flat M \to TM$. *We can associate to these data the algebra* $\mathfrak{A}_P(M) = \varinjlim \mathfrak{A}(M_i)$ *of smooth functions* $f : \varinjlim f_i =: M \to \mathbb{R}$ *which is a set of smooth functions* $f : M \to \mathbb{R}$ *whose differential* df *is a section of* $T^\flat M \to M$ *which satisfies the assumptions of Definition 7.2.*

Consider the bilinear form $\{.,.\}_P = \varinjlim \{.,.\}_{P_i}$.

Of course, if $f = \varinjlim(f_i)$ *(resp.* $g = \varinjlim(g_i)$*) then* $f_i = f \circ \varepsilon_i$ *(resp.* $g_i = g \circ \varepsilon_i$*) and so* $df_i = T^*\varepsilon_i(df)$ *(resp.* $dg_i = T^*\varepsilon_i(dg)$*). Thus we have:*

$$\{f,g\}_P \circ \varepsilon_i = df_i\left(T\varepsilon_i\left(P(dg)\right)\right) = df_i\left(P_i(dg_i)\right) = \{f_i, g_i\}_{P_i}. \tag{7.55}$$

Thus from Remark 7.13 we have
$\{f \circ \varepsilon_i, g \circ \varepsilon_i\}_{P_i} = \{f,g\}_P \circ \epsilon_i$.
Since $\varepsilon_i = \varepsilon_j \circ \varepsilon_i^j$, *it follows that*

$$\{f,g\}_P = \varinjlim\{f_i, g_i\}_{P_i}.$$

Now, as each Poisson bracket $\{.,.\}_{P_i}$ *satisfies the Jacobi identity, the same is true for* $\{f,g\}_P$ *and so* $(M, \mathfrak{A}_P(M), \{.,.\}_P)$ *is a partial Poisson manifold. Finally, (7.55) means that* ε_i *is a Poisson morphism.*

7.7 Notes

Poisson geometry has been for a long time a framework for various areas of Mathematics, Mechanics and mathematical Physics ([Lic77], [Wei83], [AMR88], [Olv93]).
An approach to this geometry in the Banach setting is relatively recent and can be found in the paper [OdRa03]. In this situation, there exist Poisson brackets which are not given by Poisson tensors as proved in [BGT18]. An extension to the framework of some locally convex topological vector spaces has been realized in [NST14].
In the article [PeCa19], one can find results about projective and direct limits of Banach partial Poisson structures.

8

Integrability of Distributions and Foliations

The core notion of integrability of weak distributions is studied in this chapter where the link with involutivity, invariance and Lie invariance in the Banach framework is pointed out.
This situation is illustrated with Banach-Lie algebroids orbits and symplectic foliations associated with partial Poisson manifolds.
One then considers the integrability of some projective limits (resp. direct limits) of distributions for which a criterion is proved.

8.1 On the Problem of Integrability of a Distribution

A *foliation* of a manifold M is a partition of M into weak submanifolds, called the *leaves* of the foliation. Foliation theory was motivated for solving problems in dynamical systems. It was initiated by the works of H. Poincaré, I. Bendixon. This theory was intensively developed from 1950 by C. Ehresmann, G. Reeb, A. Haefliger, S.P. Novikov, W. Thurston, P. Molino, A. Connes among many others. The theory of foliations has now become a research topic by itself in differential geometry and has wide applications in various areas of Mathematics and Physics, such as, for instance, dynamical systems, symplectic theory, Poisson structure, variational problems with constraints, integrable systems, quantum gravity,...

From an infinitesimal point of view, in a foliated manifold M, for each point $x \in M$ let $\mathcal{D}_x$ be the tangent space of the leaf through x. If the dimension of $\mathcal{D}_x$ is constant, the foliation is called *regular*, otherwise, the foliation is called *singular*. For a regular foliation, the collection $\{\mathcal{D}_x\}_{x \in M}$ defines a sub-bundle $\mathcal{D}$ of TM and, for any local vector fields tangent to $\mathcal{D}$, their Lie bracket is also tangent to $\mathcal{D}$.
Conversely, a *distribution* on a smooth manifold M is an assignment $\mathcal{D} : x \mapsto \mathcal{D}_x \subset T_xM$ on M, where $\mathcal{D}_x$ is a subspace of T_xM. A finite-dimensional distribution $\mathcal{D}$ is called

(1) *smooth* if, for any $x \in M$, $\mathcal{D}_x$ is generated by the local vector fields which are tangent to $\mathcal{D}$ around x;

(2) *integrable* if, for any $x \in M$, there exists a weak immersed submanifold

DOI: 10.1201/9781003435587-8

$f : L \to M$ such that

$$\forall z \in L,\ T_z f T_z L = \mathcal{D}_{f(z)}$$

(3) *involutive* if, for any local vector fields X and Y on M which are tangent to $\mathcal{D}$, the Lie bracket $[X, Y]$ is also tangent to $\mathcal{D}$;

(4) *invariant* if, for any local vector field X tangent to $\mathcal{D}$, the local flow Fl_t^X leaves $\mathcal{D}$ invariant for any t for which Fl_t^X is defined.

On a finite dimensional manifold, when $\mathcal{D}$ is a subbundle of TM, the classical Frobenius Theorem gives an equivalence between integrability and involutivity. In the other case, the distribution is *singular*. Of course, the distribution $\mathcal{D}$ defined by a singular foliation on M is singular. But, since the set of local vector fields tangent to the leaves is closed under Lie brackets, then the distribution $\mathcal{D}$ is involutive. Unfortunately, the involutivity condition for a singular distribution $\mathcal{D}$ does not imply the integrability property. Hermann was the first who stated the first integrability conditions both in the smooth and the analytic cases (without proof for this last case). In [Nag66], Nagano proved that analyticity and involutivity are sufficient conditions for integrability. Later, Stefan ([Ste74]) and Sussmann ([Sus73]) independently gave necessary and sufficient conditions for the integrability of a singular distribution (see also [Bal94] for some precisions about these results). Essentially, if $\mathcal{D}$ is generated by all local vector fields tangent to $\mathcal{D}$, then $\mathcal{D}$ is integrable if and only if $\mathcal{D}$ is invariant.

In the context of Banach manifolds, the previous definition of a distribution is valid and the properties (2), (3) and (4) are also valid. The Frobenius Theorem is again true for distributions which are split subbundles of the tangent bundle. In particular, this result permits to show that in a Banach-Lie group there exists a bijection between closed supplemented Lie Banach sub-algebras of its Lie algebra and Banach-Lie subgroups (cf. [Lan95]). From 2000, the interest for singular foliations in the Banach setting was pointed out by results of A. Odzijewicz, and T.S. Ratiu and their collaborators, on Banach-Lie-Poisson manifolds (cf. [OdRa03], [OdRa08], [BeRa05], [BRT07] for instance).

However earlier, some papers ([ChSt76], [Nub92]) showed that, when a singular distribution is closed and supplemented (i.e. $\mathcal{D}_x$ is a supplemented Banach subspace of T_xM), we have equivalence between integrability and invariance.

The purpose of the following section is to give a kind of generalization of results on singular distributions of Stefan and Sussmann in the Banach framework for singular closed distribution which are not necessarily supplemented. For the following sections § 8.2, § 8.3 and § 8.4, the reader is referred to [Pel12].

8.2 Integrability and Invariance

8.2.1 Definitions and Results

Let M be a smooth connected Banach manifold modelled on a Banach space $\mathbb{M}$.
We denote by $\mathcal{A}(M)$ the ring of smooth functions on M and by $\mathfrak{X}(M)$ the Lie algebra of smooth vector fields on M.

Definition 8.1 *A local vector field X on M is a smooth section of the tangent bundle TM defined on an open set of M denoted by* $\operatorname{Dom}(X)$.

Denote by $\mathfrak{X}_{\mathrm{loc}}(M)$ the set of all local vector fields on M.
A local vector field has a flow Fl_t^X which is defined on a maximal open set $\mathcal{O}_X$ of $M \times \mathbb{R}$.

In reference to weak submanifolds of a Banach manifold (cf. Chapter 1, § 1.9), we consider weak distributions.

Definition 8.2 *A weak distribution on a smooth Banach manifold M is a map $\mathcal{D} : x \to \mathcal{D}_x$ which, for every $x \in M$, associates a vector subspace $\mathcal{D}_x$ in T_xM (not necessarily closed) endowed with a norm $|| \,||_x$ such that:*
(i) $(\mathcal{D}_x, || \,||_x)$ is a Banach space (denoted by $\tilde{\mathcal{D}}_x$);
(ii) The natural inclusion $i_x : \tilde{\mathcal{D}}_x \to T_xM$ is continuous.

Remark 8.1 *When $\mathcal{D}_x$ is closed, via any chart, we get a norm on T_xM which induces a Banach structure on $\mathcal{D}_x$. So if $\mathcal{D}_x$ is closed for all $x \in M$, the previous assumption on the Banach structure $\tilde{\mathcal{D}}_x$ is always satisfied, and we get the usual definition of a distribution on M (compare with [Glo08], [ChSt76], [Nub92]). In this last situation, we always endow $\tilde{\mathcal{D}}_x$ with this induced Banach structure and we say that $\mathcal{D}$ is* closed.

Example 8.1 *Consider the spaces of sequences ℓ^p and ℓ^∞ as defined in Chapter 1, § 1.2.*
Denote by $I_p : \ell^1 \to \ell^p$ the natural inclusion for $p > 1$ or $p = \infty$. On the Banach space l^p, $x \mapsto \mathcal{D}_x = x + I_p(l^1)$ is a weak distribution which is not closed.

Example 8.2 *Let E and F be two Banach spaces and $T : F \to E$ a continuous operator. Denote by $\hat{T} : F/\ker T \to E$ the canonical quotient bijection associated with T that is*

$$\begin{array}{ccc} F & \xrightarrow{\;q\;} & F/\ker T \\ {\scriptstyle T}\big\downarrow & \swarrow{\scriptstyle \hat{T}} & \\ T(F) & & \end{array} \qquad (8.1)$$

We can endow $T(F)$ with the structure of Banach space such that $\hat{T}$ is an isometry. On E, the assignment $x \mapsto \mathcal{D}_x = x + T(F)$ is a weak distribution. This distribution is closed if and only if $T(F)$ is closed in E.

Example 8.3 *Let $\mathrm{L}(F, E)$ be the set of continuous operators between the Banach spaces F and E. Given a smooth map $\Psi : E \to \mathrm{L}(F, E)$, we denote by Ψ_x the continuous operator associated with $x \in E$. As in (8.1) denote by $\hat{\Psi}_x$ the canonical bijection associated Ψ_x and we endow $\mathcal{D}_x = \Psi(x, F)$ with the Banach structure such that $\hat{\Psi}_x$ is an isometry. Then, $x \to \mathcal{D}_x$ is a weak distribution on E which is closed if and only if $\mathcal{D}_x$ is closed for any $x \in E$.*

Definition 8.3 *A local vector field $Z \in \mathfrak{X}_{loc}(M)$ is tangent to the weak distribution $\mathcal{D}$, if for all $x \in \mathrm{Dom}(Z)$, $Z(x)$ belongs to $\mathcal{D}$.*

Notation 8.1 *The set of local vector fields tangent to $\mathcal{D}$ will be denoted by $\mathcal{X}_{\mathcal{D}}$.*

Definition 8.4 *We say that a weak distribution $\mathcal{D}$ is generated by a subset $\mathcal{X} \subset \mathfrak{X}_{loc}(M)$ if, for every $x \in M$, the vector space $\mathcal{D}_x$ is the linear hull of the set*

$$\{Y(x), Y \in \mathcal{X}, x \in \mathrm{Dom}(Y)\}.$$

Definition 8.5 *Let $\mathcal{D}$ be a weak distribution.*

(1) An integral manifold of $\mathcal{D}$ through x is a weak immersed submanifold $f : N \to M$ such that there exists $u_0 \in N$ where $f(u_0) = x$ and $T_u f(T_u N) = \mathcal{D}_{f(u)}$ for all $u \in N$.

(2) $\mathcal{D}$ is called integrable if, for any $x \in M$, there exists an integral manifold N of $\mathcal{D}$ through x.

Definition 8.6 *A distribution $\mathcal{D}$ generated by $\mathcal{X} \subset \mathfrak{X}_{loc}(M)$ is called $\mathcal{X}$-invariant if for any $X \in \mathcal{X}$, the tangent map $T_x \mathrm{Fl}_t^X$ sends $\mathcal{D}_x$ onto $\mathcal{D}_{\mathrm{Fl}_t^X(x)}$ for all $(x, t) \in \mathcal{O}_X$.*

If $\mathcal{D}$ is $\mathcal{X}$-invariant, we simply say that it is $\mathcal{D}$-invariant.

We now introduce two notions of local triviality.

Definition 8.7 *A distribution $\mathcal{D}$ is (locally) lower trivial if for each $x \in M$, there exist an open neighbourhood V of x and a smooth map $\Theta : V \times \tilde{\mathcal{D}}_x \to TM$, called lower trivialization, such that :*

(i) $\forall y \in V, \Theta(\{y\} \times \tilde{\mathcal{D}}_x) \subset \mathcal{D}_y$;

(ii) For each $y \in V$, $\Theta_y \equiv \Theta(y, .) : \tilde{\mathcal{D}}_x \to T_y M$ is a continuous operator and $\Theta_x : \tilde{\mathcal{D}}_x \to T_x M$ is the natural inclusion i_x;

(iii) There exists a continuous operator $\tilde{\Theta}_y : \tilde{\mathcal{D}}_x \to \tilde{\mathcal{D}}_y$ such that $i_y \circ \tilde{\Theta}_y = \Theta_y$, $\tilde{\Theta}_y$ is an isomorphism from $\tilde{\mathcal{D}}_x$ onto $\Theta_y(\tilde{\mathcal{D}}_x)$ and $\tilde{\Theta}_x$ is the identity of $\tilde{\mathcal{D}}_x$.

Remark 8.2 *If $\mathcal{D}$ is closed distribution, then Point (iii) is automatically satisfied by Θ_y.*

Definition 8.8 *$\mathcal{D}$ is called (locally) upper trivial (upper trivial for short) if, for each $x \in M$, there exists an open neighbourhood V of x, a Banach space F and a smooth map $\Psi : V \times F \to TM$, called upper trivialization, such that:*

(i) for each $y \in V$, $\Psi_y := \Psi(y,.) : F \to T_yM$ is a continuous operator with $\Psi_y(F) = \mathcal{D}_y$;

(ii) $\ker \Psi_x$ *split in F;*

(iii) if $F = \ker \Psi_x \oplus S$, the restriction θ_y of Ψ_y to S is injective for any $y \in V$;

(iv) $\Theta(y,u) = \theta_y \circ \theta_x^{-1}(u)$ is a lower trivialization of $\mathcal{D}$.

In this case the map Θ is called the associated lower trivialization.

Example 8.4 *The distribution $\mathcal{D}_x = x + T(F)$, where $T : F \to E$ is a bounded operator as in Example 8.2 on E, is lower trivial.*
This distribution is upper trivial if and only if $\ker T$ *is split in F.*

Proposition 8.1 *Let $\mathcal{D} : x \to \mathcal{D}_x \subset T_xM$ be a field on M of normed subspaces. Suppose that for each $x \in M$, there exists an open neighbourhood V of x, a Banach space F and a smooth map $\Psi : V \times F \to TM$ such that for each $y \in V$:*

(i) $\Psi(\{y\} \times F) = \mathcal{D}_y$

(ii) $\Psi_y \equiv \Psi(y,.)) : F \to T_yM$ is a continuous operator such that $\Psi_y(F) = \mathcal{D}_y$

We have the following properties:

(1) $\mathcal{D}_x$ has a natural structure of Banach space (again denoted by $\tilde{\mathcal{D}}_x$) such that the canonical continuous operator $\hat{\Psi}_x : F/\ker \Psi_x \to \tilde{\mathcal{D}}_x$ is an isometry; in particular, $\mathcal{D}$ is a weak distribution.

(2) There exists a neighbourhood W of x and, for any $y \in W$, a continuous surjective operator $\tilde{\Psi}_y : F \to \tilde{\mathcal{D}}_y$ such that $i_y \circ \tilde{\Psi}_y = \Psi_y$, where $i_y : \tilde{\mathcal{D}}_y \to T_yM$ is the natural inclusion.

(3) Assume that $\mathcal{D}$ is closed and that $\ker \Psi_x$ *is supplemented (i.e. $F = \ker \Psi_x \oplus S$). Then there exists an open neighbourhood W of x such that the restriction θ_y of Ψ_y to S is injective for any $y \in W$, and then $\Theta(y,u) = \theta_y \circ \theta_x^{-1}(u)$ is* a lower trivialization of $\mathcal{D}$. *In particular, Ψ is an* **upper trivialization** *of $\mathcal{D}$*

The context of Proposition 8.1 can be found in the framework of Banach Poisson manifold $(M, \{.,.\})$ where $\Psi : T^*M \to TM \subset T^{**}M$ is the canonical morphism associated with the Poisson structure (see for instance [OdRa03] and [OdRa08]). In fact, we have the following Corollary of Proposition 8.1:

Corollary 8.1 *Let $\pi : \mathcal{E} \to M$ be a Banach fibre bundle over M with typical fibre $\mathbb{E}$ and $\Psi : \mathcal{E} \to TM$ a morphism of bundle and $\mathcal{D} = Im\ \Psi$. If $\mathcal{D}$ is closed* *and the kernel of Ψ is supplemented in each fibre, then $\mathcal{D} = Im\ \Psi$ is an upper trivial weak distribution.*

Proof of Corollary 8.1

Given $x \in M$ there exists a local trivialization of $\mathcal{E}$ on an open set V around x. So we can identify $\mathcal{E}_V$ with $V \times \mathbb{E}$. In this context, in restriction to V, the morphism Ψ can be identified, as a map $\Psi : V \times \mathbb{E} \to TM$ which satisfies assumption (i) and (ii) of Proposition 8.1.

Remark 8.3 *Proposition 8.1 seems to be a particular case of [Pel16], Proposition 2.5 in which we only assume that $\mathcal{D}_x$ is a normed space and not a Banach space in (c). The proof of this point is essentially based on Lemma 2.8 (1), where we assume that the inclusion $\iota : E_1 \to E_2$ is only continuous and not necessary closed. Unfortunately, this assertion is wrong if we do not assume the closeness of ι. Thus this lemma will be replaced by the Lemma 8.2. A correct proof will be given in the next section.*

Let $\mathcal{D}$ be a lower locally trivial distribution on M. For any lower trivialization $\Theta : V \times \tilde{\mathcal{D}}_x \to TM$ and any $u \in \tilde{\mathcal{D}}_x$, we consider

$$X(z) = \Theta(z, u). \tag{8.2}$$

Of course, we also have $X(z) = i_z \circ \tilde{\Theta}(z, u)$ where $\tilde{\Theta}(z, u) = \tilde{\Theta}_z(u)$ and X is a local vector field on M tangent to $\mathcal{D}$ whose domain is V. Moreover, the set of all such local vector fields spans $\mathcal{D}$.

Definition 8.9 *A (local) lower section of a lower locally trivial weak distribution $\mathcal{D}$ is a map of type (8.2) for any lower trivialization Θ, any $u \in \tilde{\mathcal{D}}_x$ and any $x \in M$.*

Notation 8.2 *The set of (local) lower sections of type (8.2) will be denoted by $\mathcal{X}^-_{\mathcal{D}}$.*

We then have a sufficient condition of integrability (cf. [Pel12], Proposition 2.9):

Proposition 8.2 *If a lower locally trivial closed distribution is integrable, then $\mathcal{X}^-_{\mathcal{D}} \subset \mathcal{X}_{\mathcal{D}}$ and $\mathcal{D}$ is $\mathcal{X}_{\mathcal{D}}$-invariant and in particular $\mathcal{X}^-_{\mathcal{D}}$-invariant.*

In this context, we obtain the following version of Stefan-Sussmann Theorem (cf. [Pel12], Theorem 1.)

Theorem 8.1 *Let $\mathcal{D}$ be a lower locally trivial closed distribution on a Banach manifold M.*

(1) $\mathcal{D}$ is integrable if and only if it is $\mathcal{X}^-_{\mathcal{D}}$-invariant.

(2) If $\mathcal{D}$ is integrable, on M, consider the binary relation

$$x\mathcal{R}y \Leftrightarrow \text{there exists an integral manifold } (N,f) \text{ of } \mathcal{D} \text{ with } (x,y) \in f(N)^2.$$

Then $\mathcal{R}$ is an equivalence relation and the equivalence class $L(x)$ of x has a natural structure of connected Banach manifold modelled on $\mathcal{D}_x$.
Moreover, $(L(x), i_{L(x)})$ is a maximal integral manifold of $\mathcal{D}$ in the following sense: for any integral manifold (N,f) of $\mathcal{D}$ such that $f(N) \cap L(x)$ is not empty, then $f(N) \subset L(x)$.

Remark 8.4 *As announced in Remark 8.3, Theorem 8.1 is a corrected version of Theorem 1 of [Pel12] since this proof uses Lemma 2.8 (1) and also Lemma 2.15 which can be directly deduced from Lemma 8.2. However, we give a sketch of the proof of Theorem 8.1 in the next section. All the proofs of the results included in this chapter are based on Theorem 8.1, and the proof given in [Pel12] for the corresponding results are the same. However, in order to be complete, we will give in this chapter corrected proofs for essential results of [Pel12] and generalization of some of them.*

Remark 8.5 *The closeness of $\mathcal{D}$ is not a necessary condition (cf. Example 8.1). In fact, this condition of closeness is needed to obtain Lemma 8.2 (1). But if a weak distribution is such that the the result of Lemma 8.2 (1) is true on some neighbourhood of any point of M, then the conclusions of Theorem 8.1 are also true.*

Remark 8.6 *Taking into account Remark 8.2, the property of lower triviality of a weak distribution corresponds to the usual assumptions on the distribution that we find in [ChSt76], [Nub92] when $\mathcal{D}$ is supplemented in T_xM. Thus Theorem 8.1 corresponds exactly to the main result of integrability of distributions we can find in these papers.*

8.2.2 Proofs of Proposition 8.1, Proposition 8.2 and Theorem 8.1

Since the proofs of Proposition 8.1 and Theorem 8.1 relate to common arguments, we begin by:

8.2.2.1 Proof of Proposition 8.2

Before this proof, we need the following remarkable topological property for an integral manifold of such a distribution.
Let M be a Banach manifold modelled on a Banach space $\mathbb{M}$ and an immersed submanifold N of M[1]. According to [Nee06], we say that N is an initial submanifold of M if the following property is true:

[1]That is a subset N of M which has its own Banach structure modelled on a Banach space $\mathbb{N}$ so that the inclusion of $\mathbb{N}$ in $\mathbb{M}$ is continuous and the inclusion of N in M is an immersion.

(ISM) For any map f from a Banach manifold Z to M such that $f(Z) \subset N$, then $f : Z \to N$ is smooth if and only if $f : Z \to M$ is smooth.

Lemma 8.1 *Each integral manifold N of a lower trivial closed distribution $\mathcal{D}$ is an initial submanifold of M and so is TN in TM.*

Proof of Lemma 8.1 *Under the previous context, fix some map $f : Z \to M$ such that $f(Z)$ is contained in some integral manifold N of $\mathcal{D}$. It is clear that the property* **(ISM)** *is a local problem.*

Note that, for any $x \in N$, T_xN is modelled on the Banach space $\mathcal{D}_x$. Since Banach spaces $\mathbb{M}$ and $\mathbb{N}$ are convenient spaces, according the definition of smoothness in the convenient setting (cf. [KrMi97]), it is sufficient to prove the following property:

(ISMloc) *For $z \in Z$ and any smooth curve $\gamma :]-\epsilon, \epsilon[\to Z$ for ϵ small enough, $f \circ \gamma$ is a smooth curve in N if and only if it is a smooth curve in M.*

We set $x = f(z) \in N$. Consider a lower trivialization $\Theta : V \times \mathcal{D}_x \to TM$. Since by assumption of a lower trivialization, $\Theta_y : \mathcal{D}_x \to \mathcal{D}_y$ is an isomorphism for all $y \in V \cap N$, thus z and so x being fixed, without loss of generality, we may assume that V is an open neighbourhood of $0 \in \mathbb{M}$. Moreover, we identify $\mathcal{D}_x$ with a Banach subspace $\mathbb{D}$ of $\mathbb{M}$. Then, after shrinking V if necessary, there exists a smooth injective immersion ϕ from an open set U of $0 \in \mathbb{D}$ into V such that $\phi(U) \subset V \cap N$ is an integral manifold of $\mathcal{D}$ through x. According to this context, for ϵ small enough, $f \circ \gamma$ is contained in $\phi(U) \subset V \subset \mathbb{M}$. Now $f \circ \gamma$ is smooth in $\phi(U)$ if and only if $\phi^{-1} \circ f \circ \gamma$ is smooth in $\mathbb{D}$. But since $\mathbb{D}$ is a Banach subspace of $\mathbb{M}$, it follows that $\phi^{-1} \circ f \circ \gamma$ is smooth in $\mathbb{D}$ if and only if $f \circ \gamma$ is smooth in $\mathbb{M}$, which ends the proof of the first part.
Under the same notations, we may assume that $M = V$ and $N = U \subset \mathbb{D}$ so that ϕ is the restriction to U of the inclusion of ι of $\mathbb{D}$ in $\mathbb{M}$ and $\iota(U) \subset V$. In this way, $T\iota$ is the map $\begin{array}{ccc} U \times \mathbb{D} & \to & V \times \mathbb{M} \\ (x,u) & \mapsto & (\iota(x), \iota(u)) \end{array}$, which implies the result for TN according to the previous argument.

Proof of Proposition 8.2
By construction, we have $\mathcal{X}_{\mathcal{D}}^{-} \subset \mathcal{X}_{\mathcal{D}}$. So it is sufficient to show that, for any local vector field X defined on an open set $V \subset M$ which is tangent to N, the flow $T_x \operatorname{Fl}_t^X(\mathcal{D}_x) = \mathcal{D}_{\operatorname{Fl}_t^X(x)}$ for any $(x,t) \in \mathcal{O}_X$ on which Fl_t^X is defined. But, from Lemma 8.1, if $X : V \to TM$ is smooth, then the inclusion $\iota : V \cap N \to V$ is smooth and if $X \circ \iota(V \cap N) \subset TN$ so is $X \circ \iota$ is a smooth map from $V \cap N$ into TN. This implies that if Y is a local vector field on $V \cap N$ which is such that $Y = \iota_*(X)$ for some smooth vector field X on V, then

$$\operatorname{Fl}_t^X \circ \iota = \iota \circ \operatorname{Fl}_t^Y$$

for all (x,t) for which $\operatorname{Fl}_t^Y(x)$ and $\operatorname{Fl}_t^X \circ \iota(x)$ are defined. Therefore, we have

$$T_y \operatorname{Fl}_t^X[(D_y] = T_y(\operatorname{Fl}_t^X \circ \iota)[T_yN] = T_y(\iota \circ \operatorname{Fl}_t^Y)[T_yN] = T_{\operatorname{Fl}_t^Y(y)}N = \mathcal{D}_{\operatorname{Fl}_t^Y(y)}.$$

Now, consider any $(z,t) \in \mathcal{O}_X$. *Denote by* $]\alpha_z, \beta_z[$ *the maximal interval on which* $\mathrm{Fl}_t^X(z)$ *is defined. Given any* $\tau \in]\alpha_z, \beta_z[$, *consider the integral curve* $\gamma(t) = \mathrm{Fl}_t^X(z)$ *for* $t \in [0,\tau]$. *By compactness of* $[0,\tau]$, *there exists a finite number of integral manifolds* $(N_1, f_1), \cdots, (N_r, f_r)$ *such that* $\gamma([0,\tau])$ *is contained in* $\cup_{i=1}^r f_i(N_i)$. *Using the previous argument, by induction, we obtain:*

$$T_z \,\mathrm{Fl}_\tau^X(\mathcal{D}_z) = \mathcal{D}_{\mathrm{Fl}_\tau^X(z)},$$

which ends the proof.

8.2.2.2 Proof Proposition 8.1

These proofs need the following Lemma:

Lemma 8.2

(1) Consider two Banach spaces E_1 *and* E_2 *and let* $\iota : E_1 \to E_2$ *be an injective continuous operator. Let* $y \to \Theta_y$ *be a smooth field of continuous operators of* $\mathrm{L}(E_1, E_2)$ *on an open neighbourhood* V *of* $x \in E_1$ *such that* $\Theta_x = \iota$. *Then there exists a neighbourhood* W *of* x *in* V *such that, for each* $y \in W$, Θ_y *is an injective operator .*

(2) Let $f : U \to V$ *be a* C^1 *map from two open sets* U *and* V *in the Banach spaces* E_1 *and* E_2 *respectively such that* $T_u f$ *is injective at* $u \in U$. *Then there exists an open neighbourhood* W *of* u *in* U *such that the restriction of* f *to* W *is injective.*

Proof *(1) Let* $y \to \Theta_y$ *be a smooth field of continuous operators of* $\mathrm{L}(E_1, E_2)$ *on an open neighbourhood* V *of* $x \in E_1$ *such that* $\Theta_x = \iota$. *Since* ι *is an injective operator implies that the set* $\mathrm{L_i}(E_1, E_2)$ *of injective operators with closed range is a non-empty set. It is well known that, in this case,* $\mathrm{L_i}(E_1, E_2)$ *is an open set of* $\mathrm{L}(E_1, E_2)$. *Therefore, if* W *is the pull-back by* Θ *of this open set* $\mathrm{L_i}(E_1, E_2)$, *then* W *is open in* V.

Since we have no precise references of this result, we will give an elementary proof of it.

ι *is an injective operator with closed range if and only it is a monomorphism, that is:*

$$\exists a > 0 : \ \forall v \in E_1, \ \|\iota(v)\| \geq a\|v\|.$$

If $\theta \in \mathrm{L}(E_1, E_2)$ *and* $\|\iota - \theta\| < a/2$, *then, for every* $v \in E_1$ *we have*

$$\|\Theta(v)\| \geq \|\iota(v)\| - \|\iota(v) - \Theta(v)\| \geq a\|v\| - (a/2)\|v\| = (a/2)\|v\|.$$

Hence Θ *is an injective operator with closed range. It follows that the open ball in* $\mathrm{L}(E_1, E_2)$ *with center at* ι *and radius* $a/2$ *is contained in the set of all closed operators with closed range.*

(2) Suppose that f is not locally injective around 0. Given any pair $(u, v) \in (B(0,r))^2$ such that $u \neq v$ but $f(u) = f(v)$, we set $h = v - u$. For any $\alpha \in E_2^*$, we consider the smooth curve $c_\alpha : [0,1] \to \mathbb{R}$ defined by:

$$c_\alpha(t) =< \alpha, f(u+th) - f(u) >$$

Of course we have $\dot{c}_\alpha(t) =< \alpha, T_{u+th}f(h) >$.
Denote by $]u, v[$ the set of points $\{w = u + th, t \in]0,1[\}$. As we have $c_\alpha(0) = c_\alpha(1) = 0$, from Rolle's Theorem, there exists $u_\alpha \in]u, v[$ such that

$$< \alpha, T_{u_\alpha} f(h) >= 0. \tag{8.3}$$

But this relation is also true for sh for any $s \in \mathbb{R}$. Thus, replacing h by $\dfrac{h}{||h||}$, we can assume in (8.3) that $||h|| = 1$.

From our assumption, it follows that, for each $n \in \mathbb{N}^*$, there exists u_n and v_n in $B(x, r/n)$ so that $u_n \neq v_n$ but with $f(u_n) = f(v_n)$. So, from the previous argument, for any $\alpha \in E_2^*$, we have

$$< \alpha, T_{u_{\alpha n}} f(h_n) >= 0 \tag{8.4}$$

for some $u_{\alpha,n} \in]u_n, v_n[$ and with $h_n = \dfrac{v_n - u_n}{||v_n - u_n||}$.

Now, for n large in enough, the map $u \mapsto T_u f$ from $B(0, r/n)$ to $\mathrm{L}(E_1, E_2)$ is Lipschitz, and so there exists $K > 0$ such that from (8.4), we get, for any $\alpha \in E_2^*$,

$$\begin{aligned} |< \alpha, T_0 f(h_n) > | &= | < \alpha, [T_0 f - T_{u_{\alpha,n}} f](h_n) > | \\ &\leq ||\alpha||.K.||u_\alpha||.||h_n|| < ||\alpha||\frac{Kr}{n}. \end{aligned} \tag{8.5}$$

Also, for n large enough, $T_u f$ is injective for any $u \in B(0, r/n)$ and so it is an isomorphism from E_1 to its range. Therefore $T_u f$ satisfies the inequalities:

$$\frac{1}{||T_u f||^{\mathrm{op}}}||v|| \leq ||T_u f(v)|| \leq ||T_u f||^{\mathrm{op}}||v||.$$

But since for all $u \in B(0, r/n)$

$$||T_u f||^{\mathrm{op}} \leq ||T_0 f||^{\mathrm{op}} + ||T_u f - T_0 f||^{op} \leq ||T_0 f||^{\mathrm{op}} + Kr = M,$$

we get

$$\frac{1}{M}||v|| \leq ||T_u f(v)|| \leq M||v|| \tag{8.6}$$

Now, from Hahn-Banach Theorem, there exists $\beta_n \in E_1^*$ *such that* $< \beta_n, h_n >= 1$ *with* $||\beta_n|| = 1$. *Since* $T_0 f$ *is injective, its adjoint* $T_0^* f$ *is surjective. Thus there exists* $\alpha_n \in E_2^*$ *such that* $T_0^* f(\alpha_n) = \beta_n$. *But the adjoint operation being an isometry, from (8.6), we have*

$$\frac{1}{M}||\alpha_n|| \leq ||T_0^* f(\beta_n)|| \leq M \tag{8.7}$$

According to (8.5), this implies

$$| < \alpha_n, T_0 f(h_n) > | \leq \frac{M^2 K r}{n}. \tag{8.8}$$

But, on the other hand, we also have

$$| < \alpha_n, T_0 f(h_n) > | = | < \beta_n, h_n > | = 1$$

which gives rise to a contradiction with (8.8).

of Proposition 8.1 *At first, for any* $x \in M$, *we have a natural Banach structure on* $\mathcal{D}_x$ *such that the natural morphism* $\tilde{\Psi}_x : F/\ker \Psi_x \to \tilde{\mathcal{D}}_x$ *is an isometry. On the other hand, take a local trivialization of* TM *on a neighbourhood* W *of* x*; so we have* $TM \equiv \mathbb{M} \times W$. *In this context, on* W, Ψ *can be identified with a smooth field of continuous operators* $\Psi_y : F \to \mathbb{M}$ *such that* $\mathcal{D}_y = \{y\} \times \Psi_y(F) \subset \{y\} \times \mathbb{M} \equiv T_y M$. *Let us consider the following commutative diagram:*

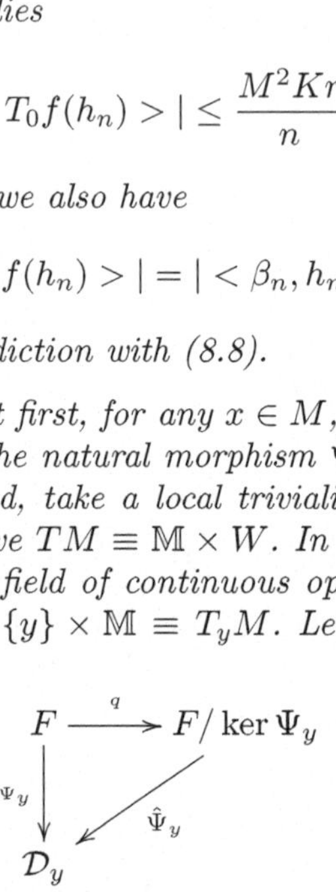

where q *is the natural projection and* $\hat{\Psi}_y$ *is the natural bijection induced by* Ψ_y. *So, if we consider the Banach structure* $\tilde{\mathcal{D}}_y$, *we get a continuous operator* $\tilde{\Psi}_y = \hat{\Psi}_y \circ q : F \to \tilde{\mathcal{D}}_y$ *so that* $\Psi_y = i_y \circ \tilde{\Psi}$.

Assume that $\mathcal{D}$ *is closed and* $F = \ker \Psi_x \oplus S$, *for some Banach space* $S \subset F$. *Let* θ_y *be the restriction to* S *of* Ψ_y *for any* $y \in W$. *Clearly,* $\theta(y, u) = (y, \theta_y(u))$ *defines a smooth map from* $W \times S$ *into* $V \times \mathbb{M} \equiv TM$ *and* $\theta_y : \{y\} \times S \to \{y\} \times \mathbb{M} \equiv T_y M$ *is a continuous operator whose image is contained in* $\mathcal{D}_y$.

Thus , $y \mapsto \theta_y$ *is a field of continuous operators from* $\mathbb{S}$ *to* $\mathbb{M}$ *so that* $\theta_y = i_y \circ \theta_y$ *for any* $y \in W$. *Since ,* $\theta_x : S \to \mathcal{D}_x$ *is an isomorphism,* θ_x *is injective, and so, from Lemma 8.2, without loss of generality, we can suppose that* θ_y *is injective for any* $y \in W$. *As* θ_y *is injective, we have* $\ker \Psi_y \cap S = \{0\}$. *It follows that* $q_1 = q_{|S}$ *is an isomorphism onto* $q(S) \subset F/\ker \Psi_y$. *Of course the restriction* q_2 *of the isomorphism* $\hat{\Psi}_y : F/\ker\Psi_y \to \mathcal{D}_y$ *to* $q(S)$ *is an isomorphism onto* $\theta_y(\mathbb{S})$ *such that* $\theta_y = q_2 \circ q_1$*. So* θ_y *is an isomorphism of* S *onto* $\theta_y(S)$*.*

Finally, the map

$$\begin{array}{rccc} \Theta: & W \times \mathcal{D}_x & \to & W \times \mathbb{M} \equiv TM \\ & (y,u) & \mapsto & \theta_y \circ [\theta_x]^{-1}(u) \end{array}$$

is clearly a lower trivialization of $\mathcal{D}$*.*

8.2.2.3 Sketch of the Proof of Theorem 8.1

Proof of Part (1). At first, according to the Proposition 8.2, integrability implies $\mathcal{X}_{\mathcal{D}}^{-}$-invariance.
So weneed to prove the converse. In fact, this proof is an adaption of arguments of Chillingworth and Stefan used in [ChSt76].

Given $x \in M$, we may assume that M is an open set of $\mathbb{M}$ and $TM \equiv \mathbb{M}\times M$. We denote by Δ the Banach space $\mathcal{D}_x$. From the property of lower local triviality, by shrinking this open set if necessary, we have a smooth field Θ_y of continuous operators from Δ to $\mathbb{M}$. Consider the family $\{X_u(y) = \Theta_y(u),\ u \in \Delta\}$ of smooth vector fields on M. By standard argument (see [ChSt76], proof of Corollary 4.2), we can choose an open ball $B(0,r) \subset \Delta$ such that the flow $\mathrm{Fl}_t^{X_u}$ is defined on an open neighbourhood W of x for all $|t| \leq 1$. We set $\Phi(t,y,u) = \mathrm{Fl}_t^{X_u}(y)$, $t \in [0,1]$, $y \in W$ and $u \in B \equiv B(0,r) \subset \Delta$ According to the following lemma (cf. [Pel12], Lemma 2.12)

Lemma 8.3 *For any smooth map* $\Phi : \mathbb{R} \times W \times B \to \mathbb{M}$*, we denote by* $D_t\Phi(t,y,u)$ *(resp.* $D_y\Phi(t,y,u)$*), resp.* $D_u\Phi(t,y,u)$ *the partial derivative of* Φ *according to the first (resp. the second (resp. the third) variable, at point* $(t,y,u) \in \mathbb{R}\times, W \times B$*.*
With these notations, $u \to \Phi(t,y,u)$ *is smooth . Moreover assume that* $T_x \mathrm{Fl}_t^{X_u}[\mathcal{D}_x] = \mathcal{D}_{\mathrm{Fl}_t^{X_u}(x)}$ *for all* t *such that* $(x,t) \in \Omega_{X_u}$ *and all* $u \in B$*, then we have:*

$$D_u\phi(t,x,u)(\Delta) \subset \mathcal{D}_{x(t)} \tag{8.9}$$

where $x(t) = \phi(t,y,u)$*.*

We are now in situation to give a sufficient condition of the existence of an integral manifold through $x \in M$:

Proposition 8.3 *Consider the map* $\Phi : B \to M$ *defined by:*

$$\forall u \in B \equiv B(0,r) \subset \Delta,\ \Phi(u) = \Phi(1,x,u) = \mathrm{Fl}_1^{X_u}(x). \tag{8.10}$$

There exists $\delta > 0$ such that $\Phi : B(0,\delta) \to M$ is a weak submanifold of M. Moreover, if we have $T_x \operatorname{Fl}_t^{X_u}[\mathcal{D}_x] = \mathcal{D}_{\operatorname{Fl}_t^{X_u}(x)}$ for all t such that $(x,t) \in \mathcal{O}_{X_u}$ and all $u \in B$, then, for $\delta > 0$ small enough, $(B(0,\delta),\Phi)$ is an integral manifold of $\mathcal{D}$ through x.

It is clear that Proposition 8.3 completes the proof of part (1) of Theorem 8.1.

We now end this subsection with the proof of the previous Proposition.

Proof *According to Lemma 8.2, it follows that, for $\delta > 0$ small enough, $(B(0,\delta),\Phi)$ is a closed submanifold of M.*

Assume now that

$$\forall t : (x,t) \in \mathcal{O}_{X_u},\ \forall u \in B,\ T_x \operatorname{Fl}_t^{X_u}[\mathcal{D}_x] = \mathcal{D}_{\operatorname{Fl}_t^{X_u}(x)}$$

From Lemma 8.3, for any $u \in B \subset \Delta$, we have:

$$D_u\Phi(\Delta) \subset \mathcal{D}_{\Phi(u)}.$$

So, it follows that, for all $u \in B$, we have

$$T_u\Phi(\Delta) \subset \mathcal{D}_{\Phi(u)}$$

Now, from the assumption of invariance, we have

$$[T_x \operatorname{Fl}_1^{X_u}]^{-1} \circ T_u\Phi(\Delta) \subset [T_x \operatorname{Fl}_1^{X}]^{-1}(\mathcal{D}_{\Phi(u)}) = \mathcal{D}_x. \tag{8.11}$$

We set $\Lambda_u = [T_x \operatorname{Fl}_1^{X_u}]^{-1} \circ T_u\Phi$ for $u \in B$. In particular, Λ_u is a continuous endomorphism of the Banach space Δ. The part (1) will be a consequence of the following Lemma which is consequence of Lemma 8.2 by passing to adjoint operator and the well known fact that the adjoint operator of an injective and closed range operator is surjective.

Lemma 8.4 *Let $\mathbb{E}_1$ and $\mathbb{E}_2$ be two Banach spaces. If the set $\operatorname{L_s}(\mathbb{E}_1,\mathbb{E}_2)$ of surjective operators in $\operatorname{L}(\mathbb{E}_1,\mathbb{E}_2)$ is non-empty, it is an open subset.*

Coming back to the proof of (1), the map $T_0\Phi$ is the inclusion map of F in F and $[T_x\phi_1^{X_0}] = Id_E$ so Λ_0 is surjective. From Lemma 8.4, for $\delta > 0$ small enough, Λ_u is surjective for all $u \in B(0,\delta)$; in particular, we get an equality

$$[T_x \operatorname{Fl}_1^{X_u}]^{-1} \circ T_u\Phi(F) = [T_x \operatorname{Fl}_1^{X_u}]^{-1}(\mathcal{D}_{\Phi(u)})$$

in (8.11), which ends the proof of Proposition 8.3.

Proof of (2). Here, we will use the notations introduced in the previous one. In particular, for any $x \in M$, we associate an integral manifold $(B(0,\delta),\Phi)$ build in Proposition 8.3. Such an integral manifold will be called a *plaque* through x.
At first, we must prove that the relation $\mathcal{R}$ is transitive. This fact is a direct consequence of the following Lemma (cf. [Pel12], Lemma 2.15):

Lemma 8.5

(1) Given any integral manifold (N, f) of $\mathcal{D}$ through $x \in M$, there exists a plaque $(B(0,\delta), \Phi)$ such that $\Phi(0) = x$ and $f^{-1}[\Phi(B(0,\delta)]$ is an open set in N.

(2) For all integral manifolds (N, f) and (N', f') through $x \in M$, then $f^{-1}[f(N) \cap f'(N')]$ (resp. ${f'}^{-1}[f(N) \cap f'(N')]$) is open in N (resp. N'). Moreover, $L = f(N) \cup f'(N') \subset M$ has a natural structure of Banach manifold modelled on $\mathcal{D}_x$ and (L, i_L) is an integral manifold of $\mathcal{D}$ through x, where i_L is the natural inclusion of L in M.

It remains to show that any equivalent class $L(x)$ of $x \in M$ carries a natural structure of connected Banach manifold modelled on $\mathcal{D}_x$. Note that $L(x)$ is the union of all the subsets $f(N)$ where (N, f) any integral manifold through x. So $L(x)$ is connected. Moreover, as in the proof of Lemma 8.5 (2) we can cover $L(x)$ by plaques and this gives rise to a natural structure of connected Banach manifold on $L(x)$. Again, $(L(x), i_{L(x)})$ is an integral manifold of $\mathcal{D}$ through x, which is maximal by construction.

8.3 Integrability and Lie Invariance

In this section, we adopt the notions used in [Ste74] and [ChSt76].

8.3.1 Case of a Lower Trivial Closed Distribution

Let X and Y be smooth vector fields on an open set U in a Banach space $\mathbb{M}$. Given any integral curve $\gamma : I = [\alpha, \beta] \to U$ of X, it is well known that the Lie bracket at some $\gamma(t)$ is given by

$$[X, Y](\gamma(t)) = \frac{d}{dt} Y(\gamma(t)) - DX(Y(\gamma(t))) \tag{8.12}$$

where DX is the differential of X.

According to the chain rule, the following definition does not depend on the choice of the local chart.

Definition 8.10 *Let X be a vector field on M and $\gamma : I = [\alpha, \beta] \to U \subset M$ an integral curve of X whose range is contained in a chart domain U of a chart (U, ϕ). Given any vector field Y along γ, i.e. $Y : I \to TM$ such that $Y(t) \in T_{\gamma(t)}M$, the Lie bracket $[X, Y]$ along γ is the vector field characterized by*

$$\phi_*[X, Y](\gamma(t)) = \frac{d}{dt} \phi_* Y(\gamma(t)) - D\phi_* X(Y(\gamma(t))).$$

Given any smooth curve $\gamma : I = [\alpha, \beta] \to M$, we denote by $T_\gamma M$ the restriction of TM to $\gamma(I)$. For any Banach space G, we denote by $L_\gamma(G, TM)$ the bundle, over γ, of morphisms between the trivial bundle $I \times G$ and $T_\gamma M$.

Let $\mathcal{D}$ be a lower trivial weak distribution on M. Consider a local vector field X and an integral curve $\gamma : [\alpha, \beta] \to M$ of X.

Definition 8.11 *An upper trivialization of $\mathcal{D}$ over γ is a smooth map $\psi : [\alpha, \beta] \to \mathrm{L}_\gamma(G, TM)$ (where G is some Banach space) such that, for each $t \in [\alpha, \beta]$, the corresponding morphism $\psi_t \in \mathrm{L}(G, T_{\gamma(t)}M)$ satisfies $\psi_t(G) = \mathcal{D}_{\gamma(t)}$.*

Definition 8.12 *Let ψ be an upper trivialization of $\mathcal{D}$ over γ. For each $v \in G$, denote by $\psi(v)$ the vector field along γ defined by $\psi(v)(t) = \psi_t(v)$. The Lie derivative of ψ by X along γ is defined by:*

$$(L_X\psi)_t(v) = [X, \psi(v)](\gamma(t)). \tag{8.13}$$

Remark that, with the previous notations, the map $v \to (L_X\psi)_t(v)$ is linear and continuous[2], so we get a smooth map $L_X\psi : I \to \mathrm{L}_\gamma(G, TM)$ so that for $t \in I$, $(L_X\psi)_t \in \mathrm{L}(G, T_{\gamma(t)}M)$.

Definition 8.13 *Let $\mathcal{D}$ be a lower trivial closed distribution.*

(1) Let $\Theta : V \times \mathcal{D}_x \to TM$, be a lower trivialization around x, and $X_u = \Theta(., u)$ a lower section on V. We say that the distribution $\mathcal{D}$ is Lie invariant by X_u, if, for any $y \in V$, we can find $\epsilon > 0$, so that, for all $0 < \tau < \epsilon$, there exist

- *an upper trivialization $\psi : [-\tau, \tau] \to \mathrm{L}_\gamma(G, TM)$ of $\mathcal{D}$ over $\gamma(t) = \mathrm{Fl}_t^{X_u}(y)$ for $t \in [-\tau, \tau]$,*
- *a smooth field of operators $\Lambda : [-\tau, \tau] \to \mathrm{L}(G, G)$ which satisfies*

$$L_{X_u}\psi = \psi \circ \Lambda \tag{8.14}$$

(2) The distribution $\mathcal{D}$ is called Lie invariant if, for any $x \in M$, there exists a lower trivialization $\Theta : V \times \mathcal{D}_x \to TM$ such that, for any $u \in \mathcal{D}_x$, the weak distribution $\mathcal{D}$ is Lie invariant by $X_u = \Theta(., u)$.

As in [ChSt76], we have the following Theorem but without the assumptions of existence of a supplement for all subspaces $\mathcal{D}_x$.

Theorem 8.2 *Let $\mathcal{D}$ be a lower trivial closed distribution.*
The following properties are equivalent:

(1) $\mathcal{D}$ is integrable;

(2) $\mathcal{D}$ is Lie invariant;

---

[2] See proof of Lemma 8.7.

(3) $\mathcal{D}$ is $\mathcal{X}_{\mathcal{D}}^{-}$-invariant.

Proof

According to Theorem 8.1, we have only to prove the equivalence (2) $\Longleftrightarrow$ (3). *Assume that $\mathcal{D}$ is $\mathcal{X}_{\mathcal{D}}^{-}$-invariant. Let $x \in M$ be a fixed point and choose a lower trivialization $\Theta : V \times \mathcal{D}_x \to TM$. Consider a lower section $X_u = \Theta(.,u)$ and $y \in V$. Note that there exists $\epsilon > 0$ such that the integral curve $t \mapsto \mathrm{Fl}_t^{X_u}(y)$ of X_u through y is defined for all $t \in]-\epsilon, \epsilon[$. Choose any $0 < \tau < \epsilon$ and set $\gamma(t) = \mathrm{Fl}_t^{X_u}(y)$ for $t \in [-\tau, \tau]$. From our assumption, we have $T_y \mathrm{Fl}_t^{X_u}(\mathcal{D}_y) = \mathcal{D}_{\gamma(t)}$. Denote by $\psi(t) = (T_y \mathrm{Fl}_t^{X_u})_{|\mathcal{D}_y}$ and set $\Gamma(t) = T_y \mathrm{Fl}_t^{X_u}$ It is clear that ψ is an upper trivialization of $\mathcal{D}$ over γ with $G = \mathcal{D}_y$. On the other hand, we have:*

$$L_{X_u}\psi = [\dot{\Gamma} - DX_u(\gamma(t)) \circ \Gamma]$$

But, we have $\dot{\Gamma} = DX_u(\gamma(t)) \circ \Gamma$ (see proof of Lemma 2.12 in [Pel12]). So we obtain $L_{X_u}\psi = 0$ on $[-\tau, \tau]$. Thus the relation (8.14) is satisfied for $\Lambda = 0$ and so we get Lie invariance for X_u

For the converse, as in [Ste74], [ChSt76] and [Nub92], we need the following result:

***Lemma 8.6** Let X be a local vector field and ψ an upper trivialization of $\mathcal{D}$ defined over an integral curve $\gamma :]-\epsilon, \epsilon[\to V = \mathrm{Dom}(X)$. Moreover we assume that, for any $0 < \tau < \epsilon$ there exists a smooth field $\Lambda : [-\tau, \tau] \to \mathrm{L}(G, G)$ such that*

$$L_X \psi = \psi \circ \Lambda.$$

Then, we have $T_y \mathrm{Fl}_t^X[\mathcal{D}_y] = \mathcal{D}_{\mathrm{Fl}_t^X(y)}$ for all $0 < |t| < \epsilon$

Now, assume that $\mathcal{D}$ is Lie invariant by X_u, let us fix some $y \in V = \mathrm{Dom}(X_u)$, and take a maximal integral curve $\gamma :]\alpha_y, \beta_y[\to V$ of X_u. Consider the set $I = \{t \in]\alpha_y, \beta_y[: T_y \phi_t^{X_u}[\mathcal{D}_y] = \mathcal{D}_{\phi_t^{X_u}(y)}\}$. This set is open from Lemma 8.6. Take a sequence (t_n) in I which converges to some $t \in]\alpha_y, \beta_y[$. From the assumption (2) of the Theorem, and Lemma 8.6 applied at the point $\mathrm{Fl}_t^{X_u}(y)$, we have $\mathrm{Fl}_s^{X_u}\left(\mathcal{D}_{\mathrm{Fl}_t^{X_u}(y)}\right) = \mathcal{D}_{\mathrm{Fl}_{t+s}^{X_u}(y)}$ for s in some neighbourhood $]t-\eta, t+\eta[$ of t. As we have some t_n which belongs to $]t-\eta, t+\eta[$, we get that I is closed. So $I =]\alpha_y, \beta_y[$ and finally we deduce that $\mathcal{D}$ is invariant by X_u.

Proof of Lemma 8.6

Let $\psi : [-\tau, \tau] \to \mathrm{L}_\gamma(G, TM)$ be an upper trivialization of $\mathcal{D}$ over an integral curve γ of X such that $\gamma(0) = y \in V$. Consider any smooth field of operators $\sigma : [-\tau, \tau] \to \mathrm{L}(G, G)$ and set $\tilde{\psi} = \psi \circ \sigma$. On a chart domain, we have

$$L_X \tilde{\psi} = \dot{\tilde{\psi}} - DX \circ \tilde{\psi} = \dot{\psi} \circ \sigma + \psi \circ \dot{\sigma} - DX \circ \psi \circ \sigma = L_X \psi \circ \sigma + \psi \circ \dot{\sigma} \quad (8.15)$$

Assume that $L_X\psi = \psi\circ\Lambda$ for some smooth field of operators $\Lambda : [-\tau,\tau] \to$ $\mathrm{L}(G,G)$. Then we have:

$$L_X\tilde{\psi} = \psi\circ\Lambda\circ\sigma + \psi\circ\dot{\sigma} = \psi\circ[\Lambda\circ\sigma + \dot{\sigma}] \tag{8.16}$$

Consider the solution (again denoted by σ) of the linear equation $\dot{\sigma} =$ $(-\Lambda)\circ\sigma$ with initial condition $\sigma(0) = Id_G$. It follows that σ is a smooth field of isomorphisms of G and in particular, for this choice of σ, we have $\tilde{\psi}(t)\,(G) = \mathcal{D}_{\gamma(t)}$ for any $t\in[-\tau,\tau]$. Moreover, from(8.16), and the choice of σ we have $L_X\tilde{\psi} = 0$. In fact the relation (8.16) does not depend on the choice of the chart, so we can get the same result even if $\gamma([-\tau,\tau])$ is not contained in a chart domain.
Thus we can assume that there exists an upper trivialization $\psi : [-\tau,\tau] \to$ $\mathrm{L}(G,TM)$ such that $L_X\psi = 0$ on γ. Again we set $\Gamma(t) = T_y\,\mathrm{Fl}_t^X$. Then $\Sigma(t) =$ $\Gamma(t)^{-1}\circ\psi(t)$ is a smooth field of continuous operators from G to $\mathbb{M}\equiv T_xM$ defined on $[-\tau,\tau]$.
On a chart domain we have

$$\dot{\psi} = \dot{\Gamma}\circ\Sigma + \Gamma\circ\dot{\Sigma} = DX(\gamma)\circ\Gamma\circ\Gamma^{-1}\circ\psi + \Gamma\circ\dot{\Sigma} = DX\circ\psi + \Gamma\circ\dot{\Sigma}$$

According to (8.12) and (8.13), on $[-\tau,\tau]$ we have

$$L_X\psi = \Gamma\circ\dot{\Sigma} \tag{8.17}$$

Now, (8.17) does not depend on the choice of the chart since Γ and $\dot{\Sigma}$ are intrinsically defined and so we have

$$(L_X\psi)_t = T_y\,\mathrm{Fl}_t^X\circ\frac{d}{dt}\{T_y\,\mathrm{Fl}_{-t}^X(\psi(t))\} \tag{8.18}$$

Therefore, even if $\gamma([-\tau,\tau])$ is not contained in a chart domain, we can obtain the same relation.
Now from our assumption $L_X\psi = 0$, as $\Gamma(t)$ is an isomorphism, we must have $\dot{\Sigma} = 0$ and so $\Sigma(t) = \Sigma(0) = \psi(0)$. We conclude that, for any $t\in[-\tau,\tau]$, we have

$$\Gamma(t)^{-1}\circ\psi(t)(G) = \psi(0)(G) = \mathcal{D}_y$$

and finally

$$T_y\,\mathrm{Fl}_t^X(\mathcal{D}_y) = \psi(t)(G) = \mathcal{D}_{\gamma(t)}. \tag{8.19}$$

Now, from the assumption in this Lemma, there exists $\epsilon > 0$ such that, we are in the previous situation for any interval $[-\tau,\tau]$ with $0<\tau<\varepsilon$. Thus (8.19) is true for any $0<|t|<\epsilon$.

8.3.2 Case of an upper trivial closed distribution

Let $\mathcal{D}$ be an upper trivial weak distribution on M. By analogy with lower sections, we introduce the notion of upper section.

Definition 8.14 *For any upper trivialization $\Psi : V \times F \to TM$ such that the associated lower trivialization is $\Theta : V \times \mathcal{D}_x \to TM$, an upper section is a local vector field Z on M defined by*

$$Z(y) = \Psi(y, u) \text{ for any } u \in F \tag{8.20}$$

Notation 8.3 *Such an upper section defined in (8.20) is denoted Z_u. The set of upper sections of $\mathcal{D}$ is denoted by $\mathcal{X}^+_{\mathcal{D}}$.*

Remark 8.7

(1) The set $\mathcal{X}^+_{\mathcal{D}}$ of upper sections generates $\mathcal{D}$.

(2) Consider any upper trivialization $\Psi : V \times F \to TM$ around x and a decomposition $F = \ker \Psi \oplus S$. We denote $\mathcal{X}_{\mathcal{D}}(V)$ the module of vector fields $X \in \mathcal{X}(M)$ whose domain contains V and which are tangent to $\mathcal{D}$ on V. The set $\mathcal{X}^+_{\mathcal{D}}(V) = \{Z_v = \Psi(., v,),\ v \in F\}$ is contained in $\mathcal{X}_{\mathcal{D}}(V) = \{X_u = \Theta(., u)\}$ and of course $\{Z_v(y),\ v \in F\}$ generates $\mathcal{D}_y$ for all $y \in V$. Moreover, if F has a Schauder basis $\{e_\alpha, \alpha \in A\}$, then the convex hull of $\{Z_{e_\alpha}(y), \alpha \in A\}$ is dense in $\mathcal{D}_y$.

(3) If $F = \ker \Psi_x \oplus S$ and if $\Theta : V \times S \to TM$ is the lower trivialization associated with the upper trivialization Ψ then each lower section $X_u = \Theta(., u)$ can be written $X_u = \Psi(., u)$ for $u \in S$. Therefore the set $\mathcal{X}^-_{\mathcal{D}}(V)$ of such lower sections is contained in $\mathcal{X}^+_{\mathcal{D}}(V)$.

(4) Consider a Banach bundle $\pi : \mathcal{E} \to M$ and $\Psi : \mathcal{E} \to TM$ a bundle morphism and $\mathcal{D} = \rho(\mathcal{E})$. According to the context of the proof of Corollary 8.1, for any given $x \in M$ we consider a local trivialization of $\tau : \mathcal{E}_{|V} \to V \times \mathcal{E}_x$ on an open set V around x. Then $\Psi_\tau := \Psi \circ \tau$ is a map $\Psi_\tau : V \times \mathcal{E}_x \to TM$ which is an upper trivialization. In this context, the set

$$\mathcal{S}^+_{\Psi_\tau} = \{\sigma(y) = \Psi_\tau(y, u), \text{ for any } u \in \mathcal{E}_x,\ y \in V\}$$

is the set of upper sections associated with the upper trivialization Ψ_τ. Therefore, if $\mathcal{S}_\Psi$ of all the $\mathcal{S}_{\Psi_\tau}$ for all trivializations τ of $\mathcal{E}$ around x and all $x \in M$, then $\mathcal{X}^+_{\mathcal{D}} = \{\Psi \circ \sigma,\ \sigma \in \mathcal{S}_\Psi\}$.

Let $\mathcal{D}$ be an upper trivial weak distribution on M and V be the domain of a chart around $x \in M$. After shrinking V if necessary there exists an upper trivialization $\Psi : V \times F \to TM$. If $F = \ker \Psi_x \oplus S$, the associated lower section Θ is a map from $V \times S$ to TM . Given any smooth function $\sigma : V \to F$, let $Z_\sigma = \Psi(., \sigma)$ be the associated vector field on V. Consider $\gamma : [-\tau, \tau] \to V$ an integral curve of Z_σ, then, Ψ_γ, defined by $\Psi_\gamma(t)(.) = \Psi(\gamma(t), .)$, is an upper trivialization of $\mathcal{D}$ along γ, according to Definition 8.10. Thus, the Lie derivative of Ψ by Z_σ along γ is $L_{Z_\sigma}\Psi_\gamma$ which we simply denoted by $L_{Z_\sigma}\Psi$. We can also define directly $L_{Z_\sigma}\Psi$ by:

$$\forall X_u = \Psi(.,u),\, \forall i \in F,\; L_{Z_\sigma}\Psi(\gamma(t),u) = [Z_\sigma, X_u](\gamma(t)) \tag{8.21}$$

Definition 8.15 *An upper trivial closed distribution $\mathcal{D}$ is called Lie bracket invariant if, for any $x \in M$, there exists an upper trivialization $\Psi : V \times F \to TM$ so that, for any $u \in F$, there exists $\epsilon > 0$, such that, for all $0 < \tau < \epsilon$, there exists a smooth field of operators $\Lambda : [-\tau, \tau] \to L(F, F)$ with the following property*

$$L_{X_u}\Psi = \Psi \circ \Lambda \tag{8.22}$$

along the integral curve $t \mapsto \mathrm{Fl}_t^{X_u}(x)$ on $[-\tau, \tau]$ of any lower section $X_u = \Theta(.,u)$.

With these definitions we have:

Theorem 8.3 *Let $\mathcal{D}$ be an upper trivial closed distribution. The following propositions are equivalent:*

(1) $\mathcal{D}$ is integrable;

(2) $\mathcal{D}$ is Lie bracket invariant;

(3) $\mathcal{D}$ is $\mathcal{X}_{\mathcal{D}}^{-}$-invariant.

Proof *According to Theorem 8.1, we have only to prove* $(1) \Longleftrightarrow (2)$. *From Lemma 8.6, property (2) of Theorem 8.3 implies that for any $x \in M$, we have $T_x \mathrm{Fl}_t^{X_u}(\mathcal{D}_x) = \mathcal{D}_{\mathrm{Fl}_t^{X_u}(x)}$ for all t such that $(x,t) \in \mathcal{O}_{X_u}$. From Proposition 8.3, $(B(0,\delta), \Phi)$ is an integral manifold through x. So* $(2) \Longrightarrow (1)$.

For the converse, we will use the following Lemma:

Lemma 8.7 *Assume that $\mathcal{D}$ is integrable. Let $\Psi : V \times F \to TM$ be an upper trivialization, and $\sigma : V \to \mathbb{F}$ a smooth map and let $X = \Psi(.,\sigma)$ be the associated vector field on V. Consider an integral curve $\gamma :]-\epsilon, \epsilon[\to V$ of X such that $\gamma(0) = x$. Then there exists a smooth field $\Lambda :]-\epsilon, \epsilon[\to L(F, F)$ such that :*

$$L_X\Psi(\gamma(t), v) = \Psi(\gamma(t), \Lambda(t)(v)).$$

Recall that, by assumption, we have a decomposition $F = \ker \Psi_x \oplus S$. So, for $\sigma(y) = (u, y)$, with $u \in \mathbb{S}$, the vector field Z_σ is the lower section X_u for $u \in S$ (cf. proof of Lemma 8.7). Therefore, clearly, from Lemma 8.7 we obtain the implication $(1) \Longrightarrow (2)$.

Proof of Lemma 8.7

Recall that we have assumed that $\mathcal{D}$ is integrable. Fix some $x \in M$. Take an upper trivialization $\Psi : F \times V \to TM$ around x. We can choose V such that

$TM_{|V} \equiv V \times \mathbb{M}$. *Recall that we have a decomposition* $F = \ker \Psi_x \oplus S$, *and* $\Theta(y,u) = \theta_y \circ [\theta_x]^{-1}(u)$ *where* θ_y *is the restriction to* S *of* Ψ_y *(see the proof of Proposition 8.1). At first, note that any lower section* X_u *associated with* Ψ *can be written* $X_u = \Psi(.,u)$ *for some* $u \in F$ *but also* $X_u = \Theta(.,u)$ *with a unique* $u \in S$. *On the other hand, from Lemma 8.5, let* $(B(0,\delta),\Phi)$ *is an integral manifold of* $\mathcal{D}$ *through* x *(associated with* Θ*). In particular, for any* $y \in \Phi(B(0,\delta) \subset M$, θ_y *is an isomorphism from* S *to* $\mathcal{D}_y$. *Given* $u \in F$, *there exists a unique* $v \in S$ *such that* $\Psi_y(u) = \theta_y(v)$. *It follows that* $F = \ker \Psi_y \oplus S$ *for any* $y \in \Phi(B(0,\delta)$.

Set $N = \Phi(B(0,\delta)) \subset M$ *endowed with its Banach manifold structure. Without loss of generality, we can identify* S *with* $\theta_x(S) = \mathcal{D}_x$ *and consider* N *as an open set of* $\mathcal{D}_x$. *Denote by* $i : \mathcal{D}_x \to T_xM \equiv \mathbb{M}$ *the canonical inclusion. We have* $T_yN \equiv \{y\} \times S$. *On* N, *each* Ψ_y *can be considered as is a smooth field of operators in* $\mathrm{L}(F,S)$. *Therefore,* $y \mapsto \theta_y$ *is a smooth field of automorphisms of* S. *We get a smooth field* $\pi_y = [\theta_y]^{-1} \circ \Psi_y$ *of operators in* $L(F,S)$ *with the following properties:*

$$\Psi_y = \theta_y \circ \pi_y \tag{8.23}$$

$$\ker \pi_y = \ker \Psi_y \pi_y(u) = u \text{ for all } u \in S \tag{8.24}$$

Take any smooth map $\sigma : V \to F$. *Then* $Z_\sigma(y) = \Psi_y \circ \sigma(y)$ *for* $y \in V$ *is a smooth vector field on* V *whose restriction to* N *is tangent to* N *and we have the following relations for all* $y \in N$

$$\Psi(y,\sigma(y)) = Z_\sigma(y) = \theta_y \circ \pi_y(\sigma(y) = \theta(y, \pi_y(\sigma(y))) \tag{8.25}$$

Consider the integral curves $\gamma(t) = \mathrm{Fl}_t^{Z_\sigma}(x)$ *defined on some interval* $]-\epsilon,\epsilon[$.
Note that γ *is tangent to* N.
For simplicity, we set $\sigma(\gamma(t)) = \sigma(t)$ *and* $P(t) = \pi_{\gamma(t)}$.
Note that, from (8.25), we have

$$\Psi(\gamma(t),u) = \theta(\gamma(t), P(t)(u)) \tag{8.26}$$

Now, from (8.21), for any $v \in S$, *we have:*

$$L_{Z_\sigma}\Psi(\gamma(t),v) = [Z_\sigma, X_v](\gamma(t)) \tag{8.27}$$

Since the restriction of Z_σ *and* X_v *to* N *are tangent to* N, *on the Banach manifold* N *the Lie bracket* $[Z_\sigma, X_v](\gamma(t))$ *belongs* $T_{\gamma(t)}N \equiv \{\gamma(t)\} \times S$. *It follows that we get*

$$L_{Z_\sigma}\Psi(\gamma(t),v) \in T_{\gamma(t)}N \equiv \{\gamma(t)\} \times S. \tag{8.28}$$

Without loss of generality, we can choose $\delta > 0$ *small enough such that on* $N = \Phi(B(0,\delta))$ *we have, for any* $y \in N$, *the following bounds:*

$$\begin{aligned}||\sigma(y)|| &\leq K \\ ||\theta(y,.)|| &\leq K \\ ||D_2\theta(y,.)[\,]|| &\leq K.\end{aligned}$$

where D_2 is the partial derivation relative to the second variable.
On the other hand, the usual expression of the Lie bracket vector fields implies:

$$\begin{aligned}[Z_\sigma, X_v](\gamma(t)) = D_2\theta(\gamma(t), v)\,&\{\theta(\gamma(t), P(t)(\sigma(t)))\} \\ - D_2\theta(\theta(\gamma(t), P(t)\sigma(t)))\,&\{\theta(\gamma(t), v)\}\,.\end{aligned} \tag{8.29}$$

From (8.29), since P is a projection, we get the majorations:

$$\begin{aligned}||D_2\theta(\gamma(t), v)\,\{\theta(\gamma(t), P(t)(\sigma(t)))\}\,|| &\leq K||v||.||\theta(P(t)\,(\sigma(t)], \gamma(t))\,|| \\ &\leq K^2||v||.||P(t)||.||\sigma(t)|| \\ &\leq K^3.||v||.\end{aligned} \tag{8.30}$$

In the second member of (8.29), the same majoration is true for

$$||D_2\theta(\theta(\gamma(t), P(t)\sigma(t)))\,\{\theta(\gamma(t), v)\}\,||.$$

Therefore we obtain:

$$||[Z_\sigma, X_v](\gamma(t))|| \leq 2K^3||v||$$

It follows that , for each $t \in]-\epsilon, \epsilon[$, the map : $v \mapsto [Z_\sigma, X_v](\underline{\gamma}(t))$ is linear continuous map from S to $\{\gamma(t)\} \times S$ and the map $t \mapsto \overline{\Lambda}(t) = \theta^{-1}_{\gamma(t)} \circ [Z_\sigma, X_{(.)}](\gamma(t))$ is a smooth map from $]-\epsilon, \epsilon[$ to $L(S, S)$. We set

$$\Lambda(t) = \bar{\Lambda}(t) \circ P(t).$$

Then $t \mapsto \Lambda(t)$ is a smooth map from $]-\epsilon, \epsilon[$ to $\mathrm{L}(F, S)$ and from (8.27) we obtain:

$$L_X\Psi(\gamma(t), v) = \Psi(\Lambda(t)(\gamma(t), v)$$

which ends the proof of Lemma 8.7.

Remark 8.8 *From the proof of Lemma 8.7, we have seen that we can choose Λ which takes values in S*

Let $\pi : \mathcal{E} \to M$ be a Banach fibre bundle over M with typical fibre $\mathbb{E}$, $\Psi : \mathcal{E} \to TM$ a morphism of bundle whose kernel is supplemented in each fibre. We denote by $\Gamma_M(\mathcal{E})$ the sheaf of local sections of $\pi : \mathcal{E} \to M$, that is smooth maps $\sigma : U \subset M \to \mathcal{E}$ such that $\pi \circ \sigma = Id_U$ where U is any open set of M. The maximal open set on which a section $\sigma \in \Gamma_M(\mathcal{E})$ is defined is called the *domain* of σ and denoted $\mathrm{Dom}(\sigma)$.

Definition 8.16 *A subset $\mathcal{S}$ of $\Gamma_M(\mathcal{E})$ is called a generating upper set of $\mathcal{D}$ if for any open set V in M and the set $\mathcal{X}_{\mathcal{S}}(V) = \{\Psi \circ \sigma,\ \sigma \in \mathcal{S}(V)\}$ contains $\mathcal{X}^+_{\mathcal{D}}(V)$.*

Of course the sheaf of sections $\Gamma_M(\mathcal{E})$ of $\mathcal{E}$ is a maximal generating upper set. We introduce the following condition **(LB)** on $\mathcal{X}_{\mathcal{S}}$ which will be used in the next theorem:

(LB) For any local section $\sigma \in \mathcal{S}$, there exists an open set $V \subset \mathrm{Dom}(\sigma)$ on which $\mathcal{E}$ is trivializable and, for any $x \in V$, we have the following property: Given any integral curve $\gamma :]-\epsilon, \epsilon[\to V$ of $X = \Psi \circ \sigma$ with $\gamma(0) = x$, there exists a smooth field $\Lambda :]-\epsilon, \epsilon[\to L(\mathcal{E}_x, \mathcal{E}_x)$ such that

$$\forall t \in]-\epsilon, \epsilon[, \forall u \in \mathcal{E}_x,\ [\Psi \circ \sigma, \Psi(., u)](\gamma(t)) = \Psi(\gamma(t), \Lambda_t(u)). \qquad (8.31)$$

Then, using Theorem 8.3, we get the following:

Theorem 8.4 *Let $\pi : \mathcal{E} \to M$ be a Banach fibre bundle over M with typical fibre $\mathbb{E}$ and $\Psi : \mathcal{E} \to TM$ a morphism of bundles with closed range and split kernel in each fibre; denote $\mathcal{D} = \mathrm{Im}\,\Psi$.*
Then $\mathcal{D}$ is an integrable distribution if and only there exists a generating upper set $\mathcal{S}$ such that $\mathcal{X}_{\mathcal{S}}$ satisfies the condition **(LB)**.
Moreover, when **(LB)** *is satisfied, then, for any complement S_x of $\ker \Psi_x$ in $\mathcal{E}_x$, we can choose Λ so that Λ is a smooth map from $]-\epsilon, \epsilon[\to L(\mathcal{E}_x, S_x)$.*

Proof *According to the context of the proof of Corollary 8.1, for any given $x \in M$ we consider a local trivialization of $\tau : \mathcal{E}_{|V} \to V \times \mathcal{E}_x$ on an open set V around x. Then $\Psi_\tau := \Psi \circ \tau^{-1}$ is a map $\Psi_\tau : V \times \mathcal{E}_x \to TM$. In this context, the set*

$$\mathcal{X}^+_{\Psi_\tau} = \{Z_u(y) = \Psi_\tau(y, u),\ \textit{for any}\ u \in \mathcal{E}_x,\ y \in V\}$$

is the set of upper sections associated with the upper trivialization Ψ_τ. By construction the union of all the $\mathcal{X}_{\Psi_\tau}$ for all trivializations τ of $\mathcal{E}$ around x and all $x \in M$ is precisely $\mathcal{X}^+_{\mathcal{D}}$.

By assumption there exists an upper generating set $\mathcal{S}$ which satisfies the condition **(LB)**. *According to Remark 8.7, $\Psi(\mathcal{S}(V))$ contains the set $\mathcal{X}^+_{\Psi_\tau}$ and so the condition* **(LB)** *is satisfied for any $\sigma \in \mathcal{X}^+_{\Psi_\tau}$. This implies that $\mathcal{D}$ is Lie invariant and so* **(LB)** *is a sufficient condition for integrability of $\mathcal{D}$ from Theorem 8.2.*

Assume now that $\mathcal{D}$ is integrable and consider an upper generating set $\mathcal{S}$ of $\mathcal{D}$ (for instance $\Gamma_M(\mathcal{E})$). Fix some section $\sigma \in \mathcal{S}$ defined on an open set V and $x \in U$. From Lemma 8.7, we get (8.31) with $\Lambda :]-\epsilon, \epsilon[\to L(\mathcal{F}_x, S_x)$ which mean that **(LB)** *is satisfied.*

Now assume that $\mathcal{D}$ is integrable. Fix some $\sigma \in \mathcal{S}$. Given any integral curve $\gamma :]-\epsilon, \epsilon[\to V$ of $X = \Psi \circ \sigma$ with $\gamma(0) = x$, assume that there exists a smooth field $\Lambda :]-\epsilon, \epsilon[\to L(\mathcal{E}_x, \mathcal{E}_x)$ such that the relation (8.31) is true. Then by Remark 8.8 we can choose $\Lambda :]-\epsilon, \epsilon[\to L(\mathcal{F}_x, S_x)$ which satisfies (8.31).

8.4 Applications to Banach-Lie Algebroids, Poisson Manifolds

8.4.1 Integrability of the Range of an Anchor

The notion of Lie algebroid is defined in Chapter 3, § 3.18 in the convenient framework. As an application of Theorem 8.4, we give a sufficient condition for the integrability of the range of the anchor of a strong partial Banach-Lie algebroid which is a generalization of Theorem 5 in [Pel12]. Note that the next Theorem is a generalization of Theorem 6.6 in Chapter 6.

Theorem 8.5 *Let $(\mathcal{A}, \pi, M, \rho)$ be an anchored bundle such that the range of ρ is closed and its kernel is split in each fibre.*

(1) If $(\mathcal{A}, \pi, M, \rho, \mathfrak{P}_M)$ is a strong partial Banach-Lie algebroid for which the set $\mathcal{S}_\rho$ (as defined in Remark 8.7 (4)) is contained in $\mathfrak{P}_M$, then the distribution $\mathcal{D} = \rho(\mathcal{A})$ is integrable.

(2) If $(\mathcal{A}, \pi, M, \rho, [.,.]_{\mathcal{A}})$ is a Banach pre-Lie algebroid[3] or a Lie algebroid, then the distribution $\mathcal{D}$ is integrable.

Proof

(1) According to Remark 8.7, (4), the set $\mathcal{S}_\rho$ is an upper generating set. So we must show satisfies the condition **(LB)** *is satisfied for $\mathcal{S}_\rho$.*

Fix some $x \in M$ and a trivialization $\tau : \mathcal{A}_V \to V \times \mathcal{A}_x$. From the assumption, the set $\mathcal{S}^+_{\rho_\tau}$ is contained in $\mathfrak{P}_M(V)$. Consider any section $\sigma \in \mathcal{X}^+_{\rho_\tau}$ and fix some $x \in V$. Again, we set $Z_\sigma = \rho \circ \sigma$ and $Z_u = \rho_\tau(.u)$ an upper section. Given an integral curve $\gamma(t) = \mathrm{Fl}_t^{Z_\sigma}(x)$ on $]-\varepsilon, \varepsilon[$, since ρ induces a sheafs Lie algebra morphims, we have

$$[Z_\sigma, Z_u](\gamma(t)) = \rho\left([\sigma, s_u]_{\mathcal{A}}\right)(\gamma(t)) \text{ where } s_u(y) = \tau^{-1}(y, u).$$

Since $\mathfrak{P}_M$ is a sheaf of Lie algebras, it follows that $[\sigma, s_u](\gamma(t))$ belongs to $\mathcal{E}_\gamma(t) \equiv \{\gamma(t)\} \times \mathcal{E}_x$ via τ. But, according to the local expression in a chart of $[.,.]_{\mathcal{A}}$ (cf. Chapter 3, Notations 3) by the same arguments as those used in the proof of Lemma 8.7, we can show that the map

$$t \mapsto [\tau(\sigma), \tau(s_u)]_{\mathcal{A}})(\gamma(t))\}$$

[3] See Chapter 3, Remark 3.12.

is a smooth field of endomorphisms of $\mathcal{E}_x$. *This implies that relation (8.31) is satisfied for the pair* $(\rho(\sigma), \rho(s_u))$. *Since this property is true for any* $x \in M$ *and any trivialization* τ *around* x, *it follows that* $\mathcal{D}$ *is integrable from Theorem 8.4.*

(2) Now, if $(\mathcal{A}, \pi, M, \rho, [.,.]_{\mathcal{A}})$ *is a Banach pre-Lie algebroid,* $\Gamma_M(\mathcal{A})$ *always contains the set* $\mathcal{S}_\rho$. *But although* $\Gamma_M(\mathcal{A})$ *is not a sheaf of Lie algebras, note that for any* σ *and* σ' *in* $\Gamma_M(\mathcal{A})$, *we have*

$$[\rho(\sigma), \rho(\sigma')] = \rho([\sigma, \sigma']_{\mathcal{A}})$$

$([\sigma, \sigma']_{\mathcal{A}}))(x)$ *belongs to* $\mathcal{A}_x$ *for any* $x \in \mathrm{Dom}(\sigma) \cap \mathrm{Dom}(\sigma')$.
From these properties the previous arguments work again and so $\mathcal{D}$ *is integrable from Theorem 8.4. If* $(\mathcal{A}, \pi, M, \rho, [.,.]_{\mathcal{A}})$ *is a Banach Lie algebroid it is in particular a Banach pre-Lie algebroid, so the previous result is in particular true.*

Remark 8.9 *Recall that in Chapter 3, §3.18.8, we have defined an equivalence relation on* M *associated with Banach Lie algebroid* $(\mathcal{A}, \pi, M, \rho, [.,.]_{\mathcal{A}})$ *in the following way:*

$x \sim y$ *if and only if there exists an* $\mathcal{A}$*-admissible curve* $\gamma : [a, b] \subset \mathbb{R} \to M$ *such that* $\gamma(a) = x$ *and* $\gamma(b) = y$.

The equivalent class of x *is called the* $\mathcal{A}$*-orbit of* x *of the convenient Lie algebroid* $(\mathcal{A}, \pi, M, \rho, [.,.]_{\mathcal{A}})$. *Since under the assumptions of Theorem 8.5, each integral manifold* L *is a closed submanifold of* M, *it follows easily that such* L *is exactly the* $\mathcal{A}$*-orbit of* x *for any* $x \in L$. *By the way, the partition of* M *in* $\mathcal{A}$*-orbits is exactly the partition of* M *by leaves of* $\rho(\mathcal{A})$.

Example 8.5 *Let* $\pi : \mathcal{A} \to M$ *be a closed Banach subbundle of* $TM \to M$, *where* π *is the restriction to* $\mathcal{A}$ *of the canonical projection of* TM *onto* M, *and the canonical inclusion* $i : \mathcal{A} \to TM$ *is a morphism bundle. Any section of* $\pi : \mathcal{A} \to M$ *induces a vector field on* M. *Assume that the set of sections* $\Gamma(\mathcal{A})$ *is stable by Lie bracket of vector fields, which means that the associated weak distribution* $\mathcal{D} = i(\mathcal{A})$ *is involutive. Then,* $\left(\mathcal{A}, i, M, [.,.]_{|\Gamma(\mathcal{A})}\right)$ *is a Banach-Lie algebroid. So it follows from Theorem 8.5 that* $\mathcal{D}$ *is an integrable distribution. Thus, we get a version of Frobenius Theorem, as we can find in [Glo08]. Thus we can consider this result as an appropriate version of Frobenius Theorem.*

Example 8.6 *Let* $\pi : \mathcal{A} \to M$ *be a Banach bundle and* $\rho : \mathcal{A} \to TM$ *an injective morphism bundle with closed range. If* $\mathcal{D} = \rho(\mathcal{A})$ *the condition* **(LB)** *of Theorem 8.4 is satisfied, then* $\mathcal{D}$ *is integrable. From the injectivity of* ρ, *it follows that we can define a Lie algebra structure on the sections* $\Gamma(\mathcal{A})$, *by:*

$$\{\mathfrak{a}_1, \mathfrak{a}_2\} = \rho^{-1}\left([\rho(\mathfrak{a}_1), \rho(\mathfrak{a}_2)]\right).$$

So, we get a Banach-Lie algebroid structure on $\mathcal{A}$.

Example 8.7 *Consider a smooth right action $\psi : M \times G \to M$ of a connected Banach-Lie group G over a Banach manifold M. Denote by $\mathfrak{g}$ the Lie algebra of G. We have a natural morphism ξ of Lie algebra from $\mathfrak{g}$ to $\mathfrak{X}(M)$ which is defined by*

$$\xi_X(x) = T_{(x,e)}\psi(0, X)$$

For any X and Y in $\mathfrak{g}$, we have:

$$\xi_{\{X,Y\}} = [\xi_X, \xi_Y]$$

where $\{.,.\}$ denotes the Lie algebra bracket on $\mathfrak{g}$ (see for instance [KrMi97], 36.12). On the trivial bundle $M \times \mathfrak{g}$, each section can be identified with a map $\sigma : M \to \mathfrak{g}$. We then define a Lie bracket on the set of sections by

$$\{\{\sigma, \sigma'\}\}(x) = \{\sigma(x), \sigma'(x)\} + d\sigma(\xi_{\sigma'(x)}) - d\sigma'(\xi_{\sigma(x)})$$

According to the triviality of $M \times \mathfrak{g}$, we get a localizable Lie bracket. An anchor morphism $\Psi : M \times \mathfrak{g} \to TM$ is defined by setting $\Psi(x, X) = \xi_X(x)$.
It follows that $(M \times \mathfrak{g}, \Psi, M, \{\{.,.\}\})$ has a Banach-Lie algebroid structure on M.
Denote by G_x the closed subgroup of isotropy of a point $x \in M$ and $\mathfrak{g}_x \subset \mathfrak{g}$ its Lie sub-algebra. Of course, we have $\ker \Psi_x = \mathfrak{g}_x$. According to Theorem 4, if $\mathfrak{g}_x$ is supplemented in $\mathfrak{g}$ for any $x \in M$ and the distribution $\mathcal{D} = \Psi(M \times \mathfrak{g})$ is closed, then $\mathcal{D}$ is integrable. In fact the leaf through x is its orbit $\psi(x, G)$.

8.4.2 Integrability of a Characteristic Distribution

We refer to Chapter 7, §7.1 and § 7.2 for the definition of a partial Poisson manifold and their properties.

Definition 8.17 *Let $(M, \mathfrak{A}_M\{.,.\}_P)$ be a Banach Poisson manifold associated with a Poisson anchor $P : T^\flat M \to M$.*
The distribution $\mathcal{D} = P\left(T^\flat M\right)$ is called the characteristic distribution.

Note that the characteristic distribution is a weak distribution which is not closed in general.

Associated with $(M, \mathfrak{A}_M\{.,.\}_P)$, we have a natural skew-symmetric bilinear form ω defined as follows:

For any α and β in $T^\flat_x M$, we have $\omega(\alpha, \beta) = \{f, g\}_P$ if f and g are smooth functions in $\mathfrak{A}(U)$ defined on an a open neighbourhood of U of x and such that $df(x) = \alpha$ and $dg(x) = \beta$ (cf. Lemma 7.3 Chapter 7). It is clear that this definition is independent on the choice of f and g).

For each x, on the quotient $T^\flat_x M / \ker P_x$, we get a skew-symmetric bilinear form $\hat{\omega}_x$.

On the other hand, let $\hat{P}_x : T^\flat_x M/\ker P_x \to \tilde{\mathcal{D}}_x$ be the canonical isomorphism associated with P_x between Banach spaces. Finally we get a skew-symmetric bilinear form $\tilde{\omega}_x$ on $\tilde{\mathcal{D}}_x$ such that:

$$[\hat{P}_x]^*\tilde{\omega}_x = \hat{\omega}_x.$$

According to [OdRa08], we give the following definition.

Definition 8.18 *A symplectic leaf of $\mathcal{D}$ is a weak submanifold $(\mathcal{L}, \iota_L)$ where $\mathcal{L} \subset M$ and $\iota_L : \mathcal{L} \to M$ is the natural inclusion with the following properties:*

(1) $(\mathcal{L}, \iota_L)$ is a maximal integral manifold of $\mathcal{D}$;

(2) on $\mathcal{L}$ we have a weak symplectic form $\omega_{\mathcal{L}}$ such that $(\omega_{\mathcal{L}})_x = \tilde{\omega}_x$ for all $x \in \mathcal{L}$

Theorem 8.6 *Let $P : T^\flat M \to TM$ be a Poisson anchor. If the range of P is closed and its kernel is split in each fibre, then the associated characteristic distribution $\mathcal{D}$ is integrable.*
Moreover, each maximal integral manifold has a natural structure of weak symplectic leaf.

For an illustration of this result, the reader will find a lot of examples of Banach Poisson manifolds in [OdRa03] and [OdRa08]. More generally, if ω is a weak symplectic form on a Banach manifold M and if $T^\flat M = \omega^\flat(TM)$ we have already seen that $P = (\omega^\flat)^{-1} : T^\flat M \to TM$ is Poisson anchor. Thus, if the distribution $\mathcal{D} = P(T^\flat M)$ is closed, then the restriction of ω on each leaf L of $\mathcal{D}$ provides L with a weak symplectic manifold structure.

Proof *At first, according to Chapter 7, Lemma 7.3, we can observe that the set*

$$\mathcal{S} = \{(df) : f \in \mathfrak{A}(U),\ U \text{ any open set in } M\}$$

is an upper generating set for $\mathcal{D}$: more precisely, given any $x \in M$, modulo any local chart around x, we can suppose that M is an open subset of $\mathbb{M}$ and so that $T^\flat M \equiv M \times \mathbb{F} \subset M \times \mathbb{M}^*$; for any $\alpha \in \mathbb{F}$ the function $f_\alpha(x) =< \alpha, x >$ is a smooth map on M such that $df_\alpha(y) = \alpha$ for any $y \in M$; so $Z_\alpha = \Psi(\alpha, y) = P(df_\alpha(y))$ is an upper section.
For any smooth local function $f \in \mathfrak{A}(U)$, we denote $X_f = P(df)$. From the Jacobi identity we have

$$[X_f, X_g] = P(d\{f, g\}) \text{ for any } f, g \in \mathfrak{A}(U) \tag{8.32}$$

According to Theorem 8.4 , to prove the integrability of $\mathcal{D}$, we have only to prove **(LB)** *for the generating upper set $\mathcal{S}$. As* **(LB)** *is a local property, again we assume that M is an open set in $\mathbb{M}$. So fix some smooth function $f \in \mathfrak{A}(U)$ and consider an integral curve $\gamma(t) = \mathrm{Fl}_t^{X_f}(x)$ through $x \in M$ defined on $]-\varepsilon, \varepsilon[$. For any $\alpha \in \mathbb{F}$, from (8.32), we have:*

$$[X_f, X_{f_\alpha}](\gamma(t)) = [P(df), P(df_\alpha)](\gamma(t)) = P(d\{f, f_\alpha\}_P(\gamma(t)))$$

But, by the same arguments as those used in the proof of Lemma 8.7 we can show that the map

$$y \mapsto [\alpha \mapsto d\{f, f_\alpha\}_P(y)]$$

is a smooth field of continuous endomorphisms of $\mathbb{F}$*. It follows that the condition* **(LB)** *is satisfied, and then,* $\mathcal{D}$ *is integrable.*

Assume now that $\mathcal{D}$ *is integrable and choose any maximal leaf* $\mathcal{L}$*. As* $T_x\mathcal{L} = \mathcal{D}_x$*, on* $T_x\mathcal{L}$ *we have the skew-symmetric bilinear form* $\tilde{\omega}_x$ *previously defined. We will show that* $\tilde{\omega}_x$ *defines a closed 2-form* $\omega_\mathcal{L}$ *on* $\mathcal{L}$*, which is a weak symplectic form.*

Fix $x \in \mathcal{L}$*. We have* $T_x^\flat M = \ker P_x \oplus S_x$*. So* $\mathcal{L}$ *is a Banach manifold modelled on* S_x*. From the definition of* $\tilde{\omega}_x$*, we have*[4]

$$\tilde{\omega}_x(\theta_x(\alpha), \theta_x(\beta)) = <\alpha, \theta_x(\beta)> . \tag{8.33}$$

As we know that θ_x *is an isomorphism from* S_x *to* $T_x\mathcal{L}$*, it follows that* $\tilde{\omega}_x$ *is a weak symplectic 2-form on the Banach space* $T_x\mathcal{L}$*.*
On one hand, since locally on each leaf, θ_y *is an isomorphism, it follows that* $\omega_\mathcal{L}$ *defined by* $(\omega_\mathcal{L})_x = \tilde{\omega}_x$ *is a smooth differential 2-form on* $\mathcal{L}$*. On the other hand, if* $i : \mathcal{L} \to M$ *is the inclusion, for any smooth function* f *defined on an open set* $U \subset M$ *which belongs to* $\mathfrak{A}(U)$*, we set* $\tilde{f} = f \circ i$*. From Lemma 8.1* $\tilde{f}$ *is smooth on* $U \cap \mathcal{L}$*. So, for any smooth functions* f*,* g *and* h *in* $\mathfrak{A}(U)$*, the Jacobi identity is satisfied for* $\tilde{f}$*,* $\tilde{g}$ *and* $\tilde{h}$ *on the open set* $U \cap \mathcal{L}$*. So, by classical arguments of Poisson bracket (see for instance [OdRa03], [OdRa08]), we get:* $d\omega_\mathcal{L}(i_*X_f, i_*X_g, i_*X_h) = 0$ *for any choice of functions* f*,* g *and* h *in* $\mathcal{A}(U)$*. So* $\omega_\mathcal{L}$ *is closed and the proof is complete.*

8.5 An Integrability Criterion for a Submersive Projective Limit of Banach Distributions

8.5.1 The Criterion and Its Corollaries

Let $\left(M_i, \delta_i^j\right)_{j \geq i}$ be a sequence of projective Banach manifolds with projective limit $M = \varprojlim M_i$ (cf. Chapter 4, § 4.6).
In order to give a criterion of integrability for projective limits of local Banach bundles on M under some additional assumptions, we need to introduce some notations.
Let ν (resp. μ) be a norm on a Banach space $\mathbb{E}$ (resp. $\mathbb{M}$). We denote by $\| \|^{\mathrm{op}}$ the associated norm on the linear space of continuous linear mappings $\mathcal{L}(\mathbb{E}, \mathbb{M})$ (cf. Chapter 1, § 1.1).

[4] θ being a lower trivialization associates to the upper trivialization defined by P

Then we have:

Definition 8.19 *Let* $\left(M_i, \delta_i^j\right)_{j \geq i}$ *be a projective sequence of Banach manifolds where the maps* δ_i^j *are submersions and* $M = \underleftarrow{\lim} M_i$ *its projective limit. A closed distribution* Δ *on* M *will be called a submersive projective limit of local anchored bundles if the following property is satisfied:*

(*) For any $x = \underleftarrow{\lim} x_i \in M$*, there exists an open neighbourhood* $U = \underleftarrow{\lim} U_i$ *of* x*, a submersive projective sequence of anchored Banach bundles* $(E_i, \pi_i, U_i, \rho_i)_{\underleftarrow{i}}$ *fulfilling the following properties for any* $z = \underleftarrow{\lim} z_i \in U$*:*

1. $\underleftarrow{\lim} \rho_i((E_i)_{z_i}) = \Delta_z$*, for any* $z \in U$*.*
2. *The kernel of* $(\rho_i)_{z_i}$ *is supplemented in* $(E_i)_{z_i}$ *and the range of* $(\rho_i)_{z_i}$ *is closed, for all* $i \in \mathbb{N}$*.*
3. *There exists a constant* $C > 0$ *and a Finsler norm* $|| \; ||^{E_i}$ *(resp.* $|| \; ||^{M_i}$*) on* $(E_i)_{|U_i}$ *(resp.* $TM_{i|U_i}$*) such that:*

$$\forall i \in \mathbb{N}, \; ||(\rho_i)_{z_i}||_i^{\mathrm{op}} \leq C, \; \forall z_i \in U_i.$$

We have the following criterion of integrability:

Theorem 8.7 *Let* M *be a projective limit of a submersive projective sequence* $(M_i, \delta_i^j)_{j \geq i}$ *of Banach manifolds and* Δ *be a local projective limit of local Banach bundles on* M*. Assume that, under the property (*), there exists a Lie bracket* $[.,.]_i$ *on* $(E_i, \pi_i, U_i, \rho_i)$ *such that* $(E_i, \pi_i, U_i, \rho_i, [.,.]_i)_{\underleftarrow{i}}$ *is a submersive projective sequence of Banach-Lie algebroids. Then the distribution* Δ *is integrable and the maximal integral manifold* N *through* $x = \underleftarrow{\lim} x_i$ *is a closed Fréchet submanifold of* M *which is a submersive projective limit of the set of maximal leaves* N_i *of* $\rho_i(E_i)$ *through* x_i *in* M_i*.*

The proof of Theorem will be given in § 8.5.4.

We have the following consequences of Theorem 8.7:

Theorem 8.8 *Let* $(E_i, \pi_i, M_i, \rho_i, [.,.]_i)_{\underleftarrow{i}}$ *be a submersive projective sequence of split Lie algebroids.*
Then, we have:

1. $\left(E := \underleftarrow{\lim} E_i, \pi := \underleftarrow{\lim} \pi_i, M := \underleftarrow{\lim} M_i, \rho = \underleftarrow{\lim} \rho_i\right)$ *is a Fréchet anchored bundle and* $\Delta = \rho(E)$ *is a closed distribution on* M*.*

2 *If* (ρ_i) *satisfies the condition (3) in Definition 8.19, then* Δ *is integrable and each leaf* L *of* Δ *is a projective limit of leaves* L_i *of* Δ_i*.*

Proof *The property (1) is a consequence of Theorem 4.16. From this property, it follows that locally* Δ *satisfies the assumptions (1) and (2) of Definition 8.19; so, if the assumption (3) is satisfied, the result is a direct consequence of Theorem 8.7.*

Theorem 8.9 *Let* $\left(M_i, \delta_i^j\right)_{j\geq i}$ *be a submersive projective sequence of Banach manifolds and* (E_i, π_i, M_i) *an involutive subbundle of* TM_i *where* π_i *is the restriction of the natural projection* $p_{M_i} : TM_i \to M_i$*. Assume that the restriction* $T\delta_i^j : E_j \to E_i$ *is a surjective bundle morphism for all* $i \in \mathbb{N}$ *and* $j \geq i$*. Then* $(E_i, \pi_i, M_i)_{\overleftarrow{i}}$ *is a submersive projective sequence of Banach bundles, and* $\left(E = \varprojlim E_i, \pi = \varprojlim \pi_i, M = \varprojlim M_i\right)$ *is an integrable Fréchet subbundle of* TM *whose each leaf* L *of* E *in* M *is a projective limit of leaves* L_i *of* E_i *in* M_i*.*

The reader will find an example in § 8.5.3 which shows that these assumptions are not necessary.

Proof *Since* $\delta_i^j : M_j \to M_i$ *is a surjective submersion, so is* $T\delta_i^j : TM_j \to TM_i$*. If* $T\delta_i^j : E_j \to E_i$ *is a surjective morphism, this implies that* $T\delta_i^j$ *is a submersion onto* E_i *and so* (E_i, π_i, M_i) *is a submersive projective sequence of Banach bundles. Let* $\iota_i : E_i \to TM_i$ *the natural inclusion and* $[\ ,\ ]_i$ *the restriction of the Lie bracket of vector fields to (local) sections of* E_i*. Then* $(E_i, M_i, \iota_i, [\ ,\ ]_i)_{\overleftarrow{i}}$ *is a Banach-Lie algebroid and since* $T\delta_i^j \circ \iota_j = \iota_i \circ T\delta_i^j$ *it follows that* $(E_i, M_i, \iota_i, [\ ,\ ]_i)_{\overleftarrow{i}}$ *is a is a submersive projective sequence of Banach-Lie algebroids. Fix some* $x = \varprojlim x_i \in M = \varprojlim M_i$*. According to Theorem 8.8, we have only to show that the condition (3) of Definition 8.19 is satisfied by* ι_i*.*
Given any norm $||\ ||^{M_i}$ *on* $T_{x_i}M_i$*, we denote by* $||\ ||^{E_i}$ *the induced norm on the fibre* $\{E_i\}_{x_i}$*, then, for the associated norm operator, we have* $||\{\iota_i\}_{x_i}||_i^{\mathrm{op}} = 1$*. So all the assumptions of Theorem 8.7 are satisfied, which ends the proof.*

8.5.2 Some Applications

8.5.2.1 Application to Submersive Projective Sequence of Banach Lie Groups

Let $\left(G_i, \delta_i^j\right)_{j\geq i}$ be a submersive a projective sequence of Banach-Lie groups where G_i is modelled on $\mathbb{G}_i$ (cf. Definition 4.12). We denote by $\mathbf{L}(G_i)$ the Lie algebra of G_i. Then $\mathbf{L}(G_i) \equiv T_eG_i$ is isomorphic to $\mathbb{G}_i$. If we set $\overline{\delta}_i^j := T_e\delta_i^j$, then each $\overline{\delta}_i^j$ is a surjective linear map from $\mathbf{L}(G_j)$ to $\mathbf{L}(G_i)$ whose kernel is supplemented.
Consider a sequence $\mathfrak{h}_i$ of supplemented sub-Lie algebras of $\mathbf{L}(G_i)$ such that the restriction of $\overline{\delta}_i^j$ to $\mathfrak{h}_j$ is a continuous surjective map. Then $\left(\mathfrak{h}_i, \hat{\delta}_i^j\right)_{j\geq i}$ is a submersive projective sequence of Banach-Lie algebra and so $\mathfrak{h} = \varprojlim \mathfrak{h}_i$ is a Fréchet Lie algebra (cf. Chapter 4). Now, from a classic result on Banach-Lie groups (cf. [Lan95]), by left translation each $\mathfrak{h}_i$ gives rise to a supplemented involutive subbundle of $\mathcal{H}_i$ of TG_i and the leaf H_i through the neutral e_i in G_i has a structure of connected Banach-Lie group, so that the inclusion

$\iota_i : H_i \to G_i$ is a Banach-Lie morphism. Note that $\mathfrak{i}_i := T_{e_i}\iota_i$ is nothing but else that the inclusion of $\mathfrak{h}_i$ in $\mathbf{L}(G_i)$ and which induces the natural inclusion $\hat{\iota}_i$ of $\mathcal{H}_i$ in TG_i.
Moreover, since $\left(\mathfrak{h}_i, \hat{\delta}_i^j\right)_{j\geq i}$ is a submersive projective sequence of Banach-Lie algebras and $\left(G_i, \delta_i^j\right)_{j\geq i}$ be a submersive a projective sequence of Banach-Lie groups, it follows that $(\mathcal{H}_i, G_i, \widehat{\mathfrak{i}}_i, [\ ,\ ]_i)$[5] is a submersive projective sequence of split Banach-Lie algebroids.

On the other hand, from Theorem 4.1, the Lie algebra $\mathbf{L}(G)$ of $G = \underleftarrow{\lim} G_i$ is $\underleftarrow{\lim}\mathbf{L}G_i$ and so $\mathfrak{h} = \underleftarrow{\lim}\mathfrak{h}_i$ is a closed supplemented Lie subalgebra of $\mathbf{L}(G)$. As in the context of the Banach setting, by left translation, $\mathfrak{h}$ gives rise to an involutive Fréchet subbundle $\mathcal{H}$ of TG which is clearly the projective limit of the submersive sequence $\left(\mathcal{H}_i, G_i, \widehat{\mathfrak{i}}_i, [\ ,\ ]_i\right)$. So, from Theorem 8.9, using the same arguments as in Banach-Lie groups, we obtain:

Theorem 8.10 *Let $G = \underleftarrow{\lim} G_i$ be a the projective limit of a submersive projective sequence of Banach-Lie groups $\left(G_i, \delta_i^j\right)_{j\geq i}$ and, for each $i \in \mathbb{N}$, consider a closed supplemented Banach-Lie subalgebra $\mathfrak{h}_i$ of $\mathbf{L}(G_i)$. Assume that the restriction of $\bar{\delta}_i^j$ to $\mathfrak{h}_j$ is a continuous surjective map. Then there exists a Fréchet Lie group H in G such that $\mathbf{L}(H)$ is isomorphic to $\mathfrak{h}$ and H is the projective limit of $\left(H_i, \delta^j_{i\,|H_j}\right)_{j\geq i}$.*

8.5.2.2 Application to Sequences of Prolongations of a Banach-Lie Algebroid over a Banach Manifold

Consider an anchored Banach bundle $(\mathcal{A}, \pi, M, \rho)$ with typical fibre $\mathbb{A}$. Let $V\mathcal{A} \subset T\mathcal{A}$ be the vertical subbundle of $p_{\mathcal{A}} : T\mathcal{A} \to \mathcal{A}$. If $\mathcal{A}_x := \pi^{-1}(x)$ is the fibre over $x \in M$, according to Chapter 6, § 6.4 and Example 6.1 (1), the prolongation $\mathbf{T}\mathcal{A}$ of the anchored Banach bundle $(\mathcal{A}, \pi, M, \rho)$ over $\mathcal{A}$ is the set $\{(x,a,b,c),\ (x,b) \in \mathcal{A}_x,\ (x,a,c) \in V_{(x,a)}\mathcal{A}\}$. It is a Banach vector bundle $\hat{\mathbf{p}} : \mathbf{T}\mathcal{A} \to \mathcal{A}$ with typical fibre $\mathbb{A} \times \mathbb{A}$ and we have an anchor $\hat{\rho} : \mathbf{T}\mathcal{A} \to T\mathcal{A}$ given by

$$\hat{\rho}(x,a,b,c) = (x, a, \rho_x(b), c) \in T_{(x,a)}\mathcal{A}.$$

From now on, we fix a Banach-Lie algebroid $(\mathcal{A}, \pi, M, \rho, [.,.]_{\mathcal{A}})$ such that the typical fibre $\mathbb{A}$ of $\mathcal{A}$ is finite dimensional. By the way, we have a Banach-Lie algebroid structure $(\mathbf{T}\mathcal{A}, \hat{\mathbf{p}}, \mathcal{A}, \hat{\rho}, [.,.]_{\mathbf{T}\mathcal{A}})$ (cf. Corollary 6.2)

We denote $(\mathcal{A}_1, \pi_1, \mathcal{A}_0, \rho_1, [.,.]_1)$ the Banach-Lie algebroid $(\mathcal{A}, \pi, M, \rho, [.,.]_{\mathcal{A}})$ over a Banach manifold $\mathcal{A}^0 = M$. Thus we have the following commutative

[5] Here $[\ ,\ ]_i$ denote again the restriction to sections of $\mathcal{H}_i$ of the Lie bracket of vector fields on G_i.

diagram:

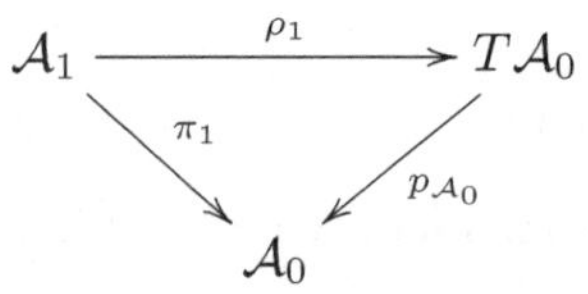

(8.34)

According to the notations of Theorem 6.5 and Corollary 6.2, we set $\mathcal{A}_2 = \mathbf{T}\mathcal{A}_1$, $\rho_2 = \hat{\rho}$, $[.,.]_2 = [.,.]_{\mathbf{T}\mathcal{A}_1}$ and $\pi_2 = \hat{\mathbf{p}}$. Then we have the following commutative diagram:

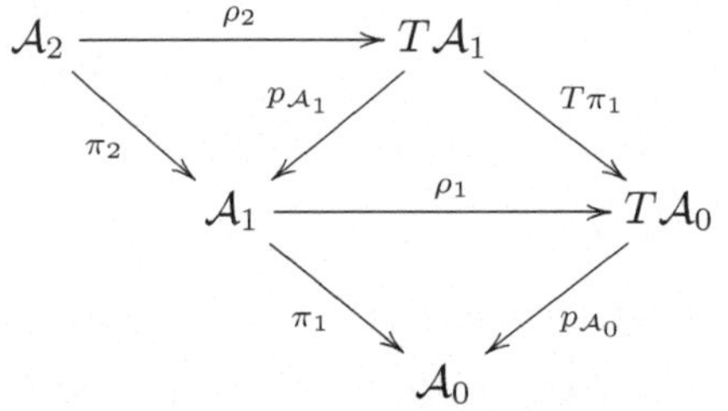

(8.35)

Fix some $x \in \mathcal{A}_0$ and a norm $|| \; ||_0$ (resp.$|| \; ||_1$) on the fibre $T_x\mathcal{A}_0 \equiv \mathbb{A}_0$ (resp. $\mathcal{A}_x = \pi_1^{-1}(x) \equiv \mathbb{A}_1$). Since the fibre $\mathbf{T}_{(x,a)}\mathcal{A}_1$ (resp. $T_{(x,a)}\mathcal{A}_1$) is isomorphic to $\mathbb{A}_1 \times \mathbb{A}_1$ (resp. $\mathbb{A}_0 \times \mathbb{A}_1$), it follows that $\sup\{|| \; ||_1 , || \; ||_1\}$ (resp. $\sup\{|| \; ||_0 , || \; ||_1\}$ gives rise to a norm on $\mathbf{T}_{(x,a)}\mathcal{A}_1$ (resp. $T_{(x,a)}\mathcal{A}_1$). Then, for the associated operator norm $|| \; ||^{\text{op}}$, we have

$$||(\rho_2)_{(x,a)}||^{\text{op}} \leq \sup(||(\rho_1)_x||^{\text{op}}, 1) \tag{8.36}$$

By induction, for $i \geq 1$, again according to notations of Theorem 6.5 and Corollary 6.2, we set
$\mathcal{A}^{i+1} = \mathbf{T}^{\mathcal{A}_i}\mathcal{A}_i$, $\rho_{i+1} = \hat{\rho}_i$, $[.,.]_{i+1} = [.,.]_{\mathbf{T}^{\mathcal{A}_i}\mathcal{A}_i}$ and $\pi_{i+1} = \hat{\mathbf{p}}$ and, as before, we have the following commutative diagram:

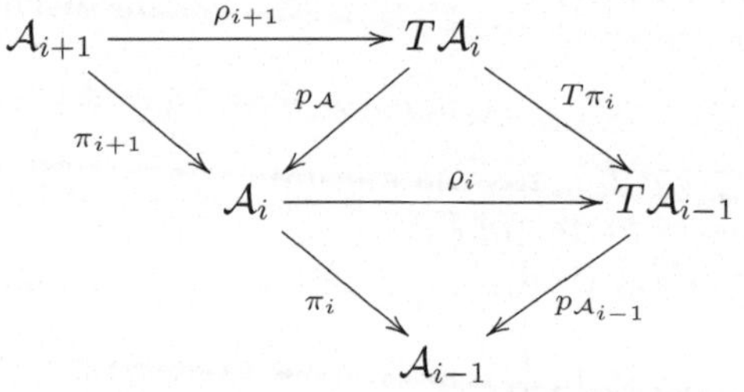

(8.37)

Also, by same arguments as for (8.36) we obtain:

$$||(\rho_{i+1})_{(x,a_1,\ldots,a_{i+1})}||^{\text{op}} \leq \sup(||(\rho_i)_{(x,a_1,\ldots,a_i)}||^{\text{op}}, 1) \tag{8.38}$$

It follows that we have a submersive projective sequence of Banach-Lie algebroids $(\mathcal{A}_i, \mathcal{A}_{i-1}, \rho_i, [.,.]_i)_{\overleftarrow{i}}$ over a submersive projective sequence of Banach manifolds $(\mathcal{A}_i)_{i\in\mathbb{N}}$ which satisfies the assumptions of Theorem 8.7. Thus, we obtain:

Theorem 8.11 *Under the previous context,*

$$(\mathcal{A} = \varprojlim_{i\geq 1}\mathcal{A}_i, \mathcal{M} = \varprojlim_{j\geq 0}\mathcal{A}_j, \rho = \varprojlim_{i\geq 1}\rho_i, [.,.] = \varprojlim_{i\geq 1}[.,.]_i)$$

is a Fréchet Lie algebroid on the Fréchet manifold $\mathcal{M}$ and the distribution $\rho(\mathcal{A})$ is integrable. Each leaf L is a projective limit of a projective sequence of leaves of type (L_i) defined by induction in the following way:
L_0 is a leaf of $\rho_1(\mathcal{A}_1)$ and if L_i is a leaf of $\rho_i(\mathbb{A}_i)$ then $L_{i+1} = (\mathcal{A}_i)_{|L_i}$.

8.5.3 A Counterexample

In this subsection we give an Example of an integrable distribution on a Fréchet bundle over a finite dimensional manifold which satisfies the assumptions (*) in Definition 8.19 but is a projective limit of a submersive sequence of Banach **not integrable** distributions.

The notion of k-jet has already been introduced in Chapter 2, § 1.12.
Let $E = M \times \mathbb{R}^m$ the trivial bundle over a manifold M of dimension n. The set $J^k(E)$ of the k-jets of sections of E over M is a finite-dimensional manifold which is a vector bundle $\pi^k : J^k(E) \to M$ and whose typical fibre is the space $\prod_{j=0}^{k} \mathcal{L}^j_{\text{sym}}(\mathbb{R}^n, \mathbb{R}^m)$ where $\mathcal{L}^j_{\text{sym}}(\mathbb{R}^n, \mathbb{R}^m)$ is the space of continuous j-linear symmetric mappings $\mathbb{R}^n \to \mathbb{R}^m$. Then each projection $\pi^l_k : J^l(E) \to J^k(E)$ defined, for $l \geq k$, by

$$\pi^l_k \left[j^l(s)(x) \right] = j^k(s)(x)$$

is a smooth surjection and is a submersion.
As seen in Chapter 4, § 4.9, $\left(J^k(E), \pi^l_k\right)$ is a submersive projective sequence of Banach vector bundles and the projective limit

$$J^\infty(E) = \varprojlim J^k(E)$$

can be endowed with a structure of Fréchet vector bundle whose fibre is isomorphic to the Fréchet space $\prod_{j=0}^{\infty} \mathcal{L}^j_{\text{sym}}(\mathbb{R}^n, \mathbb{R}^m)$.

Let s be a section of π on a neighbourhood U of $x \in M$. For $\xi = j^k(s)(x) \in J^k(E)$, the n-dimensional subspace $R(s,x)$ of $T_\xi J^k(E)$ equals to the tangent space at ξ to the submanifold $j^k(s)(U) \subset J^k(\pi)$ is an R-plane.

The Cartan subspace (cf. Chapter 2, § 1.12.3) $\mathcal{C}^k(E)$ of $T_\xi J^k(E)$ is the linear subspace spanned by all R-planes $R(s', x)$ such that $j^k(s')(x) = \xi$. So it is the hull of the union of $\left(j^k(s)\right)_*(x)(T_xM)$ where s is any local section of π around de x.

The Cartan subspaces form a smooth distribution on $J^k(\pi)$ called Cartan distribution and denoted $\mathcal{C}^k$. Then $\mathcal{C}^k$ is a regular distribution which is a contact distribution and so is not integrable (cf. [Vit17]). We have a submersive projective limit of bundles $\left(\mathcal{C}^k, T\pi_k^l, j^k(E)\right)$ whose projective limit $\mathcal{C} = \underleftarrow{\lim}\mathcal{C}^k$ is called the Cartan distribution on $J^\infty(E)$. In fact $\mathcal{C}$ is integrable (cf. [Vit17]). Note that since $\mathcal{C}^k$ is a subbundle of $TJ^k(E)$, from the proof of Theorem 8.9, the condition (3) of Definition 8.19 is satisfied.

8.5.4 Proof of Theorem 8.7

Fix some $x \in M$. According to the property (*) and the Definition 8.19, we can choose a submersive projective sequence of charts $\left(U_i, \delta_{i\,|U_j}^j\right)_{j \geq i}$ and a submersive projective sequence of Banach bundles $(E_i, \lambda_i^j)_{j\geq i}$, such that:

- $U = \underleftarrow{\lim}U_i$ is an open neighbourhood of x in M;
- If $x = \underleftarrow{\lim}(x_i)$, each U_i is the contractile domain of a chart (U_i, ϕ_i) around x_i in M_i and $(U = \underleftarrow{\lim}(U_i), \phi = \underleftarrow{\lim}(\phi_i))$ is a projective limit chart in M around x.
- The projective sequence of Banach bundles $(E_i, \lambda_i^j)_{j\geq i}$ satisfies the assumption (*) on U.

Step 1: The kernel of ρ_x is supplemented.
There exists a trivialization $\tau_i : E_i \to U_i \times \mathbb{E}_i$ which satisfies the compatibility condition:

$$(\delta_i^j \times \overline{\lambda_i^j}) \circ \tau_j = \tau_i \circ \lambda_i^j \tag{8.39}$$

where $(\mathbb{E}_i, \overline{\lambda_i^j})_{j\geq i}$ is the projective sequence of Banach spaces on which $(E_i, \lambda_i^j)_{j\geq i}$ is modeled.
Under these conditions, without loss of generality we may assume that, for each $i \in \mathbb{N}$, we have

- $x_i \equiv 0 \in \mathbb{M}_i$;
- U_i is an open subset of $\mathbb{M}_i$ and so $TU_i = U_i \times \mathbb{M}_i$;
- $E_i = U_i \times \mathbb{E}_i$.

The projection δ_i^j at point $x_j \equiv 0$ (resp. λ_i^j in restriction to the fibre of E_j over $x_j \equiv 0$) is denoted d_i^j (resp. ℓ_i^j). The morphism ρ_i in restriction to the fibre E_i over $x_i \equiv 0$ is denoted r_i and ρ_x is denoted r, so that $r = \underleftarrow{\lim}r_i$. Now, according to the context of Assumption 1 in Definition 8.19, we have the following result:

Lemma 8.8 *There exists a decomposition $\mathbb{E} = \ker r \oplus \mathbb{F}'$ with the following property:*

if (ν'_n) (resp. (μ_n) is the graduation on $\mathbb{F}'$ (resp. (μ_n) on $\mathbb{M}$) induced by the norm $|| \; ||^{\mathbb{E}_i}$ (resp. $|| \; ||^{\mathbb{M}_i}$) on $(E_i)_{x_i}$ (resp. $T_{x_i}M_i$), then the restriction of r to $\mathbb{F}'$ is a closed uniformly bounded operator according to these graduations.

Proof *At first, in such a context we have $T_{x_j}\delta_i^j \equiv d_i^j$ and the following compatibility condition:*

$$d_i^j \circ r_j = r_i \circ \ell_i^j. \tag{8.40}$$

We set $\mathbb{F}_i = r_i(\mathbb{E}_i)$ for all $i \in \mathbb{N}$. From Definition 8.19 assumption 1, $\mathbb{F}_i$ is a Banach subspace of $\mathbb{E}_i$ and there exists a decomposition $\mathbb{E}_i = \ker r_i \oplus \mathbb{F}'_i$. Thus the restriction r'_i of r_i to $\mathbb{F}'_i$ is an isomorphism onto $\mathbb{F}_i$. Now, from (8.40), we have

$$\forall (i,j) \in \mathbb{N}^2 : j > i, \; d_i^j \circ r'_j = r'_i \circ \ell_i^j.$$

But since each r'_i is an isomorphism, the restriction $(\ell_i^j)'$ of ℓ_i^j to $\mathbb{F}'_j$ takes values in $\mathbb{F}'_i$ for all $(i,j) \in \mathbb{N}^2$ such that $j \geq i$. Moreover, as δ_i^j is surjective, according to (8.40) again, this implies that $d_i^j(\mathbb{F}_j) = \mathbb{F}_i$ and so $\ell_i^j(\mathbb{F}'_j) = \mathbb{F}'_i$. Since $(\mathbb{E}_i, \ell_i^j)_{j \geq i}$ is a projective sequence, this implies that $(\mathbb{F}'_i, (\ell_i^j)')_{j \geq i}$ is a surjective projective system. The vector space $\mathbb{F}' = \varprojlim \mathbb{F}'_i$ is then a Fréchet subspace of $\mathbb{E}$.
On the other hand, let $(\ell_i^j)''$ be the restriction of ℓ_i^j to $\ker r_j$. Always from (8.40), we have

$$\forall (i,j) \in \mathbb{N}^2 : j > i, \; (\ell_i^j)''(\ker r_j) \subset \ker r_i.$$

By same argument, it implies that $(\ker r_i, (\ell_i^j)'')_{j \geq i}$ is a projective sequence and $\ker r = \varprojlim \ker r_i$. Moreover, since $\mathbb{E}_i = \ker r_i \oplus \mathbb{F}'_i$, it follows that $\mathbb{E} = \ker r \oplus \mathbb{F}'$ and also the restriction r' of r to $\mathbb{F}'$ is obtained as $r' = \varprojlim r'_i$ and r' is an injective continuous operator $r' : \mathbb{F}' \to \mathbb{M}$ whose range is the closed subspace $\mathbb{F} = \varprojlim \mathbb{F}_i$. It remains to show that r' is uniformly bounded.
According to the context of assumption 2 of Definition 8.19, there exists a constant $C > 0$ and, for each $i \in \mathbb{N}$, we have a norm $|| \; ||^{\mathbb{E}_i}$ on $\mathbb{E}_i$ and a norm $|| \; ||^{\mathbb{M}_i}$ on $\mathbb{M}_i$ such that $||r_i||_i^{\mathrm{op}} \leq C$. As $\mathbb{F}'_i$ is a closed Banach subspace of $\mathbb{E}_i$, it follows that for the induced norm on $\mathbb{F}'_i$ we have

$$\forall i \in \mathbb{N}, \; ||r'_i||_i^{\mathrm{op}} \leq C. \tag{8.41}$$

Set $\ell_i = \varprojlim \ell_i^j$ and $d_i = \varprojlim d_i^j$. By construction we have $\ell_i(\mathbb{F}') = \mathbb{F}'_i$ and $d_i(\mathbb{M}) = \mathbb{M}_i$. The norm $|| \; ||^{\mathbb{E}_i}$ on $\mathbb{E}_i$ induces a norm $|| \; ||^{\mathbb{F}'_i}$ on the Banach subspace $\mathbb{F}'_i$ and we get a natural graduation (ν'_i) on $\mathbb{F}'$ given by $\nu'_i(u) = ||\ell_i(u)||^{\mathbb{F}'_i}$ (cf. Appendix A (A.1)). In the same way, the norm $|| \; ||^{\mathbb{M}_i}$ induces a graduation (μ_i) on $\mathbb{M}$ given by $\mu_i(v) = ||d_i(v)||^{\mathbb{M}_i}$. Now (8.41) implies that r' is uniformly bounded and $r'(\mathbb{F}') = r(\mathbb{E}) = \Delta_x$ is closed by assumption. Therefore, the proof is complete.

Step 2: There exists a neighbourhood $V \subset U$ of x such that the map

$\rho' = \rho_{|U\times\mathbb{F}'}$ takes values in $\mathcal{IH}_b(\mathbb{F}',\mathbb{M})$ and is K-Lipschitz on V for some $K > 0$.

Since $E_i = U_i \times \mathbb{E}_i$, it follows that $E = U \times \mathbb{E}$ and so $\rho : E \to TU$ can be seen as a smooth map from U into $\mathcal{H}(\mathbb{E},\mathbb{M})$. Let ρ' be the restriction of ρ to $U \times \mathbb{F}'$ and so consider ρ' as a smooth map from U to $\mathcal{H}(\mathbb{F}',\mathbb{M})$. From the definition of a Finsler norm, Assumption 2 in Definition 8.19 and Lemma 8.8, the map $x \mapsto \rho'_x$ takes value in $\mathcal{H}_b(\mathbb{F}',\mathbb{M})$.
According to the notations used in the proof of Lemma 8.8, we have $\mathbb{F}' = \varprojlim \mathbb{F}'_i$ where each $\mathbb{F}'_i$ is a closed Banach subspace of $\mathbb{E}_i$. Let ρ'_i be the restriction of ρ to $U_i \times \mathbb{F}'_i$. Then we have $\rho'_i = \varprojlim \rho'_i$ and $\rho'_i = \delta_i \circ \rho \circ \lambda_i$ is a smooth map from U to $\mathrm{L}(\mathbb{F}'_i,\mathbb{M}_i)$. We denote by $||\ ||^{\mathrm{op}}_i$ is the operator norm on $\mathrm{L}(\mathbb{F}'_i,\mathbb{M}_i)$ associated with the norm induced by $||\ ||^{E_i}$ on $\mathbb{F}'_i$ and $||\ ||^{M_i}$. Now we will show:

Lemma 8.9 *There exists a neighbourhood $V_1 \subset V$ of 0 such that the map $\rho' : V_1 \to \mathcal{H}_b(\mathbb{F}',\mathbb{M})$ is Lipschitz, that is: there exists $K > 0$ such that*

$$\forall (z,z') \in V_1^2,\ \hat{\mu}_i^{op}(\rho'_z - \rho'_x) \leq K\hat{\mu}_i(z - z').$$

Proof *Let $(\hat{\nu'}_i)_{i\in\mathbb{N}}$ (resp. $(\hat{\mu}_i)_{i\in\mathbb{N}}$) be the canonical increasing graduation associated with the graduation $(\nu'_i)_{i\in\mathbb{N}}$ on $\mathbb{M}$ (resp. $(\hat{\mu}_i)_{i\in\mathbb{N}}$).*

Since the map $x \mapsto \rho'_x$ is a smooth map from U to $\mathcal{H}_b(\mathbb{F}',\mathbb{M})$ it follows that for each x the differential map $d_x\rho'$ is a continuous linear map from $\mathbb{M}$ to the Banach space $\mathcal{H}_b(\mathbb{F}',\mathbb{M})$ and so there exists $i_0 \in \mathbb{N}$ and a constant $A_x > 0$ such that

$$||d_x\rho'(u)||_\infty \leq A_x\, \hat{\nu'}_{i_0}(u) \tag{8.42}$$

for all $u \in \mathbb{F}'$ and so

$$||d_x\rho'(u)||_\infty \leq A_x\, \hat{\nu'}_i(u) \tag{8.43}$$

for all $u \in \mathbb{F}'$ and and $i \geq i_0$ and according to Remark 4.11 in Chapter 4, we set

$$||d_x\rho'||_i^{op} := \sup\{\hat{\mu}_i(d_x\rho'(u)) :\ \hat{\nu'}_i(u) \leq 1\} \leq A_x,\ \forall i \geq i_0 \tag{8.44}$$

On the other hand, for $1 \leq i < i_0$ we set $C_x^i = ||d_x\rho'||_i^{op}$. Then $d_x\rho'$ belongs to $\mathcal{H}_b(\mathbb{M}, \mathcal{H}_b(\mathbb{F}',\mathbb{M}))$ for all $x \in U$ since we have

$$||d_x\rho'||_\infty := \sup_{i\in\mathbb{N}} ||d_x\rho'||_i^{op} \leq \sup\{A_x, C_x^1, \cdots, C_x^{i_0-1}\}.$$

We set $C = ||d_0\rho'||_\infty$. By continuity, there exists an open neighbourhood V_1 of 0 such that

$$||d_x\rho'||_\infty \leq 2C \tag{8.45}$$

By choosing $K = 2C$, from the definition of $||d_x\rho'||_\infty$, that implies the announced result.

Step 3: Local flow of the vector field $X_u = \rho'(u)$.

Consider a neighbourhood V as announced in step 2. As δ_i is surjective, $V_i = \delta_i(V)$ is an open set of U_i and so we have $V = \varprojlim V_i$. For each $u \in \mathbb{F}'$, let $X_u = \rho'(u)$ be the vector field on V. If $u = \varprojlim u_i$ with $u_i \in \mathbb{F}'_i$, then $X_{u_i} = \rho'_i(u_i)$ is a vector field on V_i and $X_u = \varprojlim X_{u_i}$. Now since ρ' takes values in $\mathcal{IH}_b(\mathbb{F}', \mathbb{M})$, from our assumption, there exists a constant $C > 0$ such that $||(\rho'_i)_{z_i}||_i^{\text{op}} \leq C$ for all $z_i \in V_i$. Therefore, if u belongs to $\mathbb{F}'$ and $u_i = \lambda(u)$, then $\|X_{u_i}\|^{\mathbb{M}_i} \leq C\|u\|_i^{\mathbb{E}}$ and, from Lemma 8.9 and the definition of $\hat{\mu}_i^{\text{op}}$ we have

$$\forall\, (x_i,\, x'_i) \in (\delta_i(V))^2,\ \|X_{u_i}(x_i) - X_{u_i}(x'_i)\|^{\mathbb{M}_i} \leq K\|u_i\|^{\mathbb{E}_i}||x_i - x'_i||^{\mathbb{M}_i}. \quad (8.46)$$

Now, recall that we have provided $\mathbb{F}'$ and $\mathbb{M}$ with seminorms $(\hat{\nu'}_n)$ and $(\hat{\mu}_n)$ respectively defined by

$$\nu_n(u) = \|\lambda_i(u)\|^{\mathbb{E}_i} \text{ and } \mu_n(x) = \|\delta_i(x)\|^{\mathbb{M}_i}.$$

Since $\delta_i(X_u)(x) = X_{u_i}(\delta_i(x))$ and in this way, from (8.46), for all $n \in \mathbb{N}$, we have

$$\forall (x, x') \in V^2,\ \mu_n(X_u(x) - X_u(x')) \leq K\nu_n(u)\mu_n(x - x'). \quad (8.47)$$

Therefore, X_u satisfies the assumption of Corollary A.1. Let $\epsilon > 0$ such that the pseudo-ball

$$B_{\mathbb{M}}(0, 2\epsilon) = \{x \in \mathbb{M},\ :\ \hat{\mu}_{n_i}(x) < 2\epsilon, 1 \leq i \leq k\}$$

is contained in V and set

$$C_1 := \max_{1 \leq i \leq k} \{K\nu_{n_i}(u)\} = K \max_{1 \leq i \leq k} \nu_{n_i}(u);$$

$$C_2 := \sup_{x \in B_{\mathbb{M}}(0,\epsilon)} \left\{ \max_{1 \leq i \leq k} \mu_{n_i}(X(x)) \right\} \leq C \max_{1 \leq i \leq k} \nu_{n_i}(u).$$

By application of Corollary A.1, for any u such that $\max_{1 \leq i \leq k} \nu_{n_i}(u) \leq 1$, there exists $\alpha > 0$ such that

$$\alpha e^{2\alpha K} \leq \frac{\epsilon}{2C},$$

such that the local flow Fl_t^u is defined on the pseudo-ball $B_{\mathbb{M}}(0, \epsilon)$ for all $t \in [-\alpha, \alpha]$ for all u which satisfied the previous inequality

Note that for any $s \in \mathbb{R}$ we have $X_{su} = sX_u$. Therefore, from the classical properties of a flow of a vector field,

there exists $\eta > 0$ such that the local flow Fl_t^u is defined on $[-1, 1]$, for all u in the open pseudo-ball

$$B_{\mathbb{F}'}(0, \eta) := \left\{u \in \mathbb{F},\ :\ \ \nu'_{n_i}(u) \leq \eta,\ \ 1 \leq i \leq k\right\}.$$

(cf. for instance proof of Corollary 4.2 in [ChSt76]).

We set $B_{\mathbb{M}_i}(0, \epsilon) = \delta_i(B_{\mathbb{M}}(0, \epsilon))$. Then X_{u_i} is a vector field on $V_i = \delta_i(V)$ and

$\mathrm{Fl}_t^{u_i} := \delta_i \circ (\mathrm{Fl}_t^u) \circ \delta_i$ is the local flow of X_{u_i} which is defined on $B_{\mathbb{M}_i}(0,\epsilon)$ for all $t \in [-1,1]$ and $\mathrm{Fl}_t^{u_i}(x_i)$ belongs to V_i for all $x_i \in B_{\mathbb{M}_i}(0,\epsilon)$ and $t \in [-1,1]$ and, we have

$$\mathrm{Fl}_t^u = \varprojlim \mathrm{Fl}_t^{u_i}\,. \tag{8.48}$$

Step 4: Existence of an integral manifold.
Since $(E_i, \pi_i, U_i, \rho_i, [.,.]_i)$ is a Banach-Lie algebroid, the Lie bracket $[X_{u_i}, X_{u'_i}]$ is tangent to Δ_i and so by Definition 3.2 and Lemma 3.6 in [Pel12], we have

$$\forall t \in [-1,1],\ (T\,\mathrm{Fl}_t^{u_i})((\Delta_i)_{x_i}) = (\Delta_i)_{\mathrm{Fl}_t^{u_i}(x_i)}.$$

Therefore, according to the notations at the end of step 3, for each $i \in \mathbb{N}$, we set:

$$\forall u_i \in B_{\mathbb{F}'_i}(0,\eta) = \lambda_i(B_{\mathbb{F}'}(0,\eta)),\ \Phi_i(u_i) = (\mathrm{Fl}_1^{u_i})(0);$$
$$\forall u \in B_{\mathbb{F}'}(0,\eta),\ \Phi(u) = (\mathrm{Fl}_1^{u})(0);.$$

Lemma 8.10 $\Phi = \varprojlim \Phi_i$ *is smooth and there exists* $0 < \eta' \leq \eta$ *such that the restriction of* Φ *to* $B_{\mathbb{F}'}(0,\eta')$ *is injective and* $T_u\Phi$ *belongs to* $\mathcal{IH}_b(\mathbb{F}',\mathbb{M})$ *for all* $u \in B_{\mathbb{F}'}(0,\eta')$.

Recall that, for each $i \in \mathbb{N}$, $B_{\mathbb{F}'_i}(0,\eta'))$ is an open ball in $\mathbb{F}'_i$ and $\delta_i(V)$ is open neighbourhood of $0 \in \mathbb{M}_i$. From Lemma 8.10, since Φ is injective and each differential $T_u\Phi$ is injective, it follows that the same is true for each $\Phi_i : B_{\mathbb{F}_i}(0,\eta')) \to \delta_i(V)$. Thus we can apply the proof of Theorem 1 in [Pel12] for Φ_i. By the way, $\mathcal{U}_i = \Phi_i(B_{\mathbb{F}'_i}(0,\eta'))$ is an integral manifold of $\rho_i(\mathbb{F}'_i)$ and $(\mathcal{U}_i, \Phi_i^{-1})$ is a (global) chart for this integral manifold modeled on $\mathbb{F}'_i$. As $\Phi = \varprojlim \Phi_i$, $B_{\mathbb{F}'}(0,\eta') = \varprojlim B_{\mathbb{F}'_i}(0,\eta'))$ $\Phi_i \circ \lambda_i^j = \delta_i^j \circ \Phi_j$ for all $j \geq i$, it follows that $(\mathcal{U}_i, \delta_i^j)_{j \geq i}$ is a surjective projective sequence and so $\mathcal{U} = \varprojlim \mathcal{U}_i$ is a Fréchet manifold modeled on $\mathbb{F}'$. This last result clearly **ends the proof of Theorem 8.7**.

Proof *of Lemma 8.10. According to step 3,* Φ_i *is well defined on* $B_{\mathbb{F}'_i}(0,\eta)$ *and* $\Phi = \varprojlim \Phi_i$.
Now, for every $x \in V$, ρ'_x *belongs to* $\mathcal{IH}_b(\mathbb{F}',\mathbb{M})$ *and so is injective; after shrinking* V, *if necessary, there exists some* $M > 0$ *such that*

$$\forall x \in V,\ \|\rho'_x\|_\infty \leq M. \tag{8.49}$$

Now, by construction, we have $T_0\Phi(u) = \rho'_0(u)$ *and so* $T_0\Phi$ *is injective.*

Claim 8.1 *The map* $u \mapsto T_u\Phi$ *is a smooth map from* $B'(0,\eta)$ *to* $\mathcal{H}_b(\mathbb{F}',\mathbb{M})$.

Proof *of Claim 8.1*
We will use some argument of the proof of Lemma 2.12 of [Pel12]. We fix the index i *and, for any* $y \in B_{\mathbb{F}'_i}(0,\epsilon)$, $v \in \mathbb{F}'_i$ *we set*

$X_v(y) = \rho'_i(y, v)$
$\varphi(t, v) = \mathrm{Fl}_t^{X_v}(0)$
$A(t) = \partial_1 \rho'_i(\varphi(t, v), v)$ *(partial derivative relative to the first variable)*
$B(t) = \rho'_i(\varphi(t, v), .)$

Note that A and B are smooth fields on $[0, 1]$ of operators in $\mathcal{L}(\mathbb{M}_i, \mathbb{M}_i)$ and $\mathcal{L}(\mathbb{F}'_i, \mathbb{M}_i)$ respectively. Therefore the differential equation

$$\dot{S} = A \circ S + B$$

has a unique solution S_t with initial condition $S_0 = 0$ given by

$$S_t = \int_0^t G_{t-s} \circ B(s) ds \tag{8.50}$$

where G_t is the unique solution of

$$\dot{G} = A \circ G$$

with initial condition $G_0 = \mathrm{Id}_{\mathbb{M}}$. Given by

$$G_t = \mathrm{Id}_{\mathbb{M}} + \int_0^t A \circ G_s ds. \tag{8.51}$$

Under these notations, from [Die67], Chapter X, § 7, we have $\partial_2 \varphi(t, v)(.) = S_t$.

On the one hand, from the choice of ϵ, $\varphi(t, v)$ belongs to V_i and so from (8.45), we have $\|A(t)\|_i^{\mathrm{op}} \leq K$ for any $t \in [-1, 1]$ and from (8.49) we have $\|B(t)\|_i^{\mathrm{op}} \leq M$. Thus from (8.51), using Gronwal equality, we obtain

$$\|G_t\|_i^{\mathrm{op}} \leq e^K,$$

And so from (8.50) we obtain

$$\|S_t\|_i^{\mathrm{op}} \leq M e^K$$

This implies that

$$\|\partial_2 \varphi(t, v)\|_i^{\mathrm{op}} \leq M e^K.$$

We set $M_1 = M e^K$. Since $\Phi_i(u_i) = \varphi(1, u_i)$ it follows that:

$$\|T_{u_i} \Phi(v_i)\|^{\mathbb{M}_i} \leq M_1 \|v_i\|^{\mathbb{E}_i} \tag{8.52}$$

From (8.52) we obtain:

$$\mu_i(T_u \Phi(v)) \leq M_1 \nu_i(v). \tag{8.53}$$

But, since $T_u \Phi(\mathbb{F}') = \Delta_{\Phi(u)} \subset \{\Phi(u)\} \times \mathbb{M}$, it follows that the map $u \mapsto T_u \Phi$

can be considered as a continuous linear map from $\mathbb{F}'$ *to* $\mathbb{M}$ *which takes values in* $\mathcal{H}_b(\mathbb{F}', \mathbb{M})$. *Now as each* Φ_i *is a smooth map form* $B_{\mathbb{F}_i}(0, \eta'))$ *to* $\delta_i(V)$ *and* $\Phi = \varprojlim \Phi_i$, *this imply that* Φ *is a smooth map on* $B_{\mathbb{F}'}(0, \eta')) = \varprojlim B_{\mathbb{F}'_i}(0, \eta'))$ *to* $V = \varprojlim \delta_i(V)$ *which ends the proof of the Claim.*

End of the proof of Lemma 8.9.
At first, from Claim 8.1, the map $u \mapsto T_u\Phi$ *takes values in the Banach space* $\mathcal{H}_b(\mathbb{F}', \mathbb{M})$, *as in step 2 for* ρ, *we can show that this map is Lipschitz on* $B_{\mathbb{F}'}(0, \eta)$ *for* η *small enough. As* $T_0\Phi = \rho'_0$, *from Proposition 4.9, it follows that, again for* η *small enough,* $T\Phi$ *is injective on* $B_{\mathbb{F}}(0, \eta')$, *and we have (cf. (8.52))*

$$\forall u \in B_{\mathbb{F}}(0, \eta'),\ \forall v \in \mathbb{F}'\ \mu_i(T_u\Phi(v)) \leq M_1\nu_i(v) \tag{8.54}$$

using the fact that the range of $T_u\Phi$ *is always closed for* $u \in B_{\mathbb{F}'}(0, \eta')$. *Moreover, for* $u \in B_{\mathbb{F}'}(0, \eta')$, *as for the relation (4.8) in the proof of Proposition 4.9, we obtain:*

$$\frac{1}{\ell_u}\nu_i(v) \leq \mu_i(T_u\Phi(v))) \leq \ell_u.\nu_n(v) \tag{8.55}$$

for all $i \in \mathbb{N}$, *where* $\ell_u = \|T_u\Phi\|_\infty \leq M_1$.
Finally we obtain:

$$\forall i \in \mathbb{N},\ \frac{1}{M_1}\nu_i(v) \leq \mu_i(T_u\Phi(v)) \leq M_1.\nu_i(u) \tag{8.56}$$

Suppose that, for any $0 \leq \eta' \leq \eta$, *the restriction of each* Φ_i *to* $B_{\mathbb{F}'_i}(0, \eta)$ *is not injective. Consider any pair* $(u, v) \in [B_{\mathbb{F}'}(0, \eta)]^2$ *such that* $u \neq v$ *but* $\Phi(u) = \Phi(v)$, *we set* $h = v - u$. *For any* $\alpha \in \mathbb{M}^*$, *we consider the smooth curve* $c_\alpha : [0, 1] \to \mathbb{R}$ *defined by:*

$$c_\alpha(t) =< \alpha, (\Phi(u + th) - \Phi(u)) > .$$

Of course, we have $\dot{c}_\alpha(t) =< \alpha, T_{u+th}\Phi(h) >$.
Denote by $]u, v[$ *the set of points* $\{w = u + th, t \in]0, 1[\}$. *As we have* $c_\alpha(0) = c_\alpha(1) = 0$, *from Rolle's Theorem, there exists* $u_\alpha \in]u, v[$ *such that*

$$< \alpha, \delta_l(T_{u_\alpha}\Phi(h)) >= 0 \tag{8.57}$$

Note that, for any $t \in \mathbb{R}$, *this relation is also true for any* th. *From our assumption, it follows that, for each* $k \in \mathbb{N} \setminus \{0\}$, *there exists* u_k *and* v_k *in* $B_{\mathbb{F}}(0, \frac{\eta}{k})$ *so that* $u_k \neq v_k$ *but with* $\Phi(u_k) = \Phi(v_k)$. *So from the previous argument, for any* $\alpha \in \mathbb{M}^*$, *we have*

$$< \alpha, (T_{u_{\alpha,k}}\Phi(h_k)) >= 0 \tag{8.58}$$

for some $u_{\alpha,k} \in]u_k, v_k[$ *and* $h_k = v_k - u_k$. *From (8.58), for any* $i \in \mathbb{N}$, *any* $t \in \mathbb{R}$ *and any* $\alpha_i \in \mathbb{M}_i^*$, *if* $\alpha = \delta_i^*(\alpha_i)$, *we have:*

$$\begin{aligned} | < \alpha, T_0\Phi(th_k) > | &= | < \alpha_i, \delta_i([T_0\Phi - T_{u_\alpha}\Phi](th_k)) > | \\ &= | < \alpha_i, [T_0\Phi_i - T_{\delta_i(u_\alpha)}\Phi_i](\lambda_i(th_k))|. \end{aligned}$$

Denote by $\|.\|^{\mathbb{M}_i^*}$ *the canonical norm on* $\mathbb{M}_i^*$ *associated with* $\|.\|^{\mathbb{M}_i}$. *Then from (8.58) and since* $u \mapsto T_u\Phi$ *is* K_1*-Lipschitz on* $B'(0,\eta)$ *(for some constant* K_1*) we obtain:*

$$\begin{aligned} | < \alpha, T_0\Phi(th_k) > | &\leq (||\alpha_i||^{\mathbb{M}_i^*}.K_1.||\delta_i(u_\alpha)||^{\mathbb{M}_i}||.||\lambda_i(th_k)||^{\mathbb{E}_i} \\ &\leq (||\alpha_i||^{\mathbb{M}_i^*}.K_1.||\lambda_i(th_k)||^{\mathbb{E}_i}\frac{\eta}{k}. \end{aligned} \tag{8.59}$$

Since $u_k \neq v_k$, *there must exist at least one integer* $i \in \mathbb{N}$ *such that* $\lambda_i(h_k) \neq 0$. *Thus, by taking* $t = \dfrac{u_k - v_k}{\nu_i((u_k - v_k))}$ *in (8.59), we may assume* $t = 1$ *and* $\|\lambda_i(h_k)\|^{\mathbb{E}_i} = 1$ *In this way, for this choice of* h_k, *we have a* 1*-form* $\bar{\beta}_{k,i}$ *on the linear space generated by* $\lambda_i(h_k)$ *in* $\mathbb{F}_i'$ *such that* $< \bar{\beta}_{k,i}, \lambda_i(h_k) >= 1$ *and with* $(\|\bar{\beta}_{k,i}\|^{\mathbb{E}_i})^* = 1$. *From the Hahn-Banach Theorem, we can extend this linear form to a form* $\beta_{k,i} \in \mathbb{M}_i^*$ *such that* $< \beta_{k,i}, \lambda_i(h_k) >= 1$ *and* $(\|\beta_{k,i}\|^{\mathbb{E}_i})^* = 1$. *But since each* $T_0\Phi$ *is injective, this implies that* $T_0\Phi_i$ *is injective and so the adjoint* $T_0^*\Phi_i$ *is surjective. This implies that there exists* $\alpha_{k,i} \in \mathbb{M}_i^*$ *such that* $T_0^*\Phi_i(\alpha_{k,i}) = \beta_{k,i}$. *Thus from (8.59) we obtain*

$$\begin{aligned} 1 = | < \beta_{k,i}, \lambda_i(h_k) > | &= | < \alpha_{k,i}, T_0\Phi_i(\lambda_i(h_k)) > | \\ &= | < \delta_i^*\alpha_{k,i}, T_0\Phi(h_k) > | \leq \|\alpha_{k,i}\|^{\mathbb{M}_i^*} K_1\frac{\eta}{k} \end{aligned} \tag{8.60}$$

But, on the other hand since the operation "adjoint" is an isometry, from (8.56) we have

$$\frac{1}{M_1}\|\alpha_{k,i}\|^{\mathbb{M}_i^*} \leq \|T_0^*\Phi\alpha_{k,i}\|^{\mathbb{M}_i^*} = 1 \tag{8.61}$$

which gives a contradiction with (8.60) for k large enough.

8.6 A Criterion of Integrability for a Direct Limit of Banach Distributions

8.6.1 A Preliminary Proposition

We begin this section by the following Proposition (cf. [CaPe19]) which will be used in the proof of the theorem on the integrability of a distribution on a direct limit of Banach manifolds endowed with Koszul connections.

Proposition 8.4 *Let $(E, \pi, M, \rho, [.,.]_E)$ be a Banach algebroid. Assume that* $\ker \rho_x$ *is split and* $\mathcal{D}_x = \rho(E_x)$ *is closed in* T_xM*. Then* $\mathcal{D}$ *is an integrable weak distribution of* M*. If moreover there exists a linear connection on* E*, then there exists a non-linear connection on the tangent bundle of each leaf of the distribution* $\mathcal{D}$*.*

Proof *The first part is simply an application of Theorem 8.5. We consider a leaf* L *of* $\mathcal{D}$*. If* $\iota : L \to M$ *is the natural inclusion, it is a smooth closed immersion of* L *in* M*. Let* x *be any point of* L *and denote* K_x *the kernel of* $\rho_x : \pi^{-1}(x) = E_x \to T_xM$*. According to the assumption on* E*, we have a decomposition* $E_x = K_x \oplus F_x$*. The manifold* L *is a Banach manifold modelled on* $\mathbb{F} = F_x \equiv \mathcal{D}_x$*. Consider the pull back* E_L *of* E *over* L *via* $\iota : L \to M$*. We have a bundle morphism* $\hat{\iota}$ *from* E_L *in* E *over* ι *which is an isomorphism on each fibre. Therefore, the kernel of* $\hat{\rho} = \rho \circ \iota$ *is a Banach subbundle* K_L *of* E_L *and we have a subbundle* F_L *of* E_L *such that* $E_L = K_L \oplus F_L$*. In particular, we have an isomorphism* ρ_L *from* F_L *to* TL*. It follows that the tangent map* $T\rho_L : TF_L \to T(TL)$ *is also an isomorphism. On the other hand, according to the decomposition* $E_L = K_L \oplus F_L$*, we have also a decomposition* $TE_L = TK_L \oplus TF_L$*.*

Now, assume that there exists a non-linear connection on E *and let* $D : TE \to E$ *be the associated map connection. The map* $\hat{D}_L = D \circ \hat{\iota} \circ (T\rho_L)^{-1}$ *is smooth and maps the fibre of* $T_{(x,u)}(TL)$ *over* $(x, u) \in TL$ *into the fibre of* $(E_L)_x$ *over* $x \in L$*. As* $\hat{\iota}$ *is an isomorphism from* $(E_L)_x$ *to the fibre* E_x *over* $\iota(x)$*, it follows that* $\hat{D}_L$ *is a linear continuous map between these fibres. In particular, we can consider* $\hat{D}_L$ *as a map from* $T(TL)$ *into* E_L*. Now if* Π_L *is the projection of* E_L *on* F_L *parallel to* K_L*, the map* $D_L = \rho_L \circ \Pi_L \circ \hat{D}_L$ *defines a Koszul connection on* TL*.*

8.6.2 Criterion of Integrability for Local Direct Limits of Local Koszul Banach Bundles

Let M be a not necessarily Hausdorff convenient manifold and denote by TM its dynamical tangent bundle. A distribution Δ on M is again an assignment $\Delta : x \mapsto \Delta_x \subset T_xM$ on M where Δ_x is a subspace of T_xM. The notion of integrability and involutivity of a distribution recalled in § 8.1 can be clearly adapted to the convenient context.

We will now give a criterion of integrability for direct limits of local Koszul Banach bundles. More precisely we have:

Definition 8.20 *A distribution* Δ *on a n.n.H. convenient manifold* M *is called a local direct limit of local Koszul Banach bundles if the following property is satisfied:*

(*) *For any* $x \in M$*, there exists an open neighbourhood* U *of* x *and a strong ascending sequence of anchored Banach bundles* $(E_n, \pi_n, U_n, \rho_n)_{\underrightarrow{n}}$

endowed with a Koszul connection ∇^n such that $U = \underrightarrow{\lim} U_n$, $\underrightarrow{\lim} \rho_n(E_n) = \Delta_{|U}$ and such that E_n is a supplemented subbundle of E_{n+1}.

Remark 8.10 *In the context of paracompact finite-dimensional manifolds or Hilbert manifolds, the condition of the existence of a Koszul connection ∇^n and E_n supplemented in E_{n+1} are automatically satisfied.*

We then have the following criterion of integrability:

Theorem 8.12 *Let Δ be a local direct limit of local Koszul Banach bundles. Assume that in the property (*), there exists an almost Lie bracket $[.,.]_n$ on $(E_n, \pi_n, U_n, \rho_n)$ such that $(E_n, \pi_n, U_n, \rho_n, [.,.]_n)$ is a Banach algebroid and, over each point $y_n \in U_n$, the kernel of ρ_n is supplemented in the fibre $\pi_n^{-1}(y_n)$. Then the distribution Δ is integrable and the maximal integral manifold N through $x = \underrightarrow{\lim} x_n$ is a weak n.n.H. convenient submanifold of M which is a direct limit of the set of maximal leaves N_n of $\rho_n(E_n)$ through x_n in M_n. Moreover, each maximal leaf has the limit chart property at any point and if M is Hausdorff so is each leaf.*

Sketch of the proof *At first, for each $n \in \mathbb{N}$, we can apply the first part of Proposition 8.4. Therefore, with the notations of Property (*), if we fix some $x = \underrightarrow{\lim} x_n$, there exists a maximal integral manifold N_n of $\rho_n(E_n)$ through x_n in U_n. Recall that we have $U_n \subset U_{n+1}$ and $E_n \subset E_{n+1}$ over U_n. This implies that $N_n \subset N_{n+1}$. Since the kernel of ρ_i is split, for each $i \in \mathbb{N}$, we have a Whitney decomposition*

$$E_{i|N_i} = F_i \oplus (\ker \rho_i)_{|N_i}.$$

This implies that

$$(TN_{n+1})_{|N_n} = TN_n \oplus \rho_{n+1}(F_{n+1} \cap (\ker \rho_n)_{|N_n}).$$

Now, from Property (*) and the second part of Proposition 8.4, we have a linear connection on TN_n. Thus the ascending sequence $(N_n)_{n\in\mathbb{N}}$ satisfies the assumption of Theorem 5.8. This implies that $N = \underrightarrow{\lim} N_n$ has a structure of convenient manifold and we have $TN = \Delta_{|N}$. This means that Δ is an integral manifold of Δ through x. Moreover, N satisfies the direct limit chart property. From this construction, it follows that N is the maximal integral manifold of Δ through x.

From Theorem 8.12, one can easily obtain:

Corollary 8.2 *Let $(E_n, \pi_n, M_n, \rho_n, [.,.]_n)_{\underrightarrow{n}}$ be a strong ascending sequence of Banach algebroids provided with a Koszul connection on each E_n such that, over each point $x_n \in M_n$, the kernel of ρ_n is supplemented in the fibre $\pi_n^{-1}(x_n)$. Then $\Delta = \underrightarrow{\lim} \rho_n(E_n)$ is an integrable distribution on $M = \underrightarrow{\lim} M_n$. Moreover, for any $x = \underrightarrow{\lim} x_n$, the maximal leaf through x is a weak n.n.H. convenient* *submanifold of M and there exists a leaf N_n of $\rho_n(E_n)$ in M_n through x_n such that the sequence $(N_n)_{n\in\mathbb{N}}$ is an ascending sequence of Banach manifolds whose direct limit $N = \underrightarrow{\lim} N_n$ is an integral manifold of Δ through x such that N has the direct limit chart property at x. Moreover, if M is Hausdorff, so is each leaf.*

Now according to Remark 8.10, we also easily obtain:

Corollary 8.3 *Let Δ be a distribution on a direct limit $M = \underrightarrow{\lim} M_n$ of finite-dimensional (resp. Hilbert) paracompact manifolds. Assume that, for any $x = \underrightarrow{\lim} x_n$, there exists a sequence of finite rank (resp. Hilbert) algebroids $(E_n, \pi_n, U_n, \rho_n)_{\underrightarrow{n}}$ such that $U = \underrightarrow{\lim} U_n$, $\underrightarrow{\lim} \rho_n(E_n) = \Delta_{|U}$. Then Δ is integrable and the maximal integral manifold N through $x = \underrightarrow{\lim} x_n$ is a weak convenient submanifold of M which is the direct limit of the set of maximal leaves N_n of $\rho_n(E_N)$ through x_n in M_n. Moreover, each maximal leaf has the limit chart property at any point and is a Hausdorff convenient manifold.*

Remark 8.11 *According to the proof of Theorem 5.18 in Chapter 4, under the assumption of Corollary 8.2 or Corollary 8.3, the direct limit $\left(\underrightarrow{\lim} E_n, \underrightarrow{\lim} \pi_n, \underrightarrow{\lim} M_n, \underrightarrow{\lim} \rho_n, \underrightarrow{\lim} [.,.]_n\right)$* **is not** *a convenient Lie algeboid in general, since $\underrightarrow{\lim} [.,.]_n$ does not define a Lie bracket on all (local) sections of $\underrightarrow{\lim} E_n$ but only on direct limits of sections.*

8.6.3 Application

Consider a direct sequence of Banach-Lie groups $G_0 \subset G_1 \subset \cdots \subset G_n \subset \cdots$ such that the Lie algebra $\mathcal{G}_n$ is supplemented in the Lie algebra $\mathcal{G}_{n+1}$ for all $n \in \mathbb{N}$.
Note that this situation always occurs if for all $n \in \mathbb{N}$, each Lie group G_n is finite-dimensional or is a Hilbert Lie group. This assumption is also valid for the sequence $G_n = \mathrm{GL}(E_n)$ where $(E_n)_{n\in\mathbb{N}}$ is a direct sequence of Banach spaces such that each E_n is closed and supplemented in E_{n+1}.

Assume that for each $n \in \mathbb{N}$, we have a smooth right action $\psi_n : M_n \times G_n \longrightarrow M_n$ of G_n over a Banach manifold M_n where $M_0 \subset M_1 \subset \cdots \subset M_n \subset \cdots$ is an ascending sequence such that M_n is a Banach closed submanifold of M_{n+1}. We get a strong ascending sequence of Lie Banach algebroids $(M_n \times \mathcal{G}_n, \pi_n, M_n, \Psi_n, [.,.]_{\mathcal{G}_n})_{\underrightarrow{n}}$. Since each Banach bundle $M_n \times \mathcal{G}_n$ is trivial, we obtain a sequence of Banach-Lie algebroids with anchors $\begin{array}{rccc} \Psi_n : & M_n \times \mathcal{G}_n & \longrightarrow & TM_n \\ & (x_n, X_n) & \mapsto & T_{(x_n,e_n)}\psi_n(0, X_n) \end{array}$. Because these bundles are trivial, we get a sequence of compatible trivial Koszul connections ∇^n on $M_n \times \mathcal{G}_n$. Now, from Corollary 8.2, we obtain:

Theorem 8.13 *In the previous context, we obtain a smooth right action* $\psi = \underrightarrow{\lim}\psi_n$ *of* $G = \underrightarrow{\lim}G_n$ *on the convenient manifold* $M = \underrightarrow{\lim}M_n$*. Moreover, if the kernel of* Ψ_n *is supplemented in each fibre* $\pi_n^{-1}(x_n)$*, then the orbit* $\psi(x, G)$ *of this action through* $x = \underrightarrow{\lim}x_n$ *is a weak n.n.H. convenient submanifold of* M *which is the direct limit of the set of* G_n*-orbits* $\{\psi_n(x_n, G_n)\}_{n\in\mathbb{N}}$*. If* M *is Hausdorff, so is each orbit.*

8.7 Almost Symplectic Foliation on a Direct Limit of Poisson Manifolds

The reader is referred to Chapter 7 for notations. The following result can be found with its complete proof in [PeCa19].

Theorem 8.14 *Let* $(M_i, \mathfrak{A}(M_i), \{.,.\}_{P_i})_{i\in\mathbb{N}}$ *be a direct sequence of partial Poisson-Banach manifolds. Assume that, for each* $i \in \mathbb{N}$*, the following assumptions are satisfied:*

(i) there exists a Koszul connection on each $T^\flat M_i$*;*

(ii) over each point $x \in M_i$*, the kernel of* P_i *is supplemented in the fibre* ${p_i^\flat}^{-1}(x)$ *and the distribution* $P(T^\flat M_i)$ *is closed;*

(iii) there exists $j_i \geq i$ *such that, for any* $x \in M$*, we have for* $j \geq j_i$

$$P_j(T_x^\flat M_j) = P_{j_i}(T_x^\flat M_{j_i}).$$

Then we have:

(1) Each distribution $P_i(T^\flat M_i)$ *on* M_i *is integrable and the direct limit* $\Delta = \underrightarrow{\lim}P_i(T^\flat M_i)$ *is also an integrable distribution on* $M = \underrightarrow{\lim}(M_i)$*.*

(2) For any $x = \underrightarrow{\lim}(x_i)$*, the maximal leaf though* x *is a weak convenient manifold of* M *and there exists a leaf* N_i *of* $P_i(T^\flat M_i)$ *in* M_i *through* x_i*, such that the sequence* $(N_i)_{i\in\mathbb{N}}$ *is an ascending sequence of Banach manifolds whose direct limit* $N = \underrightarrow{\lim}(N_i)$ *is an integral manifold of* Δ *though* x*.*

(3) The natural almost symplectic structure (N, ω_N) *on a leaf* N *is such that*

$$\omega_N = \underleftarrow{\lim}(\omega_{N_i})$$

Note that the condition (iii) is always satisfied in the following cases:

- each manifold M_i is finite-dimensional;
- each bundle $T^\flat M_i$ has a finite-dimensional fibre;
- each morphism P_i has finite rank.

Sketch of Proof *We will use the notations and partial results of Chapter 7.*
At first, from Definition 7.17 (i) and notations in Proof of Proposition 7.11 (ii), since for any $i \leq j$, $M_i \subset M_j$ *we have*

$$\Delta_i^j = P_j((T^\flat_{M_i} M_j) \subset P_i(T^\flat M_i) = \Delta_i. \tag{8.62}$$

Moreover, Δ_i^j *is integrable distribution on* M_i.
For i *fixed, on* M_i, *we have a decreasing sequence of smooth distributions*[6]

$$\Delta_i = \Delta_i^i \supset \cdots \supset \Delta_i^j.$$

Note that since $\varepsilon_i^j : M_i \to M_j$ *is the inclusion and* $T^\flat_{M_i} M_j \subset T^*_{M_i} M_j$, *we have* $\varepsilon_i^j(x) = x$ *and* $T_x^* \varepsilon_i^j(\xi) = \xi_{|T_x M_i}$ *for all* $x \in M_i$ *and* $\xi \in T_x^\flat M_j$. *Therefore from Hahn-Banach theorem* $T_x^* \varepsilon_i^j$ *is surjective.*
On the one hand, since $T_x M_i \subset T_x M_{i+1} \subset \cdots \subset T_x M_j \subset \cdots \subset T_x M$, *we can choose a norm* $|| \; ||_j$ *on* $T_x M_j$ *for all* $i \leq j$ *such that* $|| \; ||_{j+1} \leq || \; ||_j$ *for all* $i \leq j$*; in particular, the operator norm of* $T_x \varepsilon_i^j$ *is bounded by* 1. *We then obtain a canonical norm* $|| \; ||_j^*$ *on* $T_x^* M$ *and so the operator norm of* $T_x^* \varepsilon_i^j$ *is bounded by* 1 *for all* $i \leq j$. *According to Property (ii) in Definition7.17 (1), all these considerations imply that the operator norm of* P_j *is bounded by the operator norm of* P_i *for* $i \leq j$. *On the other hand, if we consider the Banach quotient space* $T'_x M_j / \ker(P_j)_x$, *and so each vector space* $(\Delta_i^j)_x$ *has its own Banach space structure which is isomorphic to* $T_x^\flat M_j / \ker(P_j)_x$. *It follows that the space* $(\bar{\Delta}_i)_x$ *is provided with a Fréchet structure induced by the sequence of Banach spaces* $\{(\Delta_i^j)_x\}_{i \leq j}$. *In fact,* $(\bar{\Delta}_i)_x$ *is a Banach space.*

Fix some $x_0 \in M_i$ *and denote by* N_i^j *the maximal leaf of* Δ_i^j *through* x_0. *We then have the following sequence of weak Banach submanifolds modelled on the previous Banach structure on* $(\Delta_i^j)_{x_0}$ *and we have*

$$N_i^i \supset \cdots \supset N_i^j \supset \cdots$$

Set $\bar{N}_i = \bigcap_{i \leq j} N_i^j$. *We will show that* $\bar{N}_i$ *is a Banach manifold modelled on the Banach space* $(\bar{\Delta}_i)_x$.

Fix $x \in \bar{N}_i \subset N_i^j$. *Note that if* $(\bar{\Delta}_i)_x = \{0\}$ *then* $\bar{N}_i = \{x\}$ *and we have nothing to prove. From now on, we assume that* $\dim(\Delta_i^j)_x > 0$.

Consider a chart $(U = \underrightarrow{\lim}(U_j), \phi = \underrightarrow{\lim}(\phi_i))$ *around* x *in* M *which satisfies Property (iii) of Definition 7.17. Recall that for each* j, *if* $V_j = \phi_j(U_j)$, *we have a trivialization* $T^* \phi_j^{-1} : V_j \times \mathbb{F}_j \to T'_{U_j} M_j$. *Since* $U_i \subset U_j$ *for* $i \leq j$, *we get a trivialization* $\Theta_j : U_i \times \mathbb{F}_j \to T^\flat_{U_i} M_j$. *Now, for the anchored bundle*

[6] Recall that a distribution Δ' is contained in a distribution Δ on M if, for any $x \in M$, $\Delta'_x \subset \Delta_x$.

$(T^\flat_{M_i} M_j, M_i, P_j)$ we have $T^\flat_x M_j = \ker(P_i)_x \oplus \mathbb{S}_j$, then the vector fields $X_j(\alpha) = P_j \circ \Theta_j(.,\alpha)$ belong to $\widehat{\mathcal{M}}_{P_j}(M_i)$ for all $\alpha \in T^\flat_x M_j$. There exists a ball $B_j(0, r_j)$ in $\mathbb{S}_j$ such that

- *the map $\Phi_j(\alpha) = \phi_1^{X_j(\alpha)}(x)$ is defined for $\alpha \in B_j(0, r_j)$;*
- *there exists $0 < \delta_j \leq r_j$ such that $\Phi_j : B_j(0, \delta_j) \to M_i$ is a weak injective immersion;*
- *$\Phi_j(B_j(0, \delta_j))$ is an integral manifold of Δ_i^j through x.*

Note that, in particular, $\Phi_j(B(0_j, \delta_j))$ is an open set in N_i^j. Now from the previous choice of $(U_j, \phi_j)_{j\in\mathbb{N}}$, on U_i, for all $\alpha \in \mathbb{S}_j$, we have:

$$T^*\varepsilon_i^j \circ \Theta_j(.,\alpha) = \Theta_i(., T_x^*\varepsilon_i^j(\alpha))$$

and so we get

$$X_j(\alpha) = P_j \circ \Theta_j(.,\alpha) = T\varepsilon_i^j \circ P_i \circ T^*\varepsilon_i^j(\Theta_j(.,\alpha)) = T\varepsilon_i^j \circ X_i(T_x^*\varepsilon_i^j(\alpha)).$$

Therefore $X_j(\alpha)$ is tangent to Δ_i^j and so we have:

$$\forall \alpha \in B_j(0, r_j) \subset \mathbb{S}_j, \quad \Phi_j(\alpha) = \Phi_i(T_x^*\varepsilon_i^j(\alpha)). \tag{8.63}$$

But we have $T_x^*\varepsilon_i^j(T_x^\flat M_j) \subset T_x^\flat M_i$ and $T_0\Phi_j$ is an isomorphism from $\mathbb{S}_j$ onto $(\Delta_i^j)_x$. According to (8.63), it follows that $T_x^*\varepsilon^j_{i\,|\mathbb{S}_j}$ is a continuous injective linear map into $\mathbb{S}_i$. Again according to (8.63), we deduce that $r_j \geq r_i$ and $\delta_i \leq \delta_j$. For $i \leq j$, we set $\mathbb{S}_i^j = T^*\varepsilon_i^j(\mathbb{S}_j) \subset \mathbb{S}_i$, $\bar{B}_i^j(0, \delta_i) = T^*\varepsilon_i^j(B_j(0, \delta_j)) \cap B_i(0, \delta_i)$, $W_j = \Phi_j(B(0, \delta_j)$ and $W_i^j = W_j \cap W_i$. Then, from the previous considerations, $\mathbb{S}_i^j$ is isomorphic to $(\Delta_i^j)_x$, W_i^j is an open set in N_i^j around x and $(W_i^j, (\Phi_{i|B_i^j(0,\delta_i)})^{-1})$ is a chart for N_i^j around x. We equip $\mathbb{S}_i^j$ with the structure of Banach space such that $T_x^*\varepsilon^j_{i\,|\mathbb{S}_j}$ is an isometry when we endow $\mathbb{S}_j$ with the norm $\|\ \|_j^*$ induced from the norm $\|\ \|_j^*$ previously defined on $T_x^* M$. Then $\bar{\mathbb{S}}_i = \bigcap_{i\leq j} \mathbb{S}_i^j$ is then provided with a Banach structure. Then, from our previous construction, according to [Gal96], we obtain a Banach manifold structure on $\bar{N}_i$ modelled on $\bar{\mathbb{S}}_i$.

8.8 Notes

In finite dimension, the concept of (regular) foliation first appeared as a distinct field until in the 1940's and and really emerged from the works of Ehresmann ([Ehr51]) and Reeb ([Ree52]). The theory of foliations has now become a rich geometric subject by itself and has been marked by significant works such as Haefliger ([Hae56]), Novikov ([Nov64]), Thurston ([Thu74]), Molino ([Mol88]), Connes ([Con94]) and a long list of other authors. For a complete history and bibliography for regular foliations in finite dimension, the reader can see for instance [Hae05].

Motivated by control theory, the papers of Stefan ([Ste74]) and Sussmann ([Sus73]) on the integrability of a family of vector fields allow to define the notion of singular foliation. The characteristic foliation of a Poisson manifolds by symplectic leaves corresponds exactly to such a singular foliation (cf. [Kir76], [Lic77], [Wei83] and [Daz85]).

Note that in [Lav18], an interested reader will find a complete historical panorama on the integrability theorems of distributions on a finite-dimensional manifold.

A Frobenius Theorem in infinite dimension can be found in the context of the convenient setting ([Tei01]) and on locally convex manifolds ([Hil2000] and [Eyn14]). About first historical results on integrability of distributions on Banach manifolds, see [ChSt76], [Nub92] as mentioned in § 8.1.

A

Fréchet Spaces

In functional analysis, Fréchet spaces, named after the French mathematician Maurice Fréchet, are locally convex spaces which are complete with respect to a translation-invariant metric. The representation of Fréchet spaces as projective limits of Banach spaces is very interesting: issues arising in the Fréchet framework can be solved by considering their components in the Banach factors of the associated projective sequence. So different pathological entities in the Fréchet framework can be replaced by approximations compatible with the inverse limits, e.g. ILB-Lie groups ([Omo97]) or projective limits of Banach-Lie groups ([Gal96]), manifolds ([AbbMa99]), bundles ([Gal98], [AgSu10]), algebroids ([Cab12]), connections and differential equations ([ADGS09], [AgSu07]).

A.1 The Topology of Fréchet Spaces

Definition A.1 *A Fréchet space is a topological vector space which is locally convex, Hausdorff, metrizable and complete.*

The origin of a Fréchet space has a countable fundamental system of convex neighbourhoods: the balls of center 0 and radius $\dfrac{1}{n}$ for $n \in \mathbb{N}^*$.

Proposition A.1 *Let $\mathbb{F}$ be a Fréchet space. There exists a separating sequence of seminorms $(\nu_n)_{n\in\mathbb{N}}$ such that the topology of $\mathbb{F}$ is compatible with the metric d defined by $(\nu_n)_{n\in\mathbb{N}}$*

$$d(x,y) = \max_{n\in\mathbb{N}} \frac{1}{2^n} \frac{\nu_n(x-y)}{1+\nu_n(x-y)}.$$

Some basic examples of Fréchet spaces are listed below.

Example A.1 *Let $\mathbb{R}^{\mathbb{N}}$ be the vector space of real sequences $(a_i)_{i\in\mathbb{N}}$. Equipped with the sequence of semi norms (ν_n) where*

$$\nu_n\left((a_i)_{i\in\mathbb{N}}\right) = \sum_{i=0}^{n} |a_i|$$

$\mathbb{R}^{\mathbb{N}}$ is a Fréchet space.

DOI: 10.1201/9781003435587-A

Example A.2 *Let Ω be an open set of $\mathbb{R}^n$. We consider the space $\mathcal{C}(\Omega)$ of continuous mappings from Ω to $\mathbb{R}$. Let $(K_n)_{n\in\mathbb{N}}$ be an exhaustive sequence of compact subsets of Ω. For any $n \in \mathbb{N}$, we define the semi norm ν_n by*

$$\nu_n(f) = \sup_{x\in K_n} |f(x)|.$$

The sequence $(\nu_n)_{n\in\mathbb{N}}$ is a separating family of semi-norms which en $\mathcal{C}(\Omega)$ with a Fréchet structure, independent of the choice of the sequence $(K_n)_{n\in\mathbb{N}}$.

Any Fréchet space $\mathbb{F}$ can be realized as the limit of a surjective projective sequence of Banach spaces $(\mathbb{B}_n, \lambda_n^m)_{m\geq n}$. Following [DGV16], 2.3, we can identify $\mathbb{F}$ with the projective limit of the projective sequence

$$\left(\widehat{\mathbb{B}}_n = \{x = (x_i) \in \prod_{0\leq i\leq n} \mathbb{B}_i : \forall j \geq i, x_i = \lambda_i^j(x_j)\},\ \widehat{\lambda}_n^m = (\lambda_n^n, \dots, \lambda_n^m)\right)_{m\geq n}.$$

We denote by $\lambda_n : \mathbb{F} \to \mathbb{B}_n$ and $\widehat{\lambda}_n : \mathbb{F} \to \widehat{\mathbb{B}}_n$ the canonical surjective projections. Let $(|| \ ||_n)_{n\in\mathbb{N}}$ be a sequence where $|| \ ||_n$ is a norm on $\mathbb{B}_n$. In this way,

$$\widehat{|| \ ||}_n = \sup_{0\leq i\leq n} || \ ||_i$$

defines a norm on $\widehat{\mathbb{B}}_n$. Then

$$\hat{\nu}_n = \widehat{|| \ ||}_n \circ \widehat{\lambda}_n \text{ (resp. } \nu_n = || \ ||_n \circ \lambda_n) \tag{A.1}$$

is the semi-norm on $\mathbb{F}$ associated with the sequence $(\widehat{|| \ ||}_n)$ (resp.$(|| \ ||_n)$). Moreover, we have $\hat{\nu}_n = \max_{0\leq i\leq n} \nu_i$ and the topology of $\mathbb{F}$ is defined by $(\hat{\nu}_n)$ or (ν_n).

Definition A.2 *Let $\mathbb{F}$ be a Fréchet space and $(\nu_n)_{n\in\mathbb{N}}$ a sequence of semi-norms defining its topology.*

1. *An open pseudo-ball $B(a,r)$ in $\mathbb{F}$ is a set of type*

$$\{x \in \mathbb{F} : \nu_{n_i}(x-a) < r, \text{ for a finite sequence } n_1, \dots, n_k\}$$

2. *A closed pseudo-ball $\overline{B}(x,r)$ in $\mathbb{F}$ is a set of type*

$$\{x \in \mathbb{F} : \nu_{n_i}(x-a) \leq r, \text{ for a finite sequence } n_1, \dots, n_k\}$$

In fact, since the sequence of semi-norms $(\hat{\nu}_n)$ is increasing, such a pseudo-ball is

$$\{x \in \mathbb{F}, \text{ such that } \hat{\nu}_{n_k}(x-a) < r, \}.$$

Thus any pseudo-ball $B(a,r)$ in $\mathbb{F}$ is defined by

$$\{x \in \mathbb{F}, \text{ such that } \hat{\nu}_l(x-a) < r, \text{ for some } l \in \mathbb{N}\}. \tag{A.2}$$

The set of all such pseudo-balls generates the topology of $\mathbb{F}$.
If U be an open set of $\mathbb{F}$, we set $U_n = \lambda_n(U)$. Since λ_n is surjective, U_n is an open subset of $\mathbb{B}_n$ and we have $U = \varprojlim U_n = \varprojlim \widehat{U}_n$ if $\widehat{U}_n = \prod_{i=0}^{n} U_i$. Since U is open, for each $a = \varprojlim a_n = \varprojlim (a_0, \ldots, a_n)$ with $a_n \in \mathbb{B}_n$ and $(a_0, \ldots, a_n) \in \widehat{\mathbb{B}}_n$, we have an open pseudo-ball such $B(a, r) \subset U$.

$$\begin{aligned} \widehat{B(a,r)}_n &= \widehat{\lambda}_n(B(a,r)) \\ &= \left\{ (x_0, \ldots, x_n) \in \widehat{\mathbb{B}}_n, : ||x_i - a_i||_j < r, \ j \in \{n_i \ : n_i \leq n\} \right\} \end{aligned}$$

which is the trace of an open ball of $\prod_{i=0}^{n} \mathbb{B}_i$ on $\widehat{\mathbb{B}}_n$ and so is an open ball in $\widehat{\mathbb{B}}_n$. It follows that $\widehat{U}_n$ is an open subset of $\widehat{\mathbb{B}}_n$ and, by construction, $B(a, r) = \varprojlim \widehat{B(a,r)}_n$. Now since the linear projection $\pi_n : (x_0, \ldots x_n) \mapsto x_n$ from $\widehat{\mathbb{B}}_n$ to $\mathbb{B}_n$ is surjective, $B(x_0, r)_n = \pi_n(\widehat{B(a,r)}_n)$ is also an open ball in $\mathbb{B}_n$ and we also have $B(a, r) = \varprojlim B(a, r)_n$.

By analog arguments, $\widehat{\overline{B}(a,r)}_n = \widehat{\lambda}_n(\overline{B}(a, r))$ is a closed ball in $\widehat{\mathbb{B}}_n$ and $\overline{B}(a, r) = \varprojlim \widehat{\overline{B}(a,r)}_n = \varprojlim \overline{B}(a, r)_n$ where $\overline{B}(a, r)_n = \pi_n(\widehat{\overline{B}(a,r)})$.

As in the Banach framework, every closed subspace of a Fréchet space is also a Fréchet space.

The Hahn-Banach theorem and the open mapping theorem are still true in a Fréchet space.

The continuity of a linear map between Fréchet spaces $(\mathbb{F}_1, (p_n))$ and $(\mathbb{F}_2, (q_n))$ is checked using seminorms: A linear map $f : \mathbb{F}_1 \to \mathbb{F}_2$ is continuous if and only if, for every seminorm q_n of $\mathbb{F}_2$, there exists a seminorm p_m in $\mathbb{F}_1$ and a positive real M such that

$$\forall x \in \mathbb{F}_1, \ q_n\left(f(x)\right) \leq M \, p_m(x).$$

Remark A.1 *The dual of a Fréchet space that is not a Banach space is not a Fréchet space.*

Example A.3 *The dual of the Fréchet space* $\mathbb{R}^{\mathbb{N}}$ *is the convenient space* $\mathbb{R}^{\infty}$ *(cf. 3.10).*

The inverse function theorem is not valid in general. It becomes true in the tame Fréchet spaces framework where the tame notion is defined thanks to graduations on such spaces.

Definition A.3 *A graduation on a Fréchet space $\mathbb{F}$ is the datum of a sequence of seminorms $(\nu_n)_{n\in\mathbb{N}}$ of $\mathbb{F}$. Then $(\mathbb{F},(\nu_n)_{n\in\mathbb{N}})$ is called a graded Fréchet space. Two graduations $(\nu_n)_{n\in\mathbb{N}}$ and $(\nu'_n)_{n\in\mathbb{N}}$ are equivalent if they define the same topology on $\mathbb{F}$.*

Remark A.2 *Consider a graded Fréchet space $(F,(\nu_n)_{n\in\mathbb{N}})$. For each $n \in \mathbb{N}$, we set $\hat{\nu}_n = \sum_{i=0}^{n} \nu_n$. Then $(\hat{\nu}_n)_{n\in\mathbb{N}})$ is an increasing graduation which is equivalent to the given one. Thus, for a graded Fréchet space $(F,(\nu_n)_{n\in\mathbb{N}})$, we can always assume that the graduation $(\nu_n)_{n\in\mathbb{N}}$ is an increasing sequence.*

Definition A.4 *Let (ν_n) and (ν'_n) be two increasing graduations on $\mathbb{F}$. We say that (ν_n) and (ν'_n) are uniformly equivalent if there exists constants $C > 0$ and $C' > 0$, pairs of integers r, r' and n_0, n'_0 such that*

$$\forall n \geq n_0,\ \nu_n \leq C\nu'_{n+r}$$
$$\forall n \geq n'_0,\ \nu'_n \leq C'\nu_{n+r'}.$$

Fix some graded Fréchet space $(\mathbb{F},(\nu_n)_{n\in\mathbb{N}})$. If we denote by $\ker\nu_n$ the closed subspace $\nu_n^{-1}(0)$, then the quotient space $\mathbb{F}/\ker\nu_n$ is a normed space provided with the quotient norm $\tilde{\nu}_n$ by projection of ν_n. The Banach completion $\mathbb{B}^n$ of this normed space is called the *local Banach space associated with the semi-norm* ν_n. For $m \geq n$, we have a canonical projection $\delta_n^m : \mathbb{B}^m \to \mathbb{B}^n$ and $(\mathbb{B}^n, \delta_n^m)_{m\geq n}$ is a projective sequence of Banach spaces whose projective limit $\varprojlim \mathbb{B}^n$ can be identified with $\mathbb{F}$.

Lemma A.1 *Let $\mathbb{F}$ be a Fréchet space.*

(1) Let $(\nu_n)_{n\in\mathbb{N}}$ be a graduation on $\mathbb{F}$. Then the set

$$\mathbb{F}_b = \{x \in \mathbb{F} \ :\ \sup_{n\in\mathbb{N}} \nu_n(x) < +\infty\}$$

is a Banach subspace whose topology is the induced topology defined by $(\nu_n)_{n\in\mathbb{N}}$.

(2) Let $(\hat{\nu}_n)_{n\in\mathbb{N}}$ be the increasing graduation associated with $(\nu_n)_{n\in\mathbb{N}}$. Consider the space

$$\hat{\mathbb{F}}_b = \{x \in \mathbb{F} \ :\ \sup_{n\in\mathbb{N}} \hat{\nu}_n(x) < +\infty\}.$$

Then we have $\hat{\mathbb{F}}_b = \mathbb{F}_b$ and $\mathbb{F}_b$ only depends on the uniformly equivalence class of the increasing graduations $(\hat{\nu}_n)_{n\in\mathbb{N}}$ associated with $(\nu_n)_{n\in\mathbb{N}}$.

Definition A.5 *The space $\mathbb{F}_b$ defined in Lemma A.1 is called the bounded Banach subspace of $\mathbb{F}$.*

Remark A.3 *In [Ham82], the author defines an equivalence relation between graduations, called* tame equivalence *which is of type (A.3), with $C = C'$, $r = r'$ and $n_0 = n'_0$ but where C depends on n. In fact, two graduations which are tamely equivalent* graduations tamely equivalent *on $\mathbb{F}$ do not give rise to the same Banach spaces as show the following example: We consider*

$$\mathbb{F} := \mathcal{C}(\mathbb{R}) = \{f : \mathbb{R} \to \mathbb{R} \ : \ f \text{ is continuous } \}$$

endowed with the Fréchet topology of uniform convergence on the compact subsets of $\mathbb{R}$.
For every integer $n \geq 1$ define

$$\begin{array}{rccc} \nu_n: & \mathbb{F} & \to & \mathbb{R}_+ \\ & f & \mapsto & \sup_{t\in[-n,n]} |f(t)| \end{array}$$

and

$$\begin{array}{rccc} \mu_n: & \mathbb{F} & \to & \mathbb{R}_+ \\ & f & \mapsto & n\nu_n(f) \end{array}$$

Then one has $\nu_1 \leq \nu_2 \leq \cdots$ and $\mu_1 \leq \mu_2 \leq \cdots$, and the topology of $\mathbb{F}$ can be defined by any of the sequences of seminorms $\{\nu_n \ : \ n \geq 1\}$ and $\{\mu_n \ : \ n \geq 1\}$. Moreover note that the graduations (ν_n) and (μ_n) are tamely equivalent.

It is straightforward to check that

$$\{f \in \mathbb{F} \ : \ \sup_{n\geq 1} \nu_n(f) < \infty\} = \{f \in \mathbb{F} \ : \ \sup_{\mathbb{R}} |f(\cdot)| < \infty\}$$

which is exactly the set $\mathcal{C}_b(\mathbb{R})$ of smooth maps $f : \mathbb{R} \to \mathbb{R}$ with compact support. On the other hand,

$$\{f \in \mathbb{F} \ : \ \sup_{n\geq 1} \mu_n(f) < \infty\} = \{0\}.$$

In fact, let $f \in \mathbb{F}$ be an arbitrary function with $M := \sup_{n\geq 1} \mu_n(f) < +\infty$. For arbitrary $t_0 \in \mathbb{R}$, if $n > |t_0|$ then one has $t_0 \in [-n, n]$, hence $n|f(t_0)| \leq \mu_n(f) \leq M$. Since $n > |t_0|$ is arbitrary, we thus obtain $f(t_0) = 0$ for all $t_0 \in \mathbb{R}$.
Moreover, this example shows that the Banach space $\mathbb{F}_b$ in $\mathbb{F}$ can be reduced to $\{0\}$ and $\mathbb{F}_b$ is not defined by the topology of $\mathbb{F}$ but only by the uniformly equivalence class of a graduations.

A.2 Differential Equations on Fréchet Spaces

Differential equations (even the linear ones) need not have any solution in the framework of non-Banach infinite-dimensional locally convex topological vector spaces. This problem was in particular discussed in a famous work of Hamilton on Fréchet spaces in [Ham82]. The author gave some examples of differential equations which have no solution with some fixed initial conditions. However he proved the following result:

Theorem A.1 *Consider the differential equation $\dot{x} = F(x)$ on an open set U of a Fréchet space X. Assume that $F = Q \circ P$ where $P : U \to V$ and $Q : V \to X$ are smooth maps where V is an open set of a Banach space Y. Then, for any initial condition $x_0 \in U$, there exists a unique solution $\gamma :]-\epsilon, \epsilon[\to V$ such that $\dot{\gamma} = F \circ \gamma$ and $\gamma(0) = x_0$.*

The tame Fréchet spaces framework was developed in particular in [Ham82] where the author proved a general Nash-Moser Implicit Function Theorem. As an application of this Theorem, in [PopP92], the reader will find a version of a Theorem of existence and unicity of solutions of an ordinary differential equation in this setting. However, such a result cannot be applied to vector fields on Fréchet manifolds without additional conditions on all the transition functions associated with an atlas: These transition functions must be local diffeomophisms of the Fréchet model and respect the assumptions of such a Theorem.

In the same way, in the ILB framework developed by Omori in [Omo97], one can find a result about the existence and unicity of solutions of an ordinary differential equation. But, without the context of *strong* ILH-*Lie groups* described in this book, for the same reasons as above, this result cannot be applied to vector fields on Fréchet manifolds which are ILB manifolds without additional conditions.

In order to be complete, we will also mention an analogue result, under metric conditions in a Fréchet space, in [Glo08] (Theorem E). In this context, we have also an application to vector fields in the context of *bounded Fréchet manifolds* as developed by Eftekharinasab in [Eft15].

The theorem of existence of local solutions of a differential equation proved by C.T.J. Dodson, G. Galanis and E. Vassiliou in [DGV16], 2.4, is based on the compatibility of algebraic, topological and differential structures on Fréchet spaces with the projective limits of Banach spaces. The assumptions of this Theorem are very strong but these assumptions are invariant under any projective limit of diffeomorphisms. Therefore, this result can be applied to vector fields on Fréchet manifolds which are projective limit of Banach manifolds.

We will use the assumptions and notations of the previous subsection. The following result is in fact a reformulation in our context of Theorem 1 in [Lob07]. This formulation is explicitly used in Chapter 8, § 8.5.

Theorem A.2 *Let $\mathbb{F}$ be a Fréchet space realized as the limit of a surjective projective sequence of Banach spaces $(\mathbb{B}_n, \lambda_n^m)_{m \geq n}$ whose topology is defined* *by the sequence of seminorms $(\nu_n)_{n\in\mathbb{N}}$. Let I be an open interval in $\mathbb{R}$ and U be an open set of $I \times \mathbb{F}$. Then U is a surjective projective limit of open sets $U_n \subset I \times \mathbb{B}_n$. Consider a smooth map $f = \varprojlim f_n : U \to \mathbb{F}$, projective limit of maps $f_n : U_n \to \mathbb{B}_n$*[1]*. Assume that for every point $(t,x)) \in U$, and every $n \in \mathbb{N}$, there exists an integrable function $K_n > 0$ such that*

$$\forall\,((t,x),(t,x')) \in U^2,\ \nu_n(f(t,x) - f(t,x')) \leq K_n(t)\nu_n(x - x'). \tag{A.3}$$

and consider the differential equation:

$$\dot{x} = \phi\,(t,x)\,. \tag{A.4}$$

(1) For any $(t_0, x_0) \in U$, there exists $\alpha > 0$ with $I_\alpha = [t_0 - \alpha, t_0 + \alpha] \subset I$, an open pseudo-ball $V = B(x_0, r) \subset U$ and a map $\Phi : I_\alpha \times I_\alpha \times V \to \mathbb{F}$ such that

$$t \mapsto \Phi(t,\tau,x)$$

is the unique solution of (A.4) with initial condition $\Phi(\tau,\tau,x) = x$ for all $x \in V$.

(2) V is the projective limit of the open balls V_n of $\mathbb{B}_n$. For each $n \in \mathbb{N}$, the curve $t \mapsto \lambda_n \circ \Phi(t,\tau,\lambda_n(x))$ is the unique solution $\gamma : I_\alpha \to \mathbb{B}_n$ of the differential equation $\dot{x}_n = \phi_n\,(t,x_n)$ with initial condition $\gamma(\tau) = \lambda_n(x)$.

The following Corollary will be basic in the proof of Theorem 8.7.

Corollary A.1 *Let $U = \varprojlim U_n$ be an open subset of $\mathbb{F}$ and $X = \varprojlim X_n : U \to \mathbb{F}$ a projective limit of smooth maps $X_n : U_n \to \mathbb{B}_n$. Assume that for every $n \in \mathbb{N}$ we have*

$$\forall\,((t,x),(t,x')) \in U^2,\ \nu_n(X(x) - X(x')) \leq K_n\nu_n(x - x'). \tag{A.5}$$

For $x_0 \in U$, let $B(x_0, 2r) = \{x \in \mathbb{F},\ :\ \nu_{n_i}(x - x_0) < 2r,\ 1 \leq i \leq k\}$ be a pseudo-ball contained in U. Let us set

$C_1 = \max_{1\leq i\leq k} K_{n_i}$

$C_2 = \sup_{z\in\mathbb{R}}\left\{\max_{1\leq i\leq k}\nu_{n_i}(f(z))\right\}.$

Then for any α such that $\alpha e^{2\alpha C_1} \leq \frac{r}{2C_2}$, there exists a neighbourhood $V = B(x_0, r)$ and a smooth map $\phi_\alpha : I_\alpha \times V$ such that $t \mapsto \phi^\alpha(t,x)$ is the unique solution of $\dot{x} = X(x)$ defined on I_α with initial condition $\phi^\alpha(0,x) = x$. Moreover if $V_n = \lambda_n(V)$, consider $\phi_n^\alpha : I_\alpha \times V_n \to \mathbb{B}_n$ defined by $\phi_n^\alpha = \lambda_n \circ \phi^\alpha$; For each $z \in V_n$, the map $t \mapsto \phi^\alpha(t,z)$ is the unique solution of the differential equation $\dot{x}_n = X_n(x_n)$ defined on I_α with initial condition $\phi^\alpha(0,z) = z$

[1]This means that we have: $\forall m \geq n,\ \lambda_n^m \circ f_m = f_n \circ (Id_{\mathbb{R}} \times \lambda_n^m)$.

Remark A.4 *It is clear that the theorem of existence and unicity of solution of a ODE given in [DGV16] is a consequence of Theorem A.2 for surjective projective limits. Note that, the Theorem of existence and unicity of solutions of an ODE in [Ham82], is also a consequence of Theorem A.2 (cf. [DGV16], Remark 2.4.5).*

Remark A.5 *If* $X = \varprojlim X_n$ *is a smooth vector field defined on an open set* $U = \varprojlim U_n$ *of* $\mathbb{F}$*, which satisfies assumption (A.5), as classically, according to Corollary A.1, the map* $\mathrm{Fl}_t^X := \mathrm{Fl}^X(t, \ )$ *is* the local flow of X *that is* ϕ_t *fulfils the properties of a* 1*-parameter group:*

$\mathrm{Fl}_0^X = Id_V$

$\mathrm{Fl}_t^X \circ \mathrm{Fl}_s^X = \mathrm{Fl}_{s+t}^X$ *if* s,t *and* $s+t$ *belong to* I_α.

In particular Fl_t^X *is a diffeomorphism from* V *onto it range and its inverse is* Fl_{-t}^X. *Moreover* $\mathrm{Fl}_t^{X_n} = \lambda_n \circ \mathrm{Fl}_t^X \circ \lambda_n$ *is local flow of* $X_n = \lambda_n \circ X \circ \lambda_n$ *and we have* $\mathrm{Fl}_t^X = \varprojlim \mathrm{Fl}_t^{X_n}$.

B

Categories

The theory of categories introduced in 1945 by S. Eilenberg and S. Mac Lane ([EiMc45]) studies the mathematical structures and their relations.

Definition B.1 *A category* C *is the the datum of*

(Cat 1) *a class* $\mathbf{ob}(\mathsf{C})$ *of objects*

(Cat 2) *a class* $\mathbf{hom}(\mathsf{C})$ *of morphisms or arrows between pairs of objects: a morphism* f *from the object* X *to the object* Y *is denoted by* $f : X \to Y$ *and the set of all such morphisms by* $\mathrm{Hom}_{\mathsf{C}}(X, Y)$ *or* $\mathrm{Hom}(X, Y)$.

(Cat 3) *a partial binary operation called composition of morphisms and denoted* $\circ$ *where, for all objects* X, Y *and* Z

$$\begin{array}{rccc} \circ : & \mathrm{Hom}(X,Y) \times \mathrm{Hom}(Y,Z) & \to & \mathrm{Hom}(X,Z) \\ & (f : X \to Y, g : Y \to Z) & \mapsto & (g \circ f : X \to Z) \end{array}$$

fulfilling the properties:

(Cat 31) *the composition is associative: for any morphism* $f : X \to Y$, $g : Y \to Z$ *and* $h : Z \to T$,

$$(h \circ g) \circ f = h \circ (g \circ f)$$

(Cat 32) *for every object* X, *there exists a morphism* $\mathbf{1}_X : X \to X$ *such that all morphisms* $f : Y \to X$ *and* $g : X \to Z$, *we have* $\mathbf{1}_X \circ f = f$ *and* $g \circ \mathbf{1}_X = g$.

We give some examples of categories.

Category	Objects	Morphisms
SET	sets	functions
GRP	groups	group homomorphisms
TOP	topological spaces	continuous functions
$\mathsf{VECT}_{\mathbb{K}}$	$\mathbb{K}$-vector spaces	$\mathbb{K}$-linear maps
$\mathsf{C^*A}$	C^*-algebras	*-homomorphisms

To any category C is associated the *dual or opposite category* C^{op} with the same objects and where the morphisms are those of C reversed.

DOI: 10.1201/9781003435587-B

Definition B.2 *Let* C *and* D *be two categories. A (covariant) functor* $\mathfrak{F}$ *from* C *to* D *denoted by* $\mathfrak{F} : \mathsf{C} \to \mathsf{D}$ *is a function assigning*

— *to any object* X *of* C *an object* $\mathfrak{F}(X)$ *of* D*;*

— *to any morphism* $f : X \to Y$ *of* C *a morphism* $\mathfrak{F}(f) : \mathfrak{F}(X) \to \mathfrak{F}(Y)$ *of* C *fulfilling the following properties:*

$$\mathfrak{F}(g \circ f) = \mathfrak{F}(g) \circ \mathfrak{F}(f)$$

$$\mathfrak{F}(\mathbf{1}_X) = \mathbf{1}_{\mathfrak{F}(X)}.$$

A *contravariant functor* from C to D is, by definition, a functor $\mathfrak{F} : \mathsf{C}^{\mathrm{op}} \to \mathsf{D}$, or, equivalently, $\mathfrak{F} : \mathsf{C} \to \mathsf{D}^{\mathrm{op}}$.
So a contravariant functor $\mathfrak{F}$ reverses the direction of compositions:

$$\mathfrak{F}(g \circ f) = \mathfrak{F}(f) \circ \mathfrak{F}(g).$$

The map $(\centerdot)^\star$ which assigns to every $\mathbb{K}$-vector space E its algebraic dual space $E^\star$ and to every linear map φ its transpose t_φ is a contravariant functor from the category $\mathsf{VECT}_{\mathbb{K}}$ of all $\mathbb{K}$-vector spaces to itself seen as $(\centerdot)^\star : \mathsf{C}_{\mathbb{K}} \to \mathsf{C}_{\mathbb{K}}^{\mathrm{op}}$.

C

Sheaves Theory

We recall definitions and results used in some sections of this book. These elements come essentially from P. Schapira's notes ([Schap11]) and Vassiliou's book ([Vas05]). For more complete details and results on sheaves theory see [Godm73].

To any topological space X is associated a category TOP_X whose objets are the open sets of X and whose morphisms $\mathrm{Hom}(V,U)$ are the inclusions $\iota_V^U : V \to U$ if $V \subset U$ and $\mathrm{Hom}(V,U) = \emptyset$ if V is not contained in U.

C.1 Presheaves

Let $\mathbf{M}$ be a category. A *presheaf* $\mathfrak{P}$ on a topological space X is the datum of a functor $\mathfrak{P}$ from TOP_X to $\mathbf{M}$ such that

(PSh 1) For any object U in TOP_X, $\mathfrak{P}(U)$ belongs to $\mathbf{M}$ with $\mathfrak{P}(\emptyset) = \{0\}$.

(PSh 2) If $V \subset U$ are open sets, we have a restriction map ϱ_U^V which is a morphism from $\mathfrak{P}(U)$ to $\mathfrak{P}(V)$ and which satisfies the following axioms

(PSh 21) $\forall U \in \mathsf{TOP}_X,\ \varrho_U^U = \mathrm{Id}_{\mathfrak{P}(U)}$;

(PSh 22) $\forall (U,V,W) \in \mathsf{TOP}_X^3 : U \supset V \supset W,\ \varrho_W^V \circ \varrho_V^U = \varrho_W^U$.

Let $\mathfrak{P}$ be a presheaf on X. The elements of $\mathfrak{P}(U)$ are called the sections of $\mathfrak{P}$ over U and if $\mathfrak{s}$ belongs to $\mathfrak{P}(U)$, the element $\varrho_V^U(\mathfrak{s})$ is called the *restriction* of $\mathfrak{s}$ to V and is simply denoted $\mathfrak{s}_{|V}$.

$\mathfrak{P} = \left\{\left(\mathfrak{P}(U), \varrho_V^U\right)\right\}_{(U,V)\in\mathsf{TOP}_X^2 :\, V\subset U}$ can be seen as a double correspondence

$$\begin{array}{ccc} U & \mapsto & \mathfrak{P}(U) \\ \left(\iota_V^U : V \to U\right) & \to & \left(\varrho_V^U : \mathfrak{P}(U) \to \mathfrak{P}(V)\right) \end{array}$$

So $\mathfrak{P}$ is a contravariant functor which determines a direct limit called *stalk* of $\mathfrak{P}$ where the index set I corresponds to the topology of X directed by the relation $\preccurlyeq$ defined by

$$\forall (U,V) \in \mathsf{TOP}_X^2,\ U \preccurlyeq V \Leftrightarrow U \supset V.$$

DOI: 10.1201/9781003435587-C

For any $x \in X$, the family $\{\mathfrak{P}(U), \varrho_{VU},\ U \text{ open neighbourhood of } x\}$ is a direct system and the inductive limit

$$\mathfrak{P}_x = \varinjlim\{\mathfrak{P}(U),\ \varrho_{VU},\ U \text{ open neighbourhood of } x\}$$

is called the *stalk of* $\mathfrak{P}$ *at* x.

Let $\mathfrak{P}$ and $\mathfrak{P}'$ be two pre-sheaves on X and X' respectively, which take values in $\mathfrak{M}$. *A pre-sheaf morphism* f *from* $\mathfrak{P}$ *to* $\mathfrak{P}'$ over a continuous map $\phi : X \to X'$ is the data of family of morphisms $\{f_U : \mathfrak{P}(U) \to \mathfrak{P}'(U'), U \in \mathsf{TOP}_X, U' \in \mathsf{TOP}_X, \phi(U) \subset U'\}$ such that the following diagrams are commutative

$$\begin{array}{ccc} \mathfrak{P}(U) & \xrightarrow{\Phi_U} & \mathfrak{P}'(U) \\ \varrho_{VU}\downarrow & & \downarrow \varrho'_{VU} \\ \mathfrak{P}(V) & \xrightarrow[\Phi_V]{} & \mathfrak{P}'(V) \end{array}$$

for all open sets $V \subset U$ in X and $V' \subset U'$ in X' such that $\phi(U) \subset U'$ and $\phi(V) \subset V'$.

A *subpresheaf of* $\mathfrak{P}$ is a presheaf $\mathfrak{P}'$ such that $\mathfrak{P}'(U) \subset \mathfrak{P}(U)$ for all open set U of X

C.2 Sheaves

A presheaf $\mathfrak{P}$ on X is called a *sheaf on* X *if* $\mathfrak{P}$ if, for any open U and any open covering$\{U_i\}_{i \in I}$ of U, the following properties are satisfied:

(Sh 1) If two sections $\mathfrak{s}$ and $\mathfrak{t}$ in $\mathfrak{P}(U)$ are such that $\mathfrak{s}_{|U_i} = \mathfrak{t}_{|U_i}$ for all $i \in I$, then $\mathfrak{s} = \mathfrak{t}$.

(Sh 2) Assume that we have sections $\mathfrak{s}_i \in \mathfrak{P}(U_i)$ such that $\mathfrak{s}_{i|U_i \cap U_j} = \mathfrak{s}_{j|U_i \cap U_j}$ for all pairs (i,j) such that $U_i \cap U_j \neq \emptyset$, then there exists $\mathfrak{s} \in \mathfrak{P}(U)$ such that $\mathfrak{s}_{|U_i} = \mathfrak{s}_{i|U_i}$ for all $i \in I$.

A *morphism of sheaves* is a morphism of presheaves.

A sub-presheaf $\mathfrak{P}'$ of a sheaf $\mathfrak{P}$ is automatically a sheaf and so is called a sub-sheaf of $\mathcal{P}$.

Examples C.1 *(cf. [Schap11])*

(1) Let M be a convenient manifold. Denote by $C^\infty(U)$ the ring of smooth maps on an open set U in M. The assignment $U \mapsto C^\infty(U)$, for any open set U in M, defines a sheaf of rings on M which is denoted C^∞_M.

(2) *Let* $\pi : \mathcal{A} \to M$ *be a convenient bundle. For any open set* U *in* M*, the set of smooth sections* $\Gamma(\mathcal{A}_U)$ *is a* $C^\infty(U)$*-module. The assignment* $U \mapsto \Gamma(\mathcal{A}_U)$*, for any open set* U *in* M*, defines a sheaf of modules on* M *which is denoted* $\Gamma(\mathcal{A})$*.*

(3) *Let* $\mathfrak{P}$ *and* $\mathfrak{P}'$ *be two sheaves of modules on* M*. Then the assignment* $U \mapsto \mathfrak{P}(U) \times \mathfrak{P}'(U)$*, for any open set* U *in* M*, defines a sheaf of modules on* M *which is denoted* $\mathfrak{P} \times \mathfrak{P}'$*.*

C.3 Projective and Direct Limits of Sheaves

Let I be a directed set and let X be a topological space.
Consider a projective system of sheaves and sheaves morphisms $\mathcal{S} = \left\{ \left(\mathcal{S}_i, \delta_i^j \right) \right\}_{(i,j)\in I^2,\ j \succcurlyeq i}$. As a preliminary result about the existence of the projective limit of a projective system of differential triads[1] A *differential triad* is a triple $(\mathcal{A}, \partial, \Omega)$ where $\mathcal{A}$ is a sheaf of unital, commutative and associative $\mathbb{K}$-algebras over a topological space (X, τ), Ω an $\mathcal{A}$-module and $\partial = (\partial_U)_{U \in \tau}$ a sheaf morphism which is $\mathbb{K}$-linear and satisfies the Leibniz identity.

M. H. Papatriantafillou proved ([Pap01], Lemma 3.1.) that the projective limit of the system $\mathcal{S}$ exists in the category Sh_X of sheaves over X.
An analogous result was also proved for direct system of sheaves.

[1]The notion of *differential triad* was introduced by A. Mallios in [Mal92] as one generalization of the notion of manifold.

D

Locally Convex Vector Bundles

The definition of a locally convex manifold M modelled on a locally convex space $\mathbb{E}$ is analogous to the definition of a Fréchet manifold and so we assume that it is well known. Classically, a locally convex bundle (E, π, M) is a locally convex manifold E where π is a surjective smooth map from E to a locally convex manifold M and there exists an atlas $\mathcal{A} = \{(U_\alpha, \phi_\alpha)\}_{\alpha \in \mathbf{A}}$ on M, a locally convex space $\mathbb{E}$ and, for each $\alpha \in \mathbf{A}$, we have diffeomorphism $\tau_\alpha : \pi^{-1}(U_\alpha) \to \phi_\alpha(U^\alpha) \times \mathbb{E}$ such that :

For $U_\alpha \cap U_\beta \neq \emptyset$, the map

$$\tau_\alpha \circ (\tau_\beta)^{-1} : \phi_\beta(U_\alpha \cap U_\beta) \times \mathbb{E} \to \phi_\alpha(U_\alpha \cap U_\beta) \times \mathbb{E}$$

is written

$$(\overline{y}, \overline{u}) \mapsto \left(\tau_\alpha \circ (\tau_\beta)^{-1}(\overline{y}), \Theta_{\alpha\beta}(\overline{y})(\overline{u})\right)$$

Then $\overline{y} \mapsto \Theta_{\alpha\beta}(\overline{y})$ is a smooth map from $\phi_\beta\,(U_\alpha \cap U_\beta)$ to the locally convex space $L(\mathbb{E})$ of bounded endomorphisms of $\mathbb{E}$ taking values in the set of isomorphisms of $\mathbb{E}$.

Definition D.1 *Let E be a topological manifold and $\pi : E \to M$ a surjective map on a n.n.H. convenient manifold modelled on a convenient space $\mathbb{M}$. We say that (E, π, M) is a locally convex vector bundle over M if there exists a locally convex space $\mathbb{E}$, an atlas $\mathcal{A} = \{(U_\alpha, \phi_\alpha)\}_{\alpha \in \mathbf{A}}$ on M and, for each $\alpha \in \boldsymbol{A}$, a homeomorphism $\tau_\alpha : \pi^{-1}(U_\alpha) \to \phi_\alpha(U_\alpha) \times \mathbb{E}$ such on if $U_\alpha \cap U_\beta \neq \emptyset$, then the map*

$$\tau_\alpha \circ (\tau_\beta)^{-1} : \phi_\beta(U_\alpha \cap U_\beta) \times \mathbb{E} \to \phi_\alpha(U_\alpha \cap U_\beta) \times \mathbb{E}$$

which can be written

$$(\overline{y}, \overline{u}) \mapsto \left(\phi_\alpha \circ (\phi_\beta)^{-1}(\overline{y}), \Theta_{\alpha\beta}(\overline{y})(\overline{u})\right)$$

then the map $\overline{y} \mapsto \Theta_{\alpha\beta}(\overline{y})$ is a convenient smooth map from $\phi_\beta(U_\alpha \cap U_\beta)$ to the locally convex space $\mathrm{L}(\mathbb{E})$ which takes values in the set of isomorphisms of $\mathbb{E}$.

From this definition, it follows that E is a manifold modelled on $\mathbb{M} \times \mathbb{E}$ and $\pi : E \to M$ is convenient smooth.

DOI: 10.1201/9781003435587-D

E

The KdV Equation

The Korteweg-de Vries (KdV) equation ([KoVr1895])

$$\partial_t u = -\partial_x^3 u + 6u\partial_x u$$

is an evolution equation in 1 space dimension which was proposed as a model to describe waves on shallow water surfaces (cf [Schm04]).
This non-linear and dispersive PDE was first introduced by J. Boussinesq ([Bous1877]) in 1877 and rediscovered by D. Korteweg and G. de Vries ([KoVr1895]) in order to modelize natural phenomena discovered by J.S. Russel ([Rus1844]).

The KdV equation combines two behaviours:

1. the *dispersion process* due to the term u_{xxx}: the waves

$$u(x,t) = \exp\left(\mathrm{i}\left(kx - k^3 t\right)\right)$$

   with different frequencies disperse with different velocities;

2. the *wave breaking phenomenon* due to the non-linear term $u.u_x$: the velocity of the point of the wave is proportional to the amplitude of this point.

In general, the combination of these effects gives rise to a chaotic behaviour (cf. [Sza07]). But, in the case of the KdV equation, both of these effects compensate each other and generate solutions describing the combination of solitary waves (solitons). For more information about this type of solutions see [New85].

Let us have a look on travelling-wave solutions (cf. [MaRa99], 3.2)

$$u(x,t) = v(x + ct)$$

where $u, u_x, \ldots$ tend to 0 as $|x| \to +\infty$. For $s = x + ct$, we have:

$$u_t - u_{xxx} - 6u.u_x \;\to\; v_s^2 = v^2(c - 2v).$$

So the problem is reduced to the resolution of the ODE

$$\frac{dv}{ds} = \pm v\sqrt{c - 2v}.$$

DOI: 10.1201/9781003435587-E

We then obtain the so-called one-soliton solution of the KdV equation:

$$u(x,t) = \frac{c}{2\cosh^2\left(\frac{\sqrt{c}}{2}(x+ct)\right)}.$$

This solitary wave is a single hump traveling in time with constant speed proportional to amplitude without changing the shape.

In the figures at the end of this appendix, we show the interaction between two solitons. It is interesting to notice that each soliton recovers its shape after collision.

The KdV equation determines a non-singular algebraic hypersurface in $J^3(\pi)$ where π is the projection of the trivial bundle over the space $\mathbb{R}^2$ of independant variables (x,t) whose fibre type $\mathbb{R}$ is the space of the dependant variable u.

In [Arn66], V. Arnold suggested a general framework for the Euler equations on an arbitrary group which describes a geodesic flow with respect to a suitable one-sided invariant Riemannian metric on the group. This approach works for the Virasoro group and provides a natural geometric setting for the KdV equation (cf. [KhMi03]).

This equation was reduced in [GGKM67] to the inverse scattering problem for the Schrödinger operator $L = -\frac{d^2}{dx^2} + u(x)$[1] to solve the Cauchy problem for the KdV equation in the class of rapidly decreasing functions u.

For the periodic KdV equation with zero mean-value periodic boundary conditions (cf. [KoKu16]), we consider the Sobolev space $\mathcal{H}_\alpha$ for $\alpha \geq 1$

$$\mathcal{H}_\alpha = \left\{ u \in L^2(\mathbb{S}^1) : u^{(\alpha)} \in L^2(\mathbb{S}^1), \int_0^1 u(t)dt = 0 \right\}$$

On any space $\mathcal{H}_\alpha$[2], the KdV equation may be written as a Hamiltonian system with the Hamiltonian H_2 given by

$$H_2(u) = \frac{1}{2}\int_0^1 \left(2u^3 + u_x^2\right) dx$$

and the Poisson tensor $\frac{\partial}{\partial x}$ which defines a linear isomorphism from $\mathcal{H}_\alpha$ onto

[1] Using Miura's transformation $u = \theta + \theta_x$ and the change of variables $\theta = \frac{\psi_x}{\psi}$, one gets the linear ODE $\psi_x x + u\psi = 0$ (cf. [Gag18]).

[2] In [KaMa98], Kappeler and Makarov consider the Hilbert tower of Sobolev spaces $(H_n(\mathbb{S}^1))$ where the average $A_1 : u \mapsto \int_{\mathbb{S}^1} u(x)dx$ is a Casimir function of the Poisson tensor $\frac{\partial}{\partial x}$ and each symplectic leaf of the induced foliation is an affine space of equation $A_1(u) = c$.

$\mathcal{H}_{\alpha-1}$:

$$u_t = \frac{\partial}{\partial x}\frac{\partial}{\partial u}H_2(u)$$

It is well known after the papers of Dubrovin, Its, Lax, Matveev and Novikov ([Lax74], [NoDu74], [ItMa75], ...) that this system is integrable.

Moreover, it was shown by Kappeler and collaborators in a series of papers, starting with [Kap91], that it admits global Birkhoff coordinates.
So such an equation can be viewed as a *complete integrable system* in the following sense: there exists a sequence of action-angle variables and the system becomes linearized in these infinite set of variables (cf. [Zub97]). The dynamics of the system expressed in the action-angle variables (I, ϕ) on a separable Hilbert space (cf. [KoKu16]) is

$$\begin{cases} I_t &= 0 \\ \phi_t &= \dfrac{\partial}{\partial I}K(I) \end{cases}$$

where K corresponds to the Hamiltonian.

As most integrable Hamiltonian systems (cf. [BoBo02]), the Korteweg-de Vries equation can be written in Lax form

$$\dot{L} = [L, A]\,.$$

where

$$L = -\frac{\partial^2}{\partial x^2} + u$$

is a Strurm-Liouville operator and where

$$A = 4\frac{\partial^3}{\partial x^3} - 3\left(2u\frac{\partial}{\partial x} + u_x\right)$$

This allows to get an infinite number of first integrals for this equation.

It is well known (e.g. [FMPZ99], [MaMo84], [Olv93], [Scha80],[ZuMa91]) that this equation, as many completely integrable Hamiltonian systems arising in mechanics, mathematical physics and geometry, can be written in Hamiltonian form in two distinct ways. We then obtain a hierarchy of commuting conservation laws and Hamiltonian flows generated by the recursion operator linking both Poisson brackets.

In what follows, we give an outline of the framework of variational Lie algebroids used to describe such an evolutionary equation as in [KiVa11]. For evolutionary vector fields and their prolongations, the reader is also referred to [Olv93]. For the notion of infinite jets see Chapter 4, § 4.9.

Let $\pi : E \to M$ be a vector bundle over an n-dimensional manifold M whose fibre is m-dimensional. Denote by $\mathbf{\Gamma}(\pi)$ the $C^\infty(M)$-module of smooth sections of π.
For any $k \in \mathbb{N}$, $J^k(\pi)$ is the vector bundle of k-jets of sections of $\pi_k : J^k(\pi) \to M$. The space of infinite jets $J^\infty(\pi)$ of smooth sections of π is the projective limit of the sequence

$$M \xleftarrow{\pi} E \xleftarrow{\pi_0^1} J^1(\pi) \longleftarrow \cdots \longleftarrow J^k(\pi) \xleftarrow{\pi_k^{k+1}} J^{k+1}(\pi) \longleftarrow \cdots \tag{E.1}$$

Denote by $\varkappa(\pi)$ the $C^\infty(J^\infty(\pi))$-module of evolutionary derivations

$$\mathbf{\Gamma}(\pi_\infty^*(\pi)) = \mathbf{\Gamma}(\pi) \otimes_{C^\infty(M)} C^\infty(J^\infty(\pi)).$$

Let $\overline{\Lambda}^i(\pi)$ be $C^\infty(J^\infty(\pi))(\pi)$-module of i-th horizontal forms on $J^\infty(\pi)$. Let $\overline{H}^i(\pi)$ be the cohomology $\overline{\Lambda}^i(\pi)/\{\mathrm{im}\, \mathrm{d}_h\}$ w.r.t. the horizontal differential d_h (cf. [And92]).

Let B be the bundle over $J^\infty(\pi)$ whose 1-dimensional fibre is spanned by $d\,\mathrm{vol}(M) = dx^1 \wedge \cdots \wedge dx^n$.

Consider $\widehat{\varkappa}(\pi) = \mathrm{Hom}_{C^\infty(J^\infty(\pi))}\left(\varkappa(\pi), \overline{\Lambda}^n(\pi)\right)$.

Let $\mathbf{E}$ be the Euler variational derivative which acts on elements $\mathcal{H} = [H d\mathrm{vol}(M)] \in \overline{H}^n(\pi)$ (cf. [PrMy98]).

An antisymmetric linear operator $P : \widehat{\varkappa}(\pi) \to \varkappa(\pi)$ is called Hamiltonian if for all $\mathcal{H}_1, \mathcal{H}_2 \in \overline{H}^n(\pi)$, the associated antisymmetric bilinear bracket $\{.,.\}_P$

$$\{\mathcal{H}_1, \mathcal{H}_2\}_P = \langle \mathbf{E}(\mathcal{H}_1), P(\mathbf{E}(\mathcal{H}_2))\rangle$$

satisfies the Jacoby identity:

$$\{\{\mathcal{H}_1, \mathcal{H}_2\}_P, \mathcal{H}_3\}_P + \{\{\mathcal{H}_2, \mathcal{H}_3\}_P, \mathcal{H}_1\}_P + \{\{\mathcal{H}_3, \mathcal{H}_1\}_P, \mathcal{H}_2\}_P = 0 \tag{E.2}$$

The equivalence class of sections of B can be endowed with a structure of $\mathcal{F}$-module which can be equipped with a Lie algebra structure relatively to the bracket $\{.,.\}_P$.

We then have

$$\{\mathcal{H}_1, f.\mathcal{H}_2\}_P = f.\{\mathcal{H}_1, \mathcal{H}_2\}_P + E_{P\circ\mathbf{E}(H_1)}(f).\mathcal{H}_2$$

We then get a Lie algebroid structure over $J^\infty(\pi)$

$$\left(\overline{H}^n(\pi), \{.,.\}_P\right) \to (\varkappa(\pi), [.,.]) \tag{E.3}$$

where the anchor is $\mathfrak{a} = E_{P\circ\mathbf{E}}$.

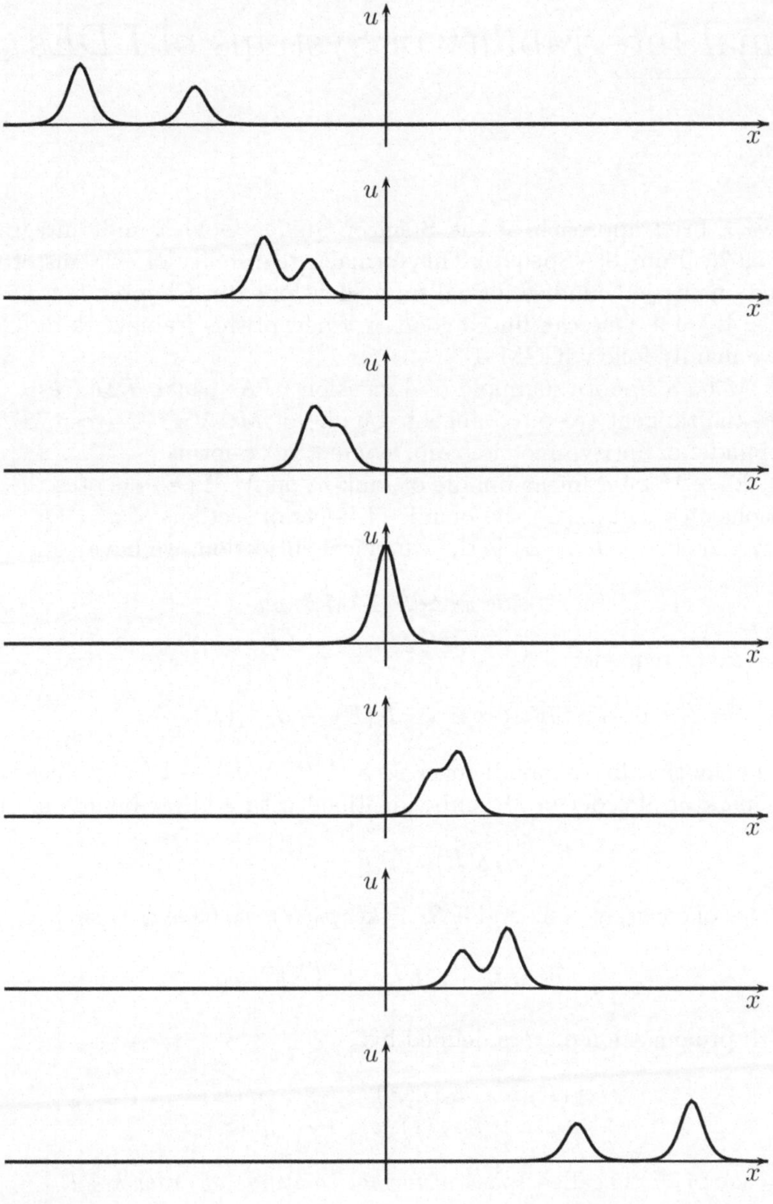

FIGURE E.1
A pair of solitons of KdV equation

F

Formal Integrability of Systems of PDEs

We give a brief approach of the Spencer-Quillen-Goldschimdt integrability (cf. [Gol67], [Pom78], [Spe69]). This formal integrability of systems of PDEs is based on the jet bundles formalism as developed in Chapter 1, § 1.12 and Chapter 4, § 4.9. One can find a good overview of this framework in [Gas85]. Here we mainly follow [CGM91].

Let M be a smooth manifold of dimension n. As usual, TM (resp. T^*M) denotes the tangent (resp. cotangent) bundle of M, $\Lambda^k T^*M$ (resp. $S^k T^*M$) is the bundle of antisymmetric (resp. symmetric) k-forms.
Let $\pi : E \to M$ be a linear bundle of rank m on M. $\mathbf{\Gamma}(E)$ denotes the space of sections of π and $J_k(E)$ the bundle of k-jets of sections of π.

If $\pi_k : J_k(E) \to J_{k-1}(E)$ is the canonical surjection, we have

$$\ker \pi_k \simeq S^k T^*M \otimes E,$$

and the exact sequence:

$$0 \to S^k T^*M \otimes E \overset{\varepsilon}{\to} J_k(E) \overset{\pi_k}{\to} J_{k-1}(E) \to 0.$$

Let F be another linear bundle over M and $P : \mathbf{\Gamma}(E) \to \mathbf{\Gamma}(F)$ a linear differential operator of order k. P can be identified with a linear bundle morphism

$$p_0(P) : J_k(E) \to F.$$

If $p_0(P)$ is of constant rank and if $R_k = \ker p_0(P)$, we have the exact sequence:

$$0 \to R_k \to J_k(E) \overset{p_0(P)}{\to} F \to 0$$

The l-th prolongation of P is defined by:

$$p_l(P) : \begin{array}{ccc} J_{k+l}(E) & \to & J_l(E) \\ j_{k+l}(s)(x) & \mapsto & j_l(Ps)_x \end{array}.$$

$R_{k+l} = \ker p_l(P)$ is called *space of formal solutions of order* $k+l$.

Definition F.1 *The operator P is said formally integrable if all the maps $\pi_{k+l} : R_{k+l} \to R_k$ are surjective for $l \geq 1$.*

In the analytic context, the formal integrability is equivalent to the existence of solutions for any 'initial conditions'.

DOI: 10.1201/9781003435587-F

Theorem F.1 *If P is an analytic linear operator of order k and formally integrable, then, for all x_0 in M and all F_0 of R_{k,x_0}, there exists a section f of E defined on a neighbourhood U of x_0 such that, on U,*

$$\left\{ \begin{array}{c} P(f) = 0 \\ (j_k f)(x_0) = F_0 \end{array} \right.$$

In order to prove the formal integrability of a linear operator, according to the definition, it would be necessary to verify an infinity number of conditions. Cartan-Kälher theorem allows to establish the formal integrability thanks to a finite number of verifications. Then we need to introduce the notion of involution. Let us consider the map ε defined by

$$\varepsilon\left((df_1)_x \bullet \cdots \bullet (df_k)_x \otimes s(x)\right) = j_k\left(f_1 \ldots f_k s\right)(x)$$

where $x \in M$, $s \in \mathbf{\Gamma}E$, $f_1, \ldots, f_k$ are germs of maps at x on M vanishing at x, and $\bullet$ stands for the symmetric product. So it is possible to calculate $\sigma_0(P)$.

Definition F.2 *We call symbol of P the map*

$$\sigma_0(P)) = p_0(P) \circ \varepsilon \sigma_0(P) : S^k T^* M \otimes E \overset{\varepsilon}{\to} J_k(E) \overset{p_0(P)}{\to} F.$$

Definition F.3 *The map $\sigma_l(P)$ defined by*

$$\sigma_l(P) : S^{k+l} T^* \otimes E \overset{\text{canonical injection}}{\to} S^l T^* M \otimes S^k T^* M \otimes E \overset{\mathrm{Id}_{S^l T^* M} \otimes \sigma_0}{\to} S^l T^*(M) \otimes E$$

is called the symbol of the l-prolongation.

In particular, $\sigma_1(P) : S^{k+1} T^* M \otimes E \to T^* \otimes F$ is defined by:

$$\forall X \in TM, \forall t \in S^{k+1} T^* M \otimes E, i_X \sigma_1(P)(t) = \sigma_0(P)(i_X t)$$

where i_X stands for the interior product by X.

Let $g_l = \ker \sigma_l$. If $\{e_1, \ldots, e_n\}$ is a basis of $T_x M$, we consider for all $j \in \{1, \ldots, n-1\}$,

$$(g_0)_{\{e_1,\ldots,e_j\}} = \left\{A \in (g_0)_x : i_{e_1} A = 0, \ldots, i_{e_j} A = 0\right\}.$$

Definition F.4 *A basis $(e_1, \ldots, e_n)$ de T_x is called quasi-regular if:*

$$\dim(g_1)_x = \dim(g_0)_x + \sum_{j=1}^{n-1} \dim(g_0)_{(e_1,\ldots,e_j)}$$

Definition F.5 *The linear differential operator P is involutive if at any point there exists a quasi-regular basis.*

We then have the fundamental Cartan-Kälher's theorem.

Theorem F.2 *Let P be a linear differential operator of order k.*
If the following conditions are fulfilled:
(i) P is involutive;
(ii) $\pi_{k+1} : R_{k+1} \to R_k$ is surjective,
then P is formally integrable.

To prove the surjectivity of π_{k+1}, we consider the following diagram of exact sequences:

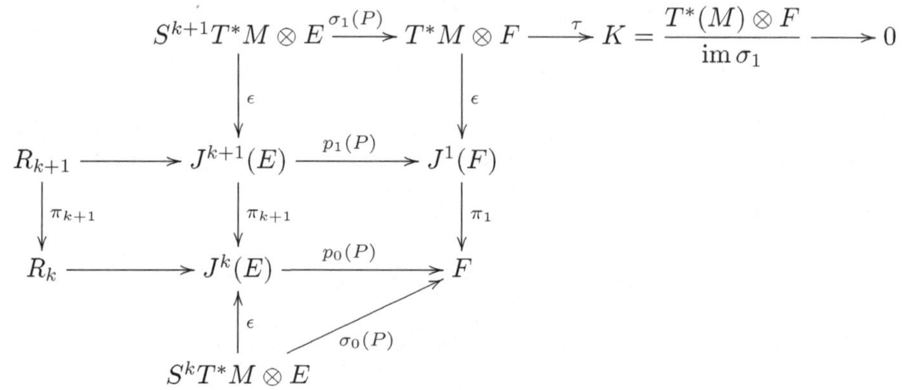

Thanks to a classical lemma of homological algebra (snake lemma), we deduce the existence of a mapping

$$\varphi : R_k \to K = \frac{T^*M \otimes F}{\operatorname{im} \sigma_1}$$

$\pi_{k+1} : R_{k+1} \to R_k$ is surjective if and only if $\varphi = 0$.
K is called the *space of obstructions*.

To determine φ, we consider a connection

$$\nabla : \mathbf{\Gamma}(F) \to \mathbf{\Gamma}\left(T^*M \otimes F\right)$$

on the bundle F. We then have:

$$p_0\left(\nabla\right) : J_1\left(F\right) \to T^*M \otimes F.$$

If $x \in M$ and $s \in \mathbf{\Gamma}(E)$ fulfils the condition $P(s)_x = 0$, one can easily see that

$$\varphi\left(j_k(s)(x)\right) = \tau\left(\nabla P(s)\right)_x .$$

To summarize, in order to show the formal integrability, one has to

1. Prove the involution of the operator P.

2. Build the mapping τ, which requires to have a good interpretation of K.
3. Build a connection ∇ on F such that, for all $x \in M$ and all smooth section s of E, one has
$$\tau\left(\nabla P\left(s\right)\right)_x = 0.$$

As an example of this theory, we consider the formal integrability of conservation laws associated with a cyclic endomorphism; a generalization of this result based on the decomposition of the associated Nijenhuis tensor in its cyclic components is given in [GrMe97].

Definition F.6 *A conservation law on a manifold M with respect with a $(1,1)$-tensor h is a smooth scalar 1-form such that*

$$\begin{cases} d\theta = 0 \\ d\left(h^*\theta\right) = 0 \end{cases}$$

where h^* is defined by

$$\left(h^*\theta\right)\left(X\right) = \theta\left(hX\right).$$

The rôle of the conservation laws in the integration of differential systems is well known (cf. [Lax57]). They constitute a fundamental tool in the study of completely integrable systems where h is the recursion operator; in [MaMo84] they are called *fundamental forms*.

Because the problem is local, the study of the existence of conservation laws is equivalent to the research of smooth functions f on M such that

$$d\left(h^* df\right) = 0. \tag{F.1}$$

Using Frölicher-Nijenhuis' notations ([FrNi56]), we can write

$$h^* df = i_h df \tag{F.2}$$

where i_h is the derivation of type i_* associated to h.
Condition (F.2) can be written as

$$dd_h f = 0 \tag{F.3}$$

where d_h is the derivation of type d_* associated to f defined as

$$d_h = i_h d - d i_h.$$

According to Frölicher-Nijenhuis' theory, to h is associated the tensor $[h,h] \in \Lambda^2 T^*M \otimes T^*M$ defined by:

$$\frac{1}{2}\left[h,h\right]\left(X,Y\right) = \left[hX,hY\right] + h^2\left[X,Y\right] - h\left[hX,Y\right] - h\left[X,hY\right].$$

Definition F.7 *A manifold endowed with a field of* $(1,1)$*-tensors* h *such that* $[h,h]=0$ *is called Nijenhuis manifold.*

Because

$$d_h^2 = d_{\frac{1}{2}[h,h]}$$

on a Nijenhuis manifold, we have $d_h^2 = 0$.
So the following proposition establishes the existence of conservation laws for some Nijenhuis manifolds.

Proposition F.1 *The differential operator*

$$\begin{array}{rccc} P: & C^{\infty}(M) & \to & \mathbf{\Gamma}\left(\Lambda^2 T^* M\right) \\ & f & \mapsto & d d_h f \end{array}$$

on a Nijenhuis manifold where h *is cyclic, is formally integrable.*

Further Studies

In this appendix, we have selected some fields of investigation linked with the frameworks we have expounded in this book[1].

(1) According to P.J. Olver ([Olv93]), *an evolutionary vector field* has a local expression of the form

$$X = \varphi^\alpha \left[x, u^{(k)}\right] \frac{\partial}{\partial u^\alpha}.$$

It can be seen as an element of $\mathbf{\Gamma}(\pi) \otimes_{C^\infty(M)} C^\infty\left(J^k(\pi)\right)$.

For example, the vector field $X = u_x \dfrac{\partial}{\partial u}$ is an evolutionary vector field where $k = 1$ (Example 4.17) which is a symmetry of the heat equation (cf. [Olv93], 5.1)

$$u_t = u_{xx}.$$

Its infinite prolongation $\mathrm{pr}^\infty(X)$ whose local expression is

$$u_x \frac{\partial}{\partial u} + u_{xx} \frac{\partial}{\partial u_x} + \cdots$$

appears as an element of $\mathbf{\Gamma}(\pi) \otimes_{C^\infty(M)} C^\infty\left(J^\infty(\pi)\right)$. Moreover, this vector field on the infinite jet bundle is a symmetry, i.e. let invariant the Cartan distribution (see Chapter 4, § 4.9.4).

In this framework, two problems of different nature appear:

(i) Existence of the flow of the prolongation of an evolutionary vector field.
Recall that the infinite prolongation $\mathrm{pr}^\infty(\pi)$ of a symmetry X of $J^k(\pi)$ has a flow obtained from the flow of X (cf. Chapter 4, § 4.9.4).

(1) Study of the structure of the tensorial product $\mathbf{\Gamma}(\pi) \otimes_{C^\infty(M)} C^\infty\left(J^\infty(\pi)\right)$.

(2) In functional analysis, a lot of spaces appear as direct limits of Fréchet structures or are modelled on such vector spaces (manifolds, Lie groups, bundles, Lie algebroids):

[1]The reader can find in the Appendix of Dodson, Galanis and Vassiliou book, a list of interesting geometrical problems in a Fréchet context (Chern-Weil theorem, holonomy theorem, ...) which could be solved using the *projective limit approach.*

- The space $\mathcal{D}$ (of test functions) on $\mathbb{R}^n$ of indefinitely differentiable functions on $\mathbb{R}^n$ with compact support is the direct limit of the Fréchet spaces $\mathcal{D}_K$ of smooth functions with support in the compact set K. It is a non-metrizable locally convex topological vector space.
- The Lie group $\mathrm{Diff}_c(M) = \varinjlim \mathrm{Diff}^\infty_{K_n}(M)$, where M is endowed with an exhaustive sequence of compact sets $K_1 \subset K_2 \subset \cdots$ with $K_n \subset \overset{\circ}{K}_{n+1}$, is modelled on the LF space of smooth vector fields with compact support (cf. [Glo07]).

The (LF)-spaces $\mathbb{F} = \bigcup_{n\in\mathbb{N}} \mathbb{F}_n$ are regular if and only if they are c^∞-completed and, in this case, the topology τ_{DL} coincides with the c^∞-topology, as in the case of LB-spaces. Moreover, a map $f : U \to Y$ from a c^∞-open set $U = \bigcup_{n\in\mathbb{N}} U_n$ of $\mathbb{F}$ in a convenient space Y is C^∞ if the maps $f_{|U_n}$ are C^∞.

(i) One can try to determine the assumptions on each structure at the level n in order to get a convenient structure on their direct limit.

(ii) The issue of existence and unicity of differential equations on direct limits of Fréchet spaces can be studied.

(3) One of the main problems in control theory is that of controllability: is it possible to drive the system from one point to another with a given class of controls? A classical result in finite-dimensional (non-linear) systems is Rashevsky-Chow theorem that gives a sufficient condition for controllability on any finite-dimensional connected manifold.
Some results have been obtained by M.K. Salehani and I. Markina in the paper [SaMa14] in the following framework: the state space M is a regular connected smooth manifold modelled on a locally convex space, the control set $\mathcal{U}$ is an arbitrary (usually closed) subset of some Euclidean space, and the dynamics is described by a map $F : M \times \mathcal{U} \to TM$ such that for each $u \in \mathcal{U}$, $F_u : M \to TM$ defined by $F_u(x) = F(x, u) \in T_xM$ for $x \in M$ is a smooth vector field. In this article, the authors obtain a generalization of Rashevsky-Chow theorem.

(4) The generalization of the finite-dimensional Poisson structures to the Banach framework gives rise to the notion of partial Poisson structure as defined in [OdRa03] by A. Odzijewicz and T. Ratiu. This generalization is extended to the convenient setting in Chapter 7.
In such a situation, the associated morphism takes values in the kinematic tangent bundle TM of a convenient manifold M can only be defined on a weak subbundle $T^\flat M$ of the kinematic cotangent bundle $T'M$.
Adapting the notion of tensors defined in the same chapter, this processus

of extension can be used for Nambu-Poisson and Jacobi structures which generalize the Poisson structure:

(i) *Nambu-Poisson structures* were introduced by Nambu in [Nam73]. In [Takh94], L. Takhtajan gave a formalism in terms of brackets on the algebra of smooth functions on a finite-dimensional manifold.
For a finite-dimensional manifold M, we denote $\mathcal{A}$ the algebra of smooth functions on M. A *Nambu-Poisson structure of order* r ($r \geq 2$) on M is a r-multilinear skew-symmetric map

$$\begin{array}{ccc} \mathcal{A}\times\cdots\times\mathcal{A} & \to & \mathcal{A} \\ (f_1,\ldots,f_r) & \mapsto & \{f_1,\ldots,f_r\} \end{array}$$

which satisfies the following properties

(L) Leibniz identity.

$$\{f_1,\ldots,f_{r-1},gh\} = \{f_1,\ldots,f_{r-1},g\}\,h + g\,\{f_1,\ldots,f_{r-1},h\}\,;$$

(FI) Fundamental identity or Filippov identity.

$$\begin{array}{rl} & \{f_1,\ldots,f_{r-1},\{g_1,\ldots,g_r\}\} \\ = & \displaystyle\sum_{i=1}^{r}\{g_1,\ldots,g_{i-1},\{f_1,\ldots,f_{r-1},g_i\},g_{i+1},\ldots,g_r\} \end{array}$$

for every $f_1,\ldots,f_{r-1},h,g_1,\ldots,g_r$ in $\mathcal{A}$.
For $r = 2$, **(FI)** is nothing but *Jacobi identity*.
In finite dimension, singularities of such structures are studied in [Duf2000], their normal forms in [DuZu99] and the deformations of the associated foliations in [MoZu20].
One can find in [CaPe21] a survey on partial Nambu-Poisson structures in the convenient setting.

(ii) The notion of *Jacobi manifold* in finite dimension was introduced independently by A.A. Kirillov in [Kir76]) and A. Lichnerowicz in [Lic78].
A finite-dimensional Jacobi manifold can be seen as a manifold M endowed with a skew-symmetric bi-differential operator on the functions space $C^\infty(M)$ of maximal order 1. So $C^\infty(M)$ is endowed with a structure of local Lie algebra, the so-called Jacobi bracket (cf [Kir76]).
In fact, a *Jacobi manifold* is a triple (M,P,X) where M is a manifold of dimension n, X is a vector field (called the Reeb field) and P is a 2-tensor such that:

$$\left\{\begin{array}{rl} [P,P] = & 2X\wedge P \\ [X,P] = & 0 \end{array}\right. .$$

This processus of extension can also be developed for

(a) *partial Dirac structures* defined on a weak Pontryagin bundle $T^{\mathfrak{p}}M = TM + T^{\flat}M$;

(b) *cometric structures.*

(5) The notion of groupoid was introduced by W. Brandt in the paper [Bra26] published in 1929. The theory of finite-dimensional Lie groupoids has received an increasing interes and has been used in various domains: in differential topology and geometry by C. Ehresman ([Ehr80]), in algebraic geometry by A. Grothendieck ([Gro61]), P. Higgins ([Hig71]) and R. Brown ([BroR87]), in analysis by G.W. Mackey ([Mckey66]) and in symplectic geometry by A. Weinstein ([Wei87]). In a series of papers, J. Pradines has extended the Lie theory from differential groups to groupoids ([Pra66], [Pra68]).
A first extension to the infinite-dimensional framework leads to consider the Banach-Lie groupoids as it was done in the paper [BGJP19].
In the continuation of this last work, we can ask how these results can be extended in the convenient setting and also how can be defined a projective (resp. direct) limit of of Banach groupoids. It might also be interesting to point out the eventual link with some differentiability questions in the theory of C^*-algebra.

(6) In finite dimension, the concept of Lie algebroid was introduced by Pradines in the 1960's ([Pra66] and [Pra68]). He associated to each Lie groupoid a Lie algebroid, and pointed out a correspondence between Lie groupoids and algebroids completely analogous to the classical theory for Lie groups and algebras and, in particular, *any Lie algebroid is isomorphic to the Lie algebroid of a Lie groupoid.*
Almeida and Molino ([AlMo85]) provided a counterexample to this Pradines assertion: they proved that a transversely complete foliation gives rise to a transitive Lie algebroid which is integrable if and only if the foliation is developable. However, there exists a wide set of finite-dimensional Lie algebroids which can be integrated by a Lie groupoid (see for instance [Daz97], [DoLa166] and [Nis2000]). A complete answer about the integrability of finite-dimensional Lie algebroid was established by Crainic and Fernandez in 2011 in [CrFe11]: there exists a local obstruction (of monodromy type) to the integrability of a Lie algebroid. Moreover, they give some geometrical characterizations of integrable Lie algebroids and they also show that any finite-dimensional Lie algebroid is "locally integrable". Recently, a new proof of this result of local integrability was produced in [CMS20].
We might look for conditions under which a Banach-Lie algebroid is integrable or locally integrable.

Bibliography

[AbPa98] M. Abate, G. Patrizio, *Isometries of the Teichmüller metric.* Ann. Scuola Norm. Sup. Pisa Cl. Sci. 26 (1998), no. 3, 437–452.

[AbTo11] M. Abate, F. Tovena, *Geometria differenziale.* Unitext, 54, Springer, Milan, 2011.

[AbbMa99] M. Abbati, A. Manià, *On differential structure for projective limits of manifolds.* J. Geom. Phys. 29 (1999), no. 1-2, 35–63.

[AbrMa78] R. Abraham, J. Marsden, *Foundations of mechanics* Second edition, revised and enlarged. Benjamin/Cummings Publishing Co., Inc., Advanced Book Program, Reading, Mass., 1978.

[AMR88] R. Abraham, J. E. Marsden, T. S. Ratiu, *Manifolds, tensor analysis, and applications.* Second edition. Applied Mathematical Sciences, 75. Springer-Verlag, New York, 1988.

[AbRo67] R. Abraham, J. Robbin, *Transversal mappings and flows.* W. A. Benjamin, Inc., New York-Amsterdam, 1967.

[AdFo03] R. A. Adams, J. J. F. Fournier, *Sobolev spaces,* Second edition. Pure and Applied Mathematics, 140. Elsevier/Academic Press, Amsterdam, 2003.

[ADGS09] M. Aghasi, C.T. Dodson, G.N. Galanis, A. Suri, *Conjugate connections and differential equations on infinite dimensional manifolds.* In: J. A. Álvarez López, E. García-Río (eds.), *Differential geometry,* World Scientific Publishing Hackensack, NJ, 2009, pp. 227–236.

[AgSu07] M. Aghasi, A. Suri, *Ordinary differential equations on infinite dimensional manifolds.* Balkan J. Geom. Appl. 12 (2007), no. 1, 1–8.

[AgSu10] M. Aghasi, A. Suri, *Splitting theorems for the double tangent bundles of Fréchet manifolds.* Balkan J. Geom. Appl. 15 (2010), no. 2, 1–13.

[AgSu13] M. Aghasi, A. Suri, *Connections and second order differential equations on infinite dimensional manifolds.* Int. Electron. J. Geom. 6 (2013), no. 2, 45–56.

[AlDu09] J-C.Álvarez Paiva, C-E.Durán, *Geometric invariants of fanning curves* Ann. Inst. Henri Poincaré, Advances in Applied Mathematics, 42-3 (2009) 290–312.

[AlMo85] R. Almeida, P. Molino, *Suites d'Atiyah et feuilletages transversalement complets.*
C. R. Acad. Sci. Paris 300 (1985) 13–15.

[AKSM18] A. Alekseev, Y. Kosmann-Schwarzbach, E. Meinrenken *Quasi-Poisson Manifolds.*
Can. J. Math. 54 (2018), no. 15.

[ABG96] W.O. Amrein, A. Boutet de Monvel, V. Georgescu, *C_0-groups, commutator methods and spectral theory of N-body Hamiltonians.* Progress in Mathematics, 135. Birkhäuser Verlag, Basel, 1996.

[AGS18] H. Amiri, H. Glockner, A. Schmeding, *Lie groupoids of mappings taking values in a Lie groupoid* arxiv.org/pdf/1811.02888

[Ana11] M. Anastasiei, *Banach Lie algebroids.* An. Ştiinţ. Univ. Al. I. Cuza Iaşi. Mat. (N.S.) 57 (2011), no. 2, 409–416.

[And92] I. Anderson, *Introduction to variational bicomplex.* In: M. J. Gotay, J. E. Marsden, V. Moncrief (eds.), *Mathematical aspects of classical field theory,* Contemp. Math., 132, Amer. Math. Soc., Providence, RI, 1992, pp. 51–73.

[ACM05] E. Andruchow, G. Corach, M. Mbekhta, *On the geometry of generalized inverses.* Math. Nachr. 278 (2005), no. 7-8, 756–770.

[Arn66] V. Arnold, *Sur la géométrie différentielle des groupes de Lie de dimension infinie et ses applications à l'hydrodynamique des fluides parfaits.* Ann. Inst. Fourier (Grenoble) 16 (1966), fasc. 1, 319–361.

[AzIo89] T. Ya. Azizov, I.S. Iokhvidov, *Linear operators in spaces with an indefinite metric.* Translated from the Russian by E. R. Dawson. Pure and Applied Mathematics (New York). A Wiley-Interscience Publication. John Wiley & Sons, Ltd., Chichester, 1989, pp. xii+304.

[BaDo02] R. Baker, C.J.L. Doran, *Jet bundles and the formal theory of partial differential equations.* In: L. Dorst, C. Doran, J. Lasenby (eds.), *Applications of geometric algebra in computer science and engineering*, Birkhäuser, Boston, MA, 2002, pp. 133–143.

[Bal94] R. Balan, *A note about integrability of distributions with singularities.* Boll. Un. Mat. Ital. A8 (1994), no. 3, 335–344.

[Bam99] D. Bambusi, *On the Darboux theorem for weak symplectic manifolds.* Proc. Amer. Math. Soc. 127 (1999), no. 11, 3383–3391.

[Ban32] S. Banach, *Théorie des opérations linéaires.* Monografje Matematyczne, tom I, Warszawa, 1932.

[Bar73] R. Barre, *De quelques aspects de la théorie des Q-variétés différentielles et analytiques.* Ann. Inst. Fourier (Grenoble) 23 (1973), no. 3, 227–312.

[Bas64] A. Bastiani, *Applications différentiables et variétés différentiables de dimension infinie.* J. Analyse Math. 13 (1964), 1–114.

[BBM12] M. Bauer, M. Bruveris, P. W. Michor, *Homogeneous Sobolev Metric of Order One on Diffeomorphism Groups on Real Line.* J Nonlinear Sci 24(5).

[Bell14] J. Bell, *Projective limits of topological vector spaces.* Department of Mathematics, University of Toronto (2014).

[Belm08] A. Belmiloudi, *Stabilization, optimal and robust control.* Communications and Control Engineering Series. Springer-Verlag, London, Berlin, 2008.

[Belt105] D. Beltiţă, *Integrability of analytic almost complex structures on Banach manifolds.* Ann. Global Anal. Geom. 28 (2005), no. 1, 59–73.

[Belt06] D. Beltiţă, *Smooth homogeneous structures in operator theory.* Chapman & Hall/CRC Monographs and Surveys in Pure and Applied Mathematics, 137, Chapman & Hall/CRC Press, Boca Raton-London-New York-Singapore, 2006.

[BGJP19] D. Beltiţă, T. Goliński, G. Jakimowicz, F. Pelletier, *Banach-Lie groupoids and generalized inversion.* J. Funct. Anal. 276 (2019), no. 5, 1528–1574.

[BGT18] D. Beltiţă, T. Goliński, A. B. Tumpach, *Queer Poisson brackets.* J. Geom. Phys. 132 (2018), 358–362.

[BPZ19] D. Beltiţă, F. Pelletier, A. Zergane, *On the smoothness of some quotients of Lie groups.* Differential Geom. Appl. **67** (2019), 101556, 22.

[BeRa05] D. Beltiţă, T. S. Ratiu, *Symplectic leaves in real Banach Lie-Poisson spaces.* Geom. Funct. Anal. 15 (2005), no. 4, 753–779.

[BRT07] D. Beltiţă, T. S. Ratiu, A. B. Tumpach, *The restricted Grassmannian, Banach Lie-Poisson spaces, and coadjoint orbits.* J. Funct. Anal. 247 (2007), no. 1, 138–168.

[BGN04] W. Bertram, H. Glöckner, K-H . Neeb *Direct limits of regular Lie groups.* Expositiones Mathematicae, (2004)

[Bil04] L. Biliotti, *The exponential map of a weak Riemannian Hilbert manifold.* Illinois J. Math. 48 (2004), no. 4, 1191–1206.

[Bir36] G. Birkhoff, *Continuous groups and linear spaces.* Rec. Math. [Mat. Sbornik] N.S., 1(43) (1936), no. 5, 635–642

[Bir38] G. Birkhoff, *Analytical groups.* Trans. Amer. Math. Soc. 43 (1938), no. 1, 61–101.

[BhVi98] K. H. Bhaskara, K. Viswanath, *Poisson algebras and Poisson manifolds.* Pitman Research Notes in Mathematics Series, 174. Longman Scientific & Technical, Harlow, 1988.

[Bla98] B. Blackadar, *K-theory for operator algebras.* Second edition. Mathematical Sciences Research Institute Publications, 5. Cambridge University Press, Cambridge, 1998.

[BCDKSTV99] A. V. Bocharov, V. N. Chetverikov, S. V. Duzhin, N. G. Khor'kova, A. V. Samokhin, Yu. N. Torkhov, A. M. Verbovetsky, A.M. Vinogradov, *Symmetries and conservation laws for differential equations of mathematical physics.* Translations of Mathematical Monographs, 182. American Mathematical Society, Providence, RI, 1999.

[Bog74] J. Bognár, *Indefinite inner product spaces.* Ergebnisse der Mathematik und ihrer Grenzgebiete, Band 78. Springer-Verlag, New York-Heidelberg, 1974.

[BoBo02] A. V. Bolsinov, A.V. Borisov, *Compatible Poisson brackets on Lie algebras.* Math. Notes 72 (2002), no. 1-2, 10–30.

[Bom67] J. Boman, *Differentiability of a function and of its compositions with functions of one variable.* Math. Scand. 20 (1967), 249–268.

[Bouc11] M. Boucetta, *Riemannian geometry of Lie algebroids.* J. Egyptian Math. Soc. 19 (2011), no. 1-2, 57–70.

[Bou67] N. Bourbaki, *Variétés différentiables et analytiques. Fascicule de résultats.* Hermann, Paris, 1967.

[Bou70] N. Bourbaki, *Éléments de mathématique. Théorie des ensembles.* Hermann, Paris, 1970, 349 pp.

[Bou71] N. Bourbaki, *Éléments de mathématique. Topologie générale.* Chapitres 1 à 4. Hermann, Paris, 1971, xv+357 pp.

[Bou06] N. Bourbaki, *Algèbre.* Chapitres 1 à 3, 2ème édition, Springer, 2006.

[Bou07] N. Bourbaki, *Espaces vectoriels topologiques.* Chapitres 1 à 5, Springer, 2007.

[Bourl19] H. Bourlès, *Fundamentals of Advanced Mathematics - 3.* ISTE-Elsevier, 2019.

[Bra26] W. Brandt, *Über eine Verallgemeinerung des Gruppenbegriffes.* Math. Ann. 96 (1926) 360–366.

[Bous1877] J. Boussinesq, *Essai sur la théorie des eaux courantes.* Mémoires présentés par divers savants à l'Académie des Sciences de l'Institut de France, XXIII (1877) 666–680.

[BroR87] R. Brown, *From groups to groupoids: A brief survey.* Bull. London Math. Soc. 19 (1987), no. 2, 113–134.

[Bru17] M. Bruveris, *Regularity of maps between Sobolev spaces.* Ann. Global Anal. Geom. 52 (2017), no. 1, 11–24.

[Bru18] M. Bruveris, *The L^2-metric on $C^k(M, N)$.* arxiv.org/abs/1804.00577

[Buc05] I. Bucătaru, *Linear connections for systems of higher order differential equations.* Houston J. Math. 31 (2005), no. 2, 315–332.

[Cab12] P. Cabau, *Strong projective limit of Banach Lie algebroids.* Port. Math. 69 (2012), no. 1, 1–21.

[Cab19] P. Cabau, *Projective limits of local shift morphisms.* Appl. Sci. 21 (2019), 69–83.

[CGM91] P. Cabau, J. Grifone, M. Mehdi, *Existence de lois de conservation dans le cas cyclique.* Ann. Inst. H. Poincaré Phys. Théor. 55 (1991), no. 3, 789–803.

[CaPe12] P. Cabau, F. Pelletier, *Almost Lie structures on an anchored Banach bundle.* J. Geom. Phys. 62 (2012), no. 11, 2147–2169.

[CaPe19] P. Cabau, F. Pelletier, *Integrability on direct limits of Banach manifolds.* Ann. Fac. Sci. Toulouse Math. (6) 28 (2019), no. 5, 909–956.

[CaPe20] P. Cabau, F. Pelletier, *Projective and direct limits of Banach tensor structures.* Differ. Geom. Dyn. Syst. 22 (2020), 42–86.

[CaPe21] P. Cabau, F. Pelletier, *A survey on partial Nambu-Poisson structures in the convenient setting.* arXiv:2112.14592 [math.DG] (2021).

[CMS20] A. Cabrera, I. M ărcut, M-A. Salazar, *On local integration of Lie brackets.* J. Reine. Angew. Math. 760 (2020), 267–293.

[Cam10] C.-M. Campos, *Geometric methods in classical field theory and continuous media.* Tesis doctoral, Universidad Autónoma de Madrid, 2010.

[Can08] A. Cannas da Silva, *Lectures on symplectic geometry.* Lecture Notes in Mathematics, 1764. Second printing. Springer-Verlag, Berlin, 2008.

[CaMa01] J.F. Cariñena, E. Martínez, *Lie algebroid generalization of geometric mechanics.* In: J. Kubarski, P. Urbański, R. Wolak (eds.), *Lie algebroids and related topics in differential geometry,* Banach Center Publ., 54, Polish Acad. Sci. Inst. Math., Warsaw, 2001, pp. 201–215.

[Caro05] N. L. Carothers, *A short course on Banach space theory.* London Mathematical Society Student Texts, 64. Cambridge University Press, Cambridge, 2005.

[Carr14] W. Cuellar Carrera, *A Banach space with a countable infinite number of complex structures.* J. Funct. Anal. 267 (2014), no. 5, 1462–1487.

[Cart71] H. Cartan, *Calcul différentiel.* Hermann, Paris, 1971.

[CaXu04] A. S. Cattaneo, P. Xu, *Integration of twisted Poisson structures.* J. Geom. Phy. 49 (2004), no. 2, 187–196.

[Cha09] M. Chaperon, *En souvenir de Paulette Libermann.* Gaz. Math. No. 122 (2009), 63–80.

[ChMa74] P. R. Chernoff, J. E. Marsden, *Properties of infinite dimensional Hamiltonian systems.* Lecture Notes in Mathematics, Vol. 425. Springer-Verlag, Berlin-New York, 1974.

[Che91] V. N. Chetverikov, *On the structure of integrable C-fields.* Differential Geom. Appl. 1 (1991), no. 4, 309–325.

[ChSt76] D. Chillingworth, P. Stefan, *Integrability of singular distributions on Banach manifolds.* Math. Proc. Cambridge Philos. Soc. 79 (1976), no. 1, 117–128.

[Choq73] G. Choquet, *Cours d'Analyse.* Tome II: Topologie. Masson et Cie Editeurs, Paris, 1973.

[ChDeDi82] Y. Choquet-Bruhat, C. DeWitt-Morette, M. Dillard-Bleick, *Analysis, manifolds and physics.* Second edition. North-Holland Publishing Co., Amsterdam-New York, 1982.

[ClGo72] R. S. Clark, D. S. Goel, *On the geometry of an almost tangent manifold.* Tensor (N.S.) 24 (1972), 243–252.

[ClGo74] R. S. Clark, D. S. Goel, *Almost cotangent manifolds.* J. Differential Geometry 9 (1974), 109–122.

[Con94] A. Connes, *Noncommutative geometry.* Academic Press, Inc., San Diego, CA, 1994, xiv+661 pp.

[CLMM09] J. Cortés, M. de León, J.C. Marrero, E. Martínez, *Nonholonomic Lagrangian systems on Lie algebroids.* Discrete Contin. Dyn. Syst. 24 (2009), no. 2, 213–271.

[CrFe11] M. Crainic, R.L. Fernandes, *Lectures on integrability of Lie brackets.* In: T. Ratiu, A. Weinstein, N. T. Zung, *Lectures on Poisson geometry,* Geom. Topol. Monogr., 17, Geom. Topol. Publ., Coventry, 2011, pp. 1–107.

[CrFa01] A.B. Cruzeiroa, S. Fangb, *Weak Levi-Civita Connection for the Damped Metric on the Riemannian Path Space and Vanishing of Ricci Tensor in Adapted Differential Geometry.* J. Funct. Anal. 18 (2001), no. 2, 681–698.

[Dah11] R. Dahmen, *Direct limit constructions in infinite dimensional Lie theory.* PhD Thesis, Universität Paderborn, 2011.

[Daz97] P. Dazord, *Groupoïde d'holonomie et géométrie globale.* C. R. Acad. Sci. Paris 324(1997), 77-80.

[Daz85] P. Dazord, *Feuilletages à singularités.* [Foliations with singularities] Nederl. Akad. Wetensch. Indag. Math. 47 (1985), no. 1, 21–39.

[DLM91] P. Dazord, A. Lichnerowicz, C.-M. Marle, *Structure locale des variétés de Jacobi.* J. Math. Pures Appl. 70 (1991), no. 1, 101–152.

[Del72] P. de la Harpe, *Classical Banach-Lie algebras and Banach-Lie groups of operators in Hilbert space.* Lecture Notes in Mathematics, Vol. 285. Springer-Verlag, Berlin-New York, 1972.

[Die52] J. Dieudonné, *Complex structures on real Banach spaces.* Proc. Amer. Math. Soc. 3 (1952), 162–164.

[Die67] J. Dieudonné, *Fondements de l'analyse Moderne.* Cahiers Scientifiques, vol. XXVIII, Gauthiers-Villars, Paris, 1967.

[Dix53] J. Dixmier, *Sur les bases orthonormales dans les espaces préhilbertiens.* Acta Sci. Math. (Szeged) 15:1-1(1953-54), 29-30.

[Dix64] J. Dixmier, *Les $C^\star$-algèbres et leurs représentations.* Cahiers Scientifiques, Fasc. XXIX Gauthier-Villars & Cie, Editeur-Imprimeur, Paris, 1964.

[Dix69] J. Dixmier, *Les algèbres d'opérateurs dans l'espace hilbertien (algèbres de von Neumann).* Deuxième édition, revue et augmentée. Cahiers Scientifiques, Fasc. XXV. Gauthier-Villars Editeur, Paris, 1969.

[DoGa04] C. T. J. Dodson, G. N. Galanis, *Second order tangent bundles of infinite dimensional manifolds.* J. Geom. Phys. 52 (2004), no. 2, 127–136.

[DGV16] C. T. J. Dodson, G .N. Galanis, E. Vassiliou, *Geometry in a Fréchet context. A projective limit approach.* London Mathematical Society, Lecture Note Series 428. Cambridge University Press, 2016.

[DoLa166] A. Douady, M. Lazard, *Espaces fibrés en algèbres de Lie et en groupes.* Invent. Math. 1, 1966, 133–151.

[DrYa05] D. Drivaliaris, N. Yannakakis, *Hilbert space structure and positive operators.* J. Math. Anal. Appl. 305 (2005), no. 2, 560–565.

[Dub72] E. Dubinsky, *Projective and inductive limits of Banach spaces.* Studia Math. 42 (1972), 259–263.

[Duf91] J.-P. Dufour, *Introduction aux tissus.* In: *Séminaire Gaston Darboux de Géométrie et Topologie Différentielle. 1990–1991.* Univ. Montpellier II, Montpellier, 1992, pp. 55–76.

[Duf2000] J.-P. Dufour, *Singularities of Poisson and Nambu structures.* Poisson geometry (Warsaw, 1998), 61–68, Banach Center Publ., 51, Polish Acad. Sci. Inst. Math., Warsaw, 2000.

[DuZu05] J.-P. Dufour, N. T. Zung, *Poisson structures and their normal forms.* Progress in Mathematics, 242. Birkhäuser Verlag, Basel, 2005.

[DuZu99] J.-P. Dufour, N.T. Zung, *Linearization of Nambu structures.* Compositio Math. 117 (1999), no. 1, 77–98.

[Ebi67] D. G. Ebin, *On the space of Riemannian metrics.* Doctoral Thesis, Massachusetts Institute of Technology, Cambridge, Mass., 1967.

[Eel66] J. Eells, *A setting for global analysis.* Bull. Amer. Math. Soc. 72 (1966), 571–807.

[Eft15] K. Eftekharinasab, *The Morse-Sard-Brown theorem for functionals on bounded Fréchet-Finsler manifolds.* Commun. Math. 23 (2015), no. 2, 101–112.

[Eft16] K. Eftekharinasab, *Fréchet Lie algebroids and their cohomology.* Armen. J. Math. 8 (2016), no. 1, 77–85.

[Eft19] K. Eftekharinasab, *On the generalization of the Darboux theorem.* Proceedings of the International Geometry Center 12 (2), 1-10 (2019).

[Ehr51] C. Ehresmann, *Sur la théorie des variétés feuilletées.* Univ. Roma Ist. Naz. Alta Mat. Rend. Mat. Appl. (5) 10 (1951), 64–82.

[Ehr53] C. Ehresmann, *Introduction à la théorie des structures infinitésimales et des pseudo-groupes de Lie.* In: *Géométrie différentielle.* Colloques Internationaux du Centre National de la Recherche Scientifique, Strasbourg, 1953, pp. 97–127.

[Ehr80] C. Ehresmann, *Oeuvres compl'etes et commentées.* (A. C. Ehresmann, ed.) Suppl. Cahiers Top. Géom. Diff., Amiens, 1980–1984.

[EiMc45] S. Eilenberg, S. Mac Lane, *General theory of natural equivalences*, Trans. Amer. Math. Soc. 58 (1945), 231–294.

[EES90] M. Elżanowski, M. Epstein, J. Śniatycki, *G-structures and material homogeneity.* J. Elasticity 23 (1990), no. 2-3, 167–180.

[Eyn14] J. M. Eyni, *The Frobenius theorem for Banach distributions on infinite dimensional manifolds and applications in infinite dimensional Lie theory* arXiv:1407.3166 (2014).

[FMPZ99] G. Falqui, F. Magri, M. Pedroni, P. Zubelli, *A Bi-Hamiltonian Theory for Stationary KdV Flows and their Separability*, SISSA 137/99/FM, 1999.

[Fau11] C.-A. Faure, *Théorie de la différentiation dans les espaces convenables*, Thèse, Université de Genève (2011).

[Fer07] V. Ferenczi, *Uniqueness of complex structure and real hereditarily indecomposable Banach spaces.* Adv. Math. 213 (2007), no. 1, 462–88.

[FeGa10] V. Ferenczi, E. M. Galego, *Countable groups of isometries on Banach spaces.* Trans. Amer. Math. Soc. 362 (2010), no. 8, 4385–4431.

[Fer02] R. L. Fernandes, *Lie algebroids, holonomy and characteristic classes*, Adv. Math. 170 (2002), 119–179.

[FeMa14] R.-L. Fernandes, I. Mărcuţ, *Lectures on Poisson Geometry.* Preprint, 2014.

[FLMR97] M. Fliess, J. Lévine, P. Martin, P. Rouchon, *Deux applications à la géométrie locale des diffiétés.* Ann. Inst. H. Poincaré Phys. Théo. 66 (1997), no. 3, 275–292.

[Flo71] K. Floret, *Lokalkonvexe Sequenzen mit kompakten Abbildungen.* J. Reine Angew. Math. 247 (1971), 155–195.

[Flo79] K. Floret, *On bounded sets in inductive limits of normed spaces.* Proc. AMS 75 (1979), no. 2, 221-225.

[Fre1906] M. Fréchet, *Sur quelques points de calcul fonctionnel.* Rend. Circ. Mat. Palermo 22 (1906), 1–74.

[Fro81] A. Frölicher, *Applications lisses entre espaces et variétés de Fréchet.* (French) [[Smooth maps between Fréchet spaces and manifolds]] C. R. Acad. Sci. Paris Sér. I Math. 293 (1981), no. 2, 125–127.

[FrKr88] A. Frölicher, A. Kriegl, *Linear spaces and differentiation theory.* Pure and Applied Mathematics (New York). A Wiley-Interscience Publication. John Wiley & Sons, Ltd., Chichester, 1988.

[FrNi56] A. Frölicher, A. Nijenhuis, *Theory of vector-valued differential forms. I. Derivations of the graded ring of differential forms.* Nederl. Akad. Wetensch. Proc. Ser. A. 59 = Indag. Math. 18 (1956), 338–359.

[Gag18] L. Gagnon, *Etude de léquation de Korteweg-de Vries en variables lagrangiennes et sa contrôlabilité, stabilisation rapide dune équation de Schrödinger et mthodes spectrales pour le calcul du contrôle optimal.* Thèse, Université Paris VI, 2018.

[Gal96] G.N. Galanis, *Projective limits of Banach-Lie groups.* Period. Math. Hungar. 32 (1996), no. 3, 179–191.

[Gal98] G.N. Galanis, *Projective limits of Banach vector bundles.* Portugal. Math. 55 (1998), no. 1, 11–24.

[Gal04] G.N. Galanis, *Differential and geometric structure for the tangent bundle of a projective limit manifold.* Rend. Sem. Mat. Univ. Padova 112 (2004), 103–115.

[GGKM67] C.S. Gardner, J.M. Greene, M.D. Kruskal and R.M. Miura, *Method for solving the Korteveg-de Vries equation.* Phys. Rev. Lett. 19 (1967) 1095–1098.

[Gas85] J. Gasqui, *Formal integrability of systems of partial differential equations.* In: N. Sanchez (ed.), *Nonlinear equations in classical and quantum field theory* (Meudon/Paris, 1983/1984), 21–36, Lecture Notes in Phys., 226, Springer, Berlin, 1985, pp. 32–36.

[Glo02a] H. Glöckner, *Infinite-dimensional Lie groups without completeness condition* in Geometry and Analysis on finite and infinite-dimensional Lie groups, Eds. A. Strasburger, W. Wojtynski, J. Hilgert and K.-H. Neeb, Banach Center Publications 55 (2002), 43–59.

[Glo02b] H. Glöckner, *Algebras whose groups of units are Lie groups.* Studia Math. 153 (2002), no. 2, 144–147.

[Glo03] H. Glöckner, *Direct limits of Lie groups and manifolds*, J. Math. Kyoto Univ. (JMKYAZ) 43-1 (2003) 1–26.

[Glo05] H. Glöckner, *Fundamentals of direct limit Lie theory.* Compos. Math. 141 (2005), no. 6, 1551–1577.

[Glo07] H. Glöckner, *Direct limits of infinite-dimensional Lie groups compared to direct limits in related categories.* J. Funct. Anal. 245 (2007), no. 1, 19–61.

[Glo08] H. Glöckner, *Stefan-Sussmann theorem for distributions of not necessarily complemented vector subspaces on Banach manifolds.* Notes, Budapest, March, 2008.

[Glo11] H. Glöckner, *Direct limits of infinite-dimensional Lie groups.* In: K.-H. Neeb, A. Pianzola (eds.), *Developments and trends in infinite-dimensional Lie theory,* Progr. Math., 288, Birkhäuser Boston, Boston, MA, 2011, pp. 243–280.

[Glo16] H. Glöckner, *Fundamentals of submersions and immersions between infinite-dimensional manifolds.* Preprint arXiv:1502.05795v4 [math.DG], 2016.

[Glo18] H. Glöckner, *Completeness of infinite-dimensional Lie groups in their left uniformity.* Canad. J. Math. 71 (2019), no. 1, 131–152.

[Glo19] H. Glöckner, *Direct limits of regular Lie groups.* Preprint arXiv:1902.06329 [math.GR], 2019.

[Godm73] R. Godement, *Topologie algébrique et théorie des faisceaux.* Troisième édition revue et corrigée. Publications de l'Institut de Mathématique de l'Université de Strasbourg, XIII. Actualités Scientifiques et Industrielles, No. 1252. Hermann, Paris, 1973.

[Gol67] H. Goldschmidt, *Integrability criteria for systems of nonlinear partial differential equations.* J. Differential Geometry 1 (1967), 269–307.

[GGU06] K. Grabowska, J. Grabowski, P. Urbanski, *Geometrical mechanics on algebroids.* Int. J. Geom. Methods Mod. Phys. 3 (2006), no. 3, 559–575.

[GrJo11] J. Grabowski, M. Jóźwikowski, *Pontryagin maximum principle on almost Lie algebroids.* SIAM J. Control Optim. 49 (2011), no. 3, 1306–1357.

[Gri72] J. Grifone, *Structure presque-tangente et connexions. I.* Ann. Inst. Fourier 22 (1972), no. 1, 287–334.

[GrMe97] J. Grifone, M. Mehdi, *Existence of conservation laws and characterization of recursion operators for completely integrable systems.* Trans. Amer. Math. Soc. 349 (1997), no. 11, 4609–4633.

[Gro61] A. Grothendieck, *Techniques de construction et théorèmes d'existence en géométrie algébrique. III. Préschemas quotients.* Séminaire Bourbaki 13e année 1960/61, no. 212 (1961).

[Gui75] V. Guillemin, *Clean intersection theory and Fourier integrals.* In: J. Chazarain (ed.), *Fourier integral operators and partial differential equations.* Lecture Notes in Math., Vol. 459, Springer, Berlin, 1975, pp. 23–35.

[GuJa04] O. Gutú, J. A. Jaramillo, *Fibrations on Banach manifolds.* Pacific J. Math. 215 (2004), no. 2, 313–329.

[Hae56] A. Haefliger, *Sur les feuilletages analytiques.* C. R. Acad. Sci. Paris 242 (1956), 2908–2910.

[Hae05] A. Haefliger, *Naissance des feuilletages d'Ereshmann-Reeb à Novikov,* In Géomtrie au XXe Siècle, 1930-2000 : Histoire et Horizons Hermann (2005). http://www.foliations.org.

[Ham82] R. Hamilton, *The inverse function theorem of Nash and Moser.* Bull. Amer. Math. Soc. **7** (1982), 65–222.

[Han71] V.L. Hansen, *Some theorems on direct limits of expanding sequences of manifolds.* Math. Scand. 29 (1971), 5–36.

[HaJo14] P. Hájek, M. Johanis, *Smooth analysis in Banach spaces.* De Gruyter Series in Nonlinear Analysis and Applications, 19. De Gruyter, Berlin, 2014.

[HaSt72] D.W. Hajek, G.E. Strecker, *Direct limits of Hausdorff spaces.* In: J. Novák (ed.), General topology and its relations to modern analysis and algebra. III. Academia, Prague, 1972.

[Har77] R. Hartshorne, *Algebraic geometry.* Graduate Texts in Mathematics, No. 52. Springer-Verlag, New York-Heidelberg, 1977.

[Hei82] R. E. Heisey, *Manifolds modelled on the direct limit of lines.* Pac. J. Math. 102 (1982), no. 1.

[Herm63] R. Hermann, *On the Accessibility Problem in Control Theory.* In: Internat. sympos. nonlinear differential equations and nonlinear mechanics. Academic Press, New York, 1963, pp. 325–332.

[HRG85] J. M. Hernando, E. Reyes, P. M. Gadea, *Integrability of tensor structures of electromagnetic type.* Publ. Inst. Math. (Beograd) (N.S.) 37 (1985), 113–122.

[Herr69] H. Herrlich, *Separation axioms and direct limits.* Canad. Math. Bull. 12 (1969), 337–338.

[Hig71] P. J. Higgins, *Notes on categories and groupoids.* Van Nostrand Rienhold Mathematical Studies, No. 32. Van Nostrand Reinhold Co., London-New York-Melbourne, 1971.

[HiMa90] P. J. Higgins, K. Mackenzie, *Algebraic constructions in the category of Lie algebroids.* J. Algebra 129 (1990), no. 1, 194–230.

[Hil2000] Hiltunen, S. A Frobenius theorem for locally convex global analysis. Monatsh. Math. 129 (2000), no. 2, 109–117.

[HSTH01] T. Hirai, H. Shimomura, N. Tatsuuma, E. Hirai, *Inductive limits of topologies, their direct product, and problems related to algebraic structures.* J. Math. Kyoto Univ. 41 (2001), no. 3, 475–505.

[HoMo07] K. H. Hofmann, S. A. Morris, *The Lie theory of connected pro-Lie groups.* EMS Tracts in Mathematics, 2. European Mathematical Society (EMS), Zürich, 2007.

[HoNe08] K.H. Hofmann, K.-H. Neeb, *Pro-Lie groups which are infinite-dimensional Lie groups.* Math. Proc. Cambridge Philos. 46 (2008), no. 2, 351–378.

[Ida11] C. Ida, *Lichnerowicz-Poisson cohomology and Banach Lie algebroids.* Ann. Funct. Anal. 2 (2011), no. 2, 130–137.

[IVV04] D. Igonin, A. Verbovetsky, R. Vitolo, *Variational Multivectors and Bracketsin the Geometry of Jet Spaces.* Proceedings of Institute of Mathematics of NAS of Ukraine 2004, Vol. 50, Part 3, 1335–1342.

[ItMa75] A. R. Its, V. B. Matveev, *Hill operators with finitely many lacunae and multisoliton solutions of the Kortewegde Vries equation.* Trudy Mat. Fiz. 23 (1975) 51–67.

[Jar81] H. Jarchow, *Locally convex spaces.* Mathematische Leitfäden. B. G. Teubner, Stuttgart, 1981.

[Jon66] W. B. Jones, *Duality and types of completeness in locally convex spaces.* Pacific J. Math. 18 (1966), 525–544. *Erratum* in Pacific J. Math. 19 (1966), 588.

[Kap91] T. Kappeler, *Fibration of the phase space for the Korteveg-de-Vries equation.* Ann. Inst. Fourier (Grenoble), 41, 1 (1991) 539–575.

[KaMa98] T. Kappeler, M. Makarov, *On the symplectic foliation induced by the second Poisson bracket for KdV.* In: *Symmetry and perturbation theory,* Quad. Cons. Naz. Ricerche Gruppo Naz. Fis. Mat. 54, Roma (1998) 135–152.

[KaPo03] T. Kappeler, J. Pöschel, *On the Korteweg-de Vries Equation and KAM theory.* In: S. Hildebrandt, H. Karcher (eds), *Geometric analysis and nonlinear partial differential equations.* Springer, Berlin, 2003, pp. 397–416.

[KaSw82] N. J. Kalton R. C. Swanson, *A symplectic Banach space with no Lagrangian subspaces.* Trans. Amer. Math. Soc. 273 (1982), no. 1, 385–392.

[Kar87] M. Karasev, *Analogues of objects of the theory of Lie groups for nonlinear Poisson brackets.* (Russian) Izv. Akad. Nauk SSSR Ser. Mat. 50 (1986), no. 3, 508–538, 638; transl in Math. USSR Izvest. 28 (1987), no. 3, 497–527.

[Kel74] H.H. Keller, *Differential calculus in locally convex spaces.* Lecture Notes in Mathematics, 417. Springer-Verlag, Berlin, 1974.

[KeTa18] M. Kerjean, C. Tasson *Mackey-complete spaces and power series —a topological model of differential linear logic.* Math. Structures Comput. Sci. 28 (2018), no. 4, 472–507.

[KhMi03] B. Khesin, G. Misiołek, *Euler equations on homogeneous spaces and Virasoro orbits.* Adv. Math. 176 (2003), no. 1, 116–144.

[Kir76] A. A. Kirillov, *Local Lie algebras.* (Russian). Uspehi Mat. Nauk 31 (1976), no. 4 (190), 57–76. *Correction* in Uspehi Mat. Nauk 32 (1977), no. 1 (193), 268. Transl. in Russian Math. Surveys **31**, 4 (1976) 55–75.

[Kis12] A.V. Kiselev, *The twelve lectures in the (non)commutative geometry of differential equations.* Preprint IHES M-12-13, 2012.

[KiVa11] A. V. Kiselev, J. W. van de Leur, *Variational Lie algebroids and homological evolutionary vector fields.* Theoret. and Math. Phys. 167 (2011), no. 3, 772–784.

[Kle62] J. Klein, *Espaces variationnels et mécanique.* Ann. Inst. Fourier (Grenoble) 12 (1962), 1–124.

[Klo11] M. Klotz, *The automorphism group of a Banach principal bundle with* {**1**}*-structure.* Geom. Dedicata 154 (2011), 161–182.

[KMS93] I. Kolář, P.W. Michor, J. Slovák, *Natural operations in differential geometry.* Springer-Verlag, Berlin, 1993.

[Kole07] K. Kolev, *Poisson brackets in hydrodynamics.* Discrete Contin. Dyn. Syst. 19 (2007), no. 3, 55–574.

[Kolm34] A.N. Kolgomorov, *Zur Normierbarket eines allgemeinen topologischen linearen Raumes.* Studia Math. 5 (1934), 29–33.

[KoKu16] E. Korotyaev, S. Kuksin, *KdV Hamiltonian as a Function of Actions*. Journal of Dynamical and Control Systems, Springer Verlag, (2016) 22 (4), 661–682.

[KoVr1895] D. J. Korteweg, G. de Vries, *On the change of form of long waves advancing in a rectangular canal, and on a new type of long stationary waves*. Phil. Mag. 39 (1895), 422–443.

[KSMac02] Y. Kosmann-Schwarzbach, K. C. H. Mackenzie, *Differential operators and actions of Lie algebroids*. In: T. Voronov (ed.), *Quantization, Poisson brackets and beyond*, ed. Theodore Voronov, Contemp. Math., 315, Amer. Math. Soc., Providence, R.I., 2002, pp. 213–233.

[KSMag90] Y. Kosmann-Schwarzbach, F. Magri, *Poisson-Nijenhuis structures*. Ann. Inst. H. Poincare Phys. Theor 53 (1990), 35–81.

[Kot69] G. Köthe, *Topological vector spaces*. I. Die Grundlehren der mathematischen Wissenschaften, 159. Springer-Verlag, New York, 1969.

[Kot79] G. Köthe, *Topological vector spaces*. II. Grundlehren der Mathematischen Wissenschaften, 237. Springer-Verlag, New York-Berlin, 1979.

[KrKe2000] I.S. Krasil'shchik, P.H.M. Kersten, *Symmetries and recursion operators for classical and supersymmetric differential equations*. Mathematics and its Applications, 507. Kluwer Academic Publishers, Dordrecht, 2000.

[KrVe11] J. Krasil'shchik, A. Verbovetsky, *Geometry of jet spaces and integrable systems*. J. Geom. Phys. 61 (2011), no. 9, 1633–1674.

[Kri82] A. Kriegl, *Die richtigen Räume für Analysis im Unendlich-Dimensionalen*. (German) [[The "right" spaces for analysis in infinite dimensions]] Monatsh. Math. 94 (1982), no. 2, 109–124.

[KrMi97] A. Kriegl, P.W. Michor, *The convenient setting of global analysis*. Mathematical Surveys and Monographs, 53. American Mathematical Society, Providence, RI, 1997.

[Krey78] E. Kreyszig, *Introductory functional analysis with applications*. John Wiley & Sons. Inc. (1978).

[KuMcK89] J. Kucera, K. McKennon, *Completeness of regular inductive limits*. Internat. J. Math. & Math. Sci. 12 (1989), no. 3, 425–428.

[Kum13a] P. Kumar, *Almost complex structure on path space.* Int. J. Geom. Methods Mod. Phys. 10 (2013), no. 3, 1220034, 7 pp.

[Kum13b] P. Kumar, *Existence of 'Darboux chart' on loop space.* Preprint arXiv:1309.2190 (2013).

[Kum15] P. Kumar (Mishra), *Darboux chart on projective limit of weak symplectic Banach manifold.* Int. J. Geom. Methods Mod. Phys. 12 (2015), no. 7, 1550072, 13 pp.

[Lan95] S. Lang, *Differential and Riemannianm Manifolds.* Graduate Texts in Mathematics, 160. Springer, New York, 1995.

[Lar19a] G. Larotonda, *Estructuras geométricas para las variedades de Banach.* Book preprint, 2019.

[Lar19b] G. Larotonda, *Metric geometry of infinite dimensional Lie groups and their homogeneous spaces.* Forum Math. 31 (2019), no. 6, 1567–1605.

[Las57] R. K. Lashoff, *Lie algebras of locally compact groups.* Pacific J. Math. 7 (1957), 1145–1162.

[Lav18] S. Lavau, *A short guide through integration theorems of generalized distributions.* Differential Geom. Appl. 61 (2018), 42–58.

[Lax57] P.D. Lax, *Hyperbolic systems of conservation laws.* II. Comm. Pure Appl. Math. 10 (1957), 537–566.

[Lax74] P. D. Lax, *Periodic solutions of the K-dV equation.* Lectures in Appl. Math. 15 (1974) 85–96.

[LMM10] M. de León, J. C. Marrero, E. Martínez, *Linear almost Poisson structures and Hamilton-Jacobi equation. Applications to nonholonomic Mechanics*, J. Geom. Mech. 2010, 159–198.

[Lesf14] A. Lesfari, *Moser's lemma and the Darboux theorem*, Universal Journal of Applied Mathematics 2 (2014), no. 1, 36–39.

[Lesl82] J. Leslie, *On the group of analytic diffeomorphisms of a compact real analytic manifold.* Trans. Amer. Math. Soc. 274 (1982), no. 2, 651–669.

[Lib55] P. Libermann, *Sur les structures presque complexes et autres structures infinitésimales régulières.* Bull. Soc. Math. France 83 (1955), 195–224.

[Lic77] A. Lichnerowicz, *Les variétés de Poisson et leurs algèbres de Lie associées.* J. Differential Geometry 12 (1977), no. 2, 253–300.

[Lic78] A. Lichnerowicz, *Les variétés de Jacobi et leurs algèbres de Lie associées*, J. Math. Pures et Appliquées **51** (1978), 453–488.

[Lie1880] S. Lie, *Theorie der Transformationsgruppen.* I. Clebsch Ann. 16 (1880), 441–529.

[Lie1888] S. Lie, *Theorie der Transformationsgruppen*, Erster Abschnitt. Unter Mitwirkung von Dr. Friedrich Engel, bearbeitet von Sophus Lie, Verlag und Druck von B.G. Teubner, Leipzig und Berlin, xii+638 pp. (1888). Reprinted by Chelsea Publishing Co., New York, N.Y. (1970).

[LiTz73] J. Lindenstrauss, L. Tzafriri, *Classical Banach spaces.* Lecture Notes in Math., vol. 338, Springer-Verlag, Berlin and New York, 1973.

[LiZh11] S.-Q. Liu, Y. Zhang, *Jacobi structures of evolutionary partial differential equations.* Adv. Math. 227 (2011) no. 1, 73–130.

[Lob07] S.G. Lobanov, *Picard's theorem for ordinary differential equations in locally convex spaces*, Izvestiya Mathematics 41(3), 465 (2007).

[MaXu94] K. Mackenzie, P. Xu, *Lie bialgebroids and Poisson groupoids.* Duke Math. J., 73 (1994), 415–452.

[Mckey66] G.W. Mackey, *Ergodic theory and virtual groups.* Math. Ann. 166 (1966), 187–207.

[McL98] S. Mac Lane, *Categories for the working mathematician.* Second edition. Graduate Texts in Mathematics, 5. Springer-Verlag, New York, 1998, xii+314 pp.

[Mak63] B. M. Makarov, *Some pathological properties of inductive limits of B-spaces.* Uspehi Mat. Nauk 18 (1963) no. 3 (111), 171–178.

[Mag78] F. Magri, *A simple model of the integrable Hamiltonian equation.* J. Math. Phys. 19 (1978), no. 5, 1156–1162.

[MaMo84] F. Magri, C. Morosi, *A geometrical characterization of integrable hamiltonian systems through the theory of Poisson-Nijenhuis manifolds.* Quaderno S/19, Università degli studi di Milano, 1984.

[Mal92] A. Mallios, *On an abstract form of Weil's integrality theorem.* Note Mat. 12 (1992), 167–202.

[Marl83] C.-M. Marle, *Lie group actions on a canonical manifold.* In: *Symplectic geometry.* Res. Notes in Math., 80, Pitman, Boston, Mass.-London, 1983, pp. 144–166.

[Marl84] C.-M. Marle, *Poisson manifolds in mechanics.* In: *Bifurcation Theory, Mechanics and Physics* . (Reidel, Dordrecht, 1984) 47-76.

[Marl08] C.-M. Marle, *Differential calculus on a Lie algebroid and Poisson manifolds.* Preprint arXiv:0804.2451v2 [math.DG], 2008.

[Marl14] C.-M. Marle, *Lie, symplectic and Poisson groupoids and their Lie algebroids.* Preprint arXiv:1402:0059v1 [math.DG], 2014.

[Mars72] J.E. Marsden, *Darboux's theorem fails for weak symplectic forms.* Proc. Amer. Math. Soc. 32 (1972), 590–592.

[Mars81] J.E. Marsden, *Lectures on geometric methods in mathematical physics.* CBMS-NSF Regional Conference Series in Applied Mathematics, 37. Society for Industrial and Applied Mathematics (SIAM), Philadelphia, PA, 1981.

[MaRa99] J.E. Marsden, T.S. Ratiu, *Introduction to mechanics and symmetry.* Second edition. Texts in Applied Mathematics, 17. Springer-Verlag, New York, 1999.

[Mar01a] E. Martinez, *Geometric formulation of mechanics on Lie algebroids.* In: M. del Olmo, M. Santander (eds.), *Proceedings of the VIII Fall Workshop on Geometry and Physics.* Publ. R. Soc. Mat. Esp., 2, R. Soc. Mat. Esp., Madrid, 2001, pp. 209-222.

[Mar01b] E. Martínez, *Lagrangian mechanics on Lie algebroids.* Acta Appl. Math. 67 (2001), no. 3, 295–320.

[Mas09] T. Masson, *Géométrie différentielle, groupes et algèbres de Lie, fibrés et connexions*, C.P.T., Marseille, 2009.

[Mau97] K. Maurin, *The Riemann legacy.* Mathematics and its Applications, 417. Kluwer Academic Publishers Group, Dordrecht, 1997.

[Meg98] R. E. Megginson, *An introduction to Banach space theory.* Graduate Texts in Mathematics, 183. Springer-Verlag, New York, 1998.

[MeSe64] N. G. Meyers, J. Serrin, *H=W.* Proc. Nat. Acad. Sci. U.S.A. 51 (1964), 1055–1056.

[Men84] A.M. Mendes-Justino, *Quelques propriétés des variétés de Poisson et des variétés de Jacobi.* Thèse 3ème cycle, Paris VI, 1984.

[Mey99] R. Meyer, *Analytic cyclic cohomology.* Preprint arXiv:math/9906205 [math.KT], 1999.

[Mica38] A. D. Michal, *Differential calculus in linear topological spaces.* Proc. Natl. Acad. Sci. USA 24 (1938), 340–342.

[MMM12] M. Micheli, P. W. Michor, D. Mumford, *Sobolev metrics on diffeomorphism groups and the derived geometry of spaces of submanifolds.* translated from Izv. Ross. Akad. Nauk Ser. Mat. 77 (2013), no. 3, 109–138; Izv. Math. 77 (2013), no. 3, 541–570

[Mil82] J. Milnor, *On infinite dimensional Lie groups.* Preprint, Institute of Advanced Study, Princeton, 1982.

[MoMr03] I. Moerdijk, J. Mrčun, *Introduction to foliations and Lie groupoids.* Cambridge Studies in Advanced Mathematics, 91. Cambridge University Press, Cambridge, 2003.

[Mol72] P. Molino, *Sur quelques propriétés des G-structures.* J. Differential Geometry 7 (1972), 489–518.

[Mol88] P. Molino, *Riemannian foliations.* Translated from the French by Grant Cairns. With appendices by Cairns, Y. Carrière, É. Ghys, E. Salem and V. Sergiescu. Progress in Mathematics, 73. Birkhäuser Boston, Inc., Boston, MA, 1988, xii+339 pp.

[MoZu20] Ph. Monnier, N.T. Zung, *Deformation of singular foliations, 1: Local deformation cohomology.* Comptes Rendus. Mathématique, Tome 358 (2020) no. 3, 273-283.

[MoZi55] D. Montgomery, L. Zippin, *Topological transformation groups.* Interscience Publishers, 1240. New York-London, 1955.

[Mos65] J. K. Moser, *On the volume elements on a manifold.* Trans. Amer. Math. Soc. 120 (1965), 286–294.

[Nag66] T. Nagano, *Linear differential systems with singularities and an application to transitive Lie algebras* J. Math. Soc. Japan 18 (1966), 398–404.

[Nam73] Y. Nambu, *Generalized Hamiltonian dynamics.* Phys. Rev. D 7 (1973), 2405–2412.

[NaBe11] L. Narici, E. Beckenstein *Topological Vector Spaces.* Pure and Applied Mathematics. Second edition (2011).

[Nas56] J. Nash, *The imbedding problem for Riemannian manifolds.* Ann. of Math. (2) 63 (1956), 20–63.

[Nee98] K.-H. Neeb, *Holomorphic heighest weight representations of infinite dimensional complex classical groups.* J. Reine Angew. Math. 497 (1998), 171–222.

[Nee05] K.-H. Neeb, *Infinite-dimensional Lie groups.* Lecture notes, Monastir, 2005.

[Nee06] K.-H. Neeb, *Towards a Lie theory of locally convex groups.* Jpn. J. Math. 1 (2006), no. 2, 291–468.

[Nee08] K.-H. Neeb, *Lie group extensions associated to projective modules of continuous inverse algebras.* Arch. Math. (Brno) 44 (2008), no. 5, 465–489.

[NST14] K.H. Neeb, H. Sahlmann, T. Thiemann, *Weak Poisson structures on infinite dimensional manifolds and Hamiltonian actions.* In: V. Dobrev (ed.), *Lie theory and its applications in physics.* Springer Proceedings in Mathematics & Statistics, 111. Springer, Tokyo, 2014.

[New85] A. C. Newell, *Solitons in mathematics and physics.* CBMS-NSF Regional Conference Series in Applied Mathematics, 48. Society for Industrial and Applied Mathematics (SIAM), Philadelphia, PA, 1985.

[Nis2000] V. Nistor, *Groupoids and the integration of Lie algebroids.* J. Math. Soc. Japan 52 (2000), no. 4, 847–868.

[Ngu97] G. Nguenchouen, *Etude des extrémales de l'énergie potentielle sur une distribution affine.* Thèse Université de Savoie, 1997.

[Nov64] S.P. Novikov, *Foliations of co-dimension* 1. (Russian) Dokl. Akad. Nauk SSSR 157 (1964), 788–790.

[NoDu74] S. P. Novikov, B. A. Dubrovin, *The periodic problem for the Kortewegde Vries and the SturmLiouville equations.* Petrovskii Seminar 6 March 1974, Uspekhi Mat. Nauk 29 (1974), no. 6, 196-197.

[Nub92] F. Nübel, *On integral manifolds for vector space distributions.* Math. Ann. 294 (1992), no. 1, 1–17.

[OkEg21] N. O. Okeke, M. E. Egwe *Characterization of Locally Convex Partial *-Algebras.* arXiv:2101.08489v1 (2021)

[OdRa03] A. Odzijewicz, T.S. Ratiu, *Banach Lie-Poisson spaces and reduction.* Comm. Math. Phys. 243 (2003), no. 1, 1–54.

[OdRa08] A. Odzijewicz, T.S. Ratiu, *Induction for weak symplectic Banach manifolds.* J. Geom. Phys. 58 (2008), no. 6, 701–719.

[OSJ14] A. Odzijewicz, A. Sliżewska, G. Jakimowicz, *Complex Manifold Structure and Algebroid of the Partially Invertible Elements Groupoid of a W*-algebra.* Geom. Meth. Phys. (2014), 143-151.

[OJS15] A. Odzijewicz, G. Jakimowicz, A. Sliżewska, *Complex Manifold Structure and Algebroid of the Partially Invertible Elements Groupoid of a W*-algebra.* J. Geom. Phys. 95 (2015), 108-126.

[OdSl16] A. Odzijewicz, A. Sliżewska, *Banach-Lie groupoids associated to of a W*-algebra.* J. Symp. Geom. 14 (2016), no. 3, 687-736

[Olv93] P.J. Olver, *Applications of Lie groups to differential equations.* Second edition. Graduate Texts in Mathematics, 107. Springer-Verlag, New York, 1993.

[Olv16] P.J. Olver, *Quantum Mathematics.* (2016).

[Omo97] H. Omori, *Infinite-dimensional Lie groups.* Translations of Mathematical Monographs, 158. American Mathematical Society, Providence, RI, 1997.

[OMY80] H. Omori, Y. Maeda, A. Yoshioka, *On regular Fréchet Lie groups. I. Some differential geometric expressions of Fourier integral operators on a Riemannian manifold.* Tokyo J. Math. 3 (1980), no. 2, 353–390. *Correction* in Tokyo J. Math. 4 (1981), no. 2, 501–503.

[OMY81a] H. Omori, Y. Maeda, A. Yoshioka, *On regular Fréchet Lie groups. II. Composition rules of Fourier Integral operators on a Riemannian manifold.* Tokyo J. Math. 4 (1981), no. 2, 221–253.

[OMY81b] H. Omori, Y. Maeda, A. Yoshioka, O. Kobayashi, *On regular Fréchet-Lie groups. III. A second cohomology class related to the Lie algebra of pseudodifferential operators of order one.* Tokyo J. Math. 4 (1981), no. 2, 255–277.

[OMY82] H. Omori, Y. Maeda, A. Yoshioka, O. Kobayashi, *On regular Fréchet Lie groups. IV. Definitions and fundamental theorems.* Tokyo J. Math. 5 (1982), no. 2, 365–398.

[OMY83a] H. Omori, Y. Maeda, A. Yoshioka, O. Kobayashi, *On regular Fréchet Lie groups. V. Several basic properties.* Tokyo J. Math. 6 (1983), no. 1, 39–64.

[OMY83b] H. Omori, Y. Maeda, A. Yoshioka, O. Kobayashi, *On regular Fréchet Lie groups. VI. Infinite dimensional Lie groups which appear in general relativity.* Tokyo J. Math. 6 (1983), no. 2, 217–246.

[Pal65] R.S. Palais, *Differential operators on vector bundles.* In: *Seminar on the Atiyah-Singer index theorem.* Annals of Mathematics Studies, No. 57 Princeton University Press, Princeton, N.J. 1965.

[Pal66] R.S. Palais, *Lusternik-Schnirelman theory on Banach manifolds.* Topology 5 (1966), 115–132.

[Pal68] R.S. Palais, *Foundations of global non-linear analysis.* W. A. Benjamin, Inc., New York-Amsterdam, 1968.

[Pal70] R.S. Palais, *Banach manifolds of fiber bundle sections*, Actes, Congrès intern. Math. Tome 2, 1970, pp. 243–249.

[Pap01] M. H. Papatriantafilou, *Projective and inductive limits of differential triads.* In: L. Kozma, P. T. Nagy, L. Tamássy (eds.), *Steps in differential geometry.* Institute of Mathematics and Informatics, Debrecen, 2001, pp. 251–262.

[Pat2000] I. Patyi, *On the* $\overline{\partial}$*-equation on a Banach space.* Bull. Soc. Math. France 128 (2000), no. 3, 391–406.

[Pee59] J. Peetre, *Une caractérisation abstraite des opérateurs différentiels.* Math. Scand. 7 (1959), 211–218.

[Pee60] J. Peetre, *Rectification à l'article "Une caractérisation abstraite des opérateurs différentiels".* Math. Scand. 8 (1960), 116–120.

[Pel12] F. Pelletier, *Integrability of weak distributions on Banach manifolds.* Indag. Math. (N.S.) 23 (2012), no. 3, 214–242.

[Pel16] F. Pelletier, *Geometrical structures on the prolongation of a pre-Lie algebroid on fibered manifolds and application to partial Finsler geometry on foliated anchored bundle.* Preprint arXiv:1412.6742v2 [math.DG], 2016.

[Pel18] F. Pelletier, *On a Darboux theorem for symplectic forms on direct limits of ascending sequences of Banach manifolds.* Int. J. Geom. Methods Mod. Phys. 15 (2018), no. 12, 1850206.

[Pel19] F. Pelletier, *Conic sub-Hilbert-Finsler structure on a Banach manifold.* In: P. Kielanowski, A. Odzijewicz, E. Previato (eds.), *Geometric methods in physics XXXVI*, Trends Math., Birkhäuser/Springer, Cham, 2019, pp. 237–244.

[Pel21] F. Pelletier, *Projective limit of a sequence of compatible weak symplectic forms on a sequence of Banach bundles and Darboux Theorem.* J. Bulsci. 168 (2021).

[PeCa19] F. Pelletier, P. Cabau, *Convenient partial Poisson manifolds.* J. Geom. Phys. 136 (2019), 173–194.

[PeSa13] F. Pelletier, R. Saffidine, *Snakes and articulated arms in an Hilbert space*, Ann. Fac. Sci. Toulouse Math. 22 (2013), no. 3, 525–557.

[PeSl14] F. Pelletier, M. Slayman, *Configurations of an articulated arm and singularities of special multi-flags.* SIGMA Symmetry Integrability Geom. Methods Appl. 10 (2014), Paper 059, 38 pp.

[Pen70] J.-P. Penot, *Sur le théorème de Frobenius.* (French) Bull. Soc. Math. France 98 (1970), 47–80.

[PeBo87] P. Pérez Carreras, J. Bonet, *Barrelled locally convex spaces.* North-Holland Mathematics Studies, 131. Notas de Matemática, 113. North-Holland Publishing Co., Amsterdam, 1987.

[PhiR40] R. S. Phillips, *On linear transformations.* Trans. Amer. Math. Soc. 48 (1940), 516–541.

[PiTa01] P. Piccione, D. V. Tausk, *On the Banach differential structure for sets of maps on non-compact domains.* Nonlinear Anal. 46 (2001), no. 2, Ser. A: Theory Methods, 245–265.

[PiTa06] P. Piccione, D. V. Tausk, *The theory of connections and G-structures. Applications to affine and isometric immersions.* XIV Escola de Geometria Diferencial. Instituto de Matemática Pura e Aplicada (IMPA), Rio de Janeiro, 2006.

[Pom78] F. J. Pommaret, *Systems of partial differential equations and Lie pseudogroups.* Mathematics and its Applications, 14. Gordon & Breach Science Publishers, New York, 1978.

[PopL11] L. Popescu , *Metric non-linear connections on the prolongation of a Lie algebroid to its dual bundle.* An. Ştiinţ. Univ. Al. I. Cuza Iaşi. Mat. (N.S.) 57 (2011), suppl. 1, 211–220.

[PoPo01] M. Popescu, P. Popescu, *Anchored vector bundles and Lie algebroids.* In: J. Kubarski, P. Urbański, R. Wolak (eds.), *Lie algebroids and related topics in differential geometry* Banach Center Publ., 54, Polish Acad. Sci. Inst. Math., Warsaw, 2001, pp. 51–69.

[PopP92] P. Popescu, *Almost Lie structure, derivations and R-curvature on relative tangent spaces.* Rev. Roumaine Math. Pures Appl. 37 (1992), no. 9, 779–789.

[Popp99] M. Poppenberg, *An application of the Nash-Moser theorem to ordinary differential equations in Fréchet spaces.* Studia Math. 137 (1999), no. 2, 102–121.

[Pra66] J. Pradines, *Théorie de Lie pour les groupoïdes différentiables. Relations entre propriétés locales et globales.* C. R. Acad. Sci. Paris Sér. A–B 263 (1966), A907–A910.

[Pra68] J. Pradines, *Troisième théorème de Lie sur les groupoïdes différentiables.* C. R. Acad. Sci. Paris Sér. A–B 267 (1968), A21–A23.

[Pre69] T. Precupanu, *Espaces Linéaries à Semi-Normes Hilbertiennes.* researchgate.net/publication

[PrMy98] A. K. Prykarpatsky, I. V. Mykytiuk, *Algebraic integrability of nonlinear dynamical systems on manifolds. Classical and quantum aspects*. Mathematics and its Applications, 443. Kluwer Academic Publishers Group, Dordrecht, 1998.

[Rat11] T. S. Ratiu, *Coadjoint orbits and the beginnings of a geometric representation theory.* In: K.-H. Neeb, A. Pianzola (eds.), *Developments and trends in infinite-dimensional Lie theory.* Progr. Math., 288, Birkhäuser Boston, Inc. Boston, MA, 2011, pp. 417–457.

[Ree52] G. Reeb, *Sur certaines propriétés topologiques des des variétés feuilletées*, Actualités Sci. Indust., no. 1183. Inst. Math. Univ. Strasbourg 11, Hermann, Paris, 1952.

[RiZa10] L. Ribes, P. Zalesskii, *Profinite groups.* Second edition. Ergebnisse der Mathematik und ihrer Grenzgebiete. 3. Folge. A Series of Modern Surveys in Mathematics, 40. Springer-Verlag, Berlin, 2010.

[RoRo64] A. P. Robertson, W. J. Robertson, *Topological vector spaces.* Cambridge Tracts in Mathematics and Mathematical Physics, No. 53. Cambridge University Press, New York, 1964.

[Rou05] J. Roubíček, *Nonlinear partial differential equations with applications.* International Series of Numerical Mathematics, 153. Birkhäuser Verlag, Basel, 2005.

[Rud91] W. Rudin, *Functional Analysis.*. International Series of Numerical Mathematics, 8 (Second ed.). New York (1991).

[Rus1844] J. S. Russell, *Report on waves.* In: *Report of the Fourteenth Meeting of the British Association for the Advancement of Sciences.* John Murray, London, 1844, pp. 311–390.

[Rya02] R. A. Ryan, *Introduction to tensor products of Banach spaces.* Springer Monographs in Mathematics, Springer-Verlag London, Ltd., London, 2002.

[Sak92] K. Sakai, *Connecting direct limit topologies with metrics on infinite-dimensional manifolds.* Compos. Math. 81, 237–245 (1992). A corringendum, Compos. Math. 102, 357 (1996)

[Sak20] K. Sakai, *Topology of Infinite-Dimensional Manifolds.* Springer Monographs in Mathematics, 2020.

[SaMa14] M. Khajeh Salehani, I. Markina, *Controllability on infinite-dimensional manifolds: a Chow-Rashevsky theorem.* Acta Appl. Math. 134 (2014), 229–246.

[Sau89] D.J. Saunders, *The Geometry of Jet Bundles.* London Mathematical Society Lecture Note Series, 142. Cambridge University Press, 1989.

[Scha80] H.H. Schaefer, *Topological vector spaces.* Springer-Verlag, Heidelberg, 1980.

[Schap11] P. Schapira, *Elementary sheaf theory for the analysts.* Lecture notes, 2011.

[Schme21] A. Schmeding, *An introduction to infinite-dimensional differential geometry,* to be published in Cambridge University Press, 2021.

[Schm04] R. Schmid, *Infinite dimensional Lie groups with applications to mathematical physics.* J. Geom. Symmetry Phys. 1 (2004), 54–120.

[Schr15] P. J. Schrader, *Existence of variationally defined curves with higher order elliptic Lagrangians.* Nonlinear Anal. 115 (2015), 1–11.

[Schr17] P. J. Schrader, *Global analysis of one-dimensional variational problems.* Bull. Aust. Math. Soc. 95 (2017), no. 1, 167–168.

[Seb79] Z. Sebestyen, *Every $C^\star$-seminorm is automatically submultiplicative.* Period. Math. Hungar. 10 (1979), no. 1, 1–8.

[Sei10] W. M. Seiler, *Involution. The formal theory of differential equations and its applications in computer algebra.* Algorithms and Computation in Mathematics, 24. Springer-Verlag, Berlin, 2010.

[SeWe01] P. Severa and A. Weinstein, *Poisson geometry with a 3-form background.* In: Y. Maeda, S. Watamura (eds.), *Noncommutative geometry and string theory.* Progr. Theoret. Phys. Suppl. No. 144 (2001), 145–154.

[SBH11] M. Sommer,K. Brazda, M. Hantel, *Algebraic construction of a Nambu bracket for the two-dimensional vorticity equation.* Phys. Lett. A 375 (2011), no. 37, 3310–3313.

[Spa14] O. Spáčil, *On the Chern-Weil theory for transformation groups of contact manifolds.* PhD Thesis, University of Aberdeen, 2014.

[Spe69] D.C. Spencer, *Overdetermined systems of linear partial differential equations.* Bull. Amer. Math. Soc. 75 (1969), 179–239.

[Sta05] A. Stacey, *The differential topology of loop spaces.* Preprint arXiv:math/0510097 [math.DG].

[Ste74] P. Stefan, *Integrability of systems of vector fields.* J. London Math. Soc. (2) 21 (1974), no. 3, 544–556.

[SiSt65] I. M. Singer, S. Sternberg, *The infinite groups of Lie and Cartan.* I. The transitive groups. J. Analyse Math. 15 (1965), 1–114.

[Smo06] N. Smolentsev, *Diffeomorphism groups of compact manifolds.* (Russian) Sovrem. Mat. Prilozh. No. 37, Geometriya (2006), 3–100; transl. in J. Math. Sci. (N.Y.) 146 (2007), no. 6, 6213–6312

[Sur17] A. Suri, *Higher order tangent bundles.* Mediterr. J. Math. 14 (2017), no. 1, 17.

[Sus73] H. J. Sussmann, *Orbits of families of vector fields and integrability of distributions.* Trans. Amer. Math. Soc. 80 (1973), 171–188.

[SzLo07] J. Szilasi, R. L. Lovas, *Some aspects of differential theories.* In: D. Krupka, D. Saunders (eds.), *Handbook of global analysis.* Elsevier Sci. B. V., Amsterdam, 2008, pp. 1069–1114, 1217.

[Swa90] R.C. Swanson, *Linear symplectic structures on Banach spaces,* Jour. Math. 10 (1990), no. 2.

[Sza07] B. M. Szablikowski, *Geometric aspects of integrable systems.* Lecture notes, Silesian University in Opava, 2007.

[Tak79] F. Takens, *A global version of the inverse problem of the calculus of variations.* J. Differential Geometry **14** (1979), no. 4, 543–562.

[Takh94] L. Takhtajan, *On foundation of the generalized Nambu mechanics.* Comm. Math. Phys. 160 (1994), no. 2, 295–315.

[TSH98] N. Tatsuuma, H. Shimomura, T. Hirai, *On group topologies and unitary representations of inductive limits of topological groups and the case of the group of diffeomorphisms.* J. Math. Kyoto Univ. 38 (1998), no. 3, 551–578.

[Tay95] J.L. Taylor, *Notes on locally convex topological vector spaces,* course notes from 1994-95.

[Tei01] J. Teichmann, *A Frobenius theorem on convenient manifolds.* Monatsh. Math. 134 (2001), no. 2, 159–167.

[ThSc91] G. Thompson, U. Schwardmann, *Almost tangent and cotangent structure in the large.* Trans. Amer. Math. Soc. 327 (1991), no. 1, 313–328.

[Thu74] W. Thurston, *A generalization of the Reeb stability theorem.* Topology 13 (1974), 347–352.

[Tor73] H. Torunczyk, *Smooth partition of unity on some non-separable Banach spaces.* Studia Mathematica (1973) 46, 41-51.

[Tre67] F. Trèves, *Topological vector spaces, distributions and kernels.* Academic Press, New York-London, 1967.

[Tro76] A.J. Tromba, *Almost-Riemannian structures on Banach manifolds: the Morse lemma and the Darboux theorem.* Canadian J. Math. 28 (1976), no. 3, 640–652.

[Tum05] A. B. Tumpach, *Structures kählériennes et hyperkählériennes en dimension infinie.* Thèse, Ecole Polytechnique, 2005.

[Tum07] A. B. Tumpach, *Hyperkähler structures and infinite-dimensional Grassmannians.* J. Funct. Anal. 243 (2007), no. 1, 158–206.

[Tum20] A. B. Tumpach, *Banach PoissonLie Groups and BruhatPoisson Structure of the Restricted Grassmannian.* Communications in Mathematical Physics 373 (2020) 795–858.

[Vai03] I. Vaisman, *Lagrange geometry on tangent manifolds.* Int. J. Math. Math. Sci. (2003), no. 51, 3241–3266.

[Vai94] I. Vaisman, *Lectures on the Geometry of Poisson Manifolds,* Birkhäuser (1994).

[Val19] F. Valencia, *Notes on flat pseudo-Riemannian manifolds,* arxiv.org/pdf/1903.08940 (2019).

[Vas05] E. Vassiliou, *Geometry of principal sheaves.* Mathematics and Its Applications (New York), 578. Springer, Dordrecht, 2005.

[Vil67] J. Vilms, *Connections on tangent bundles.* J. Differential Geometry 1 (1967), 235–243.

[Vin81] A.M. Vinogradov, *Geometry of nonlinear differential equations.* J. Soviet Math. 17 (1981), no. 1, 1624–1649.

[Vis13] A. Visintin, *Notes on Sobolev Spaces.* www.science.unitn.it/ visintin/Sobolev2013.

[Vit17] L. Vitagliano, *Differential Geometry and PDEs.* Università degli Studi di Salerno (2017).

[VonN35] J. von Neumann, *On complete topological spaces.* Trans. Amer. Math. Soc. 37 (1935), no. 1, 1–20.

[Wei71] A. Weinstein, *Symplectic manifolds and their Lagrangian submanifolds.* Advances in Math. 6 (1971), 329–346.

[Wei83] A. Weinstein, *The local structure of Poisson manifolds.* J. Differential Geom. 18 (1983), no. 3, 523–557.

[Wei87] A. Weinstein, *Symplectic groupoids and Poisson manifolds.* Bull. Amer. Math. Soc. (N.S.) 16 (1987), no. 1, 101–104.

[Wei88] A. Weinstein, *Coisotropic calculus and Poisson groupoids.* J. Math. Soc. Japan, 10 (1988) 705–727.

[Wei96] A. Weinstein, *Groupoids: unifying internal and external symmetry. A tour through some examples.* Notices Amer. Math. Soc. 43 (1996), no. 7, 744–752.

[Wei98] A. Weinstein, *Poisson geometry.* Diff. Geom. Appl. 9 (1998), no. 2, 13–238.

[WeDe73] J. C. Wells, C. R. DePrima, *Local automorphisms are differential operators on some Banach spaces.* Proc. Amer. Math. Soc. 40 (1973), 453–457.

[Wol05] J.A. Wolf, *Principal series representations of direct limit groups.* Compos. Math. 141 (2005), no. 6, 1504–1530.

[Wur95] T. Wurzbacher, *Symplectic geometry of the loop space of a Riemannian manifold.* J. Geom. Phys. 16 (1995), no. 4, 345–384.

[Yam98] A. Yamasaki, *Inductive limits of general linear groups.* J. Math. Kyoto Univ. 38 (1998), no. 4, 769–779.

[Yud16] O. Yudilevich, *The role of the Jacobi identity in solving the Maurer-Cartan structure equation.* Pacific J. Math. 282 (2016), no. 2, 487–510.

[Zar93] D. N. Zarnadze *On some topological and geometrical properties of Fréhet Hilbert spaces*, Russian Acad. Sci. Izv. Math 41 (1993), no. 2.

[Zub97] J. P. Zubelli, *Topics on wave propagation and Huygens' principle.* 21° Colóquio Brasileiro de Matemática. Instituto de Matemática Pura e Aplicada (IMPA), Rio de Janeiro, 1997.

[ZuMa91] J. P. Zubelli, F. Magri, *Differential equations in the spectral parameter, Darboux transformations and a hierarchy of master symmetries for KdV.* Comm. Math. Phys. 141 (1991), no. 2, 329–351.

Index